4th International Conference of Planning in the 2019 Era of Uncertainty

IOP Conference Series: Earth and Environmental Science
Volume 328

Malang City, Indonesia
12 - 13 March 2019

ISBN: 978-1-7138-1299-9
ISSN: 1755-1307

Printed from e-media with permission by:

Curran Associates, Inc.
57 Morehouse Lane
Red Hook, NY 12571

Some format issues inherent in the e-media version may also appear in this print version.

Institute of Physics
Dirac House, Temple Back
Bristol BS1 6BE UK

Phone: 44 1 17 929 7481
Fax: 44 1 17 920 0979

techtracking@iop.org

Additional copies of this publication are available from:

Curran Associates, Inc.
57 Morehouse Lane
Red Hook, NY 12571 USA
Phone: 845-758-0400
Fax: 845-758-2633
Email: curran@proceedings.com
Web: www.proceedings.com

TABLE OF CONTENTS

ENVIRONMENT, INFRASTRUCTURE AND INFORMATION SYSTEM

REGIONAL DEVELOPMENT AND PUBLIC POLICY

URBAN PLANNING AND DESIGN

IOP Conf. Series: Earth and Environmental Science **328** (2019) 011001 doi:10.1088/1755-1315/328/1/011001

PREFACE: The 4th International Conference in The Era of Uncertainty (ICPEU)

The honourable Vice Dean of General Affairs at Faculty of Engineering: Dr. Muhammad Ruslin Anwar, Department Head of Regional and Urban Planning, Dr. Abdul Wahid Hasyim, the honourable keynote speakers – Dr. Kakuya Matsushima from Kyoto University – Japan, Dr. ONG Ghim Ping Raymond from the National University of Singapore – Singapore , Dr. Hadid Subki from the International Atomic Energy Agency (IAEA). Vienna – Austria, Dr. Surjono from Universitas Brawijaya – Indonesia and Dr. Emil Elestianto Dardak, M. Sc. – Vice Governor of East Java Province – Indonesia, and ladies and gentlemen.

Assalamualaikum Wr Wb,

I would like to thank everyone in this room today for your enthusiasm and participation in the 4th International Conference on Planning in the Era of Uncertainty at Universitas Brawijaya

ICPEU is a biennial event that we have organized since 2013. We have focused on the discussion of sustainable development since the very beginning, as we truly believe that this topic is very relevant to what we, in the field of planning, face at the moment and in the future.

In 2013 we started off the discussion with the big question, does planning theory can be applied thoroughly in practice? It was a very memorable discussion as we explored many interdisciplinary problems and solutions in planning. We then developed our topic to sustainable development in 2015 as we considered this as an underlying topic for all relevant problems and solutions. Again, the discussion was remarkable. Papers from different countries participated at the event. In 2017, we started publishing our papers on IOP Proceedings, which is a Scopus-indexed form of publication. We published 67 papers from different universities in different countries. And we believe that this event at this moment, will be more successful than our predecessors. The topic Strengthening Rural Urban Sustainability Goals is yet another situation that we all can relate to. A problem that requires collective attention to achieve equality and sustainability in the future development.

Looking back at the events that we have organized so far, we learned that the pursuit of achieving sustainability has attracted ever-growing global interest. We have received papers from Australia, Thailand, Malaysia, Singapore, Japan, Sweden, France, and also Indonesia, discussing and showcasing multi-level efforts in achieving sustainable development from different perspectives. In short, the response that we received has been overwhelmingly positive.

I would like to thank our dearly sponsors, Universitas Brawijaya, the Faculty of Engineering, ESRI, Institut Atsiri, Persada Printing, UB Press, RRI, UB TV for their support that have allowed us to organize this event. We received a number of 163 extended abstracts for this event, and unfortunately, we could only accept a number of 76 selected ones to be published on our upcoming IOP Proceeding and Jurnal Tata Loka. Therefore, I would like to congratulate all authors that have made it this far.

To close this speech, I would like to once again welcome you to Malang, I wish you a wonderful two-day seminar, and I look forward to learning new insights. Thank you very much.

Wassalamualaikum Wr Wb.

LIST OF COMMITTEES

Steering Committee (SC)	: Dr. Ir. Budi Sugiarto Waloeya, MSP
	Dr. Ir. Surjono, MTP
	Dr. Ir. Agus Dwi Wicaksono, Lic.Rer.Reg
	Dr. Ir. A. Wahid Hasyim, MSP
	Ir. Ismu Rini Dwi Ari, MT.,Ph.D.
	Dr. Wara Indira Rukmi, ST., MT
	Gunawan Prayitno, SP., MT., Ph.D
	Dr. Eng. I Nyoman Suluh Wijaya, ST., MT
Organizing Committee (OC)	:
Head of Executive	: Dadang Meru Utomo, ST., MURP
Vice Head of Executive	: Deni Agus Setyono, ST. M.Eng
General Treasure	: Wawargita Permata W., ST., MT
	Anggi Anggraeni, S.Si
	Puji Darningsih, STP
Scientific Committee	: Dr. Septiana Hariyani, ST., MT
	Dian Dinanti, ST., MT
	Indriati Farida, SE., MM
	Dwi Maulidatuz Zakiyah, ST., MT., M. Sc
	Dr.tech. Christia Meidiana, ST., M.Eng
	Imma Widyawati Agustin, ST., MT.,Ph.D.
	Aris Subagiyo, ST.,MT.
	Kartika Eka Sari, ST.,MT.
	Wulan Dwi Purnamasari, ST., MT
	Dr. Eng. Fadly Usman, ST., MT
	Dr. Eng. Turniningtyas A.R., ST.,MT
	Adipandang Yudono. S.Si., MURP., Ph.D
	Eddi Basuki Kurniawan, ST.,MT
	Dian Indah Shofarini, ST., MT
	Wisnu Sasongko,ST.,MT
	Chairul Maulidi, ST.,MT
	AR. Rohman Taufiq Hidayat, ST., M.AgrSc
	Eka Trisye Mufida, SE
	Nindya Sari, ST.,MT
	Hastirin Widi Astuti, S.Pd, M.A (TESOL)
Technical Assistance	: Muhammad Iqbal Ashari
	Fadjar Iman Nugroho
	Regina Fortunata
	Izatul Ihsansi
	Nadya Chrisyanti Putri
	Gina Magfirah
	Adhania Nurhana
	Adelita Virenza Nurubiatmoko
	Siti Azizah Khairunnisa

IOP Conf. Series: Earth and Environmental Science **328** (2019) 011003 doi:10.1088/1755-1315/328/1/011003

Labib Eka Satria Putra
Lody Sunandar Putra

KEYNOTE SPEECH

IOP Conf. Series: Earth and Environmental Science **328** (2019) 011003 doi:10.1088/1755-1315/328/1/011003

PARALLEL SESSION

UMENTATION

Peer review statement

All papers published in this volume of *IOP Conference Series: Earth and Environmental Science* have been peer reviewed through processes administered by the proceedings Editors. Reviews were conducted by expert referees to the professional and scientific standards expected of a proceedings journal published by IOP Publishing.

IOP Conf. Series: Earth and Environmental Science **328** (2019) 012001 doi:10.1088/1755-1315/328/1/012001

Analysis Lean Construction Application to Reduce Material Waste at Bridge Construction Project

A E Sebastian[1]*

[1] Department of Civil Engineering, Universitas Pembangunan Jaya, Jl. Cenderawasih Raya Blok B7/P, Ciputat, Tangerang Selatan City, Banten, Indonesia, 15413

*email: albertus.edwarsebastian@student.upj.ac.id

Abstract. This abstract discusses the amount of concrete and steel waste that are produced at the manufacturing stage in Precast Plant as well as examines the processes that cause the generation of waste. This abstract also discusses the implementation of Lean Construction concept in bridge construction projects and the influence of this concept on waste minimization in the construction processes. The study was qualitative research using study case method and also through unstructured interviews to seek information according to the approach used in this research that is Last Planner System. This result of this research proves lean construction contribution to reduce waste.
Keywords: Material waste, bridge, precast, lean construction, last planner system

1. Introduction

The construction project involves many participants (multiparties) to carry out planned activities. Each participant interacts with each other until all scheduled work is finished. All act for their own interests, trying to optimize the use of labor, equipment, and materials they provide. For this reason, an approach is needed to bring the interests of each stakeholder in harmony with the promises to the customer.

Each project is unique in terms of design specifications, methods, administration, and people involved [1]. The method chosen to use during the construction project implementation process will provide a clear picture of the resources that must be available. In the competitive construction world, reducing costs in an effort to increase market competitiveness and profits is a common goal among all construction companies [2]. The three most common metrics for describing productivity are productivity or unit level factors (ratio of labor costs, material, and equipment to output units); labor productivity (working hour ratio to output unit), and productivity factor (schedule ratio for actual working hours).

Productivity in the construction sector experienced a negative development compared to the manufacturing industry. The level of innovation in this sector is too low, considered by many people, so that efforts are needed to improve it. Productivity is closely related to achieving a number of production units at a certain time. Productivity has implications for the duration needed. The right method means more production and or cheaper costs while still making quality as a constraint that must be fulfilled [3]. In the book Total Construction Management Project [4], it explains that planning is well-prepared before the project is implemented. In the project planning phase several main areas to be controlled are:

1. The money plan
2. The time plan

3. Quality standards
4. Material resources and delivery
5. Labor supply and productivity
6. Cashflow projections

During the process of implementing a project there are three limits that must be controlled, namely costs, quality and time of implementation. To control these constraints, implementation of planning is needed, namely: the method of implementation work, implementation budget, implementation schedule and cash budget. How much effort is used to increase efficiency, as long as the construction process takes place it will still produce waste. There is no accurate method that has been developed to measure the incidence of waste in Indonesia.

The amount of material that is wasted or the waste is the concern of construction implementers because almost all raw materials as inputs for construction are materials produced from non-renewable sources [5]. The responsibility to eliminate waste does not only depend on the Project Manager, but also clients, consultants, suppliers, foremen and workers [6].

2. Methods

This research will be that the implementation of the project must be in accordance with the plan stipulated based on the last planner system consideration. Productivity measured based on the flow of work, for this purpose, is needed in this chapter and will explain the appropriate research method to answer research questions and achieve the research objectives, namely identifying processes that produce waste (waste sources) in precast companies and evaluating the types of waste produced in bridge construction projects in precast companies by using lean construction techniques.

2.1 Research Methods

To obtain these data, the research strategy that will previously be used must be determined. Three factors that influence the type of research strategy are the types of questions that will be used, the control of the events under study and the focus on current or newly completed events (Yin, 1996). In this case the research questions related to what processes produce waste and how lean construction can reduce waste, implementation of field observations and interview interviews must be done because researchers do not have control of events but must focus on what happened during the study. The methodology used in this study covers all the issues concerned with the application of Lean Construction in reducing material waste during the construction process of the bridge project in the production phase in the precast company. In achieving the objectives of previous research, relevant data must be obtained by using a research method that refers to the research strategy suggested by Yin (1996) as shown in the table below.

Table 1. Research Strategy

Strategy	Type of Question Used	Control of Events Researched	Factors Against Current / New Events Completed
Experiment	How, Why	Yes	Yes
Survey	Who,What, Dimana, How many	No	Yes
Archival Analysis	Who, What, Where, How many	No	Yes/No
History	How, Why	No	No
Case Study	How, Why	No	No

Source: Yin, 1996

2.2 Data Analysis

2.2.1. Qualitative Data Analysis

That is a naturalistic research process that is cyclical, not linear because of its cyclical nature and, the research is done repeatedly. The number of repetition periods will depend on the level of depth and

IOP Conf. Series: Earth and Environmental Science **328** (2019) 012001 doi:10.1088/1755-1315/328/1/012001

accuracy desired, and the research will focus more on the actual problems that occur in the project. Qualitative research can be done in several codes. The process of each research period consists of.

- Research question.
- Data collection.
- Record data.
- Analyzing data.
- Making research reports

The data used in qualitative research is qualitative data, that is, data can simply be called numbers. Qualitative data has characteristics that cannot be done using mathematical operations such as addition, subtraction and multiplication.

3. Results and Discussions

3.1 Analysis of Production Flow

The design process is a refinement of the exact stage for specifications where the needs and desires that are not clearly changed into requirements, then through various number of steps, for detailed design. Simultaneously, this is the process of detecting problems and solving. The results of the design will be a benchmark for what activities should be carried out.

The Last Planner System discovered by Ballard when implementing Lean Production for the construction industry is a system that focuses on flow (Flow View). During the development of this system the aim was to shift from increasing productivity to increasing reliability / reliability of the work flow. The work flow starts with a complete design result. Furthermore, the work will be determined from the objectives of the design. Pull system is an important part of the Last Planner System.

The results of observations by researchers in the two companies of this case study, although the company indirectly knew about Lean Construction but in practice did it. Basically the concept is to streamline the work so that the process is not too much and spend a large amount of time, costs and energy. In this study, work or activities will be carried out in the production of the same girder for each order. Differences are specifications, number and time of order were made by the client. General description of the production process for PCI GIRDER in the Plant.

Precast is depicted in the Production Flow Chart as Figure 1. Each part of the production process has the same sequence of methods as follows:

1. Ironing
2. Installation of ducting
3. Installation of formwork
4. Casting
5. Demolition of formwork
6. Care

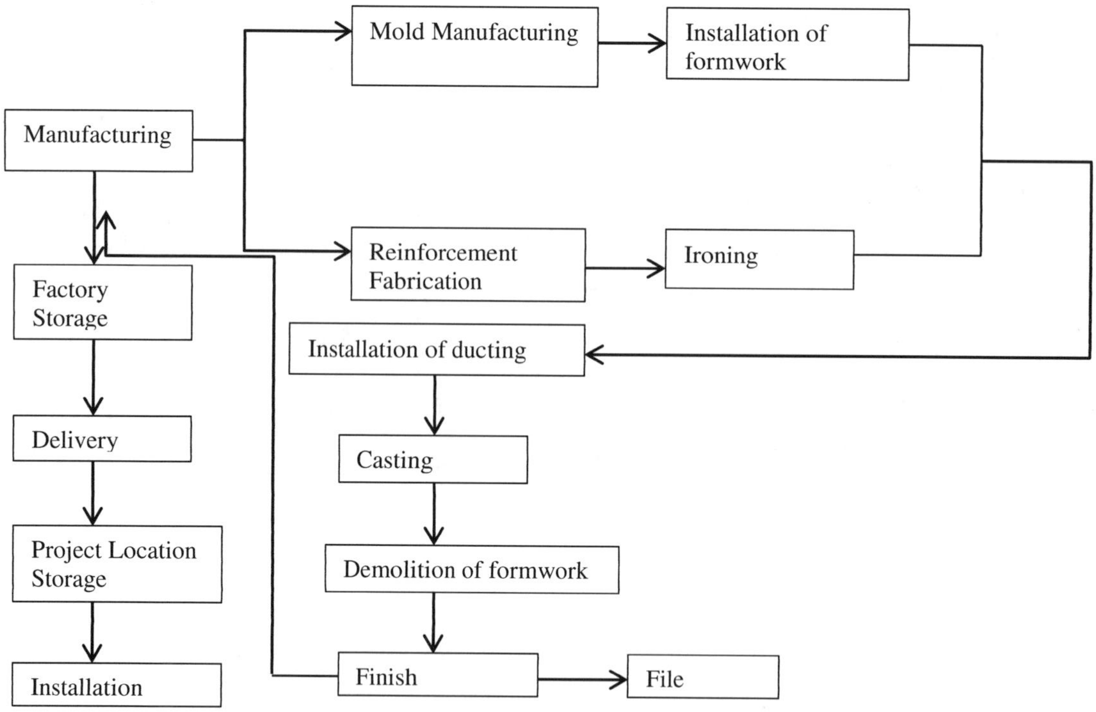

Figure 1. Girder Production Flow Chart

The above diagram explains the work done by company as a precast producer. According to the head of the production division of Company B that the scope of work carried out includes manufacturing, storage in the factory before the product is sent to the project location, product delivery and storage at the project location if the installation is not directly carried out and subsequently the installation. With the completion of the installation work, the work is considered complete. To prove the product according to the client's request, it takes as long as 180 days the maintenance period for the product installed.

3.2 Production Analysis

The construction management process is a process in which detailed design changes into construction / fabrication and plans into days of coordination and process control at the site or at the factory. The flow of material and information in girder production is done in several teams where each team does work spatially. Each team is interrelated to produce products desired by customers. This requires good teamwork and clarity of information and communication to avoid conflict which results in unfinished work.

3.2.1 Planning

Planning is done as detailed as possible until it approaches the actual work. In making a plan we must work with people who will carry out these workshops. The obstacles in the project need to be discussed together and resolved together. Planning this work is not limited to the technical part, but also all aspects involved directly or indirectly in the project.

In planning the implementation of the project, the work is divided structurally into a Work Breakdown Structure. Milestones can be divided based on the distribution of activities within the WBS. From the determination of these milestones, project implementation strategies can be carried out. Based on experience and expertise in project implementation, schedule project (Schedule master) can be arranged.

All work in the factory can be carried out according to the specifications described in the project, especially the agreed time problem. There are no significant problems that cause both parties to sit down

to discuss due to differences in perception. The success of this work is due to the fact that the factory and the client have the same perception of the goods ordered by the client.
The production process in the factory is very dependent on a production warrant that explains all information agreed to in the contract, namely costs, quality and time of execution regarding orders from clients. If the production department experiences difficulties due to lack of information from the client, the factory will immediately discuss the matter so that there is no difference in perception between the client and the adhimix.

3.2.2 PPC measurement
There is no significant problem so that the activities carried out are always completed on time. Thus, according to the Last Planner System, PPC for all activities carried out is 100%. This is not too surprising because basically precast manufacturing is a manufacturing industry not part of construction. All of them have been properly systemized so that if the productivity is small then it that should be questioned. But if the role of precast companies is different, for example, as a subcontractor who is the provider and installer of girder the possibility of planned activities that are not completed has a chance. Where in the project activities it was very crowded and filled with workers.

Table 2. PPC

Task Item	Week I	Week II	Week III	Week IV
Planned	3	3	6	6
Completed	3	3	6	6
PPC	100%	100%	100%	100%

According to experience, usually the main cause of unemployment is not equipment (Equipment). The equipment is of great potential to be the cause of incomplete work, especially cranes during lifting and transporting girders at the project site. The next problem that causes the failure of the project is erratic weather problems at the project location. This is totally out of everyone's control. Except for all the activities carried out at the factory, the weather problems, especially rain, will not stop work at the factory.

Table 3. Reasons for failure of the Plan

No	Reason for Noncompletion	Description
1	Client	controlled
2	Engineering	controlled
3	Material	controlled
4	Equipment	controlled
5	Craft	controlled
6	Pre-Requisite	controlled
7	Subcontractor	controlled
8	Plan	controlled
9	Weather	controlled
10	Other	controlled

It is difficult to determine the factors that support success in the product work done. Because the data obtained from this case study is employment data that has been completed. For this reason, the researcher conducted an interview related to the factors that influence the completion and not the completion of the planned work. Based on the experience of the head of production and the production supervisor of company A, the results are as shown in the table above.

3.2.3 Waste Analysis
Conceptualization of the design and construction process as a flow of information and appropriate materials to reduce waste by minimizing information time or material spent waiting to be used, time

IOP Conf. Series: Earth and Environmental Science **328** (2019) 012001 doi:10.1088/1755-1315/328/1/012001

spent checking information or material for conformity with requirements, time spent reworking information or material to achieve conformity and time spent on transfer. The fact that PPC achieves 100% value illustrates that to achieve the desired product, the work is done by the right person, the right time and doing the things planned

3.2.4 Concrete

The production process produces almost no waste. In the calculation of mix design, the technical section provides volume tolerance of up to 2% or in other words for 1 m^3, the more tolerable volume is 20 liters of wet concrete. But the production part requires a smaller plant tolerance of 1% so that for 1 m^3 of concrete mixture it can be tolerated by 10 liters. Because the dimensions of the girder are large, the volume is directly large, so a small percentage also results in waste that needs attention.

Waste in the production process is indeed small, except the remnants of mixer truck washing. This waste is very difficult to know how big it is. Based on experience the production of waste produced does not exceed 3% of the total volume of use of concrete mixtures. This calculation is based on the assumption that the tolerance of 2% is not used and the waste produced by mixer and cracks that occur and the possibilities of aggregates which contain mud are not in accordance with the provisions. But this waste is not just discarded because it can be used as a tool, for brick materials and heaps for homes and roads.

Technically, the quality of this waste is not the same, but the reuse of this waste, for example for bataco, provides economic value. In addition, this is the amount of waste reuse the waste discharged into landfill that will decrease so that pollution to the environment will also decrease.

To reduce the amount of waste produced from the precast production process, in addition to reuse, greater quality improvement is needed. Planning a good job, carrying out work in accordance with the plan must also be supported by better quality workforce and equipment. Increased production of labor can be done by conducting training in accordance with the focus of each work and for equipment that requires regular maintenance.

3.2.5 Iron

The use of this material in large quantities as reinforcement for each product is offered so that ironing work is not focused on one product only. Iron measurements are very important for this. The waste produced by the production process is only 1% in general. This waste is explained in Figure 1 that iron waste is produced in the iron manufacturing process. This process starts with "bestat", which is the process of determining the dimensions and shapes of iron which will be arranged as reinforcement.

Determining the dimensions and shape of this circuit is usually in manual enumeration or by using Microsoft Excel. The use of iron in a number of products causes the desired size to be varied so as to avoid cutting errors and optimize the use of iron more sophisticated. If possible by using a computer-based program that cannot cover data collection of iron stock and deductions in accordance with the provisions.

4. Conclusion

From the two implementation of case studies in precast factories it is known that by using the Lean Construction concept, the waste in the construction project can be reduced. The research was conducted at two different precast companies. The company is a company that provides precast concrete products. Customer service is done by providing precast products including:

1. Beam Column slab System
2. Facade
3. Diaphragm wall, PC Pile
4. PC Slab
5. Girder

Prestressed girder beams are precast beam structures that are used as structural elements over bridges to support gravity loads and live loads, made by monolite and segmental which consists of several small

IOP Conf. Series: Earth and Environmental Science **328** (2019) 012001 doi:10.1088/1755-1315/328/1/012001

parts during fabrication in a precast concrete plant and finishes stressed on site before erection work. From the data and discussion carried out in the previous chapter, it can be concluded;

1. During the production process of the girder, the work produces waste as discussed
   a. The process of iron fabrication in cutting iron
   b. Casting on truck mixer or concrete bucket washing
2. The lean construction system that is applied to bridge construction projects as described in the discussion gives the following results:
   a. The amount of waste material produced during the production process of Girder precast is 3% of the planned volume and 1% of the total weight of the plan.
   b. Improving production and decreasing the incidence of concrete and iron waste can be done by providing training to improve the quality of labor, maintenance of equipment and the use of computer-based programs

References

[1] Duncan ,William R.,(1996). a guide to the Project Management body of knowledge. USA: PMI

[2] Gouet , Michael C., Haas ,Carl T., & Goodrum, Paul M.(2011). Activity Analysis for Direct-Work Rate Improvement in Construction. Journal Of Construction Engineering And ManagemenT, ASCE, pp 1117-1124

[3] Nugroho, P. S.,(2012). Peningkatan Produktivitas Konstruksi Melalui Pemilihan Metode Konstruksi. Dinamika Rekayasa Vol. 8,pp 25-30

[4] Rizt, George J.,(1994). Total Construction Management Project. Mc-graw Hill Companies

[5] Ekanayake, L. L.; Ofory, George,(2000, August). Construction Material Waste Source Evaluation. Paper Presented At Proceedings: Strategies for a Sustainable Built Environment, Pretoria

[6] Alwi, S.; Hampson, K. and Mohamed, S. (2002, November), Waste in the Indonesian Construction Project. Paper Presented at Proceedings of the 1st International Conferences of CIB W107 – Creating a Sustainable Construction Industry in Developing Countries, South Africa

Swamp Land Utilization for Socio-cultural, Economic and Potential Development of Lake OPI Ecotourism in Palembang

A Husin[1*]

[1] FKIP University of Sriwijaya

*email: azizahhusin66@yahoo.co.id

Abstract. South Sumatra has a vast swamp area, now along with the development of the population the land needs an increase, so that much swamp land is converted into housing, offices, business centres, and others. One of the housing complexes that used to be a swamp is the OPI housing in Jakabaring. Amid the Jakabaring OPI housing there is a swamp that becomes a lake by the developer. This lake is quite extensive, and interesting because the environment is still beautiful and there is no construction of tall buildings except housing. The lake looks beautiful yet untouched to become an ecotourism area, so the lake has great potential to become a tourist land for the community. This study aims to find out how people utilize this tourism potential from socio-cultural and economic aspects which ultimately leads to the use of being an ecotourism area. Research uses two methods namely quantitative and qualitative descriptive. The research respondents were 40 people from the surrounding community. The research findings obtained that the community exploits the lake's potential in the aspects of social, economic, and ecotourism. The utilization that is often carried out is on the socio-cultural aspects, in the economic aspect the categories are quite utilizing, on the other hand the utilization of ecotourism aspects is still weak. For this reason, it is recommended that the regional government provide support so that the lake is maximally utilized by the community as an ecotourism area. Which ultimately prosper the community.
Keywords: swamp, flood, socio-economic, ecotourism

1. Introduction

One of the positive aspects of the extensive conversion of swamp land in South Sumatra province is to maintain an area of land converted into a potential ecotourism site. The results of the study [1] found that swamp function had caused floods everywhere and expanded and increased from year to year. This is because the swamp area has been converted into residential land, office and government centers, apartments, warehouses, shopping centers, shops, etc. without regard to the space to store water which is balanced with the area being converted. On the basis that if the building continues to pay attention to environmental aspects, it will benefit the region itself. The government should develop must pay attention to sustainable development, namely to carry out development to meet current needs without sacrificing the fulfillment of future generations' needs (According to the United Nations Brundtland Report, 1987). One factor that must be faced to achieve sustainable development is how to improve environmental destruction without sacrificing economic development and social justice Sustainable development encompasses the three spheres of development policy: economic, social and environmental protection UN documents, especially the 2005 World Summit results documents mention the three dimensions are interrelated and are the driving pillars for sustainable development. to improve the welfare of the community by utilizing natural resources in an effective manner. The importance of making development efforts by paying attention to aspects of local potential is a must [2].

Lake OPI in Jakabaring, Banyuasin district as an example of development that is concerned with the environment. Now Lake OPI attracts the attention of the surrounding community and from outside the region because of its beauty and is pleasant to visit as a place of recreation. From its own ecosystem, swamps have a high economic value, as argued by [3] Swamp land has the potential to be large enough to increase food crop production and horticulture. Of the total land area of around 13.2 million hectares, only 5 percent has been used intensively. In swamp land, agronomic activities can potentially be a source of food. The results of the study [4] the use of nature as a tourism sector have an impact on economic improvement to the people of Bajulmati Hamlet. Likewise, the results of research conducted by (Aulia & Hakim, 2017), the results of his research found that existing tourist objects are attractive to tourists, so that they can be developed further by following the principles of sustainable use of natural resources. Research conducted by [5] on Baduy communities stated that they maintain environmental preservation and develop simplicity and respect for nature. Basically swamps have economic value besides food sources, as well as a source of income, and providing employment for the community and developing local culture. . If swamps are well managed, they can even make swampland into ecotourism, so that it can bring prosperity to the community. The term "ecotourism" can be interpreted as a trip by a tourist to a remote area with the aim of enjoying and learning about the area, where the tourism pattern helps the economy of the local community and supports nature conservation [6]. Actors and experts in the field of ecotourism agreed to emphasize that the pattern of ecotourism should minimize negative impacts on the environment and local culture and be able to increase economic income for local communities and conservation values. Culture, according to Taylor in [7] is a complex totality that includes knowledge, beliefs, art, law, morals, customs, and abilities and habits that people get as members of the community which contains activities from all those elements. Social dimensions include social interaction, social processes, cooperation, social communication, and social contact. According to social interaction [8] is a dynamic relationship that brings people together with people. According to social processes [8] it is a dynamic aspect of people's lives. In it there is a process of relations between one human being and another human being. The process of the relationship in the form of social interactions that occur in everyday life. Social interaction is intended as a reciprocal influence between the two parties, namely between one individual with another individual to achieve certain goals. According to social communication [8] is a social process that provides mutual interpretation to or from the behavior of others. This study aims to explore information about how people use swamps in social, economic and ecotourism on Jakabaring OPI Lake.

2. Methods

This study uses a mixed method between quantitative and qualitative research. The researcher revealed this explorative research qualitatively [9]. The research describes the use of social culture, economic swamps, as well as seeing the utilization and development of the potential of ecotourism in the OPI Lake, Jakabaring. The technique of collecting data uses a questionnaire and makes observations directly in the lake environment. The research respondents were 40 people, with random sampling techniques. Data analysis uses descriptive percentage. Scores obtained from question items, determined by a Likert scale range of 3 namely Frequent (S) score 3, Rarely (J) score 2, and never (TP) score 1. The highest score is 120 and the lowest is 60. Data collection techniques using random. This question is addressed to people who visit Lake OPI. Whereas data collection on Lake utilization for the Development of Lake OPI Potential is carried out by observing infrastructure facilities at the location. The score interval is set to specify the category of answers for respondents with high, medium, low categories.

3. Results and Discussions

From table 1 it can be seen that people often use this lake for 72.5% social culture activities, rarely use 22.75% and never use as much as 16.25%. The category of utilization of the socio-cultural aspects of Lake OPI has a high category.

Table 1. Frequency Distribution of Utilization of Lake from Socio-Cultural Pillars

Question	Often		Rarely		Never	
	Frequency	%	Frequency	%	Frequency	%
Enjoy the fresh air from the lake with your family	35	87.5	5	13.5	0	0
Exercise with family / friends	31	77.5	7	17.5	2	5
Communicate / interact with the surrounding community	30	75	6	15	4	16
Enjoy regional snacks / culinary specialties with family / friends	20	50	18	45	2	5
Total	121	290	42	91	7	26
Average	30.25	72.5	10.25	22.75	1.75	16.25

The results obtained from the use of lakes in economic pillars are as follows: people sell food around the lake for income (income), sell services / rental entertainment water bicycle, rental seats around the lake, and sell food. It can be seen from **Table 2**.

Table 2. Frequency Distribution of Utilization of Lake OPI as Addition / Source of Revenue

Question	High		Medium		Low	
	Frequency	%	Frequency	%	Frequency	%
Selling water bicycle services	20	50	14	35	6	15
Rent a seat	17	42.5	11	27.5	12	30
Selling food and drinks	28	70	7	17.5	5	12.5
total	65	165.5	32	80	23	57.55
Average	21.6	54,2	10.67	26.67	7.6	19.17

From **Table 2** above, it can be seen that the use of Lake OPI for the source of income / economy of the community 54.2% supports income, 26.67% is rare, and does not support income 19.17%. For the utilization of the economic aspects of Lake OPI by the community, the percentage of categories in the category is quite good in utilizing for improvement and sources of income for the welfare of the community.

3.1 Results of research obtained from Environmental Conservation for Ecotourism purposes

Based on observations in the field, the following data are obtained. : The community has maintained the cleanliness of the lake and very little garbage can be seen on the lake. They provide trash bins and provide self-cleaning places such as the provision of water for washing hands and feet, toilets, not littering garbage, provide a place for selling culinary traditional and regional beverages and snacks. The community provides water bicycle rental whereas community provides a seat on the edge of the lake. All of that is provided by the community who use it to sell goods / food and services to visitors. The quality of providing facilities is still not optimal, especially the provision of makeshift toilets which is still not well managed.

3.2 Discussion

Swamp has the potential inherent in swamp ecosystems that can be developed. Land of swamp filled with water that continues to be flooded throughout the year, except in shallow swamps during drought [10, 11] Swamp also has the potential to be developed by maintaining swamp fish and plant planting both for household and agricultural needs. Swamp can also be developed into a place of recreation, creativity development, socio-cultural and economic community. In addition, it includes carrying out conservation activities. In the natural environment of the Jakabaring OPI lake swamp in South Sumatra, now, the community can utilize it through activities that are in line with the concept of sustainable development, where the community uses the natural environment for socio-cultural activities, makes new sources of income and employment, and preserves the environment. This is in line with the concept of community-based ecotourism [12], which is to promote involvement in the development of ecotourism in the region. Thus, tourism development (which is sustainable) needs to be supported by careful planning and must reflect three dimensions of interest, namely the tourism industry, environmental carrying capacity (natural resources), and local communities with the aim of improving the quality of life [13]. Environmental management policies as stated in Law No. 32 of 2009 concerning Environmental Protection and Management which instructs that the development carried out must pay attention to the environment called sustainable development as stated in article 1 number 3: "Sustainable development is a conscious and planned effort that combines environmental, social and social aspects , and the economy into development strategies to ensure the integrity of the environment as well as the safety, ability, welfare, and quality of life of present and future generations. Community-based ecotourism patterns are a pattern of ecotourism development that supports and enables full involvement of local communities in the planning, implementation and management of ecotourism businesses and all the benefits gained.

Based on the results of the research, the communities around and the Jakabaring environment have taken advantage of Lake OPI's potential for welfare, especially traders and service providers while the closest people enjoy the beauty of nature and preserve it. Here the active role of the community has gone quite well. It is based on the fact that the community has knowledge of nature and culture which is a potential and selling value as a tourist attraction, so that community involvement is absolute. The community-based ecotourism pattern understands the rights of local communities to manage tourism activities in areas that they have customarily or as managers. Activities at OPI Lake become tourist destinations both local communities and outside the region that have been useful in creating employment opportunities for local people, which reduce poverty, where ecotourism income is from tourism services by providing water bicycle rental services, traditional culinary sales area and other culinary although the community has not been able to provide other services because they have not been able to manage, do not have the knowledge to develop this business, and especially needs supports from the local government. [14] It is suggested that tourism should develop the economic viability of society, namely that tourism development must be able to create employment opportunities for all parties to engage in economic activities through a healthy and competitive economic system. The continuity of social and cultural life, namely that the development of tourism must be able to increase the role of the community in the supervision of living systems through the value system adopted by the local community as the identity of the community.

4. Conclusion

Based on the results of the study it can be concluded that the use of OPI Lake by the community as a whole has been utilized in the social, cultural and economic aspects of ecotourism. In the social cultural aspects of utilization it is grouped into the high category. The use of the income / economic aspects of the community is in the category of quite useful. However, the aspect of developing local tourism potential is still not good. The community uses Lake OPI at the level of the tourist destination and enjoyment. It is proven that tourism activities in the lake are only for the provision of simple services and the provision of simple facilities to support their economic activities. Support from related parties is needed so that the community can increase the acquisition of the benefits for tourism purposes, because it will have an impact on improving community welfare. The community uses Lake OPI for new ecotourism destinations and only to enjoy the view. It is recommended to local governments to help facilitate / support the development of local ecotourism.

References

[1] Sagala. S. D., "Alih Fungsi Lahan Rawa dan Kebijakan Pengurangan Risiko Bencana Banjir: Studi Kasus Kota Palembang). Perencanaan Tata Ruang dan Kebencanaan.," 2013.

[2] Unesco, "Sentralitas Budaya dalam Pembangunan Berkelanjutan . Retrieved February 2019, from Sentralitas Budaya dalam Pembangunan Berkelanjutan," *Unesco*, 2016. [Online]. Available: https://kebudayaan.kemdikbud.go.id/bkborobudur/mengapa-kita-harus-melestarikan-warisan-budaya/ .

[3] D. Z. Ridho, "Kegiatan Agronomis untuk Meningkatkan Potensi Lahan Lebak menjadi Sumber Pangan.," *J. Lahan Suboptimal*, vol. 2, no. 1, pp. 2252–3188, 2013.

[4] Fikri, "Pemberdayaan Masyarakat Melalui Program Ekowisata "Lepen Adventure"," *Pendidikan*, vol. 2, no. 1, pp. 1071–1076, 2017.

[5] D. Suparmini, *Pelestarian Lingkungan Masyarakat Baduy Berbasis Kearifan Lokal*. Jakarta: Universitas Negeri Jakarta, 2012.

[6] Dirjen Parawisata, "Prinsip Dan Kriteria Ekowisata Berbasis Masyarakat. (D. Pariwisata, Editor, D. Pariwisata, Producer, & Dirjen Pariwisata)," *Dirjen Parawisata*, 2009. .

[7] M. Pidharta, *Landasan Kependidikan Stimulus Ilmu Pendidikan Bercorak Indonesia*. Jakarta: Rineka Cipta, 2013.

[8] B. Baswori, *Pengantar Sosiologi*. Bogor: Galia Indonesia, 2005.

[9] Sugiyono, *Metode Penelitian Kuantitatif Kualitatif dan R&D*. Bandung: Alfabeta, 2016.

[10] D. Zoer'aini Irwan, *Prinsip-Prinsip Ekologi: Ekosistem, Lingkungan dan Pelestariannya*. Jakarta: Bumi Aksara, 2010.

[11] Sitanala Arsyad & Erman Rustiandi, *Penyelamatan Tanah, Air, dan Lingkungan*. Jakarta: Yayasan Obor Indonesia, 2008.

[12] B. Rahardjo, *RahardjoEkotourisme Berbasis Masyarakat dan Pengelolaan Sumberdaya Alam*. Bogor: Pustaka Latin, 2005.

[13] M. P. G. John Minnery, "Proceeding on the Training and Workshop On Planning Sustainable Tourism.," in *Workshop on Planning Sustainable Tourism*, 1997, p. 53.

[14] S. James J. Spillane, *Pariwisata Indonesia: Siasat Ekonomi dan Rekayasa Kebudayaan*. Yogyakarta: Kanisius, 1994.

IOP Conf. Series: Earth and Environmental Science **328** (2019) 012003 doi:10.1088/1755-1315/328/1/012003

The Performance of BRT Trans Semarang Services to Achieve Sustainable Transportation

A N Adibah[1*]

[1]University of Muhammadiyah Malang, Malang, Indonesia

*email: amalianuradibah@umm.ac.id

Abstract. Semarang provides its many important activities and movements. This city already has BRT system for public transportation, expected to make a sustainable transportation system in Semarang city. However, there are several problems, especially in service performances. The existence of BRT has not been efficient to resolve the congestion problem in the city, indicated by the low BRT load factor level. Location of bus stops are not strategic and not easy to reach, dirty, cramped, have little seatings, do not have guards, do not have feeder mode from the housing area, and the bus arrival schedule is uncertain, causing people to use private vehicles in their daily life. The main purpose of this research is to assess the service performance of BRT Trans Semarang related to safety, comfort, and regularity to achieve sustainable transportation. Conditions of BRT service performance in regard to safety level, 78% of the provision is appropriate, and 22% is not appropriate. Based on comfort level, 70% of the supply is appropriate, and 30% is not suitable. As for the regularity level, 38% of the supply is appropriate, and 62% has not been matched from the total performance indicators of BRT services that require repairs and additions. Government, thus shall improve the regularity level to enhance the load factor of BRT Trans Semarang.

1. Introduction

Congestion becomes a problem in almost all cities, even in Indonesia. Lots of private vehicles, especially cars, are able to carry a maximum of two people. The more vehicles on the road, the higher congestion that occurs. The increasing of population and size of cities, causes people to travel in greater distances [1]. However, this does not make people prefer to travel by bus compared to private vehicles, eventhough cars and motorcycle have a higher risk on the road, excluding as sustainable means of transportation [2].

Higher population and size of cities indicated the demand to initiate better city planning by improving the accessibility of the city. Increasing mix land use and density of development are also a challenge for sustainable transportation. Land use planning and sustainable transportation are interelinked in facing sustainability issues [2]. TOD is expected as an alternative to stimulate people for shifting modes [3]. TODs can also reduce private vehicle use, and congestion in the. Cities with TOD system lead to density, population, employment opportunities on public transportation, and LOS of the current road road [4].

Many efforts have been made by the government to resolve this problem. One of them is implementation of the BRT system to reduce congestion and to support the goal of sustainable transportation.

BRT system is often preferred, because BRT has fast and efficient services compared to conventional buses, with cheaper rates, emphasizing to maintain security, comfort, and safety of users [5].

Bus Rapid Transit (BRT) becomes one of the popular transportation systems that serves as a more affordable travel alternative compared to trains [6]. BRT has features, such as having large capacity, low costs, and short implementation, which can be an alternative for big cities [7].

The performance of BRT services based on the Road-based Mass Transport Minimum Service Standard [12], is divided into:

A. Safety, consisting of bus stops, bus stop supporting facilities and bus cars
B. Convenience, consisting of bus stops, bus stop support facilities and bus cars
C. Regularity, consisting of waiting time, travel speed, stopping time at the bus stop, service information, bus arrival time information, stop in and out access, information regarding which bus stop that will be passed, accuracy and certainty of bus arrival and departure schedules, information on bus trip disruption, and payment system.

The government of Semarang City also implemented this BRT system to overcome the existing congestion. In 2017, the population in Semarang City was 1,753,092 people, with the length of roads in Semarang City reaching 2,785.28 kms, the number of motorcycles as many as 80% (± 1,252,346 units), private cars as many as 18% (± 33,523 units), and public transportation as many as 2% (± 3,824 units) [13]. This indicated that the high level of population growth accompanied by the increase in the number of motor vehicle ownership. The current condition of public transport, about 90% which is not feasible, still operates [8].

The large number of private vehicles has caused traffic congestion, for example in Jatingaleh, Kaligawe Road, Majapahit Road, Bubakan Roundabout, roads around Diponegoro Tembalang University campus, and Tugu Muda Roundabout and Simpang Lima area which is in Semarang Golden Triangle area. On the other hand, the existence of public transportation does not help much to reduce the volume of traffic density. The limitations of the Trans Semarang BRT route and the absence of a separate special lane for the Trans Semarang BRT lane causes the Trans Semarang BRT rate to be hampered by other vehicles passing in its lane. Thus, the Trans Semarang BRT which should carry the rapid concept cannot be fulfilled [14]. The load factor of BRT Trans Semarang in 2017 is 47,85%, under the Minimum of Standard Service and the service was not optimal [9].

The high number of requests for the Trans Semarang BRT was not matched by the number of available Trans Semarang BRT fleets. Especially during rush hour, many passengers have to wait a long time due to the lack of Trans Semarang BRT mode that serves, as well as the operating schedules that are not in accordance with fluctuations in user demand [8]. The headway of buses is between 20-25 minutes. Location of bus stop are not strategic and not easy to reach [9], dirty, cramped, have little seating, the uncertain bus arrival schedule [10], limited service coverage of the buses, and the absence of feeder modes [11].

The low price of BRT tickets has made Trans Semarang BRT operations demanding subsidies from the government [9]. The budget that must be spent by the Semarang Government in managing the Trans Semarang BRT was IDR 73 billion in 2016 [14]. However, the users feel that the current ticket prices and the services provided by the Trans Semarang BRT are considered quite expensive. Users are willing to pay more if there is an increase in the quality of services provided by the Trans Semarang BRT. Variables that affect user satisfaction are feeders and headways. Feeders are non-BRT public vehicles integrated with the Trans Semarang BRT route [9]. Headway is affected by the difference in travel time between buses and departure interval. Departure variables between buses are influenced by total fleet, affecting the total passenger seating capacity which influences the Trans Semarang BRT load factor [9].

2. Methods

This research utilizes descriptive quantitative method by observing the condition of BRT service performance in the Semarang city, compared to the minimum standard of providing BRT performance. The data that has been collected, then compiled based on each indicators of BRT service performance for each line, and compared to the Regulation of the Minister of Transportation no. 27 of 2015.

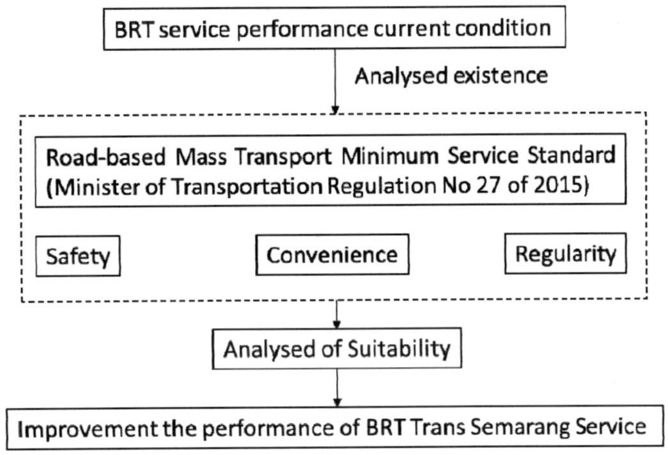

Figure 1. Methodology of the Research Study

3. Results and Discussions

3.1 Service Performance of BRT Trans Semarang

3.1.1 Safety

The assessment for BRT Trans Semarang service performance applies several safety indicators, for the bus stop, such as: lighting, security personnel, and security information. Meanwhile, for the buses themselves, the indicator consists of vehicle identity, driver identification, danger signal lights, lighting, security personnel, and window film.

Table 1. Safety Indicators Characteristic

Indicators		Existence				
		Lane 1	Lane 2	Lane 3	Lane 4	Lane 5
Bus stop	Lighting	√	√	√	√	√
	Security personnel	√	√	√	√	√
	Security information	-	-	-	-	-
BRT fleet	Vehicle identity	√	√	√	√	√
	Driver identification	-	-	-	-	-
	Danger signal light	√	√	√	√	√
	Lighting	√	√	√	√	√
	Security personnel	√	√	√	√	√
	Window film	√	√	√	√	√

Figure 2. One of the Conditions of BRT Trans Semarang Safety Indicators

Figure 3. Suitability of Safety Indicators

Based on the Regulation of the Minister of Transportation no. 27 of 2015 and from the safety indicators, the service performance for the BRT Trans Semarang supply indicated that, 78% of are suitable, and 22% are not suitable with the regulation.

3.1.2 Comfort

The assessment for BRT Trans Semarang service performance utilizes severeal comfort indicators for the bus stop, such as: lightings, temperature regulating facilities/air ventilation, floor area, and comfortable facilities for passengers' entry and exit. Meanwhile, for the buses themselves, the indicators consist of lighting, transport capacity, room temperature control facilites, cleaning facilities, floor area, and smoke prohibiton.

Table 2. Comfort Indicators Characteristic

Indicators		Existence				
		Lane 1	**Lane 2**	**Lane 3**	**Lane 4**	**Lane 5**
Bus stop	Lighting	√	√	√	√	√
	Air ventilation	-	-	-	-	-
	Floor area	√	√	√	√	√
	Entry/Exit facilities	-	-	-	-	-
BRT fleet	Lighting	√	√	√	√	√
	Transport capacity	√	√	√	√	√
	Temperature control	√	√	√	√	√
	Cleaning facilities	-	-	-	-	-
	Floor area	√	√	√	√	√
	Smoke prohibition	√	√	√	√	√

Figure 4. One of the Conditions of BRT Trans Semarang Comfort Indicators

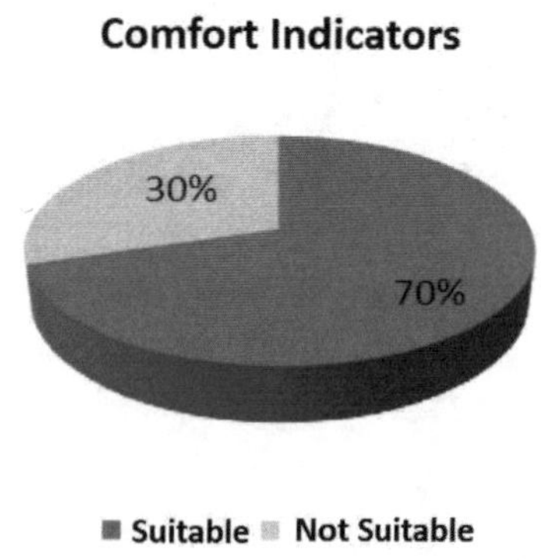

Figure 5. Suitability of Comfort Indicators

Based on the Regulation of the Minister of Transportation no. 27 of 2015 and from the comfort indicators, the service performance for the BRT Trans Semarang supply indicated that, 70% are suitable with the regulation, and 30% are not suitable with the regulation.

3.1.3 Regularity

The assessment for BRT Trans Semarang service performance applies several regularity indicators, where the indicators consist of travel time, travel speed, transit time at bus stop, service information, arrival time information, in and out access, transit information, certainty of schedule, travel disruption information, and payment system.

Table 3. Regularity Indicators Characteristic

Indicators	Existence				
	Lane 1	Lane 2	Lane 3	Lane 4	Lane 5
Travel time	√	√	√	√	√
Travel speed	-	-	-	-	-
Transit time	√	√	√	√	√
Service information	-	-	-	-	-
Arrival time information	-	-	-	-	-
In and out access	√	√	√	√	√
Transit information	-	-	-	-	-
Certainty of schedule	-	-	-	-	-
Travel disruption information	-	-	-	-	-
Payment system	√	√	√	√	√

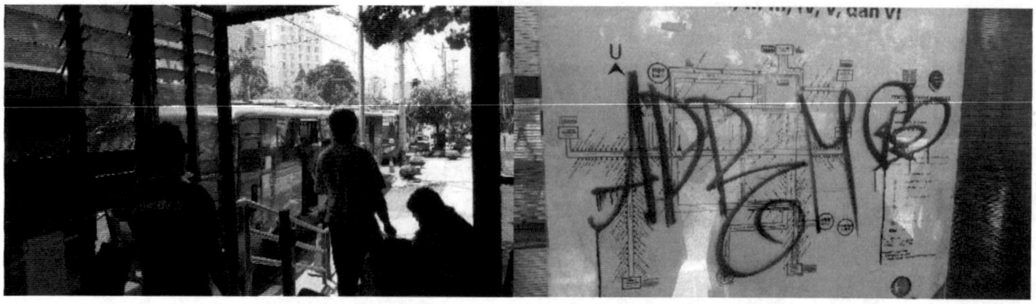

Figure 6. One of the Conditions of BRT Trans Semarang Regularity Indicators

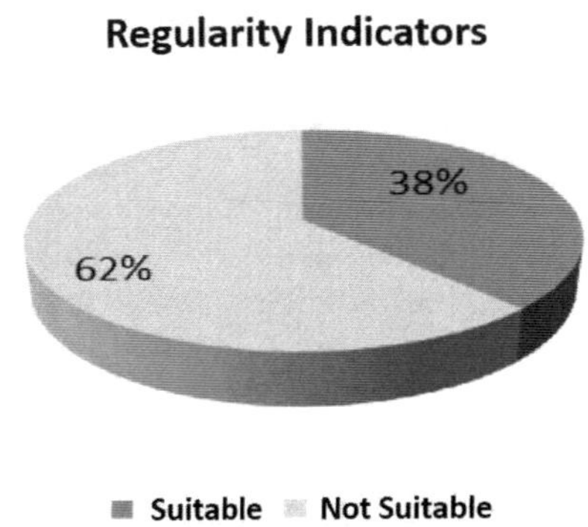

Figure 7. Suitability of Regularity Indicators

Based on the Regulation of the Minister of Transportation no. 27 of 2015 and from the regularity indicators, the service performance for the BRT Trans Semarang supply indicated that, 38% are suitable with the regulation, and 62% are not suitable with the regulation.

4. Conclusion

The assessment is based on problems related to the operational performance of BRT Trans Semarang, and the analytic result related to each influential indicator. For safety indicators it showed that 78% are suitable with the regulation. For comfort indicators, the assessment showed that 70% are suitable. However, for the assessment based on regularity indicators, it showed that only 38% are suitable with the regulation. It means that most of the service performance of BRT Trans Semarang are suitable with regulation. Yet, this condition does not cause many people to change from using private transport to BRT. The low suitability on the regularity indicator becomes a major influence towards the service performance of BRT Trans Semarang.

In addition to the service performance of BRT which deals directly with bus stops or fleets, the certainty of information regarding schedules, transit information, travel disruptions, provides important indicators for the BRT users. If compared to other cities, such as Jakarta, BRT is one of the most popular public transportation alternatives because users can estimate their travel plans, by minimizing the risks that exist. This is due to the timeliness and clear information demanded by the user. By improving the service performance of BRT, it might improve the accessibility and the mobility of the city. The high level of accessibility and mobility becomes one measurement that the city has achieved sustainable transportation.

4. 1 Recommendation

The service performance is related to the operational performance. Problem that occur in operational performance can reduce the quality of BRT Trans Semarang service perfomance. The main problem of BRT Trans Semarang is the low load factor, which can trigger the other problems. If the Government want to increase the load factor, they should consider to improve the main factors that have influence to the service performance. Based on the safety indicators, it is required to add security disturbance information stickers that contain telephone / SMS complaints, as well as to provide a driver's ID containing the driver's name and master number. Based on the comfort indicators, these are needs for additional temperature control facilities / air vents at the bus stop, an equalization of the height between the bus stop and the bus floor, and the addition of trash cans on the bus. Based on the regularity indicators, the recommendation consists of controlling the travel speed, adding service information that includes the name of the bus stop, arrival and departure schedules, routes and corridors, corridor movement, fares, service corridor route maps, and travel disruption information to determine the causes of late bus travel schedules.

References

[1] R. Iles, *Public Transport in Developing Countries*. 2005.

[2] N. S. A. Sukor and N. K. Basri, "Sustainable Transports," in *SpringerBriefs on Case Studies of Sustainable Development*, 2019, pp. 1–4.

[3] B. P. Y. Loo and K. H. Tsoi, "The sustainable transport pathway : A holistic strategy of Five Transformations," *J. Transp. Land Use*, vol. 11 No. 1, pp. 961–980, 2018.

[4] C. Shirke, G. J. Joshi, V. Kandala, and S. . Arkatkar, "ScienceDirect ScienceDirect ' Transit Oriented Development and Its Impact on Level of Service of Roads & METRO : A Case Study of Mumbai Metro Line-I ,'" *Transp. Res. Procedia*, vol. 25, pp. 3035–3054, 2017.

[5] T. K. A. Putra and A. S. Kurnia, "Analisis Preferensi Masyarakat Terhadap Bus Rapid Transit (BRT) Trans Semarang," *Diponegoro J. Econ.*, vol. 3, 2014.

[6] R. Cervero, "Bus Rapid Transit (BRT): An Efficient and Competitive Mode of Public Transport," *IURD Work. Pap. 2013-01*, no. October, pp. 1–36, 2013.

[7] W. Liu and J. Teng, "Analysis and Optimization of the Capacity Bottlenecks of Elevated Bus Rapid Transit System," *Procedia - Soc. Behav. Sci.*, vol. 138, pp. 386–393, 2014.

[8] W. Budiawan, A. Arvianto, and N. P. Adam, "Kajian Awal Simulasi Komputer Model Kebijakan Penentuan Jumlah Optimal Armada Bus Rapid Transit (BRT) Semarang," *J. SIMETRIS Vol 7 No 1*, vol. 7, no. 1, pp. 399–404, 2016.

[9] A. Susanty, S. Nugroho, and K. A. Khantari, "Penyusunan Skenario Kebijakan Untuk Pengembangan Bus Rapid Transit (BRT) Trans Semarang Dengan Pendekatan Sistem Dinamik," vol. 35, no. September 2009, pp. 17–26, 2014.

[10] A. P. Maitri, S. Sulandari, and Rihandoyo, "Kualitas Pelayanan Bus Rapid Trans Semarang (BRT) Koridor II Di Kota Semarang (Dengan Rute Terminal Terboyo Semarang-Terminal Sisemut Ungaran)," pp. 1–11, 2014.

[11] D. Suwandono, D. I. K. Dewi, Mussadun, and P. A, "Optimalisasi Jangkauan Pelayanan Halte Brt / Bus Trans Semarang," *17th FSTPT Int. Symp.*, no. August, pp. 22–24, 2014.

[12] Peraturan Menteri Perhubungan No 27 Tahun 2015. Perubahan Atas Peraturan Menteri Perhubungan No PM 98 Tahun 2013 tentang Standar Pelayanan Minimal Angkutan Orang dengan Kendaraan Bermotor Umum dalam Trayek.

[13] BPS, 2016. Kota Semarang dalam Angka 2016. Katalog 1102001.3374, 3(3), p.636.

[14] Anonimous. 2016. Anggaran BRT Trans Semarang Terus Meningkat. https://jatengekspos.com/anggaran-brt-trans-semarang-terus-meningkat/

Spatial analysis for the selection of landfill in Jepara Regency

A Nugroho[1], Suharyanto[2] and Sudarno[2]

[1] Public Works and Spatial Planning Office of Jepara Regency 59417, Central Java, Indonesia
[2] Master of Civil Engineering of Universitas Diponegoro Semarang 50275, Indonesia
E-mail: [1]gunx10@gmail.com, [2]suharyanto20@yahoo.co.id, [2]sudarno_utomo@yahoo.com

Abstract. There are three landfills in Jepara Regency, which are Bandengan, Gemulung, and Krasak. Bandengan uses a controlled landfill system and is estimated to be full in the next 1.68 years. The Gemulung and Krasak still use an open dumping landfill, and they were closed in 2013 complying the Law No. 18 of 2008. Due to the prolonging problems, it is necessary to plan a new landfill with sufficient capacity. The landfill should have sufficient capacity and meet the standard environmental requirements. The objective of this study is to examine the appropriate zones for landfill in Jepara Regency in line with the standards of SNI 19-3241-1994 and literature review. SNI uses two stages of analysis, the regional level stage, which is determining the regions considered to be feasible, i.e. inappropriate zones and stage of elimination, choosing the best site from the location that emerges from the regional stage. The analysis was conducted using the multi spatial criteria with the aid of Geographic Information Systems (GIS). From this spatial analysis, the same scoring technique was then performed which produced a map of recommendations for the various locations. The results showed that there were six feasible zones for the landfill site covering 129.53 ha or 0.13% of the total area of Jepara Regency. The Keling District was chosen as the best location for landfill among the six feasible zones.

Keywords : landfill location, spatial analysis, GIS (Geographic Information System)

1. Introduction

A landfill is a location for processing and returning waste safely to humans and the environment [1]. The selection of the landfill site is essential due to the enormous potential of environmental pollution and various technical, economic and environmental criteria which must be given proper attention.

In meeting the needs of solid waste facilities, Jepara Regency has three landfills, namely Bandengan, Gemulung, and Krasak. Bandengan is the central Landfill, while Gemulung and Krasak are the supporting Landfills. Bandengan currently uses a landfill system with a total area of 7.13 Ha. This is expected to be filled up in the next 1.68 years [2]. Gemulung landfill has an area of 0.9025 Ha while Krasak is only 0.708 Ha. Both of them use the open dumping system, and they were closed in 2013 according to Law No.18 of 2008. Due to the prolonging problems, it is necessary to plan a new landfill with sufficient capacity. The landfill should have sufficient capacity and must meet the standard environmental requirements.

The very first step of this planning is site selection. Improper site selection will lead to future environmental problems. Furthermore, if the landfill's operational activities conflict with the social interests of the surrounding community, there will be interferences [3]. In fact, to date, site selection activities are often constrained by land readiness issues. Therefore, the best option for the sites is on government lands because land readiness is the main determinant in choosing a development location.

The site selection operation is a long and complicated process. Hence, a method is needed to make the process easier and more straightforward. One method that can be successfully applied is the Geographic Information System (GIS). It can enter, store, integrate, manipulate, analyze, and display geographic reference data. The use of GIS will shorten the analysis time of various parameters of land suitability assessment for precise locations with high data accuracy [4].

Statistics say that it is necessary to conduct a research in determining the feasibility zone and alternative locations that can be recommended as a landfill in Jepara Regency. The results of the study are expected to be given due to the consideration in the preparation of new landfill planning

1.1. Problem Formulation
According to the described background, the problem statement is taken as follows
1. Which zone is suitable for use as a landfill location?
2. Which alternative sites can be recommended as the landfill locations?

1.2. Purpose and objective
The objective of this research is to select the best landfill location in Jepara. Therefore, the site should be environmentally sound and have great adequacy to be used as a facility to return waste to the environment for humans safely. Therefore, this study aims to:
1. Assess the most suitable zone for the landfill location.
2. Review alternative available sites that can be recommended as landfill.

2. Methods
2.1. Research Criteria and Data
This research begins with determining the location selection criteria according to SNI 03-3241-1994 regarding the Procedure for Selecting the location of landfill [5]. This was then followed by a literature review. The criteria based on SNI used in this study are as follows:
1. Criteria in the regional stage

This criterion is used to determine decent or inappropriate landfill zones. It consists of the following:
a. Geological factor; the geological factor consists of 2 criteria, namely 1) Not being located in the fault zone or Holocene fault. It is also not allowed to be located in the geologically dangerous zones or landslide-prone areas.
b. Hydro-geological factors which comprises of 3 criteria, which are 1) Not having a groundwater level of less than three meters, 2) not passing the soil permeability greater than 10^{-6} cm / second, 3) the distance to the source of drinking water must be greater than 100 meters downstream. In the event that there are no zones that meet the criteria above, technological input must be made.
c. Topographical factor, which means that the slope of the zone must be less than 20%
d. The distance from the airfield must be greater than 3,000 meters for turbo jet flights and must be greater than 1,500 meters for other types.
e. Must not be in a protected area/nature reserve and flood area with a 25-year return period.

2. The allowance stage criteria are used to select the best site from the output location generated based on regional criteria. These criteria can be seen in **Table 1**.

As shown in **Table 1**, SNI 19-3241-1994 is a rule used in site selection, and this was made 24 years ago (1994). For reference addition, it is vital to look at the latest literature. The literature review consists of 11 studies on the topic of choosing landfill sites. The study was conducted by 1) Colvero et al., 2018 [6], 2) Khan et al., 2017 [7], 3) Setyarini, 2016 [8], 4) Rahmat et al., 2016 [9], 5) Bahrani et al., 2016 [10], 6)

Torabi-Kaveh et al., 2016 [11], 7) El Maguiri et al., 2016 [12], 8) Djokanovic et al., 2016 [13], 9) Chabuk et al., 2016 [14], 10) Hanine et al., 2016 [15] and 11) Donevska et al., 2012 [16]. The criteria used in the study of landfill site selection are 23 as seen in **Figure 1**.

Table 1. Allowance Criteria for site selection

No.	Parameter
I	**Public**
1	Administrative Limits with Weights 5. The criteria and scores are (1) within the administrative boundary = 10, (2) outside the administrative boundary but in an integrated landfill management system = 5, (3) outside the administrative boundary and outside the management system integrated garbage = 1, (4) outside the administrative boundary = 1
2	Ownership of the land with weight 3. The criteria and scores are (1) local / central government = 10, (2) personal (one) = 7, (3) private / company (one) = 5, (4) more than one owner of rights and / or ownership status = 3, (5) social / religious organization = 1
3	Land capacity with weight 5. The criteria and scores are (1)> 10 years = 10, (2) 5 years - 10 years = 8, (3) 3 years - 5 years = 5, (4) less than 3 years = 1
4	Number of landowners with weight 3. The criteria and scores are (1) one (1) kk = 10, (2) 2 - 3 kk = 7, (3) 4 - 5 kk = 5, (4) 6-10 kk = 3, (5) more than 10kk = 1
5	Community participation with weight 3. The criteria and scores are (1) spontaneous = 10, (2) driven = 5, (3) negotiation = 1
II	**Physical Environment**
1	Land (above ground level) with weight 5. The criteria and scores are (1) the permeability score <10^{-9} cm / sec = 10, (2) the permeability score is 10^{-9} cm / sec - 10^{-6} cm / sec = 7, (3) Permeability score > 10^{-6} cm / sec reject (unless there is technology input) = 0
2	Groundwater with weight 5. The criteria and scores are (1) depth> 10 m with permeability <10^{-6} cm / sec = 10, (2) depth <10 m with permeability <10^{-6} cm / sec = 8, (3) depth = 10 m with a passing of 10^{-6} cm / sec - 10^{-4} cm / sec = 3, (4) Depth <10 m permeability 10^{-6} cm / sec - 10^{-4} cm / sec = 1
3	Groundwater flow system with weight 3. The criteria and scores are (1) discharge area / local = 10, (2) recharge area and local discharge area = 5, (3) recharge regional and local areas = 1
4	Relation to groundwater utilization with weight 3. The criteria and scores are (1) the possibility of low utilization with hydraulic limit = 10, (2) projected to be used with hydraulic limit = 5, (3) projected to be used without hydraulic limit = 1
5	Flooding danger risk with weight 2. The criteria and scores are (1) there is no danger of flooding = 10, (2) the possibility of flooding> 25 years = 5, (3) the possibility of flooding <25 years reject (unless there is technology input) = 0
6	Soil cover with weight 4. The criteria and scores are (1) sufficient soil cover = 10, (2) soil cover up to half the age of use = 5, (3) No. soil cover = 1
7	Rainfall intensity with weight 3. The criteria and scores are (1) below 500 mm per year = 10, (2) between 500 mm to 1000 mm per year = 5, (3) above 1000 mm per year = 1
8	Road to location with weight 5. The criteria and scores are (1) flat with good conditions = 10, (2) flat with bad conditions = 5, (3) up / down = 1
9	Waste transport (one road) with weight 5. The criteria and score are (1) less than 15 minutes from the garbage centroid = 10, (2) between 16 minutes - 30 minutes from the garbage centroid = 8, (3) between 31 minutes - 60 minutes from the centroid garbage = 3, (4) more than 60 minutes from the centroid garbage = 1
10	Entrance with weight 4. The criteria and scores are (1) garbage trucks not through residential areas = 10, (2) garbage trucks through medium density residential areas (≤ 300 people / ha) = 5, garbage trucks through high density residential areas (≥ 300 people / ha) = 1
11	Traffic with weight 3. The criteria and scores are (1) located 500 m from public roads = 10, (2) located <500 m in low traffic = 8, (3) located <500 m in medium traffic = 3, (4) located in high traffic = 1
12	Land use with weight 5. The criteria and score are (1) having a slight impact on surrounding land use = 10, (2) having a moderate impact on surrounding land use = 5, (3) having a large impact on land use around = 1
13	Agriculture with weight 3. The criteria and scores are (1) located on unproductive land = 10, (2) there is no impact on agriculture around = 5. (3) there is a negative influence on agriculture around = 1, (4) Located on productive agricultural land = 1
14	Protected area / nature reserve with weight 2. The criteria and scores are (1) there is no protected area / nature reserve around it = 10, (2) there is a protected area / nature reserve around it which is not negatively affected = 1, (3) There is a negative effect for protected area / nature reserve around it = 1
15	Biological with weight 3. The criteria and scores are (1) low habitat value = 10, (2) high habitat value = 5, (3) critical habitat = 1
16	Noise, odor with weight 2. The criteria and score are (1) there is a buffer zone = 10, (2) there is a limited buffer zone = 5, (3) there is no buffer zone = 1
17	Aesthetics with weight 3. The criteria and scores are (1) the hoarding operation is not visible from the outside = 10, (2) the hoarding operation looks slightly from the outside = 5, (3) the hoarding operation looks from the outside = 1

Source: *SNI 03-3241-1994*

Of the 23 criteria in **Figure 1**, there are 6 criteria that have a percentage of more than 50%. These criteria are 1) distance to road, 2) slope of the land, 3) distance to a residential area, 4) land use, 5) depth of groundwater, 6) distance to the water source. This will be included in the regional stage criteria issued by SNI to add relevance to the situation and conditions in the election. Of the 6 criteria, there are 3 that are the same as those in regional stages, namely 1) distance to rivers and water bodies, 2) depth of groundwater

IOP Conf. Series: Earth and Environmental Science **328** (2019) 012004 doi:10.1088/1755-1315/328/1/012004

level and 3) slope of the land. Only three criteria are based on the literature review added at the regional stage, namely 1) distance to the road, 2) distance to settlements and 3) land use. Hence this study established 34 criteria for site selection. They were then grouped into two, namely 12 regional criteria and 22 criteria for elimination.

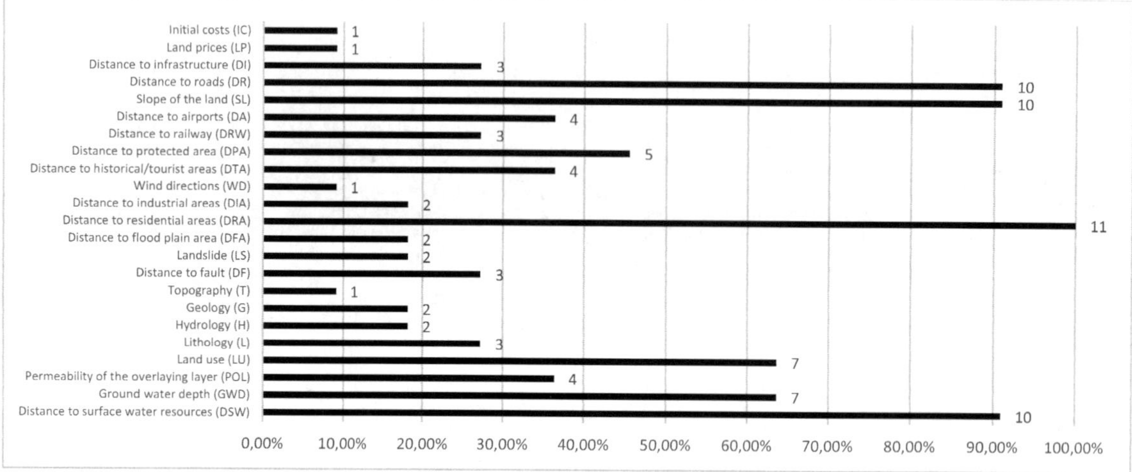

Source : Colvero et al., 2018, Khan et al., 2017, Setyarini, 2016, Rahmat et al., 2016, Bahrani et al., 2016, Torabi-Kaveh et al., 2016, El Maguiri et al., 2016, Djokanovic et al., 2016, Chabuk et al., 2016, Hanine et al., 2016 and Donevska et al., 2012

Figure 1. Amount and Percentage of Use of Landfill Site Selection Criteria

2.2. Stages of Analysis

After the relevant results were collected, data analysis was carried out. The method used was multi-criteria spatial analysis. It included the analysis of spatial data performed on many criteria using ArcMap software. Some data required preceeding process to fit the existing criteria which will eventually produce new data and the information is referred to as geoprocessing (spatial analysis). The spatial analyses used consist of buffering, overlay, query, interpolation, and the shortest path.

After obtaining the spatial analysis data, this result can be directly examined at the regional stage. At this stage, the analysis was performed on 12 criteria according to SNI and literature review, namely 1) distance to faults, 2) landslide-prone zones, 3) distance to airports, 4) protected area zones, 5) flood-prone zones, 6) soil permeability, 7) distance to river, 8) depth of groundwater table, 9) slope of land, 10) distance to road, 11) distance to settlements and 12) land use. The twelve criteria are overlaid which produce a zone map of a possible area that is not feasible for landfill. Analysis at the regional stage raised the output of prospective landfill sites. From those candidates, a report is needed at the stage of elimination. Therefore, the best output alternative location can be produced. The elimination stage has 22 criteria that will be analyzed through scoring. The total score in the allowance stage was used as the final score in ranking alternative landfill locations. The ranking was made up of alternative sites sorted from the highest to the lowest scores.

3. Results and Discussions

3.1. Regional Level Analysis

Regional stage analysis was conducted to discover the best location corresponding to SNI 03-3241-1994 and literature review. In the regional stage, spatial analysis was performed on 12 criteria as stated at the analysis stage. The 12 criteria in the overlay produced a suitable and unsuitable area zone map. To facilitate the analysis, the 12 analysed unsuitable zones were combined using union techniques. Hence, a combined zone map presented the unsuitable locations. The map of the small zone was then overlaid again with the official map so that an empty zone that was not covered by the unsuitable zone map appeared. The empty zone is a decent zone for the landfill location. The analysis process at the regional stage can be seen in **Figure 2.**

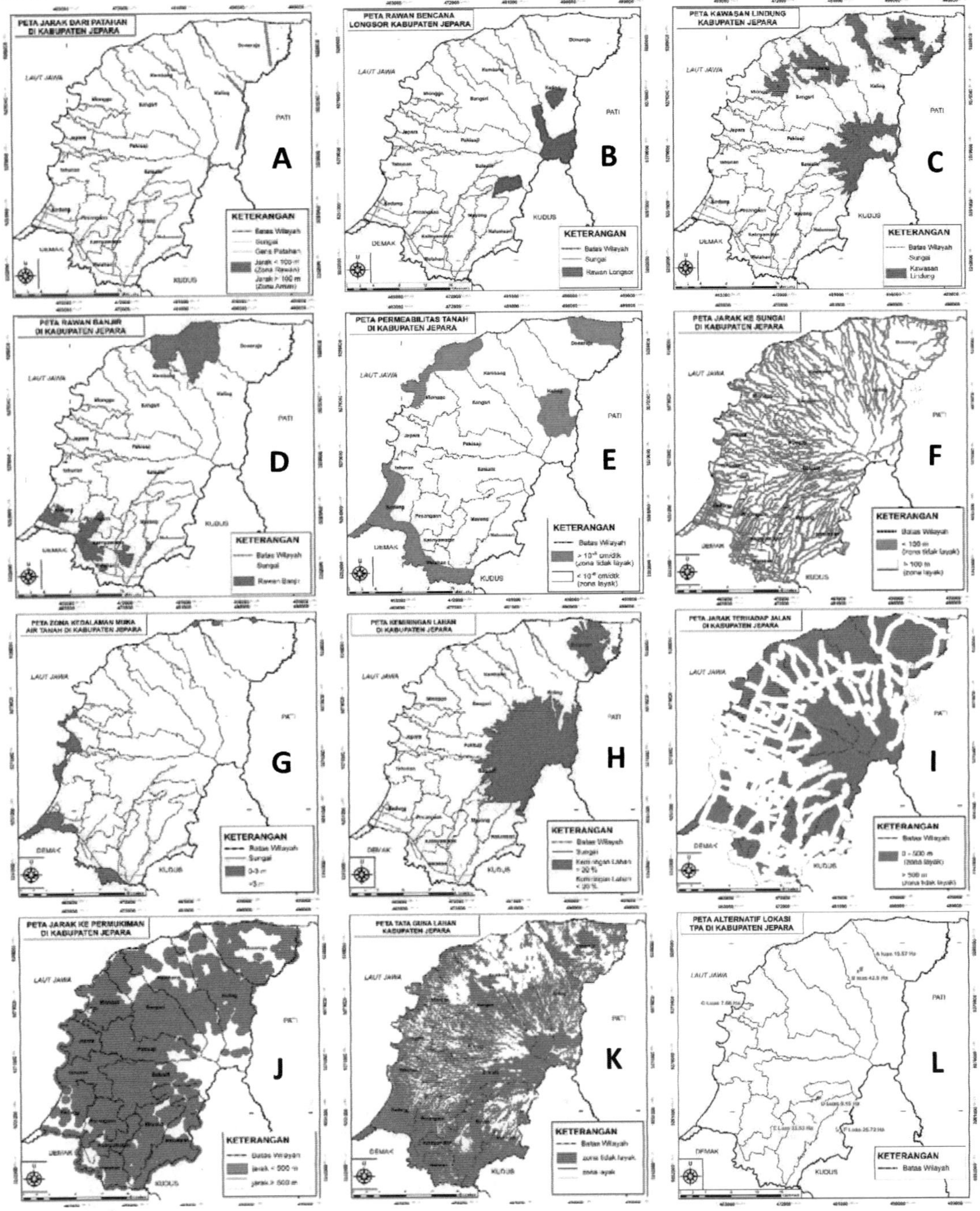

IOP Conf. Series: Earth and Environmental Science **328** (2019) 012004 doi:10.1088/1755-1315/328/1/012004

Source: Analysis of The Researcher. Information: A. Distance with Fault Map, B. Landslide Hazard Map, C. Map of Protected Areas, D. Flood-Prone Map, E. Soil Permeability Map, F. Distance to River Map, G. Map of Groundwater Face Depth Zone, H. Slope Map, I. Distance for Roads Map, J. Distance to Settlements Map, K. Land Use Map, L. Regional Phase Analysis Output Maps.

Figure 2. Analysis process of regional stages

The various locations in the feasible zone were selected as alternative points. The area of the material is one of the most essential criteria that can be used as the material for consideration in choosing an alternative. The area of a polygon in a decent zone that is quite spacious can be chosen as an alternative, while the small ones can be ignored. Of the 12 polygons in the feasible zone, there are 6 locations which are considered to be quite spacious (having an area of more than 5 hectares) and can be chosen as the alternative location. The location points chosen and determined as alternative locations can be seen in **Figure 2**. The calculator field calculated the land area of each alternative location in the table of alternative location layer attributes. From these calculations, it was known that the entire feasible zone for alternative landfill sites is 129.53 Ha or 0.13% of the total area of Jepara Regency. The output from the allowance stage analysis can be seen in **Figure 3**.

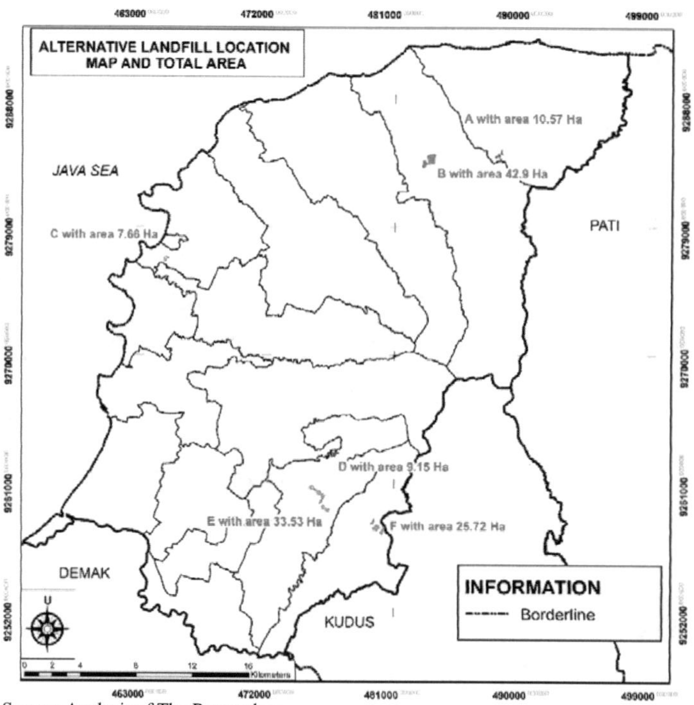

Source: Analysis of The Researcher

Figure 3. Alternative landfill location

3.2. Analysis of the stage of removal

Analysis of the allowance stage was performed to produce the most suitable location. The data on 6 alternative locations that have been generated at the regional stage was used. The elimination stage has 22 criteria that will be used for analysis. Each criterion has a different score and weight by SNI. At this stage, data has been processed and was classified according to their weight and value. This was carried out by scoring as a whole by multiplying the weight with the score of each criterion. The data obtained from the analysis are the scores of each alternative landfill location with a maximum total score of 790. The scores of each alternative result of the analysis can be seen in **Table 3**. The score can be calculated for finding the suitability of the alternative locations. The results of this scoring are used to see the level of location feasibility. From **Table 3** it can also be seen that the most suitable alternative location is location B with a score of 561 and a suitability level of 71%. Conversely, alternative locations with the lowest level of suitability (53%) are A with a score of 420. According to this rating, the most feasible alternative location recommendations appeared as the location B in Keling Village, Keling District, location E in Datar, Pule and Buaran Villages, Mayong District and location F in Ngetuk Village, Nalumsari District.

Table 2. Assessment of alternative landfill locations at the stage of elimination

No.	Criteria	Ideal score	alternative Location					
			A	B	C	D	E	F
B	Stage of Allowance							
I.	GENERAL							
1	Administrative Limits	50	50*	50*	50*	50*	50*	50*
2	Ownership of land	30	9	15*	9	9	9	9
3	Land capacity	50	5	25*	5	5	25	25*
4	Number of land owners	30	3	30*	3	3	3	3
5	Society participation	30	15*	15*	15*	15*	15*	15*
II	PHYSICAL ENVIRONMENT							
1	Soil	50	35*	35*	35*	35*	35*	35*
2	Groundwater	50	40	40	40	50*	50*	40
3	Groundwater flow system	30	30*	30*	30*	15	15	15
4	Relation to groundwater utilization	30	15	15	15	30*	30*	30*
5	Flood Hazards	20	20*	20*	20*	20*	20*	20*
6	Land cover	40	20*	20*	20*	20*	20*	20*
7	Rain intensity	30	3*	3*	3*	3*	3*	3*
8	Road to location	50	25	50*	50*	5	50*	50*
9	Garbage transport	50	15	40*	40*	40*	15	15
10	Entrance	40	20*	20*	20*	20*	20*	20*
11	Traffic	30	24*	3	24*	24*	24*	24*
12	Land use	50	25	50*	25	25	25	25
13	Agriculture	30	3	15*	3	15*	15*	15*
14	Protected Area	20	20*	20*	20*	20*	20*	20*
15	Biological	30	30*	15	30*	15	15	15
16	Noise, smell	20	10	20*	10	20*	20*	20*
17	Aesthetics	30	3	30*	3	30*	30*	30*
	Scoring Stage Score	790	420	561	470	469	509	499
	Allowance Stage Rating		6	1	4	5	2	3
	Level of Convincing Stage Suitability		53%	71%	59%	59%	64%	63%

Source: Analysis of The Researcher
Description: * sign shows the highest value among alternative landfill locations

4. Conclusion

This research was carried out to discover the most suitable landfill location in Jepara Regency as a substitute for the existing landfill. The criteria used in this research are by SNI 03-3241-1994 regarding the Procedure for Selection of Landfill Waste Sites and literature review. Hence, site selection analysis used a multi-criteria spatial analysis method, with a geographic information system device in the form of ArcGIS. The feasible zone for the landfill site has an area of 129.53 ha or 0.13% of the total area of Jepara Regency. The recommended location, ranked based on the scoring presented a result where the first location is Keling Village, Keling Subdistrict, the second location is in Datar, Pule and Buaran Villages, Mayong District, and the third location is in Ngetuk Village, Nalumsari District.

IOP Conf. Series: Earth and Environmental Science **328** (2019) 012004 doi:10.1088/1755-1315/328/1/012004

Reference

[1] Undang-Undang Republik Indonesia Nomor 18 Tahun 2008 Tentang Pengelolaan Persampahan. Jakarta

[2] Satker PPLP Provinsi Jateng 2016 *Perencanaan Teknis dan Manajemen Persampahan (PTMP) serta Rencana Teknis Rinci (RTR) Sistem Penanganan Persampahan Kabupaten Jepara Tahun 2016* (Semarang)

[3] Kementerian Pekerjaan Umum 2013 *Bahan Ajar Diseminasi dan Sosialisasi Keteknikan Bidang PLP Sektor Persampahan* (Jakarta)

[4] Mizwar A 2012 Penentuan Lokasi Tempat Pengolahan Akhir (TPA) Sampah Kota Banjarbaru Menggunakan Sistem Informasi Geografis (SIG). *EnviroScienteae Agustus 2012*:16-22

[5] Badan Standarisasi Nasional 1994 *SNI 19-3241:1994 tentang Pedoman Pemilihan Lokasi Tempat Pembuangan Sampah.* (Jakarta)

[6] Colvero D A, Gomes A P D, Tarelho, L A D C , Matos M A A D, Santos K A D 2018 Use of a Geographic Information System to Find Areas for Locating Of Municipal Solid Waste Management Facilities *Waste Management* **77** 500-515.

[7] Khan, M M U H, Vaezi M, Kumar A 2017 Optimal Siting of Solid Waste-To-Value-Added Facilities through a GIS-Based Assessment *Science of the Total Environment* **610-611** 1065-1075.

[8] Setyorini E 2016 *Analisis Spasial Multi Kriteria Untuk Pemilihan Lokasi TPA di Kabupaten Semarang* (UNDIP Semarang)

[9] Rahmat Z G, Niri M V, Alavi N, Goudarzi G, Babaei A A, Baboli Z, Hosseinzadeh M 2016 Landfill Site Selection Using GIS and AHP: a Case Study: Behbahan, Iran *KSCE Journal of Civil Engineerin* **21 (1)** 111–118.

[10] Bahrani S, Ebadi T, Ehsani H, Yousefi H, Maknoon R 2016 Modeling Landfill Site Selection BY Multi-Criteria Decision Making and Fuzzy Functions in GIS, Case Study: Shabestar, Iran *Environmental Earth Sciences* **75 (4)** 337.

[11] Torabi-Kaveh M, Babazadeh R, Mohammadi S D, Zaresefat M 2016 Landfill Site Selection Using Combination of GIS and Fuzzy AHP, a Case Study: Iranshahr, Iran *Waste Management Research.* **34 (5)** 438–448.

[12] El Maguiri A, Kissi B, Idrissi L, Souabi S 2016 Landfill Site Selection Using GIS, Remote Sensing and Multicriteria Decision Analysis: Case of The City of Mohammedia, *Morocco Bulletin of Engineering Geology and the Environment.* **75 (3)** 1301–1309.

[13] Djokanovic S, Abolmasov B, Jevremovic D 2016 *GIS* Application for Landfill Site Selection: a Case Study in Pančevo, Serbia *Bulletin of Engineering Geology and the Environment* **75 (3)** 1273–1299.

[14] Chabuk A, Al-Ansari N, Hussain H M, Knutsson S, Pusch R 2016 Landfill Site Selection Using Geographic Information System and Analytical Hierarchy Process: A Case Study Al-Hillah Qadhaa, Babylon, Iraq *Waste Management Research* **34 (5)** 427–437.

[15] Hanine M, Tikniouine O, Boutkhoum A, Agouti T 2016 Comparison of Fuzzy AHP and Fuzzy TODIM Methods for landfill Location Selection *SpringerPlus* **5** 501.

[16] Donevska K R, Gorsevsk P V, Jovanovski M, Pesevski I 2012 Regional Non-Hazardous Landfill Site Selection by Integrating Fuzzy Logic, AHP And Geographic Information Systems. *Environmental Earth Sciences* **67** 121–131.

IOP Conf. Series: Earth and Environmental Science **328** (2019) 012006 doi:10.1088/1755-1315/328/1/012006

Towards Sustainable Transportation: Identification of Development Challenges of TOD area in Jakarta Metropolitan Area Urban Railway Projects

A Rosalin[1,2]*, B Kombaitan[2], D Zulkaidi[2], P Dirgahayani[2] and I Syabri[2]

[1] Doctoral Candidate of Regional and City Planning, Bandung Technology Institute
[2] Department of Regional and City Planning, Bandung Technology Institute

*email: astriro@yahoo.com

Abstract. Transit-Oriented Development (TOD) becomes one of sustainable transportation concepts that have been implemented around the world to overcome the transport problems in certain area. This paper aims to identify and to explain general challenges encountered by Jakarta Metropolitan Area (JMA) in the process of adopting TOD. The study is expected to explain the local context discourse and important elements of the TOD area development. The study utilizes a qualitative approach, by applying both primary and secondary sources. The primary data was obtained from in-depth interview in 4 (four) Focus Group Discussions (FGD) conducted from October 2017 to July 2018. The secondary data was acquired from literature reviews, institution reports, and existing policy review. The study observed 5 (five) TOD project currently developed by the private and public sector, such as TOD Dukuh Atas, TOD Ciracas, TOD Cikoko, TOD Pasar Senen, and TOD Tanjung Barat. It is found that the challenges of developing TOD in the case studies comprise 4 (four) important elements, such as regulation, planning and design of TOD, institutional issues, and funding. In this early stage of developing TOD, Indonesia, specifically JMA, still has ample in understanding the challenges of implementing TOD.
Keywords: development challenges, Transit-Oriented Development, Jakarta Metropolitan Area

1. Introduction

1.1. Background

The concept of sustainable transportation becomes an important issue due to its relation to supply, demand, and resource management in the future. One of the concepts of sustainable transportation that has been implemented in major cities around the world is Transit-Oriented Development (TOD). This concept rests on the efficiency of mobility costs, income from property, health improvement, reduction the risks of traffic accidents, improvement of mobility for non-motorists, and decrease of energy consumption as well as adverse environmental impacts [1]. Indonesia, especially in some large cities, has adopted this concept as part of the transportation and spatial planning. The urban railway projects such as Mass Rapid Transit (MRT), Light Rail Train (LRT), and commuter rail in Jakarta Metropolitan Area (JMA) are connected to the development of TOD area projected as an integrated solution of public transportation problems, particularly for commuting journey to its satellite cities (Bogor, Depok, Tangerang, and Bekasi).

However, as an adopted concept, the planning and implementation of TOD development requires an established and well-defined policies and regulations according to the local context of a country. The process of policy adoption has the context-specific aspect and policies from different context which cannot entirely be adopted identically. Indonesia adopts TOD policy through studying the lesson of successes and failures in other countries, especially from the developed countries. The approach taken by the Government of Indonesia, in fact also occurs internationally in the context of transferable policy. In the Indonesian context, it is argued that the lack of implementing regulations and the absence of a pioneer institution leading the planning process become the two main causes of this phenomenon. Thus, policymakers tend to look for examples from other countries and to emulate policies that have been successfully adopted in their countries [2, 3, 4]. Recognizing that the policy adoption process is different from the country being emulated, understanding the specific context remains important to note. This article presents the general challenges faced by Jakarta and surrounding cities in planning the development of a TOD-based area as an effort to introduce a more specific context. It is hoped that, by understanding the challenges that arise, the development of the TOD area in the Metropolitan area of Jakarta and in Indonesia may consider these aspects more thoughtfully.

The article will be divided into 4 parts, including introduction, literature review, results and discussion and conclusions. The literature review contains a study of regulatory, planning/ design, institutional and funding aspects in developing TOD. After reviewing theory and regulation, the discussion will focus on the results and discussion consisting of challenges at the stage of regulation formulation, planning/ design, institutional and financing, concluded with recommendations.

1.2. Aims and objectives

This paper aims to identify and explain general challenges encountered by JMA in the planning process of adopting TOD. This study is expected to provide clearer direction to the government and other involved stakeholders, both from the public and private sectors, regarding the context-specific discourse and important elements of the TOD area development.

1.3. Literature review

In analyzing the theoretical point of views, the paper utilizes the findings from the study regarding the challenges in creating TOD [9], particularly in developing countries. From the varied factors, it is summarized that the challenges are classified into regulation, planning and design, institutional framework, and financing mechanism. Hence, the following literatures are referred to those four aspects. To build literature foundation, notable and important articles are gathered, synthesised, and eventually compared with the empirical cases in the result section.

1.3.1 Aspect of regulation in TOD area development

Some concerned literatures indicated that there was no specific regulation or standard applied to the TOD area, due to its innovative, contextual and growing concept. Yet, there have been general rules in understanding the thought. For instance, a regulatory concept emphasizes collaboration between levels of government, cooperation between sectors, and accommodation of all segments of society, especially in the marginalized one [9]. From this point, it can be concluded that the regulatory aspect was inseparable from the actor, such as the government apparatus and the purpose of the plan. Consequently, regulations will involve those who make the rules, regulate the subjects and produce the rules.

Supportive regulatory policies and tools will encourage the successful development of a TOD area, especially in terms of the integration of transportation and land use policies. The successful integration of land use and transit systems is visible in cities such as Hong Kong, China, Tokyo and Curitiba marked increases in local government revenues through property development and air rights

[9] due to land use and mass transportation integration. The success schemes are originated from a learning process and understanding of vision and long-term goals of TOD concept. This process requires the involvement of all actors and will be affected by the planning of a TOD area, including risk management strategies to deal with potential pitfalls or future mistakes [4]. Instrument rules about TOD are flexible and adaptable in the situations which accommodate the needs of each actor [3].

1.3.2 Aspect of planning and design in TOD area development
Aside from the integration between transportation and facilities, the design of the area becomes an important means for marketing value. With a well-planned area, it is expected that the community can contribute to change their lifestyles by using mass transportation and facilities around TOD more frequently. Furthermore, an-800-meter radius concept to plan compactness of the area enables proximity around mixed-use facilities towards the transit point. This can be done by combining different functions of activities, such as varied housing type, retail area, offices, open spaces, and other facilities aiming to provide comfort for residents and workers in using several modes of transport. In general, the development of transit points will strive to optimize the development of the area around the Rapid Transit System station through the TOD approach. The concept of TOD places transit facilities as the center of the development of the surrounding area, not only limited to the means of mass public transportation [5, 6, 7].

Theoretically, there are several TOD typologies that have been defined. Cervero divides TOD based on the stages of development, such as: new TOD, high-density TOD, and low-density TOD [6]. The characteristics of the "new TOD" are distinguished by the development of areas along with high quality and high passenger capacity corridors of public transportation. In DKI Jakarta, this type of TOD is apparently found out in the development of MRT transportation facilities with several stations along the corridor. The concept produces a hallway and is often represented as a 'node or set of nodes in the corridor', describing the development area within an 800-meter radius around the transit station. The function and density of potential development areas within a circle are differentiated from the structure outside the existing urban area. In a regional or metropolitan scale, these nodes or circles are intertwined as an infrastructure path referred as 'pearls on a necklace' [8].

The other typologies of TOD are high-density TOD and low-density TOD based on the existing conditions of the built environment. The development of TOD in high-density areas is more focused on diversification and TOD design, while in low-density areas the development of TOD is emphasized with the improvement of public transportation facilities to increase population density. Both of these concepts have been unconsciously developed by the DKI Jakarta government and its satellite cities. For the high-density, TOD example can be seen by the development of TOD Dukuh Atas located in a densely populated area with 5 modes of transportation. Whereas, low-density TOD is described as TOD in suburban environments that only consist of 1 or 2 types of public transportation. This development was seen in the construction of the Ciracas TOD.

1.3.3 Aspect of institution in TOD area development
Institutions and regulations in developing TOD are inseparable. It is revealed that institutions in TOD involve government structures, institutional governance, government initiatives (planning regulations, transit mode investments, and tax policies) [9]. Thus, the institution does not only mean institutions / organization, but also the collaboration between institutions and planning. Apparently, it is certain that institutions overshadow the actors involved, in a form of cooperation where institutions that play a role formulate policies related to TOD. In addition, the relationship between the government and the market (market) is obviously important, since the development of TOD cannot be separated from the effort to exploit land to gain profits.

The implementation of a TOD is generally pursued through cooperation between the government and the private sector as a transit agency, referring to the TOD project financing mechanism [10]. Further, the complexity of the processes that occur in the development of TODs makes achieving agreement between the actors as a matter that greatly influences the success of development. One way that can be done to minimize frictions that occur is to develop and implement strategic and coordinative policies [11]. In practice, coordinative planning is closely related to the institutional design to be implemented, where procedures, organizational structures, and regulations, encourage or prohibit an action in accordance with the objectives achieved. Referring to this understanding, it is obvious that the institutional aspects are specific but comprehensive. Specific means touching to a specific domain at each stage of planning, comprehensive that the whole process has one big umbrella which becomes a reference for more detailed stages.

In the context of practice, efforts to achieve city goals and the development of TOD cannot be separated from political aspects. It is explained that policies, regulations, norms and beliefs are seen as limit based on agreements between actors normatively and substantively [12]. While, some argue that the institutional form of the TOD area can be done by independent body, alike in Hong Kong carried out by forming an independent body where most of its shares are owned by the government. They are in charge of managing all stages of the TOD development process, starting from planning, implementing and maintaining business continuity [13].

1.3.4 Aspect of funding and financing in TOD area development

Questions in terms of funding and financing for a development are: who pays for whom? who is financing the provision of urban public transportation, and for whom? or who receives benefits from the facility? Even decision makers and finance agencies are eager to know who can afford to finance and have to finance urban public transportation, including: users, governments, companies, or business people. How should variations in funding sources be distributed and allocated to various modes of transportation? What is the most appropriate source of financing for sustainable development? How to mobilize it?

The question about urban transport is whether urban public transportation is included in public goods. Public goods are understood as goods available for consumption by everyone, without limiting who pays / finances, with non-excludable and non-rivalrous properties [14]. In the debate on this matter, a scholar describes that urban public transportation is clearly defined as a public service facility for broad public use, including the transit category and specifically the paratransit category [15].

The price or amount of financing will depend on the mode of transportation chosen. However, whatever the choice, the key lies on how to build a long-term financing condition for investment (infrastructure and rolling stock) and operations (operations, maintenance, and replacement). Therefore, all costs for long-term use must be taken into account in the strategy of choice.

However, due to the large amount of funding, conflicts usually occur to provide other public services such as health, education, access to clean water and sanitation, communication, culture, which must also be fulfilled. Therefore, it is important to consider a new mechanism of finance involving players other than the government, such as the investors from private sector, including technology providers and operators.

2. Methods

The study applies a qualitative approach, and the data is collected through primary and secondary sources. The primary data were obtained from in-depth interviews of 26 (twenty-six) interviewees from the public and private sectors. The interviews were conducted by semi-structured questions. Besides, the author also participated in 4 (four) Focus Group Discussions (FGD) from October 2017 to July 2018. Afterwards, the interview and FGD results were delivered in transcript manuscript before analyzed through thematic analysis. In doing so, the author listed a set of keywords then analyzed the result by matching the content with the keywords from interviewees' statements. To complement the data, a series of resources from literature reviews, institution reports, popular media, and existing policy reviews were acquired. The locus of study includes the 5 (five) TOD projects currently developed by the private and public sector, such as TOD Dukuh Atas, TOD Ciracas, TOD Cikoko,

TOD Pasar Senen, and TOD Tanjung Barat. In explaining the result, the study applies explanatory descriptive study in order to describe the cases empirically.

3. Results and Discussions

This chapter will explain the findings related to the challenges in the TOD planning process obtained from the results of interviews with different stakeholder groups, divided into 2 major groups, consisting of: transit agency and regional / central government, which answered questions about the regulatory, design, institutional, and financing frameworks in working on the TOD area. The results will be delivered in 4 aspects, including: regulations, planning and design, institutional and financing, which are then elaborated on by the author's thoughts.

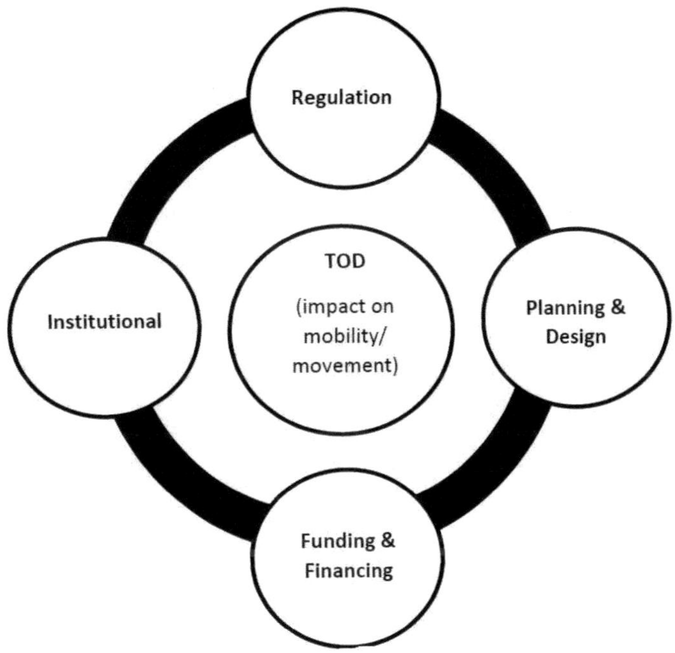

Figure 1. Challenges in developing TOD in Indonesia

In general, this study identified challenges of TOD development process that are divided into 4 (four) important elements, such as: the element of regulations, the element of planning and design of TOD area, the element of institutional issues, and the element of funding and financing.

3.1. Challenges in regulatory aspect

In general, stakeholders' views on TOD regulations are different in terms of tasks and obligations. In terms of the transit agency, the TOD regulations already existed but were not optimal. "*The TOD regulations already exist, but they are still not optimal. Regulations related to FAR are still largely unclear*" (PT MRT) while existing regulations *"are still not maximal, and they should still be developed"* (PT Adhi Karya). At present, the regulations related to TOD are still in the form of normative regulations and not yet operational, such as Minister of ATR Regulation No. 16 of 2017 concerning Guidelines for the Development of Transit Oriented Areas, with regard to incomplete regulations, excluding the scope of authority of the transit agency and an obligation of the central government. At least, one of the institutions expressed his views, *"we do not have the authority to issue policies. With regard to regulations made by other parties, we do not pay too much attention"* (PT KAI). Some of the above findings explain that inadequate regulations have become a significant challenge for transit agency and potentially hinder the TOD development process.

3.2. Challenges in planning and design aspect

Based on the results of interviews conducted, it was found that the actors' understanding of the TOD planning concept were various. For stakeholders who already have experience as developers, they value TOD as an area. Thus, the TOD planning built by the actors also processes the area around the station points into high-value strategic functions, such as mixed-use, residential, commercial, and others. Although in practice, the cultivated area is not exactly 400 or 800 meters but the embryo of regional development that is integrated with the transit point is visible. On the other hand, the TOD Dukuh Atas developed by PT MRT is still limited to the construction of transit stations.

The spokesman of PT MRT said that, "*MRT has authority related to the intensity of buildings, but does not have land that can be built*". Besides the source's limitation, another agent (PT Jakpro) also said that, "*The challenge in planning is due to an assumption that operators can manage everything related to existing land, yet the abundant makes the coordination is sometimes unclear*." The unclear regulation also perceived by Perumnas confirming that "*The absence of clear regulations made it difficult to synchronize station design. Many parties offered strange designs and developments*".

The absence of detailed regulation also leads the design process with a non-standard facility, especially the lack of placemaking design in a radius of 400 m for pedestrian or for other non-motorist' mobility and accessibility to railway stations.

3.3. Challenges in institutional aspect

In the institutional aspect, the absence of agents who lead the overall development of the TOD (leading institution) caused the parties involved to wait for instructions. The absence of an agent appointed by the government as the holder of authority in regulating the TOD also has an impact on the absence of full authority holders in maintaining TOD as a sustainable business cycle. As clarified in the interview, the role of each stakeholder was diverse, which depended on the project. PT MRT said that," *Jakarta MRT was appointed as the main operator with the second Governor Regulation, namely Governor Regulation 140 in 2013. The MRT acts as an extension of the government's involvement in managing the TOD area*". While, for PT Adhi Karya, "*Initially, Adhi planned to become the initiator of the TOD project. However, until now Adhi has been appointed as a provider of LRT infrastructure using design and built mechanism*". In this regard, PT Adhi was only appointed as provider of LRT infrastructure and was not appointed as TOD developer or manager.

The unclear constraint and role in the project also have an effect on the market mechanism where investors / business people have the lack of confidence in the political climate that is not yet established. Another basic thing is the absence of regulations that govern the process of developing TOD in detail, resulting in the rise of negotiations and improvisations between actors to facilitate development. This practice is already familiar in Indonesia, but the resulting impact weighs heavily on developers because it can increase transaction costs.

In the context of the institutional aspect, the absence of a leading institution causes a conflict of interest in which each agency acts in their own interest without optimal output. In the long run, it may stimulate the undeveloped market mechanism where the business actor is reluctant to invest and the available source remains idle.

3.4. Challenges in funding and financing aspect

The development of TOD can be an opportunity to use alternative financing mechanisms, such as Land Value Capture (LVC) by using the non-fare box component, such as: property, advertising, access fees, as company revenue. This value captures the mechanism to identify the increase in the value of a land or area due to the development of infrastructure, especially transportation infrastructure, which increases accessibility to the land [16]. LVC has been implemented by several countries to fund part or all of the TOD construction costs, such as in: Hong Kong, Japan, United

IOP Conf. Series: Earth and Environmental Science **328** (2019) 012006 doi:10.1088/1755-1315/328/1/012006

Kingdom, and Colombia. Each country applies a different LVC mechanism adapted to the economic, political and ideological systems adopted. Broadly speaking, LVC allows local governments to be able to take up the value of land caused by the construction of public infrastructure. Then, the profit is utilized to finance the construction of other public facilities / infrastructure. At present, there is no LVC mechanism that has been implemented in Indonesia. The implementation of the LVC mechanism is something that will need to be done by the government, because of the projection of land value increases in the proximity of TOD area.

Based on the interview, the value of project has not been seen vividly for a certain stakeholder. As said by PT MRT, "*Until now, the MRT has not considered financial benefits. Consideration is still only seen from the perspective of urban planning. New financial benefits will be considered when TOD will be built or developed in accordance with UDGL*". From this point of view, the return on investment mechanism remains unclear. A similar result also occurred in PT KAI, where the corporation has not thought about the funding, since they "*Don't know yet, because of* (the project) *is conducted by the different directorates".* The answer gave an insight of unsynchronized sharing information in the bodies of corporation.

Different from transit agency point of view, the central government's view represented by Jabodetabek Transportation Management Agency revealed the opportunity of obtaining profit of Floor Area Ratio (FAR). It is said that, "*FAR is still a problem. Many requests from developers, hoped that the calculation of FAR should be calculated based on area basis. But on the other hand, there must still be provisions on the limits of FAR in general, because calculations per area will cause different regulation for each region and potentially cause problems in the future.* It is understood that this stakeholder has perceived TOD as a source to gain land value through the excess allowed of FAR, yet the regulation that arranged this mechanism had not been formulated clearly. Hence, the practice of obtaining land value has not occurred in the context of case study; however, the idea has been understood conceptually.

4. Conclusion

In summary, the high quality of public transit is greatly important to the sustainability of urban development. A well-defined integration between the transit system and land use planning might encourage environmental sustainability, institution, financing and legal evolvement, as well as better quality of the urban facility. TOD more specifically - through densification approach around the station as a transit hub by placing amenities, workplaces, commercial activities and housing in close proximity becomes one of the most effective ways to achieve sustainable urban development. A well-designed property will get accessibility and agglomeration benefits.

Furthermore, the agency and all institutions involved in the TOD have pivotal roles and responsibilities in creating a supporting environment for the initiation of the TOD project. However, it is argued that provincial government is the most potential agency to manage and develop TOD project, since the development of urban dan transport project is under their expected roles by referring the decentralization concept conducted in Indonesia. However, the central government is more focused on formulating policies and regulations. The current regulations are also inadequate to overcome the problems, because it only covers normative aspects. Therefore, the detailed policies and strategies are needed to produce workable structures for all institutions involved, without draining resources or harming a certain institution. The formulation of alternative financing mechanisms is important for the sustainability of the development of TOD. To achieve that point, good collaboration among central government, local government, transit agency, developers, landowner and society are immensely required through proper and clear policy, workable working structure, as well as open dialogue among involved actors. In this early stage of developing TOD, Indonesia, specifically JMA, still has ample in understanding the challenges of implementing TOD as a stepping stone to conduct sustainable transportation in the future.

In the context of public policy, the development of TOD, with all of the promising benefits, offers the adoption of policy examples related to the integration of land use for transportation as the commonplace for government officials (as a learning material). Regardless of the good goals, the policies utilized as guidelines must be in accordance with the specific context of the location. In this point of view, several case studies of TOD area development in Indonesia, especially Jakarta Metropolitan Area (JMA) have proven this premise, by finding 4 main challenges in the development of TOD that differ from the problems of the sample countries.

References

[1] Koglin, Till & , Toth-Szabo & Varhelyi, Andras & , Angjelevska. (2011). Measuring sustainability of transport in the city – development of an indicator-set. 10.13140/RG.2.1.1451.0240.

[2] Tan, W., Bertolini, L., Janssen-Jansen, L., (2014). Identifying and Conceptualising Context-Specific Barrier to Transit-Oriented Development Strategies: The Case of the Netherlands. TPR. Vol. 85 No. 5.

[3] Thomas, R., Pojani, D., Lenferink, S., Bertolini, L., Stead, D., & van der Krabben, E. (2018). Is transitoriented development (TOD) an internationally transferable policy concept? Regional Studies. DOI: 10.1080/00343404.2018.1428740

[4] Dabrowski, M., Musiakowska, I., & Polverari, L. (2018). Introduction: drawing lessons from international policy-transfer initiatives in regional and urban development and spatial planning. Regional Studies, 1-4. DOI: 10.1080/00343404.2018.1462490

[5] Calthorpe, Peter. (1993). The Next American Metropolis: Ecology, Community, and the American Dream. New York: Princeton Architectural Press.

[6] Dittmar, Hank dan Ohland Gloria. (2004). *The New Transit Town: Best Practices in Transit Oriented Development*. Washington, D.C.: Island Press.

[7] Transit Cooperative Research Program (2007), Report 95, *Transit Oriented Development: Traveler Response to Transportation System Changes* ,The Federal Transit Administration, Washington, D.C.

[8] Cervero, R. 1998. The transit metropolis: a global inquiry. Washington, D.C.: Island Press.

[9] Suzuki, H., Cervero, R., & Iuchi, K. (2013). *Transforming Cities with Transit*. Washington DC: World Bank.

[10] Cervero, R., & Kockelman, K. (1997). Travel demand and the 3Ds: Density, diversity, and design. *Transportation Research Part D: Transport and Environment*, 199-219.

[11] Alexander, E. R. (2007). Institutionalist Perspective on Planning: Why, Where? How? In Verma, Niraj. Institutions and Planning. Elsevier. Amsterdam.

[12] Tan, W.W.Y.G.Z. (2013). Pursuing Transit-Oriented Development: Implementation through Institutional Change, Learning and Innovation. Dissertation. Amsterdam Institute for Social Science Research (AISSR).

[13] Tang, BS, Chiang, YH, Baldwin, AN dan Yeung, CW (2004), *Study of the Integrated Rail-Property Development Model in Hong Kong, Study Report for MTRC*, Research Center for Construction dan Real Estate Economics of the Hong Kong Polytechnic University

[14] Cornes, Richard and Sandler, Todd, (1996), The Theory of Externalities, Public Goods, and Club Goods, Cambridge University Press.

[15] Vuchic, V. R. (2001). Urban Public Transportation Systems. Transportation Engineering and Planning, Vol. I.

[16] Mathur, S., & Smith, A. (2013). Land value capture to fund public transportation infrastructure: Examination of joint development projects' revenue yield and stability. *Transport Policy, 30*, 327–335. https://doi.org/10.1016/j.tranpol.2013.09.016.

Performance Evaluation and Crack Repair in Building Infrastructure

A S Dermawan[1*], S M Dewi[2], Wisnumurti[2] and A Wibowo[2]

[1] Achmad Yani Banjarmasin University, Departement of Civil Engineering, Banjarmasin, Indonesia/Doctoral Student Brawijaya University, Departement of Civil Engineering, Malang, Indonesia
[2] Brawijaya University, Departement of Civil Engineering, Malang, Indonesia

*email: as_dermawan@yahoo.com

Abstract. The implementation of building infrastructure development to improve people's lives is important. The use of reinforced concrete structures has increased very rapidly. But many buildings are failed and are not safe because of overloading, changes in function and configuration of buildings, poor maintenance, and natural disasters. Many methods developed to reduce the damage caused by earthquakes so that the development of methods to strengthen and repair building structures is needed to anticipate damage to buildings. This study focuses on beam-column joint cracks with simple building characteristics, column size 20/20 and beam 15/20 (monolithic) with concrete quality fc'= 21 MPa, so it is necessary to evaluate and repair beam-column joint cracks in buildings because of overload. The method used was experimental and testing in a laboratory with a single static load and Ultrasonic Pulse Velocity (UPV) test to determine the density after repair. Test specimens are SK (control-specimens) and SK-repairing. The results showed that the beam-column joint performance before and after repairs had relatively the same flexural strength of 1,669 kg and 1,686 kg, while the stiffness was lower at 14.48%. The velocity after repair average is 3,327.5 m/s (medium-category). The elasticity average modulus of the test is 20.43 GPa.
Keywords: buildings, beam-column joints, cracks, repairs, UPV

1. Introduction

Materials that are often used in the world of construction are reinforced concrete due to its good durability, ease of implementation and it has been used for years to build various structures (houses, high-rise buildings, bridges, and dams). As a developing country which is developing massive infrastructure, the use of reinforced concrete structures has also increased very rapidly. The view that concrete is a durable and maintenance-free construction material has changed in recent years. It is possible to achieve the reinforced-concrete performance as good as expected [1]. Thousands of reinforced concrete structures are built every year throughout Indonesia. There are also a large number of poor concrete structures that become unsafe due to inadequate detailed structural design, poor construction and quality of maintenance, overloading, chemical attack, steel reinforcement, foundation, abrasion, fatigue effects, atmospheric effects, abnormal flooding [2], changes in building functions, changes in configuration and natural disasters that will affect the durability of concrete structures.

IOP Conf. Series: Earth and Environmental Science **328** (2019) 012007 doi:10.1088/1755-1315/328/1/012007

Many general problems related with repairs amerge along with various solution methods, specifications and regulations. Broken concrete, concrete surface preparation, cleaning and or steel reinforcement replacement, surface inspection, and selection of appropriate repair materials depend on the level of damage and damage conditions. Damage due to debonding too quickly is a significant problem for repairs. Structural failure in building infrastructure occurs due to less efficient bonding or due to the nature of the concrete substrate, and the repair material is not suitable. The better the materials with additives, the higher the strength of the structure [3]. The International Concrete Repair Institute has shown several patterns of premature failure due to the nature of the repair material that does not match and mentioneds properties for repair materials [4].

One way to investigate the performance of concrete testing is Non-Destructive Test (NDT) to test the elastic modulus in concrete using the Ultrasonic Pulse Velocity method [5, 6, 7]. The workings of NDT is by calculating the ultrasonic wave velocity needed for propagation in concrete and calculating the elastic modulus. The stiffness of elasticity in concrete affects NDT results. The UPV wave velocity will decrease in concrete which has poor compaction or damage to material grains.

Many research have been carried out on beam-column joint performance [8, 9], and structural performance of reinforced concrete structures using the latest repair materials where the bond between concrete and steel has disappeared, but the use of UPV to check the results of concrete reinforcement repairs has not been studied. Therefore, this study presents a study of the flexural performance of reinforced concrete column beams which are repaired using flowing concrete repair (grout) mortars.

2. Methods

The building samples for this study are simple buildings (2 levels) with 20/20 size columns and 15/20 size of beams with concrete quality of fc'= 21 Mpa. The laboratory specimens use a beam-column joint that has a rectangular cross-section with a column size of 200x200 mm with a length of 1,000 mm, and a 150x200 beam with a length of 1,000 mm. Concrete cover of 25 mm, column and beam has longitudinal reinforcement of 4 D13. The size of beam stirrups is Ø8 - 100 mm in mid-span and Ø8 - 50 mm in edge span, column stirrups are Ø10 - 100 mm in edge span and Ø10 - 50 mm in mid-span. The procedure for casting monolithic specimens, loading specimens until they reach the yielding point and repairing test specimens is based on ACI 224-07 criteria [12], and according to the requirements we then use Sikagrout 215 New which has a compressive strength of 65 N/mm^2 (28 days), flexural strength > 6 N/mm^2, expansion 0.3-1.4%, adhesion strength > 1.5 N/mm^2 and elastic modulus of 2600 N/mm^2 and SikaBond NV for connecting old and new concrete.

Materials for reinforced concrete include portland cement ex. Gresik cement, Lumajang sand and broken stone with a maximum aggregate size of 20 mm with a cement water factor (FAS) of 0.65 to produce the desired concrete quality of 21 MPa. The use of standard cylindrical samples to determine the compressive strength of concrete and Young's modulus of elasticity at 28 days. The test on specimen based on design criteria according to ACI 318-14 and ACI 352R-02 [10, 11]. The testing of each sample uses a static load with a 50 kg load interval until it reached collapse. The measurement of deflection on the beam uses LVDT (Linear Variable Displacement Transducer). Illustration of the experimental set up beam-column joint test specimens can be seen in **Figure 1**.

Figure 1. The Experimental Test Set up

UPV (Ultrasonic Pulse Velocity) testing of structures is used to determine concrete density quality and how deep the crack based on wave propagation speed. The working principle of the UPV test equipment is to produce and deliver ultrasonic waves in the form of pulses/pulses into concrete, and the average travel time of the wave from the starting point to the end point through concrete. Waves will take longer if there is an air cavity inside the concrete. UPV consists of two transducers and a Pundit Touchscreen which function to operate the transducers function.

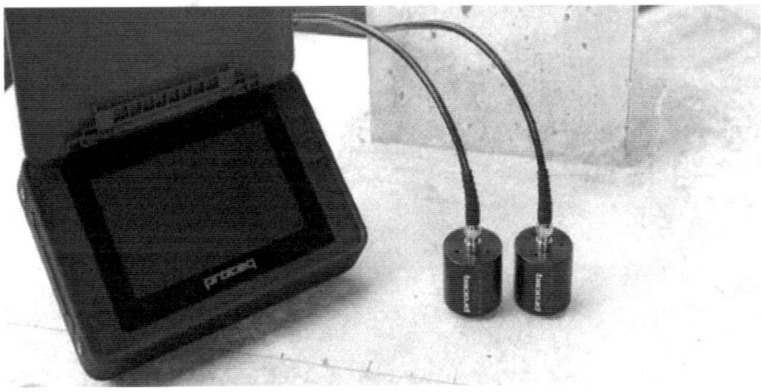

Figure 2. The Ultrasonic Pulse Velocity (UPV)

The concrete quality category with UPV is based on the wave velocity because the principle of the UPV test is to determine the time of propagation of ultrasonic waves in concrete. The better the concrete density, the shorter the time needed by the wave to propagate from the transmitter to the receiver. If the concrete has a lot of delimitation (cavity), then the waves will require a longer propagation time.

Table 1. Classification of the Quality of Concrete Based on Wave Velocity

Wave propagation speed	Concrete Quality
> 4500 m/s	Excellent
3500 – 4500 m/s	Good
3000 – 3500 m/s	Medium
< 3000 m/s	Doubtful

IOP Conf. Series: Earth and Environmental Science **328** (2019) 012007 doi:10.1088/1755-1315/328/1/012007

3. Results and Discussions

3.1. SK Specimen

The specimen control (SK) is a column of monolithic specimens before repair. From the results of a single static load, it can be seen that horizontal bending occurs when the load is 1,150 kg with a deflection of 1.98 mm on the beam on the four grid. The results of the stiffness values in the first crack in the beam-column joint are 581 kg / mm. At a load of 1,200 kilograms with a deflection of 2.18 mm, horizontal bending slits occur in the four and six lattices spaced 20 cm and 30 cm from the beam-column joints. Cracks increase in length at a load of 1,400 kg and produce a deflection of 3.38 mm. Then flexural shear cracks are started at 1,450 kilograms with a deflection of 3.74 mm. At 1,500 kilograms, crack openings occur in the meeting area of the beam and column joints, but cracks in other areas are no longer increasing. The crack opening in the beam-column joint increases in length and width until the peak load reaches 1,686 kg and the deflection is 6.08 mm. The concrete spalling enlarges towards the compression area until the test stops when the load drops to 33% of the peak load. The results of the crack pattern in the SK specimens is shown in **Figure 2**.

Figure 3. The final results of crack propagation on the SK specimen (front, back, and side view)

3.2. SK-R Specimen

The SK-R specimen is a specimen beam-column that is repaired after being damaged by grouting (Sikagrout 215 New). The cracks of horizontal bending began in the beam on the grid seven with a deflection of 0.66 mm when the load was 700 kg. The value of stiffness when the first crack occurs is 1,063.8 kg/mm. At a load of 1,200 kilograms with a deflection of 1.59 mm, there is horizontal bending crack of one and two grid with distances from 5 cm and 10 cm from the beam-column joint. At a load of 1,500 kg with a deflection of 2.91 mm the two grid cracks increase in length and width, the crack six grid increases in length and the spalling of concrete begins on the compressed of the beam zone. The flexural shear crack started when the load was 1,550 kg with a deflection of 4.08 mm at a capacity of 1,600 kg the area of the beam and column. When these openings occur, crack in other areas no longer develop. The beam column intersection opening increases in length and width until the peak load is 1,686 kg with a deflection of 8.07 mm. Concrete spalling extends to the compression area has dropped to 38% of the peak load. The crack pattern beam-column joint on the SK-R specimen are shown in **Figure 4**.

IOP Conf. Series: Earth and Environmental Science **328** (2019) 012007 doi:10.1088/1755-1315/328/1/012007

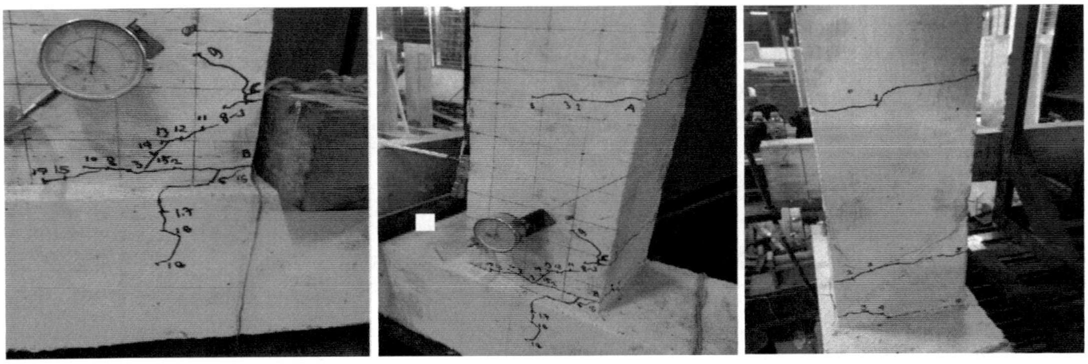

Figure 4. Crack propagation on the SK-R specimen

3.3. The results test of a single static load and discussion

The results of testing a single static load on SK specimen and SK-R specimens are shown in **Figures Figure 5,** and **Figure 6**:

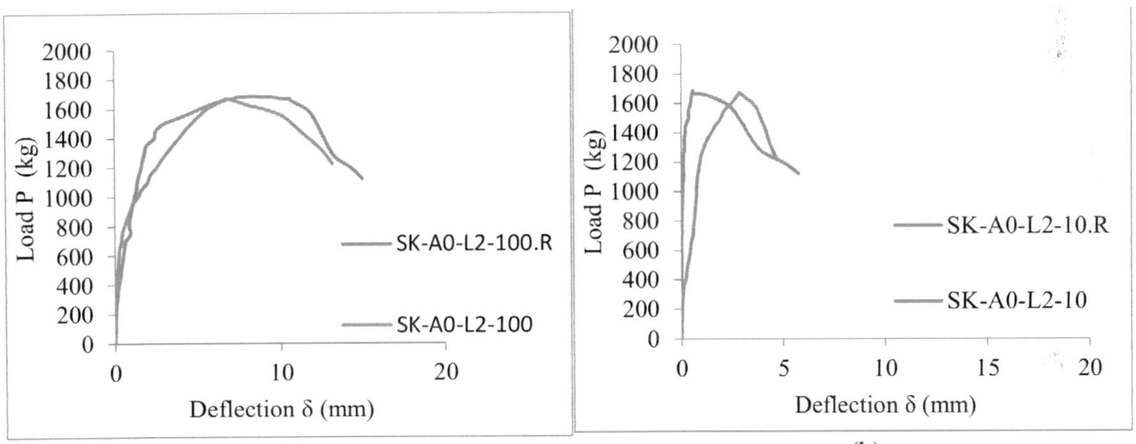

(a) (b)

Figure 5. The relationship between deflection and load (a) X = 100 cm (deflection 100 cm from the column); (b) X = 10 cm (deflection measurement of 10 cm from the column)

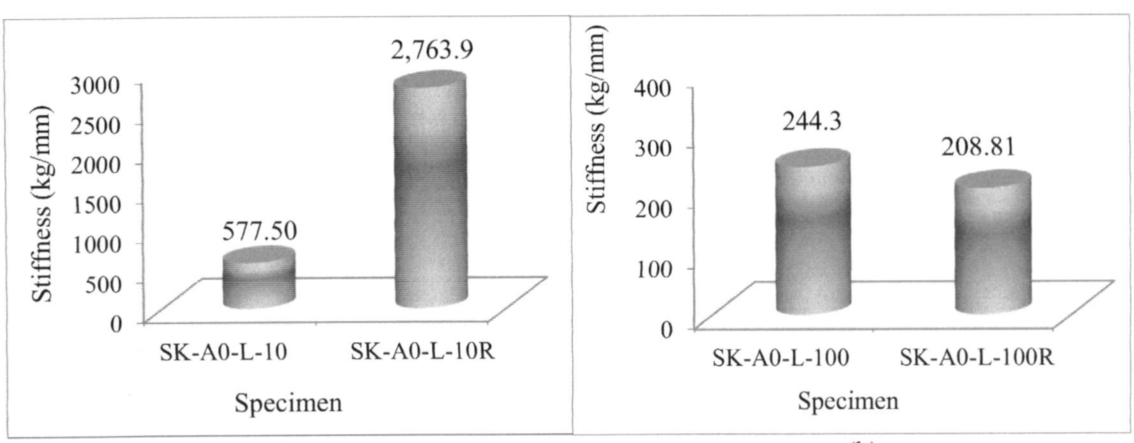

(a) (b)

Figure 6. The behavior of beam-column joint stiffness (a) X = 10 cm; (b) X = 100 cm

Figure 5 (a) shows the results of the behavior comparison between load and deflection for deflection measurement of 100 cm from the column, X = 100 cm before 1,669 kg and 6.83 mm; after 1686 kg and 8.07 mm. Meanwhile, Figure 5 (b) shows a comparison of the behavior between the load and the deflection for a deflection measurement of 10 cm from the column, X = 10 cm before 1,669 kg and 2.89 mm; after 1,686 kg and 0.56 mm. **Figure 5(a)** and **5(b)** aim to see deflection in plastic joints. **Figure 6 (a)** shows the stiffness behavior of the higher / more rigid beam connection with the first stiffness of 577.50 kg / mm and the end of 2,763.9 kg/mm with a difference of 79.11% (measurement of deflection 10 cm from the column, X = 10 cm). **Figure 6 (b)** shows the result of the beam-column joint behavior for deflection measurement of 100 cm from the column (X = 100 cm) which decreases the first stiffness of 244.36 kg/mm and the tip of 208.81 kg/mm with a difference of 14.48%.

3.4. The result of the UPV test

The results of UPV testing on SK-R specimens are shown in **Figures 7** and **Figure 8**:

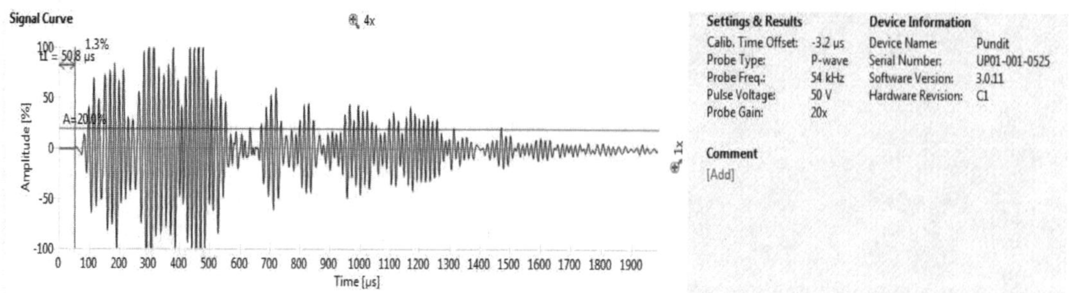

Figure 7. The UPV test result on SK-R specimens (a distance of 0.200 m)

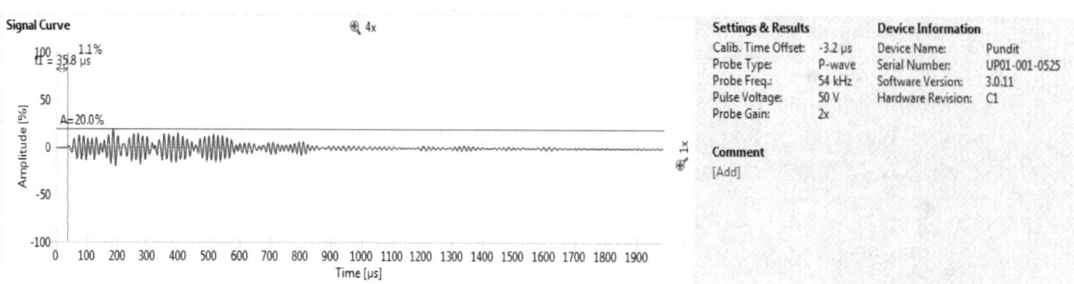

Figure 8. The result of the UPV test on SK-R specimens (a distance of 0.150 m)

Table 2. Recapitulation of UPV testing

No	Number	Date & time	Distance (m)	Total velocity (m/s)	Correction	Average quality
1	SK-A0-L2-R	12/06/2018 12.34 PM	0.200	3080	1.30	Good
2	SK-A0-L2-R	12/06/2018 12.27 PM	0.150	3575	1.30	Medium

From **Figure 7, 8** and **Table 2**, it can be seen that the concrete quality in the beam-column joint structure has an average wave velocity of 3,327.5 m/s (medium category).

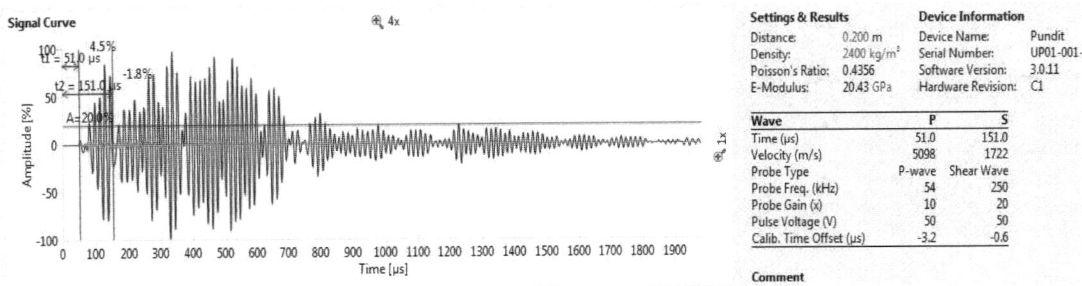

Figure 9. The result of the UPV test on SK-R specimens (E-modulus)

Figure 8 shows the results of the average modulus of elasticity of the test are 20.43 GPa.

4. Conclusion

The test results show that the repair on a joint column can restore the beam-column joint performance but it decreases stiffness performance. This performance improvement is a good bond between old concrete and concrete grouting repairs. The beam-column joint performance before and after repairs have relatively the same flexural strength of 1,669 kg and 1,686 kg, while the stiffness increases/more rigid at 79.11% (X = 10 cm). The result of the beam-column joint behavior for deflection measurement of 100 cm from the column (X = 100 cm) decreases the first stiffness of 244.36 kg/mm and the tip of 208.81 kg/mm with a difference of 14.48%. The UPV test which obtained wave propagation speed on the beam-column joint structure after retrofitting has an average velocity of 3,327.5 m/s (medium category), and the average modulus of elasticity of the test was 20.43 GPa. Repairing damage to building infrastructure must be done immediately so as not to increase the level of damage to be fatal and require a significant cost.

References

[1] Kovács T 2000 Deterioration of Reinforced Concrete Repair Mortar Layers. *Periodica Polytechnica Ser. Civ. Eng.* 44: 197–206

[2] Kenai, S. and Bahar, R.(2003). Evaluation and Repair of Algiers New Airport Building. *Cement and Concrete Composites*, 25: 633-641

[3] Dermawan A S, Fauzan A, and Yuniar D 2016 High Quality Concrete With Addition Of Viscocrete-10, Silica Fume And Quartz Powder. *Sem.Nas.Ris.Ter.1.* Poliban ISSN 2541-5670 (Online). D9-D17

[4] ACI International, BRE, Concrete Society and ICRI 2003 *Diagnosis of Deterioration in Concrete Structures* Concrete Repair Manual, 2nd Edn, Vol. 1. pp. 988-1027

[5] Chandrappa A K and Biligiri K P 2016 Influence of mix parameters on pore properties and modulus of pervious concrete: an application of ultrasonic pulse velocity. *Mater.Struct.,* vol. 49, no. 12, pp. 5255–5271

[6] Lu X, Sun Q, Feng W and Tian J 2013 Evaluation of dynamic modulus of elasticity of concrete using the impact-echo method. *Constr. Build. Mater.*, vol. 47, pp. 231–239

[7] Barazanchy D, Roth W and Giurgiutiu V 2018 A non-destructive material characterization framework for retrieving a stiffness matrix using bulk waves. *Composite Structures*. Vol.185 pp. 27-37

[8] Yuliati N C E, Dewi S M, Wibowo A and Wisnumurti 2018 Experimental Investigation of crack on a Non-Monolithic Exterior Beam-Column Joint Of Reinforced Concrete Under Static Load, *Intern.J.of Civ.Eng and Tech. (IJCIET)*, vol. 9 IAEME publication. pp. 2929-2937

[9] Shahbazpanahi S, Hejazi F, Paknahad M, Rahimipour A, and Nassimi M R 2018 Modelling Crack Propagation in RC Beam-Column Joints. *Teh. Vjesn.* Vol. 25, no. 4, pp. 1183–1189

IOP Conf. Series: Earth and Environmental Science **328** (2019) 012007 doi:10.1088/1755-1315/328/1/012007

[10] ACI Committee 318. 2014 *Building Code Requirements for Structural Concrete (ACI 318M-14)* American Concrete Institute
[11] ACI Committee 352. 2002 *Recommendations for Design of Beam-Column Connections in Monolithic Reinforced Concrete Structures* American Concrete Institute
[12] ACI Committee 224. 2007 *Causes, Evaluation, and Repair of Cracks in Concrete Structures*. American Concrete Institute

IOP Conf. Series: Earth and Environmental Science **328** (2019) 012008 doi:10.1088/1755-1315/328/1/012008

Toward Inclusive Development Through Smart Economy in Malang Regency

A Yudono[1], **Dias Satria**[2], **and Angga Erlando**[3]

[1] Department of Urban and Regional Planning, Faculty of Engineering, Brawijaya University
[2] Department of Economics,Faculty of Economics and Business, Brawijaya University
[3] Department of Economics, Faculty of Economics and Business, Airlangga University

adipandang@ub.ac.id
dias.satria@ub.ac.id
erlandoangga@gmail.com

Abstract. The concept of smart economy is used as an innovation in the economic field based on industry 4.0 and uses a different network system. Bakici*et al.* (2013) said smart economy includes several points of discussion including the formation of innovation clusters, mutual cooperation between companies, research institutions, and citizens whose role is to develop and promote innovation through the concept of smart economy [1]. Furthermore, the concept of smart economy is used as a form of sustainable economy, by using several definitions and / or other terms such as “Green Economy” and “Green Industry”.They are used to describe effective economics in a more modern way. UNEP defines the Green Economy as an economy that can produce "improved human welfare and social justice, and can significantly reduce the risk of environmental and ecological damage” [2]. The industry's revolutionary 4.0 Era, which has become a major issue throughout the world, has encouraged Malang Regency to play an active role in regional management using a citizen centric approach. These dimensions are divided into two fundamental considerations, namely the process of innovation, and the ability of competitiveness that is useful for improving the regional economy better and smarter. Based on the issue of economic challenges that occur in Malang Regency, smart economy as a part of smart city element is the focus that will be developed. It consists of 3 (three) important fields including: enterprise and innovation, productivity, then local and interconnectedness.

1. Introduction

The term Industry 4.0 was born from the idea of the fourth industrial revolution marked by the rapid development of sensor technology, interconnection, data digitalization, and data analysis to integrate all these technologies into various industrial fields [3]. The impact of Industry 4.0 also entered the realm of the Indonesian government, even those in the regions since the Industry 4.0 provides a great opportunity to streamline the function and role of government organizations in carrying out their duties [4]. Moreover, the currently centralised policies provide flexibility for local governments to create and develop extensive innovations. Local governments in this case represented by each Local Government Organisation (*Organisasi Perangkat Daerah*, OPD), are faced with the demand to adapt and accelerate the emergence of efficient and effective services. For example, with rapid IT development it can be an opportunity to accelerate e-governance implementation, digitizing data and information such as: e-budgeting, e-project planning, system delivery, administration, e-controlling, e-reporting to e-monev, and other custom applications. In this series of digital information and communication platforms, it will be more efficient and effective if the system built can integrate with one another [4].

The management of economic resources in a region of a country or region (district / city) has now been faced with the demands of innovation to realize efficiency and effectiveness. The ever expanding flow of globalization, accompanied by increasingly sophisticated technological developments in the industrial revolution era 4.0, has led to the emergence of the digital economy era, artificial intelligence, big data, robotics, and the like, as a new phenomenon of disruptive innovation. Today, business contestation is not only limited to the scope of resource allocation for profit but also innovation, competitive ability, and the use of technology, information, and communication (internet of things) [5].

Malang Regency Government covers the areas that have 5 (five) elements of business development services in the current digital era. Government service providers consist of various OPD that have digital-based services, both with their own governance system (Figure 1), and which are managed based on synergies with the Provincial Government, Central Government and Bank Indonesia programs, such as: Information Center for Strategic Food Prices (PIHPS) and (Information System for Availability and Development of Prices of Basic Commodities in East Java (SISKA PERBAPO) [6].

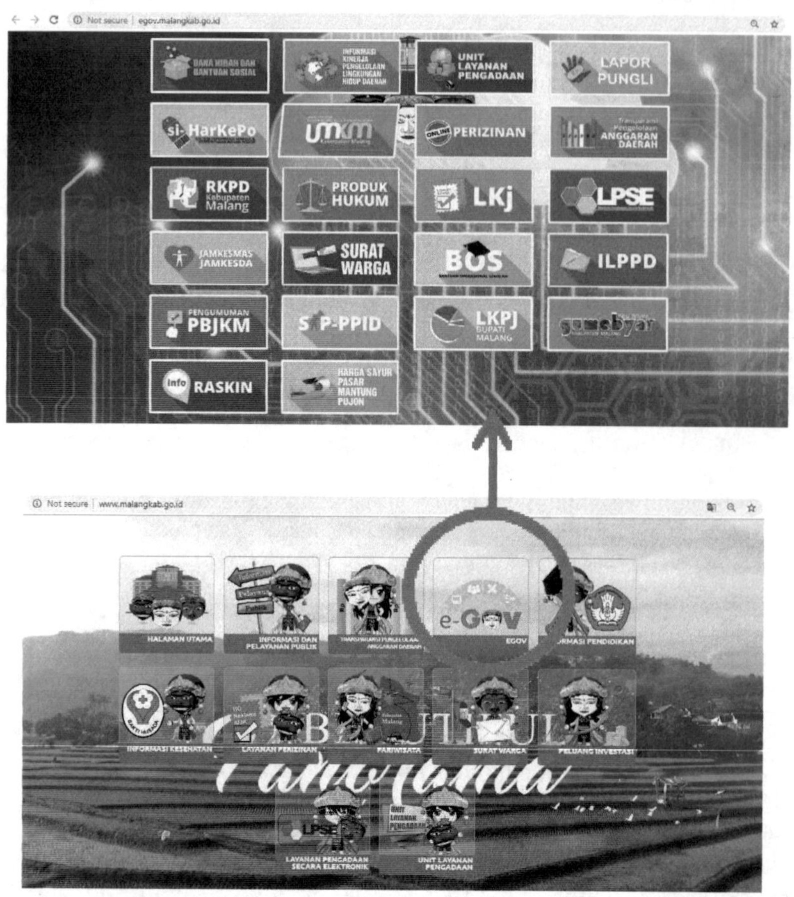

Figure 1. The Digital-Based Government Service System Applied in Malang Regency
Source: http://www.malangkab.go.id/

Based on Figure 1, Malang Regency has tried to access government services by utilizing website-based digital media, especially for e-government services, in which there are 18 service menu items that are displayed attractively. However, it is truly unfortunate that there is a fundamental weakness in the implementation of the government's digital-based website services. First, the system built in it does not yet have a road map that is integrated between one service and another. Second, the parties (OPD) have not made an open system that allows data integration and decision making that is right for policy makers. Third, the form of basic capital in the form of existing digital services has not been managed and developed continuously by the government. Therefore, it is necessary to make the development of digital services that are mutually integrated and sustainable in order to be efficient and effective.

2. Method

The research method used in this study is a mix-method approach, which is a combination of qualitative and quantitative research in exploring the potential implementation of smart economy in Malang City. The reason for using the Mix-method approach is being able to optimize studies in obtaining answers to the potential implementation of an innovation in an area [7] [8] [9]. Furthermore, Bryman (1998) suggests there is strong reason, the mix-method approach can optimize the answers to a problem statement in the study, namely Qualitative and quantitative are given balance substances [10].

Mapping the smart economy potential in Malang Regency in this paper adopt from Bruneckiene and Sinkiene (2014) with the following indicators:

1. Building a competitive industrial ecosystem (industry) through certain industry leading sectors that are integrated between primary industries (e.g. agriculture, fisheries, livestock and others), secondary industries (e.g. manufacturing, processing, packaging, etc.), and tertiary industries (e.g. regional product markets);
2. Realizing people's welfare, by: (a) developing a program to improve people's welfare through increasing household income; (b) program to increase employment absorption; (c) community economic empowerment program;
3. Building an ecosystem of financial transactions by: (a) building an ecosystem of digital financial transactions to ensure smooth payments to the public having less cash; (b) realizing a bankable society with access to capital, and; (3) realizing the digital economy ecosystem by encouraging the e-commerce industry and market place. [11]

3. Results and Discussions

Considering, the number of population in Malang Regency each year continues to grow, so that it raises a variety of complex interests. The concept of a smart city (smart city) which has become a major issue throughout the world, encourages Malang Regency to encourage active management in the area using centric citizens.

IOP Conf. Series: Earth and Environmental Science **328** (2019) 012008 doi:10.1088/1755-1315/328/1/012008

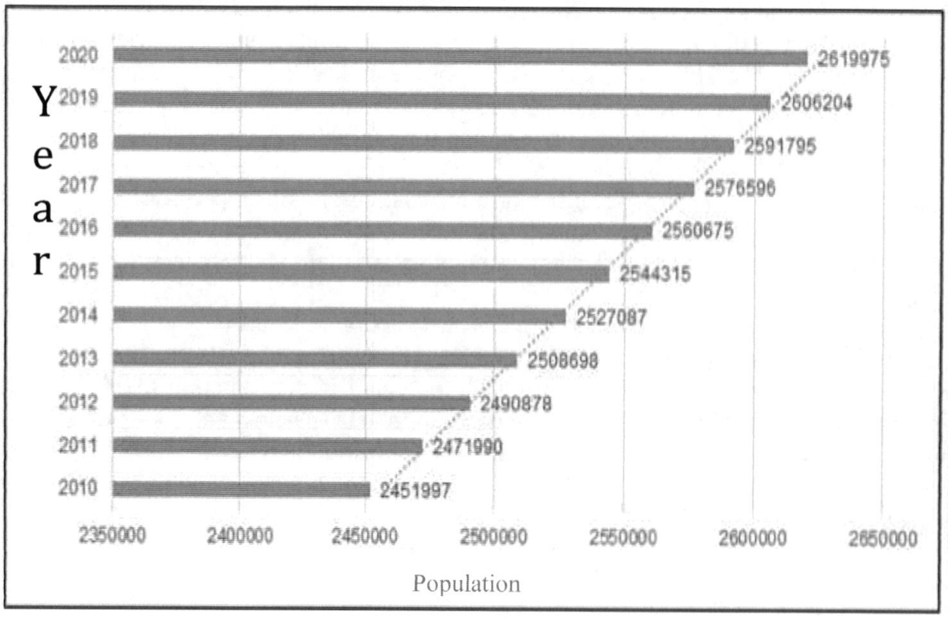

Figure 2. The Growth of Population in Malang Regency
Source: Malang Regency Bureau of Statistics, 2018 [12]

One dimension of smart city that is currently a concern in Malang Regency is the dimension of the smart economy. These dimensions are divided into two fundamental considerations, namely the process of innovation, and competitiveness that is useful for improving the regional economy in a better and smarter way. The focus on the dimensions of the smart economy is also inseparable from the issue of strategic economic challenges that have occurred in the Regency in recent years.

First, the trend of declining growth trends and dynamic (unstable) price fluctuations (inflation).

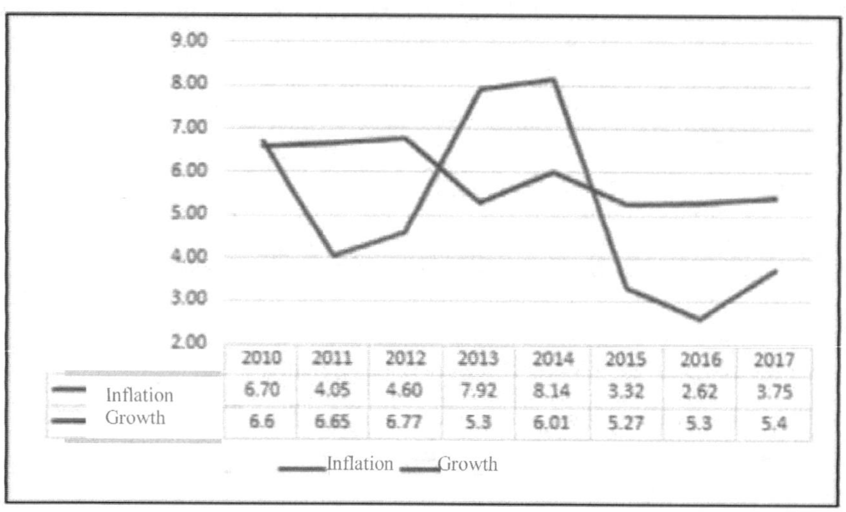

Figure 3. The Economic Growth and Inflation in Malang Regency
Source: Malang Regency Bureau of Statistics, 2018 [12]

Second, the inconsistency of efforts to reduce income inequality that occurred in Malang Regency

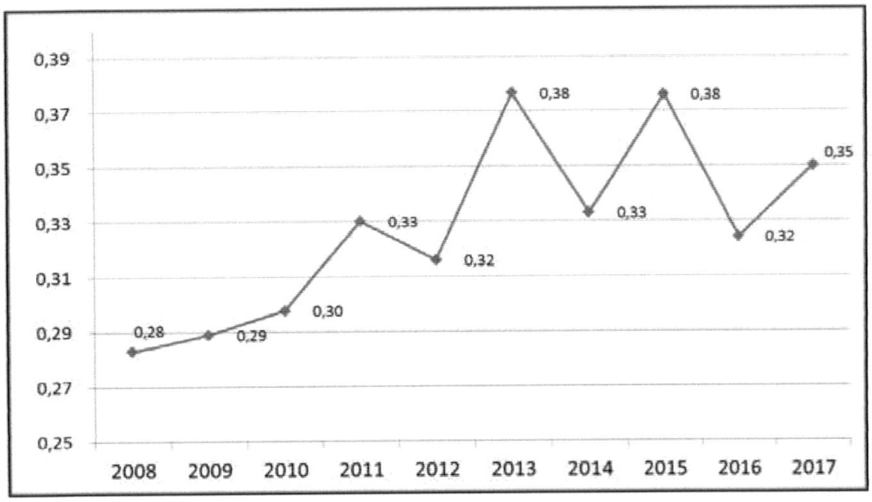

Figure 4. Gini Ratio in Malang Regency
Source: Malang Regency Bureau of Statistics, 2018 [12]

Third, the level of unemployment and poverty are inconsistent and still quite high.

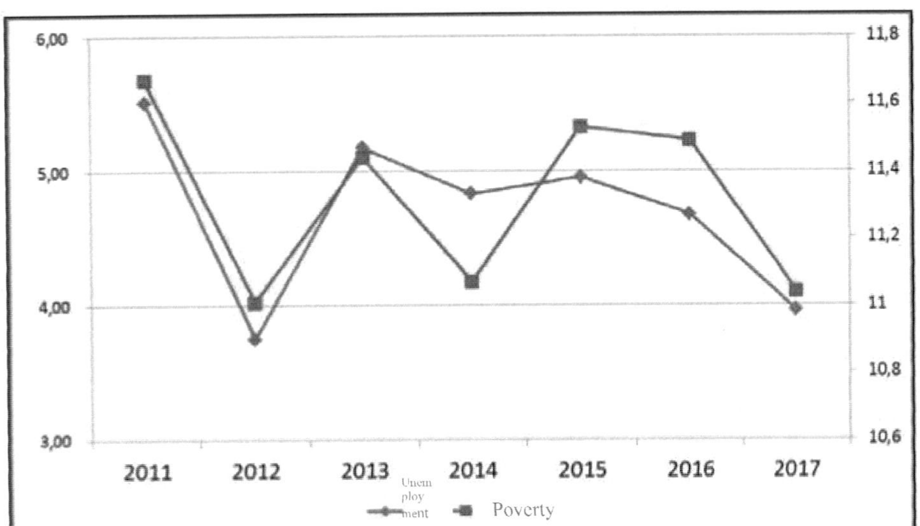

Figure 5. Unemployement and Poverty in Malang Regency
Source: Malang Regency Bureau of Statistics, 2018 [12]

Based on the issue of economic challenges that occur in Malang Regency, the smart economy as part of the smart city element is the focus that will be developed. Consisting of three (3) important fields including: enterprise and innovation, productivity, and local and interconnectedness. These three fields are substantial pillars for forming the smart economy going forward. Then the Indonesian Ministry of Communication and information interpreted these three things into the Indonesian context into three elements including: (1) industrial ecosystems; (2) improving community welfare, and; (3)

financial transaction ecosystem. This is intended to realize economic ecosystems in regions that are able to meet challenges in a disruptive information era and demand a rapid level of adaptation at this time. Afterwards, the smart economy Malang Regcncy can be formulated detailed to be

A. Building a Competitive Industrial Ecosystem (Industry)
- Building regional industry competitiveness in certain industry leading sectors that are integrated between primary industries (eg agriculture, fisheries, livestock, etc.), secondary industries (eg manufacturing, processing, packaging, etc.), and tertiary industries (eg regional product markets tourism services).

B. Realizing People's Welfare (Welfare)
- Developing community welfare improvement programs through increasing household income (income)
- Program to increase employment absorption
- Community economic empowerment program (empowerment).

C. Building a Financial Transaction Ecosystem (Transaction)
- Building an ecosystem of digital financial transactions to ensure smooth payments to the public that are less cash
- Realizing a bankable society and having access to capital
- Realizing the digital economy ecosystem by encouraging the e-commerce industry and market place

Furthermore, Malang Regency, has substantially encouraged their citizens to succeed in the context of the three smart economy elements compiled by the Indonesian Ministry of Communication and Information. In addition to the development of web digitization-based services, the Malang Regency Government has conducted: the implementation of price control and the Regional Inflation Control Team (Tim Pengendali Inflasi Daerah, TPID), access and financial literacy and Team for Accelerating Regional Financial Access (Tim Percepatan Akses Keuangan Daerah, TPAKD), tourism development based on e-tourism, e-commerce chosen by the citizens independently and directly for business development in various platforms, and so on.

Smart Economy Mapping in Malang Regency has a direction to develop three main fields including: (1) Trade as a form of realizing welfare of the people; (2) Investments which contain the ecosystem of financial transactions, and; (3) Industries through the realization of industrial competitiveness in industrial ecosystems. Meanwhile, the basis brought as the fundamental concept can be seen in terms of enterprise and innovation which means promoting innovation-based business development, roductivity that means increasing production capacity effectively and efficiently, and local and interconnectedness that are oriented on the synergy of information connection in a bottom up manner. The three main fields have designed developed thinking from various platformsconsisting of 6 (six) core program activities, which will later become a special dashboard (Figure 6).

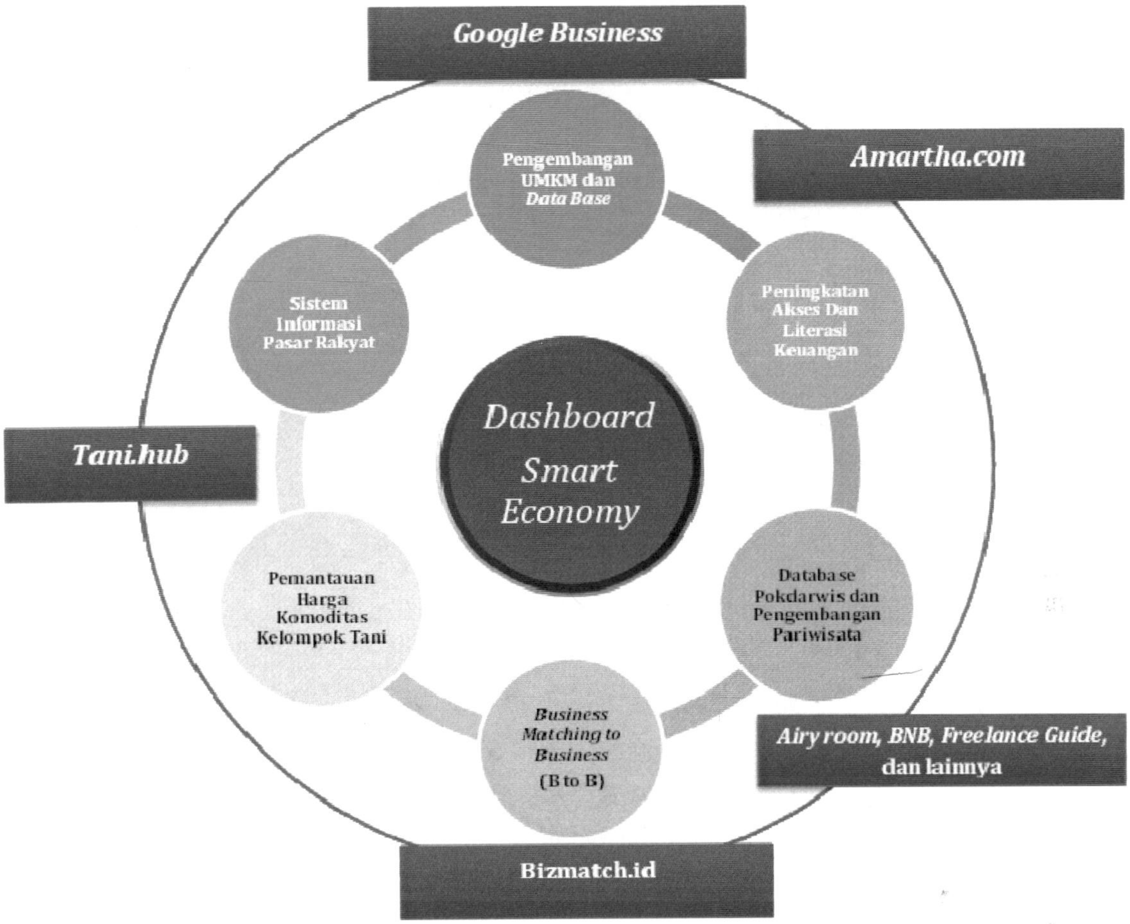

Figure 6. Smart Economy Dashboard in Malang Regency

Source: Authors' Illustration, 2018

4. Conclusion

The government of Malang Regency has began initiate the smart city agenda partially and focused on implementing the smart economy. The smart economy as a form of understanding opens wide access to information to increase opportunities for economic activity effectively, simultaneously reducing costs more efficiently. Now, the agenda has just applied in Malang Regency by forming six priorities for developing smart economy. Consists of: (1) Farmer Group Commodity Price Monitoring; (2) People's Market Information System; (3) Development of MSMEs and Data Base; (4) Increased Financial Access and Literacy; (5) Pokdarwis Database and Tourism Development, and; (6) Business Matching to Business (B to B).

The six smart ecoonomy implementation models use a web-based approach in the hope that they can build industrial ecosystems that are competitive, realize people's welfare, and build the ecosystem of modern transactions. The six models of implementation of smart economy are also formulated based on considerations. In accordance with the characteristics and specific needs of the region, the opportunities for success are implemented which are maximum, objective, targeted, and effective.

References

[1] Bakıcı, T., Almirall, E., & Wareham, J. (2013). A smart city initiative: the case of Barcelona. *Journal of the Knowledge Economy*, 4(2), 135-148.

[2] Davies, A. R., & Mullin, S. J. (2010). Greening the economy: interrogating sustainability innovations beyond the mainstream. *Journal of Economic Geography*, 11(5), 793-816.

[3] Drath, R., & Horch, A. (2014). Industrie 4.0: Hit or hype?[industry forum]. *IEEE industrial electronics magazine*, 8(2), 56-58

[4] The Indonesian Ministry of Communication and Information Technology (2017). *Buku Panduan Penyusunan Masterplan Smart City 2017* Guidence to Create Smart City Masterplan 2017 in Indonesia.

[5] Heng, S. (2014). Industry 4.0: Huge potential for value creation waiting to be tapped. *Deutsche Bank Research.*

[6] The Government of Malang Regency (2018), *Road Map of Smart Economy in Malang Regency.*

[7] Tashakkori, A., and C. Teddlie, eds., (2003). *Handbook of mixed methods in social and behavioral research.* Sage Publications, Thousand Oaks, CA.

[8] Creswell, J. W. (2003). *Research Design: Qualitative, quantitative, and mixed method approaches*, (2nd edition), Sage Publications, Thousand Oaks, CA.

[9] McDougall, K., Rajabifard, A., Williamson. I.P., (2007), *A Mixed-Method Approach for Evaluating Spatial Data Sharing Partnerships for Spatial Data Infrastructure Development.* In Onsrud, H., (ed), 2007, Research and Theory in Advancing Spatial Data Infrastructure Concepts, ESRI Press, Redlands, USA

[10] Bryman, A. (1998). *Quantity and Quality in Social Research.* Taylor and Francis, New York, USA.

[11] Bruneckiene, J., & Sinkiene, J. (2014). Critical analysis of approaches to smart economy. *In 8 th International Scientific Conference "Business and Management* (pp. 15-16).

[12] The Government of Malang Regency (2018), *Malang Regency Bureau of Statistics 2018*

IOP Conf. Series: Earth and Environmental Science **328** (2019) 012009 doi:10.1088/1755-1315/328/1/012009

Analysis of Kendal Ferry Terminal Performance

Adistirani[1,2*], B Riyanto[2], and B H Setiadji[2]

[1]Public Work and Spatial Design Office of Kendal Regency, Central Jawa, Indonesia
[2]Master of Civil Engineering, Diponegoro University, Semarang 50275, Central Java, Indonesia

*email: adistirani@gmail.com

Abstract. Kendal Ferry Terminal which has been operating since 2016 grows continuously in line with public information. As a unit of Tanjung Emas Port, Semarang, based on Masterplan of Tanjung Emas Port, the initial plan was intended as a terminal for "sharing" function. This function urged Kendal Transportation Agency to measure its existing performance to continue improving its performance. This research applied descriptive-quantitative methods that emphasized observation by examining natural conditions, systematically described as a fact in quantitative form. It was started by analyzing the existing demand and supply which increased about 55% in the second year of operation, followed by predicting the passengers' trend for ten years to reach the assumption of 30% Semarang-Kumai passengers distracted to Kendal-Kumai route, even though it was only departure. Finally, the research analyzed the existing facilities and performance, where the result showed that 65% from indicators stated that the performance is inadequate. The recommendations to improve the performance and to prepare receiving a "sharing" function are provided by increasing ship schedules and supply such as the depth of port pool, adding facilities such as passenger lounge, and realizing connectivity of transportation modes with Kedungsepur BRT lines.
Keywords : Sea transportation, Performance, Kendal Ferry Terminal

1. Introduction

Indonesia is famously recognized as one of the largest archipelagic countries with its body of waters reaching two-thirds of the country's total area with diverse geographical conditions. With these conditions, Indonesia needs various modes of transportation, such as land, water, and air transportation. Water transportation seems to be the suitable mode of transportation for areas that are often traversed by water such as Indonesia, in the form of rivers, lakes, and seas. Kendal Regency becomes one of the regencies/cities in Central Java Province that began to develop water transportation by developing a port that called Kendal Ferry Terminal. Although, there is still one DLkr (Regional Work Area) and DLkp (Regional Interest Area) of Tanjung Emas Port, Semarang that is in accordance with Minister of Transportation Regulation Number PM 18 of 2013 concerning the Master Plan of the Tanjung Emas Port in Semarang [1]. Thus, the operation is still under the Tanjung Emas Port.

Kendal Ferry Terminal began to operate in 2016, and since then, the demand for passenger increased significantly, about a 50% increase in second-year operation [2]. This increase indicated a very good achievement considering that Tanjung Emas Port will be focused as a commercial port. With the large goal to divert Semarang–Kumai passengers and freight transports at Tanjung Emas Port to Kendal–Kumai route at the Kendal Ferry Terminal, and based on Master Plan of the Tanjung Emas Port, the Transportation Agency as the operator of Kendal Ferry Terminal needs to measure its performance to prepare more demands. In accordance with Perdana, 2018 [3], it was stated that in addition to develop the economic growth of Kendal Regency, Kendal Ferry Terminal also aimed to support Tanjung Emas Port.

Supported by the potential of location which is positioned on one path with two Semarang industrial areas, such as Candi Industrial Zone and Wijayakusuma Industrial Zone, and adjacent to the Kendal Industrial Area (KIK), the readiness to accept more crossing loads is required from Kendal Ferry Terminal. In addition to the location, the support from a major land transportation path (Semarang-Kendal Pantura arterial road) is also a great potential for the Kendal Ferry Terminal. Coupled with the construction of the Semarang–Batang toll way, which is expected to increase the crossing load of the Kendal Ferry Terminal and assuming 30% Semarang–Kumai passenger and freight transport will divert into Kendal–Kumai route, it is justifiable that there is a demand in the increase of Kendal Ferry Terminal's performance.

This study aims to analyze the existing performance of Kendal Ferry Terminal and to predict the demand level of the Kendal Ferry Terminal in ten years to reach the 30% assumption of Semarang–Kumai passenger which will be diverted.

2. Methods

The methodology of the Analysis of Kendal Ferry Terminals Performance is carried out through descriptive - quantitative approach. This method emphasized field observations with natural conditions to describe the facts in a systematic way. With a quantitative approach, the research was carried out by focusing on the numbers associated with natural phenomena captured at the time of observation, with a framework as seen in **Figure 1**:

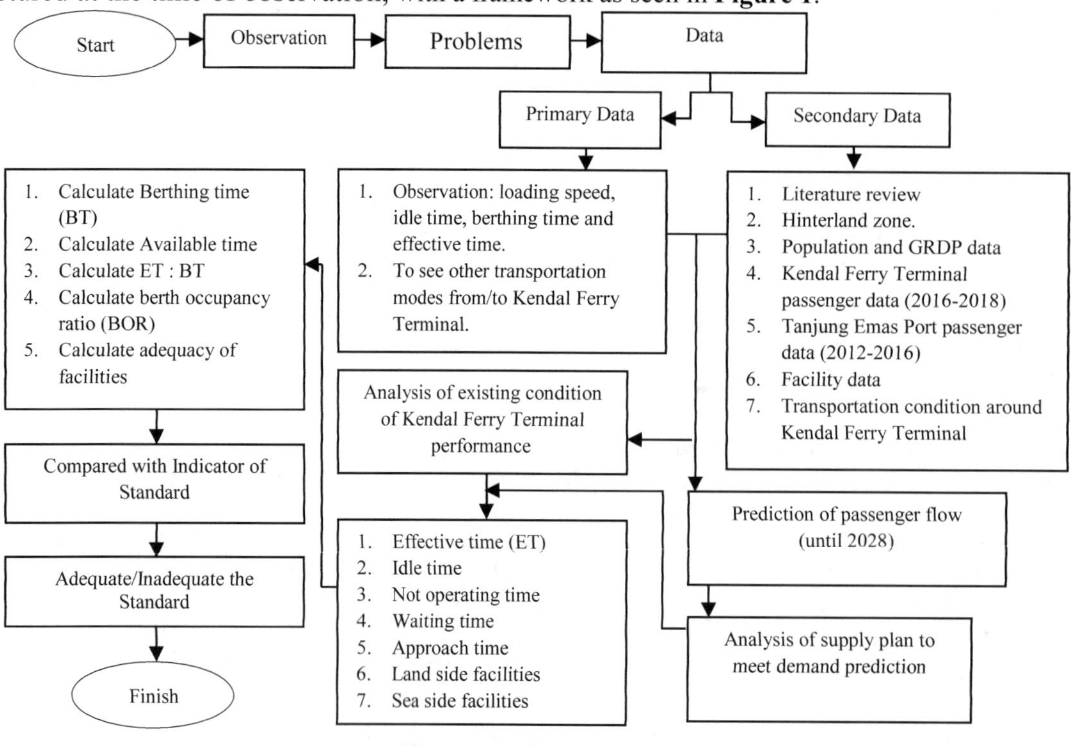

Figure 1. The Framework

Existing data were analyzed to determine the existing performance and then are compared with service performance indicator standards according to Director General of Sea Transportation regulation Number UM.002/38/18/DJPL.1/2011 Dated on December 5th, 2011 concerning Port Operational Service Performance Standards [4].

3. Results and Discussions

3.1. Analysis of Demand and Supply Existing

The existing demand of Kendal Ferry Terminal (from 2016 to October 2018) is depicted in the **Table 1**:

Table 1. Kendal Ferry Terminal Existing Demand

No	Year	Departure		Arrival	
		Passenger	Freight Transport	Passenger	Freight Transport
1.	2016	3.262	707	2.019	353
2.	2017	6.419	1.988	3.077	518
3.	2018	5.042	1.446	1.799	313

Source: Kendal Transportation Agency, 2018

With this level of demand, Kendal Ferry Terminal only has one dock, port pool and one Ro-Ro ship with a capacity for 390 passengers and 25 freight transportations as a supply. **Table 1** indicates the level of departure demand which is ± 50% higher than arrival demand. This means that Kendal Ferry Terminal needs to improve its supply and performance. Based on passenger perceptions of the existing facilities and services through the questionnaire, it is apparent that passengers demand an increase in service and additional facilities despite the good current operation.

3.2. Analysis of Demand Prediction for Ten Years

The analysis will be calculated by using Compounding Interest method. First, passenger data is compiled by predicting the empty months because of docking and by predicting normal passenger's conditions during Eidul Fitri, which will be different from the existing data in **Table 1**. The data is presented in the **Table 2**:

Table 2. Processed Demand Data of Kendal Ferry Terminal

No	Year	Departure		Arrival	
		Passenger	Freight Transport	Passenger	Freight Transport
1.	2016	2.079	764	664	369
2.	2017	2.787	2.034	1.365	533
3.	2018	3.450	1.920	952	374

Source: Kendal Transportation Agency 2018, processed

The formula is

$$Yn = Yo \times (1 + r)^{N}\ [5] \quad (1)$$

Yn : Planning year's data

Yo : This year's data

r : Growth rate, with formula : $r = \frac{\Delta Y}{\Delta Yo} \times 100\% = \frac{Y1 - Yo}{Yo} \times 100\%$;

N : Planning year

1) Departure

a. Passenger

Growth rate levels are obtained from the three districts with the most passengers based on questionnaire results, which are Tegal[6] district, Banjarnegara[7] district, and Temanggung[8] district. The level of population growth rates is r = 2,31%, with Yo = 3.450 where the prediction result is depicted in **Figure 2**:

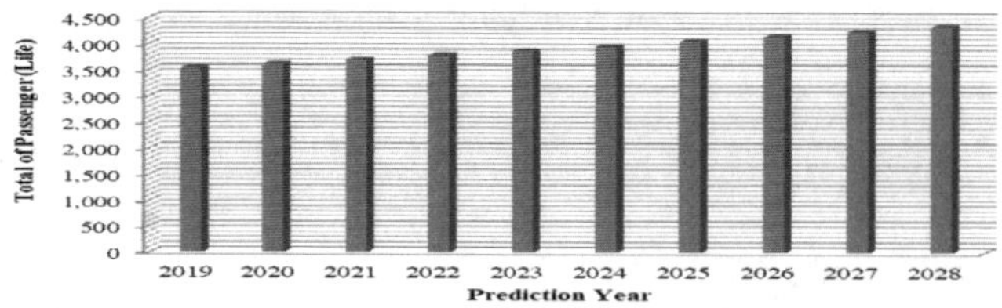

Figure 2. Departure Passenger Prediction Chart

b. Freight Transport

For freight transport, the level of GRDP growth rates used are from Tegal district, Temanggung district, and Semarang city[9]. The level of GRDP growth rates is r = 6,12%, with Yo = 1.920 where the prediction result is illustrated in **Figure 3**:

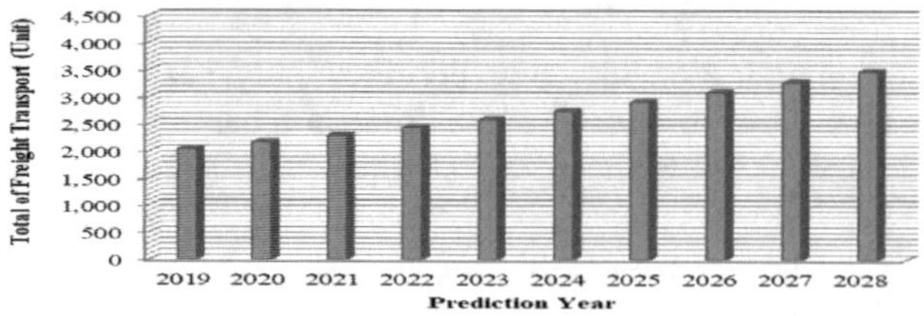

Figure 3. Departure Freight Transport Prediction Chart

2) Arrival

a. Passenger

Similar to departure, the analysis used population growth data from Tegal district, Banjarnegara district, and Temanggung district. The level of population growth rates is r = 2,31%, with Yo = 952 where the prediction result is indicated in **Figure 4**:

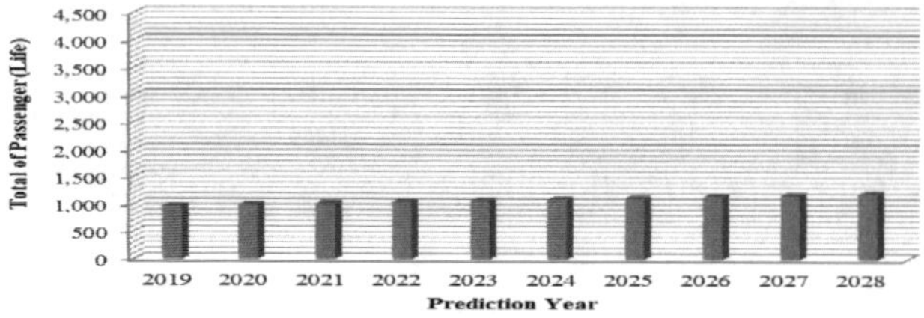

Figure 4. Arrival Passenger Prediction Chart

b. Freight Transport

The level of GRDP growth rates used in the freight transport arrival prediction are from Tegal district, Temanggung district, and Semarang city. The level of GRDP growth rates is r = 6,12%, with Yo = 374 where the prediction result is illustrated in **Figure 5**:

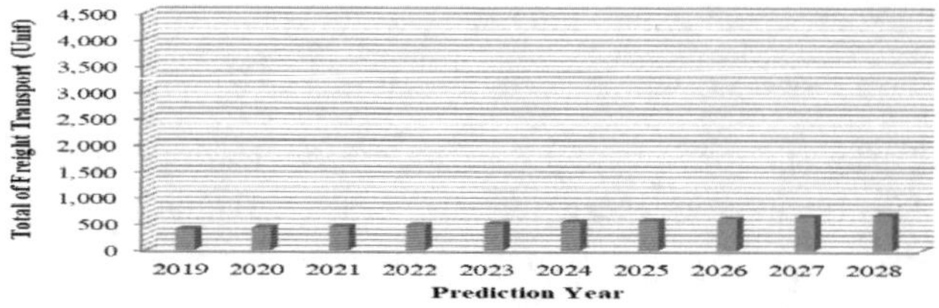

Figure 5. Arrival Freight Transport Prediction Chart

IOP Conf. Series: Earth and Environmental Science **328** (2019) 012009 doi:10.1088/1755-1315/328/1/012009

If the prediction result is compared with assumption where 30% passenger of Semarang–Kumai route in Tanjung Emas Port (first, Semarang-Kumai passenger will be predicted also), it will be diverted to Kendal Ferry Terminal. Thus, the result is presented in **Table 3**:

Table 3. Comparison of Demand Prediction with 30% Semarang–Kumai Passenger

Departure						Arrival					
Year	30% Assumption Semarang-Kumai will Distract		Year	Kendal Ferry Terminal		Year	30% Assumption Semarang-Kumai will Distract		Year	Kendal Ferry Terminal	
	P	FT		P	FT		P	FT		P	FT
2019	16.103	1.963	2019	3.530	2.038	2019	17.675	2.080	2019	974	397
2020	16.540	1.917	2020	3.611	2.162	2020	18.186	2.288	2020	996	422
2021	16.989	1.873	2021	3.695	2.295	2021	18.712	2.517	2021	1.020	442
2022	17.451	1.828	2022	3.780	2.436	2022	19.253	2.769	2022	1.043	474
2023	17.925	1.785	2023	3.868	2.585	2023	19.810	3.045	2023	1.067	503

Source: Tanjung Emas Port 2018, processed

Table 3 informs that:

1) On departure, the total freight transport demand in Kendal Ferry Terminal can achieve the
assumption. If the diversion occurs, Kendal Ferry Terminal must immediately prepare itself to receive that demand by adding supply, alike large capacity ships.
2) On arrival, the prediction of demand level in Kendal Ferry Terminal is still far from the assumption that 30% of Semarang–Kumai will be diverted. This is caused by the fact that demand level in Kendal Ferry Terminal was already low from the beginning and probably interrupted by another factor in Kumai Port.
3) For the passengers, the demand level on departure and arrival is still low. It was caused by the excess of the Kendal Ferry Terminal which made the low freight transport operation cost if it passes Kendal Ferry Terminal rather than from Tanjung Emas Port Semarang, which was not likely felt by passengers. Therefore, the number of passenger demand is still low.

3.3. Analysis of Kendal Ferry Terminal Existing Performance

The calculation of the existing performance will follow the port's operational performance indicators and calculation of the facilities area.

For facilities, the calculation resume is presented in **Table 4**:

Table 4. Resume of Kendal Ferry Terminal Facilities Analysis

No.	Facilities	Calculation Standard for Kendal Ferry Terminal	The Existing Condition in Kendal Ferry Terminal	Explanation	Action Plan
	Land Side Facilities				
1.	Office Area	A = a1 + a2 + a3 + a4+ a5 = 1.500,8 m²	1.000 m²	Inadequate	Adding the terminal building area up to 3.900 m² in accordance with the Medium Term Plan of the Kendal Ferry Terminal in the Tanjung Emas Masterplan Attachment.
2.	Parking Area	For freight transport: A = a*n*N*x*y For passenger pick up: A = a*n1*N*x* y*z*⅟n2 A = 3.139 m²	20.000 m²	Adequate	-
3.	Prayer Facility Area	60 m²	56,25 m²	Inadequate	a. Adding equipment such as prayer mats. b. Routine cleaning and maintenance. c. Although it is sufficiently extensive, the prayer area at the Kendal Ferry Terminal is

No.	Facilities	Calculation Standard for Kendal Ferry Terminal	The Existing Condition in Kendal Ferry Terminal	Explanation	Action Plan
	Land Side Facilities				
					planned to be developed up to 270 m².
4.	Health Facility Area	60 m²	24 m²	Inadequate	a. Placement of health workers who are always stationary in the health room. b. Additional stock of medicines, minimum first aid c. Increase the area of health facilities up to 176 m² in accordance with the Tanjung Emas Masterplan Attachment.
5.	Generator Area	150 m²	36 m²	Inadequate	Increase the generator area to 850 m² in accordance with the Tanjung Emas Masterplan Attachment.
	Sea Side Facilities				
1.	Dock Length	$A \geq 1{,}3L$ $A = 84{,}83$	110 m	Adequate	a. Still needs an addition of one dock for Ro-Ro ship in preparing for additional ships or new routes. b. In accordance with the Tanjung Emas Masterplan Attachment, an additional 139 m for the new dock will be built.
2.	Berthing Area	$A = 1{,}8L*1{,}5L$ $= 11.496$ m²	1.595 m²	Inadequate	Area development needed to meet the standards are as planned.

Table 5. Resume of Kendal Ferry Terminal Facilities Analysis (Sequel)

No.	Facilities	Calculation Standard for Kendal Ferry Terminal	The Existing Condition in Kendal Ferry Terminal	Explanation	Action Plan
3.	Port Pool Area	Available	Available	Adequate	a. Deepen of the port pool, from -5 LWS to -6 LWS to -10 LWS based on the results of total passenger and freight transport prediction, and the assumption of the diversion of 30% of Tanjung Emas Port passengers. b. Conduct environmental assessment before deepening the port pool to find out whether the characteristics of the waters in the Kendal Ferry Terminal allow for -10 LWS depths.
4.	Shipping Line Width	$W = 9B + 30$ m $= 156$ m	90 m	Inadequate	Increase the width of the shipping channel up to 176 m according to the Tanjung Emas Masterplan Attachment.
5.	Ship Landing Area	$A = N * \Pi * R^2$ $= 49.259$ m²	150.000 m²	Adequate	-

Source: Analysis, 2018

For performance matter, it consists of several aspects, including:

A. Waiting Time/WT

Kendal Ferry Terminal waiting time is around 30 (thirty) minutes. It is adequate according to the standard which is 1 (one) hour.

B. Approach Time/AT

Kendal Ferry Terminal does not offer this service, Kalibodri ship captain that lead the Kendal–Kumai route always moves the ship on his own without ship guide helps where the process needs ± 15 minutes.

C. Effective Time compared with Berth Time (ET:BT).

Berthing Time is total time of ship mooring, with variables consisting of:

1) Effective time, based on observation and second data from 2016 to 2018 is presented in the **Table 6**:

Table 6. Effective Time of Kendal Ferry Terminal

No	Year	Effective Time
1.	2016	122 Hours
2.	2017	210 Hours
3.	2018	146 Hours

2) Idle time of Kendal Ferry Terminal is between 0,5–1 hour and it is adequate.
3) Nonoperation time of Kendal Ferry Terminal in 2016–2018 is presented in the **Table 6**:

Table 7. Nonoperating Time of Kendal Ferry Terminal

No	Year	Not Operating Time
1.	2016	2469 Hours
2.	2017	3467 Hours
3.	2018	3417 Hours

Therefore, the berthing time of Kendal Ferry Terminal is presented in the **Table 8**:

Table 8. Kendal Ferry Terminal Berthing Time

No	Tahun	Berthing Time
1.	2016	2591 Hours
2.	2017	3677 Hours
3.	2018	3563 Hours

From that data, the Effective Time compared by the Berth Time (ET:BT) is presented in the **Table 9**:

Table 9. ET : BT of Kendal Ferry Terminal

No	Year	Effective Time	Berthing Time	ET : BT
1	2016	122	2,591	$\frac{122}{2591}$ x 100 = 4.69%

Table 10. ET : BT of Kendal Ferry Terminal (Sequel)

No	Year	Effective Time	Berthing Time	ET : BT
2	2017	210	3,677	$\frac{210}{3677}$ x 100 = 5.71%
3	2018	146	3,563	$\frac{146}{3563}$ x 100 = 4.08%

D. Berth Occupancy Ratio/BOR

BOR calculation is adapted to the type of pier of the Kendal Ferry Terminal, divided into one/several anchored vessels [10]. With available time in **Table 11**, the BOR is presented in **Table 12**:

Table 11. Kendal Ferry Terminal Available Time

No	Year	Available Time
1.	2016	1982 Hours
2.	2017	4014 Hours
3.	2018	2935 Hours

Table 12. Kendal Ferry Terminal BOR

No	Year	BOR
1.	2016	$BOR = \frac{2591}{1982} \times 100\% = 130{,}7\%$
2.	2017	$BOR = \frac{3677}{4014} \times 100\% = 91{,}6\%$
3.	2018	$BOR = \frac{3563}{2935} \times 100\% = 121{,}4\%$

E. Operation Tools Readiness

Based from the observation, Kendal Ferry Terminal did not use tools to assist the operation. Therefore, this indicator cannot be calculated.

The analysis of the existing performance of Kendal Ferry Terminal is depicted in the **Table 13**:

Table 13. Existing Performance Based on Several Indicators

No.	Indicators	Standard	Kendal Ferry Terminal	Explanation	Action Plan
1.	Waiting Time	Maximum 1 Hour	0,5 Hour	Adequate	To maintain a good waiting time, in line with increasing the number of ships and routes, the terminal required facilities such as: - Procurement of guiding ships. - Additional docks to reduce ship queues.
2.	Approach Time	Maximum 1 Hour	(None)	Inadequate	Add a guiding ship as a navigation tool to support the safety, efficiency, and comfort of ship service.
3.	Effective Time/Berthing Time	Minimum 70%	4,08 – 5,71%	Inadequate	a. Increase the depth of the port pool so that the ship does not need to wait for the tide in order to depart. b. Add departure schedule from 1 time per week to 2 times per week so that the ship does not wait too long in the mooring. It will increase non-operating time and berthing time. For the long term by looking at the prediction results, the schedule is planned to be 3-4 times per week.
4.	BOR To the post that has 2 piers: According to UNCTAD According to Director General of Sea Transportation regulation	Maximum 50% Maximum 70%	91,6 – 130,7%	Inadequate	The BOR value is influenced by the available time which is greater than the spent time. This can be reduced by: - Adding supply of ships that operate in Kendal Ferry Terminal is added along with predictions of additional passengers and freight transport. It can be the same route or new route ships. - Repairing supply, port pool depth.

Table 14. Existing Performance Based on Several Indicators (Sequel)

No.	Indicators	Standard	Kendal Ferry Terminal	Explanation	Action Plan
5.	Tools Operation	Minimum 70%	(None)	Inadequate	a. Medium-term plan consist of, procurement of guiding ships. b. Long-term plan consist of, procurement of waiting vessels.

Source: Analysis, 2018

4. Conclusion

In the end, the result of this study indicates that the demand prediction of Kendal Ferry Terminal reaches the assumption that 30% of Semarang–Kumai passengers at Tanjung Emas Port will divert to Kendal Ferry Terminal, especially on the departure side. The result of the study is still far away from the assumption, but for the freight transport, this effort is achievable. This is due to the potentials of Kendal Ferry Terminal's location, which is near the Pantura line, Semarang-Batang toll road, and Kendal and Semarang industrial area.

Kendal Ferry Terminal must improve its performance to receive diverted passengers and freight transports from Tanjung Emas Port in Semarang. This is due to the calculation of performance based on the indicators and facilities area, in which Kendal Ferry Terminal performance is still below the standard. It can be achieved by adding facilities that can support Kendal Ferry Terminal's performance including the passenger lounge, the depth of the pool, the new dock structure, and in order to attract

IOP Conf. Series: Earth and Environmental Science **328** (2019) 012009 doi:10.1088/1755-1315/328/1/012009

passengers, Kendal Transportation Agency must be objective by connecting it with other transportation modes, such as Kedungsepur BRT line and Damri bus.

References

[1] Transportation Minister 2013 *Peraturan Menteri Perhubungan Nomor PM 18 Tahun 2013 Tentang Rencana Induk Pelabuhan Tanjung Emas Semarang* (Ministry of Transportation, Jakarta)

[2] Dinas Perhubungan Kabupaten Kendal 2016 *Profil Pelabuhan Kendal* (Kendal)

[3] Perdana, AP 2017 *Pelabuhan Kendal Akan Bertaraf Internasional* able to access at https://kompas.id/baca/nusantara/2017/12/29/pelabuhan-kendal-akan-bertaraf-internasional/

[4] Sea Transportation Director General 2011 *Keputusan Direktur Jenderal Perhubungan Laut Nomor UM.002/38/18/DJPL.1/2011 Tentang Standar Kinerja Pelayanan Operasional Pelabuhan* (Ministry of Transportation, Jakarta)

[5] Ahlina AHN 2015 *Analisa Proyeksi Penumpang Kapal Cepat Palembang – Muntok* Mini Skripsi Sekolah Tinggi Transportasi Darat (Bekasi)

[6] Tegal Central Bureau of Statistics 2018 *Data Penduduk dan Data PDRB* able to access at http://www.tegalkab.bps.go.id/

[7] Banjarnegara Central Bureau of Statistics 2018 *Data Penduduk Berdasarkan Jenis Kelamin* able to access at http://www.banjarnegara.bps.go.id/

[8] Temanggung Central Bureau of Statistics 2018 *Data Penduduk dan Data PDRB* able to access at http://www.temanggungkab.bps.go.id/

[9] Dinas Komunikasi, Informasi, Statistik dan Persandian dan BPS Kota Semarang 2018 *Analisis Ekonomi Regional Kota Semarang 2016* (Semarang)

[10] Rumambi CC, Sendouw, Manoppo MRE 2016 Evaluasi Kinerja Operasional Pelabuhan Bitung *Jurnal Tekno Vol.14/No.66/Desember 2016 ISSN:0215-9617* pp.1-11 (Sam Ratulangi University, Manado)

The Influence of The Distribution of Public Green Space on The Health of The Residential Environment in Balikpapan City

Ariyaningsih[1], A N Dewanty[1], and M Ulimaz[1]
[1] Department of Urban and Regional Planning, Institut Teknologi Kalimantan

*email: ariyaningsih@itk.ac.id

Abstract. The population of Balikpapan City has increased along with economic growth. The increasing population has resulted in the densification of the population and settlements which. That are fast and uncontrolled in parts of the city. The increasing demand for space, especially for settlements and built-up land, has an impact on the deterioration of the quality of the environment, especially the public green space. Balikpapan is one of the cities that despite extensive protected forests but still has problems providing public green open spaces. One of the impacts of the lack of green space in urban areas is that the community has a limited place to socialize with the surrounding environment. Furthermore the limited green open space in urban areas can cause pollution. This has resulted in decreasing public health. The analytical method used is path analysis technique. The type of data used in this path analysis is cross section data. The results show the contribution of green space distribution in disease is 2.9% with the highest proportion of being diarrhoea. In addition, the distribution of green space affects the number of infant mortality rate is 5.5% while the rest are contributions from other variables not included in the study.
Keywords: environmental health, green space, green open space, path analysis, public health

1. Introduction

The population of Balikpapan City has constantly increased, in 2010, Balikpapan had 560,781 people, and in 2015, it increased to 615,574 people [2]. This causes a large space requirement for cities. The increasing demand for space, especially for settlements and built-up land, has an impact on the deterioration of the quality of the environment, especially the public green open space. Spatial plans (RTRW) that have been made are not able to prevent land conversion in urban areas so that the Green Open Space (RTH) is increasingly threatening and cities are not comfortable to move [4]. Balikpapan City is one of the cities that despite extensive protected forests but it still has problems providing public green open spaces. This is because the function of protected forests is not for the conservation of the local environment of the city of Balikpapan but rather the buffer function in the wider surrounding area [9]. Thus, protected forests are not part of the provision of RTH in Balikpapan City.

The existence of open green space aims to make land available as water absorbing area. In the urban aspect, green space can keep the balance between natural environment and man-made environment for people's needs. There are many functions which green space provides, such as ecology, social, and aesthetic that can give amenities [4]. One of the impacts of the lack of availability of green open space in urban areas is that the community has a limited place to socialize with the surrounding environment.

Furthermore, limited RTH in urban areas can cause pollution so that it can result in decreasing public health. Therefore, the function of public green space has a positive influence on the environmental health of the surrounding community, especially those with access to adequate public green space that tends to be healthier physically than those who do not [16].

With an inadequate area and distribution of open green space, the city of Balikpapan has a tendency to have a poor health level compared to the surrounding area. Based on the health profile data of East Borneo Province in 2015, Balikpapan City had a high morbidity rate among other cities / regencies in East Borneo including TB morbidity of 409 (the second highest), the highest number of dengue cases, highest pneumonia cases and diarrhoea cases around 133 cases. Many studies have proven that public green open space can affect the health of surrounding communities.

The city of Balikpapan is the most populous city in the province of East Borneo. This of course makes the need for green open space bigger than other cities / districts in the province of East Borneo. Seeing the importance and magnitude of the impact of the availability of public green space in urban areas, the availability of public green open space is still limited in Balikpapan City. There needs to be a strategy to fulfil public green open space that fits the needs of the people of Balikpapan, especially in improving environmental health. Based on the background described previously, it is necessary to study the influence of the distribution of public green space on the level of health of the residential environment in Balikpapan City.

2. Methods

This type of research is quantitative descriptive. According to [12] descriptive quantitative research methods are used to describe or explain events or events that occur at this time in the form of meaningful numbers. In this study, this trait was used to analyse the influence of the distribution of public green space on the quality of environmental health with statistical analysis techniques. The analysis used is path analysis. The type of data used in this path analysis is cross section data.

3. Results and Discussions

3.1 Overview Public Green Space and Distribution of Public Green Open Space in Balikpapan City

The existence of a Green Public Open Space in Balikpapan City is still very varied, such as urban forests, city parks, green lanes, cemeteries, protected forests, and others. One of the green open spaces currently maintained is the Wain River Protected Forest (HLSW) in which this area is a River Watershed (River Watershed) with Bugis Sub Watershed. This area is maintained with the aim of protecting the Wain River and Bugis Bay, as well as for the preservation of clean water sources for the people of Balikpapan City. Based on the survey results, the results of the distribution of public green open spaces in Balikpapan City are as follows:

Table 1. Distribution of Public Green Open Space in Balikpapan

Sub District	Area (m^2)			
	Park	City Forest	Green Lane	Cementery
East Balikpapan	0	545000	0	50492
South Balikpapan	18906	63240	182000	62000
Balikpapan City	19068	305805	73200	8000
Balikpapan Centre	4183	534172	12000	17500
West Balikpapan	0	0	5000	41900
North Balikpapan	675	34172	7060	90450
Total	**42832**	**1482389**	**279260**	**270342**

Source: Primary Survey, 2018

IOP Conf. Series: Earth and Environmental Science **328** (2019) 012010 doi:10.1088/1755-1315/328/1/012010

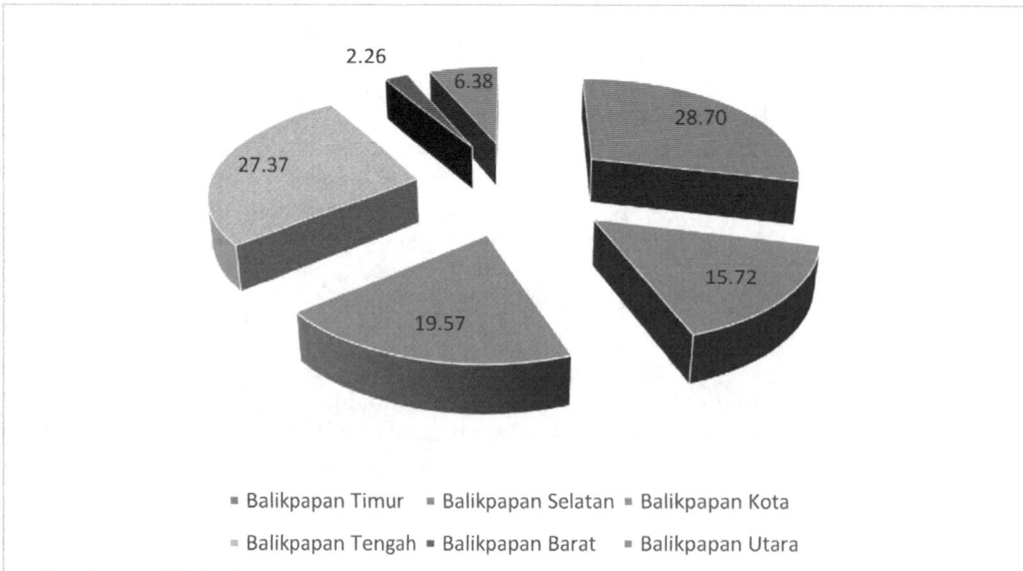

Figure 1. Distribution of Public Green Open Space in Balikpapan

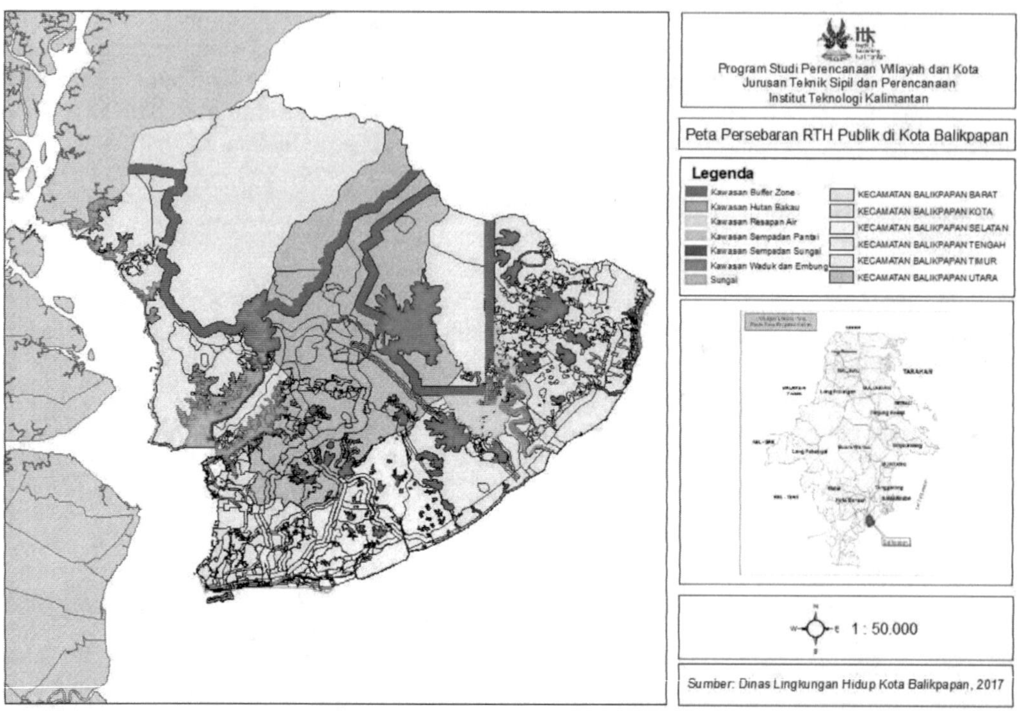

Figure 2. Maps of Open Green Space Distribution in Balikpapan

The Influence of The Distribution Of Public Green Space On The Level Of The Health Of The Settlement Environment Of Balikpapan.

A. Number Of Occurrences Of Disease

Based on the data obtained, the number of the occurrences of diarrhoea is the highest among other diseases. The number of diarrheal diseases is the highest in all sub-districts with the highest number of occurrences Balikpapan City Sub District (2444 people) with a population of 625,968 people. Whereas the number of dengue cases is still low in all sub-districts. The following is a detailed description of the number of disease occurence in all sub-districts:

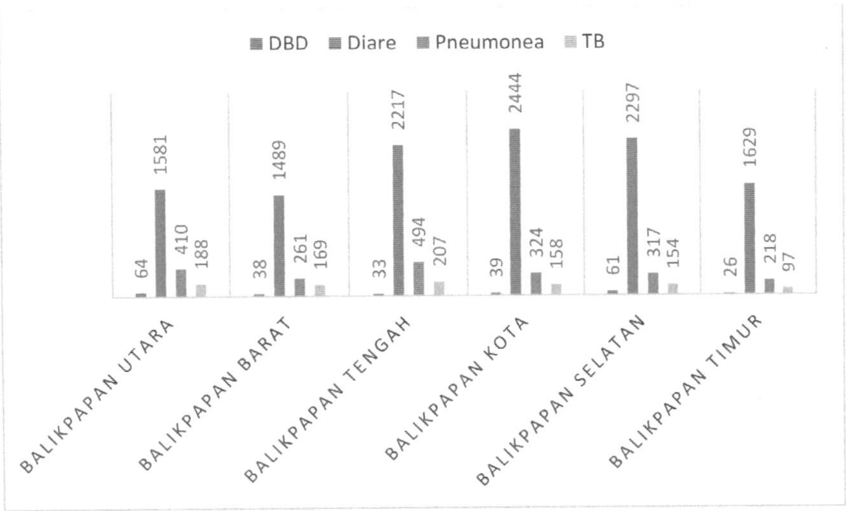

Figure 3. Number of Events and Distribution of DHF, Diarrhea, Pneumonea, and TB in 2017

B. The Number of infant mortality rates

The number of infant mortality rates is one indicator that can be used to measure the level of health of the community. The high rate of infant mortality rates can be assumed that the health condition of the community is low. The following is a description of the existing condition of the number of babies who die in Balikpapan:

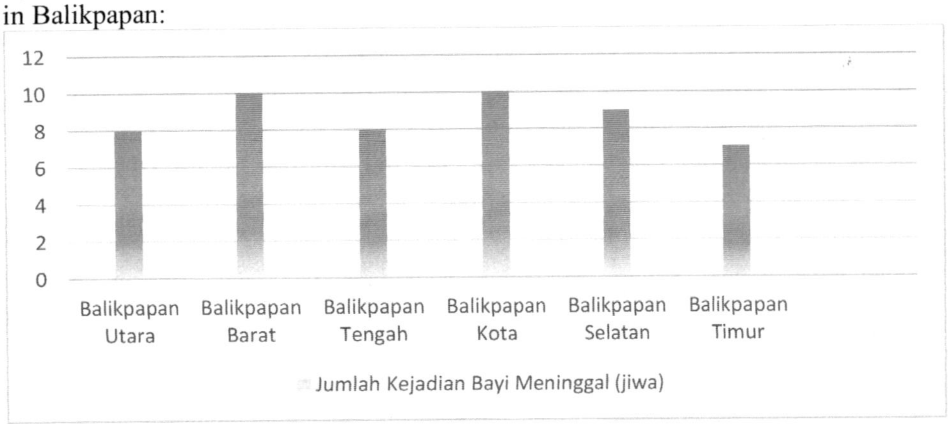

Figure 4. Number of Babies Died

Based on data on the number of diseases, the number ofinfant mortality rate , and ambient temperature, it can be seen that these data are dependent variables, while the distribution of RTH is an independent variable. The distribution of green space is a variable (x) while the number of diseases, infant mortality rate are variable (y).

Table 2. Disease Variable Regression Results

Model Summary

Model	R	R Square	Adjusted R Square	Std. Error of the Estimate

1	.170a	.029	-.214	532.654

a. Predictors: (Constant), Sebaran RTH

Coefficientsa

Model		Unstandardized Coefficients		Standardized Coefficients	t	Sig.
		B	Std. Error	Beta		
1	(Constant)	2359.245	427.418		5.520	.005
	Sebaran RTH	.000	.001	.170	.344	.748

a. Dependent Variable: Penyakit

Based on the value of the regression output in the Coefficients table, it is known that the significance value of the disease variable is 0.748. The magnitude of the R square value found in the Model Summary table is 0.029. This shows that the contribution of the influence of green space distribution to the type of disease is 2.9% while the rest is the contribution of other variables not included in the study. Then to find out which types of diseases are affected by the distribution of RTH, a regression analysis is carried out again with the following results

Table 3. The results of SPS analysis of Sub Disease Variables

No	Diseases	R square	Standardized Coefficients	Signifikansi
1.	DBD	0.15	0.121	0.819
2	Diarrhea	0.00	0.016	0.976
3	Pneunomia	0.263	0.513	0.298
4	TB	0.338	0.581	0.226

Source: analysis results, 2018

From the table above, it is known that the most significant disease affected by the spread of green space is diarrheal disease, then dengue disease, followed by pneumonia and finally TB. It is also known from the R square value that the contribution of the influence of green space distribution to TB disease is 33.8% and the pneumonia disease is 26.3% followed by DHF is 15%.

Table 4. Influence of RTH Distribution to Number of Babies Died

Model Summary

Model	R	R Square	Adjusted R Square	Std. Error of the Estimate
1	.236a	.055	-.181	1.316

a. Predictors: (Constant), Sebaran RTH

Coefficientsa

Model		Unstandardized Coefficients		Standardized Coefficients	T	Sig.
		B	Std. Error	Beta		
1	(Constant)	8.226	1.056		7.790	.001
	Sebaran RTH	1.274E-006	.000	.236	.485	.653

a. Dependent Variable: Bayi Meningal

Based on the value of the regression output in the Coefficients table, it is known that the significance value of the variable number of infants dying is 0.653. The value of the R square found in the Model Summary table is 0.055. Thisindicates that the contribution of the influence of green space distribution to the type of disease is 5.5% while the remainder is the contribution of other variables not included in the study.

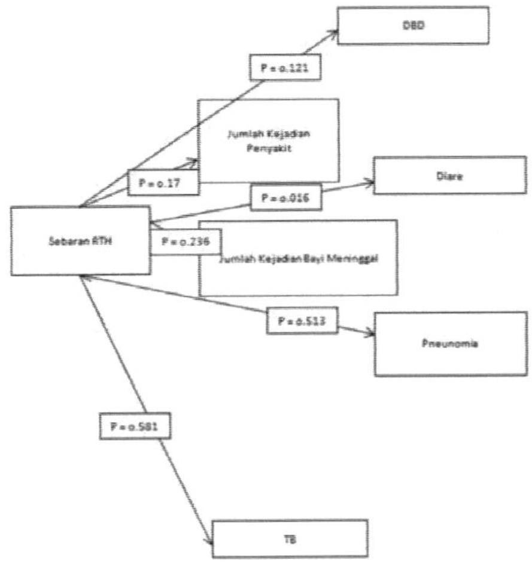

Figure 5. Path Coefficient

4. Conclusions

The existence of a Green Public Open Space in Balikpapan City is still varied, such as urban forests, city parks, green lanes, cemeteries, protected forests, and others. The variable of the influence of the distribution of public green space on the level of the health of the settlement environment of Balikpapan is "The number of Occurrences Of Disease" and "number ofinfant mortality rate". The number of disease includes DBD, diarrhoea, pneumonia, and TB. The study shows that the contribution of the influence of green space distribution to the type of disease is 2.9%. In addition, this research indicates that the contribution of the influence of green space distribution to the type of disease is 5.5% while the remainder is the contribution of other variables not included in the study.

References

[1] Badan Perencanaan dan Pembangunan Kota Balikpapan. 2012. *Materi Teknis RTRW Kota Balikpapan tahun 2012-2032.*

[2] BPS Kota Balikpapan. 2016. *Balikpapan Dalam Angka Tahun 2015.*

[3] Bewu, M. F., & Setiawan, R. P. 2015. *Penyediaan Ruang Terbuka Jikau Berdasarkan Kebutuhan Oksigen di Kota Malang. Jurnal Teknik ITS Vol.4 No.2, C-98*

[4] Dwihatmojo, Roswidyatmoko. 2013. *Pemanfaatan citra quickbird untuk identifikasi ruang terbuka hijau kawasan perkotaan.* Jurnal Seminar Nasional Pendayagunaan Informasi Geospatial untuk Optimalisasi Otonomi Daerah ISBN: 978-979-636-152-6. Pusat Pemetaan Tata Ruang dan Atlas, Badan Informasi Geospasial.

[5] Dinas Kesehatan Provinsi Kalimantan Timur. 2016. *Profil kesehatan Provinsi Kalimantan Timur Tahun 2015.* Samarinda: Dinas Kesehatan Provinsi Kalimantan Timur

[6] Hadari, Nawawi. 2003. *Metode Penelitian Bidang Sosial.* Yogyakarta. Gramedia Widiasarana Indonesia

[7] Hakim dan Utomo. 2004. *Komponen Perancangan Arsitektur Lansekap.* Jakarta: Bumi Aksara Kementerian Pekerjaan Umum. Peraturan Menteri No. 5 Tahun 2008 tentang Penyedian Ruang Terbuka di Kawasan Perkotaan. Indonesia: Kementerian Pekerjaan Umum

[8] Purnomohadi, Ning. 2006. *Ruang Terbuka Hijau Sebagai Unsur Utama Tata Ruang Kota* . Jakarta:Direktorat Jenderal Penataan Ruang Kementrian Pekerjaan Umum

[9] Sineri, A. Silas, Mahmud. 2014. *Fungsi Kawasan dan Strategi Pengelolaan Hutan Lindung Wosi Rendani Kabupaten anokwari.* Jurnal AGRIFOR Vol. XIII No. 2 Oktober 2014 hal. 131-140.

IOP Conf. Series: Earth and Environmental Science **328** (2019) 012010 doi:10.1088/1755-1315/328/1/012010

[10] Sumarmi. 2006. *Disertasi: Partisipasi Masyarakat dalam Pengelolaan RTH Permukiman di Kota Malang.* Malang: Universitas Brawijaya
[11] Sugiyono. 2009. *Statistika Untuk Penelitian.* Bandung: CV. Alfabeta

[12] Sugiyono, 2012. *Metode Penelitian Kombinasi.* Bandung: CV. Alfabeta
[13] Sunyono, Danang. 2012. *Model analisis jalur untuk riset ekonomi.* Bandung: CV. Yrama Widya *Undang-undang Republik 26 Tahun 2007 tentang Penataan Ruang*
[14] Mooi, E. , Sarstadt, M. 2011. *A concise Guide to Market Research : The Process, Data And Methods Using IBM SPSS Statistics.* Chapter 9 : Cluster Analysis p.237-284. Berlin : Springer-Verlag. http://www.springer.com/978-3-642-12540-9
[15] Murthy, P. Rama. 2007. *Operations Research.* New Delhi : New AgeInternational Limited Publisher
[16] Zhu, Yuemei, Ding, Junxiang, Zhu, Qing, Cheng, Yang, Ma, Qiuchen, Ji, Xuze. 2017.The Impact of Green Open Space on Community Attachment-A Case Studi of Three Communities in Beijing

IOP Conf. Series: Earth and Environmental Science **328** (2019) 012011 doi:10.1088/1755-1315/328/1/012011

Study of Intermittent-Irrigation Technique to Cropping Patterns and Water Availability at Pemali Kanan Irrigation Area in Brebes Regency

B B Kusumo[1,2*], Pranoto S A[2], and I Sriyana[2]

[1] Office of Water Resources Management and Spatial Planning of Brebes Regency, Brebes Regency, Central Jawa, Indonesia
[2] Master of Civil Engineering, Diponegoro University, Semarang 50275, Central Java, Indonesia

*email : bayu.bbk@gmail.com

Abstract. Paddy farming is closely related to irrigation water supply since it is a kind of plant that has a high water requirement. Therefore, in major areas in Indonesia, paddy is not recommended to be grown in dry season. In Brebes Regency, many farmers still pick paddy for all three planting seasons that they suffer from insufficient of irrigation water supply. The objectives of this study is to find out how water availability compared with water requirement at Pemali Kanan Irrigation Area in both periods (present time and 20 years to come) and is intermittent irrigation technique required to solve this water supply problem. This research is conducted by calculating water availabilty using FJ Mock Method, generating water availability for the upcoming 2 decades using Thomas-Fiering Method and analysing water requirements for all 13 cropping patterns. Output of the analysis shows that water availability of Pemali River in both periods are statistically identical. Meanwhile, the average water requirements are mostly negative. It means that water requirements for all 13 cropping patterns could be fulfilled by rainfall. The conclusion is that in most 2-weekly periods, rainfall could provide adequate water supply for all areas. Intermittent irrigation technique will only be required in a certain period that is the 3rd planting season which lasts from Juni to October.

1. Introduction

Water requirement in paddy farming is an essential element farmers should consider in order to succeed in their business. Water availability must be in adequate amount as well so that it can supply all the areas. Most of Indonesian areas have high rainfalls, 2.800 mm/year in average. Unfortunatelly, they are accumulated only in short periods, approximately 4 to 9 months a year. As the result, there are quite long dry-periods every year[1]. This condition shows that there is imbalance between water availability and water requirement [2].

One method in paddy farming commonly implemented to manage water requirements and water availability is cropping pattern. This effort requires farmers to arrange what kind of plant to grow and when to start planting so that the pattern still corresponds with water availability. Farmers should be aware of months in which water supply are lower than water requirements. They should not grow plants with high water requirements, such as paddy, within these low-water months. This way, farmers will not suffer from lack of irrigation water.

In northern part of Brebes Regency, in the dry season which last from April to October, not less than 25.180 hectare land is threathened by drought[1]. Therefore, The Government of Brebes Regency through its agencies always urges farmers especially the ones whose lands are located in the northern part of Brebes Regency to obey and implement the cropping patterns that have been set by regent decree. This cropping pattern issue has become much more sensitive since many farmers still choose paddy as the commodity to grow in the dry season. Therefore, most of them complain about insufficient water supply. Their lands were not properly watered so that the plants didn't grow well and the productions were lower than they used to be.

In terms of paddy farming, there are many kinds of advanced farming methods which had been introduced these recent years[3]. Besides providing scientific breakthroughs in farming, those methods also offer techniques to gain efficiency in water irrigation. One of the most popular farming methods which was once widely used in Indonesia is System of Rice Intensification (SRI). The principles of System of Rice Intensification (SRI) method are to do more eco-friendly farming, to minimize external input (e.g. irrigation water), to apply local wisdom, and to reduce chemicals consumption (e.g. fertilizer and pesticide)[4]. SRI methods would also offer greater paddy yields along with lower production costs due to lower consumption of seeds, chemical fertilizers and perticides. These factors would generate higher profits to farmers[5]. SRI method contains a specific technique to deal with water requirements called Intermittent Irrigation[4, 5, 6]. Unlike conventional farming method, the intermittent irrigation applies periodical watering instead of continuous watering technique.

Regarding with this water availability issue, a study is required to figure out which cropping pattern has the least water requirement. The result of this study would helps The Government of Brebes Regency to develop and to implement more suitable cropping patterns which correspond with water availability in Pemali Kanan Irrigation Area, and throughout Brebes Regency. How much the urgency of intermittent-irrigation technique implementation to save irrigation water in Pemali Kanan Irrigarion Area will also be observed..

1.1. Problem Formulation

Based on that background, the major issues in this study are as follows :

1. How is water availability compared with water requirements in Pemali Kanan Irrigation Area at present time and in the next 20 years ?
2. How much is water requirement for various cropping patterns in Pemali Kanan Irrigation Area?
3. What are farmers' opinion regarding advanced farming method like SRI Method ?

1.2. Purposes and objectives

The main objective of this study is to figure out which cropping pattern has the least water requirement so that it can be suited with the water irrigation availability in Pemali Kanan Irrigation Area. Therefore, this study is aimed to :

1. Analyze water availability in Pemali Kanan Irrigation Area either at present or in 20 years to come.
2. Analyze which cropping pattern and planting-season-commencement combination that require the least irrigation water whether it is applying conventional method or intermittent irrigation technique.

2. Methods

2.1. Materials

This research is conducted in Pemali Kanan Irrigation Area. Its area is 7.788 hectare and lies on 9 sub-districts in northern part of Brebes Regency. Pemali Kanan Irrigation Area is a part of Pemali Hilir Irrigation Area, a large irrigation area in the downstream of Pemali River whose area is 25.180 hectare [7].

Regarding the local farmers, by means of questionnaires, this study would also gain information from farmers who have an experience in applying System of Rice Intensification (SRI) Method, to find out what they consider about this advanced farming method. The respondents are choosen among farmers who have implemented System of Rice Intensification (SRI) method themselves or they have been given such training regarding System of Rice Intensification (SRI).

Based on location, respondents live in the southern area of Brebes Regency which are Banjarharjo, Bantarkawung, Bumiayu, Tonjong and Sirampog sub-districs. All of them were chosen as demplot (a kind of farm land for new-methods trial) when System of Rice Intensification (SRI) was introduced to and implemented in Brebes Regency for the first time in 2005.

2.2. Research Framework

Once all data are collected, a sequences of analysis is carried out. The first step to be taken is analyzing the existing water availability of Pemali River using FJ Mock method. This stage will also need a series of analysis to figure out what is the potential evaporation in Pemali River based on weather and climatology data.

Next step is to use the existing water availability data as an input to generate numbers of water availability of Pemali Kanan in the upcoming 20 years. Thomas Fiering Method[8] will be utilized in this step..

In order to find out how the condition of water requirements in Pemali Kanan Irrigation Area is, a number of cropping pattern simulations are carried out. These simulations involve not less than 13 cropping patterns which are formed by 4 kind of crops : paddy, corn, soybean and red onion. Technically, these simulations are conducted by switching the commencement of planting seasons for every single cropping pattern based on a 2-weekly period. It means, every cropping pattern would have 24 alternatives for a one-year period. That way, various water requirements will be obtained. After that, each of these numbers are compared to figure out which cropping pattern (along with the planting-seasons-commencement combination) that needs the least irrigation water. This step is based on the facts that incompatibility of water requirements to water availability is caused by cropping pattern violation which surely will cause loss of productivity.

In the end, questionnaires are used to get information about how farmers' opinion about certain farming methods, namely as System of Rice Intensification (SRI) method, which contains this intermittent irrigation technique. This information would be useful to discover why such a developed farming method could not last long in terms of their implementation.

3. Results and Discussions

3.1. Water Availabilty

The analysis ouput shows that for present time, water availability in Pemali River is quite high. The highest debit could reach 250,31 m^3/sec (took place in Februay II, 2009), while the average debit is 49,75 m3/sec. For certain periods, during dry seasons which last from June to October, Pemali River's water availability is recorded low that it sometimes is incapable of supplying the entire area.

This condition is similiar with the ones in the upcoming 20 years. The output of Thomas-Fiering Method, Pemali River's maximum water availability is recorded as high as 246,54 m3/sec (took place in February II, 2030) and the average debit is 48,05 m3/sec. Furthermore, statistically those two data has an identical pattern. This similiarity also proven by paired T-Test. The paired T-Test result shows that, the historical and the generated data come from the same population. **Figure 1** and **Figure 2** show how historical and generated debit are compared in terms of Average Debit and Standard Deviation.

3.2. Water Requirements

Based on the simulation, it shows that for all 13 cropping patterns, the average water requirements are relatively low compared to Pemali River's water availability. However, in certain months (from June to October) and in certain occasions such as Land Preparation phases, water requirments are recorded to be a little bit higher.

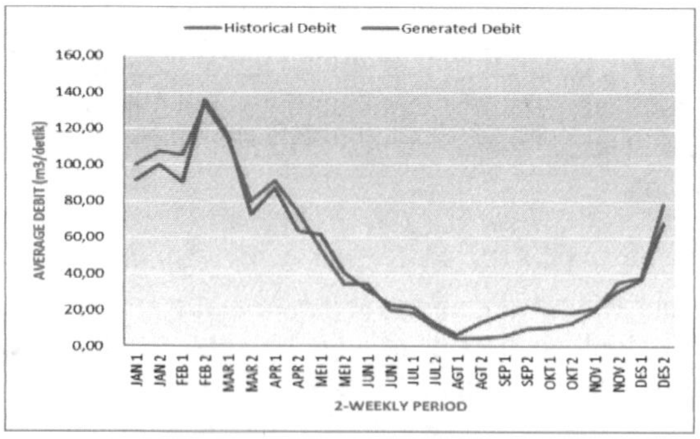

Figure 1. Average Debit Comparison

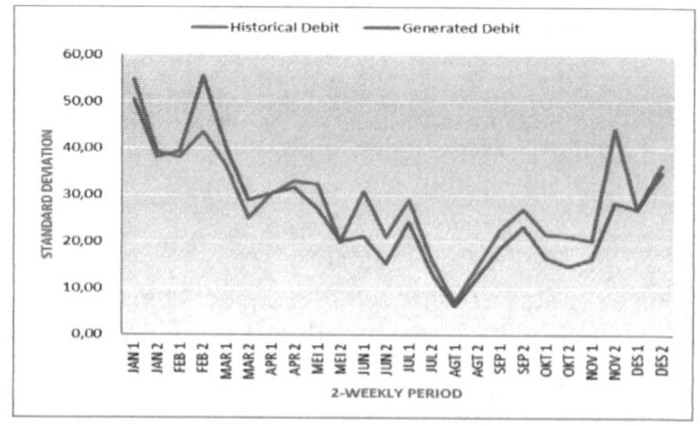

Figure 2. Standard Deviation Comparison

Table 1 shows the recap of all of the alternatives which has the least water requirements for each cropping patterns. This recap shows the water requirements of DI Pemali Kanan whose area is 7.788 hectare when conventional irrigation is applied. **Figure 3** and **Figure 4** show the comparison of water requirements (when conventional irrigation is applied) to both dependable existing debit and generated debit. For generated debit, year 2035 is taken as the example since it is the year in which the generated debit has the least average among others.

Table 1. Recap of Conventional Irrigation Water Requirements

No..	POLATANAM	SAT	JAN I	JAN II	FEB I	FEB II	MAR I	MAR II	APR I	APR II	MEI I	MEIII	JUN I	JUN II	JUL I	JUL II	AGT I	AGT II	SEP I	SEP II	OKT I	OKT II	NOV I	NOV II	DES I	DES II	RATA
1	PD - PD - CR	m3/det	0,02	6,95	9,70	0,00	3,41	5,69	6,20	5,58	1,93	0,55	14,34	15,91	8,29	10,30	11,70	11,70	12,45	6,25	2,64	27,65	2,82	3,85	5,05	2,06	7,29
2	PD - PD - SB	m3/det	3,00	1,84	4,39	1,44	2,90	0,00	1,04	11,52	13,00	6,26	8,63	10,07	9,93	8,97	5,70	2,71	4,56	7,20	8,54	8,45	5,36	2,87	12,26	9,85	6,27
3	PD - PD - ON	m3/det	3,00	1,84	4,39	1,44	2,90	0,00	1,04	11,52	13,00	6,26	8,63	10,07	9,93	8,97	5,70	2,71	4,56	6,32	6,47	6,90	5,47	5,06	12,26	9,85	6,18
4	PD - CR - CR	m3/det	5,19	1,68	4,39	1,12	0,00	0,00	17,08	1,39	4,08	5,54	5,37	6,71	4,69	26,33	6,14	7,56	10,08	10,20	9,37	8,07	0,00	13,43	12,26	2,71	6,81
5	PD - CR - SB	m3/det	5,19	1,68	4,39	1,12	0,00	0,00	17,08	1,39	4,08	5,54	5,37	6,71	4,69	26,33	6,64	8,18	10,04	9,28	6,74	4,31	0,00	13,43	12,26	2,71	6,55
6	PD - CR - ON	m3/det	5,19	1,68	4,39	1,12	0,00	0,00	17,08	1,39	4,08	5,54	5,37	6,71	4,69	26,33	5,89	6,46	8,54	9,39	9,11	0,00	0,00	13,43	12,26	2,71	6,31
7	PD - SB - CR	m3/det	5,19	1,68	4,39	1,12	0,00	0,00	17,08	1,82	4,60	5,51	4,76	4,99	3,43	26,33	6,14	7,56	10,08	10,20	9,37	8,07	0,00	13,43	12,26	2,71	6,70
8	PD - SB - SB	m3/det	5,19	1,68	4,39	1,12	0,00	0,00	17,08	1,82	4,60	5,51	4,76	4,99	3,43	26,33	6,64	8,18	10,04	9,28	6,74	4,31	0,00	13,43	12,26	2,71	6,44
9	PD - SB - ON	m3/det	5,19	1,68	4,39	1,12	0,00	0,00	5,79	1,82	4,60	5,51	4,76	4,99	3,43	26,33	5,89	6,46	8,54	9,39	9,11	0,00	0,00	13,43	12,26	2,71	5,73
10	PD - ON - CR	m3/det	12,39	0,00	0,00	0,00	0,64	3,84	1,76	0,00	0,00	14,09	14,34	7,96	10,52	10,17	11,70	11,39	6,25	2,77	28,13	3,29	0,00	0,00	0,00	0,00	5,80
11	PD - ON - SB	m3/det	12,39	0,00	0,00	0,00	0,00	1,46	0,00	0,00	0,00	14,09	14,34	7,96	10,52	10,17	11,70	11,39	6,25	2,77	28,13	3,29	0,00	0,00	0,00	0,00	5,60
12	PD - ON - ON	m3/det	1,15	1,84	2,09	9,18	0,00	1,22	0,93	3,42	4,99	0,00	0,00	0,00	15,73	15,01	9,61	11,79	12,77	12,66	11,71	3,06	0,00	20,29	1,60	0,00	5,79
13	PD - PD - PD	m3/det	10,31	0,00	2,20	1,44	3,24	5,35	3,25	11,52	13,00	6,26	6,27	7,72	8,04	7,56	5,74	17,05	17,80	10,82	13,10	11,61	8,86	7,97	0,82	9,85	7,91

KETERANGAN :
: LAND PREPARATION (LP)
: PADDY (PD)
: CORN (CR)
: SOY BEAN (SB)
: ONION (ON)
: FALLOW

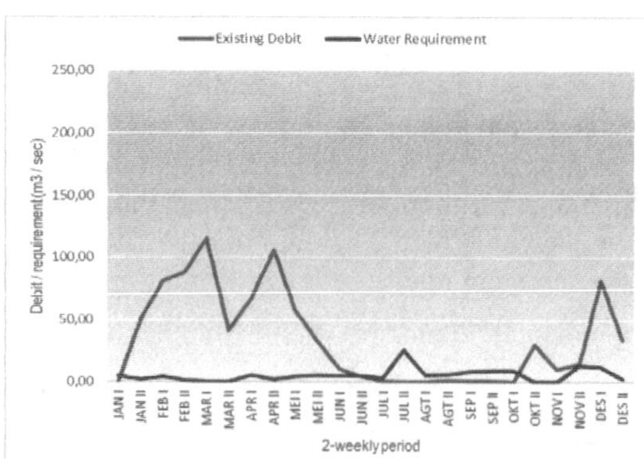

Figure 3. Comparison of Water Requirement (Conventional) to Existing Debit

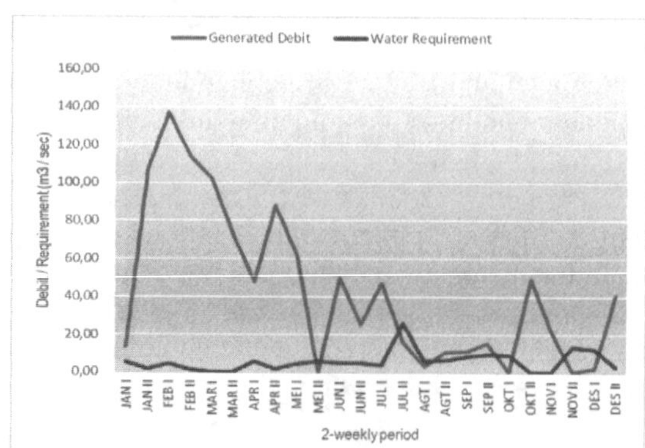

Figure 4. Comparison of Water Requirement (Conventional) to Generated Debit

Meawhile, the further calculation done regarding to intermittent irrigation shows that the proper implementation of intermittent-irrigation technique could save irrigation water to a remarkable number.

During land preparation period, intermittent irrigation could save up to 25,81% irrigation water. In cultivating period, this irrigation technique could help farmers to save irrigation water up to 37,54%. The recap of intermittent-irrigation water requirements is shown in **Table 2**.

Table 2. Recap of Intermittent-Irrigation Water Requirements

No.	POLA TANAM	SAT	JAN I	JAN II	FEB I	FEB II	MAR I	MAR II	APR I	APR II	MEI I	MEIII
1	PD - PD - CR	m3/det	0,01	5,16	7,19	0,00	2,13	3,55	3,87	3,48	1,20	0,34
2	PD - PD - SB	m3/det	1,87	1,15	2,74	0,90	1,81	0,00	0,65	8,55	9,65	3,91
3	PD - PD - ON	m3/det	1,87	1,15	2,74	0,90	1,81	0,00	0,65	8,55	9,65	3,91
4	PD - CR - CR	m3/det	3,24	1,05	2,74	0,70	0,00	0,00	12,67	0,87	2,55	3,46
5	PD - CR - SB	m3/det	3,24	1,05	2,74	0,70	0,00	0,00	12,67	0,87	2,55	3,46
6	PD - CR - ON	m3/det	3,24	1,05	2,74	0,70	0,00	0,00	12,67	0,87	2,55	3,46
7	PD - SB - CR	m3/det	3,24	1,05	2,74	0,70	0,00	0,00	12,67	1,14	2,87	3,44
8	PD - SB - SB	m3/det	3,24	1,05	2,74	0,70	0,00	0,00	12,67	1,14	2,87	3,44
9	PD - SB - ON	m3/det	3,24	1,05	2,74	0,70	0,00	0,00	4,30	1,14	2,87	3,44
10	PD - ON - CR	m3/det	9,20	0,00	0,00	0,00	0,40	2,40	1,10	0,00	0,00	10,45
11	PD - ON - SB	m3/det	9,20	0,00	0,00	0,00	0,00	0,91	0,00	0,00	0,00	10,45
12	PD - ON - ON	m3/det	0,72	1,15	1,30	6,81	0,00	0,77	0,58	2,14	3,12	0,00
13	PD - PD - PD	m3/det	7,65	0,00	1,37	0,90	2,02	3,34	2,03	8,55	9,65	3,91

No.	POLA TANAM	SAT	JUN I	JUN II	JUL I	JUL II	AGT I	AGT II	SEP I	SEP II	OKT I	OKT II	NOV I	NOV II	DES I	DES II	RATA
1	PD - PD - CR	m3/det	10,64	11,80	5,18	6,43	7,31	7,31	7,77	3,90	1,65	20,51	1,76	2,40	3,15	1,29	4,92
2	PD - PD - SB	m3/det	5,39	6,29	6,20	5,60	3,56	1,69	3,38	4,49	5,34	5,28	3,35	1,79	9,10	7,31	4,17
3	PD - PD - ON	m3/det	5,39	6,29	6,20	5,60	3,56	1,69	3,38	3,95	4,04	4,31	3,41	3,16	9,10	7,31	4,11
4	PD - CR - CR	m3/det	3,36	4,19	2,93	19,53	3,84	4,72	6,30	6,37	5,85	5,04	0,00	9,96	9,10	1,69	4,59
5	PD - CR - SB	m3/det	3,36	4,19	2,93	19,53	4,15	5,11	6,27	5,80	4,21	2,69	0,00	9,96	9,10	1,69	4,43
6	PD - CR - ON	m3/det	3,36	4,19	2,93	19,53	3,68	4,04	5,34	5,87	5,69	0,00	0,00	9,96	9,10	1,69	4,28
7	PD - SB - CR	m3/det	2,97	3,12	2,15	19,53	3,84	4,72	6,30	6,37	5,85	5,04	0,00	9,96	9,10	1,69	4,52
8	PD - SB - SB	m3/det	2,97	3,12	2,15	19,53	4,15	5,11	6,27	5,80	4,21	2,69	0,00	9,96	9,10	1,69	4,36
9	PD - SB - ON	m3/det	2,97	3,12	2,15	19,53	3,68	4,04	5,34	5,87	5,69	0,00	0,00	9,96	9,10	1,69	3,86
10	PD - ON - CR	m3/det	10,64	4,97	6,57	6,35	7,31	7,11	3,90	1,73	20,87	2,05	0,00	0,00	0,00	0,00	3,96
11	PD - ON - SB	m3/det	10,64	4,97	6,57	6,35	7,31	7,11	3,90	1,73	20,87	2,05	0,00	0,00	0,00	0,00	3,84
12	PD - ON - ON	m3/det	0,00	0,00	11,67	11,14	6,01	7,36	7,97	7,91	7,32	1,91	0,00	15,05	1,00	0,00	3,91
13	PD - PD - PD	m3/det	3,92	4,82	5,02	4,72	3,59	12,65	13,21	6,76	8,18	7,25	5,53	4,98	0,51	7,31	5,33

KETERANGAN :
: LAND PREPARATION (LP)
: PADDY (PD)
: CORN (CR)
: SOY BEAN (SB)
: ONION (ON)
: FALLOW

Figure 5 and **Figure 6** show that water requirement in which intermitent-irrigation technique is applied tends to be lower than the ones with conventional irrigation. Those figures show that 2035-debit, which has the lowest average, sometimes is incapable of supplying the entire DI Pemali Kanan area. As shown in **Figure 4** and **Figure 6**, even when intermittent-irrigation is applied, there are still spots (representing certain 2 -weekly periods) where the Pemali Kanan's debit are lower than its water requirments. From these facts, there are two conclusions. The first is that even though water requirements of DI Pemali Kanan is too low compared to Pemali river's debit, in certain 2-weekly periods, Pemali river's debit is somehow lower and inadequate to supply the entire DI Pemali Kanan. That would need a special treatments related to both cropping patterns and water supplies. Even when intermittent-irrigation technique is applied and water requirement is lower than it used to be, in certain period, this low-debit circumtances has successfully triggered complaint from farmers regarding to water supplies in their farm land.

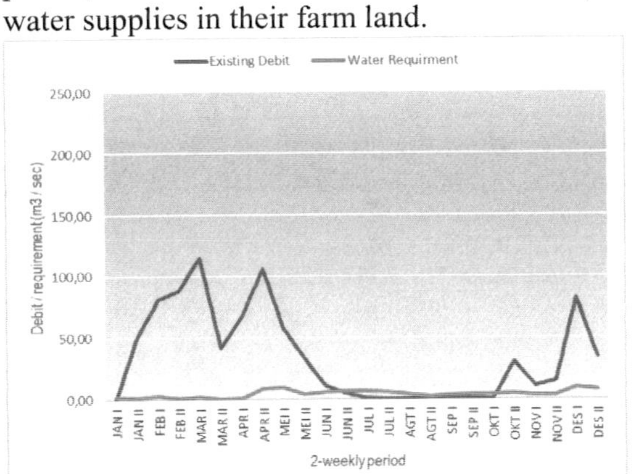

Figure 5. Comparison Water Requirement (Intermittent) to Existing Debit

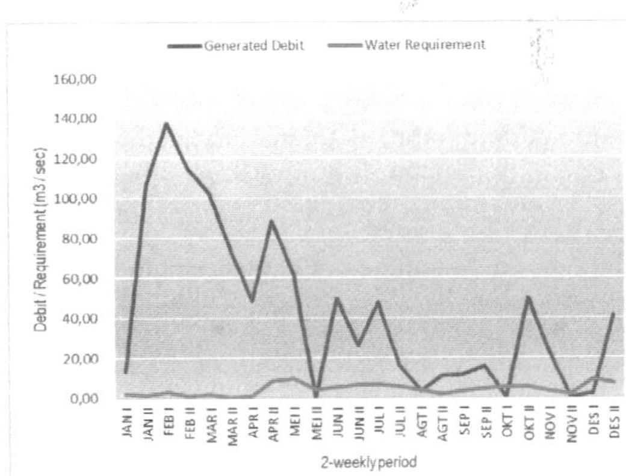

Figure 6. Comparison of Water Requirement (Intermittent) to Generated Debit (2035)

The second thing to consider is that in Pemali Kanan Irrigation Area due to its low water availability, it is not recommended to grow paddy within dry seasons which lasts from Juni to October or what so called 3rd-planting-season. Farmers is highly recommended to grow other kind of plants whose water requirement is lower than paddy (e.g corn or soy bean).

3.3. Farmers Opinion

Based on the quiestionnaires, as shown in **Figure 7**, it can be discovered that 24 out of 35 repondents (equal to 68,57%) prefer the System of Rice Intensification (SRI) method than the conventional farming method. Most of respondents are familiar with this System of Rice Intensification (SRI) so that they could give an objective preference regarding this option. They prefer to choose System of Rice Intensification (SRI) as they think that this advanced farming method has some visible advantages, such as : better quality and higher yields, more efficient in terms of irrigation-water consumption and less seeds needed. All of those advantages are considered to have positive impact to their productivity and surely would also give more benefits (especially in terms of financial aspects) to farmers.

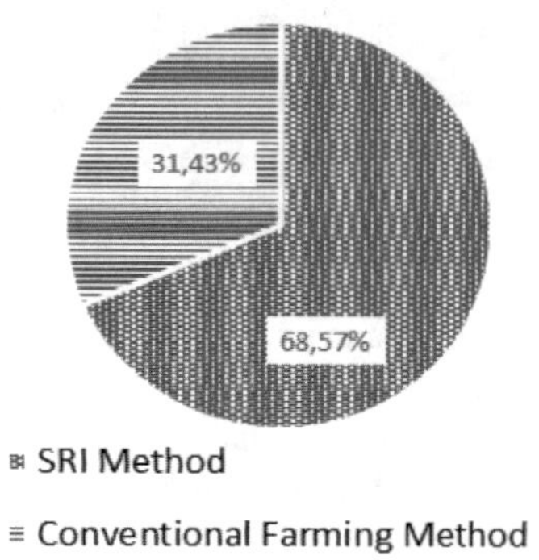

Figure 7. Farmers' Farming Method Preference

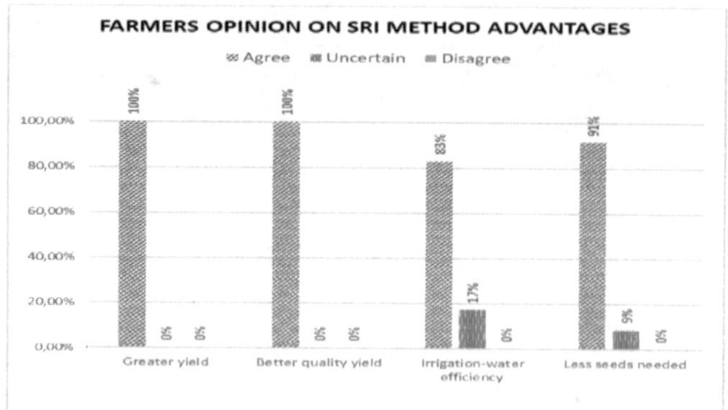

Figure 8. Farmers' Opinion on SRI Method

As shown in **Figure 8**, beside of its greater productivity and better quality, efficiency in water consumption has successfully attracted farmers to choose SRI as their method to grow paddy. Not less than 83% of respondents thought that it becomes more difficult to obtain a stable and adequate amount of irrigation water. Therefore, in addition to gain more financial benefit, farmers also consider more efficient ways in order to cut their expenses.

However, although farmers have recognized that System of Rice Intensification (SRI) method is more suitable, more efficient, more profitable and more environmentally friendly, they decide to return to conventional farming method as there are many factors that made System of Rice Intensification (SRI) seems to be difficult to implement.

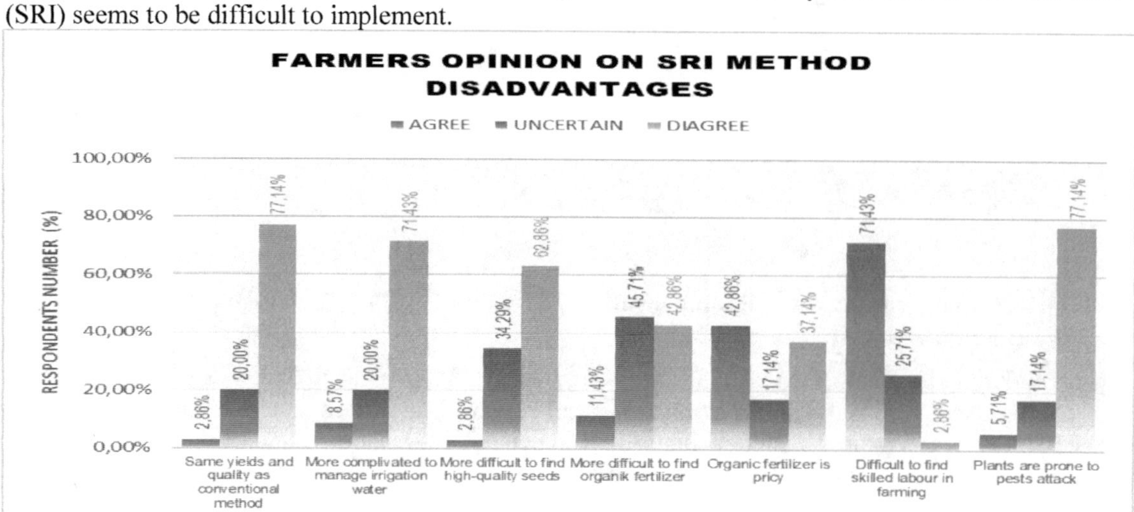

Figure 9. Farmers' opinion on SRI Method's Disadvantages

As shown in **Figure 9**, among 7 factors that could possibly be the most anticipated by farmers, lack of skilled agricultural-labourers is the factor that could be the one which cause that farmers to switch from System of Rice Intensification (SRI) method to conventional farming method which they have applied over the last decades. By this point, the question of why farmers No. longer implementing the System of Rice Intensification (SRI) method has found the answer. Shortage of skilled agricultural labourers has become more reasonable since in the System of Rice Intensification (SRI) method, there are techniques that require special skills and personal ability.

4. Conclusion

In the end, the result of this study shows that water availability of Pemali River should be more than adequate to fulfill water requirement of crops at Pemali Kanan Irrigation Area, either at present or in the next 20 years. Only in certain months, which are June to October, and in certain years, the water availability is lower than water requirement of Pemali Kanan Irrigation Area. This situation will cause some of the farm lands will not get water supply in such an adequate amount.

Water requirements of all 13 cropping patterns are varied, with a maximum number of 28,31m^3/sec which took place in October II in Land Preparation phase. The average water requirement of all 13 cropping pattern is 6,41 m^3/sec. Among all 13 cropping patterns along with their 24 commencement alternatives, formation of paddy – onions – soy bean is considered as the best cropping pattern in terms of water consumption since it has the least average of water requirement (5,60 m3/sec).

Related to the huge amount of water availability of Pemali River, The Government of Brebes Regency should consider the best way to utilize that potentials so that surplus of water could be useful during dry season (which lasts from June to October) when water availability of Pemali Kanan river is not capable to supply the entire field, especially the northern area of Brebes Regency.

Regarding the farming method, SRI method, specifically the intermittent irrigation technique, is proven to be suitable to be implemented in dry seasons. The farmers have realized that SRI Method offers many advantages compared to conventional farming method. Those farmers also had experienced in applying SRI Method in the past. However, nowadays due to many obstacles in SRI Method (e.g. lack of skilled agricultural-labourers and expensive organic fertilizer), they decide to go back to conventional farming method. The govenment agencies should consider ways to build farmers' awareness to start implementing farming methods which are more environmentally-friendly such as SRI method. Furthermore, supports from the Government of Brebes Regency in various ways could provide confidence to farmers to implement such a new farming methods.

References

[1] Sumarno, Wargiono J, Kartasasmita U G, Hasanudin A, Soejitno, Ismail I G 2008 Anomali Iklim 2006/2007 dan Saran Kebijakan Teknis Pencapaian Target Produksi Padi *Jurnal Iptek Tanaman Pangan*, 3(1), 69-97
[2] Nurrochmad F 2007 Kajian Pola-Hemat Pemberian Air Irigasi Forum Teknik Sipil 12(2) 517-529
[3] Suara Merdeka News 2018 *25.180 Ha Lahan Pertanian di Brebes Rawan Kekeringan* able to access at https://www.suaramerdeka.com/news/baca/104507/25180-ha-lahan-pertanian-di-brebes-rawan-kekeringan.
[4] Bupati Brebes 2017 *Peraturan Bupati Brebes nomor 080 tahun 2017 tentang Pedoman Pengaturan Pola Tanam dan Tata Tanam di Kabupaten Brebes 2017-2018* (Brebes, Central Java)
[5] Sato S, Hasan M 2007 Water Saving for Paddy Cultivation Under The System of Rice Intensification (SRI) in Eastern Indonesia *Jurnal Tanah dan Lingkungan*, 9(2), 57-62
[6] Sato S, Yamaji E, Kuroda T 2011 Strategies and engineering adaptions to disseminate SRI methods in large-scale irrigation systems in Eastern Indonesia *Paddy Water Environ*, 9, 79-88
[7] Syam M 2006 Kontroversi System of Rice Intensification (SRI) di Indonesia *Jurnal Iptek Tanaman Pangan*, 1, 30-40
[8] Balitbang Pertanian 2015, *Panduan Teknologi Budidaya Padi System of Rice Intensification*

(SRI), (Badan Penelitian dan Pengembangan Pertanian, Ministry of Agriculture, Jakarta)

[9] Regazzoni O, Sugito Y, Suryanto A 2013 Irigasi Berselang (Intermittent Irrigation) pada Budidaya Padi (Oryza Sativa L.) Varietas Inpari-13 dalam Pola SRI (System of Rice Intensification) *Jurnal Produksi Pertanian*, 1(2), 42-51

[10] Puteriana S A, Harisuseno D, Prayogo T B 2016 Kajian Sistem Pemberian Air Irigasi Metode Konvensional dan Metode SRI (System of Rice Intensification) pada Daerah Irigasi Pakis Kecamatan Pakis Kabupaten Malang *Jurnal Teknik Pengairan*, 7(2), 236-247

[11] PT. Yodya Karya 2010 *Laporan Detail Desain Rehabilitasi DI Pemali Kanan (7.909 ha)* (Semarang).

[12] Pratiwi, B S, Sachro S S, Suharyanto 2017 Pembangkitan Data Debit dan Skenario Pola Tanam Daerah Irigasi Embung Suruhan *Media Komunikasi Teknik Sipil*, 23 (1), 29-37

[13] Chahayati C, Sutrisno 2014 Pengaruh Debit Air terhadap Pola Tanam pada Baku Sawah di Daerah Irigasi Kebonagung Kabupaten Sumenep Jurnal MITSU Media Informasi Teknik Sipil UNIJA, 2(2), 30-39

IOP Conf. Series: Earth and Environmental Science **328** (2019) 012012 doi:10.1088/1755-1315/328/1/012012

Interaction Model Between Land Use and the Road's Level of Service in the Corridor of Lawang to Singosari

B S Waloejo[1], I WAgustin[2], W Ramadhan[3]
[123]Faculty of Engineering, University of Brawijaya, Jl Mayjen Haryono 167 Malang Indonesia

*email: budieswe@ub.ac.id; immasaitama@ub.ac.id

Abstract. Demand for transportation infrastructure is growing along with the increasing urban and economic population (Min Guo, 2014). The increase in urban movement activities often exceeds the capacity of transportation infrastructure systems which creates congestion points. Malang to Surabaya movement is the dominant movement with the largest number of movements totalling 10-20 million movements in a year and can reach 94,254 vehicles every day. Based on the RTRW Malang 2010-2030, Malang City as the core city has the main satellite area namely Lawang and Singosari Subdistricts resulting in the high intensity of these activities. The purpose of this study is to identify the characteristics of road, to formulate a model of generation/attraction of land use in the road, and to formulate the effect of land use activities on the road's level of service. This research was carried out on the Corridor of Lawang - Singosari segment 1. We descriptively described the road characteristics and for analyzed quantitatively, correlation analysis and multiple linear regression analysis and other quantitative analyzes were used to evaluate and formulate the model and its impact on the road's level of service.
Keywords: land use, land use models ,road's level of service

1. Introduction

The primary arterial road network that connects Surabaya - Pandaan - Lawang - Singosari - Malang is a primary national road that connects the two national activity centres that increase vehicle movement between regions (continuous current). In addition to vehicles between regions, based on the RTRW Malang District in 2010 - 2030, Malang City as the core city has the main satellite area namely Lawang District and Singosari District. This also concerns the movement of vehicles related to activities within the region. This area in its existing conditions has various types of land use, trade and service land use for Lawang Market and Singosari Market; use of educational land; industrial land use and warehousing; use of military land, and others. The land use is centered along the Corridor of Lawang - Singosari so as to produce vehicles generated from land use activities that also add vehicles to produce roads.

This study was conducted to see the proportion of the amount of continuous traffic flow and local traffic flow so that the effect of land use can be calculated along the Corridor of Lawang - Singosari on the road's level of service.

2. Methods

This research is descriptive research with quantitative considerations related to this study measuring variable one with other variables and supported by statistics of connectedness. Descriptive research in this study aims to get an overview or characteristic related to land use on the road's level of service in the Corriodor of Lawang - Singosari.

Data collection methods used: Primary survey consists of; field observations, interviews and questionnaires. Secondary surveys consist of; literature survey, agency survey and comparative study. The analytical method used: includes: descriptive analysis: analysis of road characteristics, analysis of land use characteristics, while for evaluative analysis: level of service analysis, road capacity analysis, traffic flow analysis, correlation analysis, and multiple linear regression analysis.

2.1 Research Variables

Based on the type of generation / attraction models commonly used is the mathematical model by taking the form of a correlation between variables in land use or characteristics of the location (as explanatory variables) with the amount of generation / attraction (as the response variable). Preferred forms of the equation is a regression equation, as it is considered to have the level of reliability that can be measured, so that it can be seen how far the model produced can be used and level of reliability can also be known.

Explanatory Variables

As candidate variables in this research are:

- Housing; X_1 = Building Area, X_2 = Tot. Floor, X_3 = Tot. Family members, X_4 = Tot. Rooms, X_5 = Tot. Family income, X_6 = Tot. Ownership of Vehicles.
- Education; X_7 = Tot. Students, X_8 = Tot. Lecturer/Employee, X_9 = Tot. Room Class, X_{10} = Land Area, X_{11} = Building Area
- Offices; X_{12} = Tot. Employee, X_{13} = Tot. of Visitors, X_{14} = Land Area, X_{15} = Building Area.
- Hospital/Clinic; X_{16} = Tot. Doctors/Employee, X_{17} = Tot. Daily patient, X_{18} = Patient Rooms Area, X_{19} = Parking Area, X_{20} = Building Area.
- Pharmachy; X_{21} = Tot. Employee, X_{22} = Tot. Visitors, X_{23} = Building Area, X_{24} = Parking Area.
- Worship : X_{25} = Building Area, X_{26} = Land Area, X_{27} = Parking Area, X_{28} = Tot. Floor, X_{29} = Tot. Visitors.
- Industrial and Warehousing; X_{30} = Tot. Building Area, X_{31} = Parking Area, X_{32} = Tot. Employee, X_{33} = Tot. Visitors, X_{34} = Tot. Work Shift, X_{35} = Tot. Delivery Frequency.
- Trade and Services :
  - Vehicle Repair and Showroom; X_{36} = Tot. Floor, X_{37} = Tot. Employee , X_{38} = Parking Area, X_{39} = Building Area, X_{40} = Tot. Visitors
  - Mini Market; X_{41} = Tot. Floor, X_{42} = Tot. Employee , X_{43} = Parking Area, X_{44} = Building Area, X_{45} = Tot. Visitors
  - Printing and Convection; X_{46} = Tot. Floor, X_{47} = Tot. Employee , X_{48} = Parking Area, X_{49} = Building Area, X_{50} = Tot. Visitors
  - Bistro and Restaurant; X_{51} = Tot. Floor, X_{52} = Tot. Employee , X_{53} = Parking Area, X_{54} = Building Area, X_{55} = Tot. Visitors
  - Convenience Store; X_{56} = Tot. Floor, X_{57} = Tot. Employee , X_{58} = Parking Area, X_{59} = Building Area, X_{60} = Tot. Visitors
  - Clothing and Accessories Shop; X_{61} = Tot. Floor, X_{62} = Tot. Employee , X_{63} = Parking Area, X_{64} = Building Area, X_{65} = Tot. Visitors
  - Cellphone and Pulses Store; X_{66} = Tot. Floor, X_{67} = Tot. Employee , X_{68} = Parking Area, X_{69} = Building Area, X_{70} = Tot. Visitors
  - Food Shack; X_{71} = Tot. Floor, X_{72} = Tot. Employee , X_{73} = Parking Area, X_{74} = Building Area, X_{75} = Tot. Visitors

IOP Conf. Series: Earth and Environmental Science **328** (2019) 012012 doi:10.1088/1755-1315/328/1/012012

- Building, Furniture and Electronic Stores; X_{76} = Tot. Floor, X_{77} = Tot. Employee , X_{78} = Parking Area, X_{79} = Building Area, X_{80} = Tot. Visitors
- Other Services; X_{81} = Tot. Floor, X_{82} = Tot. Employee, X_{83} = Parking Area, X_{84} = Building Area, X_{85} = Tot. Visitors

3. Results and Discussions

3.1 Characteristics of the Road Network

Types of roads; Corridor of Lawang – Singosari segment 1 is a two-way split (4/2D) with a road width of 12.5m, with a hot mix asphalt pavement good condition. Shoulder of the road is on the left and right, 2.5m. In some places many are used by vendors and on-street parking. Based on the above information in mind; C_O = 7.600 pcu/hour, FC_{PA} = 1, FC_{HS} = 0.99, and FC_{LJ} = 0.96, so it can be calculated that the actual capacity of the Corridor of Lawang – Singosari amount to each is 7.296 pcu/hour.

3.2 Characteristics of the Land Use

Land use along the Corridor of Lawang – Singosari segment 1 consists of housing which is 12%, trade and services are 70%, education is 1%, worship is 1%, Industrial and warehousing is 5%, and health is 2%. The total land use along the Corridor of Lawang – Singosari segment 1 is 156 buildings.

Table 1. Amount of Land Use on Corridor of Lawang – Singosari Segment 1

No	Land Use	Population
1	Housing	19
2	Trade and Services	109
3	Education	2
4	Offices	13
5	Worship	2
6	Industrial and Warehousing	9
7	Health	3
	Total	**157**

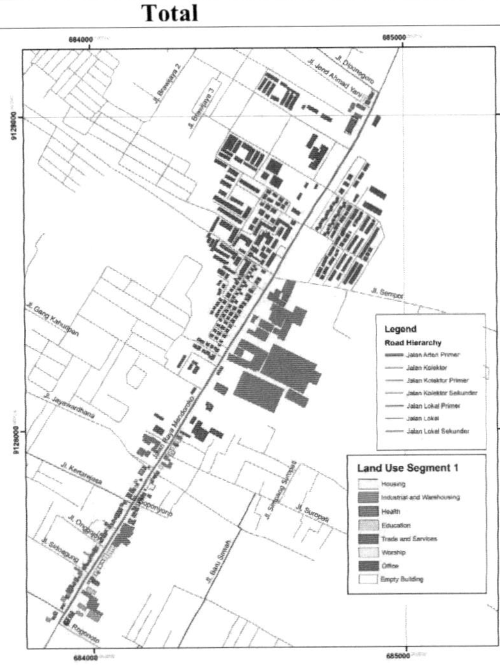

Figure 1. Land Use of Region Research

IOP Conf. Series: Earth and Environmental Science **328** (2019) 012012 doi:10.1088/1755-1315/328/1/012012

3.3 Characteristics and Movement Generation / Attraction Land Use

- Housing; Land use of housing in the Corridor Lawang – Singosari there is a total movement every day which is 112,48 pcu/day for attraction and generation movements. for the Corridor of Singosari – Lawang, the highest generation of travel occurs at 06.00 – 07.00 which is 29,34 pcu/hour with a ratio (e) of 0.26.
- Education; Land use of education in the Corridor of Lawang – Singosari there is a total movement every day which is 328 pcu/day for attraction and generation movement. For the Corridor of Singosari – Lawang, the highest generation of travel occurs at 15.00 – 16.00 that is 69,6 pcu/hour with a ratio (e) of 0.21.
- Offices; Land use of offices in the Corridor of Lawang – Singosari there is a total movement every day that is 2.379 pcu/day for attraction and generation movements. For the Corridor of Singosari – Lawang, the highest generation of travel occurs at 13.00 – 14.00 that is 373 pcu/hour with a ratio (e) of 0.16.
- Health; Land use of pharmacy In the Corridor of Lawang – Singosari, there is a total movement every day is 150 pcu/day for attraction and generation movements. For the Corridor of Singosari – Lawang, the highest generation of travel occurs at 17.00 – 18.00 that is 27 pcu/hour with a ratio (e) of 0,18.
- Trade and Services; Land use of trade and services in the Corridor of Lawang – Singosari in total movement every day is 9.363,6 pcu/day for attraction and generation movements. For the Corridor of Singosari – Lawang, the highest travel generation occurs at 16.00 – 17.00 that is 1.307 pcu/hour with a ratio (e) of 0.14.
- Industrial and Warehousing; Land use of industrial and warehousing in the Corridor of Lawang – Singosari in total movement every day is 1,152 pcu/day for attraction and generation movements. For the Corridor of Singosari – Lawang, the highest travel generation occurs at 08.00 – 09.00 which is 197.5 pcu/hour with a ratio (e) of 0.17.
- Worship; Land use of worship in the Corridor of Lawang – Singosari is a total movement every day is 135.6 pcu/day for attraction and generation movement. For Corridor of Singosari – Lawang, the highest travel generation occurs at 17.00 – 18.00 that is 28,9 pcu/hour with a ratio (e) of 0.21.

3.4 Modelling Land Use in Region Research

- Housing; Results of linear regression analysis with stepwise method, can know the model equations trip generation that occurs in housing, with the model equations $\mathbf{Y_{Housing} = 1{,}258 + 0{,}013\ (X_1) + 0{,}61\ (X_3) + 0{,}212\ (X_6)}$, wherein X_1 = variables explanatory, in the form of building area , X_3 = explanatory variables, such as number of family members, and X_4 = explanatory variables, such as number of vehicle ownership.
- Education; Results of linear regression analysis using stepwise method, then the model equations of school is; $\mathbf{Y_{school} = -8{,}762 + 0{,}111\ (X_7) + 0{,}163\ (X_{11})}$, wherein X_7 = explanatory variables, such as number of students and X_{11} = explanatory variables in the form of building area.
- Offices; Results of linear regression analysis using stepwise method, the equation is a model of Offices; $\mathbf{Y_{Offices}= -64{,}733 + 1{,}257\ (X_{13}) + 0{,}240\ (X_{15})}$ wherein X_{11} = explanatory variables, such as number of visitors and X_{13} = explanatory variables, such as building area.
- Health (Pharmacy); Results of linear regression analysis using stepwise method, the equation is a model of Pharmacy; $\mathbf{Y_{Pharmacy} = -11{,}700 + 1{,}501\ (X_{22}) + 0{,}114\ (X_{23})}$, wherein X_{22} = explanatory variables, such as number of visitors and X_{23} = explanatory variables, such as building area.
- Worship; Results of linear regression analysis using stepwise method, the equation is a model of Worship; $\mathbf{Y_{Worship}= 6.943 + 0.549\ (X_{28}) + 0.029\ (X_{27})}$, wherein X_{27} = explanatory variables, such as parking area, and X_{28} = explanatory variables, such as total visitors.

- Industrial and Warehousing; Results of linear regression analysis using stepwise method, the equation is a model of Industrial and Warehousing; $\mathbf{Y_{Industrial\ and\ Warehousing} = 5{,}104 + 0{,}736\ (X_{31}) + 1{,}832\ (X_{34})}$, wherein X_{31} = explanatory variables, such as total employee, and X_{34} = explanatory variables, such as total delivery frequency.
- Trade and Services; Results of linear regression analysis using stepwise method, the equation is a model of Trade and Services.
  - Vehicle Repair and Showroom; Results of linear regression analysis using stepwise method, the equation is a model of Vehicle Repair and Showroom; $\mathbf{Y_{Repair\ and\ Sell/Buy\ Vehicle} = 0.863 + 0.401\ (X_{39}) + 0.549\ (X_{40})}$, wherein X_{39} = explanatory variables, such as building area and X_{40} = explanatory variables, such as total visitors.
  - Minimarket; Results of linear regression analysis using stepwise method, the equation is a model of Mini Market; $\mathbf{Y_{Minimarket} = 6.631 + 0.176\ (X_{44}) + 0.442\ (X_{43}) + 0.362\ (X_{45})}$, wherein X_{43} = explanatory variables, such as parking area, X_{44} = explanatory variables, such as building area, and X_{45} = explanatory variables, such as total visitors.
  - Printing and Convection; Results of linear regression analysis using stepwise method, the equation is a model of Printing and Convection; $\mathbf{Y_{Printing\ and\ Convection} = 15.565 + 0.424\ (X_{48}) + 0.541\ (X_{50})}$, wherein X_{48} = explanatory variables, such as parking area, and X_{50} = explanatory variables, such as total visitors.
  - Bistro and Restaurant; Results of linear regression analysis using stepwise method, the equation is a model of Bistro and Restaurant; $\mathbf{Y_{Bistro\ and\ Restaurant} = -8.276 + 0.479(X_{53}) + 0.539\ (X_{55})}$, wherein X_{53} = explanatory variables, such as parking area, and X_{55} = explanatory variables, such as total visitors.
  - Convenience Store; Results of linear regression analysis using stepwise method, the equation is a model of Convenience Store; $\mathbf{Y_{Convenience\ Store} = -20.532 + 0.735\ (X_{59}) + 0.423\ (X_{60})}$, wherein X_{59} = explanatory variables, such as building area, and X_{60} = explanatory variables, such as total visitors.
  - Clothing and Accessories Shop; Results of linear regression analysis using stepwise method, the equation is a model of Clothing and Accessories Shop; $\mathbf{Y_{Clothing\ and\ Accessories\ Shop} = -19.270 + 0.587\ (X_{64}) + (0.520\ (X_{65})}$, wherein X_{64} = explanatory variables, such as building area, and X_{65} = explanatory variables, such as total visitors.
  - Cellphone and Pulses Store; Results of linear regression analysis using stepwise method, the equation is a model of Cellphone and Pulses Store; $\mathbf{Y_{Cellphone\ and\ Pulses\ Store} = 19.281 + 1.419\ (X_{68}) + 0.155\ (X_{70})}$, wherein X_{68} = explanatory variables, such as parking area, and X_{70} = explanatory variables, such as total visitors.
  - Food Shack; Results of linear regression analysis using stepwise method, the equation is a model of Food Shack; $\mathbf{Y_{Food\ Shack} = 2.676 + 0.569\ (X_{74}) + 0.176\ (X_{75})}$, wherein X_{74} = explanatory variables, such as building area, and X_{75} = explanatory variables, such as total visitors.
  - Building, Furniture and Electronic Stores; Results of linear regression analysis using stepwise method, the equation is a model of Building, Furniture and Electronic Stores; $\mathbf{Y_{Building,\ Furniture\ and\ Electronic\ Stores} = -7.934 + 1.284\ (X_{79}) + 0.304\ (X_{80})}$, wherein X_{79} = explanatory variables, such as building area, and X_{80} = explanatory variables, such as total visitors.
  - Other Services; Results of linear regression analysis using stepwise method, the equation is a model of Other Services; $\mathbf{Y_{Other\ Services} = -30.860 + 0.708\ (X_{85}) + 0.300\ (X_{84})}$, wherein X_{84} = explanatory variables, such as building area, and X_{85} = explanatory variables, such as total visitors.

Table 2. Summary of Model Generation / Attraction by Land Use

Land use		Trip Generation
Housing		$Y = 1,258 + 0,013\ (X_1) + 0,61\ (X_3) + 0,212\ (X_6)$
		Trip Attraction
Education		$Y = -8,762 + 0,111\ (X_7) + 0,163\ (X_{11})$
Offices		$Y = -64,733 + 1,257\ (X_{13}) + 0,240\ (X_{15})$
Health	Pharmacy	$Y = -11,700 + 1,501\ (X_{22}) + 0,114\ (X_{23})$
Worship		$Y = 6.943 + 0.549\ (X_{28}) + 0.029\ (X_{27})$
Industrial and Warehousing		$Y = 5,104 + 0,736\ (X_{31}) + 1,832\ (X_{34})$
Trade and Services	Repair and Sell/Buy Vehicle	$Y = 0.863 + 0.401\ (X_{39}) + 0.549\ (X_{40})$
	Minimarket	$Y = 6.631 + 0.176\ (X_{44}) + 0.442\ (X_{43}) + 0.362\ (X_{45})$
	Printing and Convection	$Y = 15.565 + 0.424\ (X_{48}) + 0.541\ (X_{50})$
	Bistro and Restaurant	$Y = -8.276 + 0.479(X_{53}) + 0.539\ (X_{55})$
	Convenience Store	$Y = -20.532 + 0.735\ (X_{59}) + 0.423\ (X_{60})$
	Clothing and Accessories	$Y = -19.270 + 0.587\ (X_{64}) + (0.520\ (X_{65})$
	Cellphone and Pulses Store	$Y = 19.281 + 1.419\ (X_{68}) + 0.155\ (X_{70})$
	Food Shack	$Y = 2.676 + 0.569\ (X_{74}) + 0.176\ (X_{75})$
	Building, Furniture and Electronic Store	$Y = -7.934 + 1.284\ (X_{79}) + 0.304\ (X_{80})$
	Other Services	$Y = -30.860 + 0.708\ (X_{85}) + 0.300\ (X_{84})$

3.5 Volume Vehicle Movement

3.5.1 External of Volume Vehicle Movement

External vehicle movement consists of alleyway movement and through movement.

Table 3. Volume Through Movement on the Corridor of Lawang – Singosari Weekdays Segment 1

Time	MC (pcu/hr)	LV (pcu/hr)	HV (pcu/hr)	Total Volume (pcu/hr)
06.00-07.00	1.581,25	2618	423,60	4.622,85
07.00-08.00	1.451,00	2874	385,20	4.710,20
08.00-09.00	1.236,75	2705	327,60	4.269,35
13.00-14.00	6.58,75	2292	308,40	3.259,15
14.00-15.00	8.19,25	2257	320,40	3.396,65
15.00-16.00	7.92,25	2546	332,40	3.670,65
16.00-17.00	9.52,25	2361	331,20	3.644,45
17.00-18.00	1.265,50	2577	441,60	4.284,10
18.00-19.00	1.120,75	2493	450,00	4.063,75
total	**9.877,75**	**22723**	**3320,40**	**35.921,15**

Table 4. Volume Alleyway Movement on the Corridor of Lawang – Singosari Weekdays Segment 1

Time	Volume Out (pcu/hr)	Volume Entry (pcu/hr)	Total Volume Entry and Out (pcu/hr)
06.00-07.00	196.70	105.90	302.60
07.00-08.00	201.85	120.25	322.10
08.00-09.00	148.15	133.30	281.45
13.00-14.00	139.30	155.55	294.85
14.00-15.00	136.75	121.45	258.20
15.00-16.00	121.45	148.25	269.70
16.00-17.00	135.40	174.10	309.50
17.00-18.00	175.10	175.55	350.65

Time	Volume Out (pcu/hr)	Volume Entry (pcu/hr)	Total Volume Entry and Out (pcu/hr)
18.00-19.00	192.25	179.75	372,00
total	**1446.95**	**1314.10**	**2.761,05**

3.5.2 Internal of Volume Vehicle Movement

Internal vehicle movement is volume of vehicles from generation and attraction of land use.

Table 5. Volume Vehicles Movement From Land Use on the Corridor Singosari - Lawang Segment 1

Time	Vehicles Movement Volume From Type of Land Use (pcu/hour)							Total Vol. From Land Use (pcu/hr)
	Housing	Trade and Services	Worship	Education	Office	Health	Industrial and Warehousing	
06.00-07.00	29.34	368.06	4.67	68.66	204.27	2.15	54.82	451.02
07.00-08.00	11.00	993.95	6.63	60.51	329.08	8.84	150.76	771.08
08.00-09.00	3.18	1233.72	3.93	11.86	323.32	17.68	197.55	839.62
13.00-14.00	10.27	1202.99	11.79	12.60	278.95	13.14	88.87	1288.50
14.00-15.00	5.87	1079.84	21.13	18.77	320.56	20.06	109.68	1237.56
15.00-16.00	15.89	1068.79	24.07	69.65	333.84	21.97	135.96	1394.82
16.00-17.00	22.25	1307.70	13.51	58.04	307.02	22.45	190.88	1080.36
17.00-18.00	9.78	1212.93	28.99	20.01	168.42	26.99	134.06	950.78
18.00-19.00	4.89	895.62	20.88	7.90	113.54	16.72	89.42	708.28

Based on the results, volume generation/attraction of vehicle movement based on land use on the Corridor of Lawang-Singosari is the lowest, occurring at 06.00-07.00 is 451.02 pcu/hour, while the volume generation/attraction of vehicle movement based on the highest land use occurs at 15.00-16.00 which is 1.394,82 pcu/hour.

3.6 Model Of Interaction Land Use – Road's Level of Services

Total volume of vehicles movement/hour on the Corridor of Lawang - Singosari will be interacted with the road capacity in order to determine the road's level of service, where the road's level of service is:

$$VCR = \frac{V_{internal} + V_{eksternal}}{C} \quad (1)$$

The Following is the total volume of vehicles movement/hour on the Corridor of Lawang - Singosari weekday segment 1:

Table 6. Total Volume Vehicles Movement on the Corridor Lawang – Singosari Segment 1

Time	Volume from Landuse (pcu/hr)	Volume from alleyway (pcu/hr)	Volume from through traffic (pcu/hr)	Total Volume (pcu/hr)
06.00-07.00	451.02	302.60	4622,85	5.657,42
07.00-08.00	771.08	322.10	4710,20	6.593,08
08.00-09.00	839.62	281.45	4269,35	6.342,02
13.00-14.00	1.288.50	294.85	3259,15	5.172,61
14.00-15.00	1.237.56	258.20	3396,65	5.230,76
15.00-16.00	1.394.82	269.70	3670,65	5.610,53
16.00-17.00	1.080.36	309.50	3644,45	5.875,81
17.00-18.00	950.78	350.65	4284,10	6.235,93
18.00-19.00	708.28	372,00	4063,75	5.584,72

Table 7. Result Interaction Land use-Road's Level of Services on the Corridor of Lawang – Singosari

Time	Total Volume Vehicle Movements (pcu/hour)	Road Capacity (pcu/hour)	Ratio V/C	LOS
06.00-07.00	5.657,42	7.296	0.78	D
07.00-08.00	6.593,08	7.296	0.90	E

Time	Total Volume Vehicle Movements (pcu/hour)	Road Capacity (pcu/hour)	Ratio V/C	LOS
08.00-09.00	6.342,02	7.296	0.87	E
13.00-14.00	5..172,61	7.296	0.71	C
14.00-15.00	5.230,76	7.296	0.72	C
15.00-16.00	5.610,53	7.296	0.77	D
16.00-17.00	5.875,81	7.296	0.81	D
17.00-18.00	6.235,93	7.296	0.85	E
18.00-19.00	5.584,72	7.296	0.77	D

The method used to calculate the road's level of services on the Corridor of Lawang – Singosari is Model of Interaction Land Use – Road's Level of Service, shown from 06.00 – 19.00 has road's level of services which varies from C to E. When morning peak hours, the volume of vehicle movement occurs at 07.00-08.00 which is 6.593,08 pcu/hour then there is a decrease in volume until noon and the volume experiences a peak hour again at 17.00-18.00 in the afternoon which is 6.235.93 pcu/hour.

4. Conclusion

Based on the information above, it is known: C_O = 3,800 veh/hour, FC_{PA} =1, FC_{HS} = 0,99, and FC_{LJ} = 0,96, so it can be calculated that the actual capacity on Corridor of Lawang - Singosari amounted to 7.296 pcu/hour. Land use along on Corridor of Lawang - Singosari segment 1 consists of housing which is 12%, trade and services is 70%, education is 1%, worship is 1%, Industrial and warehousing is 5%, and health is 2%. total land use along on Corridor of Lawang - Singosari segment 1 is 157 buildings. The total volume of vehicles movement/hour on the Corridor of Lawang - Singosari weekday segment 1 will be interacted with the road capacity in order to determine the road's level of service, where there is an interaction between land use and road's level of service by the mean of road network. This is evidenced by the contribution of the volume of vehicle movements from land use which also affects road's level of service.

References

[1] Salim N 2001 *Analisis Kebutuhan Model Interaksi Tata Guna Lahan dan Transportasi Kotamadya Bandung* (Bandung: ITB)
[2] Michael Iacono and D L G 2008 *Models of Transportation and Land Use Change: A Guide to the Territory* vol 22 Issue 4
[3] Dinas Perhubungan 2015 *Peninjauan Kembali Tataran Transportasi Lokal Kab.Malang* (Kabupaten Malang: Dinas Perhubungan)
[4] Dinas Perhubungan and LLAJ 2016 *Feasibility Studi Pembangunan Jembatan Timbang Arah Malang* ke Surabaya (Provinsi Jawa Timur: Dinas Perhubungan Dan LLAJ)
[5] Waloejo, Budi Sugiarto, Surjono and Sulistio, Harnen 2012 *The Influence of Trip Attraction on the Road's Level of Service (LOS) at Traditional Market Land Use J. Appl. Environ. Biol. Sci.* 2 p 92-96
[6] Waloejo, Budi Sugiarto 2017 *Road Network – Land Use Interaction Model: Malang City in Indonesian Case* (IOP Conf. Ser.: Earth Environ. Sci.) 70 012005
[7] Waloejo, Budi Sugiarto 2018 *Estimated Travel Time Between The City (Case Study: Waru Street-Mojokerto)* (MATEC Web Conf.: ISTSDC 2017) vol 181
[8] Waloejo, Budi Sugiarto 2018 *Controlling Core Urbanism in the Corridor of Surabaya – Sidoarjo Using the Road's Level of Service Approach* (Proc. Int. Conf. on The Bienalle ICIAP)

Effectiveness and Efficiency as a Tool of Measurement of Waste Bank's Performance

C Meidiana[1], Y A Pamungkas[1], and K E Sari[1]

[1] Department of Regional and Urban Planning, Faculty of Engineering, Universitas Brawijaya

*email: c_meideiana@ub.ac.id

Abstract. The objectives of the study are to measure the efficiency and effectiveness of waste bank as an informal institution in waste reduction. The research was conducted in five waste banks located in Batu District, Batu City. Survey concerning waste bank performance was conducted for about 5 months. The questionnaires were distributed during the survey for effectiveness assessment, while in-depth interview was conducted on waste bank staffs for attaining waste bank characteristics. Overall Equipment Effectiveness (OEE) and Elasticity index were used to measure the waste banks' effectiveness and efficiency respectively. The waste bank performance was determined by these two parameters. The results showed that all waste banks have low effectiveness indicated by the value of OEE lower than 60%. Meanwhile, the elasticity index varied among the waste banks from 0.57 to 2.74 representing low or high efficiency. Thus, the waste banks' performance in Batu District was either low or moderate depending on the combination of the two above parameters.
Keywords: Waste-Bank Performance, Effectiveness, Efficiency.

1. Introduction

Local government is the main authority in dealing with solid waste management effectively and efficiently from the residents [1]. The Local Authorities have to emphasize more on waste reduce household waste [2]. One of the informal waste sectors that are able to reduce waste is waste bank.

Waste bank is a collective waste management focusing on waste separation and recycle aiming to increase waste value and to involve community. It is one of the efforts to reduce domestic waste. Some previous studies showed that it may attract public involvement for its promising profit and its ability to reduce waste transporting to the landfill [3, 4]. Waste management through waste bank can reduce the negative impact of landfill for its methane emission to the atmosphere and infiltration of leachates into the soil. However, most existing waste bank lack the capability to operate the business professionally in order to optimize the profit for the members. This had an effect seeing the low performance of the waste bank. [5] Explained that performance is the achievement of an organization based on its responsibility that is carried out efficiently and effectively.

Furthermore, [6] defined waste bank's efficiency as a ratio of marginal product value to average waste production, while waste bank's effectiveness is a measurement representing the target achievement set by the waste bank indicated by OEE value. The same approach was also used by [7]. Waste bank's efficiency and effectiveness was used to measure waste bank's performance according

[6, 8]. Waste bank's performance may also be measured using the ratio between waste volume treated in WB with total volume of municipal solid waste [9].

2. Methods

Analysis of *Overall Equipment Effectiveness* (OEE) and elasticity index are used to measure waste bank's effectiveness and efficiency respectively. The independent variables required for both measurements are presented in **Table 1.**

Table 1. Variables used in the measurement

No.	Variables	Reference
1.	Income	[6]
2.	Number of active members	
3.	Availability	
4.	Quality	
5.	Performance	[7]

Effectiveness is measured based on the OEE value showed in **Table 2**. The value indicates that the higher the OEE value is, the better the effectiveness of waste bank will be.

Table 2. Reference for OEE Value indicating Effectiveness

OEE Value	Remarks	Effectiveness
If OEE= 100%	**Outstanding**. WB generates significant impact to programs, rapid performance, no downtime.	High
If 85% ≤ OEE <100%	**Good**. The scope of service is national wide.	
If 60% ≤ OEE <85%	**Moderate**. WB's performance is average but there are spaces for improvements.	Moderate
If 40% ≤ OEE <60%	**Marginal**. WB has low score but it can be improved easily through direct measurements.	Low
If OEE< 40%	**Low**. WB has very low score and requires much efforts to improve it.	

Source: [6]

Efficiency is measured as elasticity index calculated on monthly basis. The value is compared with indicators value presented in **Table 3**.

Tabel 3. Elasticity index of waste bank

Elasticity index	Remarks	Efficiency
> 1	More members of WB will increase the profit for its capability to generate more outputs	High
= 1	Average productivity of members are maximum that the condition can be maintained to maintain the sustainability of the program	Moderate
< 1	Members are required to be more empowered to sustain or to improve the average productivity towards the program	Low

Source: (Kristina, 2014)

Combining the value of effectiveness and efficiency generates WB's performance which are categorized into 9 groups presented in **Table 4** [9].

Table 4. Matrix for measurement of waste bank's performance

Efficiency / Effectiveness	High Efficiency	Moderate Efficiency	Low Efficiency
High Effectiveness	High Performance	High Performance	Moderate Performance
Moderate Effectiveness	High Performance	Moderate Performance	Low Performance
Low Effectiveness	Moderate Performance	Low Performance	Low Performance

Source: Sudibyakto & Priatmodjo (2016)

3. Result and Discussion

3.1. Waste bank's efficiency

WB's efficiency is reflected by its monthly elasticity calculated using income and number of active members. The result of the calculation is showed in **Table 5**.

Tabel 5. Value of waste bank's efficiency

Waste Bank	Elasticity Index	Remarks	Efficiency
Kartini Mandiri	0.57	E < 1	Low
Saras Asri	1.13	E > 1	High
Cahaya	0.96	E < 1	Low
RW 03 Krajan	2.74	E > 1	High
Kampung Damai	1.20	E > 1	High

According to **Table 5**, Kartini Mandiri and Cahaya waste bank have value of E lower than one (E<1), indicating that its members should be more empowered to maintain or to increase their average productivity in WB's programs. Low efficiency is caused by low income and instable number of members. For example, low elasticity of Kartini Mandiri was caused by significant decrease of member's marginal product. The calculation showed that the addition of one active member will decrease waste value of Rp 64,700 and Rp 64,500 in the 15th and 19th round respectively. Member's marginal product decrease leads to elasticity index to -4.26 and -4,16 in respective round. Therefore, member empowerment is necessary to increase the average productivity leading to higher efficiency. Meanwhile, Saras Asri, RW 03 Krajan, and Kampung Damai have value of E higher than one (E > 1). In this case, addition of members will profit from WB because it contributed to more output as the average productivity rose. High efficiency of the three WBs was mainly caused by steadily increasing income and number of members every round.

3.2. Waste bank's Effectiveness

Waste bank used the *Overall Equipment Effectiveness (OEE)* formula to measure WB's effectiveness that described targeted achievement of WB based on quantity, quality, and time [6]. Three parameters, *availability (A)*, *performance (P)*, dan *quality (Q)* were calculated to generate the OEE value. **Table 6** presents the result of each parameters as well the OEE value. According to **Table 2**, the effectiveness parameter was predetermined.

Tabel 6. Determination of WB's effectiveness based on OEE value

Waste Bank	A	P	Q	OEE	Effectiveness
Kartini Mandiri	92%	78%	33%	24%	Low
Saras Asri	88%	68%	50%	29%	Low
Cahaya	67%	59%	33%	39%	Low
RW 03 Krajan	100%	55%	100%	55%	Low
Kampung Damai	33%	65%	50%	24%	Low

Table 6 indicates that all WBs are categorized into WB with low effectiveness since the value of OEE is lower than 60% comprising of low and marginal group.

3.3. Waste bank's Performance

Referring **Table 5**, **Table 6**, and **Table 7** as well, WB's performance was measured and the result is presented in **Table 7**. WB in Batu district has various value of elasticity index and OEE. However, the measurement of WB performance based on these values come to the result of two main group of WBs based on its performance i.e. low and moderate. The average WB's performance is moderate.

Table 7. Matrix of WB's Performance in Batu District

Waste Bank	Efficiency		Effectiveness		Performance
	Elasticity Index	Efficiency	OEE	Effectiveness	
Kartini Mandiri	0.57	Low	24%	Low	Low
Saras Asri	1.13	High	29%	Low	Moderate
Cahaya	0.96	Low	39%	Low	Low
RW 03 Krajan	2.74	High	55%	Low	Moderate
Kampung Damai	1.20	High	24%	Low	Moderate
Mean	1.32	High	34%	Low	**Moderate**

4. Conclusion

Performance of the five WBs in Batu District was measured and the three of them have high efficiency (Sarah Asri, RW 3 Krajan, Kampung Damai) and two WBs have low efficiency (Kartini Mandiri and Cahaya). However, all WBs have low effectiveness which were caused by low availability (the actual time compared to the planned operational time), low performance (capability of WB staffs), and low quality (number of implemented programs over the planned programs). Meanwhile, high efficiency was caused by steady income and stable number of members. In conclusion, WBs in Batu district have moderate performance with high efficiency and low effectiveness.

References

[1] Sujauddin M, Huda MS, Rafiqul Hoque ATM. Household solid waste characteristics and Management in Chittagong, Bangladesh. Journal of Waste Management 2008; 28: 1688 – 1695.A reference

[2] Cole C, Osmani M, Quddus M, Wheatley A, Kay K. Towards a zero waste strategy for an English local authority. Resources, Conservation and Recycling 2014; 89: 64 – 75.Another reference

[3] Wijayanti, D. R., & S. Suryani. (2015). Waste Bank as Community-based Environmental Governance: A Lesson Learned from Surabaya. Procedia - Social and Behavioural Sciences 184, 171-179.

[4] Suryani, A. (2014). Peran Bank Sampah Dalam Effectiveness Pengelolaan Sampah (Studi Kasus Bank Sampah Malang). Jurnal Aspirasi Vol.5 No.1.

[5] Hartanto, W. (2006). Performance Pengelolaan Sampah di Kota Gombong Kabupaten Kebumen. Semarang: Magister Teknik Pembangunan Wilayah dan Kota Universitas

Diponegoro Semarang.
[6] Kristina, H. J. (2014). Model Konseptual untuk Mengukur Adaptabilitas Bank Sampah di Indonesia. *Jurnal TI UNDIP Vol IX No.1*
[7] Rahmadhani, D., Taroepratjeka, H., & Fitria, L. (2014). Usulan Peningkatan Efektivitas Mesin Cetak Manual Menggunakan Metode Overall Equipment Effectiveness (OEE) Studi Kasus di Perusahaan Kerupuk TTN. ITENAS Bandung.
[8] Suttibak, N. (2008). Assesment of Waste Recycling Performance : A Study of School Garbage Banks in Thailand. GMSARN International Journal, 83-90.
[9] Sudibyakto, & Priatmojo, A. (2016). Manajemen Risiko Bencana pada Kawasan Cagar Budaya Gunung Padang, Ciamis, Jawa Barat. Jurnal Riset Kebencanaan Indonesia Vol.2 No.1, 50-58.

Land Development Recommendation of Bunguran Island as One of the National Border Areas Based on Level of Land Suitability and Land Capability

D I Shofarini[1]*, R K Arruzzi [2], and B E N Firdauz[3]

[1] Department of Urban and Regional Planning, Universitas Brawijaya, Malang, Indonesia.
[2] Center for People and Economic Studies, Universitas Gadjah Mada, Yogyakarta, Indonesia.
[3] PT. Mitra Geotama Indonesia. Yogyakarta, Indonesia.

*email: shofarini@ub.ac.id

Abstract. According to PP no 26 of 2008 concerning National Spatial Planning documents, Bunguran Island as national border areas must be monitored and regulated as the front yard of Indonesia. This study used analytical method based on matching between land suitability criteria and land quality data that considered the most severe factors. Data were being analyzed by the criteria of land capability using the classification by The Ministry of Public Works Regulation (20/PRT/M/2007). The results obtained the form of land capability classes in Bunguran Island, which in turn resulted in land capability as the reference in determining land use directives. The direction of land capability was analyzed based on the results of land capability compared to the land use conditions that can be used as a basis for recommendations or directives for land use in Bunguran Island. The parameters used in the classification of land capabilities are slope, drainage conditions, soil permeability, soil type, texture of soil depth, and erosion data. The descriptive analysis process by data parameters were carried out according to the classification of land capabilities. The result of this research was used to compose spatial pattern planning of Bunguran Island based on the land suitability and land capability.
Keywords: Development-land-recommendation, land-suitability, land-capability

1. Introduction

In Indonesia, the majority of national border areas are being the backyard of the country with lack of development, neither in spatial nor physical development. According to PP no 26 of 2008 [1] and PP no 13 of 2017 [2] (concerning of the National Spatial Plan), Bunguran District must be monitored and regulated as the front yard of the country. Bunguran District as one of national border area, needed detailed plan to accommodate 20-year development planning (RDTR). It also must be implemented as an instrument for developing public service facilities, infrastructure, utilities, and many more. Development always raises a paradox e.g. decreasing the quality and the carrying capacity of the environment. It happened because of the improper management between natural resources and human needs. Therefore, we need to manage the land rationally [3].

The Law of Republic of Indonesia No 32 of 2009 (concerning Environmental Protection and Management) stated that environmental carrying capacity is defined as the ability of the environment to support the balancing life of human and other living things [4]. Land capability evaluation is one of the important components in land use planning process [5]. Evalution of land capability needs to be done on both land use plan within the RTRW, as well as on current land use. The results of the evaluation of land capability provides alternative land uses and limits possible uses and the management actions needed so that the land can be used sustainably in accordance with obstacles or threats.

2. Methods

This research was located in Bunguran Island. Its delineation of national border area (based on the deal of ATR Ministry and BIG of Indonesia) as PKSN Ranai was divided into 2 sides; southern Bunguran Island and Northern Bunguran Island. This study used analytical method based on the matching between land suitability criteria and land quality data that considered the most severe factors. A land use adjustment map was prepared by superimposing present composite land use map on land capability map through GIS [6]. Data were being analyzed according to the criteria of land capability which are foundation, drainage, morphology-slope, general geology including structure, shallow underwater, safe from natural disasters[7]. Every criteria has each weight. all values are multiplied by the weight of each unit of land capability classification by The Ministry of Public Works Regulation No. 20/PRT/M/2007 (Table1) [8].

Table 1. Land Capability Units's Weight

Land Capability Units	Weight
Unit of morphological land capability	5
Unit of easy to do	1
Unit of slope stability	5
Unit of foundation stability	3
Unit of water availability	5
Unit of against erosion	3
Unit of drainage	5
Unit of water disposal	0
Unit of natural disasters	5

Source: The Ministry of Public Works Regulation No. 20/PRT/M/2007

Table 2. Land Capability Classification

Amount	Land Capability Classification	Land Capability
32 - 58	Class A	Very Low
59 - 83	Class B	Low
84 - 109	Class C	Moderate
110 - 134	Class D	Rather High
135 - 160	Class E	High

Sumber: The Ministry of Public Works Regulation No. 20/PRT/M/2007

The evaluation of land use suitability forms can be classified by comparing the land form with capability class scheme with intensity and type of land use [9]. The core of land evaluation is comparing the requirements demanded by the type of land use that will be applied with the properties or quality of land. Thus, it will be known the potential of land and its carrying capacity which includes the ability class and land suitability for several types of land use [10]. The results obtained was in the form of land capability classes in Bunguran Island, which in turn, resulted in land capability as a reference in determining land use directives. The direction of land capability was analyzed based on the results of land capability compared to the existing land use conditions in Bunguran Island. The results obtained were used as a basis for recommendations or directives for land use in Bunguran District. The parameters used in the classification of land capabilities are slope, drainage conditions, soil permeability, soil type, texture of soil depth, and erosion data. From the data parameters, the descriptive analysis process was carried out according to the classification of land capabilities.

3. Result and Discussion

Based on land classification parameter's overlay, land capability in Bunguran Island was classified as somewhat high, medium, low, and very low. The high development capability covered the smallest part of the Bunguran Island, which was 27.2 ha. Areas with rather high capabilities have very small areas so that their utilization needs to be considered for prioritized activities that required land with high capability values. The land capability that was rather high is located in Cemaga Selatan Village. The other part was an area of 3255.25 ha in the Bunguran Island, which is a land that has the ability to develop land. Land capability was being spread throughout all parts of the Bunguran Island, especially in the Northern of Bunguran Island. The rest consisted of low land capability having an area of 3021.5 ha, and the very low land capability having an area of 640.0 ha which occupied a large part of the Bunguran Island, mainly occupying the Southern Bunguran Island.

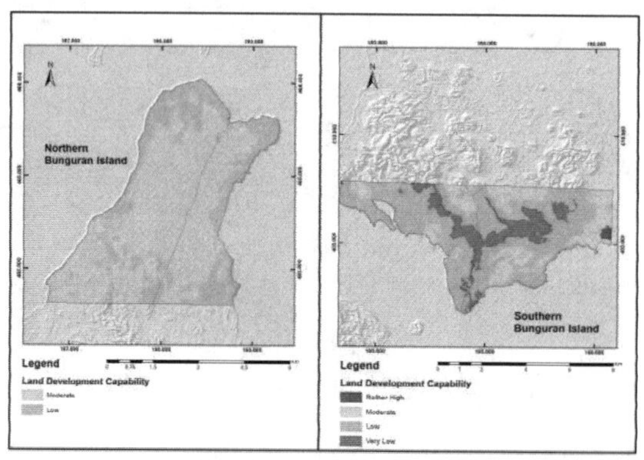

Figure 1. Land Capability Map of Bunguran Island in 2018

Table 3. Land Capability of Bunguran Island in 2018

Land Development Capability	Area
Rather High	27,25
Moderate	3.255,26
Low	3.021,26
Very low	640,01
Total	6.943,77

IOP Conf. Series: Earth and Environmental Science **328** (2019) 012015 doi:10.1088/1755-1315/328/1/012015

The result of Bunguran Island land capability was being used to plan the suitability land based on The Ministry of Public Works Regulation no. 20/PRT/M/2007. Suitability land's recommendation is divided into Land Cover Ratio Recommendation, Direction for Utilization of Raw Water, and Agricultural Spatial Arrangements, i.e.:

1) The recommendation of land cover ratio. The map of Bunguran Island's land cover ratio is divided into 4 classifications, i.e. non-building, maximum land cover of 10%, 20%, and 30%. The homogeneity of land cover ratio is defined by the slope of land in Bunguran Island. The northern Bunguran Island can be built up to a maximum of 20% land cover. The coastline has a land cover of a maximum of 10% lower than the upstream part, due to the sloping difference compared to the surrounding area. But the differences were actually not significant. The slopes on the Northern Bunguran Island coast of Buton bay are still in *landau* class, which means more intensive closure was still possible. But in the Southern Bunguran Island showed more various result based on the slope of this area. The east coast section showed the value of landing ratio stating that land cover can be built up to 30% coverage. The eastern coastal area has land slope stability and the land capability was higher than the surrounding area. The land cover ratio was expected to be smaller towards the area that is tilted towards upstream. In parts with hilly morphology, the land was directed to allocate 10% for non-buildings in mountainous areas.

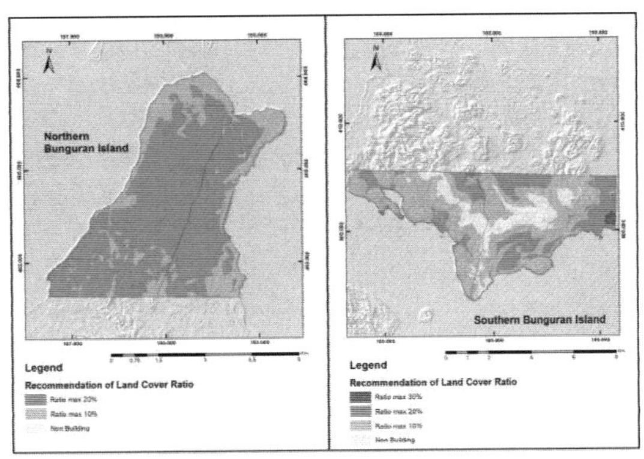

Figure 2. The Recommedation of Bunguran Island's Land Cover Ratio.

2) The recommendation of Agricultural Spatial Arrangement. Agricultural utilizations in Bunguran Island area were divided into 4 directives, namely; 1. Protected Areas, 2. Buffer Areas, 3. Annual Plant Areas, 4. Yearly / Seasonal Plant Areas. The direction of the agricultural area was strongly related to land that needed to be conserved and land that can be utilized intensively. Similar to land cover directives, agricultural directives made slope and topography as their main factors caused by the homogeneity of other factors. The northern Bunguran Island was directed to become a buffer zone and annual crop cultivation. This region is a relatively flat area but its soil conditions have a high erosion rate. Land use for annual crops was directed at areas that are flat in the upstream area, while those on steeper slopes were directed to become buffer zones with non-intensive cultivation of hard plants. Whereas the southern Bunguran Island was directed to become a protected area in the mountainous section with a buffer around it on the hilly morphology. The direction for coastal lands can be used for annual agriculture. Seasonal agriculture can be done in the eastern part of the southern village.

IOP Conf. Series: Earth and Environmental Science **328** (2019) 012015 doi:10.1088/1755-1315/328/1/012015

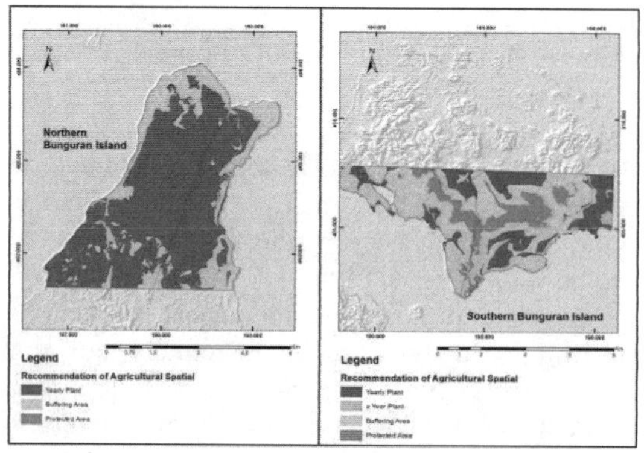

Figure 3. The recommendation of Bunguran Island's Agricultural Spatial

3) Direction for Utilization of Raw Water. The directives for raw water in Bunguran Island were divided into 3 classes, namely; 1. Good, 2. Enough, 3. Low. In general, the use of water in Bunguran Island can be utilized adequately both in the northern and in the southern regions. Water in the northern part of Bunguran Island can be used intensely in the upper reaches of this region. Whereas in the coastal part, the water was directed to be utilized with low intensity. The southern section showed a more varied class. Precisely, part of the east coast can be utilized in good intensity. The upstream was directed not to be used with high intensity. The area in Bunguran Island which has good water carrying capacity has a very small area of 27.24 ha in the southern region. An area of 3,255.09 ha spreaded in parts of Bunguran Island can also be used quite well with certain requirements to maintain the capacity of raw water. While the remaining area needed to be conserved, because of the limited water availability .

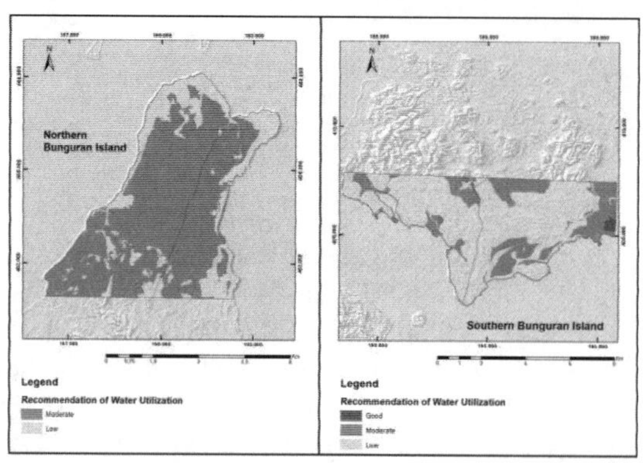

Figure 4. The recommendation of Bunguran Island's Raw Water Utilization

4) Recommendation of Building Height. The height of the building in Bunguran Island was directed to no more than 4 floors. There were two classes of building heights in the Bunguran Island, namely Non Buildings and Buildings less than 4 floors. The area in Bunguran Island which can be built to a height of less than 4 floors spanned over 3,282.5 ha. The remaining 3,661 ha was allocated to Non-Buildings. The area that is capable of being built is on the inside of the Northern Bunguran Island, where on the steep coast, it was directed not to be built. Whereas on the

Southern Bunguran Island in the coastal part, buildings with less than 4 floors can be built. The steep upstream section was directed to not built.

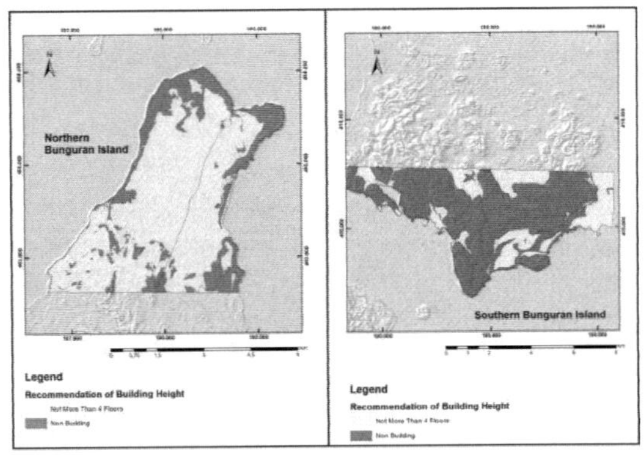

Figure 5. The recommendation of Bunguran Island's Building Height

There were five main constraints in the urban development in Bunguran Island i.e.; mountainous morphology, steep slopes, high erosion, foundation stability, and ease of work. The combined land that can be developed for urban areas in the Bunguran Island is 2,806 ha located on Northern Bunguran Island and Southern Bunguran Island. The upstream area in the Southern Bunguran Island has limited steep slopes and mountain morphology, this region also has a high erosion rate. In the Northern Bunguran Island, region steep areas are on the coast. The transitional part of the coast with hilly areas has a low foundation stability with difficult workability.

Land Capacity of Bunguran Island provided information about the ability of existing land in accommodating a number of residents and population activities. Capacity made the population in one area as part of the analysis. High capacity meant that the area can still be given additional population and community activities at a considerable level. On the contrary, low capacity of the region indicated that it is getting smaller. Unsuitability meant that the region was indeed directed to be emptied of the population. The area that is directed not to be used for population activities has the largest area in the Bunguran Island (73%), which is 5,077 ha (73%). This area is in the Southern Bunguran Island where it is located in an area with a high slope. Areas with high capacity and very high capacity occupied 8% of the Bunguran Island area located in the coastal part of the southern area and parts of northern area which showed a large capacity.

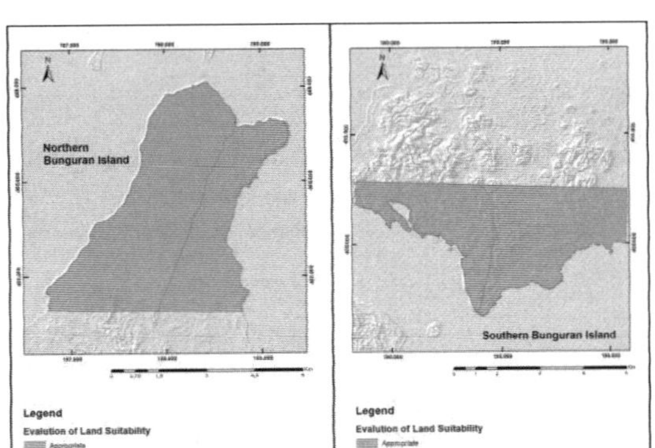

Figure 6. Land Suitability of Bunguran Island

This research was limited to just analyzing the land suitability and land capability to compose spatial pattern planning of Bunguran Island. Based on those, Bunguran Island's spatial pattern planning was divided into Protected Areas and Cultivation Area. Protected area has spacious area at 87%, which is bigger than the cultivation area. As a national border area, Bunguran Island has many potential agriculture and fisheries to support the national economy. The next development in Bunguran Island needs a spatial plan which supports agricultural and fisheries concept.

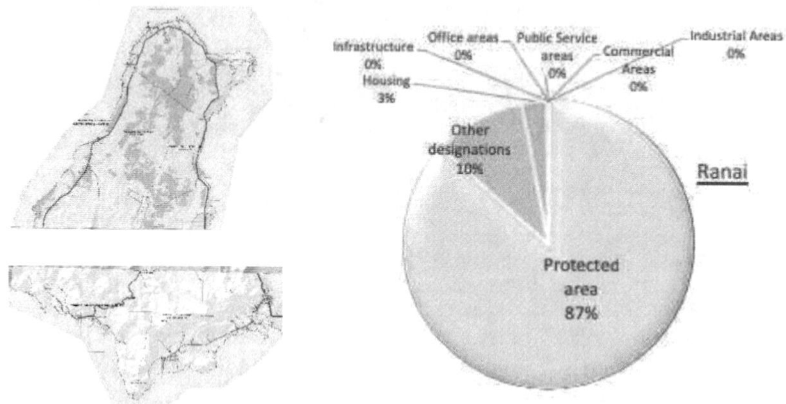

Figure 7. Land Development Recommendation of Bunguran Island

4. Conclusion

Spatial planning is a tool to make a regional regulation on detailed spatial planning and it should be an on going activity to drive the stakeholders on pursuing the goals. It also can be proposed in the annual revision agenda through zoning regulations [11]. The results of the evaluation of land capacity based on land capability and suitability are expected to be used as a basis in the formulation of the border management policy as land development recommendation of Bunguran Islan as one of national border area. The spatial pattern planning of Bunguran Island consists of 87% of protected area that cant be transferred as cultivation area. The rest of selected cultivation area in Bunguran Island can be managed and developed as settlement area, commercial area, industrial area, and more based on Agrarian and Apatial Planning Ministry Regulation no 16 of 2018 about detailed spatial planning.

Acknowledgments
The authors express their gratitude to BIG and PT Rensa who have supported in fully conceiving the Mapping of The Detailed Spatial Planning in National Border Areas in 2018 especially in PKSN Ranai section. The authors are grateful to all of those with whom they had the pleasure of working during this and other related projects.

Reference

[1] Peraturan Pemerintah no 26 Tahun 2008 tentang Rencana Tata Ruang Wilayah Nasional.
[2] Peraturan Pemerintah no 13 Tahun 2017 tentang Perubahan Atas Peraturan Pemerintah Nomor 26 Tahun 2008 Tentang Rencana Tata Ruang Wilayah Nasional
[3] Muta'ali, Luthfi. 2012. Daya Dukung Lingkungan Untuk Perencanaan Pengembangan Wilayah. Yogyakarta: Universitas Gadjah Mada.

[4] UU Republik Indonesia No 32 Tahun 2009 tentang Perlindungan dan Pengelolaan Lingkungan Hidup.
[5] Arsyad, Sitanala. 2000. Konservasi Tanah dan Air .Edisi kedua. Bogor : IPB Press.
[6] Patel, N.R. et al. 2001. Land Capability Assessment for Land Use Planning Using Remote Sensing and GIS. Agropedology, 2001. 11,1-8.
[7] Suganda, A.H., 1988. Pertimbangan Aspek Dasar Dalam Perencanaan Kota. Thesis S-2 Fakultas Pasca Sarjana ITB. Bandung (Tidak dipublikasikan).
[8] Peraturan Menteri PU No 20 Tahun 2007. Teknik Analisis Aspek Fisik & Lingkungan, Ekonomi Serta Sosial Budaya Dalam Penyusunan Rencana Tata Ruang.
[9] Simanungkalit, NM. Evaluasi Kemampuan Lahan Dan Penggunaan Lahan Pertanian Di Sub Das Gotigoti, Daerah Aliran Sungai Batangtoru, Kabupaten Tapanuli Utara. Jurnal Geografi Vol 13 No 1 Februari 2011.
[10] Hardjowigeno dan Widiatmaka. 2007. Evaluasi Kesesuaian Lahan dan Perencanaan Tataguna Lahan. Gadjah Mada University Press. Yogyakarta.
[11] Susetyo, Budi, dkk. 2014. Analisis Spasial Kemampuan Dan Kesesuaian Lahan Untuk Mendukung Model Perumusan Kebijakan Manajemen Lanskap di Sempadan Ciliwung, Kota Bogor. Makalah Ilmiah Globe. 16 No 1 Juni 2014.

IOP Conf. Series: Earth and Environmental Science **328** (2019) 012016 doi:10.1088/1755-1315/328/1/012016

Green Open Space, Towards A Child-Friendly City (A Case Study in Lembah Gurame Park, Depok City, Jakarta Greater Area, Indonesia)

E Yuniastuti[1] and H S Hasibuan[1]*

[1]School of Environmental Science, Universitas Indonesia, Indonesia

*email: hayati.hasibuan@ui.ac.id

Abstract. Green Open Space (GOS) is essential to provide a critical ecosystem service that reduces the social stress levels of a community living in densely-populated urban areas. Children, as part of an urban community, are often most affected by the lack of green space. Government seeks to fulfill these rights through its Child-Friendly City (CFC) program. This study aimed to analyze whether the existing GOS had meet the criteria of child-friendly GOS. This research was conducted in Lembah Gurame Park, as one of the urban GOS in Depok City as a part of Jakarta Greater Area. This study employed spatial analysis using GIS and conducted interviews to the visitors to find out the needs of the children on GOS. This study observed the design and management of the Lembah Gurame Park in accordance to the principal of Child-Friendly GOS including landscape, facility and utility, park furniture, flora and fauna, as well as human activities in the Lembah Gurame Park. This study found that the design and management of the GOS is below the optimum for child-friendly GOS. Many facilities and utilities are dangerous for children and very limited activities can be involved by children.
Keywords: Child-Friendly Cities, Green Open Space, Lembah Gurame Park

1. Introduction

Green open space is a space that is planned to meet the need of community interaction and joint activities, the space also serves as an active playground for children and adults [1]. The benefits of GOS are environmental benefit, economic and aesthetic benefit, as well as social and psychological benefit [2]. Green Open Space is an external public space where this type of public space is usually in the form of outside space that can be accessed by all people (public) such as city parks, squares, pedestrian paths, etc [3]. In 2011, UN-Habitat adopted resolution on sustainable urban development through access to quality urban public spaces. The proposed Sustainable Development Goals highlighted Public Space as a key and important topic. "Sustainable Development Goals, (Proposed Goal 11 Target 7 – By 2030, provide universal access to the safe, inclusive and accessible green and public spaces, in particular for women and children, older persons and persons with disabilities)".

Urban area with very rapid development often overrides the need for GOS. The number of GOS has been declining with the building and pavement. For example, Jakarta with a total area of 649.71 km2, in 1982, its GOS was around 40% of the total area. But in 2013, there was a drastic decline, leaving only 17% of the total area of Jakarta [4].

In the 1989 Convention on the Right of the Child which has been ratified by the Indonesian government through Presidential Decree No. 36 of 1990 concerning the Ratification of the Convention on The Rights of the Child, it was stated that children are those aged 18 years and under. To be able to grow and develop properly, children need an open space for playing, learning, exercising, and recreation. All those activities are very important to the practice of psychomotoric, cognitive, and affective skills. The density and spatial patterns of urban land and green spaces are considered in landscape and urban planning so that urban areas and cities can be a healthier and more comfortable urban environment to live [5].

Children growth is related to the changes in the weight, length, bone age, and the metabolic balances (body calcium and nitrogen retention). Development is the increase in skill (ability) in the structure and function of the body that is more complex in a regular and predictable pattern, as a result of the maturation process. Including also emotional development, intellectual, and behavior as a result of interaction with the environment. So it can be concluded that growth has an impact on physical aspects, while development is related to maturation of organ / individual functions. Nevertheless, both events occur synchronously in each individual [6].

The Ministry of Health of the Republic of Indonesia (2006) mentioned that the developmental aspects that can be monitored includes gross motion, subtle movements, speech and language skills, as well as socialization and independence. Three of four aspects can only be fulfilled by the presence of GOS as a place for children to practice gross motoric skills and to interact to practice language skills and psychosocial skills. Playing together is very important for children. Social forms of playing helps children to learn how to cooperate with one another and get beyond their self-centered perspective to try to see the world of the eyes of other people [7].

2. Methods

2.1 Research Area

The location of the research is at Lembah Gurame Park, the biggest park in Depok City. We selected Depok as the sample of observation, due to the majority proportion of children-age population in Depok, the suburban of Jakarta Metropolitan Area.

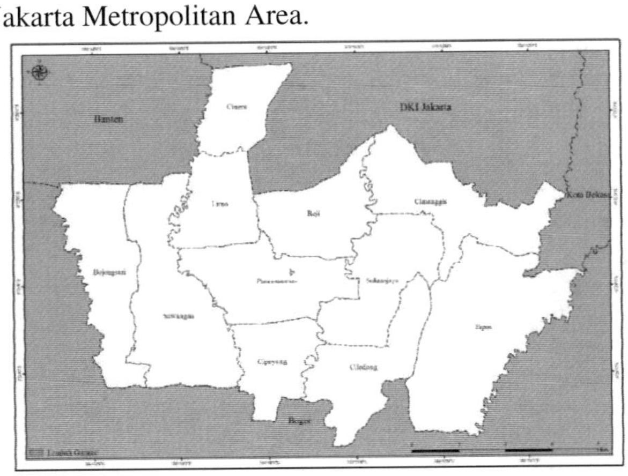

Figure 1. Studi area Depok
Source: Map of Rupa Bumi Indonesia, BIG (2016), Aerial Photo Interpretation (2018))

2.2 Indicator Observed

This research employed spatial analysis using the Geographical Information System for the distribution and the accessibility of the GOS. In accordance to the analysis of child-friendly GOS, this study observed the design of the existing GOS with the standards including the material, pedestrian path, themes, furniture, and park facilities.

IOP Conf. Series: Earth and Environmental Science **328** (2019) 012016 doi:10.1088/1755-1315/328/1/012016

2.3 Interview and questionnaire
This research conducted interviews the children-aged visitors of Lembah Gurame Park find out their motives of visiting park, the distance from their house or school, the frequency of visiting the park, the activities they do while in the park, and their opinion to the infrastructure and flora-fauna in the park. This research also observed their activities in the park.

3. Results and Discussion

3.1 Population Characteristic
Depok City is located in the south of Jakarta. As a buffer city, Depok is growing rapidly along with the development of Jakarta. Depok City consists of 11 sub-districts and 63 urban villages with a total area of 200.29 km^2 with a population of 2,254,513 or a density level of 11,256 people/km^2. As a metropolitan buffer city, settlements in Depok City are mainly residential for commuters who work every day in DKI Jakarta. As a residential area Depok City has a high number of children, where for children aged 0-19 years it reached 784,649 people or 34.8% of the total population.

Tabel 1. Children Population by age

Child age group	Total
0-4	221.122
5-9	195.593
10-14	175.931
15-19	192.003
Total	**784.649**

Source: Depok in number, 2018

3.2 Green Open Space
According to Minister of Public Works Regulation No. 05 / PRT / M / 2008 concerning Guidelines for the Provision and Use of GOS in Urban Areas, park area is classified into three classes, namely: narrow (less than 1,250 m^2), medium (1,250–9,000m^2), and wide area (more than 9,000m^2) [17]:

Table 2. Depok City Park

Classification	Name of the Park	Area (m^2)
Narrow	Bunga Pratama	326
	GTP	396
	BDN	893
	Sehat Koperasi	600
	Anantakupa	708
	15 Cinere	1000
	Dahlia	1083
	Pijar	1131
Medium	Markisa	1410
	Jembatan Serong	2500
	Lingkar UI	5000
	Lembah Leli	6000
	Jatijajar	6000
	Balaikota	7000
	Sukatani	7000
Wide	Lembah Mawar	14000
	Merdeka	15000

IOP Conf. Series: Earth and Environmental Science **328** (2019) 012016 doi:10.1088/1755-1315/328/1/012016

Lembah Gurame	36000

Depok City is a buffer zone for the capital city of DKI Jakarta. The current public green open space is only 2,015.53 hectares or 10% of public green open space out of the total area of Depok City, where it should have a public open space of 20% of the total city area [18].

3.3 Lembah Gurame Park

Lembah Gurame Park is located in Depok Jaya Village, Pancoran Mas District, Depok, with a total area of 36,000 m^2. This spacious park is located right in the middle of the Depok I Housing area. Which means, it is surrounded by residential areas that are quite dense beside the trade and education areas. The following is the map of Lembah Gurame Park and its surrounding land use.

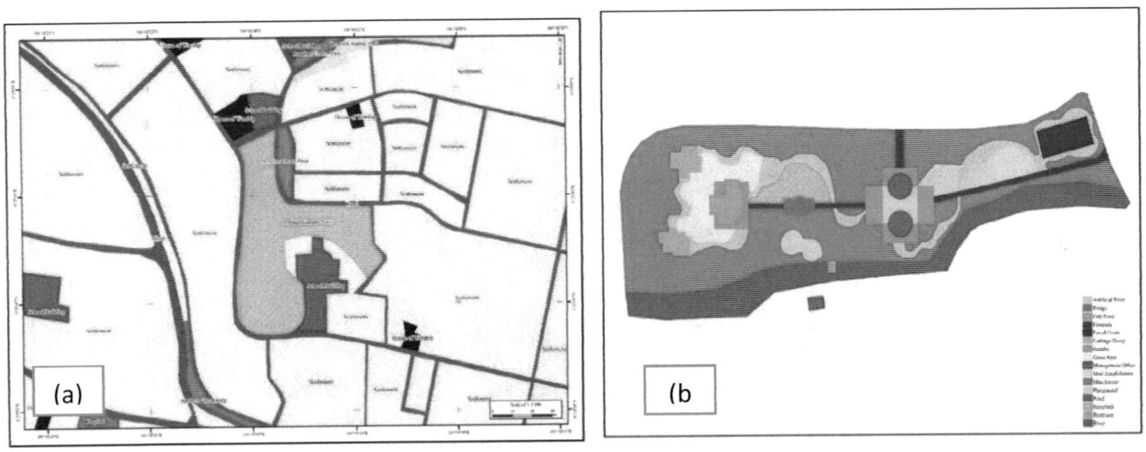

Figure 2. (a) Land use surrounding Lembah Gurame Park and (b) lay out of Lembah Gurame Park
Source: Map of Rupa Bumi Indonesia, BIG (2016), Aerial Photo Interpretation (2018)

From the picture of land use above, it can be seen that the Lembah Gurame Park is surrounded by settlements, education, and worshipping place. In the area known as Depok 1 area, this is an area developed for settlements, in other words, there are many children in this area. The location of the Lembah Gurame Park is relatively close to the settlements, so school-age children can reach it easily. Whereas for those who are located some distance away, most of them use motorcycle mode with a travel time of 5-10 minutes. The layout of the Lembah Gurame Park and pictures of infrastructures are as follows. Lembah Gurame Park is a fairly complete GOS covering ecosystems of artificial forests, pond ecosystems, and urban ecosystems where all are combined as a harmonious ecosystem. The facilities inside are quite complete starting from the game, sports, fitness, jogging track, gazebo, etc. The function of GOS must be prioritized by maintaining the existing flora and fauna well, but besides that, the function as a public space must be maximized especially as a child-friendly public space, so that the Lembah Gurame Park can play a role in supporting children's development in Depok, where to achieve this goal, there are components that must be met by green open space.

3.3.1 Landscape

The landscape in Lembah Gurame Park is very varied, there are elevations and declines in several places with conditions of paths not perfectly connected or sometimes in the form of steps that are not friendly to the disabled using wheelchairs or baby strollers. The paths are also in a damaged condition, and at some point, they need to be added handles for safety, for example on the edge of the pool. Lembah Gurame Park is crossed by an artificial river with a pond in the middle. This signifies that this area serves as a catchment area that accommodates runoff water from the surrounding area. This is very useful, though the river that crosses this area is dirty and smelly and inundated. Therefore, it is necessary to maintain the cleanliness of the river that goes through this area. Besides that, it is necessary to have a safety fence, especially in ponds with a certain depth because it will be very dangerous if children fall into the river.

3.3.2 Park Facility and Utility.
The control of the design of a children's playground was based on the function as an area of development of the creativity of the social soul, senses and self-development of children so that they can obtain pleasure. Therefore, the playground must meet the following criteria [15]:

- Ensuring safety, security and health of children to play in public spaces.
- Creating comfort and convenience for all children (healthy and with physical and mental limitations).
- Creating aesthetic visual harmony with the character of the surrounding area.
- Providing clarity about the function of the game equipment and the strength of its construction.

The facility contained in the Lembah Gurame Park in the form of sports facilities is only one futsal field, which means that it cannot accommodate the needs of residents, there needs to be additional sports fields. Playing facilities for children are quite a lot but many conditions are damaged, which is considered to be unsafe. There needs to be revitalizations and the additions of areas of dexterity and traditional games.

Figure 3. The Broken Playing Facilities
Source: Primary survey, 2018

Social facilities such as multipurpose rooms do not exist yet and need to be added so that they can be used as an indoor play area, besides as a place for community activities such as recitation, posyandu, etc. Other social facilities that need to be added are the nursing room and prayer room, so that visitors can enjoy the park to the fullest. Economic facilities do not yet exist in this region, considering that people usually buy food from street food vendors in the area, therefore, for example kiosks that sells healthy food in areas with good health and hygiene monitoring are plausible. This facility can also be managed by the surrounding community as a source income. Based on observations and interviews, the biggest problem in Lembah Gurame Park is the problem of waste, it is necessary to have good hygiene management.

3.3.3 Flora and Fauna
The existence of biotic elements, namely flora and fauna, are a must for green open space. Biotic elements in green open space generally consist of humans, plants, and animals where all three must exist and interact with each other. Urban and suburban parks can play an important role in the conservation of biodiversity [8]. The human experience is increasingly defined within an urban context. The results highlight that cities can support both biodiversity and people, but retaining these connections requires sustainable urban planning, conservation, and education focused on each city's unique natural resources [9]. In the child friendly concept, biotic elements are very important as a learning process where children can get to know common flora and fauna around them. Being able to see them directly and possibly to touch them, is of course a very good stimulus in the child's

development process. Psychologically, biotic elements such as plants and animals can provide much needed calming sensation for children who live in densely populated areas.

The existing flora consists of large trees, flowers, shrubs, grass, to water plants. The dominant trees are trembesi (*Samanea saman)* and breadfruit (*Arthocarpus altilis)* with large and shady canopy. Shrubs and flowers also thrive even though they seem untreated. There is a need for efforts to prune plants, especially branches that are fragile to make them tidier and not endangering visitors. The fauna seen in pond and river ecosystems in Lembah Gurame Park consists of Gurame (*Osphronemus goramy*), Sepat (*Trichopodus trichopterus*), Catfish (*Clarias*), and Guppy (*Poecilia reticulate*) and Frogs (*Fejervarya cancrivora*). In addition, in the trees, Sparrows (*Passeridae*), Pigeons (*Columbidae*), Lizards (*Lacertilia*), and Cats (*Felis catus*) are seen. For insects seen, there are Grasshoppers (*Caelifera*) and Butterflies (*Rhopalocera*) especially in flowering shrubs.

3.3.4 Human Activities. there are 3 (three) activities that are often done by the community in this place, which are playing, exercising, and relaxing [10]. These public parks could be an option for urban tourism both for Jakarta residence and tourists, since it is accessible as well as free of charge. With no entrace fee, public parks are suitable for every class and age [11]. Despite the rise of interest in child-friendly cities, urban planning and design is mostly adult-centered [12]. For children and adolescents, contact with nature is also associated with lessening of stress and emotional problem [13]. Quantitative evidence confirmed a positive association between presence of trails, playgrounds, and specific type of sports fields with GOS visitation and Physical Activity whereas safety and aesthetics seemed subordinate [14].

Based on the results of the interview, the travel time between visitor's house and the Lembah Gurame park varies from 5 minutes to 1 hour. Most of the visitors are residents of surrounding settlements or school students around the park, but for those with long distances usually due to the interest to gather with the community. From the results of observations and interviews, it is known that Lembah Gurame Park is one of the reference parks for several communities to gather regularly, including the Cat Lovers Community, Music and dance Community, Skateboard Community, and Volunteers of Community Book Reading in the Park. Various types of communities chose to gather here because of its location which is easily accessible and has sufficient area for creativity. Seeing the enthusiasm of the community here, the function of GOS as a public space for interaction is very vital, especially that some of the community activities can also be attended by the general public, especially children. For example, the Community who lends free reading books to children accompanied by routine storytelling visits this park every Sunday afternoon. The main purpose of children visiting GOS is for recreation, for playing and socializing. For school students, most of them come to exercise or to do school work that they do in the gazebos which are mostly available here.

The duration of the visit also varies between 1-4 hours, for parents and toddlers the duration is about 1 hour, but in adolescents it tends to be longer i.e. 2-3 hours, because of the need to exercise and do schoolwork. The longest duration is in the average, where the average reaches 4 hours due to the type of interesting activities and the same interests among them. Its location in a densely populated residential area in the middle of the city becomes a cheap and affordable recreational place for the community. Lembah Gurame Park creates a feeling of recreation, so visitors always repeat their arrival, especially at the end of the week. The function as a public space can be maximized by the addition of several public facilities, while maintaining its function as open green space. For example, among the trees, you can add healthy food kiosk facilities, waste processing training facilities, nursery facilities that can be added to the education function for the community and especially children.

4. Conclusion

In terms of accessibility, Lembah Gurame Park is very easy to reach and right in the residential area. From the results of observations and interviews, the function of the Lembah Gurame Park as GOS has been very good with complete pond and artificial forest ecosystems. But to maximize the function as a child-friendly public space, it is necessary to have more comprehensive management, both physically and non-physically in Lembah Gurame Park. There are still many facilities that are dangerous for children. Besides that, it is not supported by facilities and activities that support children's growth and development, which include cognitive, affective, and psychomotor ability. Therefore, it is necessary to add facilities and utilities that are oriented towards children's education, as well as to maximize family time in this GOS. Facilities and utilities provided must be child-friendly both in terms of material and design so as not to endanger them. The main factor that is needed is the existence of activities that attract children to come and feel at GOS, in the form of activities that have an educational element and cooperation for children. Activities must be initiated by the manager by involving children and carried out continuously with different interesting themes for children.

Acknowledgement

Funding for this study was provided by the grant of the Final Task Academic Writing issued by Directorate of Research and Public Services, Universitas Indonesia, contract number 2575/UN2.R3.1/HKP.05.00/2018.

Reference

[1] Dewi O C *et al* 2018 *Green Open Space: Awareness for Health or Sustainability? IOP Conf. Series: Earth and Environmental Science*) p 120
[2] Haq S M A 2011 Urban Green Spaces and an Integrative Approach to Sustainable Environment. (Journal of Environmental Protection vol 2) p 601-608.
[3] Carmona M et al 2003 Public Places – Urban Spaces, The Dimension of Urban Design. (Burlington: Architectural press)
[4] Budiman A et al 2014 Deteksi Perubahan Ruang Terbuka Hijau Pada 5 Kota Besar Di Pulau Jawa (Studi kasus: DKI Jakarta, Kota Bandung, Kota Semarang, Kota Jogjakarta, Dan Kota Surabaya) (Jurnal Lanskap Indonesia vol 6) 1
[5] Estoque R C et al 2016 Effects of Landscape Composition and Pattern on Land Surface Temperature: An Urban Heat Island Study in the Megacities of Southeast Asia (Science of the Total Environment)
[6] Soetjiningsih 1995 Tumbuh Kembang Anak (Jakarta: Penerbit Buku Kedokteran EGC)
[7] Hughes F P 2010 Children, Play, and Development (USA: Sage Publication)
[8] Arifin H S and Nakagoshi N 2011 Landscape Ecology and Urban Biodiversity in Tropical Indonesian Cities (Landscape and Ecological Engineering vol 7) p 33-43
[9] Aronson M F J et al 2014 A Global Analysis of the Impacts of Urbanization on Bird and Plant Diversity Reveals Key Anthropogenic Drivers (Proceedings of the Royal Society)
[10] Sutanto E and Junadi P 2018 Kalijodo Transformation in Establishment of Healthy Environment in Jakarta. (Earth and Environmental Science) p 126
[11] Adiati M P et al 2018 Public Parks as Urban Tourism In Jakarta (Earth and Environmental Science) p 126
[12] Li D et al 2018 Moving Beyond the Neighborhood: Daily Exposure to Nature and Adolescents Mood (Landscape and Urban Planning) p 173
[13] Chawla L et al 2014 Green Schoolyards as Havens from Stress and Resources for Resilience in Childhood and Adolescence (Health and Place) p 28
[14] Hecke L V et al 2018 Public Open Space Characteristics Influencing Adolescents Use and Physical Activity: A Systematic Literature Review of Qualitative and Quantitative Studies (Health and Place vol 51) p 158-173
[15] Baskara M 2011 Prinsip Pengendalian Perancangan Taman Bermain Anak di Ruang Publik. (Jurnal Lanskap Indonesia vol 3)
[16] Syahadat R M et al 2017 Identifikasi Keanekaragaman Hayati RTH Di Kota Depok (NALARs Jurnal Arsitektur Volume 17) 1

[17] Salsabila H et al 2017 Hubungan Karakteristik Lokasi Dengan Pengunjung Taman Kota Di Kota Depok (8th Industrial Research Workshop and National Seminar)
[18] Bappeda Kota Depok 2015 Rencana Tata Ruang Wilayah Kota Depok 2012-2032

IOP Conf. Series: Earth and Environmental Science **328** (2019) 012018 doi:10.1088/1755-1315/328/1/012018

Child Pedestrian Friendly Design Principle for the Settlement and Housing area

F P Makalew[1*]

[1] Civil Engineering Department, Manado State Polytechnic. Polytechnic Street Buha Manado, Indonesia

*email: fmakalew@yahoo.com

Abstract. Children pedestrian as part of sustainable transportation requires safe and comfortable area for their activity. The design principle of a good quality children pedestrian needs in-depth research. Study involving children pedestrian are related mostly to research with an adult as a major user, behaviour and pedestrian route. This study aim is to analyse the design principle of children pedestrian in settlement and housing area. The location of the study is the city of Manado, Minahasa and North Minahasa Regency. Data is collected through questionnaire and survey on field area and recording through videos and photos of activities in children pedestrian using smartphone and camera. The samples for this research are children pedestrians in selected housing and settlement areas in Winangun area, schools in Manado and North Minahasa Regency. The number of samples is 111 children between the age of 8 to 12-year-old. The result showed children preference of object and place in the walking area. Their choice for pedestrian area is also interesting in which they consider the aspect of safety, comfort and fun for doing their activities in their neighbourhood area and the surrounding school area. A specific regulation for children pedestrian is important when considering child-friendly environments.
Keywords: Children Pedestrian, Pedestrian Friendly, Settlement Area, Housing Area, North Sulawesi

1. Introduction

The available researches regarding children pedestrian and standard have limited consideration to children behavior, movement patterns and preferences. This research focuseds on design principle for children pedestrian in the area of settlements and housings in urban and rural area.

Pedestrian pathway is not always available on the street which has always been the obstacle for a comfortable walking area. Children pedestrian use all access to walk, including designated pathways and street area. They share the area with vehicles and bicycles when they walk to a certain destination. For many parts of the street, sharing areas can be a dangerous place for the users, including children. Standard that are available for pedestrian pathway has limited consideration for children as pedestrian [1]. Children are not only walking with an adult, but also with their friends or walking alone. Children also do other activities such as playing and running while walking. In designing a pedestrian area for children, adult used to be the decision maker. Children behavior have influenced the use of pedestrian area [2]. Therefore, taking children preference into account is important when designing a child-pedestrian-friendly pathway.

1.1 Children Pedestrian and Friendly Environment

Pedestrian is one of the active transportation when considering it as a sustainable transportation mode. Children pedestrian, as part of this system, has been recognized as an important role in defining pedestrian facility. Walking is a natural transportation with complex analysis [3]. Pedestrian model is based on their choice including activity, destination, mode, and route, where behaviour and interaction within pedestrian walk are important (ibid).

Based on the street classification, local street are categorized for pedestrian [1, 4]. The standard for pedestrian pathway has considered area such as housing and school where children are the most active user, though it is based on adult movement pattern. Streetscape elements are including paving, landscape planting, street lighting, street furniture and public facilities and private streetscape amenities [5]. The streetscape is a space with visual identity consisting of physical elements, such as nature elements and non-physical ones, such as artificial elements [6]. Research on children pedestrian space capacity has found that children tend to use larger space compared with standard available for adult and children due to their movement [7].

In the area of urban settlement and housing, existing researches are related to pedestrian behaviour in low-income and middle-income housing [8]. The research finding is including access and location as factors which influenced the use of facilities where the playground and local shops are placed in walking distance. In a neighbourhood plan, transportation element is recommended to include the improvement of sidewalk for pedestrian and accesses to transit [5]. Pedestrian-friendly street is related to the priority for pedestrian traffic rather than vehicle (ibid).

The concept of child-friendly environment is related to independent mobility where children have a wider area to move around [9]. The criteria for child friendly environment are affordance, the right uses of the object in which as children grow, their ability to perceive affordances is also developed (ibid). Plester et al [10], found that children can understand aerial photographs where they learn a different representation of the world including photographs. Children experience the environment from secondary sources (ibid). Children as part of the city, contribute in creating liveable and child-adult friendly city [11]. It will need an urban designer and planner to rethink and modify the participation of children in designing city (ibid).

2. Methods

This research method is a field survey with recorded video and photos, combined with a questionnaire for children between the ages of 8 to 12 years old. The samples are from 2 schools in Manado and Minahasa regency North Sulawesi and 1 neighborhood area in Winangun Manado.

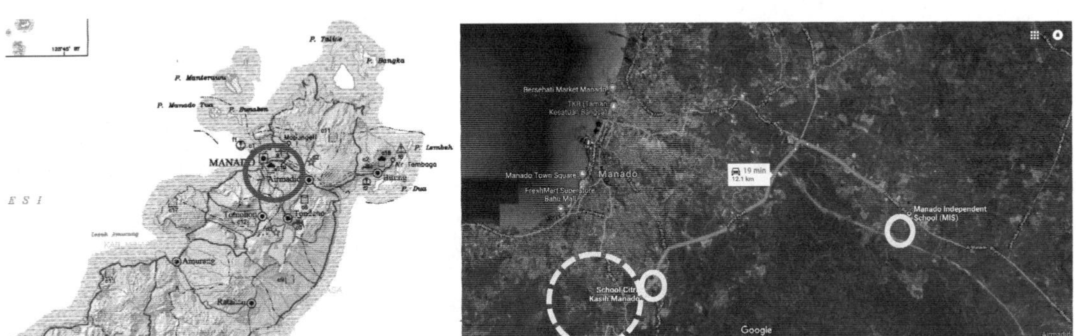

Note: O = Schools ⁛= Winangun Area

Figure 1. Map Location of Sampled Children in Manado City and Minahasa and North Regency, North Sulawesi, Indonesia

The school in the urban area is *Citra Kasih* Elementary school of Manado municipality. The school in the rural area is Manado Independent School Kolongan North Minahasa. For the neighborhood area, the samples are from children who live in the area of Winangun Manado. This area has covered an area of Manado and Minahasa regency. Children in this area live in Citraland housing area and Winangun urban and rural settlement. For children from Manado Independent School, Kolongan North Minahasa, the area of their home is varied from the city of Manado, the city of Bitung, North Minahasa and Minahasa regency. Winangun is divided into 2 major areas, Manado and Minahasa regency. Children in this group are also going to different school in Manado and Minahasa regency. The number of children is 38 children from Citra Kasih School Manado, 25 children from Winangun Area and 48 children from Manado Independent School, Kolongan North Minahasa Regency. A total number of samples are 111 children.

Sample of children is considered varying due to the characteristics of the location of their home and school and option for transportation mode choice. Children from Winangun area live in different part of the urban and rural area. Winangun area is divided into Winangun One and Winangun Two in the city of Manado and Upper Winangun in the Minahasa Regency. Upper Winangun area is considered as a rural area which tends to be rural urban area due to the increasing number of populations. The area of farming is reduced and changed into housing area. Children from Citra Kasih School mostly use cars or motorcycles which are driven by an adult to school. While children from Manado Independent school haves an option of using school bus or being drive to school by parents. The school bus picks up children at certain points where children have to walk or are driven by adults.

3. Results and Discussions

Children pedestrian area are designed for children to walk and doing their activities during their walking journey. The result shows the preference of objects and places chosen by children during their walking time along the pedestrian area and area preference as a pedestrian area based on its built environment.

For the walking area, the percentage area chosen by children is 61,6 % for neighbourhood area in settlement and housing area and 38,4% for school area. Children prefer to walk in a neighbourhood area more than school area. The reason why children chose those specific areas for walking is that they are safe, comfortable and fun. In related to the area for fun, children describe the area of the neighbourhood is fun for playing, exercising, looking at nature, exploring, walking and are good for health. While for the school area, the reasons for this choice are because it is fun for running, playing, talking, walking, relaxing, and for sport. The majority of children's choice is that the walking area is fun for many activities, including playing. Playing is the character of children when doing their activities.

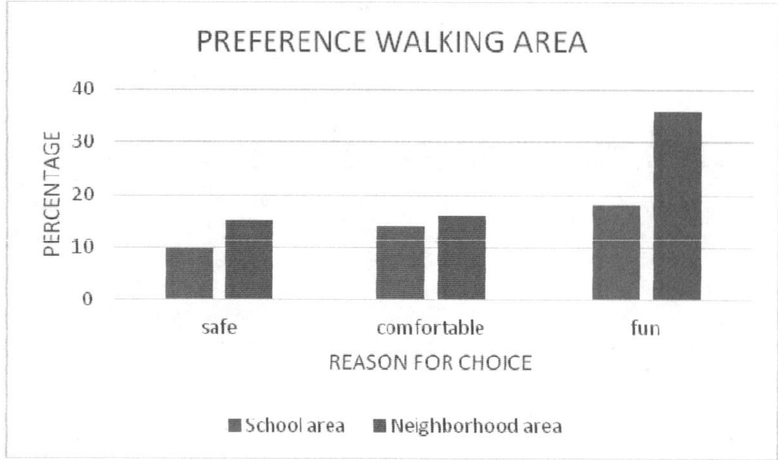

Figure 2. Children preference walking area

Moreover, object and place preference during a walking trip can be seen on the graphics as followed.

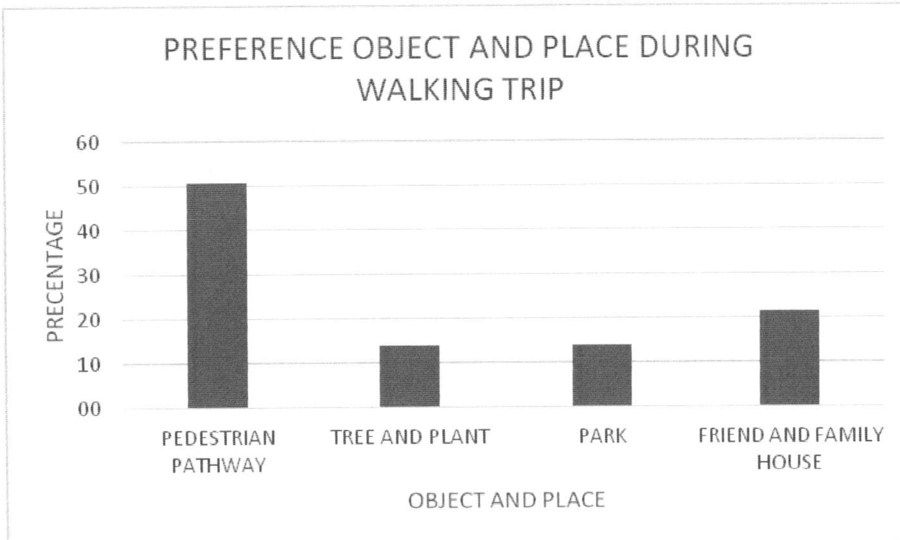

Figure 3. Children preference of object and place during the walking trip

Object and place that children chose the most is the pedestrian pathway. The next preference is friend's and family's house. This choice showed that children like to walk as well as to visit friend and family during their walking trip. In addition, when children were walking, they were also running and playing.

Children showed different responses to the pedestrian area. From 2 different photos, the first is a pedestrian pathway sharing with motorcycles in the major street with trees in bedding pots and the second is an area for pedestrian and cyclist separated from the major street with trees and plants lining along the street.

(a)

(b)

Figure 4. (a) Pedestrian sharing with motorbikes (b) Pedestrian sharing with bicycle
Source: Makalew, 2016

The pedestrian area in **Figure 4** (a) is a mix-used land area. Pedestrian pathway has always been used for parking motorbikes and street vendors. Vehicle traffic in this area is heavy due to the area being one of the major and oldest road in Manado. While **Figure 4** (b) is a pedestrian in mix-used area where the place is designed for pedestrians and cyclists at a major street in Seoul, South Korea. The preference can be seen in the figure below.

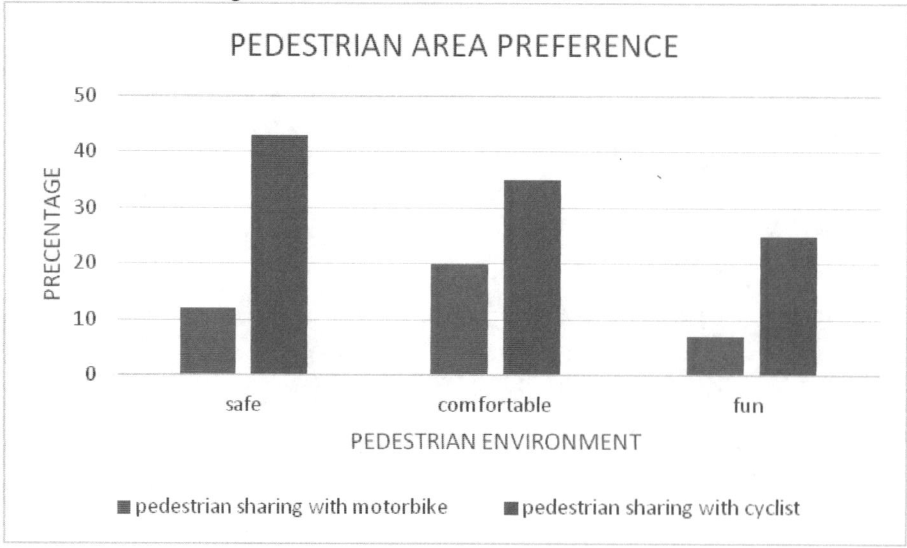

Figure 5. Pedestrian area preference

Children prefer pedestrian area in a-major road with separate pathway area where pedestrians can share with cyclists. The most chosen reason for the environment area is that the area is safely followed by comfortable and fun. Both pedestrian areas are located on a major road. The pedestrian area where pedestrian shares the pathway with motorcycles is more open to the street. There are trees along the street which are planted on bedding pot. By comparison, the second pedestrian area is designed specific for pedestrians and cyclists where the landscape consist of trees and other lower plant such as flowers.

The pedestrian area where pedestrians share the pathway with motorbikes and open to the street is considered comfortable for children to go anywhere. This area is fun for children for shopping, exploring major street, and exercise. In addition, children described the area as being noisy, polluted and having many litters.

The pedestrian area with the special landscape for pedestrians and cyclists is described by children as a safe place because it is wide. Moreover, the area is fun for many activities such as for relaxing, playing, exercise, enjoying something, sightseeing, playing bicycle, and reading a book. The area is also clean and has no trash.

It is interesting that children preference can vary based on the reasons of behind their choice. The street may be seen as dangerous for children while it is considered as fun for exploring the street and comfortable due to easy access to nearby places.

In addition, children walk in the pedestrian area beyond than the access provided. Children explore the surrounding area during their walking journey. The example, these children walked on all of the access which are available and expanded their walking area to the surrounding place. This preference can be seen in figure 6 below.

(a) A child walked in grass area in Citraland housing area

(b) A child walked by climbing to a higher area in Citraland housing area

(c) Children walked by using the street area in rural settlement Winangun area

(d) A child walkeds by exploring the natural elements in rural settlement Winangun area

Figure 6. Children walked with many activities in housing and settlement area
Source: Makalew (2016, 2018)

As can be seen in **Figure 6**, children in the housing area used the grass area for walking. This housing area only has sharing street without pedestrian pathway. Children walk on the grass area including the steep grass area to go to the next street. Citraland Manado housing area is located in an area with a different level of land with the hill and the valley. While for the rural settlement area of Winangun, children use local street and small path in the local plantation area. A child who walks with an adult is exploring the natural element surrounding the small path area.

The principle for children pedestrian considers the priority for children safety. Moreover, the pedestrian area should be comfortable and fun for children to do their activities. In order to fulfil this aspect, the quality of children pedestrian area should include facilities such as a good area of landscape and objects that support children activities. Facility for children pedestrian consists of area for relaxing, playing, exercise and sitting area. Children explore the pedestrian pathway area during their walking trip. They use natural objects as well as a man-made object while walking. Planners should have the ability to accommodate children preference for an area for many activities, as well as to create a safe place. The area of the pedestrian as children preference is part of the street that needs to consider pedestrian movement. The example of children pedestrian friendly environment can be seen on the street section and plan as followed.

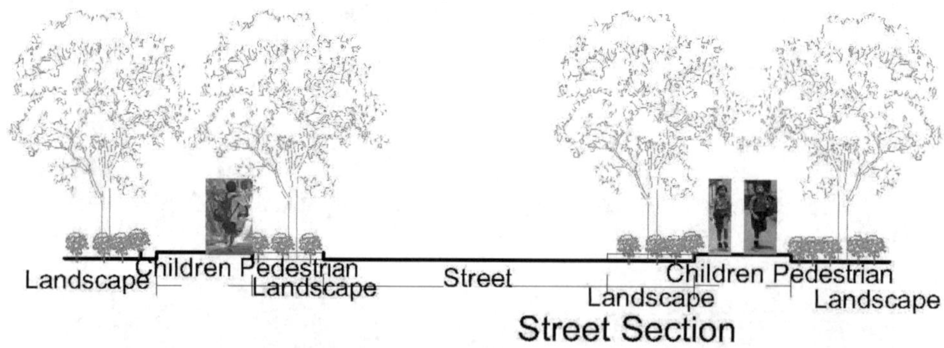

Figure 7. Street Section

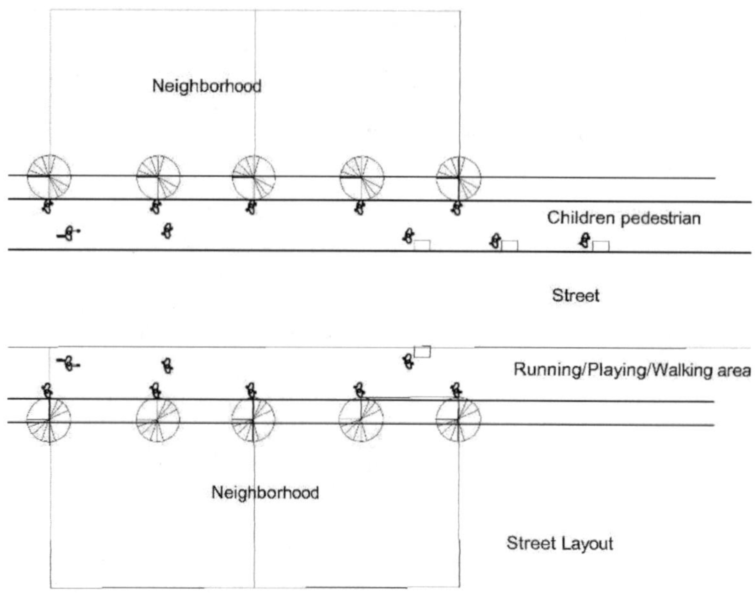

Figure 8. Street Lay Out Plan

4. Conclusion

In designing children pedestrian, it is important to provide space for their activities. The standard for children pedestrian has only considered children as part of the adult pedestrian. Neighborhood area, which is fun for doing children activities, is the most chosen preference by children for the pedestrian area. Children also have their preference as a pedestrian in terms of object and places choices and quality of the area. The area of the pedestrian pathway is the most chosen preference by children during their walking time. Moreover, friends and family house are the second most chosen preference in which shape their movement pattern on their trip. Children can describe the quality of the pedestrian based on their point of view. A good pedestrian pathway is not only for a safe and comfortable area but also fun for doing their activities such as exploring and playing. Area of walking for children are school and neighborhood where they live.

In designing pedestrian area for children, is important to look out for their needs as well as to allow them to move freely. As safety is the most important aspect for children pedestrian based on adult and children preference, the pedestrian area should be contained within the facility to support their activity. Therefore, it is crucial for further research in terms of other aspects of children preference and other places to create a more children pedestrian friendly environment.

References

[1] Ministry of Public Works. Kementerian Pekerjaan Umum. 2014. Manual for Planning, Supplying and Using Pedestrian facility in the Urban area. Pedoman Perencanaan, Penyediaan, Dan Pemanfaatan Prasarana Dan Sarana Jaringan Pejalan Kaki di Kawasan Perkotaan, Peraturan Menteri Pekerjaan Umum No: 03/PRT/M/2014

[2] Makalew, F P., Adisasmita S A., Wunas S., Hamid S. 2017. Influence of Children Pedestrian Behaviour on Pedestrian Space Usage, *IOP Conf. Ser.: Mater. Sci. Eng.* 271 012028

[3] Bierlaire, Michel and Robin, Thomas. 2009. Pedestrian Choices in Timmermans, Harry (ed), *Pedestrian Behaviour- Models, Data Collection and Applications*. Chapter 1 pp. 1-26 Emerald UK

[4] Indonesia National Standard Board (INSB) Badan Standarisasi Nasional Indonesia. 2004. SNI 03-1733-2004 *Guideline for Housing Planning in Urban Area, Tata Cara Perencanaan Lingkungan Perumahan DiPerkotaan*

[5] American Planning Association – APA. 2007. *Planning and Urban Design Standard*, John Willey and Son Inc, New Jersey

[6] Eun-Jung Ko, Bur-Deul Yoon, Sung-Won Choi and Hong-Kyu Kim. 2011. The Effects on the People's Preference on the Cityscape by the Spatial Characteristics of the Streetscape-Centered on 'Design Seoul Street', World Academy of Science, *Engineering and Technology* Vol:5 2011-11-29

[7] Makalew, Febriane Paulina., Adisasmita, Sakti Adji., Wunas. Shirly and Aly, Sumarni Hamid.,2018 Pedestrian Space Capacity and Movement Pattern For Elementary Students In Urban And Rural Area, *International Journal of GEOMATE* Vol.15 Issue 50, pp. 63 -69

[8] Azmi, Diyanah Inani & Karim, Hafazah A. 2012. A Comparative Study of Walking Behaviour to Community Facilities in Low-Cost and Medium Cost Housing, Asia Pacific International Conference on Environment-Behaviour Studies, *Procedia - Social and Behavioral Sciences* 35 (2012) 619 – 628, Elsevier B. V

[9] Kytta, Marketta. 2003. *Children in Outdoor Context, Affordances and Independent Mobility in the Assessment of Environment child friendliness*, Espoo, Thesis Chapter 4 -5 pp. 43 -94

[10] Plester, Beverly., Blades, Mark and Spencer, Christopher. 2006. Children's Understanding of Environmental representations: aerial photographs and model towns, in *Children and Their Environments- Learning, Using and Designing Spaces*. Chapter 1 pp 42-56. Cambridge University Press

[11] Francis, Mark and Lorenzo, Ray. 2006. Children and City Design: Proactive Process And The Renewal Of Childhood, in *Children and Their Environments- Learning, Using and Designing Spaces*. Chapter 4 pp 217-237, Cambridge University Press

IOP Conf. Series: Earth and Environmental Science **328** (2019) 012019 doi:10.1088/1755-1315/328/1/012019

Development of Road Network Traffic Model in Bandung City

H M Pakpahan[1*] and I Adriant[2]

[1] Transportation Management Department, Sekolah Tinggi Manajemen Logistik Indonesia, Sariasih St. No.54 Sarijadi, Phone: 022-2009562/ 2009570, Fax.: 022-2009537, Bandung 40151, Indonesia

[2] Logistic Management Department, Sekolah Tinggi Manajemen Logistik Indonesia, Sariasih St. No.54 Sarijadi, Phone: 022-2009562/ 2009570, Fax.: 022-2009537, Bandung 40151, Indonesia

*email: medipakpahan@yahoo.com

Abstract. The city of Bandung is a city with a higher number of movements; thus, it allows congestion on certain road segments. Constructing and widening new roads do not become the only solutions, but also by improving the traffic management on these roads. With the study area in some segments of Wastu Kencana Street, Cihampelas, Pajajaran, Pasir Kaliki, Kebon Kawung, and Cicendo, some of which have frequent congestion. Several alternatives have been assessed to implement traffic management policies. The solution alternatives are do nothing and do something. Do nothing alternative means the existing road conditions at the time of the study. Do something alternative consists of three traffic management scenarios: Alternative 1, changing traffic direction but not changing the existing one-way system; Alternative 2, applying a combination of one-way systems with two-way systems; and alternative 3, applying a condition of traffic management using two-way system in all roads. Whereas to evaluate network performance, the cost parameters of the trip, the average Volume Capacity Ratio (V/C), and the average speed are employed. The best policy alternative is Alternative 1 where the V/C value is 0.72, average speed of 27.57 km/ hour, and cost of travel of IDR 9,370.87 million/ hour.
Keywords: traffic management, road network, vehicle operating costs, V/C (Volume Capacity Ratio), average speed, travel costs.

1. Introduction

The city of Bandung as one of big cities in Indonesia is also inseparable from the transportation problems. At first, Bandung was only planned as a city of settlements and resorts. After the Independence of the Republic of Indonesia, this city becomes the capital of West Java Province, starting to be burdened with various functions and its population growth. Based on data from Bandung Statistics Center in 2017, there are around 2,5 million people living in an area of 167.31 km^2. This condition is exacerbated by the annual pace of road growth which is much smaller than the annual growth of vehicles. Even the width of roads in Bandung does not reach 20% of the city. In addition, geographical factors which are difficult for urban expansion become the supporting factors making traffic problems in Bandung difficult to overcome.

Solving transportation problems can be exercised from its physical aspects or traffic management application. Physical breakdown can be conducted by building a new road or by widening the available

IOP Conf. Series: Earth and Environmental Science **328** (2019) 012019 doi:10.1088/1755-1315/328/1/012019

road; therefore, it can accommodate the volume of existing traffic. Traffic management can take form in the provision of traffic signs, directional arrangements, traffic lights, etc.

Physical problem solving will cost more and requires larger area, so solving problems with traffic management offers more preferable option for large areas and more economic. Traffic management which applies to an area depends on the policy of the region which usually considers the traffic conditions of the region concerned.

The objectives in this study are:

1. To evaluate the performance of the road network in the study area under existing conditions (do nothing).
2. To evaluate the performance of the road network in the study area upon conducting any action which consists of changes in the direction of movement or in traffic management.

2. Methods

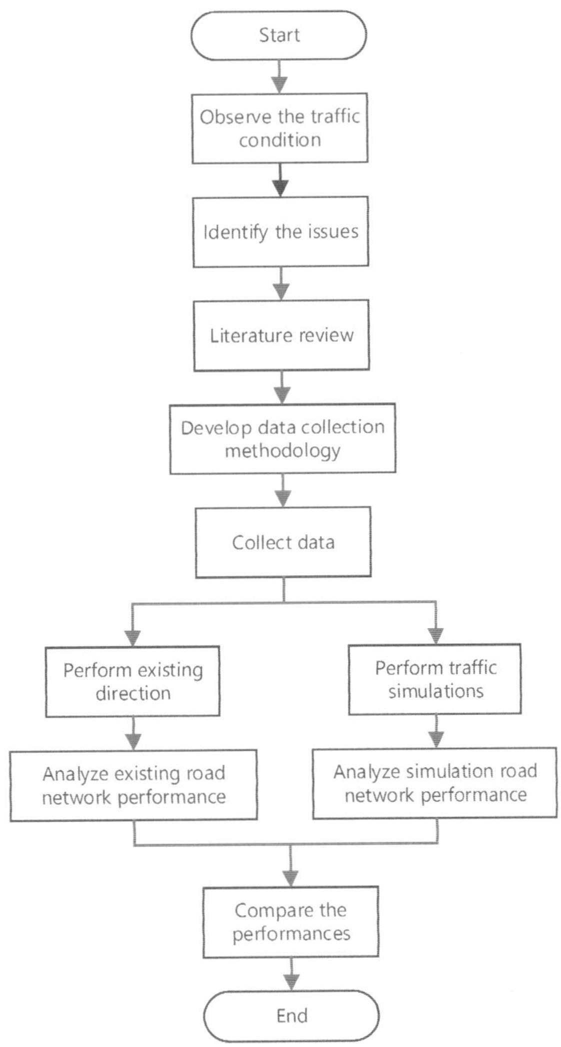

Figure 1. Methodology

2.1 Traffic Management

Traffic management is the application to control certain traffic on a road segment or several road segments in a certain area to achieve certain objectives [5]. In broadly speaking, road traffic management can be categorized into management against time and on traffic space. Basically, congestion occurs because of the movement towards the same point and at the same time. To overcome the problem of congestion, arrangements need to be made in such a way that the movement does not occur at the same time and in the same space.

2.2 Volume Capacity Ratio (V/C)
V/C is one indicator to measure road network performance [2].

$$Average\,V/C = \frac{\sum\left(V/C_{segment} \times length_{segment}\right)}{\sum length_{segment}} \quad (1)$$

Limits on the value of applied V/Cs as indicators of road network performance are presented in **Table 1Table *1*.**

Table 1.V/C and Level of Service.

Volume Capacity Ratio	Level of Road Service	Information
< 0.60	A	Smooth flow, low volume, and high speed
0.60 – 0.70	B	Stable flow, limited speed, and volume suitable for road outside the city
0.70 – 0.80	C	Stableflow, speed affected by traffic, and the volume suitable for city roads
0.80 – 0.90	D	Approaching unstable flow, and low speed
0.90 – 1.00	E	Unstable flow, low speed, and solid volume or near capacity
> 1.00	F	Obstructed flow, low speed, volume above capacity, and many stops

Source: Tamin and Nahdalina,1998 as referenced in Tamin, 2006

2.3 Average Speed
Related to the calculation of the average V/C, the speed on each road segment is obtained from the result of loading the Origin Destination Matrix on the road network database using SATURN program. The speed on each of these roads will be used to calculate the average speed in the study area, using the equation as follows:

$$Average\ Speed = \frac{\sum\left(speed_{segment} \times lenght_{segment}\right)}{\sum length_{segment}} \quad (2)$$

2.4 Travel Costs
Travel costs are the sum of two cost components: Vehicle Operating Costs and time value costs. The model used is a calculation developed by PCI as a result of a study conducted by LAPI ITB (1997)[7].

The components employed in the calculation of Vehicle Operating Costs in this study are only calculated from the cost of fuel consumption. The component of fuel consumption costs can be formulated as follows:

$$\text{Fuel Consumption Costs (FCC)} = \text{FCC}_{basic} \times \left(1 \pm \left(k_k + k_l + k_r\right)\right) \quad (3)$$

FCC_{basic}class I vehicle= $0{,}0284v^2 - 3{,}0644v + 141{,}68$
FCC_{basic} class IIA vehicle = 2,26533× (FCC_{basic} class I vehicle)
FCC_{basic} class IIB vehicle = 2,90805× (FCC_{basic} class I vehicle)
k_k = correction factor for slope
k_l = correction factor for traffic flow

k_r = correction factor for road roughness
v = vehicle speed (km/hour)

$$\text{VOC} = \sum_{i=1}^{n} \left(FCC_i \times L_i \times V_i \right) \times Rp7.800{,}00 \qquad (4)$$

VOC = Vehicle Operating Costs (Rp/year)
L_i = length of road segment i (km)
V_i = traffic volume at road segment i (pcu)
IDR7,800.00 = fuel price (pertalite) per literin 2018

Costs for time values are calculated based on the following equation:

$$C = 600 \times \text{total travel time} \qquad (5)$$

C = costs for time values
6000 = time values in Bandung

Table 2. Correction Factor of Basic Vehicle Consumption.

correction factor for negative slope (k_k)	g < -5%	-0.337
	−5% ≤ g <0%	-0.158
correction factor for positive slope (k_k)	0% ≤ g < 5%	0.400
	g ≥ 5%	0.820
correction factor for traffic flow (k_l)	0 ≤ *V/C*< 0,6	0.050
	0,6 ≤ *V/C*< 0,8	0.183
	V/C ≥ 0,8	0.253
correction factor for road roughness (k_r)	< 3 m/km	0.035
	≥ 3 m/km	0.085

g = slope
V/C = Volume Capacity Ratio
Source: LAPI-ITB (1997)

3. Results and Discussions

3.1 Review Some Related Research about Time Values

Several studies assessing time values based on Jasa Marga's formula by considering studies of the value of time have been carried out in several cities in Indonesia. The following **Table 3** are basic time values of several studies which have been conducted.

Table 3. Time Values of Several Studies.

Reference	Time Value (Rp/hour/veh)		
	Class I	**Class IIA**	**Class IIB**
PT. JasaMarga (1990-1996), Formula Herbert Mohring	12,287	18,534	13,768
Padalarang-Cileunyi (1996)	3,385 – 5,425	3,827 – 38,344	5,716
Semarang (1996)	3,411 – 6,221	14,541	1,506
IHCM (1995)	3,281.25	18,212	4,971.20
PCI (1979)	1,341	3,827	3,152
JIUTR northern extension (PCI 1989)	7,067	14,670	3,659
Surabaya-Mojokerto (JICA 1991)	8,880	7,960	7,980

Source: Highway Economics Module-Kartika, 2006

The amount of Minimum Time Value is presented by **Table 4**.

Table 4. Minimum Time Values (IDR/Hour).

Regency/City	JasaMarga			JIUTR		
	Class I	Class IIA	Class IIB	Class I	Class IIA	Class IIB
DKI Jakarta	8,200	12,369	9,188	8,200	17,022	4,246
Other than DKI	6,000	9,051	6,723	6,000	12,455	3,170

Source: Highway Economics Module-Kartika, 2006

The formula used is as follows:

$$Time\,Values = Max\{(K \times Basic\,Time\,\text{Values}); Minimum\,Time\,Values\} \quad (6)$$

While the value of K can be seen in the following **Table 5**.

Table 5. K Values for Several Cities.

No.	Regency/ City	K values
1	Jakarta	1.00
2	Cianjur	0.15
3	Bandung	0.39
4	Cirebon	0.06

3.2 Data

3.2.1 Traffic Volumes

Traffic volume surveys were carried out by collecting data with manual calculations directly on the specified road segments on weekdays and weekends. The survey of traffic volume on weekdays was carried out on Wednesday, May 16 2018.The surveys at the end of the week were conducted on Saturday, May 19 2018. The data from the survey during the morning and evening rush hours on some roads are presented in **Table 6**.

Table 6. Traffic Volumes.

Name of Roads	Survey Point	Direction of Movement	Volume (pcu/hour)	
			Weekdays	Weekend
Pajajaran 1	Kimia Farma Lab.	East- West	2,923	2,547
Pajajaran 2	Pajajaran Stadium	West- East	2,730	2,446
Cicendo	Cicendo Eye Clinic	North- South	2,846	2,632
Kebon Kawung	Ampera Restaurant	East- West	3,055	3,225
Pasir Kaliki	Jelita Parahyangan Hotel	North- South	2,390	2,568
Wastu Kencana	Kembang Market	North East–South East	4,077	3,200
R.E. Martadinata	Cibeunying Tax Office	West- East	3,251	2,749

3.3 Road Network Performance

To obtain the modeling results, the OD Matrix cannot stand alone, meaning that the OD Matrix must be charged to the road network whichhas been prepared using the iteration method to reach equilibrium conditions. Loading is carried out for three road network conditions, which are the existing road network, road network after the implementation of traffic management, and road network before the present conditions. The balance method used is User Equilibrium Method, as suitable for urban areas with high congestion levels such as Bandung. This loading process is carried out using SATURN program.

3.3.1 Do Nothing Alternative

Do nothing alternative employs the existing road network system, a one-way system. There are two one-way road loops in this condition, Pajajaran St.–Cihampelas St. – Wastu Kencana St. and Pajajaran St.–Cicendo St. – Kebon Kawung St. - PasirKaliki St. **Figure 2Error! Reference source not found.** inform details on the direction of movement on the road network of the existing study area. Each road has different characteristics, both capacity and length of segments.

3.3.2 Do Something Alternative

Do something alternative is the condition of the road network where changes are made to road arrangements by altering direction of movement, as well as road management changes into two directions on road networks in the study area to increase road network performance. In this study, three scenarios of do something alternative were carried out in an attempt to improve road performance, which areAlternative 1, Alternative 2, and Alternative 3.

A. Alternative 1

The system used in alternative 1 preserves the existing condition, a one-way management system, but some directional changes are implemented to the road network, such as part of Pajajaran St., Pasir Kaliki St., Cicendo St. and Kebon Kawung St. and removal of road dividers on some roads on Wastu Kencana St. Changes in direction in the road segment result in a change in direction on the loop of the road network. **Figure 2** presents more noticeable direction of movement in the road network of Alternative 1. The removal of road dividers on some roads on Wastu Kencana St. will increase the capacity of the road; therefore, changes to the road segments are needed.

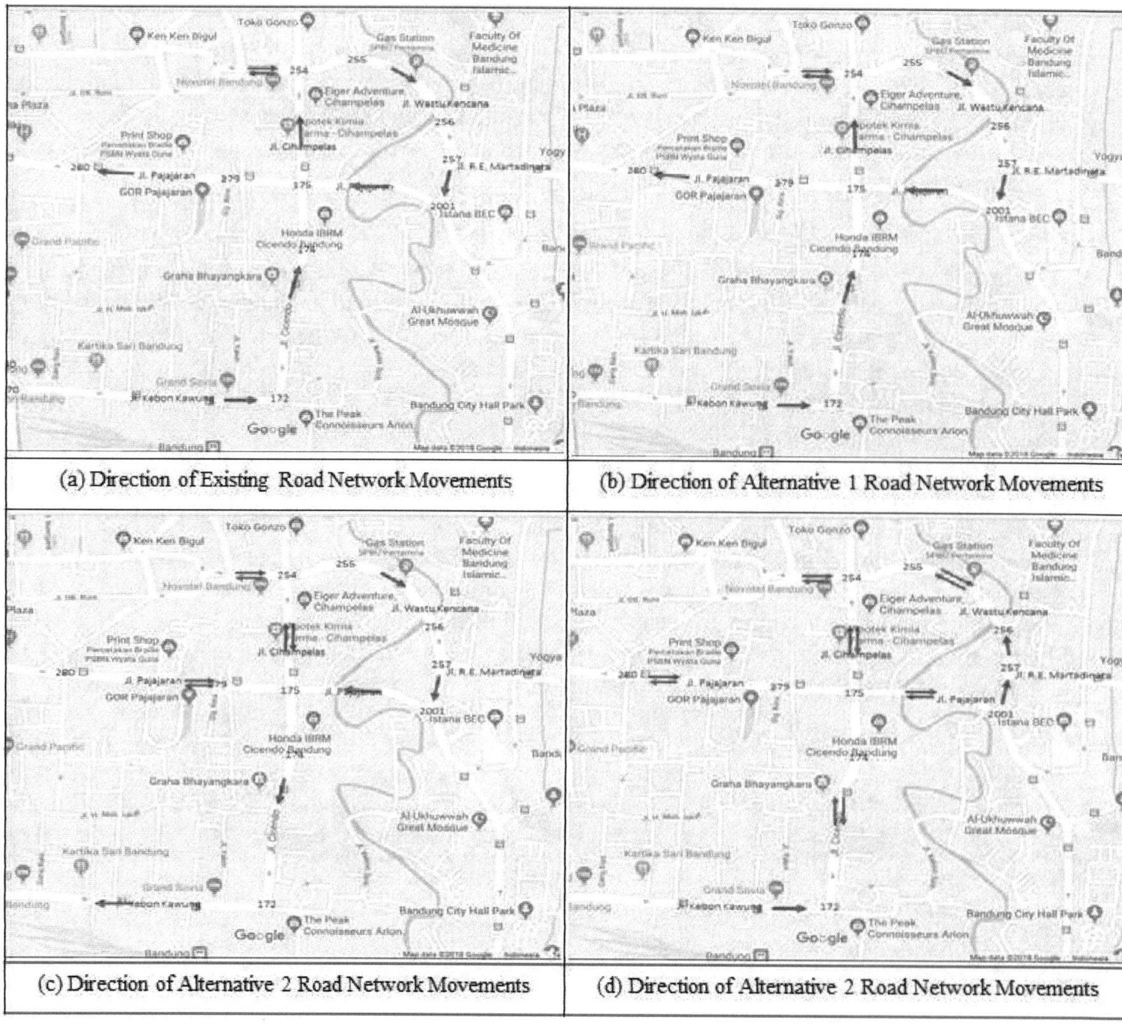

(a) Direction of Existing Road Network Movements

(b) Direction of Alternative 1 Road Network Movements

(c) Direction of Alternative 2 Road Network Movements

(d) Direction of Alternative 2 Road Network Movements

Figure 2. Direction of Road Network Movements.

B. Alternative 2

In alternative 2, there are some of the road segments in the study area modified using a 2-way system, a part of Pajajaran St. and Cihampelas St. Other road segments still use a one-way system. As in Alternative 1, Alternative 2 also removes road dividers on some roads on WastuKencana St. The direction of road network in Alternative 2 is illustrated in **Figure 2**. The system changes will change road capacity of the road segments which undergo system changes.

C. Alternative 3

In the scenario of Alternative 3, road network condition is returned as a condition before the application of one-way system (existing condition), using two-way system. All roads on the road network in this study area use a two-way system. The direction of movement can be examined in **Figure 2**. In the two-way road system, the road capacity is smaller, but more road alternatives to pass.

3.4 Analysis

To analyze the road network performance is conducted by comparing the three parameters in the study area between Alternative 1, Alternative 2, and Alternative 3 to the existing condition (do nothing). The best road network condition is the condition where the study area network has lowest travel costs, smallest average V/C, and highest average speed.

3.4.1 Average V/C

Each road network condition has a different average V/C. The average V/C value describes the state of the road network under evaluated conditions. **Figure 3** presents the average V/C value.

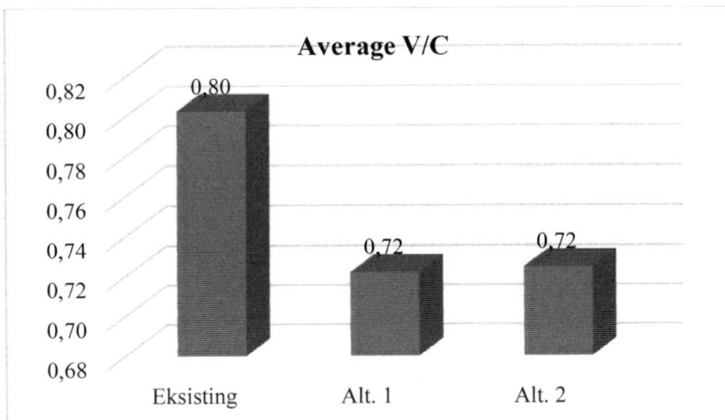

Figure 3. Average V/C Chart.

3.4.2 Average Speed

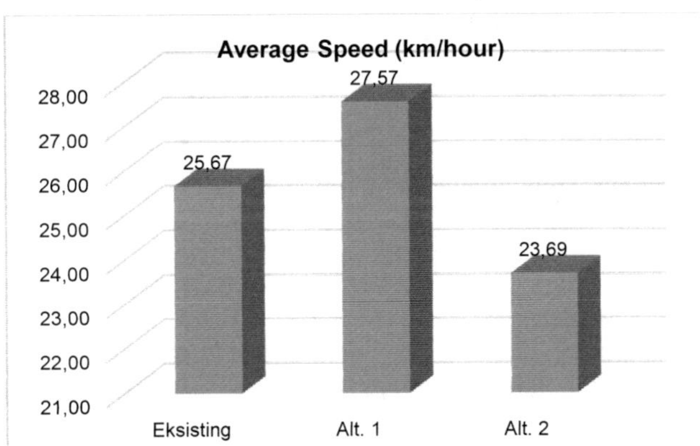

Figure 4. Average Speed Chart.

Like the average V/C, every road network condition has different average speeds. The existing condition and Alternative 1 condition have higher average speed when compared to Alternative 2 condition. It is driven by the existing road network segment, and Alternative 1 has more one-way roads which has higher design speed than the two-system direction. However, two-way system has more delays caused by setting traffic signals. The average speed is presented by **Figure 4**.

3.4.3 Travel Cost

Travel costs are the sum of vehicle operating costs and the cost of time value. Travel costs are mostlyinfluenced by the combination of V/C and speed. High V/Cs and low speeds will make traveling costs higher. The travel cost chart is presented in **Figure 5**.

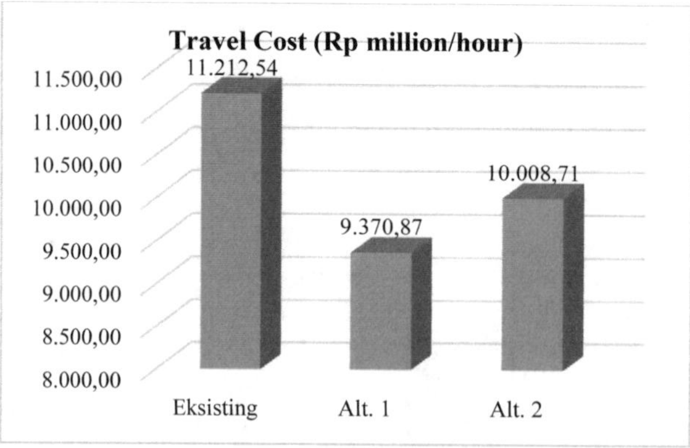

Figure 5. Travel Cost Chart.

4. Conclusions

From the analyzes performed on the predetermined parameters, the mean V/C, and the average speed, it can be concluded that the condition of the existing road network condition is unstable with the intersection only at Pajajaran St. – PasirKaliki. In Alternative 1 and Alternative 2, the study area network is in stable condition with interrupted intervals at Alternative 1 at the intersection of Pajajaran St. - Cicendo - Cihampelas, Pajajaran St. – Pasir Kaliki, Kebon Kawung St. – Pasir Kaliki, at the intersection of Kebon Kawung St. - Cicendo, at Alternative 2 at the intersection of Pajajaran St. - Cicendo - Cihampelas, Pajajaran St. – Pasir Kaliki, and Cihampelas St. - Abdul Rivai – Wastu Kencana. The road network performance of the study area on Alternative 3 is in poor condition where the performance of road segments cannot be calculated due to the length of queue at Pajajaran St.intersection - Cicendo - Cihampelas, Pajajaran St. intersection – Pasir Kaliki, and Kebon Kawung St. intersection – Pasir Kaliki. The length of queue in those intersections exceeds the length of the segment affecting to other segments.

Alternative 1 gives a better road network performance on the evaluated parameters, rather than Alternative of Do Nothing, Alternative 2 and Alternative 3. Hence, the scenario of Alternative 1 is regarded as the best solution in this study.

References

[1] Departemen Pekerjaan Umum 1997 *Manual Kapasitas Jalan Indonesia* (Jakarta).

[2] Direktorat Jenderal Perhubungan Darat 1999 *Rekayasa Lalu Lintas* (Jakarta).

[3] Hobbs, F. D. 1995 *Perencanaan dan Teknik Lalu Lintas* (Gadjah Mada University Press).

[4] Irawan, Ari 2004 *Tugas Akhir Dampak Penerapan Kebijakan Jalan Satu Arah Terhadap Sistem Jaringan Jalan di Bandung* (Bandung)

[5] Morlok, A. F., 2008. *Pengantar Teknik dan Perencanaan Transportasi* (Bandung: Erlangga)

[6] Supriatno, Novian 2004 *Tugas Akhir Pengembangan Jaringan Jalan di Kota Tasikmalaya dengan Pengaturan Parkir On Street dan Pemberlakuan One Way* (Bandung)

[7] Tamin, Ofyar.Z. 2000 *Perencanaan dan Pemodelan Transportasi* (Bandung: Penerbit ITB)

[8] Van Vliet, Dirck 2004 SATURN 10.5 *User Manual*(Leeds)

The Flexibility of Parking Space of Traditional Market in the City Center (Case Study: Pasar Gede Of Surakarta, Indonesia)

I Aliyah[1*] and S A S Aulia[1]

1 Department of Regional and Urban Planning, Faculty of Engineering, Universitas Sebelas Maret, Surakarta, Indonesia

*email: istijabatul@ft.uns.ac.id

Abstract. The economic growth in the city of Surakarta could be seen in its trading activities. The existing condition of the current parking space is located on the side of the road or on street parking. This has an impact on traffic jams during peak hours. Therefore, the purpose of this research is to discover the flexibility of parking space at Pasar Gede Trade Area. The method began with empirical data collection based on Category Based Analysis method in grouping the duration of use, area extent, and technical arrangement of parking space. The result of grouping was then analyzed in terms of temporary spatial flexibility. The results show that the use of parking space area temporarilly can be regarded as a space with the flexibility of Time Cycle and Time Management, whereas conceptually, it is due to the flexibility of Convertibility that occurs naturally from the consciousness of the users. The contribution of the theory which can be generated is the flexibility of space in the context of spatial constraints through the diversity of spatial planning techniques.
Keywords: flexibility, parking space, traditional market area, City Center

1. Introduction

Nowadays, all regulations relating to the provision of facilities of health, education, shopping and praying are allocated in consideration of the distribution of services referring to the Central Place Theory [1]. Various decisions on city planning and development are taken on the basis of the service-scale rule in accordance with population and demands of community service needs from social and economic aspects [2]. On the other side, the economic growth in Surakarta is seen in trading activities, one of them is in the biggest and oldest traditional market in Surakarta, Pasar Gede Hardjonagaro. Traditional markets are not merely seen as a space of trading activity but are viewed as an organizational system consisting of interconnected, intregated elements forming a complex unity of mutual support to each other. Traditional market is a part of urban activities having various functions. Traditional market's location occupies a certain area in the city center [3]. Pasar Gede is located in downtown of Surakarta precisely on Jenderal Urip Sumoharjo Street and RE Martadinata Street junction; thus, it has strategic location and most-visited by buyers from inside and outside Surakarta. These activities have an impact on the high demand for motor vehicle parking spaces. The existing condition of the current parking space is located on the side of the road or on street parking. This has an impact on the stagnation of regional traffic flow during peak hours. Parking errors in very limited spatial dimensions and traditional market position in the city center require a flexible management of space utilization, so that activities can be better accommodated. Planning by creating Flexible Space becomes one of the strategies to overcome the limitations of space and time [4]. Therefore, the objective of this research is to discover the flexibility of parking space at Pasar Gede Trade Area. Potential and strategic location affects the rapid development of Surakarta and has an impact on the increasing number of city dwellers, as one of the demands on the number of supporting facilities for community activities, especially in terms of transportation. Demand for parking has been growing due to the possibility to own and use private vehicles increases. In most cases, private vehicle users require two parking spaces: at the beginning and at the end of the movement [5]. Parking problems are very important to be studied deeply, as all the activities of trading areas require parking facilities, which should be adequately available due to greater volume of traffic in Pasar Gede Trade Area. The currently used parking system is inefficient as the driver is allowed to park unlimitedly and parking facilities cannot be fully used [6]. Various regulatory efforts have been made to overcome parking problems, ranging from the variety of space and parking management techniques. Hence, it is necessary to conduct a study of the flexibility of parking space in traditional market areas at the city center.

1.1 Literature Review

1.1.1 The Definition of Spatial Use Flexibility
Flexibility has been discussed since the 1950s, and at present the concept and application continues to experience renewal. Therefore, flexibility can be considered as the ability of space to provide differences in rules, configurations, and conformity [7]. Based on the Great Dictionary of Indonesian Language (2007), the flexibility of spatial use presents the characteristics of a space for various characteristics and activities, dealing with the arrangement of space by preserving the order of the building [8]. Meanwhile, Carr & Dionisio (2017) review Flexible Spaces in cities by analyzing the advantages of under-utilized, abandoned, or empty spaces, along with urban revitalization needs. Carr & Dionisio integrate the possibility of using Flexible Space as a tool in urban planning [4]. Three concepts of flexibility include: 1) Expansionibility, which is the flexibility of space that can accommodate growth through expansion; 2) Convertibility, which is the flexibility of space with a change of governance in one space; 3) Versatibility, which is the multi-functional spatial flexibility [9] and the building is said to be flexible when other activities can be used without changing its shape [10]. Temporal flexibility can be temporarily analyzed in three aspects of the temporal dimension suggested by Carmona, et. al. (2003), including: 1) Time Cycle and Time Management, emphasizing that activity is always changing according to space as well as time; 2) Continuity and Stability, stating that although the environment is always changing over time, a space is able to adapt to change; and 3) Implemented Over Time, indicating that a space can overcome all changes to the environment without much changes [11].

1.1.2 Trade Area Parking Space
According to the Decision of the Director General of Land Transportation No. 272 of 1996 on the Technical Guidelines for the Implementation of Parking Facility, the definition of parking is a non-moving condition of a vehicle that is not temporary. According to Taju, (1996) in Rinaldi Mirsa (2012), the definition of parking is a space to place a vehicle (for transport/goods or motorized/non-motorized) by stopping at a place at a certain time period. According to Joseph Dechiara & Lee Koppelmen (1975) [12] and revealed by Ririh Sudiraharjo (2004), the ways of parking placement are: 1) on street parking; and 2) off street parking. According to the status, parking is divided into five, consisting of: 1) public parking; 2)

special parking; 3) emergency parking; 4) parking park; and 5) parking lots [13, 14]. On the other hand, normatively, the parking space of the Directorate General of Land Transportation has determined the size of the parking space (PSU) that is affected by: 1) the standard vehicle dimension; 2) vehicle-free parking spaces, vehicle-free spaces are provided for lateral and longitudinal vehicles; and 3) width of vehicle door openings. The size specification of parking space requirement and parking space unit is illustrated in **Table 1** and **Table 2**.

Table 1. Parking Space Requirement Size

Designated	Unit (PSU for Cars)	Parking Space Requirement
Trade Center, consists of: • Pertokoan • Pasar Swalayan • Pasar • Shops • Supermarkets • Market	PSU / 100 m^2 effective floor area	3.5 – 7.5
Office Center, consisting of: • Non-Public Service • Public service	PSU / 100 m^2 effective floor area	1.5 – 3.5
School	PSU / College Students	0.7 – 1.0
Hotel / Place to stay	PSU / Bedroom	0.2 – 1.0
Hospital	PSU / Bed	0.2 – 1.3
Cinema	PSU / Seat	0.1 – 0.4

Source: Decree of the Director General of Land Transportation No. 272 of 1996

Table 2. Parking Space Unit Standard

Vehicle type	Parking Space Unit (m^2)
Passenger Cars, classified into:	
a. Group I	2,30 x 5,00
b. Group II	2,50 x 5,00
c. Group III	3,00 x 5,00
Bus/ Truck	3,40 x 12,50
Motorcycle	0,75 x 2,00

Source: Decision of the Director General of Land Transportation No. 272 of 1996

1.1.3 The Definition of Traditional Markets

Traditional markets especially those in urban areas have grown in Indonesia since the beginning of the emergence of settlements or kingdom. There are several definitions of traditional market based on experts' opinion, such as expressed by Rutz that traditional market as the center of socio-economic activity of democracy; thus, the pattern of economic relations occurs in traditional markets resulting in the intertwined social interaction among sellers-buyers, sellers-sellers, and sellers-distributors as a social heritage of need representation for socializing among individuals [15, 16]. Traditional markets are in kiosk, stalls and open land buildings. In the main part, there are kiosks in permanent buildings, stalls in the form of makeshift or semi-permanent buildings, and the 'oprokan' or open parts which are used by temporary sellers with a smaller area than stalls [17]. Formally, traditional market institutionally is a market built and managed by government, private, cooperative or self-supporting local community with business premises in the form of shops, kiosks, stalls and tents, or other similar terms owned/managed by small and medium scale sellers, with small business and small capital, with the process of buying and selling through bargaining [18]. Modern markets and traditional markets in social and cultural aspects have different functions and roles, depicted by the empirical facts in the field, due to similar role as the city's economic facilities.

2. Methods

Data collection methods employed in this study include Spatial Data, Parking Space Width Data, Parking Capacity Data, Parking Space Need Data, and Duration of Parking Space Use. Spatial data was obtained by conducting field observation to calculate the number of available parking space. Furthermore, it was supported by additional data in the form of basic maps and satellite imagery maps utilizing Geographic Information System (GIS). Meanwhile, Parking Space Width Database was obtained by interviewing parking attendants in charge. The interview was conducted within a period of 2 days, Monday representing weekday and Sunday representing weekend with data on the number of vehicles parked in Pasar Gede Area during the time span between 02.00 a.m. - 06.59 a.m.; 07.00 a.m. - 11.59 a.m.; and from 12:00 p.m. to 03:00 p.m. After all data was obtained, the data was recapitulated to examine whether there was the required data for analysis purposes. Parking Space Capacity is a number stating the total number of vehicles including the parking load, reporting the number of vehicles for a given period of time, in units per hour or per day [19]. The following is the formula to calculate the capacity of parking space with vehicle as the unit used.

$$\text{Parking Space Capacity (unit)} = \frac{\text{Parking Area Width (m}^2\text{)}}{\text{Parking Space Unit (PSU)}} \quad (1)$$

Meanwhile, the Data of Parking Space Requirement was collected by calculating parking space requirement with the unit in PSU (Parking Space Unit). Calculation can be performed with the following Parking Space Requirement formula:

$$\text{Parking Area Needs (m}^2\text{)} = \text{Quantity of Vehicles (unit) x Parking Space Unit (PSU)} \quad (2)$$

Parking duration covers the time span (length of time) of vehicles parked at a certain place. Parking duration can be calculated by the formula (Suwardi, 2008). The result of the subsequent grouping process was analyzed temporarily in terms of spatial flexibility.

3. Result and Discussion

3.1. Pasar Gede Constellation as Traditional Market in the City Center

Traditional market provides a place unlimited to purchase and selling activities of goods and service exchange, but also to the exchange of information and knowledge activities in traditional markets. This is in line with Geertz's theory that *pasar* is an economic order as well as a way of life. Within the scope of Javanese society, the power of economic activity is centered on traditional market. Traditional market is viewed as an interrelated and interdepent part of organizational system to form a complex and mutually supportive unity between components. Meanwhile, the components in the market system include rotation, production, distribution, transportation and transaction. In its development, traditional market reaches a wider scope as a node of regional exchange of goods and services which then grows and evokes activities in the city. The displacement of traditional market location is able to change the land use planning, road pattern, movement and pattern or type of building, circulation line distribution, and land use. The constellation of Pasar Gede to Surakarta is depicted in Figure 1 indicating that the area around Pasar Gede is a densely populated settlement, such as Pecinan (Chinese-dominated settlement).

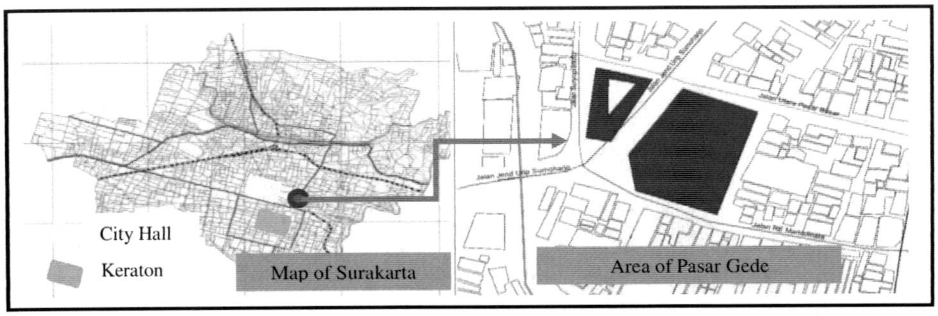

Figure 1. Constellation of Pasar Gede at Surakarta City Center

3.2. Existing Parking Area in Pasar Gede Trade Area

Pasar Gede area consists of two main buildings, each consisting of two floors and each separated by a highway. At the front of the building on the east side, there is a writing of "Pasar Gede Hardjonegoro" above the main gate as an identity which is so far maintained. The parking location in Pasar Gede of Surakarta is linear with parking placement on road side or on street parking. The parking area in Pasar Gede is illustrated in **Figure 2**.

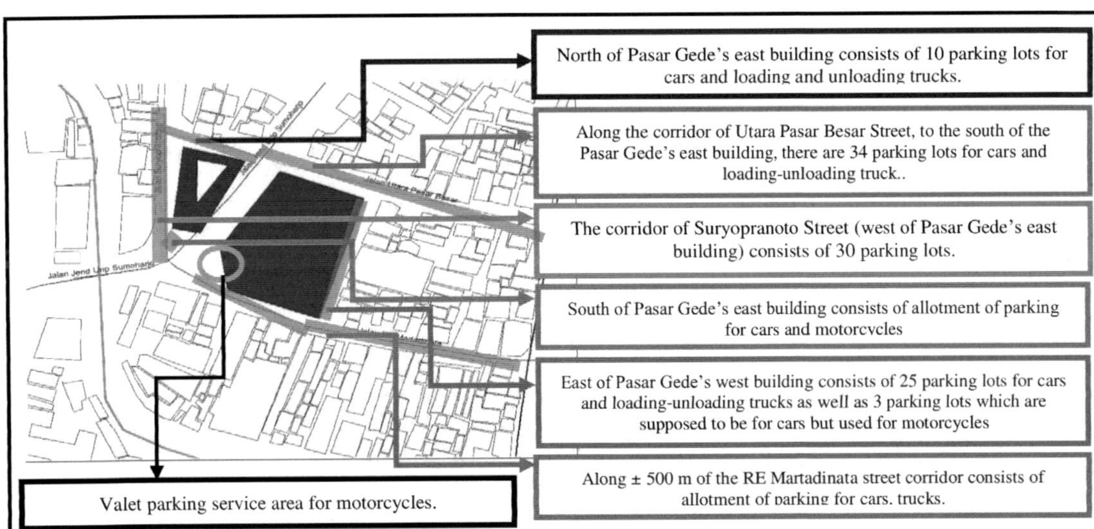

Figure 2. The Use of Parking Area at Pasar Gede

3.3. Categorization of User and Parking Space Service in Pasar Gede Area

Users of parking spaces in Pasar Gede can be classified into: a) Cars, b) Motorcycles, c) Trucks, and d) Non-Motorized Vehicles. This unfortunately has an impact on the reduced comfort of pedestrians who will use the sidewalk because the pedestrian area is reduced and sometimes pedestrians choose to walk along the shoulder of the road rather than down the sidewalk. It is different from research on Modern Market of Pasir Pengaraian City which merely classifies the type of vehicle into car types (sedan, jeep, and pick up) and motorcycle [20]. Goods loading and unloading in Pasar Gede usually take place in the early morning between 2 a.m. to 6.59 a.m. by using trucks which size ranges from small trucks to large trucks. As for the motorcycle, it refers to motorcycles with 2 or 3 wheels, and three-wheeled vehicles in accordance with the Bina Marga classification system. Parking service at Pasar Gede is illustrated in

Figure 5. In contrast to the provision of parking services in Hanoi Vietnam, the term "para-parking" is a type of parking service utilizing a private area used for semi-automatic parking. The concept of "para-parking" is expected to increase the supply of legal parking to urban areas in developing cities. Besides, parking is expected to benefit transportation planners to control unplanned land use, legalized illegal parking, parking business in urban areas with reasonable pricing schemes and effective regulations [21].

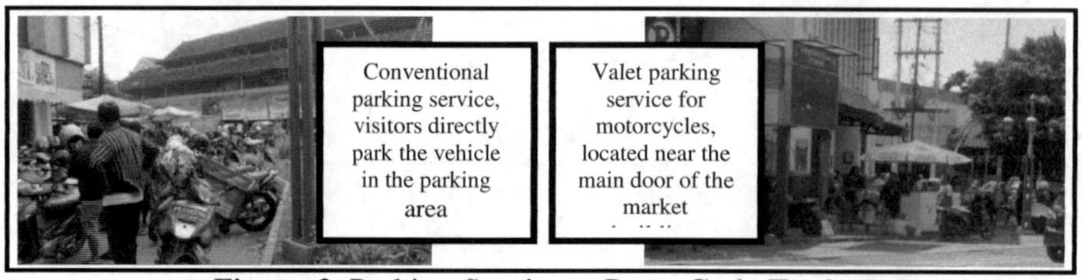

Figure 3. Parking Service at Pasar Gede Trade Area

3.4. Parking Duration in Pasar Gede Area

Parking duration is the time span or the length of time a vehicle is parked at a certain place. Parking duration is stated in minutes. The vehicle parking duration in Pasar Gede Trade Area differs from a research in Tomohon Central Trade Area which calculates the parking duration from 08.00 a.m. to 06.00 p.m. [22]. The highest average parking duration of cars during the research period occurred on Sunday at 7 a.m. - 11.59 a.m. which reached 52 minutes, while the lowest occurred on Monday at 12 p.m. to 3 p.m., reaching 36 minutes. On the other hand, the highest average parking duration of motorcycles during the research occured on Sunday at 7 a.m. - 11.59 a.m., reaching 68 minutes, while the lowest occurred on Monday at 12 p.m. to 3 p.m., reaching 42 minutes. Meanwhile, the highest average parking duration of trucks during the research occurred on Sunday at 2 a.m. - 6.59 a.m., reaching 92 minutes, while the lowest occurred on Monday at 12 p.m. to 3 p.m., reaching 79 minutes. It can be seen that of the three types of motor vehicles having the highest parking duration is truck occurring on Sunday at 2 a.m. - 6.59 a.m. reaching 92 minutes. Meanwhile, the lowest parking duration is car occurring on Monday at 12 p.m. to 3 p.m. which only reached 36 minutes. This result indicates that the truck's parking is required for urban goods shipping which is different from commuter's parking. Unlike cars, trucks usually need a place to park temporarily to load or unload goods at their destination in the central business area. Trucks take more space and time and also require a short distance to the desired destination.

3.5. Parking Space Capacity in Pasar Gede Trade Area

Parking capacity can be interpreted as the maximum number of vehicles which can be parked in a parking area under certain conditions and time. Parking space capacity is a value stating the total number of vehicles including the parking load, including the number of vehicles in a certain period of time which usually uses per hour or per day unit. The capacity of car and truck parking space can be calculated together because in Pasar Gede Trade Area there is no distinction between the parking space specifically for cars and the parking space specifically for trucks. As a result, the loading and unloading activities are done in the same parking space for cars. Furthermore, it is known that the area of parking for cars and trucks is 2196.5 m^2 with Parking Space Unit that is 11.5. Therefore, the calculation is made to determine the capacity of parking space. Meanwhile, in the case of the parking space capacity for motorcycle, it is revealed that the area of the motorcycle parking area is 253 m^2 with the Parking Space Unit that is 1.5. From the calculation, it can be seen that the capacity of parking spaces in Pasar Gede Area for cars and trucks is 191 units, while for motorcycle is 169 units. It indicates that the available parking space in Pasar Gede area can merely accommodate 191 units of cars and trucks. On the other hand, the available parking space in Pasar Gede area can only accommodate 169 units motorcycles. Thus, it shows that the availability of parking space in Pasar Gede area cannot yet meet the need of parking space there. The intensity of the demand is so high that it causes even without the edge of the road allocated for parking, the curb is still utilized as a parking space that will increase traffic congestion [23].

3.6. Parking Space Need in Pasar Gede Trade Area

The need for parking space is a condition of which the available parking area is not yet able to accommodate all vehicle units in a particular parking location. It is revealed that the availability of parking space for cars and trucks is 2196.5 m2, while the need of parking space for cars and trucks is 9180 m^2. Indeed, the available space does not fulfill the demand for parking space for cars and trucks, thus the addition of parking space of 6911.5 m^2 is needed. Meanwhile, the available parking space for motorcycles is 253 m^2 and the need of parking space for motorcycles is 1410 m^2. The available space is indeed insufficient for fulfilling the demand for parking spaces for motorcycles, hence the addition of parking space of 1157 m^2 is required. Therefore, it can be concluded that currently additional parking space is required to accommodate the number of vehicles that will be parked in Pasar Gede area with a minimum area of 8069 m^2, consisting of 6911.5 m^2 to accommodate cars and trucks, and 1157 m^2 to accommodate motorcycles. Besides adding the parking space in Pasar Gede area, a good parking policy and management can also be enforced. Parking policy is considered as one of the strongest steps to manage travel demand pattern and travel reallocation in urban areas [24, 25, 26]. Other efforts that can be approach to identify improved availability of parking facilities and better implementation of parking regulations are by encouraging easy and efficient urban traffic performance. The implementation of parking on the roadside with time variation becomes one of the ways to overcome traffic jam [27].

4. Conclusion

The existence of traditional market as a node of regional exchange of goods and services, which then grows and develops, generates various activities within a city. The existence of traditional markets in the city center is able to change land use, road patterns, movement and pattern or type of building, distribution of circulation path, and land use as parking area. The findings of this research indicate that the highest average parking duration is 52 minutes for cars, 92 minutes for trucks, and 68 minutes for motorcycles. From the analysis of parking space, it is found that the area of on street parking in Pasar Gede trade area is insufficient. Thus, it is necessary to add parking space of 8068.5 m^2, consisting of 6911.5 m^2 to accommodate cars and trucks, and 1157 m^2 to accommodate motorcycles. Meanwhile, the parking arrangement technique in Pasar Gede area is conducted conventionally and by valet parking. The use of parking space in Pasar Gede area temporarily can be regarded as a space with the flexibility of Time Cycle and Time management, belonging to the flexibility of Convertibility. On the other hand, the spatial constraints management strategy in Pasar Gede is different from the results of research conducted by Carr and Dionisio. In this case, Pasar Gede area creates the spatial use strategy in the parking area with the flexibility of Time Cycle and Time management, while Carr and Dionisio manages with Flexible Space strategy to overcome the excess space which has not been or has been likely utilized, abandoned, or left empty. The theoretical contribution that can be generated from this research is the spatial flexibility in the context of spatial constraints through the strategy of diverse spatial planning techniques.

Acknowledgments

The researcher team would like to express gratitude to the Regional and Urban Planning Department of the Faculty of Engineering, Universitas Sebelas Maret, particularly to the Regional Trade Planning Course who has facilitated this research.

References

[1] Christaller, W. (1966). Central Places in South Germany. (W. Baskin, Trans.) . New york, USA:
[2] Aliyah, I., Setioko, B., & Pradoto, W. (2016). The Roles of Traditional Markets as the Main Component of Javanese Culture Urban Space (Case Study: The City of Surakarta, Indonesia). IAFOR Journal of Sustainability, Energy & the Environment, 3(1), Pages 103 - 120.
[3] Aliyah, I., Setioko, B., & Pradoto, W. (2017). Spatial flexibility in cultural mapping of traditional market area in Surakarta (A case study of Pasar Gede in Surakarta). In City, Culture and Society, Volume 10, 41-51.
[4] Carr, J., & Dionisio, M. R. (2017, October). Flexible spaces as a "third way" forward for planning urban shared spaces. Cities, 73-82.
[5] Maternin, G. (. (2014). Trasporti e citta. Forli: Mobilita e pianificazione urbana.
[6] Sofiene Abidi, S. K. (2015). A New Heuristic for Solving the Parking Assignment Problem. In Procedia Computer Science, Volume 60, 312-321
[7] Rabeneck, A., Sheppard, D., Town, P, (1973), Housing flexibility? Architectural Design, Volume 43, Issue 11, Pages 698-727
[8] The Central Lexicographer (2007). The Great Dictionary of Indonesia. Jakarta: Balai Pustaka Publisher (in Indonesian).
[9] Rahadi, R. (2008). Architecture. Retrieved from http://www.arsiteka.com/2008/11/fleksibilitas-ruangmassa.html (Accessed 11 December 2015). (in Indonesian) .Richard Arnott, E. I. (2015). Downtown curbside parking capacity. In Journal of Urban Economics Volume 86, Pages 83 - 97.
[10] Kronenburg, R. (2007). Flexible: Architecture that responds to change. London: Laurence King Publishers.
[11] Carmona, M. (2003). Public places-urban spaces: The dimensions of urban design (1st ed.). Boston, MA: Architectural Press.
[12] Rinaldi, M. (2012). Elemen Tata Ruang Kota [Elements of Urban Spatial Planning]. Yogyakarta: Graha Ilmu.
[13] Chiara, J. D., & Koppelman, L. (1975). Urban planning and design criteria. Van Nostrand Reinhold.Wnglewoods Cliffs, N.J. Prentice Hall, Inc.
[14] Sudirahardjo, R. (2004, April). Analysis of Parking Space Requirement in Bandarjo Ungaran Market. Pilar, Volume 13, 8-24. (in Indonesian)
[15] Rutz, W. (1987). Cities and town in Indonesia. Berlin, German: Gebruder Borntraeger.
[16] Wiryomartono, B. (2000). Building art and urban building art in Indonesia. Jakarta: Gramedia Pustaka Utama (in Indonesian).
[17] Kusmawati, F. (1996). Pattern of market days in Gunungkidul Yogyakarta. Yogyakarta: Gadjah Mada University (in Indonesian)
[18] Permendagri (2007). Regulation of minister of domestic affairs No 42 in 2007 on rural village. Jakarta: Minister of Domestic Affairs of Republic of Indonesia (in Indonesian).
[19] Hobbs, F.,D., (1995), Traffic Planning and Techniques, penerjemah Suprapto T.M., Waldijono; penyunting Achmad Djunaedi, Gadjah Mada University Press, Yogyakarta
[20] Fahmi, K. (2014). Analisa Kapasitas Ruang Parkir Pasar Modern Kota Pasir Pengaraian [Analysis of Parking Space Capacity of Pasar Modern Kota Pasir Pengaraian]. JURNAL APTEK Vol. 6 No. 1, 107 - 116.
[21] Thanh, T. T., & Friedrich, H. (2017). Legalizing the illegal parking, a solution for parking scarcity in developing countries. Transportation Research Procedia, 4950–4965.
[22] Prasetiyo, J. A. (2014). Analisis Kebutuhan Ruang Parkir Pada Kawasan Pusat Perdagangan Kota Tomohon [Analysis of Parking Space Need In Tomohon City Trade Center Area]. Sabua Vol.6,

No.3, 333 - 340.
[23] Richard Arnott, E. I. (2015). Downtown curbside parking capacity. In Journal of Urban Economics Volume 86, Pages 83 - 97.
[24] André Romano Alho, J. d. (2014). Analyzing the relation between land-use/urban freight operations and the need for dedicated infrastructure/enforcement — Application to the city of Lisbon. In Research in Transportation Business & Management, Volume 11, Pages 85-97.
[25] Emmanouil Chaniotakis, A. J. (2015). rivers' parking location choice under uncertain parking availability and search times: A stated preference experiment Volume 82. In Transportation Research Part A: Policy and Practice, Pages 228-239.
[26] Mehdi Nourinejad, A. W. (2014). Truck parking in urban areas: Application of choice modelling within traffic microsimulation. In Transportation Research Part A: Policy and Practice, Volume 64, Pages 54-64.
[27] Zhengtian Xu, Y. Y. (2017). Optimal parking provision for ride-sourcing services. In Transportation Research Part B: Methodological, Volume 105, Pages 559-578.

Effect of Land Use Change to the Increasing Land-Based Emission and Urban Heat Island Phenomenon in Bandung City

I B Puspita[1*] and F A Saputra[1]

[1]National Technology Institute, Bandung

*email: iredobettie@gmail.com

Abstract. In order to fulfill the needs of built areas in Bandung encouraged land use changes in several areas. This land use change is one of the contributing factors to the increasing levels of carbon emissions in the air, which will have a further impact on the increasing urban temperature and Urban Heat Island phenomenon. Thus, this study was aimed to identify the effect of land use change on the increase of land-based emissions and temperatures in Bandung, as well as the trend in the future. This study utilized stock difference analysis method and the value of land-based emissions by using simple regression method. In addition, this study identified the effect of increase in emissions to the increase in temperature. The equation model produced from the regression was applied to identify the trend in 2031. Based on the results of the analysis, it was identified that land use change that occurred in 2003, 2011, 2014 and 2016 had an influence on increasing levels of carbon emissions in Bandung and had a significant influence on the increase in temperature. This temperature trend is projected to increase by an average of 1°C in 2031.
Keywords: land use change, land-based emission, temperature surface temperature

1. Introduction

Bandung is a significantly developed city in West Java Province since it was first planned by Thomas Karsten in 1941[1], Bandung city continues to grow in response to population growth each year. Identified in 2011 to 2015, the population of Bandung City increased by 56,152 people. This had an impact on natural land use that changed into built-up area to respond the number of population.

Land use change as a respond to urbanization might improve emissions, due to the reducing amount of vegetations [2]. The increase in emissions is related to the increase in temperature due to the amount of gas exhausts in the atmosphere resulting them to gather up which cover the Earth, reflecting heat radiation and trapping them inside. The increase in temperature will occur in urban areas due to the concentration of human's productive activities, such as trade and services, also transportation in urban areas. In addition, buildings and roads affect radiation and will often produce heat island effect, i.e. Urban Heat Island (UHI).

The UHI phenomenon also began to occur in Bandung City, indicated by the average annual temperature increase. The average annual air temperature in the Bandung city increased by 0.15 ° C, from 23.38 ° C in 2011 to 23.53 ° C in 2016[3]. The trend of increasing average temperature turned out to be in line with the land use change. From 2003 to 2014, the percentage of the built area continued to increase, which originally amounted to 72%,proceeded to develop to 88% [3]. With changes in land use, the increase of emissions in the air was projected to incline due to land use change.

Indications of the relationship between land use change, emission, and air temperature have not been specifically identified for the case of Bandung City. This study is expected to greatly contribute the intervention that will be carried out in the future spatial plan. Therefore, this study was aimed to identify the effect of land use change on increasing emissions and urban temperatures as an indication of the existence of the urban heat island phenomenon.

2. Methods

This study is an explanatory research with a causal approach, exploring the effects of land use change on the increasing land-based emissions and the UHI phenomenon, as well as providing the predictions for the future. The land uses observed in this study are the land use data in Bandung City in 2003, 2011, 2014, and 2016 in accordance with the availability of data. This study utilized the three main methods of analysis, such as: (1) method of calculating surface temperature changes, (2) methods of calculating land-based emissions, and (3) method of correlation.

2.1. Method of calculating land surface temperature changes

In this study, the phenomenon of the urban heat island was identified by using observations from Tenperature Land Surface (LST) to measure the surface heat emitted by the land. LST was measured by applying the remote-sensing techniques, estimating the brightness level of the uppermost layer in the atmosphere, which was captured based on its brightness in the lansat image [4]. The value of this LST represented the value of urban temperature by converting Digital Number to spectral radiation, which has its brightness temperature valued. LST will be measured based on Landsat 7 with band 6, conducted by using the following formulas and stages of:

1. Changing the DN values into spectral radiance by using the conversion formula on Landsat 7 as follows [5], [6]:

$$L\lambda = \left(\frac{L\lambda_{\max} - L\lambda_{\min}}{QCAL_{\max} - QCAL_{\min}}\right) \times \left(Qcal - QCAL_{\min}\right) + L\lambda_{\min} \tag{1}$$

$L\lambda$: Spectral radiance (Watts/(m2 * srad * μm))
$L\lambda_{max}$: Maximum spectral radiance in scale of $QCAL_{max}$. $L\lambda_{max}$ for Landsat 7 with band 6 is 17.040
$L\lambda_{min}$: Minimum spectral radiance in scale of $QCAL_{min}$. $L\lambda_{min}$ for Landsat 7 with band 6 is 0
$QCAL_{max}$: Maximum pixel value. $QCAL_{max}$ for Landsat 7 with band 6 is 255
$QCAL_{min}$: Minimum pixel value. $QCAL_{min}$ for Landsat 7 with band 6 is 1
Qcal : Quantized and calibrated standard product pixel values, Digital Number which used (thermal band)

2. Calculating the Brightness Temperature by converting ($L\lambda$) spectral radiance then converting it to TB value into temperature with unit Kevin (K). This used thermal constants provided in metadata landsat 7 ETM^+ files [6]:

$$T_B = \frac{K_2}{In\left(\frac{K_1}{K_2} + 1\right)} \tag{2}$$

T_B : At-satellite brightness temperature (K)
$L\lambda$: Spectral radiance (Watts/(m2 * srad * μm))

K1 : Band-specific thermal conversion constant from The metadata (K1_CONSTANT_BAND_x, where x is the Andnumber), with the constant for Band 6 0f Landsat 7 is 666,09
K2 : Band-specific thermal conversion constant from the metadata (K2_CONSTANT_BAND_x, where x is the band number), with the constant for Band 6 0f Landsat 7 is 1282,71

3. Calculating the LST by converting the estimation results of T_B, which have kelvin units to centigrade by reducing each result of the calculation with a value of 273,15.

$$T = T_B - 273{,}15 \quad (3)$$

T : Temperature in Celcius (C)
T_B : At-satellite brightness temperature (K)

2.2. Method of calculating land-based emission
Calculation of land-based emissions was carried out by identifying the value of carbon stocks from a land use. In this study, the approach applied the stock difference approach, that identifies an increase or a decrease in carbon stocks, due to changes in land cover biomass as a representation of land use change [2]. This method was carried out by calculating the total carbon stock from each type of land cover. The value of carbon stocks used in this study refers to the value of aboveground carbon stocks for each type of land cover issued by the National Development Planning Agency (Bappenas) in the framework of preparing Regional Action Plans for the reduction of Greenhouse Gases (RAD-GRK). The calculation carried out for each type of land use change, including the realization of the spatial pattern plan contained in the Bandung City Spatial Plan (RTRW) with the basic formula as follow [2], [7]:

$$EV = AD \times FE \quad (4)$$

EV : Emission Value, the amount of greenhouse gases released into the atmosphere from a landscape (ton CO^{eq})
AD : Activity Data, area of land cover changes (ha)
EF : Emission Factor, the amount of carbon stored from a biomass in land cover (ton CO^{eq})

From the results of the calculation of carbon stock values, the amount of emissions produced was gathered from the difference in the carbon stocks value from two different periods. If the value of the carbon stock difference is greater than 0 (emission value > 0), carbon release into the air was present, which means there was an increase in the level of emissions in the air. If the value of the carbon stock difference is less than 0 (emission value < 0), carbon storage or sequestration was present. If the value of the difference in carbon stock is equal to 0 (emission value = 0), then there is no change in the carbon content in the air [7].

2.3. Method of correlation
This study identified the correlation through the effect of changes in the value of land-based emission toward LST and its consequences for the UHI phenomenon. This correlation was identified by using a simple regression analysis between each change in the value of land-based emission with each change in the value of LST in a location experiencing changes in land use. Aside to identify the influence, this analysis was also utilized to build an equation model of the influence, which will be used to estimate the conditions that will occur in the future, when the spatial pattern plan for Bandung City RTRW in 2031 is realized. The hypothesis of the equation that was proven as follows:

$$Y = a + bX \quad (5)$$

Y = dependent variable of LST for each pixel land use change
X = independent variable of land-based emissions as a consequence of land use changes calculated from each pixel change in land use
a = constant
b = coefficient estimator of X

This method was carried out through 3 (three) main tests, namely: (1) Pearson correlation test, (2) F-test, and (3) t-test.

Pearson correlation was applied to find out the strength and direction of the influence between changes in land-based emission to the changes in LST. The closer Pearson correlation value to 1, the greater the relationship formed and conversely, the farther it is to 1, the smaller the relationship formed. If the correlation value is positive (+), then the direction of the relationship between changes in land-based emission are in line with changes in LST or in other words the greater the changes in land-based emission, the greater the LST changes that occur. Conversely, if the correlation value is negative (-), then the relationship between changes in land-based emission is inversely proportional to changes in LST or in other words the more changes in land-based emission that occur, the smaller the LST changes that occur.

The F-test was performed to determine whether the independent variable changes in land-based emission (X) had a joint effect on the dependent variable LST (Y). This F-test proves that the model generated from regression analysis can / cannot predict temperature conditions in 2031. The hypotheses tested are:

- Ho: Land-based emissions do not have a joint effect to predict temperature / model or the equation formed cannot be used to predict temperature.
- H1: Land-based emissions have a joint effect to predict temperature / model or the formed equation can be used to predict temperature.

The test criteria are as follows: Ho: accepted (H1 rejected) if the significance is F≥0, 05 and H1: rejected (Ho accepted) if the significance is F≤0, 05

The t-test essentially shows how far the effect of 1 (one) variable changes in land-based emission in explaining LST variables. The hypotheses tested are:

- Ho: Variable of land-based emissions have no effect on LST.
- H1: Variable of land-based emissions have the effect of predicting LST.

The test criteria are as follows: Ho: accepted (H1 rejected) if the significance is to≥0, 05 and H1: rejected (Ho accepted) if the significance is to≤0, 05

3. Result and Discussions

3.1. Temperature Changes as a consequence of Land Use Change in the City of Bandung

The results of processing by converting brightness levels to landsat 7 images into temperature values indicated that the city of Bandung had a high LST value in the city center, while for areas far from the city center tended to have a low LST. Changes in the LST of Bandung City from 2003-2016 experienced fluctuations. The highest LST is in the built area, which is marked by red, marking the heat due to the prolonged and existing heat absorption in building during the day [8].

In 2003, the highest LST was found in the northern part of Bandung; in Cibeunying Sub-district, and in the airport at Cicendo Subdistrict. On other areas, the average LST was only 25-26° C and the lowest was in the East of Bandung. In 2011, the highest LST was in the city center to the North and South with an average surface temperature of 32°C. The increase of LST from 2003 to 2011 was quite high, because almost all areas of Bandung City in 2011 had high LST. In 2014, the LST of the city of Bandung decreased to 6-14°C at the lowest, while the highest was 24-26°C. The highest LST was spread in the central part of the city to the northern part, from the District of Bojongloa Kidul to Sukasari District. In 2016, the LST of the city of Bandung experienced an increase compared to 2014. The lowest LST ranged from 13-16°C, and the highest was 27-34°C. The highest LST were only scattered in the central part of the city and in the North such as Cibeunying District, and Cicendo District. For some parts such as North Bandung, the average low LST ranged only from 13-16°C due to the large area of vegetation in the area. The distribution of LST changes in the city of Bandung is depicted in Figure 1.

3.2. *Land-Based Emission as Consequences Land Use Change*

Emission calculations were carried out by using data on land use change in 2003, 2011, 2014, 2016, along with spatial pattern of the RTRW in 2031. With these data, the carbon stock change process was carried out by using the difference approach of two different-time periods. The processing results showed that emissions due to land use change, continued to increase. This increase was influenced by the reduction in non-built land changing into built-up land, as the consequence of the increasing population.

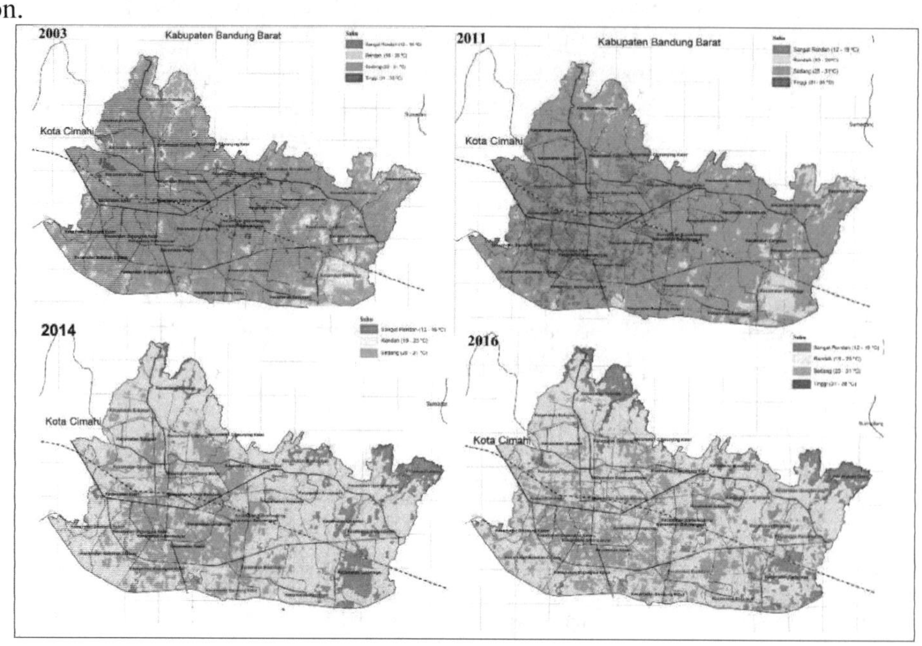

Figure 1. LST Changes in 2003-2016
Source: Analysis Result, 2018

Figure 2 presents the information that the change in total emission values is striking between the changes in 2011-2014 and 2003-2011. In 2011-2014, the largest increase of carbon stocks (sequestration) was identified, which was in contrast to the conditions in the changes from 2003-2011 producing the largest emissions from the range of years during the research. However, this condition does not work consistently, as in 2016, the drive to rebuild the city was unstoppable, with changes in rice fields, as well as in residential areas. This has an impact on the emissions resulting from changes in 2016 (increase). The fluctuations in land use change were anticipated in the spatial pattern of the Bandung City Plan (RTRW) 2031 to restore better condition of Bandung City, by producing greater sequestration. In terms of distribution, changes in the value of emissions marked land use changes in the eastern part of Bandung, such as Gedebage and Ujungberung, with a large amount of non-built land available in 2003, as target for urban development allocation (**Figure 2**).

3.3. *Corellation of Land-based Emission Change and LST Change as Concequences of Land Use Change*

Changes in land-based emissions during the period of 2003-2016 were essentially in line with the condition of the LST of the city of Bandung at the same period. This was identified from the value of emissions in 2011-2014, which resulted in the largest sequestration value with the average LST at the lowest value among all years of research. On the contrary, in 2003-2011 and 2014-2015, a large land-based emission value had a high average LST. The existence of conditions that are in line between changes in emissions was due to changes in land use with LST (indicating a relationship between the two) [8]. As for proving this corellation, a correlation analysis was performed by using a simple regression analysis between changes in land-based emission with changes in LST on each type of land use change. This analysis was carried out separately for each type of land use change to identify the linkages more specifically. From this simple regression analysis, the threetest results were obtained, including R^2, F test, and t test.

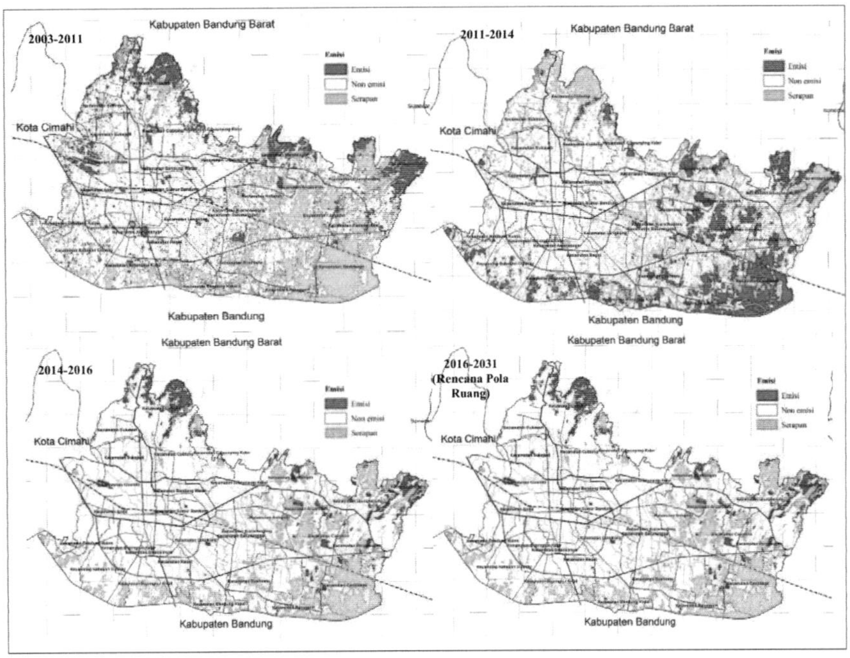

Figure 2. Distribution of Land-based Emission 2003-2031
Source: Analysis Result, 2018

From the results of the analysis, it was identified that changes in any land use change that produced a certain emission value, turned out to have an effect on the existing LST conditions with a positive relationship (**Table 3**). Even though it has a relationship, the relationship formed is relatively small when seen for each of these changes. From these results, it was identified that the value of R^2 which gave the most influence was the use of station land by 0,1%, while the value of R^2 was the smallest for sports areas of 0,001%. This identified that the change in land use into a built area such as a station had a relatively large effect compared to other land use changes with the green land cover.

Table 1. R^2 Value dan t-test for Each Type of Land Use Change

No	Changes From Using Any Land to Being	R^2	Information R^2	t-stat	Sig. t-test	Information t-test
1	Airport	0,044	Corellated	2,609	0,01	Corellated
2	Military Area	0,015	Corellated	1,909	0,057	Corellated
3	Forest	0,047	Corellated	-2,772	0,006	Corellated
4	Religion Facility	0,003	Corellated	1,856	0,064	Corellated
5	Industry	0,015	Corellated	7,161	0,000	Corellated
6	Health Facility	0,015	Corellated	2,609	0,009	Corellated
7	Sport Facility	0,001	Corellated	1,165	0,244	Not Corellated
8	Education Facility	0,012	Corellated	4,996	0,000	Corellated
9	Government Office	0,007	Corellated	2,373	0,018	Corellated
10	Trading Facility	0,005	Corellated	7,567	0,000	Corellated

No	Changes From Using Any Land to Being	R^2	Information R^2	t-stat	Sig. t-test	Information t-test
11	Residential Area	0,004	Corellated	7,286	0,000	Corellated
12	Stasion	0,116	Corellated	6,416	0,000	Corellated
13	Park	0,003	Corellated	3,656	0,000	Corellated
14	Agriculture Area	0,015	Corellated	11,798	0,000	Corellated
15	Terminal	0,057	Corellated	2,438	0,017	Corellated

Source: Analysis Result, 2018

Based on the t-test statistic related to the influence of the variable changes in land-based emission (X) and LST (Y), the value of Sig. was less than 0,05 for each land use except for sports facilities, because the use of the land hadnot changed. Based on the t test, there was a significant effect on the increase in LST (Y) in the city of Bandung, due to an increase in land-based emission (X). Each use of land except sports facilities produced changes in land-based emission (X) with the highest concentration value from the agricultural land use with a t value of 11,798 with a significant value of 0,00. Thus, it is concluded that there is an increase in land-based emission (X) from the area built into an agricultural area. Land use with the concentration of land-based emission (X) becomes the lowest for forest land uses with a t value of -2,7772. This explains that there is a change in land-based emission (X) to be sequestration due to changes in land use into forest areas.

The corellation between variables X and Y as a single unit model was understood by the F / Anova correlation test using the formula in equation (5). From the results of the analysis, it was explained that the model of the relationship between changes in land-based emission and changes in LST formed for each type of land use change were all insignificant. Of the 15 patterns of land use change that occurred, it was identified that the change became sports facilities, giving a significance value greater than 0,05, identifying that the model of the relationship equation formed cannot be used to predict conditions in the future. Meanwhile, 14 other land-use change patterns provided conditions that the model from this simple regression analysis was significant to predict future conditions (**Table 4**).

Table 2. F-test Value for Each Type of Land Use Change

No	Changes From Using Any Land to Being	F	Sig.	Information	Equation Model (Y=a+bX)	Arti Model Persamaan
1	Airport	6,808	0,01	Corellated	Y= 24,245+0,002 (x)	Every increase in emissions by 1% causes a LST increase of 0,002°C
2	Military Area	3,644	0,057	Corellated	Y= 24,667+0,044 (x)	Every increase in emissions by 1% causes a LST increase of 0,044°C
3	Forest	7,683	0,006	Corellated	Y= 22,847+-0,006 (x)	Every increase in emissions by 1% causes a LST increase of 0,006°C
4	Religion Facility	3,445	0,064	Corellated	Y= 0,659+25,273 (x)	Every increase in emissions by 1% causes a LST increase of 0,0659°C
5	Industry	51.274	0,000	Corellated	Y= 23,165+0,39 (x)	Every increase in emissions by 1% causes a LST increase of 0,39°C
6	Health Facility	6,808	0,009	Corellated	Y= 25,376+1,346 (x)	Every increase in emissions by 1% causes a LST increase of 1,346°C
7	Sport Facility	1,358	0,244	Not Corellated	Y= 24,801+0,01 (x)	Every increase in emissions by 1% causes a LST increase of 0,01°C
8	Education Facility	24,959	0,000	Corellated	Y=24,549+0,039 (x)	Every increase in emissions by 1% causes a LST increase of 0,039°C
9	Government Office	5,630	0,018	Corellated	Y=25 +0,29 (x)	Every increase in emissions by 1% causes a LST increase of 0,029°C
10	Trading Facility	57,256	0,000	Corellated	Y=24,188+0,001 (x)	Every increase in emissions by 1% causes a LST increase of 0,001°C
11	Residential Area	53,088	0,000	Corellated	Y=23,509+0,002 (x)	Every increase in emissions by 1% causes a LST increase of 0,002°C
12	Stasion	41,169	0,000	Corellated	Y=25,088+0,493 (x)	Every increase in emissions by 1% causes a LST increase of 0,493°C
13	Park	12,706	0,000	Corellated	Y=23,250+0 (x)	Every increase in emissions does not causes an increase of LST

No	Changes From Using Any Land to Being	F	Sig.	Information	Equation Model (Y=a+bX)	Arti Model Persamaan
14	Agriculture Area	139,188	0,000	Corellated	Y=22,315+0 (x)	Every increase in emissions does not causes an increase of LST
15	Terminal	5,944	0,017	Corellated	Y=24,383+0,632 (x)	Every increase in emissions by 1% causes a LST increase of 0,632°C

Source: Analysis Result, 2018

3.4. *Effect of Spatial Planning in Reducing Emissions and UHI Phenomena*

The equation model obtained from simple regression analysis on each land use change is utilized to predict the distribution of LST in Bandung in 2031 with the spatial pattern plan (the Bandung City Spatial Plan (RTRW) 2031). As for changes to sports facilities, it is assumed to remain constant because the equation model formed from the results of simple regression analysis showed no effect.

From the results of calculations with the equation formed in **Table 4**, it was identified that the average LST of Bandung in 2031 would increase by 0,92°C (close to 1°C) or at an average of 24°C. This average LST is predicted to increase compared to the average LST in 2016 (23,02 °C) due to changes in the value of emissions in vegetation area that becomes built area, resulting in high emissions of 10,235 $CO_2^{eq.}$ in which a 1% increase in emissions causes the increase in LST of 0,6°C.

When compared to the land-based emission produced, it showed that a sequestration leads to the absorption of carbon stocks with higher LST. This means that the increase in the allocation of protected areas and parks in the Bandung Spatial Plan (RTRW) 2031 has not been able to reduce the overall LST. This also means that the composition of protected areas and parks in Bandung is not enough to reduce the rising temperature of settlements as a result of the construction of the built area. The prediction of changes in LST in 2031 will occur a lot in eastern and northern part of Bandung, which currently has a tendency to change into a built up area (**Figure 3**).

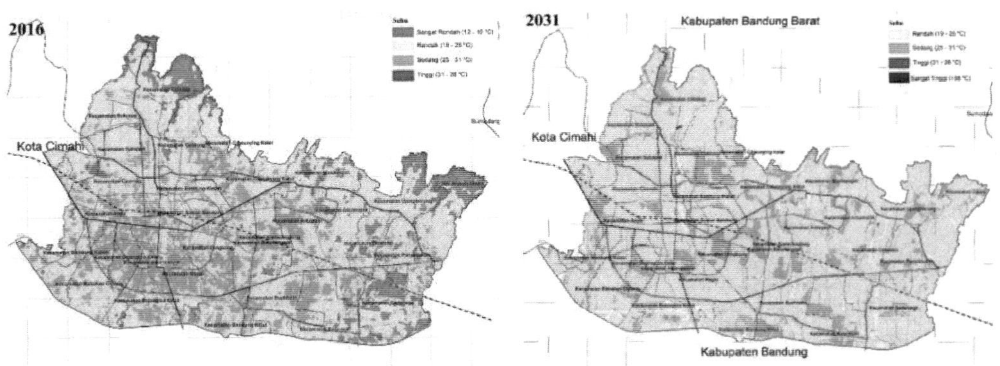

Figure 3. Comparation of LST Distribution 2016 and 2031
Source: Analysis Result, 2018

4. Conclusion

This tendency of increasing LST indicates that the UHI phenomenon occurred in Bandung. The increasing value of LST indicated the increasing value of air temperature as the existence of the UHI phenomenon in urban areas [4]. The value of this LST was converted into UHI values in urban areas. For the case of Bandung City, the increase in the UHI phenomenon was caused by the increase in land-based emission with response to the development needs of Bandung City contributing to the occurrence of the UHI phenomenon.

The Bandung City Spatial Plan (RTRW) 2031, which has allocated protected areas and parks, has not been able to control and reduce the UHI phenomenon. The allocation of sufficiently good space needs to be complemented by detailed directions on each type of land use change that can later be followed up in zoning regulations in a detailed spatial plan. The detailed form of directives included

IOP Conf. Series: Earth and Environmental Science **328** (2019) 012021 doi:10.1088/1755-1315/328/1/012021

the determination of a large green coefficient in the built area, the use of materials that can absorb heat, and the selection of the right type of vegetation for absorbing heat and absorbing emissions.

5. References

[1] Voskuil, R. P. G. A. 1996. *Bandung: citra sebuah kota: Bandoeng, beeld van een stad*: Asia Maior.

[2] Agus, F., Santosa, I., Dewi, S., Setyanto, P., Thamrin, S., Wulan, Y., & Suryaningrum, F. 2013 *Pedoman Teknis Penghitungan Baseline Emisi dan Serapan Gas Rumah Kaca Sektor Berbasis Lahan: Buku I Landasan Ilmiah*. Badan Perencanaan Pembangunan Nasional Republik Indonesia, Jakarta.

[3] Statistic Berau of Bandung City 2016 *Bandung City in Figure 2016*. BPS: Bandung.

[4] Tsou, J., Zhuang, J., Li, Y., & Zhang, Y. 2017 *Urban heat island assessment using the Landsat 8 data: a case study in Shenzhen and Hong Kong*. Urban Science, 1(1), 10.

[5] Schott, J. R., Palluconi, F. D., Helder, D. L., Hook, S. J., Markham, B. L., Chander, G., & O'Donnell, E. M. 2003 Landsat TM and ETM+ thermal band calibration AU - Barsi, J A. *Canadian Journal of Remote Sensing*, 29(2), 141-153. doi:10.5589/m02-087.

[6] Mallick, J., Kant, Y., & Bharath, B. 2008 Estimation of land surface temperature over Delhi using Landsat-7 ETM+ *J. Ind. Geophys. Union, 12(3)*, 131-140.

[7] Puspita, I. B. 2017 Mewujudkan Perencanaan Tata Ruang yang Ramah Lingkungan melalui Analisis Emisi Berbasis Lahan (Hal. 81-89). *REKA RACANA, 3*(1).

[8] Musco, F. 2016 *Counteracting Urban Heat Island Effects in a Global Climate Change Scenario*: Springer.

IOP Conf. Series: Earth and Environmental Science **328** (2019) 012022 doi:10.1088/1755-1315/328/1/012022

Expansive Clay Soil Stabilization Using White Soil Material and Sulfuric Acid Solution (H_2SO_4) For Subgrade in Godong Area - Grobogan District

I K Suwantara[1], S P R Wardani[2]*, and Y A Priastiwi[2]

[1]Student of Masters Degree, Civil Engineering, Universitas Diponegoro
[1]Advisor, Civil Engineering Masters Program, Universitas Diponegoro

*email: arswandari@gmail.com

Abstract. Expansive clay is classified as an unstable soil. Soil stabilization of expansive clay can be achieved by means of replacement, chemical, and physical methods. This research studies the stabilization of expansive clay using chemical methods, by combining white soil from NTT with H_2SO_4. This study was conducted to determine the effect of adding white soil and H_2SO_4 to soil physical and mechanical properties and to improve expansive clay. The method used physical and mechanical testing, namely: property test, CBR, UCS, and Oedometer with variations in composition with an addition of 3%, 4%, and 5% white soil from the weight of expansive clay and 5% H_2SO_4 from OMC water content at 0, 7, 14, and 28 days of aging, then compared to untreated expansive clay. The results showed that physically stabilized clay became inactive, although its plasticity and expansion rate decreased. Mechanically, there was an increase in the compressive and shear strength of 43.82%, and unsoaked CBR density of 83.25% and soaked CBR of 7.4%, which is directly proportional to the aged soil. Swell potential and swell pressure decreased by 90.71% and 65.71% respectively. The optimum composition is the composition with an addition of 3% white soil and 5% H_2SO_4 at 28 days of aging.

1. Introduction

The tropical climate causes high rainfall and a long dry shift in Indonesia so that land most areas of the country tends to have large shrinkage. This type of soil experiences significant volume changes along with changes in water content [1]. One way to design the nature of the soil to meet certain technical requirements is the soil stabilization method. Technically, soil stabilization can be achieved by mechanical, physical and chemical stabilization [2].

An example of expansive soil stabilization by means of chemical admixtures is the stabilization of soil with lime, which is widely used in road projects. For optimum results, the ratio of lime used is usually between 3% - 7%. Thomson found that with an addition of lime levels between 5% - 7%, it would produce greater strength than 3% lime [3]. In another study, the results obtained by stabilizing the soil with cement were almost the same as soil stabilization with lime [4].

White soil from Kupang city, East Nusa Tenggara has the potential to be a good soil stabilization material. The dominant chemical elements in this soil are 52% Magnesium Calcium Carbonate and 48% Magnesium Phosphate [5]. This chemical content in white soil can function as a chemical stabilization material, while the compound used as stabilization agent is sulfuric acid (H_2SO_4). Clay soil stabilization using H_2SO_4 on sub grade in Godong-Purwodadi at 50 Km mark, Grobogan showed an increase in physical and mechanical properties, while swelling pressure and swelling potential did not show a significant increase [6]. Expansive soil stabilization by adding limestone to embankment work provides a significant CBR value [7].

Expansive clay is one type of fine-grained colloidal soil formed from expansive minerals, namely: *montmorillonite*, *illite*, *kaolinite*, *halloysite*, *chlorite*, *vermiculite*, and *attapulgite* [8]. These colloidal soils have a high potential for shrinkage, in accordance to changes in water content. If the clay mineral content increases, the surface area, liquid limit, and plasticity index will increase, so the potential of the shrinkage will increase as well [9]. Test of expansive clay properties ex. Godong, Grobogan found 49,1% *PI*, with the following mechanical properties: $q_u = 4.61 – 6.87$ kg/cm^2, $C_u = 2.31 – 3.44$ kg/cm^2, *CBR Unsoaked* = 7.10–11.40%, *CBR Soaked* = 1,21–4.32%, Dry Nett (γ_d) = 1,34–1,41gr/cm^3, *Disturbed Swell Potential* = 13.96%, *Disturbed Swell Pressure* = 700 kPa, *Undisturbed Swell Potential* = 15.03%, *Undisturbed Swell Pressure* = 900 kPa [10].

Identification of expansive clay by indirect method is an empirical relationship on experience and identification of Atterberg Limit, clay content, initial moisture content, dry volume weight, and shrinkage limit [11]. Atterberg limit of soil is the main parameter in indirectly identifying expansive soils [12]. This test refers to ASTM D4318–000 standards [13]. The Plasticity Index (PI) is the difference between the Liquid Limit (LL) and the Plastic Limit (PL) expressed by equation 1. The correlation between PI value with the type and swelling soil is shown by table 1 and table 2.

$$PI = LL = PL \tag{1}$$

Table 1. PI value and type of soil [14].

PI	Properties	Soil Type	Cohesion
0	Non-Plastic	Sand	Non-Cohesive
<7	Low Plasticity	Silt	Partially Cohesive
7-17	Medium Plasticity	Silt Clay	Cohesive
>17	High Plasticity	Clay	Cohesive

Table 2. Correlation of PI and expansion rate [15].

Plasticity Index, PI (%)	Expansion Probability (%)	Expansion Rate
<18	<20%	Low
15 - 28	10 - 20	Medium
25 - 41	20 - 30	High
>35	>30	Very High

The Activity Method is intended to indirectly identify expansive clay [16]. The parameter of activity can be seen in equation 2.

$$Activity\ (A)_{remolded} = \frac{PI}{C\text{-}10} \tag{2}$$

Legends: *PI* = *Plasticity Index*, *C* = the percentage of clay escaping the filter 0,002 mm. Because the dividing coefficient in equation 2 only applies to remolded conditions, equation 3 is used for unremolded soil conditions.

$$Activity\ (A)_{unreemolded} = \frac{PI}{C\text{-}n} \tag{3}$$

Legends: n = valued at 5. The land categories of the equation are A <0.75 (inactive), 0.75 <A <1.25 (normal), A> 1.25 (active).

Unconfined Compression Test (UCS) is the amount of axial load per unit area when the specimen has collapsed or when the axial strain reaches 20%. Experiments of free compressive strength in the laboratory were carried out on remolded soil samples. The axial stress on the ground can be calculated by equation 4.

$$\sigma = q_u = \frac{P}{A} \tag{4}$$

Legend: $\sigma = q_u$ = Axial strain, P = load, A= unit area. Soil shear strength can be calculated by equation 5.

$$C_u = \frac{\sigma_1 + \sigma_3}{2} = \frac{\sigma 1}{2} = \frac{q_u}{2} \tag{5}$$

Legend: C_u *=undrained shear strength,* C_u *=undrained shear strength,* $\sigma_3 = 0$, C_u *=undrained shear strength,* $\sigma_3 = 0$, q_u = *unconfined compressive strength.* q_u *as unconfined soil strength.*

CBR is widely used for highway planning as recommended by *Nelson* and *Miller* [17]. Test of expansion of expansive clay is performed using *CBR* tools with ASTM D 1883-99 standards [18]. The expansion that occurs is expressed in percent, which is the ratio between the amount of expansion of the soil divided by the initial height of the expansive clay soil multiplied by 100%, expressed in the formula in equation 6.

$$CBR = \frac{PT}{PS} \times 100\% \tag{6}$$

Legend: *CBR* = *California Bearing Ratio*, PT = load-on-trial, PS = standard load.

Expansion potential as a percentage of the development of lateral samples soaked (saturated) with pressure is <6.9 kPa, while for compacted soils it can reach maximum density at its optimum water content [19]. Clay which contains many montmorillonite minerals has the potential to expand and is generally tested using the Loaded Swell Test method with the ASTM-D-4546-90 method B [20]. This test can be calculated using equation 7 as follows:

$$SPt = \frac{\Delta h}{h0} \times 100 = \frac{(e_{v0} - e_o)}{1 + e_0} \times 100 = \left(\frac{\gamma_{d0}}{\gamma_{dvo}} - 1\right) \times 100 \tag{7}$$

Legends: SP_t= *Swell Potential* (%), e_{v0}= *Void ratio* after expansion is stabilized at vertical pressure σ_{v0}, γ_{dv0}= dry density at void ratio e_{v0}.

2. Methods

This study is carried out by experiments. Primary data was collected from making expansive test specimens of white soil stabilized clay at 3%, 4%, and 5% of the weight of untreated expansive dry soil and stabilizer material in the form of 5% H_2SO_4. Test specimens were aged for 0, 7, 14, and 28 days to determine the change in stabilized soil, after which tests were conducted which included the test of Index Properties (Attereberg in accordance with the ASTM D4318 standard (2000), hydrometer and grain size in accordance with the ASTM D 422 standard (1963)), UCS test from Proctor samples in accordance with the ASTM D2166 standard (2016), swell potential and swell pressure in accordance with the ASTM D4546 standard (1996), and CBR in accordance with the ASTM D 1883 (2014). The analysis is conducted by comparing between existing expansive clay and expansive clay after stabilization.

3. Results and Discussions

The mixture composition is defined as follows: Composition 1, expansive clay + 3% white soil + 5% H_2SO_4 + Water. Composition 2, expansive clay + 4% white soil + 5% H_2SO_4 + Water. Composition 3, expansive clay + 5% white soil + 5% H_2SO_4 + Water.

3.1. Physical Properties of Expansive Stabilized Soils

Based on the PI values of stabilized clay at **Figure 1**(a) it shows that there is no activity of expansive clay. This is indicated by the PI value smaller than 0.75, while the untreated expansive clay shows activity (**Figure 1**(b)). The type of soil after stabilization is still classified as clay type with high plasticity and is cohesive, this is indicated by the PI value greater than 17 [15], although the PI value of stabilized soil is still smaller than the PI value of the untreated soil (**Figure 2**(a)). Viewed from the rate of expansion based on the PI value, the stabilized expansive clay is still very high with a PI value greater than 30. Nevertheless, the stabilized PI values of expansive clay soils are still smaller than those of the untreated expansive clay (**Figure 2**(b)).

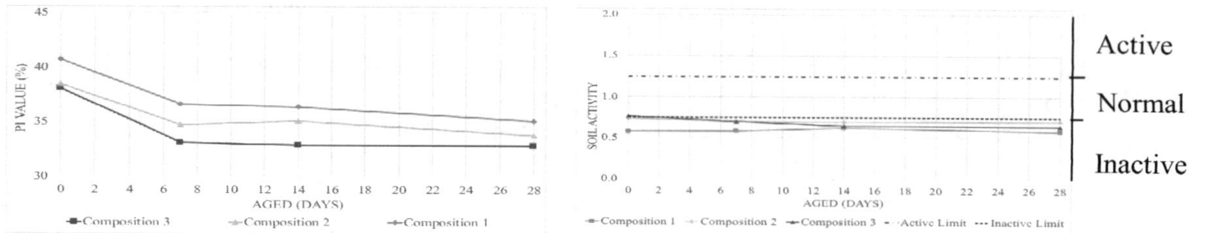

(a) (b)

Figure 1. *PI* Value of expansive clay (a), expansive clay activity based on *PI* values (b) before and after stabilization.

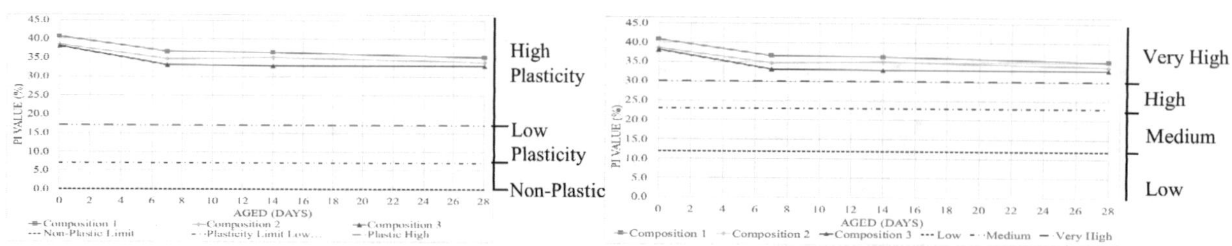

(a) (b)

Figure 2. Expansive clay types (a), rate of expansive clay soil swelling (b) based on *PI* value before and after stabilization.

3.2. Mechanical Properties of Stabilized Expansive Soil

3.2.1. Unsoaked CBR Test

CBR value in unsoaked condition with 56x blows on **Figure 3**(a) shows the highest results compared to 25x blows (**Figure 3**(b)) and10x blows (**Figure 3**(c)). Composition 1 showed the highest *CBR* value. The *CBR* value at 56x blows significantly increased in the untreated soil. The biggest increase was found at 83.25% and the smallest at 10x blows of 21.27% for Composition 1. The biggest increase was found at 76.49% and the smallest at 10x blows of 2.82% decrease for Composition 2. Meanwhile, the biggest increase was found at 74.74% and the smallest at 10x blows of 7.32% decrease for Composition 3. It was also found that the length of aging is proportional to the increase in *CBR* value. The average increase in *CBR* values in each composition was 30.24%, 34.43%, and 31.95% for Compositions 1, 2, and 3 respectively.

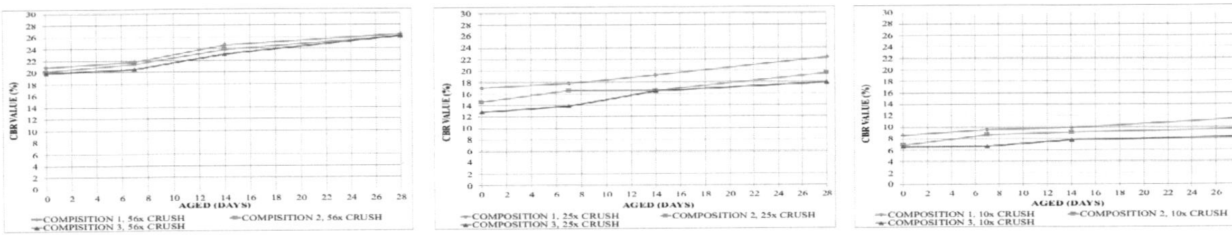

(a) (b) (c)

Figure 3. The biggest CBR value of 56x blows (a), 25x blows (b), 10x blows (c) of unsoaked stabilized clay soil aged for 0, 7, 14 and 28 days.

3.2.2. Soaked CBR Test

The *soaked CBR* result at 56x blows on **Figure 4**(a) shows the highest value compared to 25x blows (**Figure 4**(b)) and 10x blows (**Figure 4**(c)). Composition 1 shows the highest CBR value, despite the decrease in the CBR value of the existing soil. For Composition 1, the decrease is 4.86% at 56x blows and at 10x blows is 73.15%. For Composition 2, the decrease is 17.82% at 56x blows and at 10x blows is 75.46%. Whereas for Composition 3, the most the decrease is 54.17% at 56x blows and at 10x blows is 82.18%. Similar to the results of unsoaked CBR test, the results also shows that the length of soil aging is proportional to the increase in *CBR* value. The average increase in *CBR* values in each composition were 83,77%; 75,66%; dan 122,57% for Compositions 1, 2, and 3 respectively.

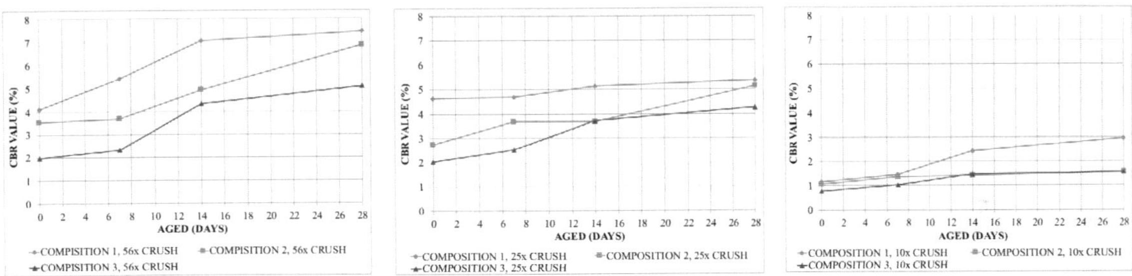

(a) (b) (c)

Figure 4. The biggest CBR value of 56x blows (a), 25x blows (b), 10x blows (c) of soaked stabilized clay soil aged for 0, 7, 14 and 28 days.

3.2.3. UCS Test

UCS Test as seen in **Figure 5**(a) on Composition 1 shows that the axial strain that occurs at 0, 7, and 14 days of aging tends to be greater than that of 28 days of aging, and that there is a significant increase in compressive strength at 28 days. UCS Test results for Composition 2, shown in **Figure 5**(b), shows a behavior that is slightly different from the graph shown in **Figure 5**(a). Greater axial strain occurs only at 0 day of aging, while the strain at 7, 14, and 28 days of aging tend to be lesser. This finding indicates increasing compressive strength at 7, 14, and 28 days of aging. A similar phenomenon is also found in composition 3 as shown in **Figure 5**(c), suggesting that the longer the pressure is, the more compressive strength is obtained, which means improved soil stability.

Based on unconfined compressive strength test shown at **Figure 5**, the maximum q_u obtained for Composition 1 is 8.741 kg/cm^2 at 28 days of aging, as seen in **Figure 6**(a). This composition tends to possess a greater q_u value compared to that of other Compositions in this study. On the other hand, the value of C_u found was half of unconfined compressive strength value, at 4.371 kg/cm^2 (**Figure 6**(b)). This value is proportional to the unconfined compressive strength, which tends to have a greater value than other compositions.

Based on the level of soil consistency [18], the q_u value of Expansive clay from the untreated soil sample of 3.909 kg/cm^2, as shown in **Figure 6**(a), is classified as a level of hard soil consistency (>3.83 kg/cm^2). This also shows that the stabilized expansive clay with q_u = 8.741 kg/cm^2>q_u untreated expansive clay soil = 3.909 kg/cm^2, which also falls into the category of hard soil.

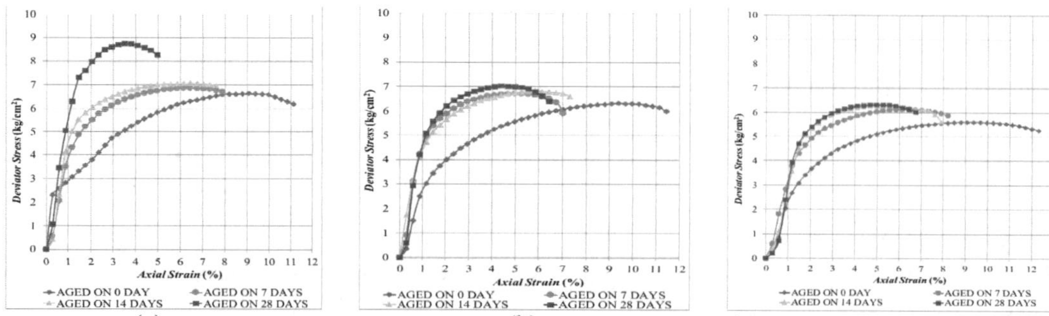

(a) (b) (c)

Figure 5. Comparison chart of axial strains with strain deviator of stabilized expansive clay with composition 1 (a); composition 2 (b); and composition 3 (c) at 0, 7, 14 and 28 days of aging.

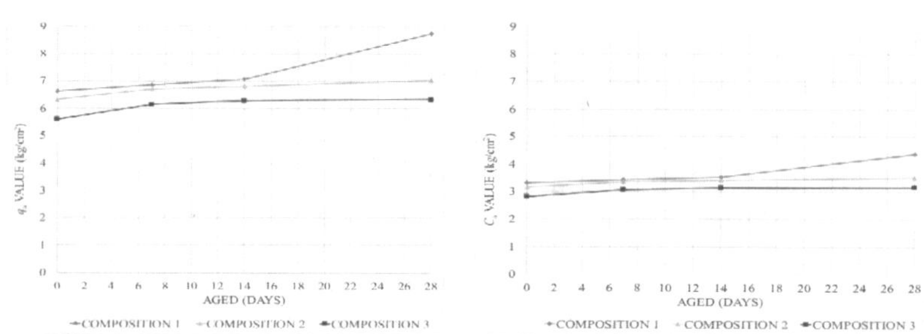

(a) (b)

Figure 6. Comparison charts of q_u value (a) and of $C_u$value (b) at 0, 7, 14 and 28 days of aging, expansive clay soils stabilized using composition 1, composition 2, and composition 3.

3.2.4. Swelling Test

Swelling test on the expansive clay soil ex. Godong, Grobogan has a fairly high expansion rate of 15.03% in undisturbed state and 13.96% in disturbed state (**Figure 7**(a)). The test results of all three Compositions indicate that all compositions lessen the potential for soil expansion compared to that of the soil in untreated state. Treatment with Composition 1 shows a gradual decrease in expansion potential from 5.92% at 0 days of aging, to 4.28% at 7 days of aging, and continues to decrease by 4.23% at 14 days of aging, until finally it reaches 2.49% at 28 days of aging. However, treatment with Composition 2 and 3 only reduced expansion potential at 0 to 14 days of aging, and an increased expansion potential was found at 28 days of aging (**Figure 7**(a)). This phenomenon may occur due to the influence of the higher levels of water absorption on the higher percentage of white soil, making the stabilized soil condense faster, which affects the potential for expansion of the soil.

The effect of the duration of soil aging which is inversely proportional to the expansion potential of the soil in Composition 1 treatment shows an improvement in soil conditions due to the effect of changes in water content. Meanwhile, the treatment of Compositions 2 and 3 can induce greater swell potential compared to Composition 1. Original expansive clay in disturbed conditions has an expansion pressure of 700 kPa and under undisturbed conditions it has an expansion pressure of 810 kPa (**Figure 7**(b)). On the other hand, the maximum expansion pressure of Composition 2 is 240 kPa (**Figure 7**(b)), indicating that the stabilized expansive clay has less pressure, so that the condition of this soil is better than the untreated expansive clay.

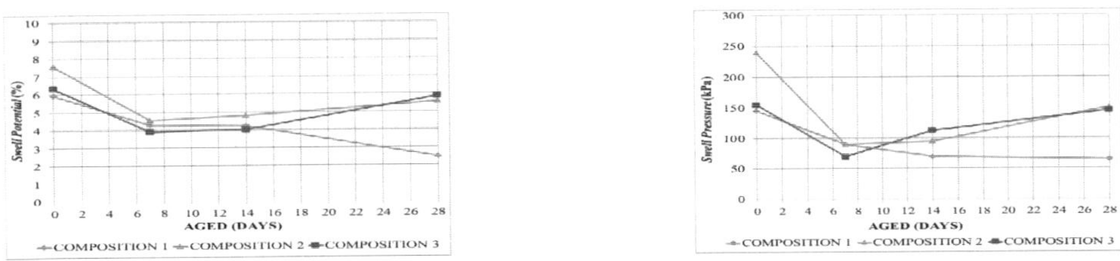

(a) (b)

Figure7. Comparison of potential expansion charts (a) and expansion pressure comparison charts (b) at 0, 7, 14, and 28 days of aging expansive clay soils are stabilized using composition 1, composition 2, and composition 3.

4. Conclusions

1. The addition of White Soil and H_2SO_4 has an effect on the physical and mechanical properties of stabilized expansive clay. The effects are making the soil inactive because the activity is smaller than 0.75, with high plasticity and is cohesive. The level of land expansion is still very high. However, the limit of plasticity and its expansion rate is 17.1% less than that of non-stabilized expansive clay. There is an increase in compressive and shear strength of 43.82%. The duration of aging affects the values of q_u and C_u, in that the longer the aging, the more increase in q_u and C_u found. There is an increase in CBR values of 83.25% in unsoaked condition and unsignificant decrease is 7.4% in soaked condition. The duration of soil aging is found to be proportional to the increase in CBR values, both in soaked and unsoaked conditions. There is a decrease in swell potential between 82.16% and 45.77%, with a decrease in comparable swell pressure in between 90.71% and 65.71% respectively. Longer duration of aging tends to lower swell potential and pressure.
2. The composition that is most suitable to increase CBR value, compressive strength, and expansive clay shear strength, and is able to reduce the percentage of expansion is Composition 1, which is made of expansive clay soil mixed with 3% white soil and 5% H_2SO_4 + Water solution (OMC condition). CBR values increased by 83.23% unsoaked and unsignificant decrease by 72.45% soaked, compressive strength and shear strength increased by 43.82%, swell potential decreased by 82.16%, and swell pressure decreased by 90.71%.

Acknowledgments
The author would like to thank the Master of Civil Engineering Program, Department of Infrastructure Engineering Management, Universitas Diponegoro who facilitated this research, the supervisors who have provided consultation for this research, technicians of the Land Mechanics Laboratory, Civil Engineering Department, Faculty of Engineering, Universitas Diponegoro for their assistance in many testing processes, fellow classmates, and the Regional Research and Development Center of Denpasar Region II for supporting this research.

References

[1] HardiyatmoHC2006Mekanika Tanah I, 4th Ed., Gadjah Mada University Press Yogyakarta.
[2] InglesOGdan Metcalf J B 1972*Soil Stabilization Principles and Practic Butterworth* Sydney, Melbourne – Brisbane ASTM International2000*Standard Test Methods for Liquid Limit, Plastic Limit, and Plasticity Index of Soils* Copyright ©ASTM D 4318 – 00, 100 Barr Harbor Drive, West Conshohocken, PA 1928-2959 United States.
[3] Thompson, M. R. (1968). "*The Significance of Soil Properties in Lime-Soil Stabilization*". Civil Engineering Studies, Highway Engineering Series No. 13, University of Illinois.
[4] Chen, F.H. (1988). Foundations on Expansive Soils. Developments in Geotechnical Engineering 54. Elsevier Amsterdam-Oxford-New York-Tokyo. Chen & Associates, 96 South Zuni, Denver, CO 80223, U.S.A.

[5] Integrated Laboratory of Universitas Diponegoro 2017 Hasil Uji Kandungan Kimia Tanah Putih Semarang.

[6] Prabandiyani SHardiyati S Muhrozi dan Pardoyo B2015Stabilisasi Tanah Lempung dengan Mengunakan Larutan Asam Sulfat (H_2SO_4) pada Tanah Dasar di Daerah Godong - Purwodadi Km 50 Kabupaten GroboganJurnal MKTSJurnal Ilmu dan Terapan Bidang Teknik Sipil Volume 21, No 1 pp 13-22.

[7] Sutikno dan DamiantoB2009Stabilisasi Tanah Ekspansif dengan Penambahan Kapur (Lime): Aplikasi pada Pekerjaan Timbunan Jurnal Teknik Sipil dan Perencanaan, Number2, Volume 11, pp 101 – 108.

[8] ChenFH1975*Foundation on Expansive Soils*Developmentsin Geotechnical Engineering, Elsevier Scientific Publication Company, New York, USA.

[9] Muhunthan B1991 Liquid Limit and Surface Area of Clays Geotechniques, Vol XLI, No 1

[10] Laboratorium Mekanika Tanah UNDIP2017 Hasil Uji *OMC* Standar Proctor, *UCS*, *CBRSwelling sample disturb* dan *undisturb* pada tanah lempung ekspansif ex. Godong KM. 49, Grobogan Semarang.

[11] SapazB2004Lateral Versus Vertical Swell Pressure in Expansive SoilsEtd.lib.metu.edu/upload/1053040/index.

[12] SudjiantoAT2015Tanah Ekspansif (Karakteristik dan Pengukuran Perubahan Volume)Graha Ilmu Yogyakarta.

[13] ASTM International. (2000). *Standard Test Methods for Liquid Limit, Plastic Limit, and Plasticity Index of Soils*. Copyright ©ASTMD 4318 – 00, 100 Barr Harbor Drive, West Conshohocken, PA 1928-2959. United States.

[14] Jumikis AR1962 Soil Mechanics. D Van Nostrand, NJSeed H B Woodward R J and Lundgren R 1962 Prediction of Swelling Potential for Compacted Clays Journal of the Soil Mechanics and Foundation Division, ASCE, vol 88, no SM3, Proc. Paper 3169.

[15] SnethenDR1975A Review of Engineering Experiences with Expansive Soils in Highway SubgradesReport No. FHWA-RD-75-48, Federal Highway Administration Offices of Research & Development, Washington, D.C. 20590 USA.

[16] Skempton A W1953 *The Colloidal Activity of Clays*Poceeding 3rd Int. Conference Soil Mechanics Found. Eng., Switzerland SnethenDR1975, A Review of Engineering Experiences with Expansive Soils in Highway Subgrades, Report NoFHWA-RD- 5-48, Federal Highway Administration Offices of Research & Development, Washington, D.C. 20590, USADasBM1998 Mekanika Tanah (Prinsip-Prinsip Rekayasa Geoteknis), Erlangga, Jakarta GourlyCSNewill D and Schreiner HD1993, Expansive Soil,TRL"sResearch Strategy, Proc. Ist Inc. Symp. Engineering Characteristics of Acid Soils, London, UK.

[17] Nelson J D and Miller D J 1992 Expansive soils—problems and practice in foundation and pavement engineering International Journal for Numerical and Analitical Methods in Geomechanics. Department of Civil Engineering, Colorado State University, Wiley, New York, 259 pp ISBN 0 471 51186 2.

[18] ASTM International2014*Standard Test Method for CBR (California Bearing Ratio) of Laboratory-Compacted Soils* Copyright © ASTM D 1883 – 99, 100 Barr Harbor Drive, PO Box C700, West Conshohocken, PA 19428-2959, United StatesLaboratorium Mekanika Tanah UNDIP2017Hasil Uji OMC Standar Proctor, UCS, CBR Swelling sample disturb dan undisturb pada tanah lempung ekspansif ex. Godong KM. 49, Grobogan.

[19] Seed, H. B., Woodward, R. J., and Lundgren, R. (1962). "*Prediction of Swelling Potential for Compacted Clays*". Journal of the Soil Mechanics and Foundation Division, ASCE, vol. 88, no. SM3, Proc. Paper 3169.

[20] ASTM International.(1996). *Standard Test Methods for One-Dimensional Swell or Settlement Potential of Cohesive Soils*. Copyright ©ASTM D4546 – 96, Committee on Standard, 100 Barr Harbor Drive, West Conshohocken, PA 19428. United States.

IOP Conf. Series: Earth and Environmental Science **328** (2019) 012023 doi:10.1088/1755-1315/328/1/012023

Analysis of Car Accident at the Location of Black-Spot and Rating for Accident-Prone Roads in Surabaya

I W Agustin [1]

[1] Department of Urban and Regional Planning, Brawijaya University, Jl. MT. Haryono 167, Malang City, East Java, Indonesia, 65145

*email: immasaitama@ub.ac.id

Abstract. Surabaya is a city that has a relatively high accident rate. As the capital of East Java Province, Surabaya has areas with diverse activities that attract movements. The large number of movements has the potential to cause traffic accidents on the roads in Surabaya. Since 2015 until 2017 there has been an increase in the number of accidents by 179 accidents. Cars became the type of vehicle with the second largest number of accidents after a motorcycle with a total of 6,666 accidents. One of the causes of accidents in Surabaya is the condition of roads that have holes and traffic violations. The main purpose of the research was to analyze car accidents at the location of black-spot and to compile accident-prone roads in Surabaya. The research used accident frequency method and accident level method. The results showed that Road segment that has the highest frequency value in the city of Surabaya is Ahmad Yani Street with a total percentage of 19.67 per cent.
Keywords: Car-accident, Black-spot, Surabaya

1. Introduction

According to the Law of the Republic of Indonesia No. 22 of 2009 on Road Traffic and Transportation, accidents are unexpected and unintentional incidents involving motor vehicles with other road users resulting in casualty and material loss. Traffic accidents can be classified into several categories: minor accidents resulting in damage to goods and vehicles, moderate accidents resulting in minor injuries and damage to property, and serious accidents resulting in death or serious injury. Traffic accidents can be caused by three factors including human factors, vehicle factors and road factors [1]. In the last three years, traffic accidents in Indonesia has been the third largest cause of death after coronary heart disease and tuberculosis based on WHO assessment [2]. Traffic accidents have different crash characteristics. Characteristics of accidents can be classified by type of accident, impact on the victim, time of incident, type of traffic accident involved. The number of vehicles involved is divided into three i.e. single accident is an accident that involves one vehicle, double accidents is an accident that involves two vehicles, and accidents that involves more than two vehicles [3].

The condition of the victim in a traffic accident is divided into three, namely the victim of minor injuries, serious injuries, and fatal / death [3]. Minor injury is a type of injury that does not require hospitalization in hospital. Severe injury is the condition of the victim who has a permanent defect and requires treatment for more than 30 days. The fatal injury / death is the condition of the victim who is confirmed to be killed by accident. Traffic accidents are categorized by day and hour of accident. Traffic accidents can be divided into weekdays (Monday, Tuesday, Wednesday, Thursday, and Friday) and weekends (Saturday and Sunday). The timing of the event can be determined at a certain time interval. The characteristics of traffic accidents based on the age of the driver indicate that about 20 per cent of drivers involved in the accident are drivers who are under the age of 17 years. Of course, this can be a matter of thinking given that new driver can have a driver's license at the age of 17 years, while quite a lot of drivers involved in accidents were under 17 years [4].

Some researchers analyze that 91.9 per cent of pedestrian accidents occur in built areas, but only 35.6 per cent of accidents are caused by pedestrians. On the street, the situation is different, in 60.8 per cent of the cases, pedestrians are the suspected offenders [5]. The survey results showed that the percentage of users who used their phones while driving is very high and for this reason it is necessary to introduce tools that can reduce the number of accidents caused by this type of driving [6]. The evidence indicating a connection between a single car accident and suicide is not correct, but the topic is feasible for future investigations and may lead to a much better understanding of the causes of car accidents and proposals to reduce their incidence [8].

2. Methods

This research was conducted in 13 streets in Surabaya. The selection of the location in this research is based on the number of accidents that occurred in the roads in Surabaya City. This research used two methods to identify accident-prone locations, namely: frequency method and accident level method.

2.1 Frequency Method

This method was used to determine the amount of accidents occurred in a year for each kilometer, so that the result will present which segment has the highest or lowest part of the accident rate (black-spot). The equation for calculating accident rate by accident frequency method (Accident Frequency Method) is:

$$AF = A/(L \times T) \tag{1}$$

AF = accident frequency (accident/km/year)
A = total of accident
L = segment length/road (km)
T = period (year)

2.2 Accident Level Method

This method combines the frequency of accidents with the presence of vehicles (ie, traffic volume) and is expressed as "accidents per million vehicles for crossing" or "accidents per million vehicle - miles of travel" for the section of the highway. The place is then ranked in order of crash rates. A highway system of 10,000 miles or less can use this method.

a. To point places:

$$Rsp = \frac{A(1.000.000)}{365(TV)} \tag{2}$$

b. For part of the road:

$$Rsp = \frac{A(1.000.000)}{365(TVL)} \tag{3}$$

Rsp = Accident level at one point (accident/million vehicles)
Rse = Accident level at a part of road (accident/million vehicle-mill)
A = Total accident

T = AADT in the research period (for intersection)
V = Sum of incoming volumes for all branches of intersection
L = Road length (mill)

3. Result and Discussion

This study used average daily traffic data or often abbreviated LHR to find out how many vehicles pass through the location of black-spot road segment in Surabaya.

3.1 Accident Characteristics

Based on Annual Traffic data, it can be seen that Mayjend Sungkono Road has the highest LHRT value, that is 8631.14 pcu/hour compared to 13 other study area roads. The result of the calculation of Accident Frequency Method (AF) was obtained from the calculation of the number of accidents in a given period of time divided by the length of road which is multiplied by the number of periods used in the calculation, then it was obtained that Mayjend Sungkono Road has the highest AF value with a value of 4.586 and it can be considered as the black-spot most vulnerable to accidents in Surabaya (**Table 1**).

Table 1. Accident Frequency Method (AF)

No	Name	Road length (km)	Period (Year)	LHRT (pcu/hour)	Total Accident 2014-2017	AF
1	Jalan Ahmad Yani	9,73	4	9675,23	83	2,132
2	Jalan Raya Mastrip	8,20	4	1845,83	66	2,013
3	Jalan Raya Darmo	5,11	4	5807,7	30	1,467
4	Jalan Raya Diponegoro	5,35	4	5713,65	31	1,450
5	Jalan Tambak Osowilangun	3,81	4	3621,88	43	2,819
6	Jalan Raya Jemursari	3,35	4	4084,2	24	1,793
7	Jalan Ngagel	2,87	4	4921,7	16	1,392
8	Jalan Tanjungsari	1,10	4		14	3,184
9	Jalan Raya Sememi Surabaya	1,40	4	4047,45	18	3,222
10	Jalan Kenjeran	3,11	4	4370,8	22	1,768
11	Jalan Kertajaya	2,63	4	6162,36	9	0,857
12	Jalan Dr.Ir H. Soekarno	4,50	4	6440,1	35	1,946
13	Jalan Citra Raya	3,06	4		12	0,979
14	Jalan Mayjend Sungkono	1,04	4	8631,14	19	4,586

The severity of car accident victims in Surabaya is dominated by casualties group with no injury condition, while the severity of serious injury is the group with the lowest percentage of victims. The severity of casualties in the group of minor injuries and serious injuries each year has an increased number of casualties, while the number of casualties in the group with no injuries and the group that died showed fluctuating number from 2014 to 2017 (**Figure 1**).

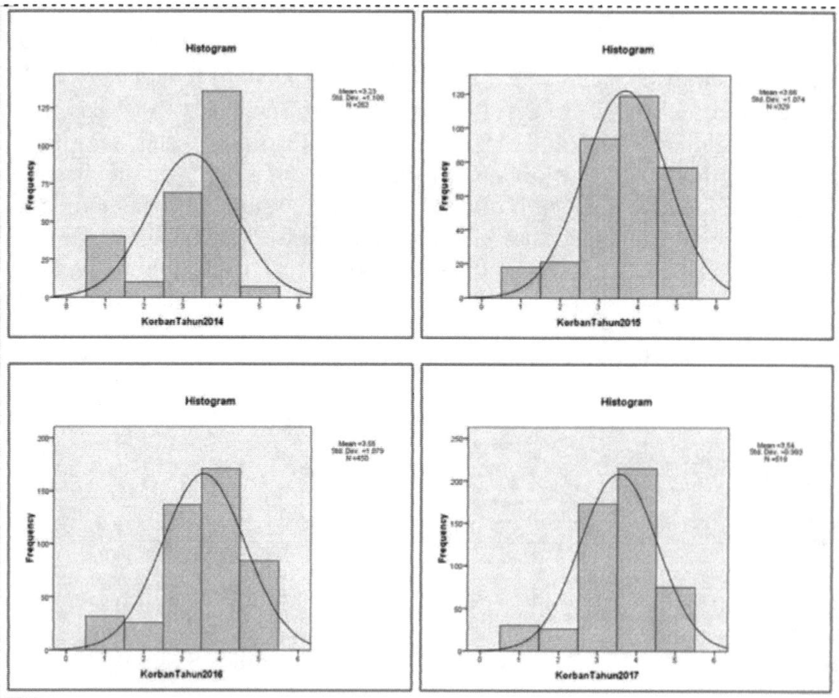

Note:
1 : Dead
2 : Serious injuries
3 : Minor injuries
4 : No injuries
5 : Unknown

Figure 1. Accident victims from 2014 to 2017

3.2 Rating of Road Accident Prone area in Surabaya

The accidents rate ranking can be determined by the Applied Equivalent (EAN) Scale approach, so that it can be known which location has the highest accident rate according to the severity level. Road segment that has the highest frequency value in the ranking table of accident-prone roads in Surabaya is at Ahmad Yani Road with a total percentage of 19.67 per cent (**Table 2**).

Table 2. Rating Road Accident Prone in Surabaya

Name		Year				Total
		2014	2015	2016	2017	
Ahmad Yani	Total Data	17	19	22	25	
	%	4,03	4,50	5,21	5,92	19,67
Raya Mastrip	Total Data	12	21	19	14	
	%	2,84	4,98	4,50	3,32	15,64
Raya Darmo	Total Data	12	10	5	3	

Name		Year				Total
		2014	2015	2016	2017	
	%	2,84	2,37	1,18	0,71	7,11
Raya Diponegoro	Total Data	11	8	6	6	
	%	2,61	1,90	1,42	1,42	7,35
Tambak Osowilangun	Total Data	8	11	18	6	
	%	1,90	2,61	4,27	1,42	10,19
Raya Jemursari	Total Data	7	6	7	4	
	%	1,66	1,42	1,66	0,95	5,69
Ngagel	Total Data	6	2	5	3	
	%	1,42	0,47	1,18	0,71	3,79
Tanjungsari	Total Data	5	2	4	3	
	%	1,18	0,47	0,95	0,71	3,32
Raya Sememi Surabaya	Total Data	4	2	7	5	
	%	0,95	0,47	1,66	1,18	4,27
Kenjeran	Total Data	4	8	5	5	
	%	0,95	1,90	1,18	1,18	5,21
Kertajaya	Total Data	4	2	2	1	
	%	0,95	0,47	0,47	0,24	2,13
Dr.Ir H. Soekarno	Total Data	5	2	10	18	
	%	1,18	0,47	2,37	4,27	8,29
Citra Raya	Total Data	1	5	6	0	
	%	0,24	1,18	1,42	0,00	2,84
Mayjend Sungkono	Total Data	3	7	5	4	
	%	0,71	1,66	1,18	0,95	4,50

4. Conclusion

The severity of car accident suspect in Surabaya City is dominated by group of suspect with minor injury condition, whereas the severity of serious injury is group with percentage of lowest suspect number. There has been an increase in the number of suspect based on the severity of minor injury group each year. The number of suspects in the group of minor injuries is in 2014 as many as 111 people, in 2015 as many as 112 people, in 2016 as many as 152 people, and in 2017 as many as 213 people. The results showed that Road segment that has the highest frequency value in the city of Surabaya is Ahmad Yani Road with a total percentage of 19.67 per cent. Specifically for the Black-spot area, things that can be conducted are installing careful warning signs at the accident-prone areas, improving and upgrading road facilities and pedestrian facilities, and making policies that can reduce the level of accidents at the site such as adding a lane to both roads, installing banners of appeal, and socializing good and proper traffic ways on the highway.

References

[1] Tahir, Anas. Studi Penyebab Kecelakaan Lalu Lintas Di Kota Surabaya. *Maktek (Majalah Ilmiah);* 2006, Vol. 8(2).

[2] Hidayati, Anisa dan Lucia Novita Henrayati. Analisis Risiko Kecelakaan Lalu Lintas Berdasar Pengetahuan, Penggunaan Jalur, Dan Kecepatan Berkendara. Jurnal Teknik Sipil; 2016, Vol. 4(2).

[3] Andriyati, Susi. Kajian Lokasi Rawan Kecelakaan Lalu Lintas di Surabaya Selatan tahun 2013. *Jurnal Teknik Sipil;* 2015, Vol. 1(1).

[4] Putri, Cahaya, Eka. Analisis Karakteristik Kecelakaan dan Faktor Penyebab Kecelakaan pada Lokasi Blackspot di Kota Kayu Agung. *Jurnal Teknik Sipil dan Lingkungan;* 2014, Vol. 2(1).

[5] Attila Glász, János Juhász. Car-pedestrian and car-cyclist accidents in Hungary, *Transportation Research Procedia;* 2017, 24 p. 474–481.

[6] G. Espinosa-Garza, I. Loera-Hernández, N. Antonyan. Process and method for the reduction of car accidents due to the while driving, *Procedia Manufacturing*; 2017, Vol. 13, p. 1360–1364.

[7] M. Pompili et al. Car accidents as a method of suicide: A comprehensive overview, *Forensic Science International;* 2012, Vol. 223, p. 1–9.

IOP Conf. Series: Earth and Environmental Science **328** (2019) 012024 doi:10.1088/1755-1315/328/1/012024

Exploring the Role of Transportation Demand Omotenashi (TDO) in the City Center Plaza of Batu

I W Agustin[1]*, R A C Adhianti[2], Fikriyah[2], N F Shakia[2], I A Maulidiah[2]

[1] Lecturer at Department of Urban and Regional Planning, Brawijaya University, Jl. MT. Haryono 167, Malang City, East Java, Indonesia, 65145.
[2]Student at Department of Urban and Regional Planning, Brawijaya University, Jl. MT. Haryono 167, Malang City, East Java, Indonesia, 65145.

*email: immasaitama@ub.ac.id

Abstract. City center plaza of Batu is one of the icons of Batu city because of its strategic location and attractive design. The large number of visitors to City center plaza of Batu makes the area have a high number of trip attractions. Therefore, it is necessary to have integrated planning, especially in the city center plaza of Batu. The concept of TDO is hospitality, familiarity, and warmth which are considered the right solution to overcome traffic problems in Batu City, especially in the city center plaza of Batu. The research used Transportation Demand Omotenashi Concept and Process Hierarchy Analysis (AHP). The results obtained in the research are in the form of strategies that will be applied to the city center plaza of Batu, namely pedestrian arrangement, pedestrianway addition, enforcement of traffic regulations, improvement of public transport services, and parking management.
Keywords: Transportation-Demand-Omotenashi, City-Center-Plaza, Batu-City

1. Introduction

Transportation problems in Indonesia arise because of an imbalance between facility transportation capacity (supply side) and the movement of people (demand side). This resulted in traffic congestion, delays, noise, and air pollution [1]. The approach that is always used by urban transportation planners is to accommodate every transportation growth in the form of increasing the capacity and efficiency of the infrastructure network system. One way that is often conducted is the implementation of management on transportation needs, also known as Transport Demand Management (TDM). The strategy that results in more efficient use of transportation resources is a general term for TDM [2]. Implementation of the TDM concept and its dependence on technology, and several previous studies have focused on definitions and policy characteristics [3,4].

TDM is one way of managing vehicle demand in the transportation system, by reducing the number of private vehicles by encouraging public transportation modes and reducing the need for expensive planned repairs, such as widening roads or new highways. Decreasing the number of vehicles can reduce emissions, increase air quality, and save energy [5]. One of the Transport Demand management (TDM) strategies is TDO (Transportation Demand Omotenashi) which uses the concept of "Omomotenashi" which is a mindset of Japanese-style prime service with the concepts of friendliness, kinship and warmth [6]. Omotenashi also means hospitality, namely the concept of hospitality in service.

There are some research related to Transportation Demand Management such as [7] examines the role of TDM policy interactions at the macro and micro level. They adopted the mode choice model of the city of Tehran. [8] explained the methodology for predicting the effects of active transportation and demand management (ATDM) measuring highway performance that can be applied using microsimulation or Highway Capacity Manual (HCM) analytical technique. [9] has hypothesized that TDM policy might offer important potential to reduce the problem of traffic congestion in Istanbul because very minimal demand-side management measures have been implemented. In this case, one goal is to investigate and test the potential of the right TDM policy that offers a solution to the problem of traffic congestion in Istanbul. Conventional economic variables are not enough to increase demand for public transport [10]. Park and Ride is one of the Transport Demand Management, the TDM scheme, which is very popular in several densely populated cities with large parking areas on the border line or suburban areas [11].

Batu City is a newly formed city in 2001 as a fraction of Malang Regency. The population of Batu City in 2017 was 203,997 people with a population density of 4,965 people / km^2 [12]. Batu City is known as one of the leading tourist cities in Indonesia because of its potential for natural beauty, and has become one of the best-selling tourist destinations in East Java Province. The city center plaza of Batu is one of the icons of Batu city, because of its strategic location (in the city center) and it has an attractive design. The large number of visitors to the city center plaza of Batu makes the area has a high number of trip attraction. This resulted in the accumulation of private vehicles, so that congestion is frequently found at several points, one of which is on Diponegoro Street. Therefore, it is necessary to have integrated planning, especially in the city center plaza of Batu. Transportation Demand Omotenashi (TDO) with the concept of hospitality, familiarity and warmth is considered the right solution to overcome traffic problems in Batu City, especially in the city center plaza of Batu. The research was conducted by taking a case study in the City Centre Plaza of Batu City. The concept of Transportation Demand Omotenashi in the research using analysis of process hierarchy (AHP).

2. Methods

2.1 Location

The research location is in Batu City which is known as one of the leading tourist cities in Indonesia because of the potential of natural beauty and become one of the best-selling tourist destinations in East Java Province. The population of Batu city is 163,393 people with a density of 806 people/km (Fig. 1).

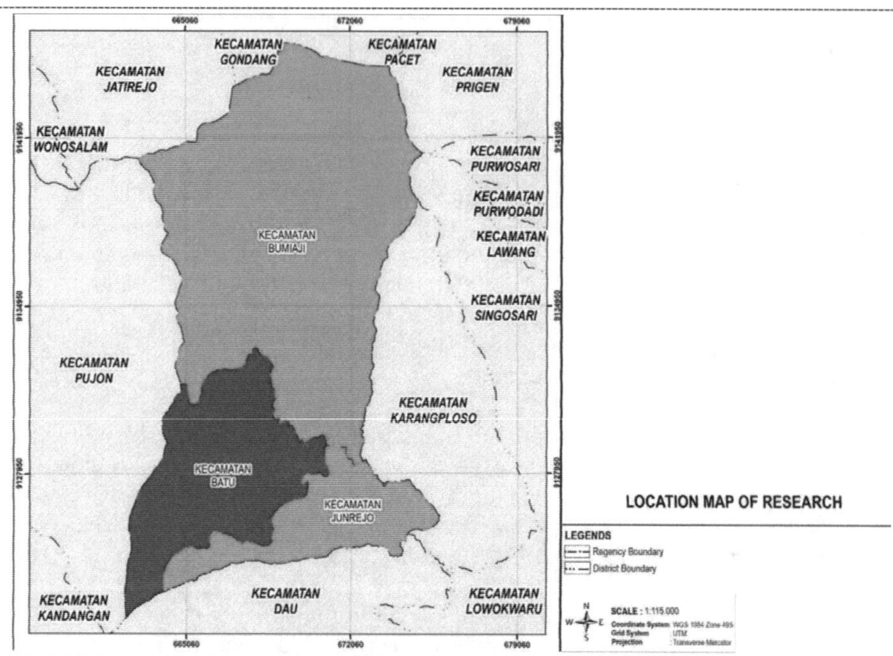

Figure 1. Research Location

2.2 Transportation Demand Omotenashi

Analysis of Transportation Demand Omotenashi (TDO) is an analysis used to assess the possibility of applying TDO in Batu City. TDO used the concept of hospitality, familiarity and warmth. The concept of hospitality is focused on the road's level of service, pedestrian way performance by using customer satisfication index. The familiarity concept focuses on parking capacity and evaluative descriptive analysis. In the concept of warmth, the analysis used is evaluative descriptive analysis [13].

2.2.1. Hospitality

A. The Road's Level of Service Analysis

The road performance analysis is carried out by measuring the road's level of service on the 29 selected road segments. The road's level of service (LoS) can be obtained from a comparison between traffic volume and road capacity [13,14].

$$\mathrm{VCR} = \frac{\mathrm{V}}{\mathrm{C}} \quad (1)$$

VCR = volume capacity ratio (service level value)
V = traffic volume (pcu/hour)
C = road capacity (pcu/hour)

B. Pedestrian-way Performance Analysis

The relationship between speed, density, and traffic flow is high, if the density of pedestrians increases, the speed of movement of pedestrians on the pedestrian-way will decrease [13,15]:

$$\upsilon = \mathrm{S} \times \mathrm{D} \quad (2)$$

υ = pedestrians flow (ped/min/ft)
S = pedestrians speed (ft/min)
D = pedestrians density (ped/ft)

C. Customer Satisfication Index Analysis

The Customer Satisfaction Index is used to determine the level of overall visitor satisfaction by looking at the importance of the product / service attributes [16].

$$\mathrm{CSI} = \frac{\sum_{i=1}^{P} \mathrm{MS_i}}{\mathrm{KS}} \mathrm{X}\ 100\% \quad (3)$$

WS_i = weigh scores
HS = maximum scale used

2.2.2. Familiarity

A. Parking Capacity Analysis

Parking capacity is how much capacity is available in the study area, at a certain time [13,14]. Parking capacity is calculated by the formula:

$$\mathrm{Capacity} = \frac{\mathrm{S}}{\mathrm{D}} \quad (4)$$

S = total official stall (stall)
D = parking duration (hour)

B. Descriptive Evaluative Analysis
Evaluative descriptive analysis aims to describe the description of the object under study and measure the success of an activity that has been carried out. Research that uses evaluative methods is a research activity that evaluates an activity / program that has a goal to measure the success of a program activity and determine the success of the program, whether it has been in line with expectations [17]. In this study evaluative descriptive analysis was used to determine the percentage of passengers who is engaged in social interaction.

2.2.3. Warmth
The warmth concept assessment uses evaluative descriptive analysis. Evaluative descriptive analysis is an analysis that aims to describe the description of the object under study and measure the success of an activity that has been carried out. In this study evaluative descriptive analysis was used to determine the physical attractiveness of the region, the social attractiveness of the region, and the historical attractiveness of the region.

2. 3 Process Hierarchy Analysis (AHP)
Process hierarchy analysis is a decision-making technique developed by Thomas L. Saaty to solve complex multi-factor problems or multi criteria into a hierarchy. In this study process hierarchy analysis was conducted to determine the priority strategies that can be done to create the TDO concept based on the variables of Hospitality, Familiarity and Warmth in the city centre plaza of Batu City (Fig. 2).

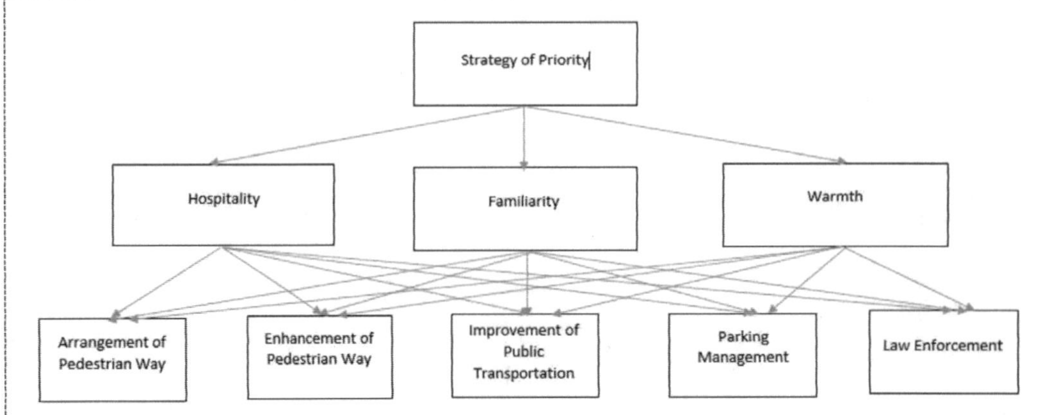

Figure 2. Process Hierarchy Analysis (AHP)

3. Result and Discussion

3.1. The Possibility of TDO in the City Centre Plaza of Batu City
An assessment of the possibility of applying the TDO in the Batu City Square was carried out by analyzing the three concepts used, namely the concepts of Hospitality, Familiarity, and Warmth.

3.1.1. Hospitality
A. The Road's Level of Services
Road performance is known after calculating the capacity and volume of traffic in the road segment under study. Analysis of road performance obtained is:

Table 1. The Road's Level of Service

Location	LoS		LoS		Hospitality Level	
	Weekday	Weekend	Weekday	Weekend	Weekday	Weekend
Ahmad Yani Road to Brantas Road	0.18	0.19	A	A	Hospitable	Hospitable
Brantas Ke Jalan Ahmad Yani	0.24	0.21	B	B	Hospitable	Hospitable
Brantas Road	0.51	0.52	C	C	Sufficient	Sufficient
Semeru Road	0.94	1.00	E	E	Unhospitable	Unhospitable
WR Supratman Road – Diponegoro Road (Right)	0.18	0.18	A	A	Hospitable	Hospitable
WR Supratman Road – Diponegoro Road (Left)	0.05	0.05	A	A	Hospitable	Hospitable
Diponegoro Road (Left to the Right)	0.34	0.42	B	B	Hospitable	Hospitable
Diponegoro Road (Right to the Left)	1.52	1.50	F	F	Unhospitable	Unhospitable
Diponegoro (Right) - WR Supratman Road	0.05	0.05	A	A	Hospitable	Hospitable
Diponegoro Road (Left) to WR Supratman Road	0.11	0.11	A	A	Hospitable	Hospitable
WR Supratman Road - Ahmad Yani Road	0.11	0.12	A	A	Hospitable	Hospitable
WR Supratman Road – Suropati Road	0.07	0.07	A	A	Hospitable	Hospitable
WR Supratman Road – Sudarno Road	0.02	0.03	A	A	Hospitable	Hospitable
Ahmad Yani Road - WR Supratman Road	0.18	0.18	A	A	Hospitable	Hospitable
Ahmad Yani Road – Suropati Road	0.09	0.10	A	A	Hospitable	Hospitable
Ahmad Yani Road – Sudarno Road	0.01	0.02	A	A	Hospitable	Hospitable
Suropati Road - WR Supratman Road	0.08	0.09	A	A	Hospitable	Hospitable

Location	LoS		LoS		Hospitality Level	
	Weekday	Weekend	Weekday	Weekend	Weekday	Weekend
Suropati Road - Ahmad Yani Road	0.07	0.08	A	A	Hospitable	Hospitable
Suropati Road – Sudarno Road	0.01	0.01	A	A	Hospitable	Hospitable
Sudarno Road - WR Supratman Road	0.07	0.07	A	A	Hospitable	Hospitable
Sudarno Road - Ahmad Yani Road	0.03	0.03	A	A	Hospitable	Hospitable
Sudarno Road – Suropati Road	0.02	0.02	A	A	Hospitable	Hospitable
Ahmad Yani Road – Brantas Road	0.19	0.22	A	B	Hospitable	Hospitable
Brantas Road – Bromo Road	0.18	0.24	A	B	Hospitable	Hospitable
Ahmad Yani Road – Bromo Road	0.21	0.27	B	B	Hospitable	Hospitable
Bromo Road – Semeru Road	0.17	0.23	A	B	Hospitable	Hospitable
Kelud Road - Semeru Road	0.11	0.12	A	A	Hospitable	Hospitable
Panglima Sudirman Road	0.80	0.99	D	E	Sufficient	Unhospitable
Pangeran Diponegoro Road	0.78	0.96	D	E	Sufficient	Unhospitable

Overall, out of 29 road segments, the city centre plaza of Batu City has 17.5% or 5 road segments that are not hospitable, and 24 road segments are hospitable for motorized vehicle drivers. Therefore, roads that are sufficient and unhospitable require transportation management strategies to prevent congestion.

B. Pedestrian Way Performance

Assessment of pedestrian way performance can show the level of performance of pedestrian facilities having a density problem.

Table 2. Pedestrian Way Performance

Location	Ratio V/C		LoS		Hospitality Level	
	Weekday	Weekend	Weekday	Weekend	Weekday	Weekend
Gajahmada Utara Road (the front of Mosque and City Centre Plaza of Batu)	1.0	1.2	A	A	Hospitable	Hospitable
Gajahmada Selatan Road (City Centre Plaza of Batu)	1.1	2.0	A	A	Hospitable	Hospitable
Sudiro Utara Road	0.7	1.2	A	A	Hospitable	Hospitable

Location	Ratio V/C		LoS		Hospitality Level	
	Weekday	Weekend	Weekday	Weekend	Weekday	Weekend
Sudiro Selatan Road (City Centre Plaza of Batu)	1.0	2.1	A	A	Hospitable	Hospitable
Kartini Utara Road (the front of Pawnshop office)	0.3	0.3	A	A	Hospitable	Hospitable
Kartini Selatan Road (the front of Slamet's Bebek Restaurant)	0.3	0.4	A	A	Hospitable	Hospitable
Ahmad Yani Road	0.9	0.9	A	A	Hospitable	Hospitable
Brantas Road	0.6	0.9	A	A	Hospitable	Hospitable
Semeru Road	0.7	0.8	A	A	Hospitable	Hospitable
Diponegoro Road	0.6	0.7	A	A	Hospitable	Hospitable
WR Supratman Road	0.7	0.8	A	A	Hospitable	Hospitable
Suropati Road	0.4	0.9	A	A	Hospitable	Hospitable
Sudarno Road	0.9	1.0	A	A	Hospitable	Hospitable
Bromo Road	0.6	0.9	A	A	Hospitable	Hospitable
Kelud Road	0.5	1.0	A	A	Hospitable	Hospitable
Munif Road	1.2	1.5	A	A	Hospitable	Hospitable
Agus Salim Road (City Centre Plaza of Batu)	0.9	1.1	A	A	Hospitable	Hospitable

The LoS A value shows pedestrians can walk freely and can determine the direction of walking at a relatively fast speed without causing interference between fellow pedestrians. The number of pedestrians who has little or large influence also, because of the effective width of the pedestrian way is not disturbed by street vendors. So, 17 pedestrian way segments fulfill the concept of hospitality.

C. Public Transport Passenger Satisfaction

Attributes to measure satisfaction level consist of five aspects, namely security, safety, affordability, and order. The following is the result of the Customer Satisfaction Index (CSI) obtained based on calculations.

Table 3. Public Transport Passenger Satisfaction

Aspect	Attribute	average value			
		Level of Importance	Level of Satisfaction	Weight Factors	Weight Score
Security	A1	4,15	3,59	7,62	0,27
	A2	4,02	1,6	7,38	0,12
	A3	4,18	2,8	7,68	0,22
	A4	3,69	2,86	6,78	0,19
Safety	B1	4,72	2,65	8,67	0,23
Comfort	C1	4,25	2,78	7,8	0,22
	C2	3,64	1,86	6,68	0,12
	C3	4,04	1,84	7,42	0,14
Affordability	D1	4,29	2,72	7,88	0,21
	D2	4,32	3,45	7,93	0,27
Regularity	E1	4,51	2,83	8,28	0,23
	E2	4,24	3,02	7,79	0,24

Aspect	Attribute	average value			
		Level of Importance	**Level of Satisfaction**	**Weight Factors**	**Weight Score**
	E3	4,41	2,88	8,1	0,23
Total		54,46	34,88	100	2,7

Based on the results of the Customer Satisfaction Index (CSI), it can be concluded that the level of customer satisfaction for service quality performance is 54%, so that customers can be said to be quite satisfied.

3.1.2. Familiarity

A. Parking Capacity

Parking capacity is the parking capacity provided by the manager. On street parking in the City Centre Plaza of Batu is located on five road segments, namely Diponegoro Road, Agus Salim Road, Munif Road, Sudiro Road, Gajah Mada Road. The location of on-street parking is located on the left side of the road as many as 10 parking lots and there are 8 units on the right side of the road (Fig. 3).

The number of on-street parking vehicles in the area of City Centre Plaza of Batu as much as 94% does not exceed the parking capacity and 6% exceeds the parking capacity. The number of parking vehicles that exceed the parking capacity is on the right side of Sudiro Road, due to the presence of dense service trade activities. In addition, on the left side of Sudiro Road there is the City Centre Plaza of Batu, so that many visitors to the City Centre Plaza of Batu their motorbikes on the on-street parking area of Sudiro Road.

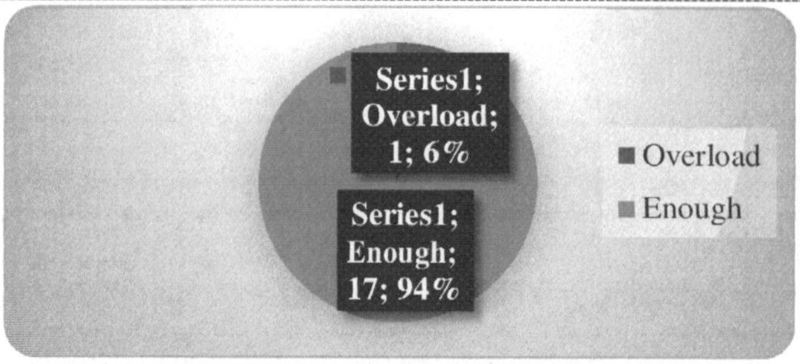

Figure 3. Parking Capacity

B. Public Transport Passenger Interaction

Social interactions that occur include two conditions, namely the existence of social contact and the occurrence of communication between passengers. Based on the results of the questionnaire, it can be seen that 68% of respondents knew new people in transportation, 60% of respondents interacted in transportation. As many as 45% of respondents boarded their own transportation, 23% of respondents boarded transportation with friends, 17% of respondents boarded transportation with children, and 15% of respondents boarded transportation with their husbands/wife.

3.1.3 Warmth

A. Regional Attraction

Ferris wheel, fountain, and apple statue are icons of the Batu City. Ferris wheel rides can be enjoyed from day to night and to ride the Ferris wheel just pay the ticket 8000 Rupiah, so it can be reached by all circles. The sparkling lights that glow the Ferris wheel and the fountain at night further add to the physical attraction of the City Centre Plaza of Batu.

B. Social Attraction
In the central area of the square there are empty spaces equipped with seating for visitors. This seat is often full of visitors who talk and rest by enjoying snacks sold on the sides of the square. Facilities in the City Centre Plaza of Batu area consist of children's playgrounds, seats, toilets and information rooms.

C. Historical Attraction
The historical attraction of the City Centre Plaza of Batu area is located in the 1967 Sticky Rice Post. This sticky post has its own historical appeal for it has been around since 1967. The building has been modest and has not changed until now, but it has always been filled with the visitor. In addition, on all sides of the square there are also locations that become tourist attractions, Laron Market (Senggol), Batu Plaza, Square Culinary Tourism, and An-Nuur Mosque.

3.2. Strategy for Application of TDO in the City Centre Plaza of Batu
Priority strategies used in implementing the TDO strategy in the City Centre Plaza of Batu area, namely the arrangement of pedestrian ways, the development of pedestrian way, enforcement of traffic regulations, improvement of public transport services, and parking management.

Based on AHP calculations, the concept of TDO can be applied in the study area but on the condition that it is able to overcome transportation problems in Batu City. The priorities of the TDO strategy based on the opinions of experts are as follows:

1. Arrangement of Pedestrian way (13.10%)
2. Enhancement of Pedestrian way (22.89%)
3. Law Enforcement (31.52%)
4. Improvement of Public Transportation (10.39%)
5. Parking Management (2310%)

Table 4. Process Hierarchy Analysis

Expert	Arrangement of Pedestrian way	Enhancement of Pedestrian way	Law Enforcement	Improvement of Public Transportation	Parking Management
R 1	0.11	0.22	0.28	0.08	0.31
R 2	0.12	0.23	0.30	0.12	0.22
R 3	0.17	0.20	0.36	0.11	0.16
Inconsistency	0.13	0.22	0.32	0.10	0.23

4. Conclusion

TDO is the right solution to overcome the problems of public transportation, parking and pedestrians in the City Centre Plaza of Batu, using the concept of hospitality, warmth and familiarityp which will be a variable to measure the possibility of applying the TDO strategy in the City Centre Plaza of Batu. It can be seen that the geometric of pedestrian ways found in the City Centre Plaza of Batu area does not meet the concept of hospitality, but 24 road segments and pedestrian ways in the area have met the concept of hospitality. The existence of public transportation in the area can encourage community interaction so that it meets the familiarity concept, while the parking characteristics found in the City Centre Plaza of Batu have not met the familiarity concept because most parking lots do not have parking lots and the number of vehicles exceeds the parking capacity. Therefore, the strategy that will be applied to the City Centre Plaza of Batu area is structuring rigid pedestrian ways, arrangement of pedestrian way, enforcing traffic regulations, improving public transportation services, and parking management.

References

[1] Pradana, M. Fakhruriza. 2017. *Analysis of Strategy Transportation Demand Management to Solve Traffic Congestion in Cilegon City*. Proceeding of The 7th Annual Basic Science International Conference-2017. Fakultas Teknik Sipil, Univeristras Sultan Ageng Tirtayasa.

[2] Litman, T. 2003. *"The online TDM encyclopedia: mobility management information gateway"*, Transport Policy, 10, pp. 245–249.

[3] Litman, T. Victoria. 2011. Transport Policy Institute, Retrieved March 11, Available on the Web: http://www.vtpi.org/tdm/tdm76.htm#_Toc252264028.

[4] Gatersleben, B. 2003. ''On yer bike for a healthy commute'', in: Human decision making and environmental perception: understanding and assisting human decision making in real-life settings, Hendrickx, L., Jager, W., Steg, L., Eds., pp. 161–182, University of Groningen, Groningen, Netherland.

[5] Ferguson, Erik. *Transportation Demand Management Planning, Development, and Implementation*. Journal of the American Planning Association; Auntumn 1990; 56, 4; ABI/INFORM Global p. 442.

[6] Kubota, Hisashi & Uemura, Takayuki & Kojo, Masafumi & Sakamoto, Kunihiro. (2006). *Transportation Demand Omotenashi (TDO)*. INFRASTRUCTURE PLANNING REVIEW. 23. 711-716. 10.2208/journalip.23.711.

[7] M. Habibian, M. Kermanshah. 2011. *Exploring the role of transportation demand management policies' interactions*. Scientia Iranica A, 18 (5), 1037–1044.

[8] Richard Dowling, Richard Margiotta, Harry Cohen, Alexander Skabardonis, Aaron Elias. 2011. *Methodology to Evaluate Active Transportation and Demand Management Strategies*. Procedia Social and Behavioral Sciences, 16, 751–761.

[9] Irfan Batur, Muammer Koc. 2017. *Travel Demand Management (TDM) case study for social behavioral change towards sustainable urban transportation in Istanbul*. Cities 69, 20–35.

[10] Hakim Hammadou, Aurelie Mahieux. 2014. *Transportation Demand Management In A Deprived Territory: A Case Study In The North Of France*. Transportation Research Procedia 4, 300 – 311.

[11] Erica Buchari. 2015. *Transportation demand management: a park and ride system to reduce congestion in Palembang city Indonesia*. Procedia Engineering 125, 512 – 518.

[12] Badan Pusat Statistik. 2018. *Kota Batu dalam Angka Tahun 2018.* Batu: Badan Pusat Statistik.

[13] IW, Agustin. 2015. Tingkat Keramahan dalam Kemungkinan Penerapan Transportation Demand Omotenashi (TDO) Di Kawasan Pusat Kota Malang. Jurnal Sosial Ekonomi Pekerjaan Umum. Vol. 7, No. 2. 80-158.

[14] Kementerian Pekerjaan Umum. 1997. *Manual Kapasitas Jalan Indonesia*. Jakarta: Kementerian Pekerjaan Umum.

[15] Box, P. C., & Oppenlander, Joseph C. 1976. *Manual of Traffic Engineering Studies.*Virginia: Institute of Transportation Engineers.

[16] Nurdiana, Iva. 2015. *Pengukuran Indeks Kepuasaan Pelanggan untuk Peningkatan Kualitas Layanan*. Universitas Kanjuruhan. Malang.

[17] Sugiyono. 2009. *Metode Penelitian Kuantitatif, Kualitatif, dan R&D*. Bandung: Alfabeta.

Fairness Of Intensity of Energy Consumption for Cooling Targets

P Satwiko*, V S C Widarji, A D Istiadji, and N Sekarlangit[1]

[1] Faculty of Engineering, Universitas Atma Jaya Yogyakarta, Indonesia

*email: prasasto.satwiko@uajy.ac.id

Abstract. In Indonesia, the auditing of building energy follows the procedure of SNI 6196-2011, while the whole building's performance evaluation uses the criteria of the GBCI. Both institutions' target is based on the intensity of energy consumption (IEC), of which 65% is the intensity of energy consumption for cooling (IECC). This research studied the IECCs of hotels in Indonesia, using a computer-simulation method with the Design Builder software. A virtual hotel was constructed based on two design scenarios, using (1) ordinary design and (2) green design as suggested by the GBCI. The two variables used were the hotel's geographical coordinates and altitude. The research found a significant potential for cooling energy savings if, in particular, the building's altitude was considered in the standards. Green buildings in lower altitudes have their IECC higher than those in higher altitudes. The highest IECC is 191.79 kWh/m².year for a green hotel in Medan, which is 35.71 kWh/m².year lower than GBCI standard. The lowest IECC is 85.39 kWh/m².year for a green hotel in Bandung, which is 142.11 kWh/m².year lower than GBCI standard. The average deviations of the IECCs from the standards of the GBCI and the ASEAN USAID were 76.25 kW/m2.year and 43.75 kW/m2.year, respectively.
Keywords: Intensity of energy consumption; energy savings; green building standard; geographical coordinate.

1. Introduction

Driven by global warming, climate change is a real problem worldwide, which threatens the sustainability of the Earth as a habitable planet for all human beings. At the Conference of the Parties 21, held in Paris in December 2015, the 195 countries that attended agreed to reduce carbon dioxide (CO_2) emissions from fossil-based fuels to maintain the CO_2 concentration in the atmosphere below 380 ppm and to limit the global environment temperature increase to 1.5°C [1]. Previously, the Intergovernmental Panel on Climate Change [2] assumed the global temperature increase to be between 1°C and 3.7°C in the twenty-first century, which would depend on future greenhouse gas emissions. Scientists agree that the continuing increase of global temperature promotes longer heat waves, more frequent droughts and torrential rains, as well as stronger typhoons. Reducing fossil-based fuel consumption is thus urgent.

Buildings consume a large percentage of energy. For example, in Indonesia, buildings use around 50% of the country's total energy and 70% of its electrical energy, which are higher than the energy used by its industries and means of transportation [3]. For buildings, air conditioning can consume more than 50% of their total energy consumption [4]. The increase in outdoor air temperature caused by global warming and urban heat islands might contribute to the increase in air conditioning installations to keep buildings thermally comfortable, as shown in the estimated 700 million air conditioners installed worldwide [5].

It is crucial to control buildings' energy consumption through better building designs, building energy management and audit, supported by realistic building standards. Countries have established their own institutions to assist architects and builders in constructing more energy-efficient buildings, such as the USA with its Leadership in Energy and Environmental Design (LEED) and Singapore with its Building and Construction Authority Greenmark [6]. In Indonesia, the Standar Nasional Indonesia regulates the procedure of the building energy audit through SNI 6196:2001, which is based on a research report by the Association of Southeast Asian Nations–United States Agency for International Development's (ASEAN–USAID) Building Energy Conservation Project. Another institution called the Green Building Council of Indonesia (GBCI) provides a Greenship (a rating tool that contains energy efficiency and conservation guidelines). The ASEAN–USAID and the GBCI target their energy audit according to the intensity of energy consumption (IEC), which is based on total building energy value [7]. The intensity of energy consumption for cooling (IECC) is 65% of the IEC the ratio of electrical energy consumption to air-conditioned floor area per year, kWh/m^2.year (**Error! Reference source not found.**).

Table 1. IEC and IECC of ASEAN–USAID and GBCI

No.	Classification	ASEAN–USAID	GBCI
		IEC/IECC (kWh/m^2.year)	
1	Offices	240/156	250/162.5
2	Hotels (apartments)	300/**195**	350/**227.5**
3	Shopping centres (commercial)	330/214.5	450/292.5

The IEC targets of the ASEAN–USAID and the GBCI are grouped according to the building classifications. Hotels fall under the commercial sector, contributing 3% of Indonesia's national energy consumption, with a growth rate of 8.6% a year. Indonesia hotels grew rapidly at 12.5% from 2007 to 2011, in response to the 9–13% tourism increase in the corresponding years (Indonesia–USAID *et al.*, 2015). The Indonesian government vigorously promotes its tourism industry and invites foreign investors [8], speeding up the development of hotels.

The cities' geographical positions and altitudes give them different climates and weather conditions, which affect the buildings' air-conditioning performances. A city's geographical position determines its climatic parameters, that is, air temperature, relative humidity and solar radiation. Its altitude affects its outdoor air temperature.

This paper reports the results of computer simulation-based experiments to answer two questions: (1) What is the energy performance of Indonesia's hotels according to their geographical positions and altitudes? (2) Are there any significant deviations between the hotels' IECCs and the targets of the ASEAN–USAID and the GBCI? The computer simulation also tested the effect of each building's orientation on its IECC, which could give architects a reasonable argument when choosing a certain orientation.

1.1 Theoretical review

A building's energy consumption is influenced by its characteristics, occupants' behaviour and other factors, including subjective comfort preferences [9][10], socio-demographic characteristics [11], cognitive factors [12] and cultural dimensions [9] [13]. The heating, ventilating and air conditioning (HVAC) system determines the rate of heat transfer in the building and contributes a significant percentage (up to 65%) of the building's energy consumption [14].

A warm-humid climate is usually characterised by warm outdoor air (25–32°C), high relative humidity (60–90%) and high prevalence of no wind. HVAC is commonly associated with cooling because heating is not needed in most areas. However, there are highlands in Indonesia with cooler outdoor air, below 25°C, where cooling is not really needed if indoor heat gain is insignificant. The hotels' indoor air temperature is commonly set far below 24°C to provide a stronger cooling sensation to their customers upon entry. However, since the building energy efficiency movement has become more popular, hotels have responded by setting their thermostats to 24°C, which is believed to be an acceptable common thermal comfort standard. Hotels located in areas with 24°C outdoor air temperature have an advantage to use this outdoor air; thus saving energy for cooling by letting in outdoor air. Using Equation (1), places with altitudes higher than 466 m above sea level are likely to have outdoor air temperatures lower than 24°C. The relationship between air temperature and altitude is expressed through Equation (1):

$$T = Tsl - 0.6h^{0} \quad (1)$$

Where:

T = the air temperature of a location at °C; Tsl = the average air temperature at sea level at °C of the same or similar latitude to the corresponding location; 0.6 = a constant; and h = the location's altitude (in 100 m). For Indonesia, the Tsl is 26.8°C.

2. Methods

This research used a computer-simulation method to analyse the energy performances of two digital models of a hotel, namely, ordinary and green. The ordinary hotel was constructed based on common materials, while the green hotel was constructed based on the green building principles. The software used for this research is called Design Builder 4.1.0.042. The energy performances of those two virtual hotels were simulated as if they had different geographical locations and altitudes. The two dependent variables were the hotels' world coordinates and altitudes. The independent variable was the IECC. Nine cities were sampled as representatives of Indonesia's cities and were simulated based on nine scenarios (**Table 2**).

Table 2. Nine representative cities

Coordinate group	Altitude group	City	Coordinates	Altitude above sea level (m)
North latitude (+23.5° to +0.5°)	Low land	Medan	03° 38' N 098° 38' E	20
	Medium land	Manado	01° 33' N 124° 53' E	475
	High land	Toba Samosir	02° 35' N 099° 27' E	1,550
Equator (+0.5° to -0.5°)	Low land	Pontianak	00° 05' N 109° 22' E	1
	Medium land	Gorontalo	00° 34' N 123° 05' E	250
	High land	Bukittinggi	00° 18' N 100° 22' E	925
South latitude (-0.5° to -23.5°)	Low land	Jakarta	06° 10' S 106° 49' E	46
	Medium land	Yogyakarta	07° 48' S 110° 21' E	210
	High land	Bandung	06° 57' S 107° 37' E	865

Source: google earth
Lowland: 0–200 m; medium land: 201–700 m; high land: above 700 m.

Latitude is distance from the north or south of the equator (an imaginary circle around the Earth halfway between the North Pole and the South Pole). It determines the angle of sunlight striking the building's envelope. Altitude, like elevation, is the distance above sea level.

The IEC targets of the ASEAN–USAID and the GBCI are for the whole electrical energy consumption. Since this research focused on the electrical energy consumption for air conditioning, the IEC targets were modified. Air conditioning contributes 65% of a building's whole energy consumption. Thus, the IEC targets of the ASEAN–USAID and the GBCI were modified to IECC, from 300 kWh/m^2.year and 350 kWh/m^2.year to 195 kWh/m^2.year and 227.50 kWh/m^2.year, respectively. The design differences between the two hotels and the parameters used in the calculation are respectively presented in **Table 3** and **Table 4**.

Table 3. Differences between ordinary and green hotels

Design	Ordinary hotel	Green hotel
Building coverage ratio	40.00%	40.00%
Floor area ratio	2.4	2.4
Layout	double loaded	double loaded
Window-to-wall ratio	6.67%	6.67%
Number of floors	6	6
Area of building	65 x 18 (m^2)	65 x 18 (m^2)
Height of building	22.0 m	22.0 m
Ground to floor height	4.0 m	4.0 m
Floor to floor height	3.6 m	3.6 m
Number of rooms	100	100
Material of roof	metal deck	metal deck – super insulated
Material of floors	concrete floor tiles	concrete floor tiles
Secondary skin	No	glass fibre reinforced concrete board 50%
Shading	No	concrete-plastered insulation
Window	single glazed	double glazed low-E
Window frame	Aluminium	aluminium
Wall	plastered brick	brick with insulation
Coefficient of performance, Air Conditioner	3.0	3.5

A healthy and comfortable hotel follows certain standards which consist of air temperature, humidity, and air exchange. Those standards can be used as a reference. The scenario for the healthy and comfortable hotel standards are presented in **Table 4.**

Table 4. Parameters and criteria for thermal calculation

Parameters	Criteria	
	Ground floor	1st–5th floors
Thermostat range	12–24°C	12–24°C
Humidity	50%	50%
Air speed	0.25 m/s	0.25 m/s
Air change rate	0.73 ach	0.50 ach
Wind sensitivity	0.25 ach	0.25 ach
Clothing	1.00 clo	1.00 clo
Number of people	126 persons	84 persons
People's activity	100 W	70 W
Sensible heat gain	13.99 W/m^2	11,84 W/m^2
Latent heat gain	2.00 W/m^2	2.00 W/m^2
Type of system	mixed mode	mixed mode
Efficiency	95%	95%
Lighting level	300 lux	300 lux
Hours of operation		
Weekdays	1 x 24 h	1 x 24 h
Weekend	1 x 24 h	1 x 24 h

3. Results and Discussions

Table 5 presents the results of the computer-based experiments. For all cities, the IECCs of ordinary hotels are higher than those of green hotels. Their deviations range from 21.4 kWh/m^2.year to 108.75 kWh/m^2.year, regardless of orientation. These findings show that the GBCI's design recommendation worked well in reducing energy consumption. However, ordinary hotels on high lands could have their IECCs lower than the ASEAN–USAID's and the GBCI's standards. Converting ordinary hotels to green hotels further drops their IECCs and therefore offer significant energy-saving potential. As expected, the IECCs of north–south-facing hotels are lower than those of east–west-facing ones.

Table 5. IECC from computer-based simulation

Coordinates	City	Altitude	Orientation	IECC from simulation		IECC target from standard	
				Ordinary hotels	Green hotels	ASEAN–USAID	GBCI
				kWh/m².year			
North latitude (NL)	Medan	Low land (L)	N–S	289.45	191.79	195.00	227.50
			E–W	301.65	192.90		
	Manado	Medium land (M)	N–S	278.62	185.13		
			E–W	293.33	186.67		
	Toba Samosir	High land (H)	N–S	188.22	98.09		
			E–W	200.38	99.85		
Equator (EQ)	Pontianak	Low land	N–S	264.53	180.13		
			E–W	284.79	183.18		
	Gorontalo	Medium land	N–S	215.76	169.83		
			E–W	249.41	174.68		
	Bukittinggi	High land	N–S	110.63	90.14		
			E–W	126.51	92.31		
South latitude (SL)	Jakarta	Low land	N–S	275.96	190.93		
			E–W	292.84	191.90		
	Yogyakarta	Medium land	N–S	212.98	159.98		
			E–W	238.55	161.66		
	Bandung	High land	N–S	106.79	85.39		
			E–W	123.84	88.01		

Figure 1 summarises north–south-oriented hotels in the nine representative cities. It shows the consistent phenomenon that hotels on high lands have lower IECCs than those on low lands. This phenomenon is expected because high lands have higher air temperature than low lands, which affects the energy consumption of the air conditioning system. Ordinary hotels on high lands (Samosir, Bukittinggi and Bandung) have IECCs slightly lower than the ASEAN–USAID standard. Even without adopting any green building guidelines, the hotels in those cities comply with the ASEAN–USAID's IECC standard. However, adopting those guidelines in the hotels significantly lowers their IECCs. This shows that there is still a high potential for energy savings if green building guidelines are applied in locations where the outdoor air temperature is comfortable.

Medan, Pontianak and Jakarta are on low lands. In those cities, green hotels can only reach IECCs slightly lower than the ASEAN–USAID's standard. For example, a green hotel in Medan has an IECC of 191.79 kWh/m².year, which is 1.79 kWh/m².year lower than the ASEAN–USAID's standard. This shows that adopting green building guidelines is important for buildings on low lands.

The cities' geographical positions do not seem to affect their IECCs (Figure 1). Even though all of the nine selected cities are within the tropical lines (23.5 NL 23.5 SL), they are not far from the equator. The northernmost city in the sample, Medan, lies at 03° 38' NL, and the southernmost city, Yogyakarta, lies at 07° 48' SL. Pontianak lies at 00° 05' NL and it is known as a hot city. However, the experiments found that ordinary and green hotels located on the equator have lower IECCs than those in other latitudes with corresponding altitudes. Hotels in Pontianak have IECCs lower than those in Jakarta.

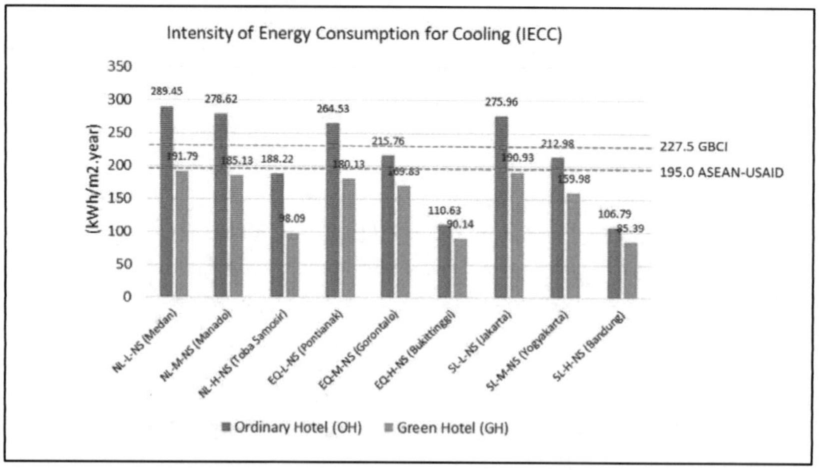

Figure 1. IECCs of hotels in the nine selected cities.

The experiments confirmed that the higher the altitude of a hotel's location is, the lower the hotel's IECC is, as higher lands have lower air temperature. Therefore, altitude has a larger effect on hotels' IECCs than geographical position (latitude). However, this finding should be interpreted carefully. It might be caused by the relatively narrow spread of the samples. The farthest sample from the equator is Yogyakarta, which is located at 07° 48' SL. Simulations with hotels located far from the equator line but still within 23° 30' NL (tropic of Cancer) and 23° 30' SL (tropic of Capricorn) should be conducted to verify whether geographical position does not affect IECC. Between the tropic of Cancer and the tropic of Capricorn, the sun passes the zenith of any location twice a year during its path, back and forth, to the north and to the south.

The simulation results show a wide range of deviations between the ordinary hotels' and the green hotels' IECCs and the ASEAN–USAID's and the GBCI's targets. The average deviation, maximum IECC and minimum IECC are calculated by using Equation (2) and presented in Table 6 (ordinary hotels) and Table 7 (green hotels).

$$d_{\bar{x}} = \frac{\sum |x_i - \bar{x}|}{n} \quad (2)$$

Where:

$d_{\bar{x}}$ = average deviation in kWh/m².year; x_i = IECC from simulation in kWh/m².year; $\bar{x}$ = IECC target in kWh/m².year; and n = sample population.

Maximum IECC = $\bar{x} + d_{\bar{x}}$, and minimum IECC = $\bar{x} - d_{\bar{x}}$.

Table 6. Ordinary hotels' IECC deviations from ASEAN–USAID's and GBCI's standards

ASEAN–USAID's IECC target	195.00 kWh/m².year
minimum IECC deviation from ASEAN–USAID's target	164.76 kWh/m².year
maximum IECC deviation from ASEAN–USAID's target	225.24 kWh/m².year
average deviation from ASEAN–USAID's target	30.24 kWh/m².year
GBCI's IECC target	227.50 kWh/m².year
minimum IECC deviation from GBCI's target	225.24 kWh/m².year
maximum IECC deviation from GBCI's target	229.76 kWh/m².year
average deviation from GBCI's target	2.26 kWh/m².year

Table 6 shows that for ordinary hotels using the ASEAN–USAID's IECC standard, the minimum and the maximum IECCs are 164.76 kWh/m².year and 225.24 kWh/m².year, respectively. With a

deviation of 30.24 kWh/m².year, the IECC for a thermally comfortable and energy-saving hotel ranges from 164.76 kWh/m².year to 195 kWh/m².year. For a thermally comfortable but not energy-saving hotel, the IECC ranges from 195.01 kWh/m².year to 225.24 kWh/m².year. The same procedure has been performed by using the GBCI's IECC standard. It is found that for a thermally comfortable and energy-saving hotel, the IECC ranges from 225.24 kWh/m².year to 227.50 kWh/m².year. For a thermally comfortable but not energy-saving hotel, the IECC ranges from 227.51 kWh/m².year to 229.76 kWh/m².year.

Table 7 shows green hotels' IECCs. Using the same procedure applied in Table 6, the green hotels' average deviation from the ASEAN–USAID's target is 43.75 kWh/m².year. Thermally comfortable and energy-saving hotels are possible with the IECC ranging from 151.25 kWh/m².year to 195 kWh/m².year. Thermally comfortable yet not energy-saving hotels are possible with the IECC ranging from 195.01 kWh/m².year to 238.75 kWh/m².year.

Table 7. Green hotels' IECC deviations from ASEAN–USAID's and GBCI's standards

ASEAN–USAID's IECC target	195.00 kWh/m².year
minimum IECC deviation from ASEAN–USAID's target	151.25 kWh/m².year
maximum IECC deviation from ASEAN–USAID's target	238.75 kWh/m².year
average deviation from ASEAN–USAID's target	43.75 kWh/m².year
GBCI's IECC target	227.50 kWh/m².year
minimum IECC deviation from GBCI's target	151.35 kWh/m².year
maximum IECC deviation from GBCI's target	303.75 kWh/m².year
average deviation from GBCI's target	76.25 kWh/m².year

Whether to adopt the ASEAN–USAID's or the GBCI's IECC standards is the hotel owner's choice. Currently, the Indonesian government does not enforce the IECC target yet even though it is urgently required. The ASEAN–USAID's IECC target is 32.5 kWh/m².year; it is lower than the GBCI's. An ordinary hotel in Gorontalo (medium land) can have its IECC comply with the GBCI's target but not with ASEAN-USAID's (see Figure 1). In other words, there is no need for hotels in Gorontalo to adopt green building guidelines if their owners refer to the GBCI. However, those hotels have significant potential for energy savings if green building guidelines are applied.

4. Conclusion

Computer simulation results found energy saving potential of green hotels.

1. Based on the results of the computer-based experiments, it can be concluded that applying a single IECC target for all hotels, which are located at different altitudes and geographical coordinates, is arguable; significant energy-saving potentials can be overlooked. Flexible IECC targets are then needed and achievable. With the current online advanced geographic information system, it is possible to easily obtain the weather data of a certain location to calculate the lowest possible IECC for green buildings. Thus, the fairness of the IECC can be achieved. The highest IECC is 191.79 kWh/m².year for a green hotel in Medan. This is 3.21 kWh/m².year and 35.71 kWh/m².year lower than, respectively, ASEAN-USAID and GBCI standard. The lowest IECC is 85.39 kWh/m².year for a green hotel in Bandung. This is 109.61 kWh/m².year and 142.11 kWh/m².year lower than, respectively, ASEAN-USAID and GBCI standard.
2. The average deviation in green building is 43.75 kWh/m².year based on ASEAN-USAID standard and 76.25 kWh/m².year based on the GBCI standard.

Further studies are needed to elaborate on the research conclusion that provides practical suggestions for more flexible and applicable IECC targets. Those studies should consider the following:

1. This research cannot firmly conclude the effect of buildings' latitude to the IECC yet. This is because the latitude of the samples only ranges from 03° 38' N (Medan) to 07° 48' S

(Yogyakarta). This is the range of hotels' latitude found in Indonesia. Further research should expand the latitude to 23°30' N and 23°30' S.

2. This research did not take each hotel's rank into account. The hotels might have a wide range of internal load that reflects their rank from one to five stars.

Acknowledgement
This research has been made possible by the support of the Master of Architecture Program, Universitas Atma Jaya Yogyakarta, which has provided all the required hardware and software.

References

[1] United Nations Framework Convention on Climate Change, "Report of the Conference of the Parties on its twenty-first session, held in Paris from 30 November to 13 December 2015," 2015.
[2] IPCC, "Climate Change 2013: The Physical Science Basis. Contribution of Working Group I to the Fifth Assessment Report of the Intergovernmental Panel on Climate Change," *Clim. Chang. 2013 Phys. Sci. Basis. Contrib. Work. Gr. I to Fifth Assess. Rep. Intergov. Panel Clim. Chang.*, p. 33, 2013.
[3] Indonesia USAID, B. Kencana, I. Agustina, R. Panjaitan, and T. Sulistiyanto, "Panduan Praktis Penghematan Energi di Hotel," p. 110, 2015.
[4] K. J. Chua, S. K. Chou, W. M. Yang, and J. Yan, "Achieving better energy-efficient air conditioning - A review of technologies and strategies," *Applied Energy*, vol. 104. pp. 87–104, 2013.
[5] L. W. . b L. W. Davis and P. J. . b P. J. Gertler, "Contribution of air conditioning adoption to future energy use under global warming," *Proc. Natl. Acad. Sci. U. S. A.*, vol. 112, no. 19, pp. 5962–5967, 2015.
[6] A. Ahankoob, S. R. Morshedi.E, and K. G. Rad, "A comprehensive comparison between LEED and BCA Green Mark as Green Building assessment tools," *Int. J. Eng. Sci.*, vol. 2, no. 7, pp. 31–38, 2013.
[7] BSNI, "Standar Nasional Indonesia (SNI) Prosedur Audit Energi Pada Bangunan Gedung," 2011.
[8] D. Prasetya and M. Rani, "PENGEMBANGAN POTENSI PARIWISATA Kabupaten Sumenep, Madura, Jawa Timur (Studi Kasus: Pantai Lombang)," *J. Polit. Muda*, vol. 3, no. 3, pp. 412–421, 2014.
[9] J. F. Busch, "A Tale Of Two Populations: Thermal Comfort In Airconditioned And Naturally Ventilated Offices In Thailand," *Energy Build.*, vol. 18, no. 3–4, pp. 235–249, 1992.
[10] N. H. Wong, H. Feriadi, P. Y. Lim, K. W. Tham, C. Sekhar, and K. W. Cheong, "Thermal comfort evaluation of naturally ventilated public housing in Singapore," *Build. Environ.*, vol. 37, no. 12, pp. 1267–1277, 2002.
[11] T. J. M. Camilleri, N. Isaacs, A. Pollard, A. Stoeklien, J. Tries, *Energy Use in New Zealand Households. Report on Aspects of Year Four of the Household Energy End-use Project (HEEP), BRANZ*, vol. 122. 2003.
[12] L. Lutzenhiser, "A question of control: alternative patterns of room air-conditioner use," *Energy Build.*, vol. 18, no. 3–4, pp. 193–200, 1992.
[13] G. Prins, "On condis and coolth," *Energy Build.*, vol. 18, no. 3–4, pp. 251–258, 1992.
[14] M. Hart and R. De Dear, "Weather Sensitivity In Household Appliance Energy End-Use," *Energy Build.*, vol. 36, no. 2, pp. 161–174, 2004.

ICT-Based Innovation in the Smart City Masterplan and Its Relation to Regional Planning

R Rachmawati[1*]
[1] Department of Development Geography, Faculty of Geography, Universitas Gadjah Mada, Indonesia

*email: rinirachma@ugm.ac.id

Abstract. Masterplan is one of the planning models implemented in Indonesia, including the smart city masterplan. In contrast to the characteristics of the masterplan in general that focuses on products on the physical/spatial the design of smart city masterplan, it is more oriented to strategic planning as an effort to achieve smart city goals. To support this, the smart city masterplan is dominated by ICT-based programs. This paper describes the kinds of smart innovations and its relation with regional planning. The study area consists of two districts which are included in the 100 smart city movements in Indonesia in 2018, namely Blora Regency and Kendal Regency. Commonly smart cities are emphasized in city, but this paper discusses its implementation in regency.
Keywords: ICT, Smart City, Masterplan, Regional Planning

1. Introduction

Indonesia is currently active in implementing various programs and activities to make smart cities come true. One of the programs is the Movement of 100 Smart Cities in Indonesia. It was initiated by the Ministry of Communication and Informatics. There are 100 cities and regencies altogether that are prepared to be smart cities. One of the steps to realize smart cities is by designing the Masterplan of Smart City. Therefore, not only cities are encouraged to be involved in the Movement of 100 Indonesian Smart Cities, but regencies are also encouraged to do so. In relation to the regional characteristics, regencies consist of city areas and mostly rural areas. Therefore, the priority of the development and implementation of a smart city is certainly different.

Smart city must include six dimensions to be called smart, namely smart governance, smart people, smart economy, smart mobility, smart living and smart environment [1]. Meanwhile, the design of Masterplan of Smart Cities in the Movement of 100 Indonesian Smart Cities that is coordinated by the Ministry of Communication and Informatics, the dimension of being smart must include smart governance, smart society, smart economy, smart branding, smart living, and smart environment [2].

In contrast to the characteristics of the masterplan in general that the focus on products on the physical /spatial (city seen physically) and in the process containing problem seeking-programming-designing (action), the design of smart city masterplan is more oriented to strategic planning as an effort to achieve smart city goals. To support this, the smart city masterplan is dominated by ICT-based programs.

This paper is aimed at describing ICT-based applications forwarded by regencies. Especially from Blora and Kendal Regency. Both are included in the Movement of 100 Indonesian Smart Cities. Further discussion on ICT-based applications is related to the need of region. This paper also presents the result of identification related to the applications mostly needed and applied in supporting smart regency and regional planning as well.

2. Result and Discussions

2.1 The Concept of Smart City and its Implementation

A smart city can be defined as a city with well-planned performance designed for future period in economic field, society, administration, mobility, environment, and other aspects of life. It is built based on smart combination of endowment and activities of its subjects that are self-reliant, aware, and self-conscious [3]. It is also the one that tries to make itself smarter with the indications of being more efficient, sustainable, fair, and proper to live in [3]. In studying and selecting the concept and designing a smart city, it is necessary to adapt it with the formulated vision and missions of a city. Therefore, it is possible that the implementation of smart city will be different from one city to another. There are several basic components to be a smart city [4]. They are, among others, 1) a city holistic horizon, determining parameter of Smart City for a city, region, or clusters. Smart city takes holistic horizon from all people's activities in all areas, including city administration, schools, hospitals, infrastructure, resources, and businesses. This needs to be implemented to determine which purpose to achieve. 2) Information and Communication Technology. The use of technology to improve the rate of liveability, workability, and sustainability in a city is to connect and integrate various components of one administration to another (G2G), citizens (G2C), and businesses as well (G2B). The purpose for the government is to obtain the latest evaluation on how a city can function well at any time, prepare technology to anticipate and solve an incident. 3) Data Collection, Dissemination, and Analysis. The use of connected system is to collect data and analyse the present condition of a city. Therefore, it is necessary to have hardware like smart meter or traffic censor to obtain data, and related software to run and filter information. After all information is collected, the data from the censor are communicated through cable or wireless network. The data are then analysed to find out what is happening at present and to anticipate what may happen next.

Several important steps need to be taken to implement to be a smart city [5]. They are; studying the concept and adapting to the city condition, identifying the need and priority of city problems, the smoothness of city operation, and involvement of stakeholders in developing smart city. In relation to identifying the need and priority of city problem, which problems are considered urgent and serious in a city and regency and that it is able to implement in a smart way and therefore they can be identified specifically. That way, a city or regency is expected to be able to focus on solving the problem. In solving a city problem, it is expected to do the planning and implementation of the concept of smart city. After that, it is necessary to search stakeholders to realize those plans. This is a good way to make a well-planned governance. A city is impossible to do all the implementations, operations, and maintenance of city components as it takes a complicated analysing ability. Meanwhile, there are four models for the involvement of stakeholders in developing smart city; First, Building Own model, developing and operating by itself, meaning that city government becomes the main contractor in providing smart city services so that the operation and maintenance of the city is under the control of the city government. Second, Transfer model. In this case, in developing and operating a city, the government let private sectors develop and operate the infrastructure of smart city. Third, Open Business model, meaning that the city government makes a bid for some projects to private sectors. In this case, the city government involves firms that meet the requirements to develop city infrastructure and provide city services under the guidance and regulations designated by the city government. Fourth, Public Private Partnership model. In this case, a city is cooperating totally with private sectors in developing smart city, starting with planning, implementing, and evaluating. Start-up project is implemented to initiate the development of smart city, starting with small scale to large one to avoid the failure of project in developing smart city.

2.2 The Use of ICT and Applications to Support Smart Cities

Smart city is related to spaces for living together among the people who make use of technology, considering that the condition is changing fast as well as the aspects of economic, social, and environmental sustainability [6]. A smart city can also be defined as the one where information and communication technology (ICT) strengthens the freedom to speak and accessibility to public information services [3].

Various innovations in using ICT have been operated to support the application of a smart city. There have been many countries that are successful in implementing smart city. Smart cities do not only deal with the use and development of technology, but they also deal with how the people in a city manage their environment in a smarter way. This is a part of smart city itself. However, it is inevitable that technological aspect also plays an important role in applying smart city.

An example of applying an application is the Smart City Masterplan of Kendal Regency [7]; this is Smart Neighbourhood Security Program. It is a program intended to encourage people to involve themselves in keeping public order, people protection e.g. an obligation to report if a family has some guests over for at least 24 hours, Panic Button (ambulance, natural disaster, crime acts, fire, and any other disturbance), video and audio calls, and SDK. Meanwhile, the aim of the implementation of this program is to provide facility for people to make a fast and responsive report that can be accessed in 24 hours online. The innovation and smart value of this program is a genuine one from this city, the one that is not applied in other cities. It is a program combining system/application, infrastructure, and human resources.

The second application of Kendal Regency is online health service, an android-based application as the result of the development and innovation of the application SIAGA 119. This application is really useful to answer the need of the people for health services and to give fast and real-time response, and to provide fast service for victims because of an incidence. Therefore, there are always officers on duty for 24 hours. The features and ways of health service application to work are; Google Auth for verifying account and minimizing fake information (hoax), One Push for connecting an officer at the call centre of health service online SIAGA as well as transmitting signal GPS in order that the call centre officer can rush to the scene spot accurately, reporting an incident by sing photos or other documentations accompanied with the coordinate of the one reporting it, tracking the location of the one reporting, tracking vehicles (ambulance, police, fire brigade), chatting with officers on WhatsApp, and transmitting signal GPS as well, providing information on the availability of hospital ward, providing information on the availability of blood stock at PMI (Indonesian Red Cross). The information on the closest health service to the users is also equipped with the distance information to the health service, information or news of public health, and information connected to health services in other related regencies. For further development, this application is, in fact, also useful as a reference system for mothers and children. The result of using this application is expected to be able to increase the number of people accessing health and referral services and it results in the implication of the decreasing rate of sickness and mortality, including the mortality rate of mothers and children, as well as improving the rate of life expectancy.

The following is an example of the Smart City Masterplan of the Regency of Blora [8] in which this is Online Admission System for the Information on the Availability of Beds in hospital. The information technology on this health services like hospital can be realized in the form of Hospital Management Information System called SIMRS. Since Android technology has been familiar to people's daily life, the development of SIMRS based on android is one the forms of innovation that is properly used to improve services to the hospital users. Hospitals often receive complaints from people about refusing a patient in a hospital because there is no more bed available. It is expected that the availability of beds in a hospital can be displayed and accessed by people since there has been transparency from the hospital. In addition, since there in no a special social medium from the hospital to receive complaints from people and slow response from the hospital, they often share their complaint in public social media.

Since the application available is the one that is based on android, those who have no android-based smart phone can still access health services by using queue machine manually through admission office and SMS on the information of the availability of beds in admission section. In this case, officers are required to keep updating data of inpatients on SIMRS because the information of the available beds on application has been integrated to SIMRS so that the information about the available beds can be real time.

The realization of the application of the Smart City concept includes integrating the use of ICT in various aspects of urban development [9]. This also applies to the district. Not all elements in Smart City can be implemented perfectly for the entire city, but needs to look at the conditions, potentials and problems of each [10]. Likewise the application applied to support smart city/smart regency must be adjusted to the needs of the region and OPD (regional implementing organizations)

Along with the increasing use of ICT-based services, it is predicted that in the future ICT will be a determining factor in the decentralization of economic services [11]. The use of ICT is allegedly also reducing movement [12]. Associated with ICT-based government services will also reduce population movements because services can be accessed via the internet. In addition, services can be accessed through sub-districts that are connected virtually with the government at the district level so that people do not need to go to government services in the city centre. In this case decentralization is realized in the form of services that are not only accessible at the centre of government at the district level but shift to lower levels in the sub-district.

3. Conclusion

In making masterplan of smart city, the understanding of the characteristics of the area is very important. Each city has its own characteristics, the same thing applies to regency. In terms of the difficulty in making masterplan of smart city, the main problem is synchronizing between the program and the budget in the masterplan with the Medium Term Development Plan (RPJM). The preparation of this masterplan needs to be synchronized with regional development plans. In addition, the main obstacle in implementing smart city depends on the head of the region/city as well as the human resources of the local government. The head of the region/city should have a strong motivation in planning and handling the implementation of smart city supported by good human resources staff.

In the framework of implementing Smart City in Indonesia, many ICT-based applications have been made to support various dimensions of being smart. In the future, the availability of human resources for the government offices dealing with operating such applications especially in each OPD needs to been taken care of. The digital literacy of the community is also needs improving. This is because they are also a part of stakeholders who make use of the applications available. ICT-based applications need to be designed and implemented well adapted to the needs of the region.

References

[1] Giffinger, R., Fertner, C., Khamar, H., Kalasek, R., Pichler-Milanovic, N., Meijers, E. 2007. *Smart Cities-Ranking of European Medium-Sized Cities*. Final Report.

[2] Kementerian Komunikasi dan Informatika (Kemenkominfo). 2018. *Guideline Masterplan Smart City Gerakan Menuju 100 Smart City*. Direktorat Jenderal Aplikasi Informatika.

[3] Nam and Pardo, 2011, Conceptualizing Smart City with Dimensions of Technology, People, and Institutions, The Proceedings of the 12th Annual International Conference on Digital Government Research, Center for Technology in Government, University at Albany, State University of New York, U.S., *Dg.o'11*, June 12–15, 2011, College Park, MD, USA.

[4] Wharton, K. 2016. *Smart Cities: The Economic and Social Value of Building Intelligent Urban Spaces*. Knowledge Wharton. Philadelphia

[5] Brown, J. 2015.*7 Steps on the Road to Becoming a Smart City*. Majalah Government Technology. http://www.govtech.com.

[6] Branchi, P. E., Fernandez, C., & Matias, I. R. 2014. Analysis Matrix for Smart Cities. *Future Internet*, 61-75.

IOP Conf. Series: Earth and Environmental Science **328** (2019) 012026 doi:10.1088/1755-1315/328/1/012026

[7] Kabupaten Kendal. 2018. Masterplan Smart City. Gerakan 100 *Smart City* di Indonesia-Kemenkominfo.

[8] Kabupaten Blora. 2018. Masterplan Smart City. Gerakan 100 *Smart City* di Indonesia-Kemenkominfo.

[9] Rachmawati, Hapsari, A.A., Cita, A.M. 2018. Virtual space utilization in the Digital SMEs Kampongs: Implementation of Smart City and Region. *Human Geographies – Journal of Studies and Research in Human Geography*. Vol. 12, No. 1.

[10] Rachmawati, R. 2018. Pengembangan Smart Village untuk Penguatan Smart City dan Smart Regency. *Jurnal Sistem Cerdas*. Asosiasi Prakarsa Indonesia Cerdas (APIC). Volume 01 No 02 ISSN : 2622-8254 Hal : 12 - 18

[11] Rachmawati, R., Rijanta, R., Djunaedi, A. 2015. Location Decentralization Due to The Use of Information and Communication Technology: Empirical Evidence from Yogyakarta, Indonesia. *Human Geographies-Journal of Studies and Research in Human Geography*, Volume 9, Issue 1, May.

[12] Rachmawati, R. and Rijanta, R. 2012. Population Mobility and Urban Spatial Structure: Does the Use of Information and Communication Technology Matter? *Regional View*, Japan, Number 25,March, Page 9-19.

Planning for Urban Green Area and Its Importance for Sustainability: The Case of Jakarta

R Setiowati[1]*, H S Hasibuan[1], R H Koestoer[1], and R Harmain[1]

[1] School of Environmental Science, University of Indonesia, Jl. Salemba Raya, Kenari, Senen, Jakarta, 10430, Indonesia

*email: retnodki@gmail.com

Abstract. Population growth and rapid development in Jakarta have caused a reduction in the amount of Green Open Space (GOS) in urban areas. Based on data in 2017, the public GOS managed by the Forestry Agency is only 4.65% and the private GOS can not be counted until now. Although the target of public GOS achievement is 20%, but the green plan is only 7,514.08 ha. The aim of this study is to analyze the extent and distribution of GOS and compare it with green plan. This study used the method of Geographic Information Systems to determine the extent and distribution by comparing the GOS area in year 2011 and 2018. The next step is reviewing the green plan zone according to Jakarta Spatial Planning 2030. Based on the results of this study, it can be concluded that the area of GOS in 2011 was 3,816.34 ha (5.84%) and it decreased in 2018 to 3,473.94 ha (5.31%). The distribution of GOS spreads in 5 administrative cities, with the largest area being in the East Jakarta. The result of the comparison of GOS in 2018 with the green plan shows that there is a lack of green areas of 4,040.13 ha.

1. Introduction

Most urban areas are home to more than half of the world's population [1]. More than half of the world's population lives in urban areas and by 2050 it is estimated that two out of three people live in cities in less developed countries [1]. Today many megacities are formed, namely cities with a population of more than 10 million, even up to 30 million [2]. Growing development in urban areas is often at the expense of green landscapes that are converted into concrete and asphalt buildings causing limited water absorption into the soil [1]. Population growth in urban areas has been the cause of environmental problems [3].

Sustainable development is the middle way to promote development that is adapted to the sustainability of landscape ecology [4]. Sustainability indicators for urban development must include more parameters of public space and urban green open space (GOS), as well as an index that reflects the satisfaction and perception of its citizens towards their environment [5]. Urban planners must consider green space management in a variety of ways to meet the needs and expectations of all population segments (children, families, the elderly, etc.) [5].

Green city is a city that has reached and moved towards long-term environmental sustainability [6]. In the future, city planners face difficulties in choosing between protecting green cities, urban economic growth, and social justice [7]. Greater London has become a green city where around 40–

50% of its land area is GOS and has been made using the UK Government Generalized Land Use Database [8]. The existence of small parks scattered in the city is important to form space and habitat [2].

Law Number 26 of 2007 concerning Spatial Planning mandates the availability of open space for at least 30% of the total area of the city where 20% is public GOS and 10% is private GOS. Jakarta faces the problem of land availability related to development activities that take place intensively so that the built area tends to grow rapidly. Besides the problem of land availability, the price of land in Jakarta is also one of the obstacles in the provision of GOS. In addition, there is a potential for GOS that can be developed under the authority of the Central Government.

The masterplan covers local area authority and regulates land use and infrastructure patterns throughout the region through zoning or land allocation maps where the plan regulates all or part of the local area authority and shows zoning of specific location details for buildings, land use and infrastructure [9]. Based on Regional Regulation Number 1 of 2008 concerning the Regional Medium-Term Development Plan (RPJMD) in 2007-2012, Jakarta was directed to compile the GOS Masterplan. However, until now Jakarta does not have the masterplan.

The topic of the problem of GOS lately has become important because along with the increasing population, the number of available GOS is also reduced. At present the problem of GOS limited area becomes the most urgent issue in urban planning, especially in Jakarta. Based on data in 2017, the public GOS managed by the Forestry Agency is only 4.65% (3,080.89 ha). Although the target of public GOS achievement in Jakarta is 20%, but the green plan zoning stipulated in Regional Regulation 1 of 2012 is only 11.7% (7,749.36 ha).

The study aims to analyse the need of GOS's planning for urban sustainability. The objective of this study is to investigate the planning document stipulated the GOS Masterplan, about how the plan of GOS in Jakarta is. Based on the background that has been described before, the research problems is, how many is the number and how is the distribution of GOS in Jakarta?

2. Methods

2.1. Study area

Jakarta Province is a special area that functions as the Capital of the Republic of Indonesia and at the same time as an autonomous region at the provincial level.Geographically, the total area of Jakarta is 7,660 km^2, with a land area of 662 km^2 (including 110 islands scattered in the Kepulauan Seribu) and an area of the ocean of 6,998 km^2. In terms of government administration, Jakarta Province is divided into 5 (five) administrative cities and 1 (one) administrative district namely South Jakarta, East Jakarta, Central Jakarta, West Jakarta, North Jakarta, and Kepulauan Seribu Regency. In detail, the administrative boundary map of Jakarta Province can be seen in **Figure 1**.

2.2. Method

The method of this study used Geographic Information Systems (GIS) to determine the extent and the distribution by comparing the GOS area in year 2011 and 2018. The next step is comparing and reviewing the green plan zone according to Jakarta Spatial Planning 2030. The framework of this study can be seen in **Figure 2**.

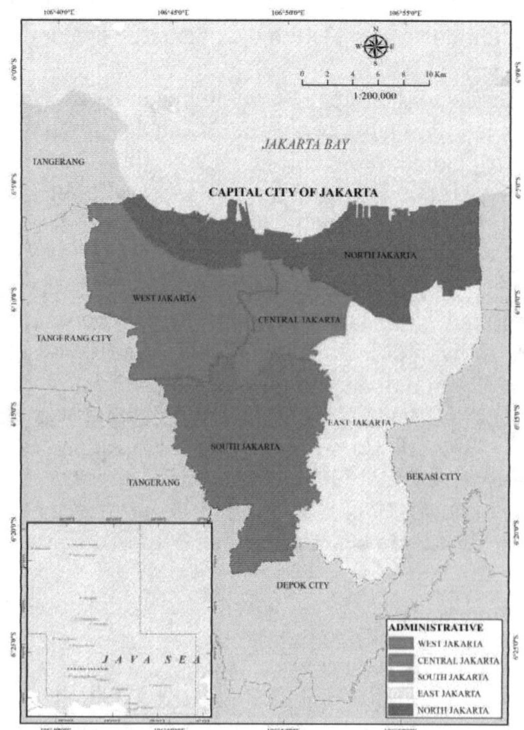

Figure 1. Map of administration of Jakarta Province

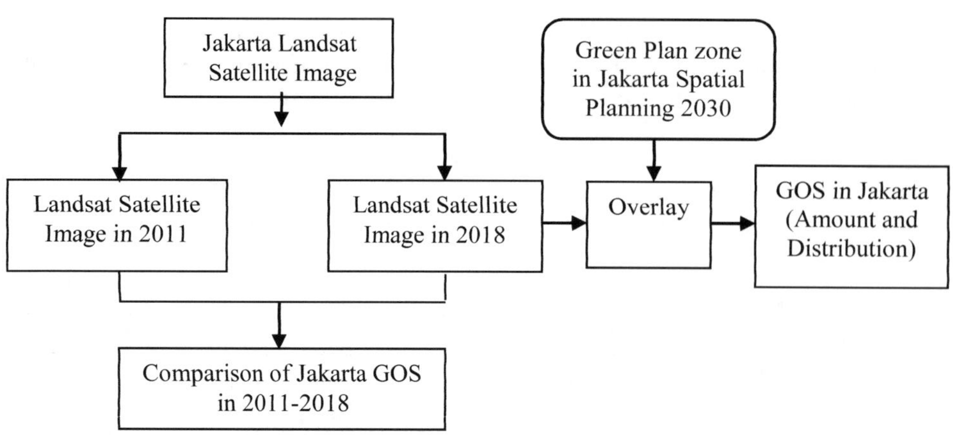

Figure 2. Framework of this study

3. Result and Discussions

3.1. Land Use/Cover Change (Built-Up Area Development)

Land use change is generally affected by both natural factors, such as geographical and soil characteristics, and socioeconomic factors, such as population growth, development plans, land use control, zoning and other related laws [13]. From 1995 to 2012, the built up area in Jabodetabek grew rapidly. Jakarta is dominated by built land represented by the designation of buildings, road infrastructure and other infrastructure [13].

Based on the Academic Study of Jakarta Spatial Planning 2030, satellite image interpretation provides information that around 66.62% of Jakarta's mainland area is built land, while 33.38% can be interpreted as land that has not been built, among which is still natural, such as urban forest, lane green, cemeteries, agricultural land, parks, vacant land and more.The allocation for housing occupies the largest proportion, which is around 48.41% of the main land area of Jakarta, followed by the allocation of industrial buildings, offices and trade with around 15, 68%.

3.2. Green plan based on spatial plan Jakarta 2030

Jakarta has a green plan including public and private GOS which is outlined in Regional Regulation Number 1 of 2012 concerning the Spatial Planning Jakarta 2030. The percentage of the green plan amounts to only 11.4% (7,514.08 ha). Jakarta's green plan can be seen in Figure 4. Referring to Chapter V of the General Policy Direction of Regional Development for Environmental Affairs in the Field of Environment in Regional Regulation Number 1 of 2008 concerning Regional Medium Term Development Plans (RPJMD) for 2007 - 2012, the governmentis directed to prepare GOS Masterplan. In addition to these directives, referring to Article 79 of Regional Regulation 1 of 2012, Jakarta is also directed to make Governor's regulation on GOS and its development. Green plan zoning can be seen in figure 2.

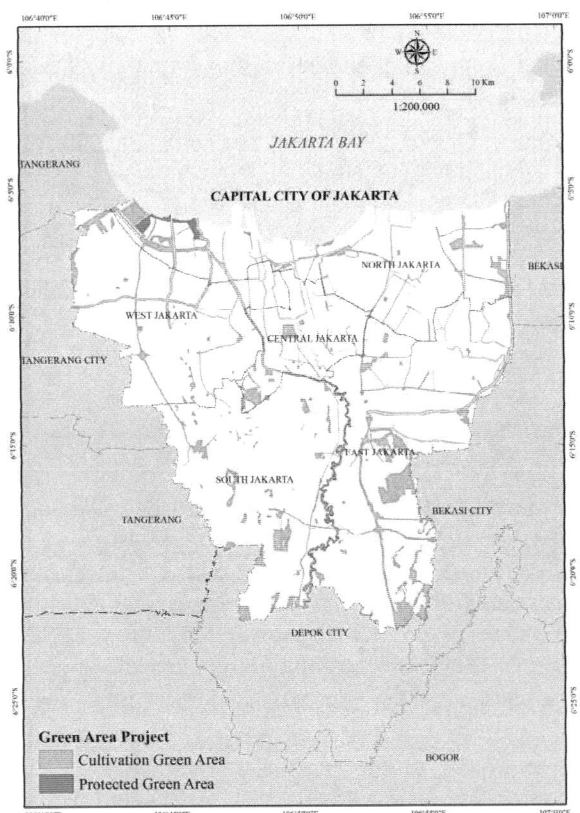

Figure 3. Green plan in Jakarta
Source: Regional Regulation No. 1 of 2012 [10]

Although the drafting of the GOS Masterplan has been mandated according to the regulation, until now the government does not have the masterplan nor the Governor's regulation on GOS and its development. The absence of the GOS Masterplan has caused not to have applicable and measurable long-term guidelines and plans in achieving the 30% GOS target by 2030. Based on Law Number 26 of 2007 concerning Spatial Planning, the provision of GOS based on the area in urban areas is a minimum of 30% consisting of 20% of public GOS and 10% of privat GOS. The government also targets the same amount of GOS achievement as Act No. 26 of 2007 as outlined in Part Two of Spatial Planning Policy Article 6 paragraph (5) letter b of Regional Regulation 1 of 2012 concerning Regional Spatial Planning (RTRW) Jakarta 2030. Although the target of public GOS achievement in Jakarta is 20%, but the green plan zoning stipulated in Regional Regulation 1 of 2012 is only 11.4% (7,514.08 ha).

3.3 GOS in Jakarta

Jakarta has no definite data related to the number of public owned and managed GOS. The government also does not have GOS Private achievement data in Jakarta. There are differences in the number of GOS by several units in Jakarta. Data from Forestry Agency in 2017 stated that data on the management of Public GOS in Jakarta up to semester 1 of 2017 was only 4.65%. This is different from the data held by Environmental Agency in 2017 stating that the achievement of Jakarta's GOS in 2015 was 11.97%. Meanwhile, based on the data from Public Works, Spatial and Land Affairs Agency, the number of GOS in Jakarta in 2018 is 7.89%. GOS achievement data by agency in Jakarta can be seen in **Table 1**.

The realization of the GOS is still far from the target of 20% of public GOS fulfillment by 2030. In addition, this difference is a problem related to the determination of the base line of the GOS achievement in Jakarta. Each agency in Jakarta has a different calculation method in calculating the number of GOS. GOS base line is needed in determining future GOS achievement targets and it must be included in the GOS masterplan.

Table 1. GOS Achievement by each agency in Jakarta

No.	Agency	Wide (ha)	Percentage (%)	Information
1.	Forestry	3,080.89	4.65	GOS Public (Year 2017)
2.	Environmental	7,927.83	11.97	GOS Public & Private (Year 2015)
3.	Public Works, Spatial and Land Affairs	5,096.69	7.89	GOS Public (Year 2018)

Source: Researcher, 2018

Based on the results of the GIS analysis,the width of GOS in Jakarta in 2011 and 2018 can be shown in **Figure 4** and **Figure 5**. The results of comparison of land use found that there was a decrease in the number of GOS in Jakarta as many as of 342.40 ha in a period of seven years. The decrease in the number of GOS in Jakarta was mostly found in the green line as many as 255.11 ha and followed by cemetery as many as 148.71 ha. The decrease in GOS in the green line was caused by the construction of infrastructure and public facilities such as MRT, LRT, and pedestrian arrangement. In addition, there was also an addition of city/environment park zones of 44.88 ha. The details of comparisons for each of the GOS's zone in year 2011 and 2018 are shown in **Table 2**.

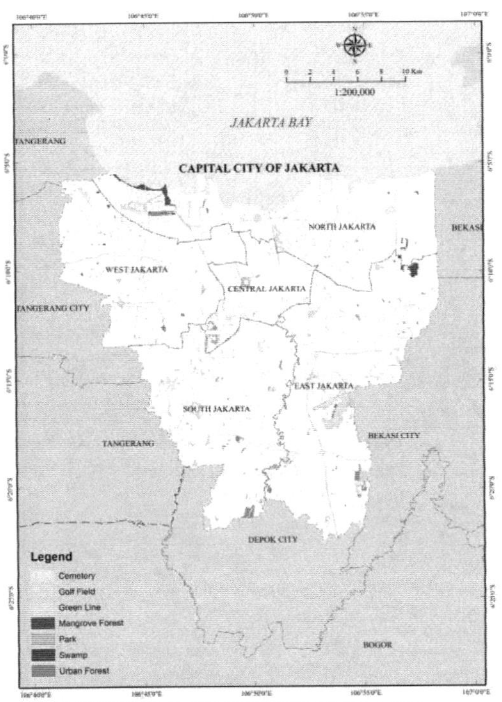

Source: Researcher, 2019

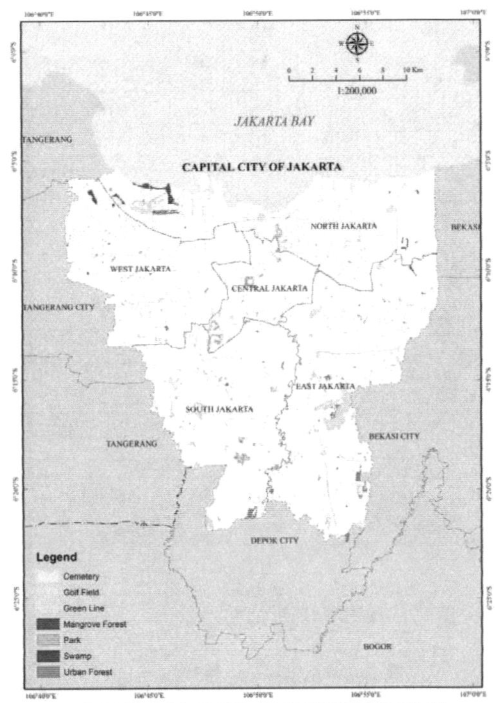

Source: Researcher (2019)

Table 2. Comparison of the 2011 and 2018 green zones

No.	Zone	Sub Zone	Code	2011 (ha)	2018 (ha)	Ratio (ha)
1.	City forest	City forest	H1	211.13	235.13	23,99
2.	City/environment park	City/environment park	H2	580.04	624.92	44.88
3.	cemetery	Cementary	H3	513.87	365.16	(148.71)
4.	Green line	Green line	H4	1,812.99	1,557.88	(255.11)
		Green high voltage	H5			
		Green railroad tracks	H6			
5.	Recreational green	Recreational green	H7	698.3	690.86	(7.44)
6.	Open green cultivation	Open green cultivation	H8			
	Deviation			3,816.34	3,473.94	(342.40)

Source: Researcher, 2019

Map of GOS in year 2018 has overlaid with a green plan map on the Jakarta Spatial Planning 2030. The comparison of the existing GOS in 2018 in Jakarta is still far from the target of the green plan which is 4,040.13 ha. The biggest difference is in the green line zone, consist of 2,974.64 ha and followed by city/environment parks with 1,069.73 ha. Urban forests in Jakarta have exceeded from the existing targets in green plan. It is found that urban forests in Jakarta have not all been included in the green plan listed on the Jakarta Spatial Planning 2030. Details of the comparison of each GOS zone in year 2018 with the green plan of the Jakarta Spatial Planning 2030 can be seen in **Table 3**.

Table 3. Comparison of the 2011 and 2018 green zones

No.	Zone	Sub Zone	Code	Green Plan (ha)	2018 (ha)	Ratio (ha)
1.	City forest	City forest	H1	17.58	235.13	217.55
2.	City/environment park	City/environment park	H2	1,694.65	624.92	(1069.73)
3.	Cemetery	Cemetery	H3	605.87	365.16	(240.71)
4.	Green line	Green line	H4	3,818.46	1,557.88	(2,260.58)
		Green high voltage	H5	445.79		(445.79)
		Green railroad tracks	H6	268.27		(268.27)
5.	Recreational green	Recreational green	H7	663.46	690.86	27.4
6.	Open green cultivation	Open green cultivation	H8			
	Deviation			7,514.08	3,473.95	4,040.13

Source: Researcher, 2019

4. Conclusion
The base line data on GOS is needed as a basis for making GOS masterplan. The absence of GOS masterplan has no applicable and measurable guidance and long-term plan in achieving 30% GOS target by 2030. The target of green public open space in Jakarta is 20%, but the green plan zonation stipulated in Jakarta Spatial Plan 2030 is only 11.4%. GOS in urban area is one of the important elements of the urban landscape for sustainability.

Acknowledgment
Funding for this study was provided by the grant of the Final Task Academic Writing issued by Directorate of Research and Public Services, University of Indonesia.

References
[1] Miller GT and Spoolman SE 2012*Environmental Science Fifteenth Edition.* Cengage Learning. Boston. USA
[2] Cunningham WP and Cunningham MA 2008 *Environmental Science A Global Concern Tenth Edition* (New York: Mc Graw Hill Higher Edition)
[3] Pebley AR 1998 Demography and the Environment.*Demography*, Vol. 35(4) 377-389
[4] Enger ED and Smith B F 2009*Environmental Science: A Study of Interrelationship Twelfth Edition* (New York: Mc Graw Hill Higher Edition)
[5] Chiesura A 2004*Landscape and Urban Planning* 68 129–138
[6] LindfieldM and Steinberg F 2012.*Green Cities*.Urban Development Series Asian Development Bank. Manila. Filipina
[7] Campbell, S2016 The Planner's Triangle Revisited: Sustainability and the Evolution of a Planning Ideal that Can't Stand Still *J. Am. Plan. Assoc* 82 388–397
[8] PocheeHand Johnston I2017 Understanding design scales for a range of potential green infrastructure benefits in a London Garden City*Building Serv Eng Res Technol*38728–756
[9] Albrechts L 2004 Environment and Planning B: Planning and Design 31 743-758
[10] Regional Regulation No. 1 of 2012 about Jakarta Spatial Plan 2030

Optimizing Building Planning and Design to Control External Energy Loads

R Y Sandi[1,2*] and M Ismail[2]

[1]Magister Student at School of Environmental Science of University of Indonesia.
[2]Architecture graduated of University of Hasanuddin

*email: rezayuniarsandi@yahoo.com

Abstract. In Indonesia, the building sector is responsible for 50% of the total energy consumption from all sectors and more than 70% of overall electricity consumption. The highest energy loads of the office sector is around 57% of the building and 50% of the energy loads on the building belongs to the air conditioning. This study attempted to optimize planning and design, especially through controlling the external energy loads (heating) that enter the building envelope (wall and vents). Case study was conducted in Makassar at Residential and Building Development Information Center (PIP2B Building). The method utilized OTTV (Overall Thermal Transfer Value) consisting of planning and building design analysis especially accounting for heat energy entering the building through windows and vents of the building. According to the Ministry Public works regarding OTTV, the standard for building the energy conservation was 45 W/m² but the standar was increased to 30-35 W/m². From the results of OTTV calculations, an external energy load entered into the building through the building envelope (walls and windows). The result of OTTV calculation shows that some sides of building have an increased amount of energy that went through the wider windows. Through optimizing and planning the windows area, thedecreasing external energy (heat energy) that entered from the outside is encouraged.
Keywords: Optimizatizing, Planning, Design, Energy, Building

1. Introduction

Energy needs are a very noticeable issue; therefore, some strategies are required to control energy consumptions from various sectors of energy use especially in the building sector (about 40% of total energy consumption in the world) [4, 20, 22]. In Indonesia, the building sector is responsible for 50% of energy consumption from all sectors and more than 70% of the overall electricity consumption [8, 12]. Office sector consumes about 57% of the energy, which are the highest energy loads compared with other building sector and the air conditioning's loads which is 50% of the energy loads in the office building [13]. There are energy saving programs in the building sector which was a program launched by the world [3, 7, 17, 18], that aims for the efficiency and conservation of energy use. From the building sector, the office sector has the largest energy loads of the office sector at around 57%. Therefore, it requiress control in reducing the energy loads. Energy loads is divided into 2 parts, including external loads (solar radiation that enters the building) and internal loads (human heat and equipments). There are several ways for measuring and reducing the energy consumption in building, such as: thermal insulation, sensitivity and optimization, life cycle analysis, technical and economic analysis of renovation of the existing buildings, strategies to control heating/cooling, ventilating, air-conditioning (HVAC) and lighting system [9]. This study evaluates the external loads that entered the building from the planning and design of the PIP2B office through the building envelopes with OTTV (Overall Thermal Transfer Value) analysis. The standards and procedures for planning the energy conservation in buildings and the standar of buildings thermal condition are compiled by the Ministry of Public Works, where OTTV on the walls of a buildings must not exceed 45W / m²[16], but their values are now reduced to 30-35 W/m² [10].

1.1 Optimization Planing and design

Design in architecture becomes a strategy in solving problems that uses the ability of creativity, art and science to get a solution to a problem [21]. Design involves a complex activity that produces various and varied forms from each of the different designer backgrounds [5]. The purpose of architectural design as a basic course is to build the environment and human's relationship with the environment [14]. The design of a building is expected to reduce the amount of energy consumption, along with increasing window ratio settings and visible transmission with a higher impact on energy consumption level [1]. The window to wall ratio (WWR) also influences the intensity of energy use (EUI) in buildings [19], especially with the fact that increased the electricity usage at around 40-50% forair conditioner loads [11]. Facade planning in every building orientation should consider more specifically on natural ventilation [15], to control the increasing external energy.

1.2 Case Study

The case study is located at the Residential and Building Development Information Center (PIP2B) in Makassar. The research location is on Jl. Batara Bima KM 16 Kec. Biringkanaya Makassar South Sulawesi. It is a business and industrial area in Makassar with a total area of 1,185.74m², ground floor area of 403.61 m² and the 2nd floor area of 324.24m². The building utilizes red brick plaster with thickness of 144 cm that is coated with shiny white paint; the roof uses zinc with red wavy pattern.

1.3 Existing Condition

Building materials have different absorption values (α) depending on its material type red brick plaster with 144 cm thickness with shiny white paint) [6], wall transmission value of 3,64 W/m², and windows transmission value of 2.94. The shading coefficient values (SC) is 0, 61 W/m², (it can be found from glass fabric), depending on the kind of glass. Solar factor (SF) depends on city that the building is located.

Table 1. The value of equivalent temperature difference

Wall Construction	Mass/area (kg/m²)	ΔTeq
Low	<125	15
Medium	126-95	12
High	>195	10

Table 1. The value of equivalent temperature difference (ΔTeq) depends on the kind of wall material and its areal density, such as: mass/area (kg/m²).

IOP Conf. Series: Earth and Environmental Science **328** (2019) 012028 doi:10.1088/1755-1315/328/1/012028

Table 2. The value of solar factor

Orientation	North	North East	East	South East	South	South West	West	North West
Solar Factor	37	150	154	125	105	176	243	211

Table 2. Solar factor value took place in Makassar city

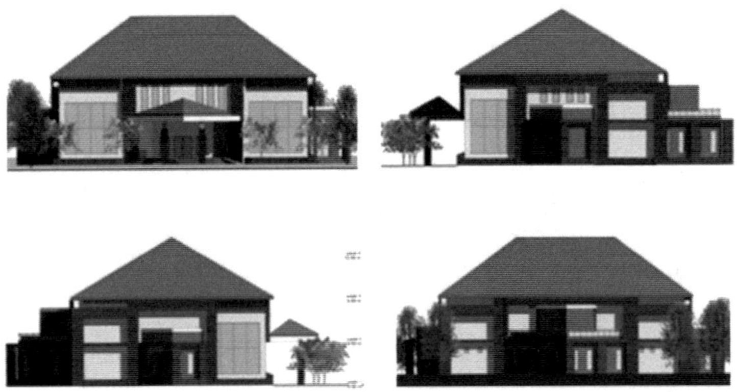

Figure 1. Each View of Building the PIP2B office

2. Methods

The research method applies the OTTV analysis (Overall Thermal Transfer Value). It is a method of which the (quantitative) calculation is carried out by theoretically determining the amount of heat load that will enter through the building envelope (Heryanto, 2004). The OTTV formula that was used to calculate the heat entering through wall and glass conduction and glass radiation is as follows:

$$OTTV = (Wall\ Conduction + Glass Conduction + Glass\ Radiation) \quad (1)$$

$OTTV$: $\alpha\{U(1-WWR)\}.\Delta Teq + (Uf.WWR.\Delta T) + (SC).(WWR).(SF)$

OTTV : Total heat transfer value on the outer wall at a certain orientation, W / m²

α : Absorption of solar radiation on the surface of the wall

Uw : Wall transmission, W/m²degC

Ug : Window transmission, W/m²degC

WWR : Comparison between window area and outer wall area, such as: window-to-wall ratio.

ΔTeq : Equivalent temperature differences between the inside and outside

ΔT : Difference in planning for outside and inside temperatures within 5degC

SF : Solar radiation factor, W / m²

SC : Shade coefficient of penetration system (openings)

To calculate the average OTTV every 1m² on the entire wall surface, the equation is as follows:

$$OTTV = \frac{(OTTVu \times An) + (OTTV \times Ae) + (OTTVs \times As) + (OTTVb + Aw)}{(Au + At + As + Ab)} \quad (2)$$

OTTV : Heat transfer value for all outer walls, W/m²

OTTVn: OTTV value on the north side outer wall, W/m²

An : The total area of the outer wall and the north side window, W/m²

OTTVe : OTTV value on the east side outer wall, W/m²

Ae : Total area of the outer wall and east side window, W/m²

OTTVs : OTTV value on the south side outer wall, W/m²

As : Total area of the outer wall and south side window, W/m²

OTTVw: OTTV value on the west side outer wall, W/m²
Aw : Total area of the outer wall and west side window, W/m²

2.1. Analysis of External Energy Loads on buildings
Analysis of external energy loads on the buildings used OTTV on the existing building envelopes (windows and walls). Firstly, a partial OTTV calculation was performed from each orientation (8 orientations) of the sheath; and secondly, the average OTTV that entered every 1m² of wall and window surfaces was calculated.

2.2. Strategy optimizing external energy loads on building planning and design.
The results of the OTTV calculation from each orientation of the building envelope showed results that indicate whether any of them do not meet the OTTV standards (more than 35 W/m²).

3. Result and Discussions

The analysis of the calculation result of external energy loads was calculated by using OTTV to find out the external energy that entered the building through the building envelope. Analysis of External Energy Expenses on buildings.

Before calculating the OTTV for each building envelope, the first step is conducted to look at the absorption value, the wall transmittance value and the equivalent temperature of difference value.

3.1. Determining the value of sun absorption on the surface of the building (α),of all sides of the building are the same)
Building materials use brick walls (αw = 0.89) with shiny white paint with a value of αp = 0.30
α : (αw + αp) / 2
α : (0.89 + 0.30) /2 = 0.595

3.2. Determining the wall transmitance value
The thickness of plastered bricks on both sides are 114 mm with a U value of 3.64 / m²degC

3.3. Determining the ransmittance value of the wall with specifications of plastered bricks with both sides are 114 mm thick with a value of U = 3.64 / m²degC (attachment 2).

$$P = Mass\ /Volume \tag{3}$$

Mass : (2400kg / m³) x (1m³ x 1 m³ x 114m³)
Mass : 273.6 kg
From the results of the weight of the wall of 273.6 kg the value of ΔTeq = 10K (10 degC) is obtained. (Table 1).

3.4. Calculating the OTTV value of each building envelope orientation (W / m²)

Table 3. The value of OTTV each for side of walls

Elevation	Area m² (A)	Window Area m²	WWR %	SF	SC	α	Uw/Ug	ΔT/ΔTeq (degC)	OTTV (W/m²)
Northeast Walls	192.2922	58.0704	30.1	150	0.61	0.595	3.64/ 2.94	5/10	47
East Walls	20.2975	3.116	15,3	154	0.61	0.595	3.64/ 2.94	5/10	34.7
Southeast Walls	144.0021	27.2467	18,9	125	0.61	0.595	3.64/ 2.94	5/10	34.4

IOP Conf. Series: Earth and Environmental Science **328** (2019) 012028 doi:10.1088/1755-1315/328/1/012028

Elevation	Area m² (A)	Window Area m²	WWR %	SF	SC	α	Uw/Ug	ΔT/ΔTeq (degC)	OTTV (W/m²)
South Walls	64.6245	9.566	14.8	105	0.61	0.595	3.64/ 2.94	5/10	29.7
Southwest Walls	136.4172	3.93	2,8	176	0.61	0.595	3.64/ 2.94	5/10	24.2
West Walls	64.6245	6.4	9.9	243	0.61	0.595	3.64/ 2.94	5/10	35.3
Northwest Walls	144.0021	27.2467	18,9	211	0.61	0.595	3.64/ 2.94	5/10	44.4
North Walls	20.2947	1.8087	9,2	137	0.61	0.595	3.64/ 2.94	5/10	28.2
Total	**786.55075**	**137.3834**	**119,9**	-	**0.61/0,47**	**0,595**	**3.64/ 2.94**	**5/10**	**265.6**

Table 3. The results of OTTV for each of the building envelope orientation are obtained. The OTTV standard is more than 35 W / m² on the northeast wall (47 W/m²), the west wall (35.3 W/m²), and the northwest wall (44.4 W/m²).

3.5. OTTV Calculation value of each building envelope orientation (W / m²)

$$OTTV = \frac{(OTTVtl \times Atl) + (OTTVt \times At) + (OTTVtg \times Atg) + (OTTVs + As) + (OTTVbd \times Abd) + (OTTVb \times Ab) + (OTTVbl \times Abl) + (OTTVu \times Au)}{(Atl + At + Atg + As + Abd + Ab + Abl + Au)}$$

$$OTTV = \frac{(47 \times 192.2) + (34.7 \times 20.2) + (34.4 \times 144) + (29.7 \times 64.6) + (24.2 \times 136.4) + (35.3 \times 64.6) + (44.4 \times 144) + (28.2 \times 20.2)}{(192.2 + 20.2 + 144 + 64.6 + 136.4 + 64.6 + 144 + 20.2)}$$

$$OTTV = \frac{(9{,}033.4 + 700.9 + 4{,}953.6 + 1{,}918.6 + 3{,}300.8 + 2{,}273.3 + 6{,}393.6 + 569.6)}{(192.2 + 20.2 + 144 + 64.6 + 136.4 + 64.6 + 144 + 20.2)}$$

$$OTTV = \frac{29{,}143.8}{786{,}2} = 37.07\ W / m^2$$

From the calculation of the average OTTV for each m², result shows value of 37.07 W/m², which is greater than the standard set (more than 35 W / m²).

3.6. Strategies in optimizing external energy loads on building planning and design

The strategy to reduce external energy load is by decrease the WWR in each of building's envelope orientation where the value is more than 35W / m² (i.e. northeast wall, west wall, and northwest wall) which used the OTTV equation to obtain the following results.

3.6.1. The strategy reducing the external energy loads on the northeast wall

Table 4. Northeast Wall OTTV values

Elevation	Area m^2	Window Area m^2	WWR %	SF	SC	α	Uw/Uf	ΔT/ΔTeq (degC)	OTTV (W/m^2)
Northeast Walls	192.2922	58.0704	30.1	150	0.61	0.595	3.64/ 2.94	5/10	47
Northeast Walls	192.2922	48.05	25	150	0.61	0.595	3.64/ 2.94	5/10	42.63
Northeast Walls	192.2922	38.44	20	150	0.61	0.595	3.64/ 2.94	5/10	38.21
Northeast Walls	192.2922	34.59	18	150	0.61	0.595	3.64/ 2.94	5/10	36.45
Northeast Walls	192.2922	30.75	16	150	0.61	0.595	3.64/ 2.94	5/10	34.69

Table 4. The northeast wall shows the results of WWR 16% reducing the external energy load entering through the building envelope of 34.69 W/m^2 in the standard set (35 W/m^2).

3.6.2. The strategy reducing the external energy loads on the west wall

Table 5. The value of OTTV of the Western Wall

Elevation	Area m^2 (A)	Window Area (m^2)	WWR %	SF	SC	α	Uw/ Uf	ΔT/ΔTeq (degC)	OTTV (W/m^2)
West Walls	64.6245	6.4	9.9	243	0.61	0.595	3.64/ 2.94	5/10	35.3
West Walls	64.6245	6,3	9.8	243	0.61	0.595	3.64/ 2.94	5/10	31.2

Table 5. The west walls's WWR were reduced by 0.1% (6, 3 m^2). The reduction lead to the decrease of external energy load that enters the building envelope by 31.2 W/m^2, which is under specified standard (35 W/m^2).

3.6.3. The strategy reducing the external energy loads of the northwest wall

Table 6. The value of OTTV Northwest Walls

Elevation	Area m^2 (A)	Window Area (m^2)	WWR %	SF	SC	α	Uw/Uf	ΔT/ ΔTeq	OTTV (W/m^2)
Northwest Walls	144.0021	27.2467	18.9	211	0.61	0.595	3.64/ 2.94	5/10	44.4
Northwest Walls	144.0021	20.16	14	211	0.61	0.595	3.64/ 2.94	5/10	38.25
Northwest Walls	144.0021	17.28	12	211	0.61	0.595	3.64/ 2.94	5/10	35.77
Northwest Walls	144.0021	15.84	11	211	0.61	0.595	3.64/ 2.94	5/10	34

Table 6. The northwest walls (Table 3.4) through the reduction WWR to 11% (15, 84 m^2) decreases the external energy load that enters the building envelope by 34 W/m^2 which is under the specified standard (35 W/m^2).

3.6.4. Average OTTV calculation in every m^2 on the surface of walls and windows

$$OTTV = \frac{(OTTVtl \times Atl) + (OTTVt \times At) + (OTTVtg \times Atg) + (OTTVs + As) + (OTTVbd \times Abd) + (OTTVb \times Ab) + (OTTVbl \times Abl) + (OTTVu \times Au)}{(Atl + At + Atg + As + Abd + Ab + Abl + Au)}$$

$$OTTV = \frac{(34.6 \times 192.2) + (34.7 \times 20.2) + (34.4 \times 144) + (29.7 \times 64.6) + (24.2 \times 136{,}4) + (31.2 \times 64.6) + (34 \times 144) + (28.2 \times 20.2)}{(192.2 + 20.2 + 144 + 64.6 + 136.4 + 64.6 + 144 + 20.2)}$$

$$OTTV = \frac{(6{,}650 + 700.9 + 4{,}953.6 + 1{,}918.6 + 3{,}300.8 + 2{,}009 + 4{,}896 + 569.6)}{(192.2 + 20.2 + 144 + 64.6 + 136.4 + 64.6 + 144 + 20.2)}$$

$$OTTV = \frac{24{,}998.5}{786.2} = 31{,}07$$

From the results of OTTV, the average yield of each m² is 31.07 W/m², where this is in accordance with the standard (not more than 35 W/m²). Therefore, the profitable strategy in the Planning and Design aspect which decrease the external energy loads of the building is executed through the reduction of window area or WWR (Windows Wall Ratio).

4. Conclusion

The reduction of external energy (heat energy) can be controlled by optimizing planning and design of a building. OTTV strategies were applied with changes in windows-wall-ratio (WWR). Initially, the Residential and Building Development Information Center (PIP2B) had an average OTTV of 37, 07 W/m², which was above the standard (35 W/m²). There was a lot of external energy (heat energy) that entered the inside of building from Northeast widows (47 W/m²), West windows (35.3 W/m²), and Northwest (44.4 W/m²), because the windows area is wider than the other side. For achieving the standard (35.3 W/m²), there should be a decrease in the in the amount of energy with regards to the windows-wall-ratio. Northeast windows area should be 16% in order to reduce the external energy load (34.69 W/m²) that entered through the building envelope. West windows area were set at 9.8% in order to achieve the size of 31.2 W/m², while 11% of WWR was implemented on the northwest windows area in order to reach the size of 34 W/m². The reduction of each sides of the windows (Northeast, West, and Northwest) can decrease the average OTTV by 31, 07 W/m² which is below the specific standard.

References

[1] Anthoilah, R and Totok R B (2014). Optimization Natural Lighting in Working Space by Setiing Window and Wall. *Journal of Teknik Pomits vol , No. 1.*

[2] Badan Standarisasi Nasional. 2001 SNI 03-2396-2001 Procedures for designing building with natural system.

[3] Baubiology, (1975). Study about the building impact of human health by Anton Scheneider Germany

[4] Bauer, Michael, Peter Mosle and Michel Scwarz. (2007). Green Building Guidebook for Sustainable.

[5] Cavallin, Humberto. (2009). Caliphs, Intentionalities, and the Design Process: *Journal Architecture-Design Methods- Inca Struture*, Kassel University Press.

[6] Cv. Karya Prima. 2010. Report of building project residential and building development information center.

[7] Declaration for Environmentally Symbiotic Housing, 1997; The Association of ESH. Tokyo

[8] Gunawan, Billy 2012. Energy efficiensy manual book for building design in Indonseia: Energy efficiency and Conservation house.

[9] Hamzah, B., A. Kusno, and R. Mulyadi. (2018). "Design of energy efficiency and the thermally comfortable air-conditioned university classrooms in the tropics." *International Journal of Sustainale Energy*: DOI: 10.1080/14786451.2018.1539394

[10] Heryanto, Sani. 2004. Architecture and energy efficiency building. *Jurnal Ilmiah Arsitektur UPH Vol 1 No. 1.*

[11] Loekito, Sandra (2006). Conservation of Energy Analysis through Building Sheaths', *Journal of Civil Engineering Dimension,* Vol. 8, No, 2, 93-98.

[12] MEMR. 2012. Study of supply and demand energy. Center Mismistry of energy and mineral resources.

[13] OECD. IEA and UUROSTATS (2005) Manual statistic energy. Available online http://www.iea.org/about/copyright.asp.

[14] Saghafi, Mahmoud R, FarhangMozaffar, Seyed Moosavi. (2015). *Teaching Methods in Architectural Dessig Basic.* CiênciaeNatura, Santa Maria, v. 37 Part 1 2015, p. 379−387.

[15] Setiani, A.N, Anis R, and Reska R (2017). Evaluation of Overall Thermal Transfer Value (OTTV) Calculation in Building Construction. *Journal of Universitas Muhammadiyah Palembang.* ISSN 2580-1155.

[16] SNI 03-6398-2000 Energy conservation of buiding envelope.

[17] The Architecture Charter for Global Environment and its significance, (1990). On Rion Declaration on Environment and Development. Rio de Janiero.

[18] UIA / AIA World Congress of Architect, 1993 ACE Declaration on Architecture and Sustainability Adopted by the General Assembly of the ACE Final.

[19] Vidiyanti, Christy (2015). Building Retrofit Assesment as Reducing Efforts Operational Energy Consumption. *Journal of Architecture, Building and Environment.*

[20] WEC. 2016. *World Energy Issues Monitor.* London, UK: Published by The world energy council.

[21] Wardah, Khalil. (2016). *Design Process and Strategic Thinking in Architecture,* Proceedings of 2016 2nd International Conference on Architecture, Structure and Civil Engineering.

[22] Yang, Liu, Haiyan Yan, and Joseph C. Lam. 2014. "Thermal Comfort and Building Energy Consumption Implications A Review." Applied Energy 115: 164–173

IOP Conf. Series: Earth and Environmental Science **328** (2019) 012030 doi:10.1088/1755-1315/328/1/012030

Natural Lighting System to Provide Visual Comfort in Library Reading Room at Universitas Budi Luhur, Jakarta

S Kurniasih [1*] and O Saputra[1]

[1]Universitas Budi Luhur

*Email: sri.kurniasih@budiluhur.ac.id

Abstract. The general purpose of natural lighting is to produce quality of light that is efficient and minimizes direct glare, and excess light level ratios. The natural light that enters through a window can come from several sources, namely direct sunlight, clear sky, clouds or reflections of the lower surface and surrounding buildings. The light from each source varies not only from the amount and heat it carries, but also on other qualities, such as colour, spread, and savings. Light is the most important part of human life, to get the right lighting in a space, the right lighting system is needed according to people needs so that visual comfort can be achieved. The library is an important means to support teaching and learning activities and in supporting the implementation of the Three Principles of Higher Education and, the library should be the centre of attention, especially in its lighting system. This study applies a quantitative analysis method with a descriptive approach where the measurement of natural lighting levels and calculation of the dimensions of shading devices are carried out as part of the research stage. The independent variables in this study are the use of shading devices, while the dependent variables observed are natural lighting levels and natural lighting factors. The results of this study are natural lighting systems in the form of design solutions that allow it to apply in the 3rd-floor library reading room at Budi Luhur University.
Keyword: daylighting; visual comfort; natural lighting system; shading devices; library

1. Introduction

The library is one of the workplaces where most of its activities rely heavily on the work of one of our five senses, namely the eye. Therefore, good lighting in the library will increase comfort in working for employees and students. Good lighting includes natural lighting and artificial lighting. The existence of libraries in higher education is an important means to support learning and teaching activities and in supporting the implementation of the Three Principles of Higher Education so that, the library must be the centre of attention.

Visual comfort is the need for a good level of lighting in a room. Good lighting is lighting that can meet the needs of its users, related to the types of activities carried out in the space. To achieve visual comfort in a room, it is necessary to adjust the intensity of the incoming light. Based on SNI 03-6197-2000 on Energy Conservation in Lighting Systems, the average lighting level in the library reading room is 300 lx. The following is the level of lighting in accordance with SNI for energy conservation in the lighting system [1];

Table 1. Average lighting levels recommended colour rendering and temperature

The Function of the room	Illumination Level (Lx)	Colour Rederation group	Colour temperature		
Educational institutions					
Classroom	250	1 or 2		•	•
Library	300	1 or 2		•	•
Laboratory	500	1		•	•
Studio of Architecture	750	1		•	•
Canteen	200	1	•	•	

Source: Indonesian National Standards on energy conservation in lighting systems

Based on the results of measurements of the intensity of natural lighting obtained the average value of natural lighting intensity in the 3rd-floor library reading room of Budi Luhur University is 663 lx, this indicates that the average lighting power that occurs exceeds the SNI standard. The problem that arises in the library on the 3rd-floor is the occurrence of glare that can disrupt library users' comfort, this glare is influenced by the position of the library which is located on the 3rd floor with a floor to floor height of 11.10 m with that height without a shade tree barrier on the outside of the building [2].

The design of a natural lighting system needs to know the availability of natural light received at a local location. The general purpose of natural lighting is to produce quality light that is efficient and minimizes direct glare, reflective layers, and excess light level ratios [3]. The basic strategy of natural lighting can be imagery, natural lighting transfer and view taking [4].

Based on the results of the measurement of the intensity of the lighting, it is necessary to have a daylighting system strategy that can be applied to achieve visual comfort in the library reading room. The object of the research is the library reading room on the 3rd-floor, this is due to the glare that disturbs the reader in the space. The scope of this study is a daylighting system strategy that can be applied to window openings in the 3rd-floor library of Budi Luhur University, Jakarta.

2. Methods

The research method used in this study is a quantitative research method with a descriptive approach. The quantitative research method is one type of research whose specifications are systematic, planned and structured clearly from the beginning to the making of the research design. Research with a descriptive approach aims to describe the object of research or the results of research [5]. The application of quantitative methods is by measuring the intensity of natural lighting as data to analyze the visual comfort level in the 3rd-floor library reading room and calculating the dimensions of shading devices. While the application of the descriptive approach is done by designing shading devices as a natural lighting system strategy in the openings of the 3rd-floor library reading room window. The scope of this research is visual comfort.

The research steps carried out were a theoretical study of the level of natural lighting and shading devices, measuring the intensity of lighting and calculating dimensions of elements of shading devices, designing alternative lighting systems, analyzing and synthesizing results and making conclusions.

3. Results and Discussions

Based on SNI 03-6197-2000 on Energy Conservation in Lighting Systems, the average lighting level in library reading rooms is 300 lx. The conditions for measuring the level of lighting in the reading room of the Budi Luhur University library are:

1. Library reading room on the 3rd floor with window openings facing East and West.
2. Measurements are made on each reading room table with a height of 75 cm from the floor.

3. Measurements are made in the morning at 09.00 a.m., at noon at 11.00 a.m. and in the afternoon at 3:00 p.m. when the sky is clear.
4. Measurements are carried out in March, April, and May.
5. Measurements are made on the level of lighting in the reading room with conditions without artificial lighting.

The measurement results of the average illumination of the reading room in the 3rd-floor library of Budi Luhur University:

Table 2. The average illumination of the library reading room without artificial lighting
The condition of the room with daylighting

Location	Time: 09.00 am Illumination (Lx)	Time: 11.00 am Illumination (Lx)	Time: 15.00 am Illumination (Lx)	Average Illumination (Lx)
3rd-floor reading room	294	462	1232	**663**

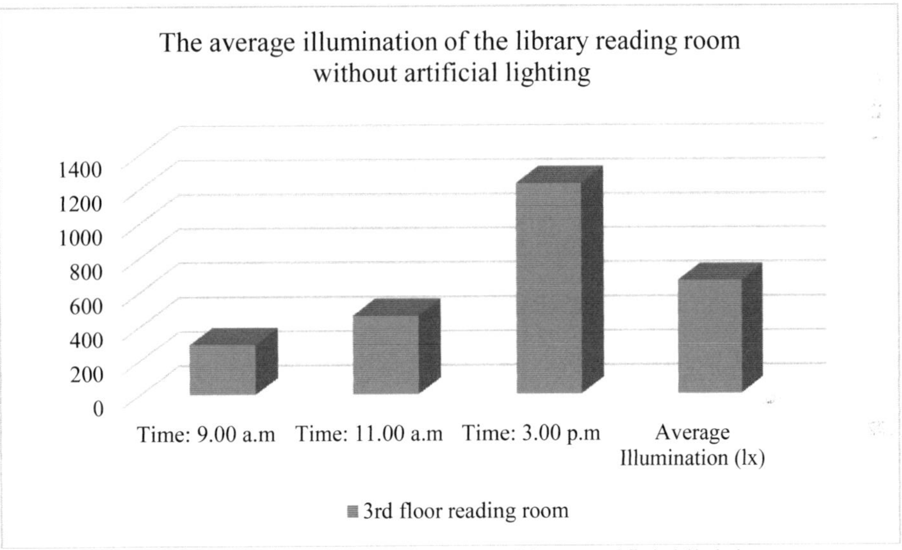

Figure 1. Graph of illumination without artificial lighting

Based on the results of these measurements it can be seen that the average intensity value in the library of the 3rd-floor is 663 lx with conditions using daylighting. The lighting intensity exceeds the lighting intensity value required in SNI 03-6197-2000, giving rise to glare problems that can disturb the comfort of reading in the room.

The following are some lighting system strategies in the form of shading devices [6] that can be applied in the 3rd-floor library to reduce the lighting that goes into the reading room. The design of shading devices is a design strategy for daylighting systems that can reduce glare problems that occur in the 3rd-floor library reading room [7].

IOP Conf. Series: Earth and Environmental Science **328** (2019) 012030 doi:10.1088/1755-1315/328/1/012030

Table 3. Various shading devices

No	Description	Best orientation	Information
1	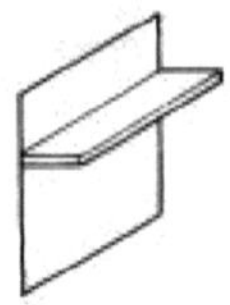 *Overhang Horizontal panel*	South, East, West	• Hot air traps • Can be loaded by wind
2	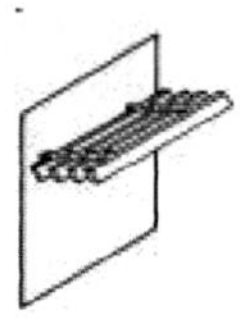 *Overhang Horizontal louvers in a horizontal plane*	South, East, West	• Free air movement • Small wind loads
3	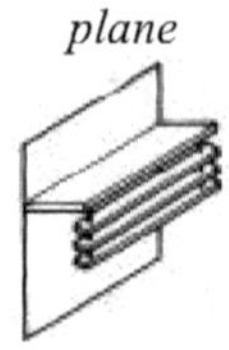 *Overhang Horizontal louvers in a vertical plane*	South, East, West	• Reducing the length of the overhang • Vision limitation • There are decorative lines on the window
4	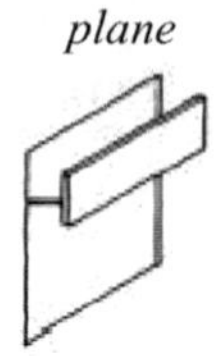 *Overhang Vertical panel*	South, East, West	• Free air movement • Vision limitation
5	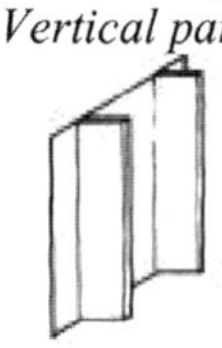 *Vertical fin*	South, East, West	• Vision Limitation • For the northern facade in only hot climates
6	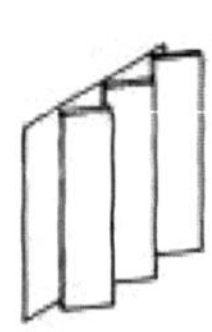 *Vertical fin slanted*	East, West	• Tilt to the North • Significantly limits vision

No	Description	Best orientation	Information
7	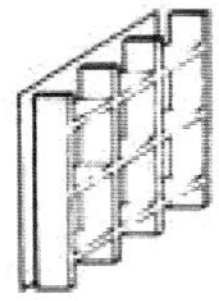 *Eggcrate*	East, West	• For very hot climates • Very limited vision • Hot air traps
8	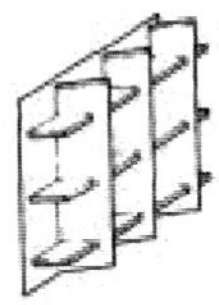 *Eggcrate with slanted fins*	East, West	• Tilt to the North • Very limited vision • Hot air traps • For very hot climates

Source: Lechner, Heating, Cooling and Lighting, 2001

Based on some of the designs of the shading devices, the design of shading devices chosen and used to overcome glare problems in the 3rd-floor library is horizontal panel overhang and vertical fin. The selection of the selected shading devices design is based on the orientation of the building that is facing East-West and the criteria for its use.

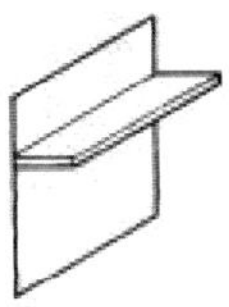

Overhang Horizontal Panel

Vertical Fin

Figure 2. Recommended shading device

Here are the alternative designs of shading devices as a strategy for designing natural lighting systems to overcome glare problems that occur in the library on the 3rd floor of Budi Luhur University;

a. *Shading Devices* type *Horizontal Overhang*
   - Alternative 1 with overhang width = 100 cm

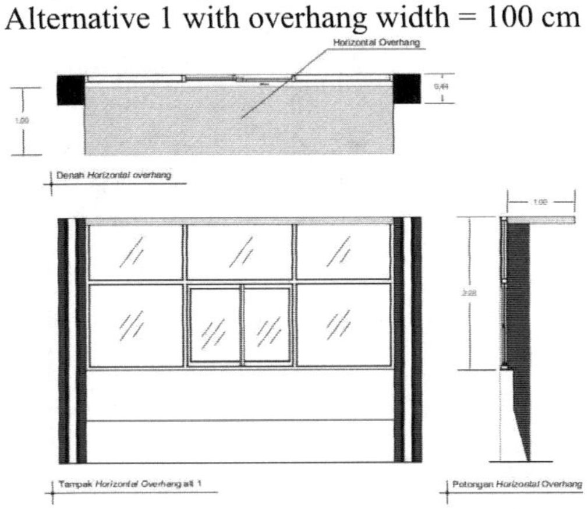

Perspective

IOP Conf. Series: Earth and Environmental Science **328** (2019) 012030 doi:10.1088/1755-1315/328/1/012030

- Alternative 2 with overhang width = 150 cm

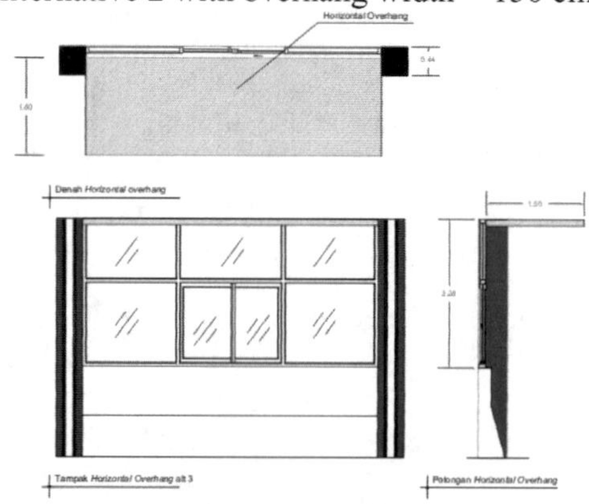

Perspective

- Alternative 3 with overhang width = 67 cm

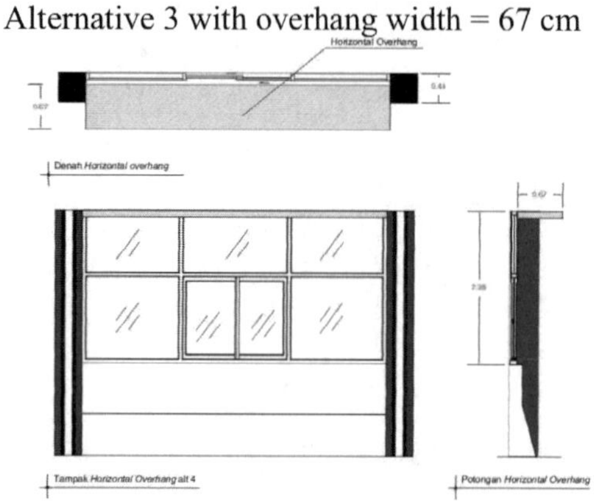

Perspective

- Alternative 4 with overhang width = 200 cm

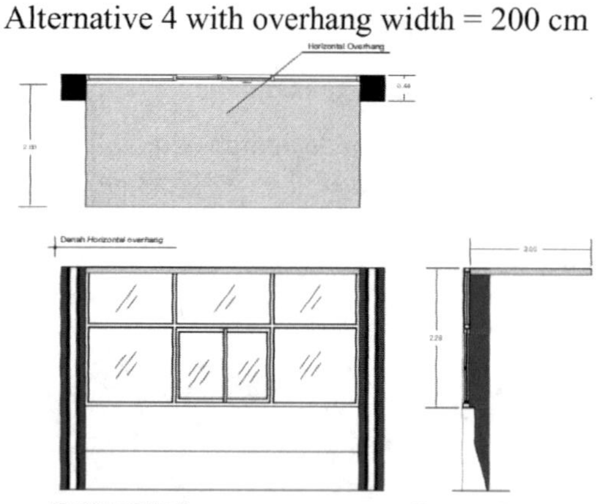

Perspective

b. Shading Devices Vertical Fin type
Based on the dimensions of shading devices, the width of the vertical fin is 50 cm.

IOP Conf. Series: Earth and Environmental Science **328** (2019) 012030 doi:10.1088/1755-1315/328/1/012030

- Alternative 1 with a distance of louvers every 50 cm

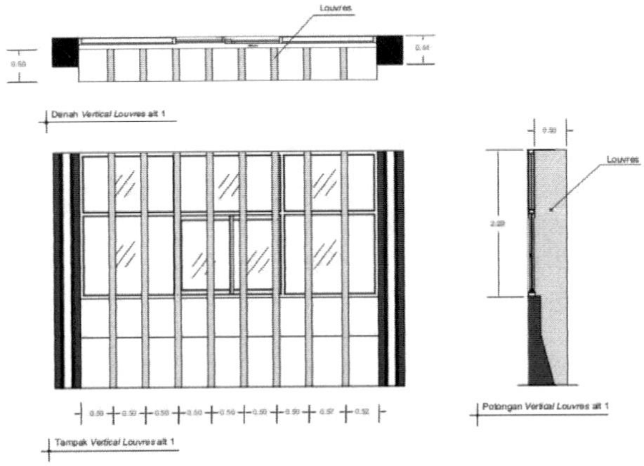

Perspective

- Alternative 2 with a distance of louvers every 100 cm

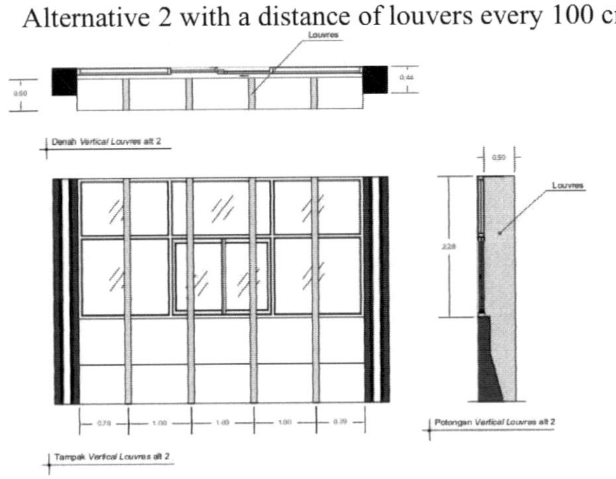

Perspective

- Alternative 3 with a distance of louvers every 150 cm

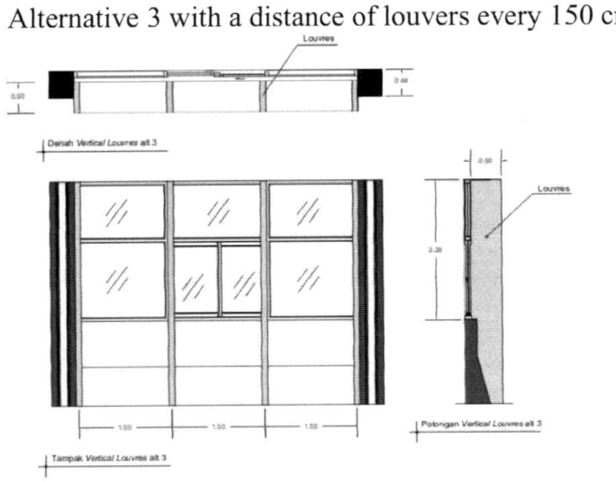

Perspective

IOP Conf. Series: Earth and Environmental Science **328** (2019) 012030 doi:10.1088/1755-1315/328/1/012030

4. Conclusion

The natural lighting system design strategy that can be a design solution to reduce glare problems is to use horizontal type shading devices and vertical fin with louvers so that the sunlight entering the reading room can be reduced.

References

[1] S. N. Indonesia and B. S. Nasional, "Konservasi energi pada sistem pencahayaan," 2011.
[2] S. Kurniasih, "Evaluasi Tingkat Pencahayaan Ruang Baca pada Perpustakaan Universitas Budi Luhur, Jakarta," Jakarta, Indonesia, 2018.
[3] Soegijanto, *Bangunan di Indonesia dengan Iklim Tropis Lembab Ditinjau dari Aspek Fisika Bangunan*. Jakarta, Indonesia: Direktorat Jenderal Pendidikan Tinggi, Departemen Pendidikan dan Kebudayaan, 1998.
[4] P. I. Principles, "Part i: principles."
[5] Sugiyono, *Metode Penelitian Kuantitatif, Kualitatif.* Bandung, Indonesia: Alfabeta, 2013.
[6] Norbert Lechner, *Heating, Cooling, Lighting Metode Desain untuk Arsitektur*, Kedua. Jakarta, Indonesia: Rajagrafindo Persada, 2001.
[7] W. I. Rachel Felicia, Jusuf Thojib, "Pencahayaan perpustakaan.pdf," *J. Mhs. Jur. Arsit.*, vol. 4 No. 4, 2016.

Sustainable Coastal Protection in Pasuruan City : Jetty Structure in Rejoso Estuary

A L Rakhmawati

Coastal Engineering. Ocean Engineering Department. Sepuluh Nopember Institute of Technology. Surabaya. Indonesia.

*Email : annisaklaila@gmail.com

Abstract. Rejoso Estuary has an unstable river mouth and threatens the existence of residents in the surrounding area. The rainy season, marked with high water discharge and river water overflows, affects flood in residential areas located around the coast. It happens due to sediment deposits formed closes the flow of the river to the sea. This study aims to find out an alternative, by processing data such as location maps, wind speed data, tidal data, current velocity, and sediment transport. From the calculation results, the researcher obtained the amount of longshore sediment transport amounting to 205837.7 m3 / year with the rate of perpendicular sediment transport of the coast which was 677915.27 m3 / year. The jetty is planned as the main alternative in this study to concentrate the flow of river water on a predetermined path to the sea, and to anticipate sedimentation due to sediment transport along the coast. The jetty was planned to be installed on the left and right of the estuary with a depth of 1.2 meters and length 25 meters. In this study, the jetty was considered quite effective to protect the estuary from the backwater effect and sedimentation which could cause flooding in the city area.
Keywords: *estuary, jetty, sediment*

1. Introduction

East Java is one of the provinces that having several coastal areas, both in the north and the south. This coastal area is rich in natural resources with economic value, used for ports, tourism, etc. It is possible for the beach to experience various damage caused by natural factors, as well as factors from coastal area users. One of the estuaries in East Java is Rejoso which is located at ± 3000 meters from the center of Pasuruan and has a river mouth known as the Rejoso River, located on the border of the villages of Rejoso and Lekok.

In the rainy season, high water overflow occurs due to the obstruction of the flow of river water to the sea, caused by sediment transport along the coast settling right at the mouth of the river, in which the river water overflows, threatening existing facilities and infrastructure in the Pasuruan area.The longshore sedimentary material is driven by waves, entering and settling in the estuary, later causing then settles and causes the sand tongue to occur. Thus, it is necessary to plan the appropriate protections to withstand sediment transport along the coast to prevent deposited sediment in the estuary.

IOP Conf. Series: Earth and Environmental Science **328** (2019) 012031 doi:10.1088/1755-1315/328/1/012031

Based on the background as outlined above, this study questions about (1) Reviewing area in the Rejoso River Pasuruan estuary, (2) Calculating sediment transport (3) Installing jetty planning analysis.

This study aims to plan coastal buildings to reduce sediment deposition caused by longshore sediment transport to protect the settlements around the estuary from sea water waves. With the existence of the safety building, it is able to anticipate the sedimentation that occurs due to the influence of sea waves. While the benefits of this study are (1) Providing an understanding of coastal protection buildings, especially building jetty, (2) Understanding the character of waves, tides and sedimentation in the estuary of the Rejoso River. (3) Revealing the effects of making a coastal protection in the form of a river in the Rejoso estuary area, (4) Providing a reference in learning about coastal protection buildings.

1.1 Literature Review

1.1.1. Beach Overview

There are two terms about coastline in Indonesian which are often confused with their use, such as the coast and shore. Coastal is a land area on the edge of the sea exposing the influence of the sea such as tides, sea breeze and seepage of sea water. While the shore is an area on the edge of the water which is affected by the highest tide and lowest tide, the coastline is the boundary line between the land and sea water, where the position is not fixed and can move according to the tides and coastal erosion that occurs.

A. Wave. Waves in the sea can be generated by wind (wind waves), attraction of the sun and moon (tides), volcanic eruptions or earthquakes in the sea (tsunami), and so forth. Waves can generate energy to form the coast, giving rise to the flow and transport of sediments in a perpendicular direction and along the coast, causing forces acting on coastal structures. Tides are also an important factor because it can cause strong currents, especially in narrow areas, for example in bays, estuaries and river estuaries. In addition, tide and low tide elevation is also important for planning coastal buildings.

B. Small Amplitude Wave Theory. In general, waveforms in nature are very complex and difficult to describe mathematically because non-linear, three-dimensional and has a random shape (a wave series has a different height and period). Some theories only describe simple waveforms with natural wave approaches. The simplest theory discusses the theory of small amplitude waves introduced by Airy in 1985, often called as Airy waves.

C. Hindcasting Waves. Hindcasting waves are an upcoming wave forecasting technique that uses wind data in the past. Wind data can be used for estimate the height and period of waves in the sea due to the occurrence of ocean waves most affected by the wind. In the wind blast area, an event of wind will transfer energy to water. The initially calm water surface will be disrupted and small wave ripples arise above the water surface, as a result of tension caused by the speed of the wind on the sea surface. With increasing wind speed, ripples will become large and eventually will form waves. The longer the wind blows, the bigger the waves are formed. Hindcasting waves will produce an estimate of the height (H) and period (T) of the wave due to the presence of wind with a certain magnitude, direction, and duration.

D. Calculation of Fetch. Fetch is a sea wave generator area which is limited by the land surrounding the sea. In the analysis of wave forecasting, the initial work is performed to know the duration of the fetch from each direction of the wind that might generate waves.

IOP Conf. Series: Earth and Environmental Science **328** (2019) 012031 doi:10.1088/1755-1315/328/1/012031

E. Surface Wind Estimation. Wind data used for wave forecasting is data at sea level at the generation site. The first step in analyzing wind data, is conducted get the value of Wind Stress Factor (UA), to be used in conducting wave forecasting.

F. Formation of Deep Sea Waves. In limited fetch conditions, the wind blows constantly long enough for the wave height to reach equilibrium at the end of the fetch. The duration and the wave conditions are limited by the length of the wind blowing.

G. Wave Deformation. In wave propagation from deep waters to shallow waters, the wave will experience a change in height, direction, speed and wavelength caused by silting and refraction process wave. Refraction and the effect of siltation, diffraction, wave reflection, and breaking waves will determine the wave height and pattern (shape) of the wave crest line somewhere in the coastal area.

H. Sedimentation and Erosion. River estuaries can be divided into three groups depending on the dominant factors that influence it. In an estuary, these three factors work simultaneously, but usually one of them has a more dominant influence than the other. Sedimentation is an unexpected deposition process that occurs outside the coastline (river mouth).

I. Dominated by Sea Waves. This type of estuary is characterized by large sediment transports along the coast each year. In this type the estuary is usually covered by a sand tongue with sedimentation patterns. Large waves on sandy beaches can cause sediment transport (sand), either in a perpendicular direction or parallel / along the coast. Of the two types of the transport, the longshore sediment transport is the most dominant formation. Sediment transport can move into the mouth of the river as the wave conditions are calm, settling the sediment. The larger the wave, the greater the sediment transport occurs.

J. Dominated by the River. This estuary type characterized by river discharge each year is large enough resulting in debit as the main parameter of the estuary formation in the sea with relatively small waves. The river carries sediment transport from the uppermost large enough.

K. Tidal Dominant. This type of estuary is characterized by considerable tidal fluctuations. If the tidal height is large enough, the volume of tide entering the river is very large. The water will accumulate with water from upstream of the river. At low tide, the huge volume of water flows out over a period of time depending on the type of tide. Thus, the current velocity during low tide is large enough, which is potential to form a river mouth.

L. Coastal Sediment Transportation. Coastal sediment transport is sedimentary motion in the coastal area caused by waves and currents. Coastal sediment transport results in changes in shoreline both in the form of accretion and erosion. Coastal sediment transport can be in the form of longshore sediment transport and perpendicular sediment transport of the coast. Longshore sediment transport has a longshore average direction; whereas, perpendicular coastal sediment transport has an average direction perpendicular to the coast.

M. Longshore Sediment Transportation. Long shore transport is caused by turbulent sediments when the wave breaks, then moves are carried away and the components of the waves are parallel to the coast. At a point on the beach, the sediments come and go away (transported). If the transported sediments are greater than the sediments that come, beach erosion will occur. Beach upright sediment is divided into two parts of : the transport of sediments to the offshore direction such as those that occur during a storm, and heading towards the coast (onshore) as occurs during waves. Sediment transport towards the deep sea occurs quickly and sediment will form a bar that will function to reduce wave.

1.2. Coastal Protection Technical Review.
The main purpose of coastal protection and protection is to secure and protect: a.) People who live along the coast from the threat of waves or floods; b.) Public facilities along the coast; c.) Coastal areas against the threat of coastal erosion; d.) Pollution of the coastal environment that can damage coastal biota.

A. Protection with Soft Structures. Nature generally provides an effective natural coastal protection mechanism. The examples of protection with soft structures include sand nourishment, beach recycling (beach passing), and beach drainage (beach drains).

B. Protection with Hard Structures. Protection of the beach with a harder structure consists of protection along the coast, which is perpendicular to the beach. Protecting the coastline for example breakwaters, seawalls and revetments becomes a coastal protection commonly seen in Indonesia. This building consists of structures that are directly attached along the coastline, or near and parallel to the coast for coastal protection from wave attacks. Whereas protection is perpendicular to the beach, such as in groin and jetty.

C. Jetty building. Jetty is a beach perpendicular building that is placed on one or both sides of a river mouth which functions to prevent river overflows by coastal sediment deposits. During the deposition process it is usually accompanied by turning the river mouth in the same direction as the direction of sediment transport along the coast. Prevention estuarine closure is divided into countermeasures for ship traffic (long jetty) and countermeasures for mouth closure which causes flooding (short jetty).

1.3. Building Type

A. Fix / Rigid Structure. The types of immovable construction that have a massive, structure and the advantages of ease and speed in installation, cheaper construction prices and lower maintenance costs. Meanwhile, the shortcomings lie in a more complicated planning procedure, if a sudden and total disaster occurs, it is difficult to repair [1].

B. Flexible Structure. The types of movable construction which have advantages and conveniences in planning. The structure is relatively simple, a high stability factor, because it can absorb most of the wave energy that hits the surface of the building, and the building is still functioning despite heavy damage, and is easy to repair. Meanwhile, the shortcomings lie in the availability of material (rock material) in a large amount of volume for the required diameter and quality (usually requires large quantities of large rocks) [1].

C. Detailed Construction. The construction part of the guide building and the wave barrier at the outlet of the river estuary can be broadly divided into two main parts, including the main construction part and the protective construction part.

D. Calculation of Weight of Protective Stone Grains. The head part of the building requires a heavy protective stone larger than the building's arm. This is because the building head can receive wave attacks from various directions. The KD value for the building head part is smaller than the KD value in the building's arm. The weight of protective stone is calculated by the formula of Hudson and Jackson 1962 [3].

On a sandy beach, the making of jetty which scours far enough into the sea can cause obstruction of sediment transport along the coast. As a result, sediments that move from the left will be blocked by jetty, leading to precipitation in the area. The area to the right, the waves that come in an angle to the coastline cause a current along the coast. The current can carry sediment. However, there is no sediment supply in this area, because the sediment moving from the left is blocked by the building. As a result, the beach on the right side of the jetty will experience erosion.

2. Methods

2.1 Data Collection Survey
Secondary Data includes:
- The last 11 years wind speed data (2001-2011) from the Meteorology, Climatology and Geophysics Agency (BMKG).
- Aerial photos from Google Earth 2010 software.
- World map of Microsoft Encarta 2009 software.

Primary data: The survey was conducted by visual observation and interviews with the people who live around the estuary of the Rejoso river.

2.2 Secondary Data Analysis

2.2.1 Wind Data. For wave forecasting, wind data is required. The data are from a minimum of daily data in 10 years, and the wind data are obtained in the last 11 years. The wind data is taken from BMKG. Wind data consist of speed, direction and duration. Wind speed is expressed in units of knots, where 1 knot = 1 nautical mile / hour, 1 nautical mile = 6080 feet (feet) = 1853.18 m and 1 knot = 0.515 meters / second. Wind direction is expressed in 8 directions of the wind (North, Northeast, East, Southeast, South, Southwest, West, and Northwest).

2.2.2 World map. In wave forecasting, it is necessary to understand the effective fetch length of each main direction. To do fetch measurements, Microsoft Encarta software is utilized to have world map facilities and distance gauges. To facilitate the depiction, the researcher utilize AutoCad software.

2.2.3 Aerial Photos. An aerial photograph of the Google Earth software is required to determine the measured shoreline position.

2.3 Primary Data Analysis

2.3.1 Damage Assessment and Coastal Protection Technical Planning. Based on the survey conducted, it is apparent that the problems that occur around the mouth of the Rejoso river which are often subjected to sand deposition.

2.3.2 Flow Measurement Data. Current measurement is intended to get an overview of ocean currents in locations, such as the dominant speed and direction of the current.

3. Results and Discussions

3.1 Frequency Analysis
Based on the analysis of the frequency distribution of wind speeds that have been carried out, it is revealed that in January, February, March and April the dominant speed is West and North. The biggest wind speed in the West is in January with a percentage of 51.3% dominantly occurring between 1-5 knots, followed by the second largest wind speed in April with a percentage of 32% dominant occurring between 1-5 knots. Whereas the largest wind speed in the North direction is in March with a percentage of 39.7% dominantly occurring between 1-5 knots followed by the second largest wind speed in February with a percentage of 37.1% dominant occurring between 1-5 knots.

IOP Conf. Series: Earth and Environmental Science **328** (2019) 012031 doi:10.1088/1755-1315/328/1/012031

3.2 Calculation of Broken Waves

The calculation is conducted to determine the value of H'o and Hb: (From table A-1 Bambang Triatmodjo "Ports" get H / Ho 'values based on the value of d / Lo) [2].
d / Lo = 0.0453; H / Ho '= 1.042
Then: H'o = Ho / (H / Ho ') = 1.0703 / 1.042
a. 1,027 m
Thus:
H'o / gT2 = 1.0703 / (9.81.376292) = 0.0077
The value of Hb is obtained from the plot between the value of H'o / gT2 and the slope of the beach (m) from the graph of "Determination of the Height of Broken Wave (page 96), Coastal Engineering, Bambang Triatmodjo " [2].
The plot on the graph for H'o / gT2 = 0.0077 and m = 0.2, obtained Hb / Ho = 1.025 and Hb = 1.053 m. Thus: Hb / gT2 = 1.053 / (9.81. 3.76292) = 0.0076, obtained db / Hb = 0.82 and db = 0.863 m

3.3 Analysis of Sediment Transport

3.3.1 Longshore Sediment Transportation. Calculation of longshore sediment transport is carried out by using the CERC method. The parameters are obtained based on previous calculations. By accumulating the amount of sediment transport every month for one year, the recapitulation of total longshore sediment transport is 205837.70 m3 / year from the North and Northwest, but the transportation of sediments is more dominant, from the North. The northern direction of sediment transport is 203650.17 m3 / year, with the largest sediment transport occurring in December, which is 32508.37 m3 / year, and the lowest is in March which is 218.11 m3 / year. Whereas the northwest direction of sediment transport is 2187.53 m3 / year, with the largest sediment transport in April, which is 503.71 m3 / year and the lowest is in August.

3.3.2 Upright Straight Beach Sediment Transportation. From the calculation of sediment transport perpendicular to the coast, by accumulating the amount of sediment transport every month for one year, the total recapitulation of upright coastal sediment transport is 677915.27 m3 / year originating from the North, with the largest sediment transport in January, which is 78042.98 m3 / year and transportation the lowest sediment in July is 31097.98 m3 / year.

3.4 Building Layout

To determine the location of the safety building, building layout is performed to understand the wave conditions and coastal currents at the location of the study. To prevent the accumulation of sediments in the estuary area, it is necessary to make jetty on both sides of the river mouth, and add two series of jetty on the east coast of the river estuary, and to anticipate damage to the coastal area of Estuary Rejoso, considering that there are settlements along the coast.

4. Conclusion

At the mouth of the Rejoso river, sand deposition forms when waves and dominant currents deposit sediments form a longitudinal plain and result in the closure of river mouths. Since the research location included sandy beaches, the closure of the estuary resulted in sand spit in the estuary, the estuary type was dominated by ocean waves and wind.

- The maximum height and wave period that occurred in 2001 to 2011 was in February 2001, with a height of 1,070 m and a period of 3,763 seconds originating from the North. Likewise, the dominant wave originates from the North.
- By using the CERC (Coastal Engineering Research Center) method, the maximum longshore sediment transport rate is obtained, which is equal to 205837.70 m3 / year originating from the North.

IOP Conf. Series: Earth and Environmental Science **328** (2019) 012031 doi:10.1088/1755-1315/328/1/012031

Besides, the maximum perpendicular beach sediment transport rate is also obtained, which is equal to 677915.27 m3 / year from the North.

• The alternative problem solving for the Rejoso river estuary is conducted by using a protective beach in the form of jetty, because the area is not the port area where the ship is anchored. There are only small boats that are often used by fishermen for their survival. Thus, the construction is easier and more economical.

• Jetty is built on both sides of the river mouth, coupled with two jetties on the coast east of the river mouth to prevent the accumulation of sediment due to sediment transport which is much greater than the river discharge.

• In the jetty in the river estuary, overtopping is expected to occur, especially when the storm takes place to be considered for planning the type of construction, although overtopping is not too large. A damage inventory study in the Rejoso area is needed to find out more about coastal damage that will occur in a certain period of time.

References

[1] Jatmoko H 2003 *Konstruksi Jetty* (Jakarta : Departemen Permukiman dan PrasaranaWilayah)
[2] Triatmodjo B 1999 *Teknik Pantai* (Yogyakarta : Beta Offset)
[3] CERC 1984 *Shore Protection Manual (SPM '84)* (Washington : US Army Coastal Engineering, Research Center)
[4] Kementerian Pekerjaan Umum 2010 Pedoman Perencanaan Teknis Pengamanan Pantai. (Jakarta : Ministry of Public Work)

IOP Conf. Series: Earth and Environmental Science **328** (2019) 012032 doi:10.1088/1755-1315/328/1/012032

Community service facilities and infrastructure improvement in remote areas of East Kalimantan (Case study : Conflict in Batoq Kelo Village, Mahakam Ulu Regency)

A R T Hidayat[1]* and R A Vidianti[2]

[1]Urban and Regional Planning Department, Brawijaya University
[2]Graduate student of Urban and Regional Planning Department, Brawijaya University

*Email : a.r.taufiq.h@ub.ac.id

Abstract. Conflict at the local level is one of the implications of regional division. Mahakam Ulu Regency is a new autonomous region which is a division of Kutai Barat District, East Kalimantan Province which was established in 2013. Mahakam Ulu Regency is located in the border area with the State of Malaysia and is located in a remote area of East Kalimantan province. one of which was faced was conflict in fulfilling the needs of village community service facilities and infrastructure. this research explores the condition of facilities and infrastructure in Batoq Kelo village which is divided into two residential development areas by involving community participation. the approach used is Community Needs assessment (CNA) with an effort to identify needs with involvement from the community. The results of the analysis show that there are differences in the needs of facilities and infrastructure in the two residential areas of Batoq Kelo Village. directed at the two residential areas to develop BUM Desa as an effort to improve the government's financial capacity in the administration of the government and can increase people's income by involving the economic activities of the community and the potential of the Batoq Kelo Village.
Keywords: conflict, remote area, community services, infrastructure, community needs assesment.

1. Introduction

In Indonesia, village at least has 3 different names. One of them is a "Kampung". Based on Law No. 6 concerning Villages in 2014 [1], the village has the opportunity to develop its territory in accordance with local customs and culture. Of course the construction does not conflict with applicable regulations. The law also explains that in regulating villages it has the principle of diversity. the existence of diversity triggers conflict. Conflict in society is mainly caused by how to respond to differences that arise, both differences in interests, perceptions, identities, jobs and positions. Another factor is the existence of interest groups, organizational institutions, and social classes in society that do not always have the same and harmonious interests [2].

Conflict at the local level is one of the implications of regional sparation [3]. Mahakam Ulu Regency is a new autonomous region which is the division of West Kutai Regency, East Kalimantan Province which was established in 2013. Mahakam Ulu Regency is located in the border region with the State of Malaysia. as a new district and is located in a remote area of East Kalimantan province. The remote area is defined as a remote (rural/settlement) location within an area [4]. One of the problems experienced is the still difficult accessibility between regions within the Mahakam Ulu District or other districts. The main transportation between villages in Mahakam Ulu Regency is using water transportation. As new autonomous region, they are facing conflict in meeting the needs of village facilities and infrastructure. This study aims to determine the needs of village facilities and infrastructure that have experienced local conflicts in Mahakam Ulu District, namely Batoq Kelo Village

1.1 Literature Review

The village is defined as an association of human beings who live together and have an organization in a government that is supported by its own rules of regulation, the village is under the village leadership who is elected for joint decisions [5]. In planning an area arrangement (as an example of a village) one that is needed is support for government policies, one of which includes the development of infrastructure conditions.

Infrastructure has an important influence in improving the quality of life and human well-being[6]. one of the constraints of the government in the effort to develop village infrastructure is the rejection of the community due to the mismatch between the infrastructure built and their needs, hence the participation of the community in the planning process is an effective way [7]. A person or community will participate in rural development efforts (village infrastructure) influenced by a belief in a situation of interdependence, and a network of social organizations that facilitate cooperation for mutual benefits [8]**Error! Reference source not found.**.

According to Suriana [9], the existence of community participation can increase the independence of the community. It can accelerate the rural development. The benefits are as follows:

- Participation broadens the knowledge base
- Participation helps build communication transparency and power relations between stakeholders
- Participation can improve solutions based on local understanding and knowledge
- Participation will encourage local ownership, commitment. Engagement can help create sustainable results by facilitating community ownership of the program.

Social conflict is a conflict between individuals and groups in society to fight over and contrast various things that are considered true or valuable. Todd stated[10], the conflict is divided into 3 stages, namely:

- Pre-conflict, interpreted as a condition based on someone or group dissatisfaction
- Conflict, interpreted as a situation where the parties involved are aware of the existence of such dissatisfaction
- Post conflict, defined as a situation where the conflict is declared public (in public) and involves a third party

Villages have various conflicts. The conflict included conflict with rural communities with higher levels of government and conflicts among village communities. The conflict can be interpreted positively or negatively. If interpreted positively, conflict encourages people to improve conditions to be able to resolve the conflict. Conflict can be interpreted negatively. Conflict can disrupt community activities. Often conflicts are interpreted in negative dimensions. Because conflict has a greater negative impact on society. Moreover, the conflict was completed in a relatively long period of time. Rural conflicts can disrupt village development. So conflicts in the countryside need to be resolved quickly. Settlement capital is available, namely the social bond of the community. This becomes the main tool in resolving conflicts.

In this study aims to find out the needs and desires of the community. The Community Needs assessment approach was used. Based on the Participant Workbook: Community Needs assessment [11], Community needs assessment has a definition as an effort to identify the needs of the community in providing fun services to those needs with involvement from the community. with this community needs assessment approach, it provides a brief overview to researchers or stakeholders regarding current policies, systems and strategies for changing the local environment and helps identify areas that need to be improved. With this data, the community can create strategies to make positive and sustainable changes in the community (in this study the villagers of Batoq Kelo village)

2 Methods

The study site area was Batoq Kelo village. Batoq Kelo village is one of the villages in Long Bagun district with the condition of the development area divided into two different locations. The flow in this study begins by identifying the existing conditions and problems at the study location, the next stage of data collection. Data collection is done by primary and secondary surveys. in this study also conducted a Forum Group Discussion (FGD) with local communities and extracted information related to plans to increase the availability of village facilities and infrastructure based on community perceptions. after the existing condition data and planned fulfillment of needs are collected, an analysis of needs and desires is carried out in the availability of village facilities and infrastructure. The purpose of this research is determining the needs of facilities and infrastructure using community participation approach. The variables of this reserach are as follows:

Tabel 1. Research variables

Variables	Sub variables
Community services facilities	• The number and condition of educational facilities; • The number and condition of health facilities; • The number and condition of worship facilities; • The number and condition of Government Community Service and Public Services; • The number and conditions of commercial services facilities.
Infrastructure	• Road conditions; • Condition of transportation infrastructure (docks); • The condition of clean water infrastructure; • Conditions for sanitation infrastructure and waste; • Telecommunications infrastructure conditions; • Condition of electricity infrastructure.
Needs and desires of the community.	• Existing conditions of village facilities and infrastructure; • Consideration of important village leaders (key person); • Assessment of village facilities and infrastructure in Forum Group Discussion (FGD).

The approach in this study uses community needs assessment (CNA). Figure 1 shows the step of CAN that consists of 5 steps. Facilitator must identify community teams as well as keypersons, develop a team strategy, define scope and location, and collect basic data related to the issue.

To obtain the data, a primary survey was carried out by using interviews, field observations and questionnaires which then the results of the field conditions were planned with community participation in Forum Group Discussion (FGD) activities. namely the system or customary norms that apply. at the time of the FGD a comparison was made between the existing conditions of the village infrastructure. Preparation of plans for improvement and improvement of the availability of village facilities and infrastructure based on community perceptions and consideration of important village leaders (keyperson). The CNA in this study did not establish an action plan. Because the final results of this study are used in the preparation of a comprehensive village development master plan. Before hand, this reseach investigating historical profile to understand how the conflict was emerged.

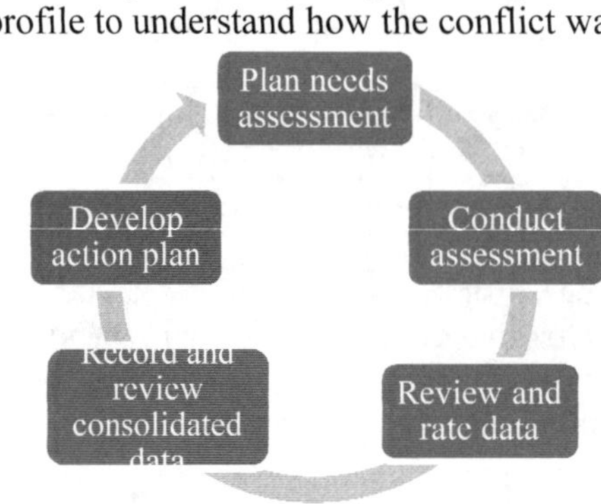

Figure 1. Community needs assessment [11]

IOP Conf. Series: Earth and Environmental Science **328** (2019) 012032 doi:10.1088/1755-1315/328/1/012032

3 Results and Discussions

3.1 Current condition of Batoq Kelo Village

Mahakam Ulu Regency was established in 2013. It was part of West Kutai Regency. Mahakam Ulu Regency consists of 5 sub-districts, namely Long Apari District, Long Pahangai, Long Bagun, Laham, and Long Hubung. Mahakam Ulu Regency consists of 49 villages with the regency capital, Ujoh Bilang. The area of Mahakam Ulu Regency is ± 15,315 km2. The boundaries of the Mahakam Ulu Regency are as follows:

Long Apari District is the largest district in the Mahakam Ulu Regency. Batoq Kelo Village is one of the villages in Long Bagun District, Mahakam Ulu Regency. Batoq Kelo Village is divided into two residential development areas, Batoq Kelo Ulu at upper stream of Mahakam River and Batoq Kelo Ilir at lower stream of Mahakam River. Those are sparated 20 Km. Due to this separation, it elicited conflict in improving community service facility and infrastructure. In the otherhand, government has limited budget.

Batoq Kelo village can be accessed by land and river routes. The area of Kampung Batoq Kelo is 1,940 Km2 which is divided into 4 neighborhoods (i.e. Rukun Tetangga or RT). RT 01 and 02 are in the ilir section, namely in Long Bagun Tengah and RT 03 and 4 in the upper part. In the village of Batoq Kelo ilir close to the capital of the Regency, namely Kampung Ujoh Bilang. the condition of the conflict in the development of the Kampung Batoq Kelo has the following history.

Tabel 2. Historical profile of Batoq Kelo Village

Year	Events
1912-1982	in 1912 about ± 40 people from the Kalteng tribe (the Udanum tribe, the Kahajan tribe and the Siyaang tribe) came to East Kalimantan (East Kalimantan) precisely to the Batoq Kelo. Then, they formed settlements and planted crops in Batoq Kelo Village. until in 1982 the movement of the Batoq Kelo settlement began to continue to Long Bagun, precisely in Long Bagun Tengah
1982-1994	Due to the access of the upper Kampung Batoq Kelo to the center of the activity at that time in Long Bagun or Kampung Ujoh Bilang. The purpose of the resettlement was because the Batoq Kelo Ulu area was difficult to reach by the government and where at that time Batoq kelo was still within the administrative area of West Kutai Regency. After considering some options, the people of Batoq Kelo preferred to move and to settle in Long Bagun which was right in the middle of the village of Long Bagun ilir and Long Bagun ulu.
2003	In 2003, Batoq Kelo Ilir community slowly re-opened residential and agricultural land in the Batoq Kelo Ulu. Precisely, it is located between across sub-watershed Tebaq Kecil Village and Tebaq Besar Village. the reason the people of Batoq Kelo ilir re-cleared agricultural land and the construction of settlements in Batoq Kelo Ulu was a response to the impact of rapidly increasing population growth and even greater land needs. so the need for land to build new settlements has dropped dramatically. Batoq Kelo Ilir people, one by one, are reopening residential land in Batoq Kelo Ulu. But there are a number of villager who remain reside at Batoq Kelo Ilir.
2015-2018	In 2015, Village office proposed adding 2 neighborhoods unit (RT) in Batoq Kelo Ulu namely RT 03 and RT 004. RT 01 and RT 02 were in Batoq Kelo ilir. Mahakam Ulu Regency government responded and agreed to the proposal. Until 2018, 80% of development is focused on Batoq Kelo Ulu. And 20% development was carried out in Batoq Kelo Ilir. Batoq Kelo Ilir received a lower percentage because villagers can access neighboring village community service facilities, Ujo Bilang which is the Regency capital.

The majority of the population lives in Kampung Batoq Kelo Ilir (neighborhood unit or RT 01 and RT 02) which account for 66% of the total population of Batoq Kelo Village. Due to separation of the settlement development areas, it inflicted an internal conflict.

3.2 Community services and infrastructure

Community services are the most important supporting tools in social activities or economic activities. Facilities can also be said as facilities that function for the implementation and development of economic, social, cultural life in the form of commercial facilities or workplaces, educational facilities, health facilities, worship facilities, government facilities and public services [12]. the condition of facilities in the research area can be explained by the following tables.

Tabel 3. Public services at Kampung Batoq Kelo

Community services facilities	Kampung Batoq Kelo Ulu	Kampung Batoq Kelo ilir
Education facility	2 units. Fair condition but need to be repaired	-
Health facility	3 units. Limited equipment	2 units, Limited equipment
Government Community Service and Public Services	4 units	3 units
Worship Community Service	2 units	1 unit
Commercial facilities	1 unit	7 units

Tabel 4. Infrastructure at Batoq Kelo Village

Infrastructures	Kampung Batoq Kelo Ulu	Kampung Batoq Kelo ilir
Road	main road mainly without pavement and muddy during rainy season	One concret pavement road
Dock	1 unit at fair condition	1 unit at fair condition
Sanitation	5 units public toilet in fair condition	Every household has sanitation system
Clean water	Community piped clean water system. It has spring as water source	60% of Household own well 40% of household using river as water source
Electricity	Household scale electricity source. Only used at 18.00 to 24.00	24 hours/day. Often Blackedout
Telecommunication	No signal	Fair signal is avaliable

3.3 Analisis Community needs assessment (CNA)

To find out the needs of Batoq Kelo village facilities and infrastructure, the Community Need Assessment approach was carried out by involving the people of Batoq Kelo Village through Focus Group Disscussion. Researchers were taking role as a facilitator in the forum so that the community provides an assessment of the things that are felt necessary for the development of Batoq Kelo Village facilities and infrastructures At this stage, the researchers were able to identify the current conditions community facilities and infrastructure.

During the FGD, the community was quite enthusiastic in compiling plans for community service facilities and infrastructure. Community engagement in planning can accelerate rural development. The community can participate in the stages of planning, implementation and evaluation of development program programs [13]. The forum is conducted in the villager house.

Participants identify the current condition and elicited development plan. Then, they assessed their option and classified into needs and desires. The assessment was carried out by considering existing conditions of community service faclities and infrastructure and opinions of important village person (i.e. keyperson).

Based on the results of the analysis, facilities and infrastructure in Batoq Kelo ilir village are more complete. This is because the location of the settlement is close to regency capital, Kampung Ujoh Bilang. Educational facilities, Batoq Kelo Ulu requires the construction of a kindergarten building. Because, until now, it is still integrated with SD buildings. Development of Junior high school building is considered as desires because the population is still low and the capacity is sufficient. In the Batoq Kelo Ilir, there are currently no educational facilities. However, the community does not need new formal education facilities. Because students go to school that located in Long Bagun Ulu villages and Ilir. The reading park development plan is still included in the desire category. Because people don't need it. It has an impact on the limited opportunities of school-age children to access science. But the community considers the number of visitor/user still low. Moreover, students attend school in the district capital which has a large collection of books.

Health facilities In Batoq Kelo Ulu and Batoq Kelo Ilir require the addition of complementary facilities. An example is for handling dental health. The existing condition of health equipments are insufficient and only capable of serving minor ailments. Although the distance between the village and the district capital center is relatively affordable, community health center/*Puskesmas* as a first aid facility requires qualified equipment to treat patients with various illnesses. While the community said that the addition of pharmacy facilities and midwife practices was included in the category of desires. Each village has at least a midwife. But the community considers the practice of midwives not a necessity because of the distance of the center to the district capital relatively close. So for emergencies condition, the community goes to health facilities in the regency capital.

The community does not need the addition of government service facilities and public services. the security post unit is classified as desire. At existing condition of the security post is sufficient because the area of the village settlement is relatively narrow. The community needs repairs to Customary building/*lamin adat* buildings, adding cultural fields around custom lamin buildings. *lamin adat* is a cultural activity facility which is expected to become a tourist attraction in Batoq Kelo Village. In both residential areas requires developing BUMKampung as a business institution that can encourage economic productivity of the community. It is expected that with BUMKampung, it can support community economic activities such as the sale of products from the Batoq Kelo Village. The community expects BUMKampung to run several types of businesses. Those are social business / serving, rental business, and trade

Regarding the village infrastructure development plan, there are still residential areas with road conditions in the form of land pavement and in poor conditions. Due to limited funds to improve road quality. Both villages need to improve the quality of residential roads by concrete road pavement. Good road conditions will facilitate community accessibility. The community needs to improve the quality of the dock as a water transportation infrastructure. The community considers that the current dock is still sufficient to serve their activities. Their dock is very traditional. The dock is made of wood. Different results were pointed out by the Batoq Kelo ulu community. They need to repair road access to the pier. The community determines these needs because access to the pier often hampers community activities. Access to the pier is still soil pavement.

The differences between the villages of Batoq Kelo Ulu and Ilir are teleocommunication and electricity infrastructure. Access to the two infrastructures in Batoq Kelo ulu is still lacking. The village of Batoq Kelo does not yet have access to telecommunications. It is necessary to develop Base transceiver station (BTS) in Batoq Kelo Ulu. Information and Communication Technology infrastructure provides benefits to its users both from socio-economic, social psychology, and socio-cultural aspects. Rural telecommunications development has a high and significant benefit to rural development, socio-economic and sufficient but significant benefits in the social psychological and socio-cultural aspects.

Electricity Infrastructure, in the existing condition of electricity fulfillment in the village of Batoq Kelo Ulu, is still lacking and can only serve at certain hours, this limits community activities at night. Collaboration with the government program in the form of electricity is included in the village and the Development of BUMKampung in diesel power plant management in Batoq Kelo Ulu. By knowing the needs and desires of village facilities and infrastructure, it can be an input for the government in making the priorities of village development. With the formation of community participation in the village planning process, communication was also formed between the community as the party who felt the impact of the development and the government as a policy maker [14].

The impact of the conflict that arose in the village was the need for community service facilities and infrastructure. The needs of Batoq Kelo Ilir is lower than Batoq Keloq Ulu because the availability of community service facilities and infrastructure are supported by the next village. Whereas batoq kelo ulu requires a lot of additions because this part of the village stands alone. If villagers need supporting facilities and infrastructure, they must go to the regency capital. Due to the distance from the regency capital is quite close, the people of Batoq Kelo Ulu tend to separate from Batoq Kelo Village and testablish new village.

The results show that Batoq Kelo Ulu needs to build new kindergarten, building quality improvement, and complementary facilities for elementary school, repairing or rebuilding village offices, repairing *Lamin adat* buildings, adding cultural fields around custom lamin buildings, improving the quality of facilities for worship and commercial facilities. In the aspect of infrastructure, it is prioritized for repairing the road environment and improving the quality of access to telecommunications and electricity.

In contrast to Kampung Batoq Kelo Ilir, the need for facilities and infrastructure is more for improving health facilities at community health service, repairing mosques, and adding worship equipment, adding shop / service units to sell daily necessities. In the aspect of infrastructure, needs additional residential roads with cement road pavement and communal clean water supply that managed by the BUMKampung, and increasing clean water piping coverage area and quality of paping system.

In Batoq Kelo village, it was directed develop BUMKampung in both settlement areas. This strategy aims to improve the government's financial capacity and villagers' income by involving the economic activities of the community and the potential of the Batoq Kelo village.

The conflict that occurred in Batoq Kelo Village is a rare problem. These problems emerged because of the process of sustaining community livelihoods. This conflict can be considered as natural process in human settlement development. However, these conditions have a negative impact on infrastructure development. Because it is separated into two parts, then from a legal point of view, the construction of facilities and infrastructure will experience legal issues. From the point of view of community development, the separation of the villages showed a lack of community harmony and formed two different communities, even mutually hostile.

Conflict in Batoq Kelo shows problems in terms of village development. one of the problems was supported by the local government by giving permission to open new settlements in the area outside the village administration. however, the tragedy carried out by the regional government shows a careful attitude in resolving the conflict. Mahamkam Ulu district government gets more financial assistance from the central government because of its status as a new autonomous region and border region. Therefore financial support for the development of areas that experiencing conflict is available.

The construction of the two groups of Kelo Batoq Villages requires higher costs than they should because the location of the two villages is separated. Settlement facilities and infrastructure for the community must be able to meet all the people who live within the scope of their territory. if the location of the village is far away, the construction of facilities and infrastructure will focus on the village and cannot support one another. villages or villages in Kalimantan usually have one residential center so that the construction of facilities and infrastructure can be utilized effectively. In addition, it can save village fund allocation for other activities.

4 Conclusions

The results show that there is a big problems to fulfilling community service facilities and infrastructure. First, in Batoq Kelo village is located at remote area that needs high financial support to improve their facilities and infrastructure. Second, Batoq Kelo Village has 2 sattlement areas that need to be met their facilities and infrastructure. As response, Batoq Kelo Ilir only receive 20% budget for improving community service facilities and infrastructure. Because it receive support from Ujoh Bilang Village which is Regency capital. 80% of development is concentrated in Batoq Kelo Ulu. Because, it needs to improve their facilities and infrastructure without support from neighbour villages.

The results also show that settlement facilities are complete but in some facilities have conditions that need quality improvement. the main problems of infrastructure in the villages of Batoq Kelo Ulu and Ilir are different. Batoq Kelo Ulu is lack access to telecommunications, electricity and roads. while in Batoq Kelo needs quality improvement of community service facilities and infrastructure.

Acknowledgment
This research was funded by the government of Mahakam Ulu Regency.

References

[1] Undang Undang Republik Indonesia nomber 6 Year 2014 regarding Village
[2] Surbakti, Ramlan 1992 *Memahami Ilmu Politik Gramedia* Widya Sarana Jakarta
[3] Riadi, Slamet 2007 *Analisis Pemekaran Wilayah Dan Potensi Konflik Di Kabupaten Donggala* Jurnal Academica Vol. 1, No. 2 ISSN 1411- 3341
[4] Macintyre, R and Macdonald, J 2011 *Remote from What? Perspectives of Distance Learning Students in Remote Rural Areas of Scotland International* Review of Research in Open and Distance Learning Vol. 12.4
[5] Sumpeno, W 2011 Perencanaan Desa Terpadu. Aceh : Reninforcement Action and Development. second edition
[6] Atmaja, H Kurniadi and Mahalli, K 2015 *Pengaruh Peningkatan Infrastruktur Terhadap Pertumbuhan Ekonomi Di Kota Sibolga* Jurnal Ekonomi dan Keuangan Vol. 3 no. 4 ISSN: 2303-3525
[7] Nur, F 2006 *Partisipasi Masyarakat Dalam Proses Pembangunan Infrastruktur Desa Studi Kasus : Program Alokasi Dana Desa di Desa Bialo Kabupaten Bulukumba.* Makkasar : Universitas Hasannudin
[8] Putnam, R D 1993 *The Prosperous Community: Social Capital and Public Life* The American Prospect
[9] Suriana 2009 *Analisis Keberlanjutan Pengelolaan Sumber Daya Laut Gugus Pulau Kaledupa berbasis partisipasi masyarakat* Bogor : Institut Pertanian Bogor
[10] Todd, Harry F. and Nader, Laura. 1978 *The disputing process, law in ten societies* New York : Colombia University Press
[11] Centers for Disease Control and Prevention (CDC) 2013 Community Needs Assessment Atlanta:GA
[12] Sutikno R 2011 *Perencanaan Infrastruktur Permukiman Perkotaan* Malang : Universitas Brawijaya
[13] Laksana, N Setyasa 2013 *Bentuk-Bentuk Partisipasi Masyarakat Desa Dalam Program Desa Siaga di Desa Bandung Kecamatan Playen Kabupaten Gunung Kidul Provinsi Daerah Istimewa Yogyakarta* Jurnal Kebijakan Dan Manajemen Publik Vol. 1 No 1 ISSN 2303-341X
[14] Laily Elida I N 2015 *Partisipasi Masyarakat Dalam Perencanaan Pembangunan Partisipatif* jurnal Kebijakan dan Manajemen Publik Volume 3, Nomor 2

IOP Conf. Series: Earth and Environmental Science **328** (2019) 012033 doi:10.1088/1755-1315/328/1/012033

Transborder Movements of Borderland People in Indonesia – Papua New Guinea (Case Study: Iwur District, Pegunungan Bintang Regency)

A Subagiyo[1], U S Adam[2] and A D Wicaksono[3]

[123] Urban and Regional Planning Department, Universitas Brawijaya, Indonesia 65145

*Email : arissubagiyo@ub.ac.id

Abstract. The border area between Indonesia and Papua New Guinea (PNG) is mostly defined based on geographic coordinates without considering cultural boundaries, which has resulted in high intensity of interaction and trans-border movements in the area. Trans-border movements in the border area of Indonesia and PNG, specifically in Pegunungan Bintang Regency border area, are classified as illegal movements. This is due to lack of border area facilities to accommodate these movements. There are varieties of purposes of trans-border movements in this area, which are classified into 4 categories. They are economic purposes, cultural purposes, social purposes, and political purposes. Circular or temporary movement is a trans-border movement category based on the maximum of 6 months of stay in the destination country. There are three factors that influence the borderlanders on conducting trans-border movements, namely individual characteristics factors, social-economy factors, and cultural factors. The movements are mostly conducted for economic purposes in addition to the influence of individual characteristics factors.
Keywords: border area, transborder movement

1.Introduction

Borderland is an 'arena' for trans-border crimes which also represents dynamic history in cultural and socio-economic exchange between two countries at the same time [1]. According to [2], trans-border agreement between two countries represents unity of borderland people in socio-economic and socio-cultural aspects.

Although borderlines are drawn precisely on maps and both countries have a bilateral treaty, in fact, the practice is quite different. No matter how many customs officials are placed in border area, how clear the borders are agreed, and how many watchtowers are built, people will ignore borderline whenever it suits them [3].

According to [4], migration is one of the borderland problems between Indonesia and Papua New Guinea (PNG). Understanding determinants and processes of trans-border movements in migration is needed for the development of the borderland area. The civilians' responsiveness to the different situation and condition of both countries causes the migration to occur. There are many factors which cause the existence of both legal and illegal trans-border movements between Papua and PNG [5].

Trans-border movements also exist in Pegunungan Bintang Regency. Pegunungan Bintang Regency belongs to Papua Province, consisting of 34 districts and 8 of them are adjacent to PNG. According to [5], Papuan of both Indonesia and Papua New Guinea shared the common traditions and habits. They had a good relationship with one another before the concept of country and constitution was applied. Since then, nationalism and security level in the borderland have been diminished.

Cultural factor such as Ulayat land and administrative boundary between the tribes of Pegunungan Bintang Regency and Papua New Guinea, social factor e.g. family, economic factor e.g. trade, and security factor like separatism and limited security in the borderland, have a huge impact on the movements in the Pegunungan Bintang Regency borderland. Trans-border movements in Pegunungan Bintang Regency are mostly illegally conducted, which are dominated by social-cultural activities like visiting families, marriage, hunting, and traditional ceremonies as well as economic activities like farming, shopping, and trading.

Related study on trans-border movements in borderland is needed, especially for those that happen in Pegunungan Bintang Regency in order to sustain the persistence of borderland in Pegunungan Bintang Regency as the exit-entry point.

2.Research Method

This research employs quantitative method. Quantitative research is one of the scientific research types which is systematic both in terms of its parts and relationship among them by using mathematical model [6]. This type of research uses measurable data and objective statistics data.

Unit of analysis is used as a unit of research object. Unit of observation is a source from unit of analysis and it is a part of unit of analysis which is used to draw a conclusion. Unit of analysis and unit of observation that are used in this research are:

1. Unit of analysis: people who live in the borderland between Indonesia and Papua New Guinea.
2. Unit of observation: people in the borderland (whose citizenships are of Indonesian and Papua New
Guinea) that lived in Iwur District during the research (in 2016).

2.1.Data Collection

This research used primary and secondary data collections methods. Primary data was obtained through direct observations; it is in the form of raw data. The primary data collection was divided into three methods; interview, questionnaire, and direct observation. Interviews were given to concerned parties like headmen and security officers i.e. Army, while questionnaires were given to the civilians belonged to the unit of observation. In the other hand, observation was aimed to identify the condition of facilities and infrastructures in the field of research.

2.2.Sampling

According to [7], sample is a part of population. The sampling technique used in this research is purposive technique with sample of people aged 15 and above because trans-border movements were mostly performed by people who are in their productive age, 15-64 years old [8]. The data from interviews and questionnaires was generalized to estimate the condition of population. This research used [9] sampling method to determine the interview sample size. Based on the result of sample calculation, the sample consisted of 208 people.

2.3.Identification of Trans-border Movement in Borderland

The identification of trans-border movements in the borderland was done by decrypting the data from questionnaires and interviews. Then, the data was described using descriptive statistics. This identification was aimed to find out the purpose, frequency, and type of movements, along with the living period in the destination country and how long the movements were performed.

2.4. Binary Logistic Regression Analysis

Regression analysis is an analysis to predict the impacts of independent variables toward the value of dependent variable. In this research, the dependent variable is dichotomy variable which shows someone's decision to do a trans-border movement. Thus, binary logistic regression analysis is used in this research. Whereas, the independent variables were divided into three factors, namely individual characteristic factor, social-economic factor, and cultural factor.

3. Results and Discussion

3.1. Types of Determining the Borderland in Indonesia – PNG
There are two types of determining the borderland in Papua Island, which is divided into two zones (Indonesia and Papua New Guinea). They are:

1. Natural Border, (the deepest point of Fly River) exists between Boven Digoel regency and Merauke regency in Papua as well as the Western Province zone in PNG.
2. Artificial Border, consists of border stakes built along the island following the astronomical line on 141° 01'00"BT from the north Papua to the south on the natural border, which then continued to the south passed the meridian on 141 ° 01' 10" BT.

Although the demarcation of Indonesia-PNG boundary between Indonesia and PNG has been arranged and has come to a decision that the borderline of Papua-PNG is +820 Km restricted by 52 border pillars, according to BNPP data on 2015, the borderland between Papua-PNG is not optimal and unfinished. Due to the landlords' ownership problem of tribes in borderland, determining the borderline of country, especially Papua-PNG needs further consolidation with tribes living at the borderlines.

The major problem on the long borderline is the difficulty of both countries in doing maintenance of the border stakes. One of the efforts to maintain the Papua-PNG border stakes was done by Indonesia by the end of 2016. Through this operation, Digi village, which administratively belongs to Indonesia but is subjected to PNG government, was discovered.

Ever since, borderland people were classified into two categories, transnational borderlanders or respondents who performed trans-border movements in 2016, and national borderlanders or those who did not perform trans-border movements in 2016.

In conclusion, the societies who are categorized as trans-national borderlanders are more than the national borderlanders. The result of this survey showed that society in the borderland of Indonesia – PNG, especially in Iwur District had an intense relationship with the society in the other side country. Approximately, 30% PNG civilians are female (transients), who became national borderlanders as they did not do any trans-border movements in 2016 and had also been resided in Indonesia for more than 6 months. The reason for this is that those people had married to the natives whose citizenship of Indonesian and live in Iwur District.

Iwur District is adjacent to 3 districts and 2 districts in PNG. Based on the result of the interview, Iwur District has 11 villages, but only 10 villages are registered in Pegunungan Bintang Regency.

3.2.Transborder Movements in Iwur Borderland

3.2.1.*Characteristic of movement lane*
Based on the survey results, there are three preferred routes from Iwur District to PNG or vice versa. They are Marantikin village in Tarub District (on the east of Iwur District), Ok Tedi track (one of the mining area in PNG), and the last is a track passed by people of Digi village.

3.2.2.*Respondent Characteristics*
Based on the interview results, trans-border movements were commonly performed by a group of 3 to 10 people who came from several families. All of the movements were done without passing the border posts. People usually walked from one village to another through several lanes. From the research samples of 204 people, 123 people are the doers of trans-border movements. Those samples consisted of 138 Indonesians and 66 PNG people. The movements were frequently performed by males, either Indonesian or PNG.

All respondents of the interview live in Iwur District. According to the education level, most of the residents are uneducated. According to the occupation, 172 respondents are farmer, 21 respondents are housewives, 1 respondent is a priest, 1 respondent is a nurse, 3 respondents are state civil apparatus (ASN), and 6 respondents are unemployed. Based on the survey results, 30 respondents that consist of 20 Indonesians and 10 PNG are OSEA members who did trans-border movements.

3.2.3.Trans-border Movements Characteristics.

During 2016, 123 respondents had done 451 trans-border movements. Depending on their length of stay, the movements are divided into two, migration and circular [10], while according to the purpose, they are classified into 4 [11], which are:

1. Economic purpose. It consists of three purposes, namely shopping, trading, and farming.
2. Social purpose. It consists of two purposes, namely visiting families and getting facilities of public services such as health and education.
3. Cultural purpose. It consists of three purposes, namely marriage, hunting, and traditional ceremony.
4. Political purpose, which is organization activity.

Farming and shopping which belong to economic purpose have the greatest impact on the trans-border movements in Iwur District borderland, while political purpose like organization activity has the least. Movements with the purpose of farming were mostly done in the harvest season, whereas movements for shopping were mostly done approaching the religious feast days such as Easter, Christmas, New Year, as well as the harvest season. Trans-border movements were generally temporary, i.e. 1 to 30 days. The permanent movements consisted of 6 movements. The journey variously lasted for 3 days to a week. Most of the respondents did trans-border movements for 10 to 12 days. Beside the purpose and duration of the movements, the movement frequency was also one of the factors being studied. The amount of trans-border movements done in 2016 was more than 8 times. In general, the doers were those who moved 3 up to 4 times within a year.

3.3. Analysing the Factors Affecting Borderland Interaction using Binary Logistic Regression Analysis

1.

The factors affecting borderland interactions in Iwur District border were analysed using binary logistic regression with dummy variables. It aimed to identify the factors that affected the trans-border movements in Pegunungan Bintang Regency, particularly in Iwur District. Here are dependent and independent variables used in binary logistic regression with dummy variables.

Y = 0, do not perform any migration
= 1, perform migration
JK1 = 1, if the respondent is male
= 0, if the respondent is female
Age = age of the respondent (years old)
SP1 = 1, if the respondent is married
= 0, if the respondent is not married
PT1 = 1, if the respondent is uneducated
= 0, if others
KO1 = 1, if the respondent is a part of organization
= 0, if the respondent is not a part of organization
KW1= 1, if the respondent is Indonesian
= 0, if the respondent is Papua New Guinean
Income = the respondent's average income per month
P1 = 1, if the respondent is employed
= 0, if the respondent is unemployed
TU1 = 1, if the respondent possesses a customary land in other countries
= 0, if the respondent does not possess a customary land in other countries
AS1 = 1, if the respondent comes from Murop tribe
= 0, if others
KA1 = 1, if the respondent participates in communalism
= 0, if the respondent does not participate incommunalism

3.4.Logistic Regression Similarity

Based on the results of analysis using SPSS 23, the similarity model formed is as the following:

$$\pi(x)=\frac{exp(5{,}222-0.103\,Usia\,Responden+2{,}074\,Kewarganegaraan+0{,}207\,Status\,Pekerjaan-0{,}820\,mengikuti\,adat)}{1+exp(5{,}222-0.103\,Usia\,Responden+2{,}074\,Kewarganegaraan+0{,}207\,Status\,Pekerjaan-0{,}820\,mengikuti\,adat)}$$

The use of the obtained model was simulated for the people (the samples) that did the movements and those who did not. Those who moved were simulated using dummy variable code 1 on each independent variable or on the annotated variable:

1. The respondent is at the age of 25
2. The respondent is an Indonesian
3. The respondent has an occupation
4. The respondent has desire for communalism

The motionless samples were simulated using dummy variable code of 0 on each independent variable or on the annotated variable:

1. The respondent is at the age of 68
2. The respondent is a PNG citizen
3. The respondent has an occupation
4. The respondent has desire forcommunalism

The calculation results for the simulation were 0,984 for the samples who did the movements and 0,069 for those who did not. The prediction results between code 0 (not doing trans-border movements) and code 1 (doing trans-border movements) are limited to 0,5 (cut value 0,5). If the result is more than 0,5 so, that samples are categorized as performing trans-border movements, while if it is less than 0,5 so, that samples are categorized as not performing trans-border movements.

4. Conclusion

According to the results and discussion, there are 6 conclusions that can be drawn, which are:

1. Based on the interview with people who live in borderland of Pegunungan Bintang Regency, it was found that 60% of the people are Transnational Borderlanders and 40% of them are National Borderlanders. It showed that the borderlanders of Indonesia – PNG, especially those in Iwur District bond with the people of the adjacent country. To conclude, Iwur District as a private location whose function as an exit-entry point, is expected to accommodate the activities of trans-border movements.
2. Based on the survey results, there were three preferred routes from Iwur District to PNG or the other way around. Those are Marantikin village in Tarub District (on the east of Iwur District), Ok Tedi track (one of the mining area in PNG), and the last is the track passed by people of Digi village. The route selection depended onthe proximity of the exit-entry point and the destination.
3. Based on the interview results, trans-border movements were commonly performed by a group of 3 to 10 people that came from several families. All of the movements were done without passing the trans-border posts. People usually walked from one village to another through several lanes for 3 days to a week.
4. Most respondents are farmers at the age of 30-34 and residents at the age of 35-39. Economic purpose like farming and shopping greatly drove trans-border movements. Most movements for farming were in the harvest season, while for shopping were close to religious feast days such as Easter, Christmas, New Year, and in the harvest season.
5. The amount of trans-border movements done by borderland people in Iwur District in 2016 is 8 times. It was dominated by people who moved 3 up to 4 times a year.
6. Based on the previous researches, there were 12 variables that drove people to do trans-border movements, namely gender, age, marital status, occupation, citizenship, educational level, separatist membership, income, customary landlord status in the adjacent country, tribe and communalism. Based on these results of regression analysis, there are 4 variables that significantly affect the movements (B value< Sig. 0,05). Those variables are age, citizenship, occupation, and communalism.

References

[1] Blum, C. 2014.Cross-Border Flows between Nigeria and Benin. Nigeria. Friedrich-Ebert Stiftung.

[2] Hadiwijoyo, S. S. 2009. Batas wilayah negara Indonesia: dimensi, permasalahan, dan strategi penanganan: sebuah tinjauan empiris dan yuridis. Yogyakarta. Gava Media.

[3] Baud, M. & Van Schendel, W. 1997.Toward a Comparative History of Borderlands. Journal of World History Vol. 8, No. 2, Fall, 1997. (pp. 211-242). Hawai'i. University of Hawai'i Press.

[4] Bandiyono, S., Raharto, A., Djohan, E., Yogaswara, H., & Hargiono, S. 2004.Mobilitas

[5] Setyawan. 2009. Mengelola Perbatasan Indonesia – Papua New Guinea dengan Pendekatan Keamanan Non-Tradisional. Jurnal Mengelola Perbatasan Indonesia di Dunia Tanpa Batas: Isu, Permasalahan, dan Pilihan Kebijakan: 149 . Yogyakarta. Graha Ilmu Yogyakarta.

[6] Creswell, J. W. 2003.Research design: Qualitative, quantitative, and mixed methods approaches (2nd ed.).Thousand Oaks. CA: Sage.

[7] Arikunto, S. 1998.Prosedur Penelitian: Suatu Pendekatan Praktek. Jakarta. Rineka Cipta

[8] Tjiptoherijanto, P. 2001.Proyeksi Penduduk, Angkatan Kerja, Tenaga Kerja, dan Peran Serikat Pekerja dalam Peningkatan Kesejahteraan. Majalah Perencanaan Pembangunan Edisi 23 Th. 2001.

[9] Godden, B. 2004. Sample Size Formula.Chicago: William Godden Website. http://williamgodden.com/samplesizeformula.pdf/. (diakses tanggal 1 Oktober 2017).

[10] Mantra, I. B. 2000. Demografi Umum. Yogyakarta. Pustaka Pelajar.

[11] Gunawan, T., Sukwardjono, Sukoco, M., Sudarsono A. & Suwadi. 2007.Fakta dan Konsep Geografi. Jakarta. Interplus.

IOP Conf. Series: Earth and Environmental Science **328** (2019) 012034 doi:10.1088/1755-1315/328/1/012034

Sustaining Innovation on Public Health Service by Building Creative Culture in Rural Areas: Case on Program of *Pemburu-Bumil-Resti* at Banyuwangi Regency - Indonesia

A Y Asmara[1] and A Y S Rahayu[2]
[1]Graduate Student of Public Administration and Policy at University of Indonesia (UI),Indonesian Institute of Sciences (LIPI)
[2]Professor of Public Administration and Policy at University of Indonesia (UI)

*Email: a.yuka.asmara@gmail.com

Abstract. Banyuwangi Regency, located in East Java Province has been popular several breakthroughs in public sector as initiated and promoted by Local Government of Banyuwangi. Innovation on public sector is aimed to deliver public services to Banyuwangi's local people. One of prominent innovation forms is on public health sector called as The *Pemburu-Bumil-Resti* Program, aimed to minimize the number of baby and mother mortality rate in Banyuwangi Regency. It is necessary for pregnant women with potential for high-risk birth especially for those occupying in rural areas. The program is successfully considered to improve local people health quality in rural areas, as appreciated by national and international organizations. Unfortunately, this program is still dominantly conducted by Health Agency of Banyuwangi Government at recent time. This study is post-positivistic approach by using case study method conducted in rural areas in Sempu District – Banyuwangi Regency. The research is on an ongoing study starting from 15 January to 15 February 2019. The temporary research finding states that The *Pemburu-Bumil-Resti* program should be reinforced through building creative culture on rural community to improve participation and awareness of this program and to sustain it in the next period of leadership in Banyuwangi Regency.

Keywords: Innovation, Health Service, Culture, Rural Areas, Banyuwangi

1. Introduction

1.1 Background

Banyuwangi Regency, located in East Java Province, becomes one of innovative local governments in Indonesia. Local Government of Banyuwangi (Pemkab-Banyuwangi) has received many innovatie and public awards at national and international level. There are about 50 exciting awards during leadership period of Mr Abdullah Azwar Anas as a Regent of Banyuwangi (2010 year-present) [1, 2, 3]. Banyuwangi Regent government has breakthrough in delivering public service, promoting new approach in delivering public health service to local people whom they inhabit in rural areas. Local people in rural areas, notably context in Indonesia, is often characterised as limited knowledge, depending on farming as main livelihood, and restricted in using other languages [4]. This phenomenon leads to importance of affordable and equitable access for local people in rural areas such as in urban areas to obtain public health service.

Banyuwangi Regent government has initiated and implemented new ways in delivering public health service for villagers in order to get affordable and equitable access. In this regard, Health

Agency of Banyuwangi Government *(Dinkes-Kabupaten-Banyuwangi)* is the most responsible actor to create easier, faster, cheaper, and more gratifying services for them. One of innovative programs on public health service is *Pemburu-Bumil-Resti,* a part of larger program called *Sakinah*. The program is aimed to look for pregnant women with potential for high-risk birth. It is totally facilitated by *Dinkes-Kabupaten-Banyuwangi*, and is technically conducted by mobile vegetable sellers which all of them are Banyuwangi's women [3, 5]. In rural areas of Banyuwangi, and some rural areas in East Jatim Province, the mobile vegetable sellers are usually called as *mlijo*. Actually, they do not only sell vegetables, but also sell various daily cooking needs such as meat, seasoning, food,beverage, and many more. It is firstly implemented in rural areas at *Sempu* District. In this area, it has been successfully implemented because it can press the number of baby and mother mortality rate until zero number in 2014-2015[6]. In 2017-2018 period, the number of baby and mother mortality rate is still zero, and Local Government of Banyuwangi receives the international awards of Open Government Partnership (OGP) on Asia-Pacific Regional Meeting in Seoul, South Korea [7].

To maximize the performance of this program, *Dinkes-Kabupaten-Banyuwangi* equips *mlijo* with smartphone and pulse, buckets containing information about pregnant women health, and boots [8]. Involvement of *mlijo* to find out pregnant women with potential for high-risk birth who live in rural areas is a reasonable way. They can sell vegetables to specific areas where there is no any public transport to achieve the location. Besides, this way assists to reduce burden at *Dinkes-Kabupaten-Banyuwangi* in delivering public health services. In this context, role of Health Agency of Banyuwangi Government at Sempu District *(Puskesmas-Sempu)* is very dominant in promoting program of *Pemburu-Bumil-Resti*, even *The Puskesmas-Sempu* is the main actor to stop number of baby and mother mortality rate in rural areas. However, the program of *Pemburu-Bumil-Resti* is adequately susceptible in that the program, which is highly relied on political situation at Local Government of Banyuwangi. Anggraini et. al, pointed out that to make robust local people in maintaining and improving the program, becomes a serious challenge regarding with low awareness of Banyuwangi's villagers about the importance of health [6]. The relevant studies indicate that involvement of local people or communities and related actors/stakeholders in promoting innovation in health public service is totally required [9, 10].

Active involvement of villagers at Sempu District to sustain *Pemburu-Bumil-Resti* program as an innovative program is actualized by building creative culture. According to Burdon et al. creative culture is one of pivotal factors in responding a change in an organization. In this term, organization is local community or local people occupying in rural areas at Sempu District, called as rural community [11]. Innovative culture is a precondition concerning how an innovation is mainly rooted from aggreement and cooperation of all actors, rural community, government, and supporting organizations. In rural area, building creative culture is not easy, OECD and Clavel note that building rural area provides a challenge requiring coordination and cooperation of across sectors at inter-disciplinary field. Certainly, it also needs mutual understanding among actors to achieve rural development goal [12, 13]. Therefore, this study is focused on building creative culture on three main actors such as: rural community at Sempu District, Health Agency of Banyuwangi Government at Sempu District *(Puskesmas-Sempu)*, and mobile vegetable sellers at Sempu District *(mlijo)*.

1.2 Problem Statement and Research Goal

Building creative culture by involving rural communities, *Puskesmas-Sempu*, and *mlijo* in sustaining innovative program of *Pemburu-Bumil-Resti* at Sempu District in Banyuwangi Regency requires attention on two things. Firstly, it is important to know perception of those actors about innovation on health service in rural coomunities. Secondly, It is necessary to identify builder factors of mutual interaction of those actors in delivering program of *Pemburu-Bumil-Resti* at recent time. In this regards, there are two main research questions that will be addressed on this study such as:

1. How is the perception of rural communities, *Puskesmas-Sempu*, and *mlijo* about program of *Pemburu-Bumil-Resti* at Sempu District in Banyuwangi Regency ?

2. What are builder factors of mutual interaction of *Puskesmas-Sempu* in collaboration with *mlijo* and rural communities to deliver program of *Pemburu-Bumil-Resti* at Sempu District in Banyuwangi Regency ?

This research goal is primarily aimed to answer both questions as above, including:

1. To describe the perception of rural communities, *Puskesmas-Sempu*, and *mlijo* about program of *Pemburu-Bumil-Resti* at Sempu District in Banyuwangi Regency.
2. To identify builder factors of mutual interaction of *Puskesmas-Sempu* in collaboration with *mlijo* and rural communities to deliver program of *Pemburu-Bumil-Resti* at Sempu District in Banyuwangi Regency.

1.3 Conceptual Review

1.3.1 Innovation on public sector organization

Innovation on public sector organization (PSO) is relatively new compared to innovation on business sector organization (BSO), early adopted by developed countries in Continental Europe and Anglo Saxon on 1980s [14]. In end of 1990s and in early 2000s, the utilization of PSO is widely applied in Asia countries and other developing countries to respond global challenges and public demands [15]. Basicaly, innovation definition either on PSO or BSO is similiar in terms of newness. According to Clavel, innovation is the implementation of a novelty or newness stemmed from an idea, process or technology. It is also a novel manner to produce several benefits for social, economic, and others such as the stakeholders and for community [13]. Innovation is not occuring in a vacum, but it is the result of a interaction process among various actors.

Alike BSO, innovation on PSO comprises newness on: product and services, service delivery way or process, organization, and marketing or diffusing [16]. In this context, innovation is primarily focused on newness on public service delivery way or process, notably on public health service sector. Despite the urgency to create innovation on public service delivery, it is often necessarily needed to make an innovative organization in making newness on it. It is important that each innovation on PSO is often as result of another innovation on previous newness. Innovation on PSO has striking differences to BSO. Windrum and Hartley provide at least two domains of innovation on PSO which are: 1) presence of political factor, 2) achievement not only for economy benefits, but also for socio benefits or may be political benefits for politicians. It is important to be noted that innovation on PSO is not simple as on BSO [17, 18], Lynn Jr. points outthat innovation PSO is a complex process involving various key actors to get new ideas and to get approval from politicians or stakeholders [19]. Based on specific characteristic of innovation on PSO, the researchers add that innovation on PSO is characterized by inclusivity, where all of related consumers (public) can access result of innovation.

1.3.2 Rural area and rural development

Rural area is a place where a population density is accounted under 150 dwellers per km^2, it is classified in thinly populated area. It is based on rural definition coined by Local Administrative Level 2 (LAU2), as part of the OECD survey [20]. European Commission defines rural areas is *"a spatial phenomenon that extends across regions, landscapes, natural areas, agricultural land, villages, and other larger urban centres, pockets of industrialisation and regional centres. It encompasses a diverse and complex economic and social fabric. It is the home of a great wealth of natural and cultural resources and traditions. It is becoming more important as a place for relaxation and leisure activities"* [21 p. 1]. Both definition are widely developed and utilized by countries in Europe. Definition of rural area is not single definition that can be applied at each country. Definition of rural area conveys various terms acccording to characteristics at each country. A particular definition should be developed at each country based on several measures such as: population size, socioeconomic structure of population, industrial area, and so on [20]. Ní Fhlatharta and Farrell and Vicol et al. reveal that traditional agriculture sector as main livelihood of villagers has currently shifted to service and industrial sector, especially on agriculture-based industry [23, 24]. According to Gilaninia, rural development is defined as "maturity of though and hard work and determined action of villagers to improve their community, through self-education, self-help and cooperation; thus, rural development can be considered as development of economic, social and culture"[25].

In Indonesia case, rural area is also known as village, a place where local dwellers and head of village can meet together to discuss and solve all affairs in this area [4]. Rural or village has recently had self-government since 2014, in which village government is able to arrange appropriate programs,in line with villagers needs. Even, there is special village funding allocated by central government to support rural development [22]. Defining a place as a rural area in Indonesia also considers local values and main economy sources in that place. There is not only specific limitation of population number occupying a village of rural, but the geographical factor becomes an influencing factor in determining term of village in Indonesia. Surely, Indonesia has many terms of rural which is different at each region, such as: nagari in West Sumatera Province and Desa in East and Central Java Province. Alike other countries, rural areas of Indonesia is growing due to globalization effect and increasing needs of villagers. Many new business units are developed by local people to improve local well-being, attempting to secure local values or culture that shape local behaviour of them.

1.3.3 Creative culture

Innovation practice on both BSO and PSO are not linear process, involvement and collaboration among various actors and stakeholders are essential to materialize it. Rather, innovation is built by robust ecosystem by involving active participation ranging from senior manager, innovation manajer [26] and individu or participants as beneficiaries of innovation [27]. Emergence of innovation is inseparable from culture embedded in an organization. Culture is often deemed as development base of a society, as a foundation to generate a particular ecosystem. Culture is defined as all activity of society, as important sector facilitating innovation to resolve problems or yield development in other fields, or even on culture field itself [28]. Culture is necessary to make innovation workable that supports innovation by not hindering culture.

Creative culture is a popular term in BSO environment, but it is rarely faced in PSO environment. Howkins defines creative or creativity as intellectual property, not private activity, spiritual matter, and explicited in public sphere as well [29]. Landry and Bianchini and Stamm state that creativity is stage of ideas generated, while innovation is stage of ideas implemented and used by consumers [30, [31]. In another meaning, innovation is a result of creativity to directly support innovation practice [11]. The creative culture definition becomes a pivotal factor in growing innovation in PSO. Building creative culture in PSO is very complex process due to existence of political factor in building creative culture and presence of various goals of beneficiaries.

2. Methods

This research is post-positivitism approach using two main concepts as a research analysis framework such as innovation and rural development. Both concepts are used to analyze innovative culture builders in sustaining innovation on public health service in 2 (two) rural areas at Sempu District in Banyuwangi Regency, East Java Province. The time period of research was from 15 January to 15 February 2019. Case studybecomes the main research method by collecting and compiling descriptive data derived from key informants (Table 1.). Case study isutilized to describe field phenomenon at whole which is inseparable. By this method, the study will yield descriptive qualitative data which is analyzed by using both concepts [32]. Thus, perception and valuation of researcher(s) in vieweing phenomena must be framed by a research guidance to keep a research focused on goal(s). The research site is conducted in two villages at Sempu District – Banyuwangi Regency covering Sempu Village and Jambewangi Village. Both villages are chosen because they have extreme difference in population number and public health facilities. Sempu is the smallest population area (5.441 people), while Jambewangi is the largest population area (20.740 people). In addition, Sempu village has 8 health service units for mother and baby *(Posyandu),* while Jambewangi village has 33 units of *Posyandu* [33].

IOP Conf. Series: Earth and Environmental Science **328** (2019) 012034 doi:10.1088/1755-1315/328/1/012034

Table 1. Key Informants List

Type of Informants	Number
Head and official staffs of *Puskesmas-Sempu*	3
Local communities in two villages: Sempu village and Jambewangi village at Sempu District as representative informants*	2
Women as *mlijo* at each village *	2
Women as target groups of program in two villages: Sempu village and Jambewangi village at Sempu District	2
Midwife at each village	2
Village government at each village	2
An expert innovation on public sector organization (PSO) and An expert in rural areas	1
Total	**14**

Note: * at each village by at least 1 informant

This study also applies a literatures review including relevant official documents, scientific articles, books, and other literatures related to innovation, culture, rural areas, health service, and Sempu District. To validate the field data findings, the collected data will be triangulated by means of discussion with the expert(s) understanding deeply about innovation in public sector and rural area development to minimze subjective interpretation conducted by researchers. However, intrepretation of researchers cannot be fully omitted in this study due to nature of research on case study method.

3. Results and Discussions

3.1 Public Health Service in Banyuwangi Regency

Banyuwangi Regency is 5.782,50 km^2 width with total population of 1.692.324 people and population density of 292,66 persons/km^2 in 2017 year [34]. To deliver public health service to local people in Banyuwangi Regency, Local Government of Banyuwangi has formulated and implemented public health programs to fulfill health need of Banyuwangi's local people.

3.2 The Program of Pemburu-Bumil-Resti and Its Effect

Pemburu-Bumil-Resti program becomes a specific part of *Sakinah* program to stop the number of baby and mother mortality rate in rural areas. The *Sakinah* program has been nominated as Top 33 Public Service Innovation by Ministry of Civil Aparratus Empowerment and Bureaucracy Reformation (KemenPAN dan RB) in 2015 [35]. In period of 2014-2015, the program could press the number of baby and mother mortality rate until zero number [6]. Indeed, until 2017-2018 period, the number of baby and mother mortality rate is still zero [7]. In 2016 year, *Puskesmas-Sempu* receives an international awards from GIZ, a non-profit organization which is operated by Germany's institution [35], and in 2018 year, this program has been appreciated on the international award of Open Government Partnership (OGP) held on Asia-Pacific Meeting in Seoul, South Korea [7].

3.3 Profile of Sempu Village and Jambewangi Village at Sempu District – Banyuwangi Regency

Sempu is a district consiting of 7 villages including: Sempu (width 4,9 Km^2), Tegalarum (5,77 Km^2), Jambewangi (15,59 Km^2), Temuasri (5,15 Km^2), Karangsari (12,7 Km^2), Temuguruh (8,5 Km^2), and Gendoh (6,1 Km^2). Total population of Sempu district is 72.323 people divided into 36.000 men and 36.323 women. Most of local people work as farmer (16.825 persons), followed by worker in processing-industry (4.414 persons), entrepreneur or worker on trade, restaurant, and hotel (9.058 persons), and as worker or owner on transportation, warehouse, and communication (875 persons). On health sector, there are 134 health units to deliver public health service to local people supported by health workers as accounted to 50 persons comprised by doctors, dentists, nurses, and midwifes. Besides, there are traditional health assistants who assist birth process for pregnant women (we call as *dukun bayi*) accounted to 30 persons to assist in keeping people health through massage (we call as *tukang pijat*) accounted to 68 persons [33].

3.4 Perception of Rural Communities, Agency of Puskesmas-Sempu, and Mlijo about Program of *Pemburu-Bumil-Resti*

Three actors such as: rural communities, *Puskesmas-Sempu*, and *Mlijo* have various perception to respond program of *Pemburu-Bumil-Resti* at Sempu District. It is apparent that common perception towards such actors remains good indicating that this program is positively worthwhile for local people occupying in rural areas at Sempu District – Banyuwangi Regency, especially for pregnant women. Based on their perception, this program is not seperated from real action of staffs at *Puskesmas-Sempu* in collaboration with *mlijo* and rural communities. Each actor also states that the program of *Pemburu-Bumil-Resti* will face serious challenge at future when there is local political leader election by 2020 in Banyuwangi Regency. Generally speaking, at local government level in Indonesia, each top local leader is substituted by another top local leader, followed with substitution at line of middle and low manager. Automatically, the program is also substituted with new program initiated by new top local leader. In this case, a stable ecosystem in a supporting frame for sustaining the program of *Pemburu-Bumil-Resti* is necessarily demanded.

3.5 Builders of Creative Culture in Sempu District

Building good and positive ecosystem to support the program of *Pemburu-Bumil-Resti* becomes one of the wishes perceived by rural communities, agency of *Puskesmas-Sempu, and Mlijo.* At recent condition, mutual interaction of those actors to conduct the program of *Pemburu-Bumil-Resti* has been totally triggered by Agency of *Puskesmas-Sempu* approximately during 3 years. This mechanism is susceptible in political condition uncertainity by 2020 where it will be top political leader substitution in Banyuwangi Regency. To foster a sustainable program suc as *Pemburu-Bumil-Resti*, stable ecosystem is neccesariy created not merely by Agency of *Puskesmas-Sempu,* but also by collaborating several key actors such as rural comunities, *mlijo,* academics, and related actors of Banyuwangi Regency in promoting sustainability of this program. Rural communities that are fully understanding the real situation of village, is expected as main trigger in sustaining this program, understanding deeply about number, address, condition, and other update data regarding with pregnant women with potential for high-risk birth in villages.

4. Conclusion

In Sempu district, three actors such as rural communities, *Puskesmas-Sempu*, and *Mlijo* have responded positively towards the program of *Pemburu-Bumil-Resti* at recent time. In the future, they consider sustainbility of this program by 2020 due to political leader election. Therefore, it is required a stable ecosystem to foster sustainable program during the political uncertainity era at local level. Building creative culture is a must to implement the program of *Pemburu-Bumil-Resti* sustainably. To foster a sustainable program, rural communities role should be improved to contribute on bargaining position towards local government agencies and political leaders.

References

[1] Pemkab Banyuwangi. 2013. Banyuwangi Raih gelar Travel Tourism Award 2013 "The Most Creative". (Banyuwangi receives Travel Tourism Award 2013 "The Most Creative"). Retrieved from *https://www.banyuwangikab.go.id/berita-daerah/banyuwangi-raih-gelar-travel-tourism-award-2013-the-most-creative.html (21 November 2018). [In Indonesian].*

[2] Pemkab Banyuwangi[a].2016. Laporan Kinerja Pemerintah Kabupaten Banyuwangi Tahun 2015. (Performance Report of Local Government of Banyuwangi in 2015 Year). Retrieved from *https://banyuwangikab.go.id/new/media/doc/edoc/lakipbanyuwangi2015.pdf (21 November 2018). [In Indonesian].*

[3] Pemkab Banyuwangi. 2018. Laporan Kinerja Instansi Pemerintah Kabupaten Banyuwangi Tahun 2017 (LKjIP). (Performance Report of Local Government of Banyuwangi in 2017

Year) Retrieved from https://kinerja.banyuwangikab.go.id/dok/lkjipbanyuwangi2017.pdf (21 *November 2018). [In Indonesian].*

[4] Alamsyah, M. Nur. 2011. Memahami Perkembangan Desa Di Indonesia *[Understanding Rural Development in Indonesia]. Jurnal ACADEMICA Fisip Untad 3 (2):647-660 [In Indonesian].*

[5] Pemkab Banyuwangi[b]. 2016 (23 Desember). Di Banyuwangi Tukang Sayur Plus Jadi Pemburu Ibu Hamil. (In Banyuwangi, the vegetable sellers are also the searcher of pregnant women). Retrieved from https://www.banyuwangikab.go.id/berita-daerah/di-banyuwangi-tukang-sayur-plus-jadi-pemburu-ibu-hamil.html (10 *Desember 2018). [In Indonesian].*

[6] Anggraini, Mita Dian; Murdyastuti, Anastasia; dan Wasiati, Inti. 2017. Inovasi Pelayanan SAKINA di Puskesmas Sempu Kecamatan Sempu Kabupaten Banyuwangi *(SAKINA Service Innovation at Sempu Public Health Center, Sempu District, Banyuwangi Regency). E-SOSPOL IV (2):75-81. [In Indonesian].*

[7] Detik News. 2018. Program Pemburu Bumi di Banyuwangi Raih Penghargaan Dunia. (Pemburu-Bumil Program has been appreciated at international level). *Retrieved from* https://news.detik.com/berita-jawa-timur/d-4291866/program-pemburu-bumil-di-banyuwangi-raih-penghargaan-dunia (25 *January 2019) [In Indonesian].*

[8] Kabar Banyuwangi. 2016 (December 24). Cooperate Mlijo So Hunter High Risk Pregnant Women. Retrieved from https://www.kabarbanyuwangi.info/en/gandeng-mlijo-jadi-pemburu-ibu-hamil-risiko-tinggi.html. *(25 January 2019).*

[9] Scheer, Wilma van der; Noordegraaf, Mirko; and Meurs, Pauline. 2011. Innovating Entrepreneurship in Health Care: How Health Care Executives Perceive Innovation and Retain Legitimacy. In Victor Bekkers, Jurian Edelenbos and Bram Steijn (Eds). *In Innovation in the Public Sector Linking Capacity and Leadership. (pp. 176-194)*. Palgrave Macmillan, Hampshire, UK.

[10] Vickers, Ian; Lyon, Fergus; Sepulveda, Leandro; and McMullin, Caitlin. 2017. Public Service Innovation and Multiple Institutional Logics: The Case of Hybrid Social Enterprise Providers of Health and Wellbeing. *Research Policy 46: 1755–1768.*

[11] Burdon, Stephen; Al-Kilidar, Hiyam; and Mooney, Grant. 2013. Evaluating an organisation's Cultural Readiness for Innovation. *International. Journal of Business Innovation and Research 7(5):* 572-589.

[12] OECD. 2014. *Innovation and Modernising the Rural Economy*. OECD Publishing. http://dx.doi.org/10.1787/9789264205390-en.

[13] Clavel, Danièle. 2014. *Knowledge and Rural Development-Dialogue at the Heart of Innovation. London*: Springer. DOI 10.1007/978-94-017-9124-3.

[14] Osborne, Stephen P. 2013. A Services-Influenced Approach to Public Service Innovation? In Stephen P. Osborne and Louise Brown (Eds.), *Handbook of Innovation in Public Services (pp. 60-71).* Cheltenham, UK: Edward Elgar.

[15] Valkama, P., Stephen, J. B., & Ari-Veikko, A. (2013). Contexts and Challenges of Organizational Innovation in Public Services. In Pekka, Valkama; Stephen, J. Bailey and Ari-Veikko, Anttiroiko (Eds.), *Organizational Innovation in Public Services Forms and Governance (pp. 3-12).* Hampshire, UK: Palgrave Macmillan.

[16] Gault, Fred. 2018. Defining and Measuring Innovation in All Sectors of the Economy. *Research Policy, 47: 617–622.*

[17] Windrum, P. 2008. Innovation and Entrepreneurship in Public Services. In Paul Windrum and Per Koch (Eds.), *Innovation in Public Sector Services: Entrepreneurship, Creativity and Management (pp. 3-20)*. Cheltenham, UK: Edward Elgar.

[18] Hartley, J. 2013. Public and Private Features of Innovation. In Stephen P. Osborne and Louise Brown (Eds.), *Handbook of Innovation in Public Services (pp.44-59).* Cheltenham-UK: Edward Elgar.

[19] Lynn, Jr., L. E. (2013). Innovation and Reform in Public Administration: One Subject or Two? In Stephen P. Osborne and Louise Brown (Eds.), *Handbook of Innovation in Public Services (pp. 29-43).* Cheltenham, UK: Edward Elgar.

[20] Dijkstra, Lewis and Poelman, Hugo. 2014. A Harmonised Definition of Cities and Rural Areas: The New Degree of Urbanisation. Regional Working Paper 2014 WP 01/2014. A series of short papers on regional research and indicators produced by the Directorate-General for Regional and Urban Policy. European Commission. *http://dx.doi.org/10.1787/9789264205390-en.*

[21] Tapiador, Francisco J. 2008. *Rural Analysis and Management: An Earth Science Approach to Rural Science.* Berlin Heidelberg: Springer-Verlag.

[22] Suwandana, Endan; Mutaqin, Agus Zaenal; Rostiawati, Enong; and Oktaviana, Oki. (2015). Kajian Penguatan Lembaga Kolaboratif dalam Penguatan Desa Inovatif di Provinsi Banten. (Collaborative Institution Reinforcement Study in Reinforcing Innovative Village in Banten Province) *National Proceedings of Pertemuan Ilmiah Tahunan (PIT) Nasional ke – 2 Ikatan Widyaiswara Indonesia (IWI) Provinsi Banten Pandeglang, pp. 453-467 [In Indonesian].*

[23] Ní Fhlatharta, Aoife M. and Farrell, Maura. (2017). Unravelling the Strands of 'Patriarchy' in Rural Innovation: A Study of Female Innovators and Their Contribution to Rural Connemara. *Journal of Rural Studies 54:15-27.*

[24] Vicol, Mark; Neilson, Jeffrey; Hartatri, Diany Faila Sophia; Cooper, Peter. 2018. Upgrading for whom? Relationship Coffee, Value Chain Interventions and Rural Development in Indonesia. *World Development 110: 26–37.*

[25] Gilaninia, Roghayeh. (2015). Village, Villagers and Rural Development. *Singaporean Journal Of Business Economics, And Management Studies, Vol.3, No.6, pp. 39-44.*

[26] Mensch, Gerhard O.. 1985. Perspective: Get Ready for Innovation by Invasion. *Journal of Production Innovation Management 4:259-265.*

[27] Sun, Jun. 2016. Tool Choice in Innovation Diffusion: A Human Activity Readiness Theory. *Computers in Human Behavior 59: 283-294.*

[28] Jelinčić, Daniela Angelina. 2017. *Innovations in Culture and Development - The Culturinno Effect in Public Policy.* Cham, Switzerland: Palgrave Macmillan. DOI 10.1007/978-3-319-52721-5.

[29] Howkins, John, 2007. Enhancing Creativity. Retrieved from *http://www.india-seminar.com/2005/553/553%20john%20howkins.htm.*

[30] Landry, Charles and Bianchini, Franco. 1995. *The Creative City.* London-UK: Demos.

[31] Stamm, Bettina von. 2003. *Managing Innovation, Design and Creativity.* Chichester, West Sussex –England: John Wiley & Sons Ltd.

[32] Yin, Robert K. 2003. *Case Study Research: Design and Methods.* London-UK: SAGE Publications.

[33] BPS Kabupaten Banyuwangi. 2017 Kecamatan sempu dalam angka 2017. (Sempu District in Figure of 2017 Year). Retrieved from *https://banyuwangikab.bps.go.id/publication/download.html (28 January 2019) [In Indonesian].*

[34] Pemkab Banyuwangi. 2019. Kependudukan-Data Persebaran Penduduk (Population-Data and Populatiom Deployment. Retrieved from https://www.banyuwangikab.go.id/profil/kependudukan-dan-naker.html *(25 January 2019). [In Indonesian].*

[35] Pemkab Banyuwangi[c]. 2016. LKPJ Bupati Banyuwangi Akhir Tahun Anggaran 2016. (Performance Report of Local Government of Banyuwangi in 2016 Year) Retrieved from *https://banyuwangikab.go.id/new/media/doc/edoc/LKPJ%20ATA%202016.pdf (21 November 2018). [In Indonesian].*

The Analysis of Landslide Vulnerability in Settlement of Citizenship Association VI, Sukorejo Village, Semarang

B Rahman[1*], R Fimanasari[1] and U C Sari[2]

[1]Urban Regional and Planning, Sultan Agung Islamic University, Semarang, Indonesia
[2]Civil Engineering Department, Diponegoro University, Semarang, Indonesia

*Email: bobyrahman@unissula.ac.id

One of the disasters that must be noticed in mountains and hills is landslide. This is especially if the area is used as settlement. Semarang as metropolitan city with various topography is not separated from the use of its hilly area into residential areas. One of them is in Gunungpati District. However, this utilization of land can cause this area prone to landslide. Sukorejo Village, especially in Citizenship Association (CA) VI, is the landslide prone areas in Gunungpati District, Semarang. Almost every year, Citizenship Association (CA) VI of Sukorejo Village experiences landslide causing damage to infrastructures. Therefore it is needed to analyse the landslide vulnerability in Citizenship Association (CA) VI of Sukorejo Village. Thus, the vulnerability of the area is evaluated as disaster mitigation in accordance with physical, social, economic and institutional criteria. This study utilized quantitative method by scoring analysis to measure the level of landslide vulnerability. The vulnerabilities which were analysed were physical, social, economic, and institutional. The results of this study indicate that settlement in Citizenship Association (CA) VI of Sukorejo Village has high physical vulnerability, moderate to high social vulnerability, moderate economic vulnerability and low institutional vulnerability to disaster.
Keywords : Landslides, Settlements, Disasters Vulnerability

1. Introduction

Landslide fatalities are concentrated in less developed countries, where there is little investment to understand the dangers and risks associated with landslides, mainly due to lack of appropriate resources [1]. This condition indicates that the concentration of handling disaster problems is indeed appropriate to be considered in developing countries..

According to statistical data from the National Disaster Management Agency [2], there have been more than 4,607 landslide cases in Indonesia in the past 10 years. Based on the data, it was also stated that around 24,407 people were victims. Therefore, with an increase in the incidence of landslides every year, it is necessary to determine the level of vulnerability of a region to landslides to make prevention and protection efforts, including the strength of the buildings.

One reason for the high number of cases and landslide victims is the need for settlements. Settlement becomes one of the main human needs as a place for housing and space in carrying out its life activities. The rapid growth rate of the population is very influential on the demand for settlements and housing needs. In fact, the population of residential land is greater than the availability of land to settle (demand is greater than supply) [3] due to. The development of the population in settlements is dynamic while the existing land area is permanent, so it cannot compensate for the growing needs of settlements. This imbalance of needs affects the compulsion of the community to settle on existing land, even though the land is not feasible to be built in a disaster-prone area [4].

The resilience of the people to choose living in disaster-prone areas can be lately set in making decisions, whether they remain in the area or choose to stay [5]. The choice to stay and settle in a disaster-prone location is influenced by certain factors, one of which is to anticipate potential disasters that will occur with the "resilience" that is owned by the community. On the other hand, the settlement of the community can also be influenced by the socio-economic conditions of the community that do not allow them to move.

The case study in this paper was in the settlement of Sukorejo Village of Semarang City, A settlement located in an area prone to landslides in the city of Semarang. Sukorejo Village is included in the zone of high ground movement, namely an area that has a high degree of vulnerability for the occurrence of soil movements [6].

1.1 Literature Review

Landslides are a geological event caused by the movement of rock or soil mass which can be defined as the displacement of the material forming a slope, which is in the form of original rock or embankment material that moves according to the earth's gravitational force [7]. Landslides are due to several factors such as nature and humans. Thus, it is necessary to consider these aspects in identifying vulnerabilities and disaster mitigation efforts. Physical conditions are due to earthquake prone zones, hilly terrestrial forests with high rainfall intensity, where some areas in Java are prone to landslides [8].

Financial capability and knowledge of the community about landslides also affect vulnerable areas [9]. Financial capability will help people adapt to upgrade buildings and regions, while knowledge capabilities will help people adapt to behaviour and actions. This paper analyzes both aspects, such as the physical and human aspects (social, economic and institutional) in terms of vulnerability, in order to provide input on mitigation efforts from both aspects.

2. Methods

This study utilized a quantitative research method with a weighting or scoring approach. The aspects that would be reviewed were aspects of the physical vulnerability of the area, economic vulnerability, social and institutional vulnerability. These three aspects have their own parameters.

2.1 Social Vulnerability Parameter

The physical vulnerability of the land can be determined based on weighting parameters of daily rainfall intensity, slope, soil type, land movement and land use. Variables are divided into 3, which are: low vulnerability (score 1), moderate vulnerability (score 2) and high vulnerability (score 3). The weighting parameters of land physical vulnerability are depicted in Table 1.

Table 1. Weighting parameters of Physical Vulnerability in Citizenship Association (CA) *VI*

Parameter	Variable	Score	Source
Daily Rainfall Intensity	Low< 20,7 mm/hr	1	[10]
	Moderate 20,7-27,7 mm/hr	2	
	High 27,7– 34,8 mm/hr	3	
Slope	0-15%	1	
	15%-25%	2	[10]
	>25%	3	
Soil Type	Low sensitive: Alluvial, gleisol, planosol, G, laterite groundwater, latosol Less sensitive : Aluvial, glei, planosol, hodromorf kelabu, laterite, latosol	1	[10]

Parameter	Variable	Score	Source
	Rather sensitive : Brown forest soil, caleie brown, mediteran	2	
	Sensitive : Andosol, laterite, grumosol, podsol, regosol, litosol, organosol, renzina	3	
Land Movement	Low<0,0016 Mild 0,016-0,081 High>0,081	1 2 3	[11]
Land Use	Habitable Settlement Non-habitable Settlement	1 3	[12]

2.2 Social Vulnerability Parameter
Social vulnerability can be determined based on weighting parameters of population density, percentage of population of vulnerable age groups and average level of education of the household head. Social vulnerability variable is also divided into 3, which are low vulnerability (score 1), moderate vulnerability (score 2) and high vulnerability (score 3). The parameters of weighting social vulnerability are presented in Table 2.

Table 2. Weighting parameters of Social Vulnerability in Citizenship Association (CA) *VI*

Parameter	Variable	Score	Source
Population density	Low (<150 person/ha) Moderate(100-200 person /ha) High (>200 person /ha)	1 2 3	[13]
Percentage of population of vulnerable age groups	Low < 5 % Moderate 5 % - 10 % High>10%	1 2 3	[14]
The percentage of the average level of education of the household head	Low (College) Moderate (High School) High (Secondary School, Primary School/Ungraduated/Uneducated)	1 2 3	[15]

2.3 Economic Vulnerability Parameters
Economic vulnerability is identified based on the weighting of the parameters of the number of poor people, the percentage of unemployment, and the percentage of decent homes. Variables are divided into 3 as two previous vulnerability reviews, which are low vulnerability (score 1), moderate vulnerability (score 2) and high vulnerability (score 3). The economic vulnerability weighting parameters are depicted in Table 3.

Table 3. Parameters of Weighting Economic Development in *Citizenship Association (CA) VI*

Parameter	Variable	Score	Source
Amount of Poor Population	Low < 10 % (below hard core poverty)	1	[16]
	Medium 10%-20% (hard core poverty)	2	
	High >20% (Absolute)	3	
Percentage of unemployment	Low < 20 %	1	Interval Class Analysis
	Medium 20 % - 40 %	2	
	High >40%	3	
Percentage of non-liveable houses	Low< 25 %	1	[17]
	Medium26 % - 74%	2	
	High>75%	3	

2.4 Institutional Vulnerability Parameters
The vulnerability parameter of the institutional aspect is assessed based on the interval level [18], which are: Low Vulnerability = has mitigation institutions as well as disaster management and has a village disaster risk reduction forum, score 1; Moderate vulnerability = only has one of the regional disaster management agencies or village disaster risk reduction forums, score 2; High Vulnerability = There are no regional disaster management agencies and village disaster risk reduction forums, score 3.

2.5 Vulnerability Score Calculation
The three aspects of vulnerability are calculated in a score to get the vulnerability. To divide the vulnerability class score, the interval is calculated between the highest score and the finest score divided by the number of interval classes (high, medium and low vulnerability).

$$Vul\ nerability = \frac{Total\ of\ High\ Score - Total\ of\ Low\ Score}{Interval\ Class} \tag{1}$$

Table 4. Score Guideline of Physical, Social, Economic and Institutional Vulnerability

Vulnerable Aspects	High	Moderate	Low
Physical	>13	12-9	< 9
Social	> 7	7-6	< 6
Economic	> 7	7-6	< 6
Institutional	3	2	1

3. Results and Discussions
Sukorejo Sub district has experienced several landslides every year which causes damage to infrastructure. In 2014, Sukorejo experienced a landslide caused by a downturn in heavy rains that occurred in the Semarang City area for a long time. This incident resulted in 39 houses being victims of severe damage and some of them actually collapsed by the brunt of the land [19]. Then, in 2015, it experienced a landslide triggered by heavy rain causing several houses damaged by landslides [20]. Furthermore, in February 2017, Sukorejo Village again experienced a landslide at 9:30 p.m. The landslide that occurred was a sinking gutter because it was not strong enough to hold the ground after the area had been rained for 30 minutes [21].

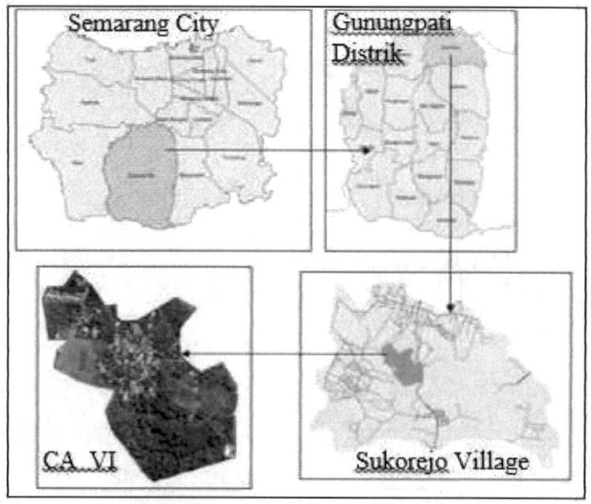

Figure 1. Research Location

In April 2017, landslide occurred in Sukorejo Village [22]. The landslide was quite severe causing the road to collapse and leaving only a few meters wide of the road. Some locations of landslide points in Sukorejo Village are presented in Figure 2.

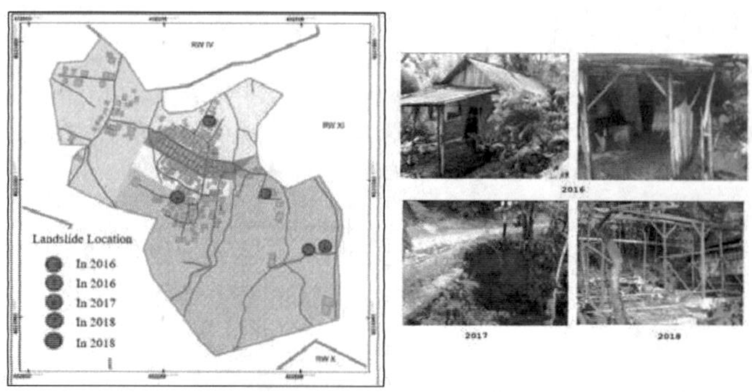

Figure 2. Landslide Location

The disasters that occurred in Sukorejo Village did not force the community to leave the settlements they occupied, although the settlements often experienced landslides. The local government has offered to relocate them to Rusunawa at Kaligawe, thus they still leave their settlements [23].

3.1. Physical Vulnerability Calculation

Physical conditions at the location have similarities, rainfall intensity of 27.7 mm / day –34.8 mm / day resulting in a score of 3, slope with 15% -25% or score 2, dark brown Mediterranean soil type or score 3, ground motion> 0.08 or score 3, and conformity to the spatial plan that has not matched the designation with a score of 3. The results of the physical vulnerability calculation score indicate a classification of scores less than 9 which means low vulnerability, 9-12 is a moderate vulnerability and more than 12 is high vulnerability. According to that result, Citizen Association (CA) VI of Sukorejo Village has a high physical vulnerability to land.

3.2. Social Vulnerability Calculation

Based on the social vulnerability weighting parameters and the weighting results in Citizen Association (CA) VI of Sukorejo Village (Table 3 and Table 4), the results of the calculation of social vulnerability scores indicate a classification of less than 6 indicating low vulnerability, 6-7 is a moderate vulnerability, greater than 7 is a high vulnerability. Some spots of Citizenship Association (CA) VI of Sukorejo Village have a high social vulnerability, such as in Neighbourhood Association (NA) 2 and 3.

Table 5. Social Vulnerability Score Results

CA VI	Social Vulnerability						Total Score	Classification
	Population Density		*Vulnerable Group*		*Education Level of Household Head*			
	Existing	*Score*	*Existing*	*Score*	*Existing*	*Score*		
NA 1	200	2	9,3	2	Primary School	3	7	Moderate Vulnerability
NA 2	238.2	3	9.4	2	Primary School	3	8	High Vulnerability
NA 3	71.1	1	11.9	3	Primary School	3	8	High Vulnerability
NA 4	16.5	1	11.7	3	Primary School	3	7	High Vulnerability
NA 5	64	1	9.1	2	Primary School	3	6	Moderate Vulnerability
NA 6	28.7	1	13.1	3	Primary School	3	7	Moderate Vulnerability

3.3 Economic Vulnerability Calculation

Based on the economic vulnerability weighting parameters above, the weighting results in Citizenship Association (CA) VI of Sukorejo Village is presented in Table 6. The results of the calculation of economic vulnerability scores indicate a classification less than 5 indicaring a low vulnerability, where 5-7 is a moderate vulnerability and more than 7 is a high vulnerability. Some spots of Citizenship Association (CA) VI of Sukorejo Village have a high economic vulnerability, such as in Neighbourhood Association (NA) 1 and 4.

Table 6. Economic Vulnerability Score Results

CA VI	Economic Vulnerability						Score	Classification
	Total of Poor Populatin		*Unemployed Rate*		*uninhabitable house*			
	Existing	*Score*	*Existing*	*Score*	*Existing*	*Score*		
NA 1	50.85%	3	29.7	2	80.43%	3	8	High Vulnerability
NA 2	32.04%	3	17.5	1	80.39%	3	7	Moderate Vulnerability
NA 3	36.30%	3	15.9	1	80.00%	3	7	Moderate Vulnerability
NA 4	46.85%	3	24.2	2	79.41%	3	8	High Vulnerability
NA 5	29.55%	3	16.5	1	79.07%	3	6	Moderate Vulnerability
NA 6	48.60%	3	13.3	1	80.65%	3	7	Moderate Vulnerability

IOP Conf. Series: Earth and Environmental Science **328** (2019) 012035 doi:10.1088/1755-1315/328/1/012035

3.4 Institutional Vulnerability Calculation

The vulnerability of institutions is related to the role and function of institutions, where both the community and government are able to manage the risk of landslides. Institutional vulnerability Citizenship Association (CA) VI of Sukorejo Village has a low level of vulnerability; this can be seen from the possession of an agency that manages landslide disaster risks and discussion forums.

Based on field reviews and study of [24], the institutional role in managing landslide disaster risk has been very good. This can be seen from the existence of the Disaster Preparedness Village allowing the community to participate in the implementation of disaster management. This can create a disaster response of community to minimize the impact of disasters; to increase the ability of the community to deal with disasters before, during and after a disaster.

In addition, there are institutions such as the Semarang City Regional Disaster Management Agency which have 3 functions in disaster management, such as in emergency and logistics, prevention and preparedness, and rehabilitation and construction. There is also a Community Service Activity from the academy which holds a Landslide Disaster Mitigation Training that is aimed for women community in Citizenship Association (CA) VI.

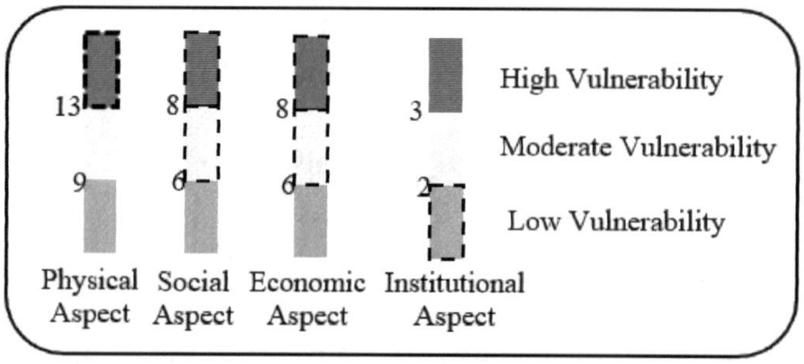

Figure 3. Position of Landside Vulnerability Each Aspect in Settlement of Citizenship Association VI, Sukorejo Village, Semarang

4. Conclusion

The conclusion of this study is the settlement of Citizenship Association (CA) VI in Sukorejo Village which has high physical vulnerability. This is because the physical condition of the area is on a slope with high rainfall intensity. The settlement of Citizenship Association (CA) VI in Sukorejo Village has high and moderate social vulnerabilities. In this study, population density and population structure according to education level became one of the factors that drive the high social vulnerability in the study area.

Citizenship Association (CA) VI in Sukorejo Village has medium economic vulnerability. The large numbers of uninhabitable homes in the region, poverty and unemployment rates are included from influencing factors. This must be prevented because many people live in a disaster area thus, it is necessary to have an element of economic readiness in the community to form resilience to disasters. Citizenship Association (CA) VI in Sukorejo Village has low institutional vulnerability; the existence of Disaster Prepared Village has become one of the efforts to realize the disaster response community.

The suggestions are provided to address the majority of high vulnerability in physical, social and economic conditions and settlement locations that are not in accordance with the designation of Semarang City and Regional Spatial Plan in 2011-2031, relocating the people to safety and spatial suitability settlement. The mitigation option in the existing settlement activities is to provide physical strength of the area such as environmental engineering for safeguards because this aspect becomes a high vulnerability at the study area. Another mitigation option is to provide socialization and training related to disaster mitigation to increase community sensitivity and awareness to reduce aspects of social and economic vulnerability.

References

[1] D Petley 2012 Global patterns of loss of life from landslides *Geology* vol 40 no 1 pp 927–930.

[2] BNBP 2019 Bencana Alam Di Indonesia Tahun 1998 S/D 2018.

[3] M Arief and B Pigawati 2015 Kajian Kerentanan di Kawasan Permukiman Rawan Bencana Kecamatan Semarang Barat, Kota Semarang *J. Tek. PWK* vol 4 no 2 pp 332–344.

[4] M Usamah, J Handmer, D Mitchell, and I Ahmed 2015 Can the vulnerable be resilient? Co-extence of vulnerability and disaster resilience: informal settlements in the Philippines *Int. J. disaster risk Reduct* vol 10 no 1 pp 178–189.

[5] M. Macchi *et al* 2008 *Indigenous and Traditional Peoples and Climate Change* IUCN

[6] Sugalang and Siagian 1991 *Peta Zona Kerentanan Gerakan Tanah, Jawa Lembar Semarang dan Magelang Skala 1:100.000* Bandung: Direktorat Geologi Tata Lingkungan

[7] B Nurlaksito and B Legowo 2012 Identifikasi Bidang Gelincir Pemicu Bencana Tanah Longsor Dengan Metode Resistivitas 2 Dimensi Di Desa Pablengan Kecamatan Matesih Kabupaten Karanganyar *Indones J Appl Phys* vol 2 no 1 pp 51–60.

[8] A Saputra, J Sartohadi, D S Hadmoko, and C Gomez Geospatial Assessment of Coseismic Landslides in Baturagung Area *Forum Geogr* vol 29 pp 99–113.

[9] Widjonarko and H Wijaya 2014 Pemetaan Potensi Bencana Longsor Di Kelurahan Kembang Arum *Geoplanning* vol 1 no 2 pp 93–101.

[10] SK Kementan Republik Indonesia 1980 Tentang Kriteria Dan Tata Cara Penetapan Hu Tan Lindung.

[11] K Sugianti and D Mulyadi 2014 Pengklasan Tingkat Kerentanan Gerakan Tanah *Ris. Geol. dan Pertamb* vol 24 no 2 pp 93–104.

[12] Badan Perencanaan Kota Semarang 2011 Rencana Tata Ruang Wilayah Kota Semarang

[13] Kementrian Pekerjaan Umum 2004 SNI 03-1733-2004 tentang Tata Cara Perencanaan Lingkungan Perumahan.

[14] P Setyaningrum and S R Giyarsih 2012 Identifikasi Tingkat Kerentanan Sosial Ekonomi Penduduk Bantaran Sungai Code Kota Yogyakarta Terhadap Bencana Lahar Merapi *J Bumi Indones* vol 1 no 3 pp 261–269.

[15] I N Dewi, Zulkarnain, and R K S Utami 2012 Faktor - Faktor Penyebab Anak Lulusan Sd Tidak Melanjutkan Ke Sltp Di Desa Marga Batin Kecamatan Waway Karya Kabupaten Lampung Timur Tahun 2009-2012 *Jurnal Penelit Geogr* vol 2 no 6 pp 1–9.

[16] D Wahyudi and T W Rejekingsih 2015 Analisis Kemiskinan di Jawa Tengah *J Econ* vol 2 no 1 pp 1–15.

[17] Kementrian Pekerjaan Umum 2016 Pedoman Penyusunan RP2KPKP.

[18] M Habibi and I Buchori 2013 Model Spasial Kerentanan Sosial Ekonomi Dan Kelembagaan Terhadap Bencana Gunung Merapi *J Tek PWK* vol 2 no 1 pp 1–10 2013.

[19] Jateng Tribunews.com 2014 Longsor, Puluhan Rumah di Trangkil Gunungpati Ambruk dan Rusak Parah

[20] Berita Jateng 2015 Longsor Timpa Tiga Rumah di Gunungpati.

[21] Jateng.Tribunews.com 2017 BREAKING NEWS: Longsor di Sukorejo Gunungpati, Tak Ada Korban Jiwa.

[22] Radarsemarang.com 2017 Jalan Longsor, Sekolah Terancam

[23] Liputan6.com 2014 Puluhan Korban Longsor Sukorejo Akhirnya Direlokasi.

[24] R Firmanasari 2017 Karakteristik Ketahanan Masyarakat Di Kawasan Rawan Longsor (Studi Kasus : RW IV Kelurahan Sukorejo Kecamatan Gunungpati) Universitas Islam Sultan Agung.

IOP Conf. Series: Earth and Environmental Science **328** (2019) 012036 doi:10.1088/1755-1315/328/1/012036

Urbanization Phenomena of Small Inland City: Case of Parakan, Central Java

B Septiarani[1], W Handayani[2], and S Ma'rif[3]

[1] Department of Civil and Planning, Vocational School, Diponegoro University, Semarang, Indonesia
[2,3] Department of Urban and Regional Planning, Faculty of Engineering, Diponegoro University, Semarang, Indonesia

*Email: bintang.septiarani18@pwk.undip.ac.id

Abstract. Urban development in Indonesia has been notably figured out with the development of coastal area as the place to provide regional-international trade having the control of labour. Hence, coastal cities and the surrounding areas are rapidly growing to urbanized area. However, rapid development with limited carrying capacity and climate related issues of coastal area has become the challenges of their development. Meanwhile, the growth of inland city also started to gain some attentions. In Central Java, for instance, the emerging of inland city centre has become a new phenomenon in urbanization. Parakan, one of the inland cities in Central Java selected as the study area due to the growing of rural-urban characteristic. This new inland city centre is a part of Temanggung Region, Central Java strategically located in the national road of Semarang to Purwokerto. The intention of this paper is to elaborate the growth of inland city in Central Java, especially in Parakan. The GIS is utilized to analyse the spatial changes in Parakan meanwhile the economic aspects are analysed to find out the indication of shifting economic activities within the city. From the analysis, it is found that changes in population appear in urban area as one of indicators of urbanization in Parakan inland city. As a consequence, the economic activities and the GDP also experience changes due to the growing of urbanized area. The growing of Parakan has been proven that the urbanization phenomenon in Central Java occurs in the inland city inferring that the urbanization pattern in Indonesia, specifically in Central Java, is not only limited to the growth of the coastal cities.

1. Introduction

The urbanization has been understood as the common phenomenon of urban development. According to the United Nation, the world urban population is expected to increase from 3.6 billion in 2011 to 6.3 billion in 2050 [1] which means that the growing urban population will reach 72% by 2050. Urbanization can also be simply understood as a demographic phenomenon [2]. However, the demographic phenomenon led to broader issues regarding with physical and environmental changes causing land use transformation. Land prices and land availability are the common factor behind land use transformation. Unbalance functional changes from farm use to non-farm uses emerge as the result of rapid modification and urbanization. Thus, the urbanization also leads to the environmental degradation in the new urbanized area.

In Indonesia, especially in Central Java, urbanization was started with the development of coastal city and the area surrounding it. For instance, the major big city in Central Java is Semarang City

IOP Conf. Series: Earth and Environmental Science **328** (2019) 012036 doi:10.1088/1755-1315/328/1/012036

which is also the coastal city. Despite the rapid growth of Semarang as a coastal city, coastal area of Semarang faces its own vulnerability. In contrast with the development, coastal areas are mostly prone to the environmental hazard such as climate change, which would affect the urban development. Most climate-related disasters, such as floods, tidal flooding, and windstorms, affect coastal areas [3]. Hence, the emerging of inland city development should be encouraged to bridging the balanced urban development of coastal area and other urbanized area surround them [4].

Parakan as one of inland cities in Central Java is a unique sample of urban development. The development of this small city is influenced by the existence of national road to the tourism area and the plywood industry. This strategic position has implications for the role of the city as a service center for the region behind it, where the urban area of Parakan tends to be a distribution city for the other cities as the provider due to insufficient sources of primary raw material, for which the other regions acts as the provider.

Growth and development of a region can also cause changes in spatial use of the area. Hence, this paper aims to further elaborate the urbanization of inland city in Central Java, especially in Parakan through analyzing the growth of the built up area, the changes of activities, growth of population in urban area, and the changes of GDP. By understanding the urbanization characteristic of the area, it is beneficial for the future development of Parakan as the growing inland city.

2. Methods

2.1. Study Area

The study area of this research is Parakan urban area. The urban area of Parakan includes Parakan Sub-district as the center of economic and government of Temanggung Regency. This sub-district is located on the slopes of Mount Sindoro-Sumbing and the urban area of Parakan sub-district is crossed by the route from Wonosobo Regency to Yogyakarta City/Semarang City and crossed by the route from Yogyakarta City to the Pantura Line/Jakarta. Urban and non-urban region in Parakan are identified through the latest data from the Central Bureau of Statistic (BPS) [5].

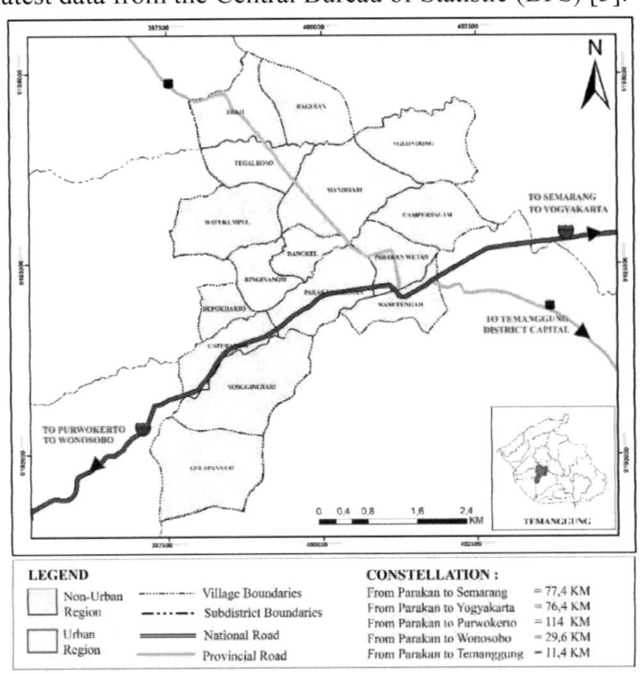

Figure 1. Study Area of Parakan

2.2. Data Collection

The data for this research was spatial data and socio-economic data of Parakan. The spatial data consist of the built up area maps from the year of 1990 to 2016 which were derived from the Google Maps. The maps are from the year of 1990, 2000, 2010 and 2016 of Parakan indicating the built up area. Besides, the data of rural-urban status from the Central Bureau of Statistic in the year of 1990, 2000, and 2010 are also utilized to analyze the growth of Parakan inland city. Meanwhile, economic data are derived from the Central Bureau of Statistic.

2.3. Methods

There are two methods to further elaborate the urbanization in Parakan, through the Geographical Information System (GIS) analysis and from the analysis of economic and activities aspect. The series of maps are processed in GIS to see the changing of built up area. Meanwhile, descriptive analysis was applied to analyze the changing economic activity and the growth of GDP in Parakan.

3. Results and Discussions

3.1. The growth of Parakan

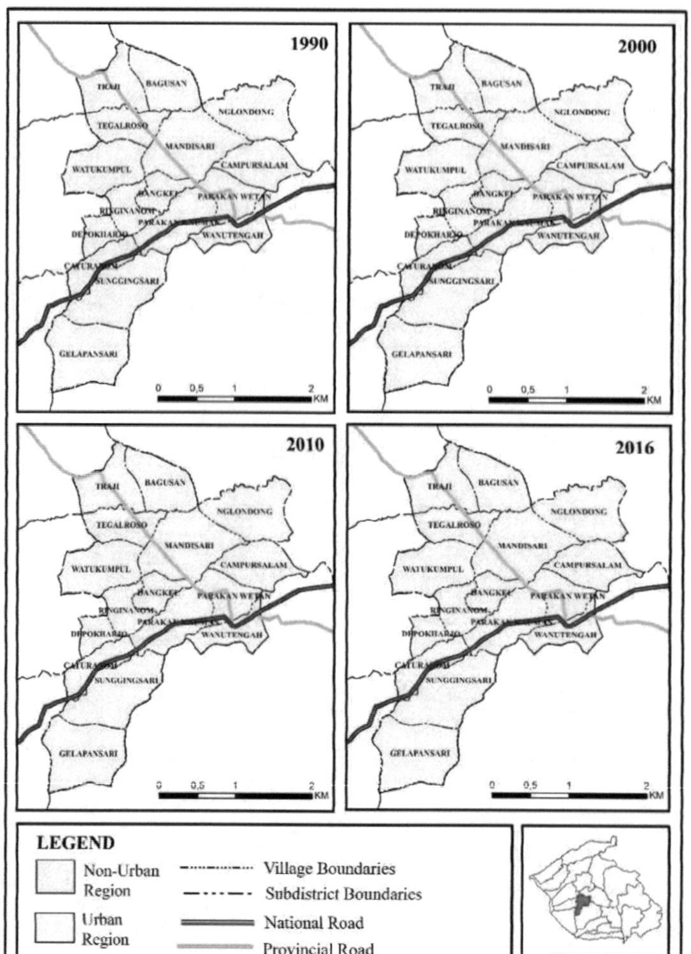

Figure 2. Urban and Non-Urban Region of Parakan from 1990-2016

The number of population in Parakan in 2017 was 52.346 with total of 42.089 people living in urban area meaning that more than 80% of people in Parakan are living in urbanized area. Figure 2 illustrates the number of growing urban village of Parakan. Based on the rural-urban data of BPS, the number of village with urban status are growing from only three villages in 1990 to eleven villages in 2016. The change of status from rural to urban mostly occurs in the village directly crossing to the national and provincial road. There are additional 8 villages which were adjanted to the national and provincial road that changed from rural to urban villages within 1990 to 2016.

Table 1. Built up Area in Parakan (ha)

Year	Built up Area (ha)
1990	230,74
2000	292,33
2010	347,51
2016	406,45

In line with the growing number of urban status of village in Parakan, the growing number of urbanized area in Parakan is also supported with the growth of the built up area (as referred to Table 1). By viewing the maps of built up area in Parakan from time to time, the growing of built up area as seen in Figure 3, is mostly along the national and provincial road. Those roads are the route of Wonosobo Regency to Yogyakarta City/Semarang City as well as crossed by the route from Yogyakarta City to the Pantura Line/Jakarta. Those roads attract many passerby hence the route connects the tourism activity in Wonosobo Regency, Borobudur Temple (Magelang, Central Java), Yogyakarta City and alsoto other cities like Purwokerto City to Semarang as the capital city of Central Java.

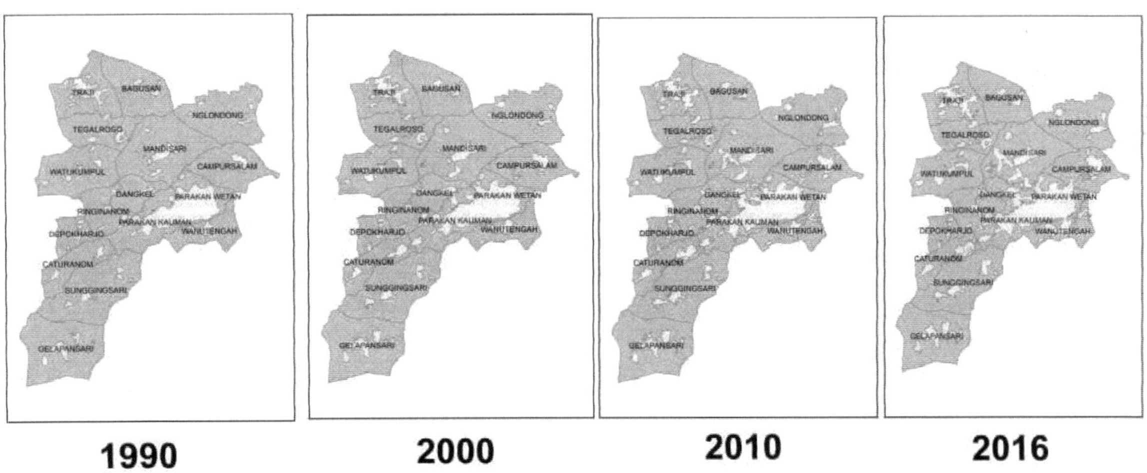

Figure 3. Built-up area Changes of Parakan Inland City

3.2. Activities and GDP Changes of Parakan

The existence of national and provincial road in Parakan is essential to the urbanization in Parakan. Figure 4 indicates the land use changes along the national road. It is apparent on Figure 4 that the lands use along the national road covers industrial cluster, central market and shop house. The growing of small industrial area is spreading along the road to Purwokerto and Wonosobo Regency. These are the changing land use from agricultural area to the industrial area although the numbers are not quite high. However, it is obvious that the changes are potentially getting wider along the road. Meanwhile, there is also growing of shop house along the national road, owned by private household which find the opportunity in selling goods along the strategic place. The national road crossing Parakan is considered as strategic place for the growing of shop house due to the numbers of passersby of the tourist from and to the nearby tourism area which are in Dieng Plateu (Wonosobo Regency), Borobudur Temple (Magelang Regency), and tourism city like Yogyakarta. The shop houses in Parakan are mostly selling the traditional goods such as local souvenir and traditional snacks made by local people in Temanggung Regency especially from Parakan.

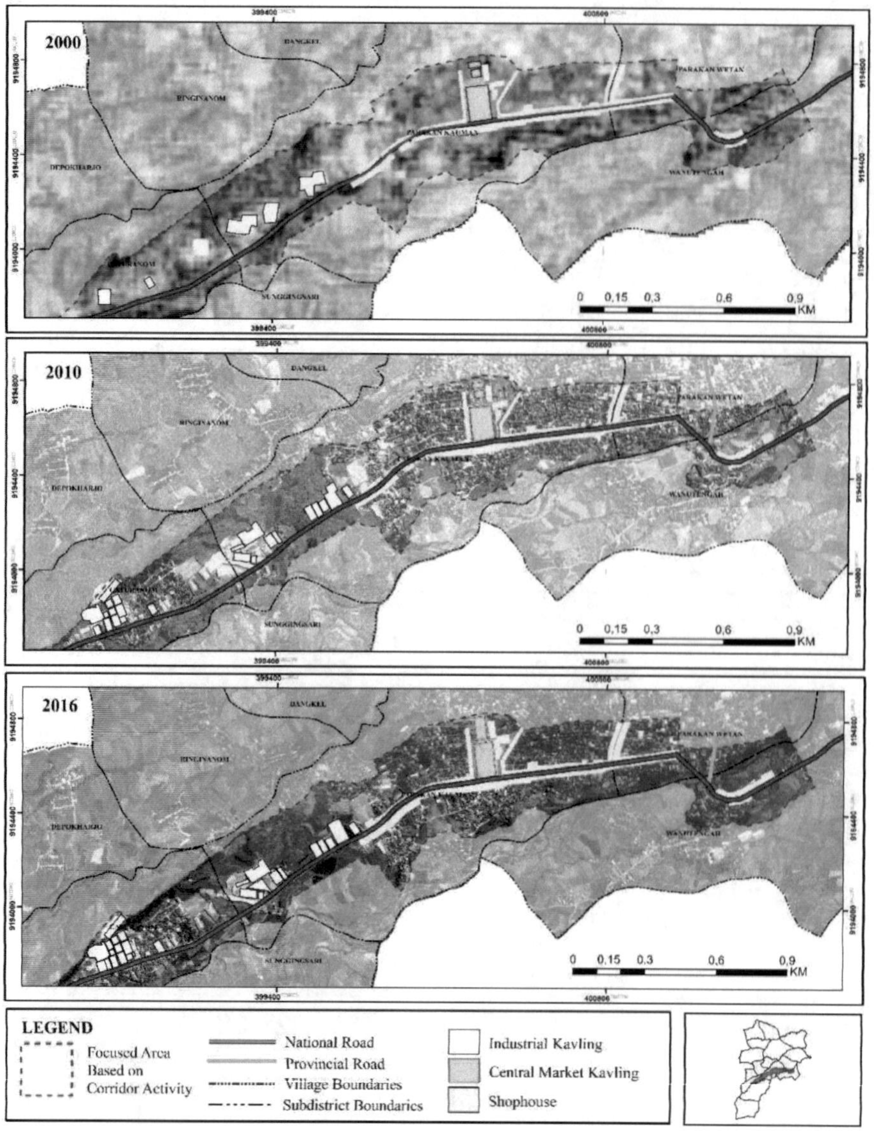

Figure 4. Spatial Changes alongside the National Road in Parakan

From the socio-economic aspect, there are also the shifting of economic sector within 2010-2014. Table 1 illustrates the percentage of economic sector from Gross Domestic Product (GDP) of Parakan. Table 2 depicts that the agricultural sector is slightly decreasing from 17,33% in 2010 to 12,98% in 2014. In contrary, the processed industry sector has been growing from 23,95% in 2010 to 25,55% in 2014. In line with the processed industry, the building/construction as well as trading, restaurant and hotels have been growing more than 1% from 2010 to 2014. The percentage indicates that the agricultural sector in Parakan slightly decreases to the secondary and tertiary economic activities such as industry, trading, restaurant and hotels (Figure 5).

Table 2. The Percentage of GDP for each sectors in Parakan, 2010-2014

Economic Sector		2010	2011	2012	2013	2014
Primary	Agricultural	17,33	14,14	14,13	13,82	12,98
	Mining	0,55	0,51	0,46	0,46	0,44
		17,88	**14,65**	**14,59**	**14,28**	**13,42**
Secondary	Processed Industry	23,95	25,03	25,57	25,48	25,55
	Gas, Water and Electricity	1,09	1,12	1,12	1,12	1,11
	Building/construction	4,48	4,54	4,71	4,73	4,95
		29,52	**30,69**	**31,4**	**31,33**	**31,61**
Tertiary	Trading, Restaurant and Hotels	23,31	23,82	23,93	23,64	24,14
	Transport and Communication	10,09	10,35	9,87	10,02	10,09
	Finance, Rental and Service Company	6,3	6,82	6,74	7,11	7,31
	Services	12,91	13,67	13,48	13,62	13,43
		52,61	**54,66**	**54,02**	**54,39**	**54,97**

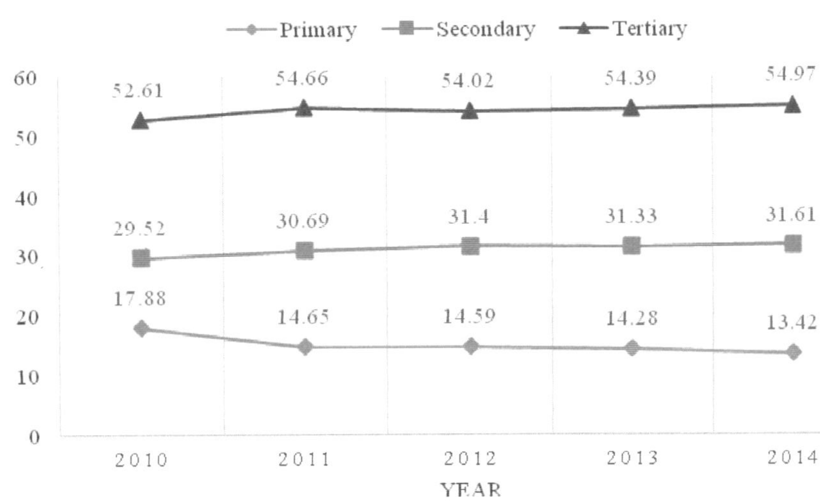

Figure 5. GDP Percentage for Each Economic Sector

4. Conclusion

This paper presents the findings that the urbanization occurs in the small inland city. Many scholars have been proved that the urbanization in Indonesia occurs in the coastal area [6][7]. However, the elaboration of the growth of the built up area, the changes of activities, growth of population in urban area, and the changes of GDP in Parakan indicate that the urbanization phenomenon occurs in the small inland city such as in Parakan, at Central Java.

IOP Conf. Series: Earth and Environmental Science **328** (2019) 012036 doi:10.1088/1755-1315/328/1/012036

Acknowledgement
The authors would like to thank Diponegoro University for administering the funding assisting the team to conduct proper research activities. However, all contents of this publication are solely the responsibility of the authors/researchers.

References

[1] U. Nations, "World Urbanization Prospects The 2011 Revision," 2012.
[2] W. Handayani, "Rural-Urban Transition in Central Java: Population and Economic Structural Changes Based on Cluster Analysis," pp. 419–436, 2013.
[3] I. Rudiarto, "A Regional Perspective on Urbanization and Climate-Related Disasters in the Northern Coastal Region of Central Java , Indonesia," *land*, vol. 7, no. 34, 2018.
[4] J. S. Setyono, H. S. Yunus, and S. R. Giyarsih, "THE SPATIAL PATTERN OF URBANIZATION AND SMALL CITIES DEVELOPMENT IN CENTRAL JAVA: A CASE STUDY OF SEMARANG-YOGYAKARTA-SURAKARTA REGION," *Geoplanning*, vol. 3, no. 1, pp. 53–66, 2016.
[5] Central Bureau of Statistic, *Kecamatan Parakan dalam Angka*. BPS, 2018.
[6] UNFPA, *UNFPA Indonesia Monograph Series: Urbanization in Indonesia*. UNFPA, 2015.
[7] J. S. Setyono, W. Handayani, I. Rudiarto, and L. Esariti, "Resilience Concept in Indonesian Small Town Development and Planning : a Case of Lasem , Central Java," in *E3S Web of Conferences*, 2018, vol. 68, pp. 1–8.

Effects of Coal Freight Road Policy Reforms on Transportation and Environmental Economics

D Yuniar[1], A Wicaksono[2], L Djakfar[3] and A Efendi[4]

[1]Department of Civil Engineering, Achmad Yani Banjarmasin University, Banjarmasin, Indonesia/ Doctoral Student Brawijaya University, Departement of Civil Engineering, Malang, Indonesia
[2,3]Department of Civil Engineering, Brawijaya University, Malang, Indonesia
[4]Department of Statistics, Brawijaya University, Malang, Indonesia

*Email: dewiyuniar_81@yahoo.co.id

Abstract. Since 2017, the application of the Provincial Government of South Borneo No.3 of 2012 regarding the use of public and particular roads for transportation of mining products and plantation companies has a positive impact on changes in transportation and environmental economics with the principle of benefit, justice, and interests of the nation. This study analyzes the effects of policy changes on coal transportation routes on transportation (traffic congestion) and environmental economics (air pollution, noise, loss value). The research method involved literature reviews and field surveys. The results indicated that after the 2017 policy, there was a decrease in the volume of coal transportation passing public roads by 92.4%, a decrease in speed and traffic congestion of 13% and 0.2 points. The noise level decreased to 4.87 dB (A). CO2 emissions reduced by 44.30%, CO by 48.82%, HC by 46.22%, Nox by 22.03%, SOx by 22.01% and PM10 by 25.66%., the value of losses due to the burden of vehicle emissions decreased by 24.3%. Visually, pedestrians and road users become more comfortable and safe in traffic. The government is expected to support and supervise particular roads for coal transportation to avoid the unexpected effect on roads.

Keywords: policy, coal transportation, economy, environment

1. Introduction

Coal provides a sizeable financial contribution to the region but raises problems from the transportation side from the stockpile to the port. According to Law No. 4/2004 clause 91, the public infrastructure facilities applicable and beneficial for mining purposes [1] granting the permits for coal transportation entrepreneurs to pass public roads in several districts/cities. Thousands of coal transport trucks move from the mine area to the port or factories having a payload of more than eight tons with modified tubs [2], disturbing road users, pedestrians and surrounding community. Regarding transportation, the traffic volume and congestion increase, speed decreases and inconveniences of road users exacerbate the existence of coal transportation. Regarding environmental economics, air pollution from traffic and coal dust, the air temperature increases, noise levels and vehicle emission load the increase.

In 2008 South Kalimantan Province issued regional regulation No.3 of 2008 and revised Regional Regulation No. 3 of 2012 about the use of public and particular roads for the transportation of mining products and the results of plantation companies. According to clause 3 (1), every transport of mining products and the effects of plantation companies are prohibited from passing public roads; (2) Every mining product and yield of a plantation company must transport through a particular road determined by the Governor [3, 4]. Dispensation for two years gives an opportunity for mining companies to build specific roads for coal transportation. The provincial government policy requires District/City supporting systems because it involves intergovernmental relations and building harmonization among stakeholders to accelerate regional socio-economic development. Dissemination of information in

October, November and December 2016 was declared to transport coal to not cross public roads. In January 2017, the integrated team installed a portal on the Banjarmasin-Marabahan national road as a form of law enforcement for coal transportation that does not comply with the regulations. The mixed team oversaw the policy consisting of elements from the South Kalimantan Regional Police, the South Kalimantan Provincial Transportation Agency, Korem 101/Antasari, and the Public Works Center.

The policy is successful if it had a good impact based on mutual interests. Therefore, it is essential to analyze the changes in coal freight policy on transportation (congestion, volume, and speed of traffic) and environmental economics (air pollution and noise levels), for providing road users a comfortable and safe experience.

1.1 Related Literature

1.1.1 Study of the policy of transporting coal, transportation, and the environment

Various strategies of coal transportation present the legal foundation such as Law Number 4 of 2004 about mining of minerals and coal, article 91 concerning RI Law Number 4 of 2009 about Mineral and Coal Mining, that IUP and IUPK Holders can use public infrastructure and facilities for mining purposes after fulfilling the provisions of legislation, as well as Regional Regulation Number 3 of 2008 and Number 3 of 2012 [1, 3, 4] . Coal transportation must go through a particular road that was built by the mining company [5]. In the study of transportation and environmental economics about coal in the world [6], several legal foundation are presented such as in, the Law as Number 22 of 2009 concerning traffic and road transportation rules (clause 210) [7], and Regulation of the Minister of Environment Number 12 of 2012 concerning guidelines for controlling emissions in the region [8].

1.1.2 Motor vehicle emissions, air pollution, loss and noise values

Emissions are substances, energy and other components produced from an activity that enters into ambient air which has potential as a pollutant element [8]. Air pollution threatens public health in various cities in Indonesia, especially large cities such as respiratory diseases, hypertension, impaired kidney function, decreased intellectual ability of children, coronary heart disease, and premature death. The components of air pollution mainly sourced from transportation activities in Indonesia are carbon monoxide (CO), Nitrogen oxides (NOx), Sulfur dioxide (SO2), Hydrocarbons (HC), PM10 and other particles. Similar research on the level of dust around the truck and the highest load of deposited coal is at a distance of one meter from the roadside which is 267.74 tons/km^2/month, and the lowest is 200 m from the roadside which is 19.895 tons/km^2/month. Dust levels at all sampling points (South Borneo) are above the environmental quality standard of 10 tons/km^2/month [8] [9]; the influence of PM10 and PM2.5 on health [10]. NOx affects the respiratory system, SO2 can worsen respiratory or cardiovascular diseases, and CO is poisonous and in high concentrations can lead to unconsciousness and even death [11]. Other related studies include, impact of traffic congestion on health [12]; the environmental effects of changing speed limits (a quantile regression approach) [13]; the effect of economic policy instruments in promoting environmentally sustainable transport [14]. Increased traffic density causes vibrations and noise along the roadside. In theory, if humans are exposed to the intensity of sound in the numbers 55-65 dB(A), it can cause narrowing of the arteries and an increase in heart rate frequency. If it persists, it can increase the chances of heart and blood vessel pain. Noise is an unexpected sound from a business or activity at a certain level and time that can disrupt human health and environmental comfort [15].

2. Methods

The research was carried out on public roads that were crossed by coal transportation in the Banjarmasin City, Indonesia. The research method was based on the literature review (2012 traffic data) and field survey for eighteen hours of observation in 2012 and 2017.

3. Results and Discussions

3.1. Traffic congestion

Calculated by applying the road capacity (C) (smp/hour) formula [16],

$$C(smp/jam) = C_0 \; x \; FCw \; x \, FCsp \, x \, FCsf \; x \, FCcs \tag{1}$$

With Co (basic capacity for certain conditions, smp/hour); FCw (Traffic Line Adjustment Factor); $FCsp$ (Direction Separation Adjustment Factor); $FCsf$ (Adjustment Factor Side Barriers); $FCcs$ (City Size Adjustment Factor); Q (traffic flow capacity, smp/hour). The degree of saturation (DS) [16]:

$$DS = \frac{Q}{C} \tag{2}$$

The results of road capacity analysis and degree of saturation before and after policy implementation are explained in Figure 1, with Co = 2,900 smp/hour; FCw = 1.07; $FCsp$ = 1; $FCsf$ = 0.99; $FCcs$ = 0.94 so that C = 2,887.7 smp/hour. The equivalent factor of a passenger car (emp) is a light vehicle (LV) = 1.0; heavy vehicles (HV) = 1.2 and motorbikes (MC) = 0.25 [16].

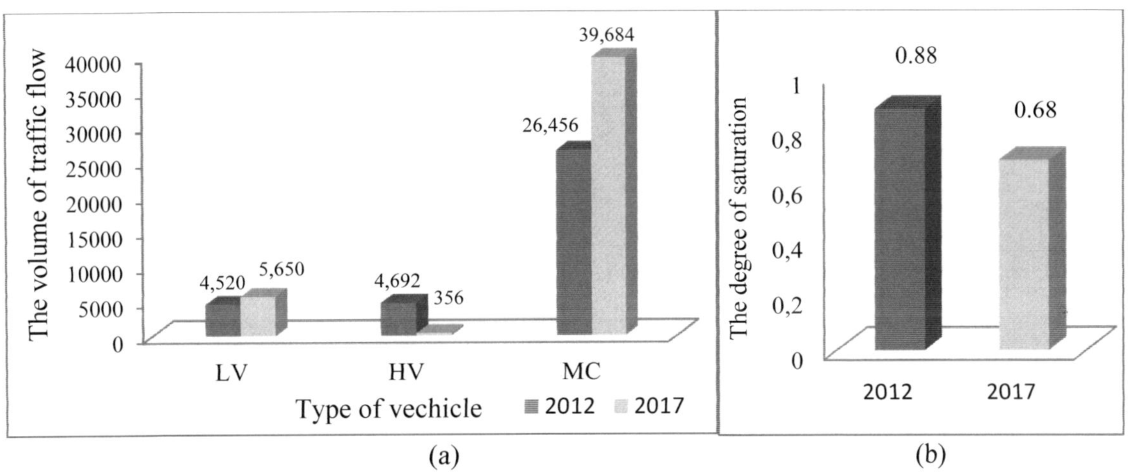

Figure 1. The volume of daily traffic flow (a) and the degree of saturation of peak hours (b) between 2012 and 2017

Figure 1 (a) illustrated the different volume of daily traffic flows. The number of heavy vehicles (trucks) drastically reduced by 92.4%, the number of motorbikes increased after the policy of banning entry of trucks crossed public roads, indicating that motorcyclist began to comfortably and safely cross the street. Figure 1 (b) demonstrates the degree of saturation before policy (2012) 0.88 > 085 and failling to 0.68 at peak hours (2017). The average speed of traffic was 23 km/h in 2012, and was 36 km/h in 2017.

3.2. Analysis of vehicle emission loads with Indonesia's emission factors

Vehicle emission load calculations utilize the emission load calculation method [17], through a Tier II approach, with available data in Indonesia.

$$E = Vehicle \; Volume \; x \; VKT \; x \, FE \; x \; 10-6 \tag{3}$$

note: *E*: Emission load (ton/year), Vehicle Volume: Number of vehicles (vehicle/year), *VKT*: Total trip length traveled (km), *FE*: Emission factor (g/ km/vehicle).

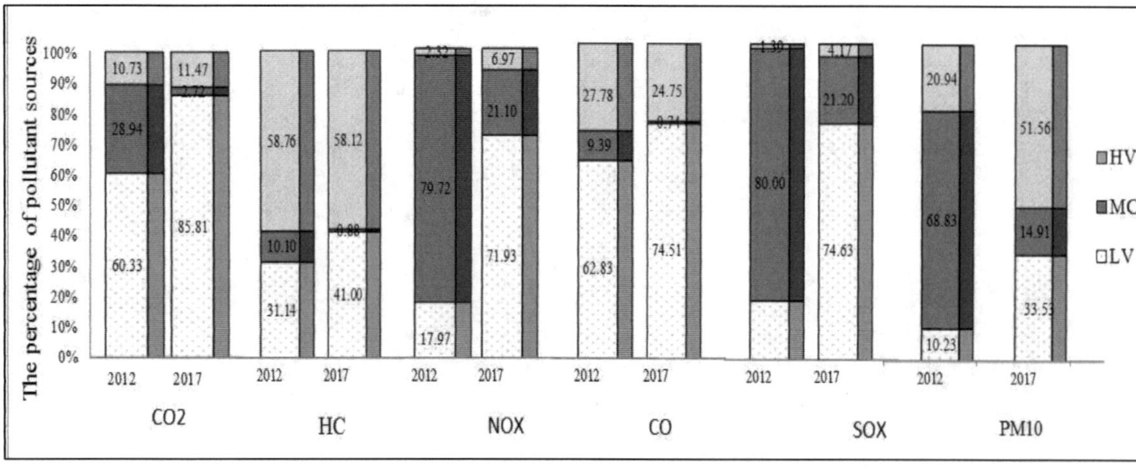

Figure 2. The percentage of pollutant sources for each type of vehicle

The results of vehicle emission loads of all vehicles in the road segment are presented by using Indonesian emission factor data [17]. The source column diagram of each type of vehicle is a light vehicle (LV), heavy vehicle (HV) and motorcycle (MC). Figure 2 indicates that heavy vehicles look significantly decreased by an average of 93% of the value of the post emission policy load (2016). CO2 emissions decreased by 44.30%, CO by 48.82%, HC by 46.22%, NOx by 22.03%, SOx by 22.01% and PM10 by 25.66%. The result illustrates that this government policy can significantly reduce air pollution from transportation.

3.3. Analysis of the value of losses due to vehicle emissions

Vehicle emissions loss due to increased vehicle volume and congestion assuming costs are based on 2005 pollutant costs from research ($/ton): CO = $ 205/ton, PM10 = $ 3.17/ton, CO2 = $ 205/ton, SO2 = $ 1000/ton, Nox = $ 934/ton and HC = $ 44/ton [18]. The $/tonne emission costs are converted into Indonesian currency (IDR), of which 1 $ = Rp. 13,500 (January 2017) and the value of losses due to the burden of emissions as presented in following Table 1:

Table 1. Value of losses due to vehicle emissions per year (Rupiah)

Pollutants	Total price (2012)			Total price (2017)		
	LV	HV	MC	LV	HV	MC
CO	285,061,072	136,758,511	50,695,127	322,522,897	10,227,666	43,092,765
HC	675,753	219,169	1,275,022	764,559	16,391	1,083,816
NOX	10,310,039	45,748,186	1,330,325	11,664,952	3,421,339	1,130,826
PM10	31,877,515	4,765,263	14,095,964	36,066,757	356,377	11,982,100
SOX	527,932	2,269,173	39,292	597,312	169,703	33,400
CO2	1,826	12,281	3,737	2,066	918	3,176
'otal		585,666,186			443,137,019	

Table 1 indicates that the value of losses due to vehicle emissions by enforcing policies was able to cut losses by 24.3% and percentage of heavy vehicles to the total price of the emission load decreased significantly by 93% and the difference in price was by IDR. 142,529,167.

3.4. Noise level
The level of noise intensity has an impact on humans such as physiological/psychological disorders, communication disorders, and deafness. Noise level analysis results before policy (2012) are 74.75 dB (A) > 70 dB (A) (unsafe), while after the policy (2017) are 69.88 dB (A) < 70 dB (A) (safe for trade and services) [23].

3.5. Strategy for controlling pollution of coal transportation
Control strategies to cut pollution include green belts, periodic road spraying, periodic truck emission checks and coal packing rules on trucks must follow regulations to reduce coal spills and dust on the passed road, changes in fuel type and a shift in types of vehicle modes.

4. Conclusion
Implementation of the South Kalimantan Provincial Regulation No. 3/2012 acts not only as transportation policy but also as part of environmental, economic policies and effectiveness in reducing congestion, pollution and material losses including coal transportation across public roads and forcing employers to build special roads. This policy also enables road users and pedestrians to be more comfortable and safe. The impact that occurred was the volume of heavy vehicles (trucks) decreased by 92.4%, the degree of saturation dropped to 0.68 at peak hours, the difference in average speed was by 13 km/hour. The percentage of heavy vehicles to the total price of the emissions load decreased significantly by 93%. The noise level decreased by 4.87 dB (A). CO2 emissions decreased by 44.30%, CO by 48.82%, HC by 46.22%, Nox by 22.03%, SOx by 22.01% and PM10 by 25.66%.

4.1 Recommendations
In issuing the policies, relations among the Regional Government, the private sector and civil society must be constructive and collaborative by prioritizing the principles of participation, transparency, and accountability during the government process. Thus local governance is expected to sustain the program. The government needs to support and supervise special roads for coal transportation to avoid unexpec effects on the road.

References

[1] Law of the Republic of Indonesia Number 4 of 2004 Mining of minerals and coal
[2] Yuniar D and Susanto H 2015 Coal stockpile volume deviation analysis with surpac software version 6.3 and manually. *J.Polhasains* 3 (1). **26-31**
[3] Regional Regulation of South Kalimantan Province Number 3 of 2008 Regulations on the Use of Public Roads and Special Roads for the Transport of Mining Products and Results of Plantation Companies. Banjarmasin
[4] Regional Regulation of South Kalimantan Province Number 3 of 20012 Amendment to Number 3 of 2008. Banjarmasin
[5] Yuniar D and Fatihin H 2016 Identification of road damage and handling repairs on the mine road. *J. Polhasains.* 4 (1), **34-40**
[6] Huaman N, Ruth and Xiu T 2014 Energy-related CO2 emissions and the progress on CCS projects: A review. *Renewable and Sustainable Energy Reviews.* 31, **368-385**
[7] Law of the Republic of Indonesia Number 22 of 2009 Traffic and Road Transportation
[8] Regulation of the Minister of Environment Number 12 of 2010 *Implementation of Air Pollution Control in the Region.* Jakarta
[9] Pratiwi T, Junaidi and Ali Z *2017* Effect of pollutant source distance on sulfate content (SO_4) In deposited dust along the coal hauls road. *J. Kesehatan Lingkungan* Vol. 14 No. 2 July 2017. **533-540**
[10] WHO 2011 *Burden of disease from environmental noise.* Copenhagen: World Health Organisation

[11] Neidell M 2004 Air pollution, health, and socioeconomic status: the effect of outdoor air quality on childhood asthma. *J. Health Econ*. 23, **1209–1236**
[12] Currie J and Walker R 2011 Traffic congestion and infant health: evidence from E-Zpass.Am. Econ. *J. Appl. Econ*. 3, **65–90**
[13] Bel G, Bolance C, Guillen M and Rosell J *2015* The environmental effects of changing speed limits: a quantile regression approach. *Transport Res. Transport Environ*. 36, **76–85**
[14] Elvik R and Ramjerdi F 2014 A comparative analysis of the effects of economic policy instruments in promoting environmentally sustainable transport. *Transp. Policy*. 33, **89–95**h
[15] State Minister of Environment, 1996. *Standard Noise Level, Decree of the State Minister of Environment* Number: Kep-48/MENLH/1996/25 November 1996, Jakarta
[16] Department of Public Works 1997 *Indonesian Road Capacity Manual (MKJI)*. Directorate General of Highways, Jakarta
[17] Regulation of the Minister of Environment Number 12 of 2010 Implementation of Air Pollution Control in the Region. Jakarta
[18] Victoria Transport Policy Institute 2011 *Transportation Cost and Benefit Analysis II.*Australia

Geographical Pattern of Economic Activities: an Evidence from Large and Medium Manufacturing Industries in Indonesia

E Santoso[1], M Ismail[2], I Noor[3] and M Ekawaty[4]

[1] Doctoral Student in Economic Science, Brawijaya University, Indonesia
[2,3,4] Lecturer at Economic Department, Brawijaya University, Indonesia

*E-mail: santoso_edy@yahoo.com

Abstract. The aim of this paper is to analyze geographical pattern of economic activities from an evidence of large and medium manufacturing industry in Indonesia. By utilizing data for manufacturing industry for the 2010–2015, the exploratory spatial data analysis is applied to identify geographical distribution of economic activities. The result indicates that the spatial distributions of manufacturing industries are unevenly distributed indicating the clustered phenomenon pattern. A high concentration of manufacturing industries is depicted in several provinces, as well as in Java Island. The Moran's *I* tend to decrease along the period of study. From the LISA, is apparent that there was a statistically significant high clustering. Only some provinces indicate significant clustering, such as in west Java, Banten, DKI Jakarta, and Central Java. The result of LISA indicates a positive spatial autocorrelation in research area revealing the existent cluster of economic activities. This paper attempts to contribute theoretical thought as well as empirical notes on spatial concentration in Indonesia. The researchers utilize spatial approach by taking spatial dependence among region units explicitly into account in peeling phenomenon of geographical concentration of manufacturing industry in Indonesia.
Keywords: *geographical pattern, manufacturing industries, spatial dependence, Indonesia*

1. Introduction

The most characteristic of spatial distribution of manufacturing industry is focused on concentration and unevenness. The concentration of manufacturing industry indicates that industrialization is a selective process and occurs only in certain cases when viewed in geographical terms. The noticeable examples are viewed in manufacturing belt in the United States [1] and in similar industrial spatial concentrations which are also found in Axial Belt industrial estate in UK and Ruhr manufacturing belts in Germany [2]. Similar phenomena also can be found in Indonesia, for more than the last three decades which unevenly distributed. Indonesia's manufacturing industry has tended to be spatially concentrated in Java since the 1970s [3]. According to [4], the manufacturing industry was spatially concentrated in Java since the 1970s, although in the late 1980s it was shifted to Sumatra and Bali.

The spatial distribution of economic activity in Indonesia demonstrates the presence of inequality and concentration. The distribution of large and medium manufacturing industries is not evenly distributed and concentrated in a few regions. For 1990, the number of large and medium industries in Java reached 79.95 percents of the total industries in Indonesia. In 2010, the composition of the companies number changed to 83.65 percents in Java and 16.35 percents for outer Java. Thus, for the last 25 years (1990-2014) the Indonesian manufacturing industries was still concentrated in Java.

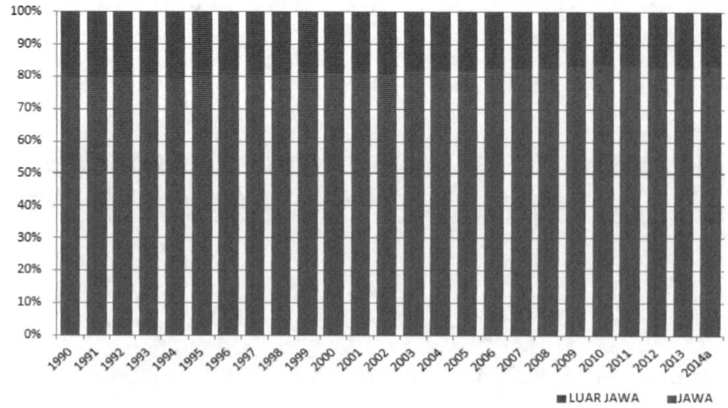

Source: Large and Medium Annual Industry Survey BPS, 2015

Figure 1. Distribution of Large and Medium Industries by Region, 1990–2014

Several reasons are blamed for the unbalance distribution. For the manufacturing industry, the locations of industry are determined by any resources, such as: wage costs, transportation costs and market access, and the externalities of spatial concentration associated with saving localization and saving urbanization (so-called agglomeration economy). Uneven distribution of the resources leads to disparities in industrial concentrations among regions. Thus, it gives the explanation on the concentration of industry occurring in certain areas only. Hence, the spatial pattern of economic activity, especially manufacturing industry, has become an interesting phenomenon to be analyzed.

Spatial concentration is the grouping of industry and economic activity located in a particular region spatially [5]. Spatial dependence might arise due to the existence of technology spillovers, labor and non-labor migration, commodity flows, and other spatial interactions that benefit the common interest [6]. The spatial concentration of economic activity can be understood by regional economic theory especially from the location theory approach. The purpose of the location theory is to determine the reasons and explain why certain factors is important for one industry but not for other industries [7]. It also involves the principle of substitution, in which the industry selects a location from some alternative location, which in economic theory, is similar to the problem of how to replace labor production factors with capital or land and vice versa. Many factors are considered in relation to the determination of the appropriate location for the industry. These factors vary depending on the peculiarities of an industry [8].

The main objective of this paper is to analyze spatial distribution of manufacturing industries. The analyses would be focused on spatial exploratory analysis of Indonesia's manufacturing industries. Spatial exploratory analysis of industry location will aid in capturing important intangible aspects concerning spatial dependence. The majority of empirical work which relates to the concentration of the economic activities is still based on calculations of basic statistical measurements where the geographical properties of the data do not play any role. This paper considers spatial interaction among region units by taking spatial dependence explicitly into account. This spatial analysis will also aid in identifying phenomenon pattern for the distribution of manufacturing industries expressed which is clustered or randomly distributed. This objective is achieved by utilizing Global Moran's I Statistic and Local Indicators of Spatial Association (LISA) statistics.

The rest of this study is organized as follows: the second section discusses about the data and the applied methodologies. The third section analyzes the main expected result of the study. Finally, the last section provides conclusion of the study.

2. Methods

This study analyses the geographical pattern of economic activity especially for the spatial distribution of large and medium manufacturing industries by using the provincial data set from 2008 to 2015 for 33 provinces in Indonesia. In practice, there have been 34 provinces in Indonesia including several newly formed provinces being added to the former ones during decentralization era. In this case, data from newly formed provinces are combined with those of their original provinces.

For exploratory spatial data analysis, Moran's I was applied to determine the general level of spatial autocorrelation in the data. Theoretical models usually recognize the existence of spatial autocorrelation between regions; empirically, Moran's I becomes a popular tool for measuring spatial autocorrelation, which is an index of linear association between a set of spatial observations x_i x_j, and a weighted average w_{ij} of their neighbors [9]. For the number of establishments y, Moran's I is:

$$I = \left(\frac{n}{\sum_i \sum_j w_{ij}} \right) \text{X} \left(\frac{\sum_i \sum_j (y_i - \mu)(y_j - \mu)}{\sum_i (y_i - \mu)^2} \right) \quad (1)$$

Where w_{ij} indicates elements of the spatial weight matrix W (distance/contiguity weight matrix) between two points (i and j), μ the mean of all y observations, and $i, j = 1\ldots$ n. This tool measures spatial autocorrelation (feature similarity) based not only on feature locations or attribute values alone but on both feature locations and feature values simultaneously. A set of features and an associated attribute, evaluates whether the pattern expressed is clustered, dispersed, or random. The tool calculates the Moran's I value and a Z score evaluates the significance of the index value. In general, a Moran's I value near +1.0 indicates clustering while an index value near -1.0 indicates dispersion. In the case of the Spatial Autocorrelation tool, the null hypothesis states that "there is no spatial clustering of the values". If the Z score is large (or small) enough to such that it falls outside of the desired significance, then the null hypothesis can be rejected. If the null hypothesis is rejected, the next step is to inspect the value of the Moran's I. If the value is greater than 0, the set of features exhibits a clustered pattern. If the value is less than 0, the set of features exhibits a dispersed pattern. To execute the data, GeoDa software is applied and designed to be a software tool that facilitates the exploration and analysis of geospatial data [10].

3. Results and Discussions

The main objective of this paper is to analyze the distribution of Indonesian manufacturing industries geographically. There are two expected results to be achieved in this paper, which are: to uncover the existence of spatial dependence related to the distribution of manufacturing industries in Indonesia and to contribute significantly to industrial cluster in Indonesia. The expected results are discussed as follows:

3.1. Spatial dependence of the distribution of manufacturing industries

The spatial distribution of manufacturing industries for the year 2015 is presented in figure 2. The figure illustrates that the distribution of Indonesia's manufacturing industries is unevenly distributed for all 33 provinces. Mostly, the manufacturing industries are concentrated in Provinces which are located in Java Island. In addition, this spatial concentrated was also caused by the infrastructure and labor accumulated in Java [11]. It implies that the Indonesia's manufacturing industries shows a spatial pattern and it is not independently distributed over space. Economic activity tends to be geographically clustered. Moran Model allows recognizing the existence of spatial dependence between regions. To prove that the Indonesia's manufacturing industries are not independently distributed over space, Global Moran's I is applied. This index analyzes the spatial dependence (autocorrelation) among region related to the distribution of manufacturing industries. The high value of the Moran's I indicates the presence of spatial interactions resulting in spatial dependencies among regions. Spatial dependence is measured as the existence of statistical dependence in a collection of random variables, each of which is associated with a different geographical location.

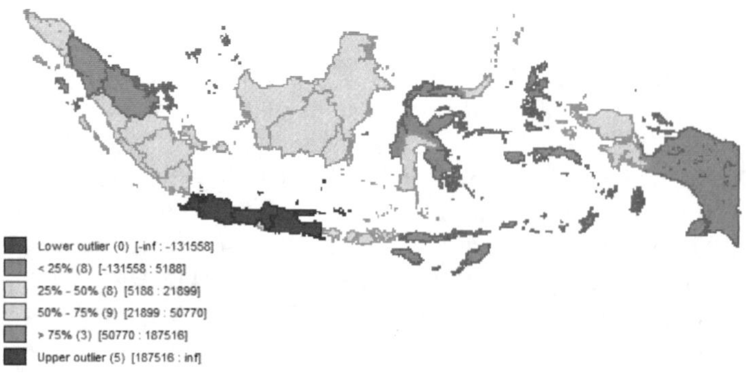

Figure 2. The Spatial Distribution of Indonesian Manufacturing Industries, 2015

For Moran's *I* indicate positive spatial correlation, showing that province with a high or low numbers of manufacturing industries are similar to their neighboring province. Conversely, a negative and significant value for Moran's *I* indicate negative spatial correlation, showing that province with a high or low number of manufacturing industries are unlike their neighboring province [12]. The Moran's *I* statistic is equal to 0.623179 for the year of 2015, indicating a strong positive spatial relationship (autocorrelation) and the phenomenon pattern expressed which is clustered. To prove its significance, pseudo p-value in the randomization is utilized for reporting p-value, which is $\rho = 0.001$ (proving that it is definitely significant). The size of the Moran's *I* also indicates that the Indonesian manufacturing industry is distributed in a clustered pattern.

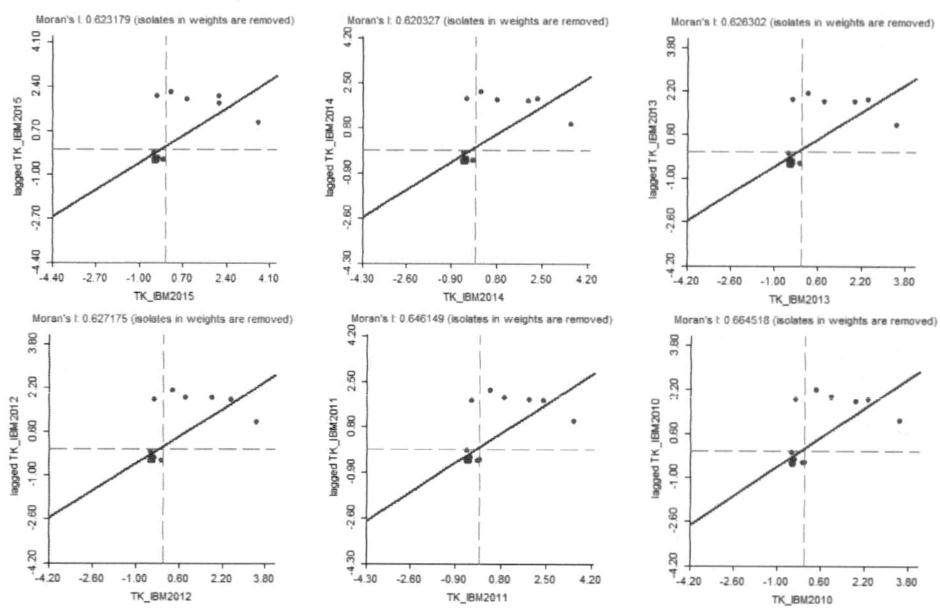

Source: calculated by author

Figure 3. Global Moran's *I* Scatter Plots for Manufacturing Industries, 2010-2015

By comparing the value of Moran's *I* in a series, we could know the trend of spatial dependence is presented among region. Whether the spatial dependence among province getting strengthened or weakened for along study period. For example, the Moran's *I* in 2008 was 0,6728, in 2010 was 0,6645, reached its maximum value in 2007 was 0,6898 and in 2015 amounted to 0,6232 (authors calculation). That the spatial distribution of manufacturing industry for along periode of study has a tendency to weakening. Figure 4 shows the global Moran's *I* statistic for spatial distribution of manufacturing industries in Indonesia for the year 2007 to 2015 using the spatial weight matrix. The values of global Moran's *I* statistic show a downward trend from 2007 to 2015 and significant positive spatial autocorrelation. The results suggest that the locations of province with high or low the number of manufacturing industries are clustered, and the spatial pattern has a developing trend of decrease spatial autocorrelation. Its mean, the spatial distribution of manufacturing industries are still clustered but the degree of spatial dependence among region are going down. This was due to a series of deregulation by government corcerned with industrial development [4]. So, the spatial distribution of manufacturing industry more evenly distributed.

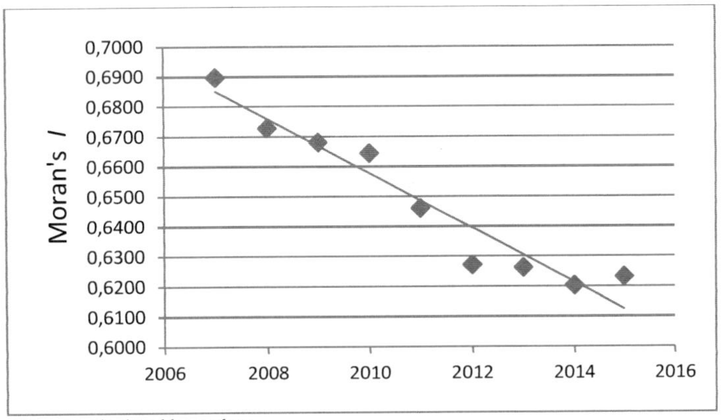

Source: calculated by author

Figure 4. Trend of Moran's *I* statistics for Manufacturing Industries (2007-2015)

When looking on formation of spatial clusters in detail local spatial autocorrelation methods such as LISA [13] – the local equivalent of Moran's *I* – can be used. The mapped results of this analysis answer not only the question of whether any spatial pattern across studied area exists but also where the clusters are, what they look like, etc [14].

3.2. Local Indicator of Spatial Accociation (LISA)

An important component of geographical analysis is a focus on the local, refrected in new methods to deal with local spatial autocorrelation [15]. Specifically, considerable interest has been devoted to local indicators of spatial association (LISA) since the original LISA framework was outline in Anselin [13, 16]. The Moran's *I* is a global statistic. So, we know there is clustering going on in the research area, but we do not know exactly where clustering exist. To investigate further, we would do a LISA. This test not only tests for regional clustering, it can also show the presence of significant spatial clusters or outlier. LISA statistics was developed by [13], which measures the contribution of individual region to the global Moran's *I* statistic. Local Moran's *I* useful for showing places where significant spatial autocorrelation exists.

IOP Conf. Series: Earth and Environmental Science **328** (2019) 012038 doi:10.1088/1755-1315/328/1/012038

Figure 5. LISA Cluster Map for Indonesian Manufacturing Industries, 2015

LISA cluster map indicates locations with a significant Local Moran statistic. The LISA cluster map shows that how the attribute of industries clusters. The red color show the tracts where the province with a high rate cluster surrounded by province with a high rates cluster too (High-High), and blue shows the province with a low rate cluster surrounded by province with a low rates cluster too (Low-Low). There is even mix of high-low, for example the pink color. So, there is a cluster of high industries surrounded by cluster of lower industries. The HH and LL locations suggests clustering of similar values (positive spatial correlation), whereas the HL and LH locations indicate spatial outliers (negative spatial correlation) [13]. A positive and high autocorrelation is found in some provinces really show significant clustering. For the year of 2015, A high cluster (HH) industries occurs in four provinces, these provinces are West Java, Banten, DKI Jakarta, and Central Java. These provinces are have significant contribution to the global Moran's I statistic. For East Java, indicates not significant high clustering due to the spatial distribution of manufacturing industries spread from the northern region of the southern region (called the north-south corridor).

4. Conclusion

This study analyzes the geographical changes in spatial pattern of economic activities in Indonesia. The paper deals with the spatial distribution of large and medium manufacturing industries which examines whether a phenomenon pattern expressed is clustered, dispersed, or random using data set from 2010 to 2015 across 33 provinces in Indonesia. In this paper we contribute to the empirical literature on the spatial distribution of manufacturing sector by taking spatial dependence among region units explicitly into account. The Moran's I was employed to characterize the spatial distribution of manufacturing industries. The result shows a strong positive spatial relationship (autocorrelation) in research area and shows that the phenomenon pattern for the distribution of manufacturing industries expressed is clustered, and a high concentration of manufacturing industries is seen in Java Island. For along the period of study the Moran's I are tending to decrease. Its mean, the domination of Java Island as the centre of economic activities in Indonesia is no longer. From the LISA, we can ascertain that there was statistically significant high clustering in Indonesia. Only some provinces really show significant industrial.

The findings are not only strengthened the regional economic theory especially the economic geography theory, but also important to develop a policy recommendation due to industrial development planning. As a policy implication, the result showed that spatial dependence across regions was matter and in other side will be beneficial for the economic growth of the regions. It implied that all regions need to cooperate due to economic integration.

For further research, it would be interesting to develop model take into account some variables which determinant to spatial distribution of manufacturing industries. Hence, the model can be able to explain factors influencing industrial location.

References

[1] Thomas J H and Stevens J J 2003 *Spatial Distribution of Economic Activities in Nort America. Handbook of Regional and Urban Economic: Cities and Geograpgy* vol **4** (Amsterdam: North-Holland)
[2] Hayter R 1997 *The Dynamic of Industrial Location: The Firm the Factory and the Production System* (Chichester: John Wiley)
[3] Azis I J 1994 *Ilmu Ekonomi Regional dan Beberapa Aplikasinya di Indonesia* (Jakarta: Lembaga Penerbit Fakultas Ekonomi Universitas Indonesia)
[4] Kuncoro M 2012 *Ekonomika Aglomerasi: Dinamika dan Dimensi Spatial Kluster Industri di Indonesia* (Yogyakarta: UPP STIM YKPN)
[5] Fujita M, Krugman P and Venables A 1999 *The Spatial Economy : Cities, Region, and International Trade* (Cambridge: MIT Press)
[6] Rey S J and Janikas M V 2005 Regional Convergence, Inequality and Space *Journal of Economic Geography* vol **5** pp 155-176
[7] Greenhut M L 1982 *Plant Location in Theory and in Practise, the Economic of Space - The Theory of Least-Cost Location*, Chapter 1 (California: Greenwood Press)
[8] Sambidi P R 2007 *Spatial Econometric Analysis of Agglomeration Economies Associated with The Geographical Distribution of The US Biotech Industry* A Dissertation The Graduate Faculty of The Louisiana State University and Agricultural and Mechanical College
[9] Moran P A P 1950 Notes on Continous Stochastic Phenomena *Biometrika* vol **37** pp 17-23
[10] Anselin L 2017 *The GeoDa Book Exploring Spatial Data* (Chicago: GeoDa Press LLC)
[11] Wahyudin 2004 *Industri dan Orientasi Ekspor: Dinamika dan Analisis Spasial* (Surakarta: Universitas Muhammadyah Press)
[12] Pacheco A I and T J Tyrrell 2002 Testing Spatial Patterns and Growth Spillover Effect in Clusters of Cities *Journal of Geographic Systems* vol **4** pp275-285
[13] Anselin L 1995 Local Indicator of Spatial Association-LISA *Geographical Analysis* vol **27** pp 93-115
[14] Netrdova and Nosek 2010 Regional and Spatial Concentration of Socio-economic Phenomena: Empirical Evidence from The Czech Republic *Ekonomicky Casopis* vol **58** pp 344-359
[15] Anselin L 2017 *A Local Indicator of Multivariate Spatial Association: Extending Geary's c* Center for Spatial Data Science (Chicago: University of Chicago)
[16] Anselin L 1996 The Moran Scatterplot as an ESDA tool to Assess Local Instability in Spatial Association In Fischer M, Scholten H, and Unwun D Editor *Spatial Analytical Perspective on GIS in Environment and Socio-Economic Sciences* pp 111-125 (London: Taylor and Francis)

Determining the Location of Shelters for Tsunami Evacuation Based on Service Area Analysis in Paseban Village, Kencong Sub-District, Jember District

F Usman[1] and I C Sari[2]

[12]Department of Urban and Regional Planning, Faculty of Engineering, University of Brawijaya, Malang, East Java 65145, Indonesia

*Email : fadlypwk@ub.ac.id; cahayaindah360@gmail.com

Abstract. In each earthquake and tsunami event, efforts to evacuate to a safe place are always the right choice. This is because after the earthquake, the damage caused is usually quite massive. There is little time to react in an evacuation effort to move to a place safe from the reach of the tsunami wave. This study aimed to identify areas affected by the tsunami and the number of people exposed. The data will be used to select the temporary evacuation points that can be used to protect oneself from tsunami waves. The evacuation point was done by the scoring method. The determining of evacuation points took into account the parameters of building functions, distance from shoreline, building area, altitude, number of building floors, distance from the road, and its capacity. The results of this study indicates that, the determining of evacuation sites greatly impacts the area and service areas for Temporary Evacuation Shelter (TES)

Keywords: Tsunami, Temporary Evacuation Shelter (TES), Geographic Information System (GIS)

1. Introduction

Indonesia is an archipelago that has various types of disasters and also with varying levels of risk in each region, one of which is East Java Province. Karangkates Meteorology and Geophysics Agency (BMKG), Malang, noted that eight districts in the southern region of East Java are prone to tsunamis because they are categorized into zone I (high) of being prone to earthquakes. East Java is an area where several areas have potential for disaster-prone areas, one of which is Jember District which has an area of 3,092.34 Km^2 [1]. Jember District is one of the regencies in East Java which is included in the disaster-prone zone. Based on the East Java Province Disaster Hazard Index Map which was sourced from the National Disaster Management Agency post, Jember District has a high level of disaster vulnerability which is included in the high classification [2].

Based on the 2015-2019 National Agency for Disaster Management Strategic Plan, Jember District is a city in Indonesia which ranked 10th as a city with a high disaster risk class with a disaster risk index score worth 219.2. According to the Regional Disaster Management Agency (BPBD) of Jember, almost all potential disasters have the potential to occur in the area of Jember District. Data at the Jember Regional Disaster Management Agency (BPBD) noted that 19 of the 31 sub-districts in Jember were disaster-prone areas being vulnerable to floods, landslides, tornadoes and tsunamis. Based on the Jember Regency Spatial Plan (RTRW) for 2015-2035, Kencong, Gumukmas, Puger, Ambulu,

Wuluhan and Tempurejo are the sub-districts in Jember District which are included in tsunami-prone areas (article 38 section 8) [3].

Historically, the South Coast of Java Island had experienced a tsunami that originated from the Indian Ocean, there was one in 1994 in East Java and 2006 in West Java. The tsunami disaster on June 12, 2006 had a 7.7 scale righter earthquake power, the height of the 4.6 meter tsunami run-up resulted in 668 fatalities. The biggest impact due to the tsunami that hit the coast in Jember occurred was when an earthquake accompanied by a tsunami struck Banyuwangi on June 2, 1994 with a magnitude of 7.2 righter scale, resulting a 13-meters-high tsunami run-up which caused 238 deaths and 400 other injuries and experienced material losses [4].

The vulnerability of Jember Beach to tsunami can be estimated from the large distribution of shallow depth centers located on the seabed around the coast. This is supported by the fact that the south coast of Java is directly facing a subduction zone at the base of the Indian Ocean. This zone is a series of ring of fire which results in a high intensity of the geological process in the form of seafloor earthquakes and areas that have very high seismic activity which has the potential to trigger tsunamis. Paseban Beach is a beach in Kencong Subdistrict, which is 460 meters from the settlement and physically located between two rivers, namely the Bondoyudo River and the Tanggul River. Coastal Kencong Subdistrict includes a moderate tsunami vulnerability zone with inundation heights of 1 to 3 meters and a tsunami landing distance reaching 980 m from the coastline where it will inundate settlements around the coast [5]. Based on the existing information and data, this study was carried out as a mitigation effort against the threat of the tsunami disaster by evaluating the available temporary evacuation shelters (TES). The evacuation shelters considered data from the tsunami affected area obtained from the Regional Disaster Management Agency of Jember District.

2. Methods

In this study, several stages of analysis were carried out, specifically related to tsunami disaster mitigation strategies by carrying out evacuations. The first stage was spatial analysis related to the use of existing land in the study area and tsunami-affected areas based on data from the BPBD of Jember Regency. From the two spatial data above, this research can find out the extent of the affected area, and which areas that will not be affected by the tsunami.

There are two methods of evacuation of the tsunami threat; horizontal and vertical. Horizontal Evacuation is the evacuation by going away from the coast towards the tsunami-safe area (recommended) on foot, instead of driving. This method requires evacuation pathways that utilize roads in the city, and several major roads in residential areas. Vertical evacuation is the act of saving oneself toward higher grounds. There were neither a hill (natural shelter) nor a multi-storey building [1]. Because the study area has no hills at all or has very flat topography, then the option is to build more of the higher-than-three-story buildings, or utilizing the existing multi-story buildings as vertical building. There are several naming for vertical evacuation building, namely Tsunami Evacuation Buildings (TEBs) Vertical Evacuation Structures. In Jember District, it is called the TES or Temporary Evacuation Shelter. It is called temporary, because it is expected to accommodate refugees, within 2-3 hours, before the refugees are evacuated to the tsunami evacuation centre in the safe zone. In addition to it, the standard equipment of vertical building shelters are also not available on the TES, therefore the refugees are left standing.

Table 1. Score for Shelter Location

Building Condition	Criteria	Score
Location from the road	Road side	1
	Intersection of local road	2
	Main road side	3
	Intersection of main road	4
Number of floors	One floor	1
	Two floors	2

Building Condition	Criteria	Score
	Three floors	3
	> Three floors	4
Volume capacity	<100 peoples	1
	100-500 peoples	2
	500-1000 peoples	3
	>1000 peoples	4
Function of buildings	Residential	1
	Health facilities	2
	Government or Private Office	3
	School, Religious	4

Source : Agus Suharyanto, 2012 (Journal of Environmental Science)

The next step was to weight the distribution of Temporary Evacuation Shelter (TES). Weighting of temporary evacuation shelter is very important, this was done to find out how feasible the building or place is available as a temporary evacuation site based on parameters of building functions, distance from shoreline, building area, altitude, number of building floors, location from the road and capacity. Table 1 showed the scoring parameter used in the study. Then the analysis was conducted to the service area using the Network Analyst GIS. By using ArcGIS software, this study can find out the coverage area of each evacuation site based on the rational distance from the residential area to the Temporary Evacuation Shelter (TES).

3. Results and Discussions

3.1 General Description of Jember District Tsunami Disaster

Based on the Regional Disaster Management Agency of Jember District, there are 10 types of potential disasters in Jember, namely floods, flash floods, extreme weather, extreme waves and abrasions, earthquakes, forest and land fires, droughts, volcanic eruptions, landslides, and tsunamis. The potential area and class of potential disasters are presented in **Table 2.** below.

Table 2. Potential Hazard Areas and Classes of Jember District

Disaster Type	Hazard	
	Area (Ha)	Class
Flood	168,646	Medium
Flash floods	19,918	High
Extreme weather	215,236	High
Extreme waves and abrasion	2,417	High
Earthquake	311,006	Medium
Forest and land fires	157,106	High
Drought	311,006	Medium
Tsunami	**5,196**	**High**
Raung Volcanic Eruption	8,063	Low
Lamongan Volcanic Eruption	644	Low
Landslide	101,184	**Tinggi**

Source: Disaster Risk Study Document of Jember Regency 2017-2021 (BPBD)

IOP Conf. Series: Earth and Environmental Science **328** (2019) 012039 doi:10.1088/1755-1315/328/1/012039

3.2 Spatial Analysis

3.2.1 Land Use

Paseban Village is one of the coastal villages in Kencong District, Jember District. The land use in the Coastal Areas of Paseban consists of settlements (residential), fishponds, farm (agriculture/ rice fields) and garden (horticulture). The largest land use in Paseban Village is used as a residential area which spans to 262.06 Ha and rice fields which spans to 354 Ha. (Source: Kencong Sub-District in Figure 2018)

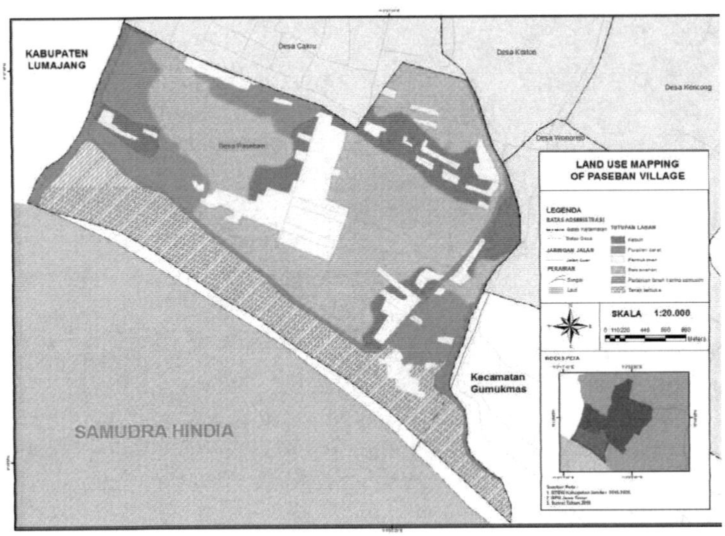

Figure 1. Land Use Mapping of Paseban Village

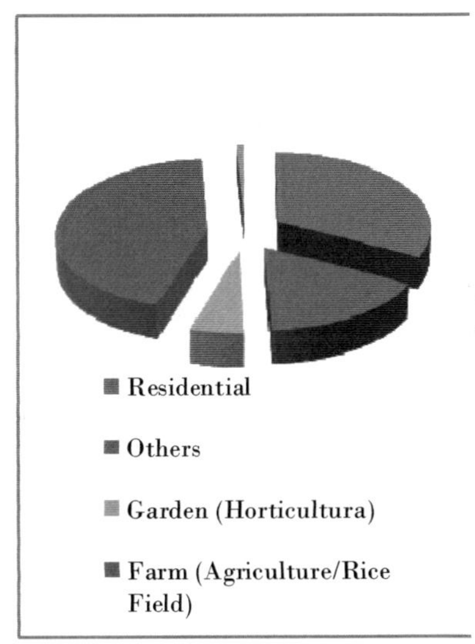

Figure 2. Diagram of Land Use Percentage

3.2.2 Tsunami Affected Area

The tsunami hazard assessment produced the area exposed to tsunami hazards. The selection of area of exposure to tsunami hazards used measurement parameters based on the methodology of disaster risk assessment Regulation of the Head of National Disaster Management Agency (BNPB) Number 2 of 2012. The tsunami hazard of each sub-district in Jember District was based on the Disaster Risk Study Document of Jember Regency 2017-2021, which is shown on the **Table 3.** Below.

Table 3. Tsunami Hazard Potential and population affected on each Sub-District in Jember District

Sub-District	Hazard		Affected Population (People)
	Area (Ha)	Class	
Ambulu	899	High	2.076
Gumukmas	1.693	High	6.963
Kencong	**680**	**High**	**1.536**
Puger	1.243	High	14.813
Tempurejo	476	High	21
Wuluhan	205	High	-
Jember Regency	5.196	Tinggi	25.409

Source: Disaster Risk Study Document of Jember Regency 2017-2021 (BPBD)

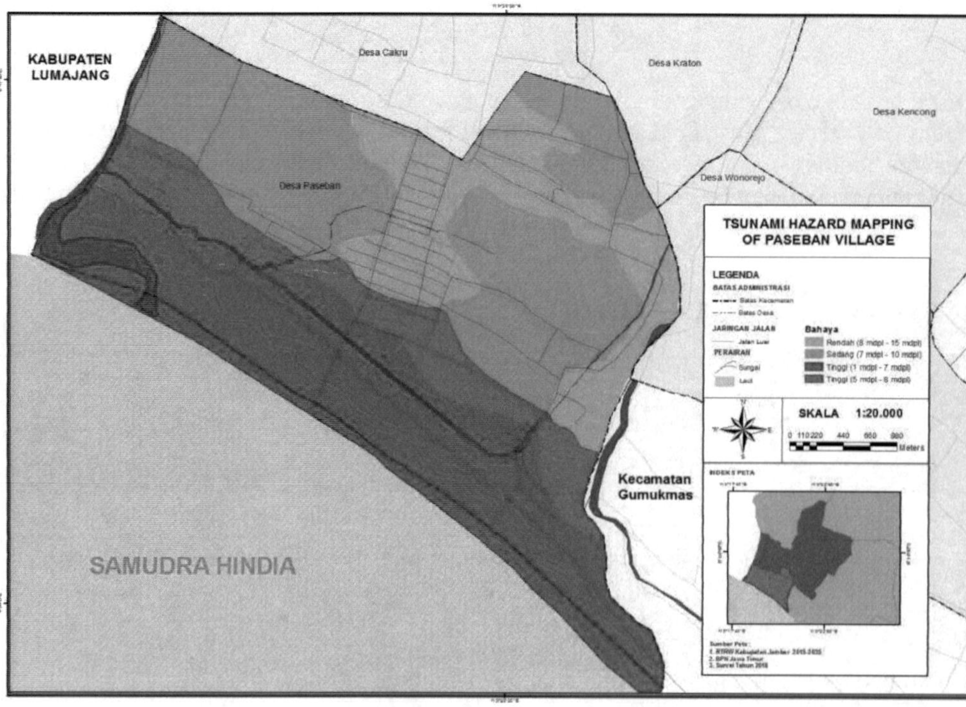

Figure 3. Tsunami Hazard Mapping of Paseban Village

3.3 Evacuation Shelter

Planning and selecting the temporary evacuation shelter (TES) in Paseban Village, Kencong Subdistrict, in Jember District considered the land use prioritized the form of public facilities and paid attention to their capacity. The determined evacuation point must be able to serve and accommodate according to the space requirements of the refugees that need to be accommodated. The refugee space needs are obtained from the estimated number of affected residents multiplied by the minimum space requirement per person, i.e. 1.64 m^2 (*Source: The American National Red Cross Evacuation Shelter Standard*).

Table 4. Location of Shelters in Paseban Village, Kencong Sub-District (Existing Condition)

Code	Coordinate		Function of Buildings	Building Area (m^2)	Number of Floors	Volume Capacity (People)	Altitude (mdpl)	Distance From Coastline
	Latitude	Longitude						
A1	8.3068699	113.3293015	Religious	678,66	2	414	9 mdpl	1500 m
D1	8.3083147(-)	113.3289783	Residential	156,71	2	191	9 mdpl	1450 m
B1	8.3084404 (-)	113.3343125	Private Office	1199,37	1	731	10 mdpl	1700 m
A2	8.3041395(-)	113.3195825	Religious	308,95	1	188	6 mdpl	1200 m
D2	8.3003225(-)	113.3361113	Residential	299.11	2	364	12 mdpl	2500 m
C1	8.3157015(-)	113.3431933	School	1132,95	1	691	8 mdpl	1600 m
A3	8.3147142(-)	113.3470574	Religious	424,65	1	259	8 mdpl	2000 m
C2	8.3077805(-)	113.3265559	School	921,65	1	562	8 mdpl	1300 m
C3	8.2990081(-)	113.3367419	School	1255,10	1	765	12 mdpl	2900 m
A4	8.2959956(-)	113.3305574	Religious	341,69	1	208	12 mdpl	2600 m
C4	8.3131646(-)	113.3482255	School	1243,31	1	758	10 mdpl	2200 m

Selection of evacuation points considered accessibility aspect in the form of ease of access of the locations, where temporary evacuation points are located close to the main roads in Paseban Village, Kencong District. In addition, temporary evacuation sites are in safe areas, or outside of the tsunami hazard potential area with high classification. Based on tsunami evacuation planning guidelines, the height of the rescue building on the recommended evacuation zone has a building height of more than 3 meters [9].

Table 5. Selection of Location of Shelters in Paseban Village, Kencong Sub-District (Scoring)

Code	Name (Picture)	Score							Total of Score	Classification
		A	B	C	D	E	F	G		
A1	Baitussalam Mosque	4	2	2	2	2	1	4	17	Middle
D1	Rumah Masyarakat	1	1	2	2	2	1	2	11	Unsuitable
B1	Paseban Village Hall	3	4	1	3	3	2	3	19	Middle
A2	Al Mubarokah Mosque	4	1	1	2	1	1	2	12	Unsuitable
D2	Rumah Pasar Baru Kencong	1	1	2	2	4	4	3	17	Middle
C1	Paseban 2 Elementary School	4	4	1	3	2	1	2	17	Middle
A3	Baitul Amin Paseban Mosque	4	1	1	2	2	2	2	14	Unsuitable
C2	Paseban 1 Elementary School	4	3	1	3	2	1	2	16	Middle
C3	MI Muhammadiyah 02	4	4	1	3	4	4	4	24	Unsuitable
A4	Nur Islam Mosque	4	1	1	2	4	4	4	20	Suitable
C4	MI Assuniyah	4	4	1	3	3	3	2	20	Suitable

A = Scoring for Function of Buildings
B = Scoring for Building Area
C = Scoring for Number of Floors
D = Scoring for Volume Capacity
E = Scoring for Altitude
F = Scoring for Distance from Coastline
G = Scoring for Distance from the Road

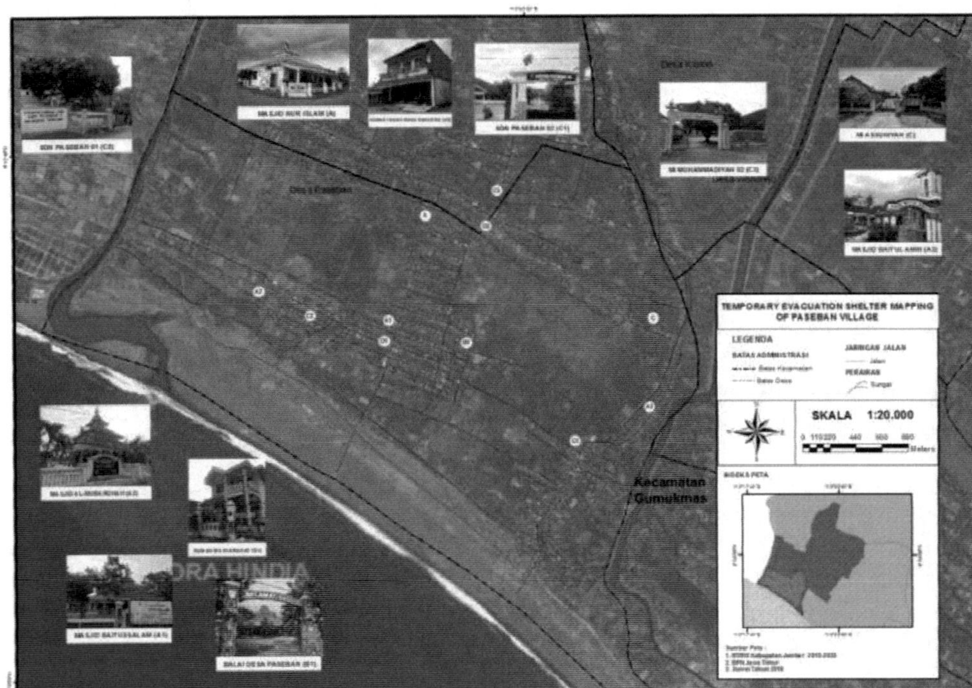

Figure 4. Photo mapping of Temporary Evacuation Shelter (TES) in Paseban Village

3.4 Service Area

The service area aimed to find out the minimum area for the community to able to reach the temporary evacuation point (shelter area) that is closest to the residence by walking within a certain range. Data needed for service area analysis included the distribution of temporary evacuation points, road network data, and settlement distribution of Paseban Village.

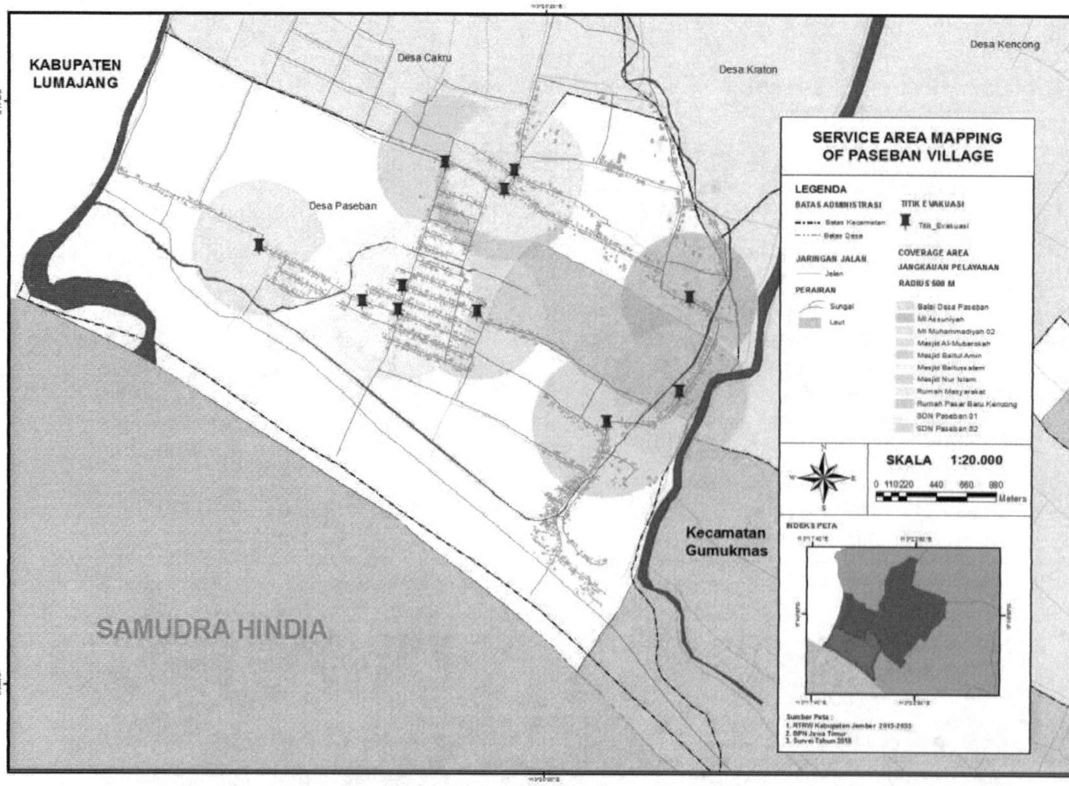

Figure 5. Service Area Mapping of Paseban Village

4 Conclusion

Based on the results of the analysis of the current conditions of each shelter, it can be seen that the distribution of shelters cannot provide services to all residential areas in Paseban Village, Kencong Subdistrict, especially residential areas around the coastal area. Some of technical advices need to be done, so that potential buildings for temporal shelters can be added to serve the entire residential areas that have the potential to be affected by tsunami.

References

[1] Ashar, Faisal *et all.* 2014. The Analysis of Tsunami Vertical Shelter in Padang City. Journal Science Direct. Procedia Economics and Finance 18 (2014) 916 – 923. Published by Elsevier B.V.

[2] Suharyanto, Agus et all. 2012. Predicting Tsunami Inundated Area and Evacuation Road Based On Local Condition Using GIS. IOSR Journal of Environmental Science, Toxicology and Food Technology (IOSR-JESTFT) ISSN: 2319-2402, ISBN: 2319-2399. Volume 1, Issue 4 (Sep-Oct. 2012), PP 05-11. www.iosrjournals.org.

[3] Rencana Strategis Badan Nasional Penanggulangan Bencana Tahun 2015-2019

[4] Putturuhu, Ferad. 2015. *Mitigasi Bencana dan Penginderaan Jauh.* Yogyakarta : Graha Ilmu

[5] Maemunah, Sulaeman dan Robiana. 2011. *Jurnal Lingkungan dan Bencana Geologi.* Identifikasi Potensi Kerawanan Tsunami di wilayah Kabupaten Jember, Jawa Timur. Vol. 2 No. 2 Hal 141 – 152.

[6] Rencana Tata Ruang Wilayah (RTRW) Kabupaten Jember Tahun 2015-2035

[7] Peraturan Kepala Badan Nasional Penanggulangan Bencana Nomor 02 Tahun 2012 tentang

Pedoman Umum Pengkajian Risiko Bencana
[8] Disaster Risk Study Document of Jember Regency 2017-2021 (Badan Penanggulangan Bencana Daerah Kabupaten Jember)
[9] Tsunami Refuge Planning, Sea Defense Consultant, 2007

Outer Island Development in Crossborder Cooperation Framework: Cooperation or Exploitation?
(Study Case: Natuna Island in Singapore – Johor- Riau Cooperation)

H M Putri[1] and W Salim[2]

[12]Department of Regional and Urban Planning, School of Architecture, Planning and Policy Development, Bandung Institute of Technology, Indonesia

*Email: hertria.maharani@gmail.com

Abstract. The development of the outer islands has not been widely discussed in literatures. As a peripheral area, its location which are remote and intersects with border areas of other countries caused the form of development to be different from the development in the mainland. One form of development on the outer island can be linked with cross-border cooperation. This study argued that based on the Natuna Island's case study, trans-border development through cooperation between countries is a promising form for the development on the outer islands. But exploitation will occur if the government does not guard the process of development of the outer islands in the trans-border cooperation. The flow of goods that are not accommodated with appropriate access can make the outer islands being only as suppliers of goods that are not recognized by formal data. It caused their existence not being considered important in the cooperation between countries.

Keywords: *Trans-border Development, SIJORI, Natuna, Periphery*

1. Introduction

Indonesia as an archipelago has many outer islands that needs to be developed. In the Presidential Decree No. 6 of 2017, the government established 111 outer small islands to be the focus of attention in the context of the development of Republic of Indonesia (NKRI). However, the form of development of the outer island region was still very limited in the discussion of regional planning literature, so it required deeper exploration of the concept of the outer island development that can be applied specifically to the nation's island.

The concept of regional development has been based on neo-classical economics theories which emphasize economic growth and commodity competitiveness, thus potentially increasing the disparity between regions. Development programs initiated by the government, mostly emphasized on macro-economic growth and often ignored large gaps between locations that are favored by investors and not based on location theory, which only cause exploitation of natural resources for peripheral regions (3).

The island, as the area that need to be developed, has different spatial conditions than the mainland. The island is a combination of smallness, remoteness, and oceanic (Kakazu, 2011), so it is necessary to understand the concept of regional development with these characteristics. Some of the outer islands also intersect directly with other countries, which can change the shape of the development. The outer islands are not only the peripheral areas of the closest region but also for other countries.

With the characteristics of the location, the use of cross-border development as a framework for developing the outer islands might be allowed. Borders can be a fence that separates one area from another and makes it difficult to cooperate, but on the other hand it can also be a zone of connectivity and cooperation (10). The aims of this study are looking at cross-border cooperation in the development of the outer island region and knowing the role of the outer islands in trans-border cooperation to create a balanced regional development.

2. Methods

2.1. Object, Place and Time of Research

The object of the study focused on the activities of the fisheries business in the studied location. The fisheries business is considered as an economic activity that is quite influential after the gas industry and had a significant role in the SIJORI agreement. This research was conducted in Natuna Regency, Riau Islands Province, Indonesia. While the time of research was in October 2017.

2.2. *Research Methods*

This study used descriptive methods that aim to provide a general, systematic, factual and actual description. Descriptive methods were carried out using a qualitative approach and supported by quantitative data. The implementation of the research at the location of the study was by survey techniques. Survey technique is a research that takes samples from one population and uses questionnaires as the main data collection tool (16).

3. Results and Discussions

3.1. Concept of Outer Islands Development

Regional development are the step to improve the socio-economic life of the community, to reduce disparity between regions, and to manage environmental sustainability (14). Regional development is indispensable because of the socio-economic, cultural and geographical conditions that are very different from one region to another and are an effort to harmonize natural resources, human beings and technology, taking into account environmental carrying capacity (Zen, 1999 in (15)).

The concept of regional development is largely based on the development of the neo-classical regional economy which only prioritizes economic growth. But if one is talk about developing the archipelago, of course it will have a distinctiveness from the mainland countries. Especially when talking about the outer islands that have elements of remoteness, smallness, and oceanic (Kakazu, 2011) as well as their spatial contact with other countries. From the various concepts of regional development offered, the development of the outer islands would be more appropriate if this study uses a core-periphery perspective, because the development of the outer islands in various parts of the world originated from the understanding that the outer islands are peripheral regions (9). The peripheral area is a lagging region that hopes that the economic activities that occur in its core region will cause what is called by Hirscman (1958) as trickle down effects.

In developed countries that have outer islands such as Japan and the Philippines, they tried to build their peripheral regions for both economic and security purposes. The national government of Japan created a good business climate so that investors are willing to build infrastructure and some other forms of investment in the outer islands and balanced with the readiness of local communities, both local entrepreneurs and local governments (9). Usually, the national government builds the outer islands as a destination and as a place for the development of several renewable energy programs. That is, their national governments have given an important role and clear division of regional functions to the outer islands.

The concept of that kind of development is in accordance with the concept of functional integration, an alternative development approach that was created intentionally at various centers of growth because of its complementary functions. The concept of peripherals that are functionally integrated have strong bond going forward and backward from a production process.

3.2. Cross-border Cooperation: Singapore - Johor - Riau (SIJORI)

In functionally integrating between regions, one form of development that can be initiated is cross-border cooperation. Trans-border development was considered as a development framework to build an economy between border regions (6). The purpose of developing a region with trans-border development is the utilization of potential and resource management possessed by each region, so that it will complement each other and be integrated. This integration is not only because it has the same economic needs but also due to the closeness of the culture given by the adjacent spatial conditions. It is hoped that later in the long-term perspective, integration that will occur will foster regional and local economies so that the region will develop and become independent.

In cross-border cooperation between regions, the scope is no longer a cross-border area, but trans-border or between cross-border areas (Baranyi, 2007). Cross-border areas then become the arena of economic, social and political cooperation owned by different countries (Engl, 2009). Usually, the initial growth of closeness between border areas is based on the process of moving goods, due to the increasingly-open access to transportation and the movement of people between regions. This background also happened to Singapore, Malaysia, and Indonesia.

Singapore has transformed into a highly-developed region with a high level of economic growth. As a result of Singapore's very limited land area, this caused land prices in Singapore to become very expensive. In addition to land prices, labor costs also increased. The average income from the manufacturing sector in 1989-1990 increased 12%, making foreign companies evaluate their investment in Singapore. Increased wages that are not offset by increasing productivity and increasingly expensive land prices make companies think of relocating their businesses to other regions with cheaper production costs (8).

Singapore, which has skilled workers and excellent infrastructure, began to look for other areas as a form of expansion of its business. Its proximity to Johor and Riau, prompting Singaporean Prime Minister Goh Chok Tong to think of building cooperation that allows industrial relocation, given the availability of sufficiently large and relatively inexpensive land in the two regions (13). Johor had a half-trained staff and Riau had a low wage for workers. Finally an agreement was reached regarding the specialization of the division of roles in shaded economic activities in the 1989 SIJORI agreement. The SIJORI agreement was expected to be "The Triangle of Growth" with the aim of covering; infrastructure development; development of natural and human resources; and the development of industries for export interests beyond national borders.

3.3. Implementation of SIJORI Cooperation: Unbalanced Regional Interactions

The division of work that occurred in the Growth Triangle can be seen from the core-periphery perspective. Singapore has good infrastructure, financial resources, technology and transportation, making it the core region. The Johor area was devoted to the development of economic zones and research activities (R & D). While for Riau Province, it was divided into Batam as a high value-added manufacturing area, Bintan for tourism activities and labor intensive manufacturing industries, while small islands such as Natuna, Anambas, and other islands were initially only for nature resources such as mineral, fisheries and agriculture.

SIJORI's cooperation has resulted in several changes, including increasing foreign investments and the population due to migration. In 1990, Johor received foreign investment from Singapore, Japan, and Taiwan worth USD 800 million / year, an increase of four times compared to the SIJORI agreement (7). Batam also experienced an increase in investment flows of USD 230 million / year. Economic growth in both regions continues to increase (18).

The positive impact was seen by the government both Malaysia and Indonesia as a success of mutual coherence. Finally the government submitted 3 Malaysian states and 6 other provinces in Indonesia to be included in the growth triangle. But it becomes ineffective because the division of roles was not very clear (13). This was compounded by the occurrence of crises in both Indonesia and Malaysia which eventually slowed the economic pulse in the triangle area and changed relations between countries after the crisis ended due to changes in political order (12). Each country considered economic activity in the triangle area to be limited to business-to-business which is nothing more than a form of trade and cross-border business. In fact, Batam was only used as an entry point for goods to be exported and industries with foreign capital that are very dependent on foreign market demand, without having a unique value of competitiveness. In the 2012-2017 period, 170 were closed and foreign companies withdrew their capital.

The success of the growth triangle that seems stalled, showed that there were neglected sustainability aspects, especially political will sustainability. SIJORI is currently considered no longer developing, according to SIJORI observer Prof. Dr. Ir Bun Yamin Ramto SE, as a result of policy makers and economic actors in the three regions being less proactive in accommodating common interests in such cooperation (11). Growth Triangle must be driven by strong political commitment and manifested in concrete action activities. The sense of dependency must be owned by each country as something complementary so that the need to remain partnered remains high. The concept initiated by Martinez regarding interactions in the border region can help see the trans-border cooperation between Singapore, Malaysia, and Indonesia, as shown below:

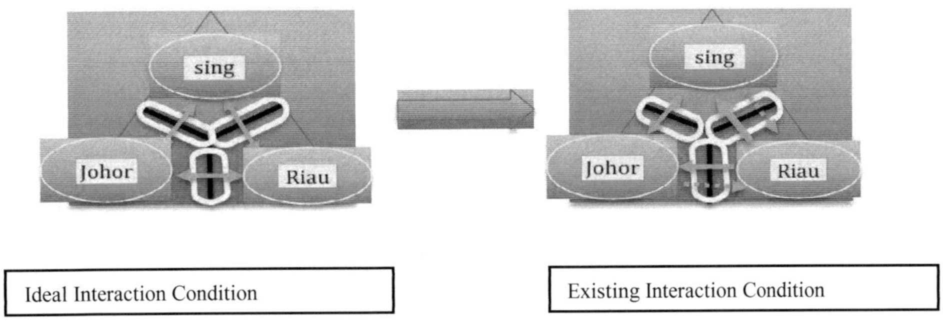

Figure 1. Interaction in the Border Region (Constructed from Martinez, 1994)

The implementation of the current SIJORI trans-border development has not been as expected. Ideally, the economies of countries that join functionally, and the free movement of people and goods across the border should no longer be restricted. Both countries can be seen as a single social system (Hardi, 2004). Between Singapore, Malaysia, and Indonesia, a maximum integration process was not established because it seemed that only the Indonesian side needed the presence of Singapore and Malaysia as a market. The supply of goods from Indonesia to the two countries was very good but the reciprocity was not felt by the Indonesian side. This failure was also one of the weak political will of the government to jointly make trans-border cooperation into mutually beneficial cooperation. Participation of every stakeholder at national, regional, and local levels is needed to realize the long-term goals of neighboring relations within a sustainable development framework (Anderson, 2003 in (5)).

3.4. Natuna's role in the Development of SIJORI Cooperation.

The SIJORI cross-border cooperation has affected the development of the outer islands. The economic activity that occurred in Batam as one of the cores of Natuna Island, also caused economic writhing on Natuna Island. The island can no longer be isolated and tries to find a way to get into the construction of its closest area. One of the potentials of Natuna Island is fishery commodities. The economic structure of Natuna Regency based on the business field is indeed dominated by the Mining and Excavation sector, especially the Oil and Gas sub-sector. The share of GRDP in the sector in 2015 was more than 75%. While the agricultural sector share including the fisheries and forestry sector was very low (less than 10%). However, if the Oil and Gas sector is issued, the role of the fisheries sector is very dominant (more than 40%). This shows that the fisheries sector can be a mainstay in the economic development of Natuna Regency (19).

The fisheries commodity offered by Natuna Regency as the outer island, turns out to have the potential to be sold in Singapore and Malaysia. Singapore is a country that is growing with tourism (bringing in many tourists and the tourism industry) and the lifestyle of people who are turning to healthy lifestyles. In 2016, Singapore imported fishery products amounting to USD 1.07 billion or 193,700 tons from China (US $ 139 million or 20,543 tons, Malaysia (US $ 135 million or 43,812 tons), Indonesia (US $ 110 million or 31,995 tons) and Vietnam (US $ 104 million - 32,599 tons), Norway (US $ 75 million - 8,685 tons and (Japan (US $ 58 million or 2,776 tons). While Malaysia, imports fish worth USD 885 million in 2016. Malaysian fish suppliers come from China at USD 234 million, Indonesia is USD 171 million and Thailand is USD 81 million (17). This shows that Singapore and Malaysia's needs for fish are quite high, thus becoming a promising market.

From a number of fish that flew to Singapore and Malaysia, there was an important role from Natuna Island. A number of fish were the catches of Natuna fishermen who were taken to Singapore and Malaysia through the cities of Batam, Pemangkat, and Tanjung Pinang . But what later became the reason for the role of the outer islands not seen in trade, especially fish in the SIJORI context was, that there was no recorded fish volume from Natuna Regency to Singapore, but Singapore fish traders already had accomplices in Batam. This had implications, not only does Natuna Island's role not being recognized, but also the pricing that was very determined by traders in Batam and finally, the farmer shares of Natuna fishermen were very small. This form of trading shows the periphery-core hierarchy as follows:

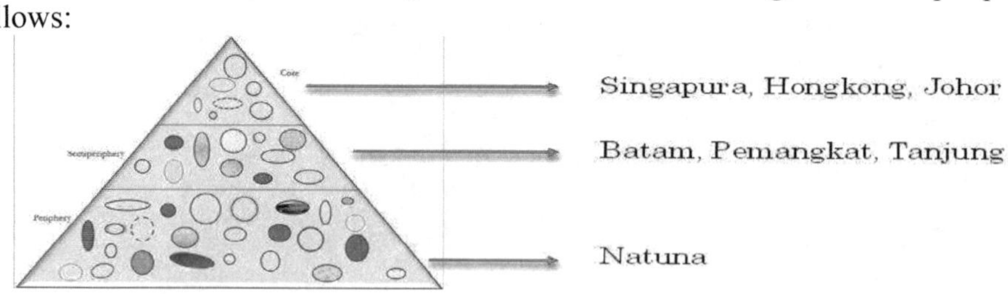

Chase-Dunn and Hall (1997)

Figure 2. Core-Periphery Structure in SIJORI

Many peripheral regions eventually continued to be left behind because the core area only extracted peripheral regions according to their needs (4). This was the case in Natuna because it was not guarded by regulations and stakeholders (national and regional governments). Natuna is the bottom periphery of the existing regional hierarchy. The form of polarization in the SIJORI triangle area can cause two effects, namely what Myrdal (Streeten, 1990) calls a backwash effect (increasingly marginalizing the periphery) and spread effects (encouraging the development of peripheral areas).

Table 1. Analysis of Backwash Effect and Spread Effect of Natuna Regency

No	Effect	Condition
Backwash Effect		
1.	Human resource flow	Mobility of residents from the Natuna region to Batam. This shows that there is no human resource development in the periphery region
2.	Flow of financial resources	Unbalanced investment flows, lack of physical development on Natuna Island
3.	Flow of raw materials	Natuna extracts its fisheries resources for the business interests of other regions (Batam, Tanjung Pinang, Pemangkat, Hongkong, Singapore)
4.	Information resource flow	Business actors (Natuna fishermen and agents act as price takers that have been set in Singapore as "receptors")
5	Power flow	In the Sijori context, power was held by the Riau Provincial Chamber of Commerce and trade representatives from Singapore and Malaysia
Spread Effect		
1	Trade network	The opening of a wider market outside the Natuna area
2	Transportation facilities	The construction of several means of transportation to facilitate the movement of goods (raw material in the form of fish)

From the table above, it can be seen that in the context of transborder cooperation, the form of regional development carried out more often creates a backwash effect that can actually be improved by stakeholders. Government escort and the participation of the role of the local community are expected to minimize the backwash effect caused by transborder development in the SIJORI region.

4. Conclusions

The border area consists of a group of people that need to be directed, controlled, and protected because it is full of different interests that must be protected politically. Trans-border development carried out through cooperation is a promising form for the construction of the outer islands. The presence of the government as a regulator and staff member in cooperation with trans-border area is very vital. The role of the government as a regulator and provider of infrastructure must be felt by the outer island communities in order to create a good business climate.

Exploitation will occur if the governments are insensitive and do not guard the process of developing the outer islands in the framework of trans-border cooperation. Even the flow of goods that are not accommodated with appropriate access can make the outer islands only as suppliers of goods that are not recognized, which make their existence not considered as important. Outer islands such as Natuna, which have primary products as important needs of neighboring countries, should be given access to directly export their fish to core countries so that the bargaining position of the outer islands will be more visible. In addition, given the opportunity to export without having to sell it to Batam, it will increasingly pump up the local economy which is expected to bring local entrepreneurs on the outer islands.

Technically, it needs to be understood that the division of roles in such cooperation must be clear, including the role of the outer islands, and is important so that complementary integration can occur between regions. The division of labor must be stated in an agreement integrity pact as part of the cooperation that must be fulfilled. The steps that can be taken by the government are: 1). Evaluation and negotiation with partner countries related to trans-border cooperation between Singapore, Johor, and Riau by including Natuna's role in the framework of cooperation.The division of roles in the development of cooperation needs to be followed up in regional policies including the ease of direct export of fish to Singapore and Malaysia;3). The need for integration of plans and policy documents as financial plan and legal umbrella in each Regency / City should be adjusted.

References

[1] Andrew K. Copus (2001) *From Core-periphery to Polycentric Development: Concepts of Spatial and Aspatial Peripherality, European Planning Studies, 9:4, 539-552, DOI: 10.1080/71366649*

[2] Baranyi, B. (2007): Dimensions of Border Crossing in Hungary, Dialog - Campus Publisher, Budapest - Pécs, 2007

[3] BPS Provinsi Riau, 2016., *Riau Dalam Angka.*

[4] Busega, I.,& Postoiu, C. (2015). *Inter-Regional Disparities in the European Union. Romanian Review of Regional Studies*. Journal of the Centrefor Regional Geography, 11(1), 3-14

[5] Chase Dunn & Hall (1997). *Rise and Demise: Comparing World-Systems.* Boulder, CO: Westview Press.

[6] Dastanka, Alena A dan Chuprys, Volha I (2014) *Euroregions As A Part Of Trans-Border Cooperation Of Belarus: Legal And Sociological Aspects*. Regional Formation and Development Studies, No. 2 (13) ISSN 2029-9370

[7] Ehlers, N. – Buursink, J. (2000): *Binational Cities: people, Institutions, and Structures*. In: M. van der Velde, H. van Houtum: Borders, Regions, and People / European Research in Regional Science 10. London: Pion Ltd.

[8] Engl, A. (2009). Territorial cooperation in Europe: Coordinated strategy or lost in confusion? In Innovation for Good Local and Regional Governance – A European Challenge, Institute of Governance Studies (pp. 1–21). University of Twente Enschede.

[9] Hardi, T. (2004): Formation of Cross-border Regions, Hungarian Science, p. 991

[10] Hirschman, A. (1958) The Strategy of Economic Development. Yale University Press, New Haven

[11] Hutchinson, Francis E (2015). Minor Images in Different Frames? Johor, The Riau Islands and Competition for Investment From Singapore. Singapore: Institute of Southeast Asia Studies.

[12] Kakazu, Hiroshi (2011) Challenges and Opportunities for Japan's Remote Islands. Eurasia Border Review < Voices from Asia's "Periphery" > Vol2 No 1

[13] Lee Tsau Yuan (1991). *Growth Triangle: The Johor – Singapore – Riau Experience*, Singapore: Institute of Southeast Asia Studies

[14] Martínez, Oscar, (1994) Border People: Life and Society in the US-Mexico Borderlands, United States, University of Arizona

[15] Matsumura, Yuko; Miyoshi, Emako (2018), Forces in the development of remote islands in Japan
: A case study of local energy enterprises in Tsushima Island. Osaka Human Sciences. 4 P.61-P.83. DOI : 10.18910/68280

[16] Ratti, R. – Reichman, S. (1993): *Theory and practice of transborder co-operation*. Basel: Helbing&Lichtenhahn
[17] Sjamsir Sjarif, (2000). Kerja Sama SIJORI Gagal. Riau Post: 29 September 2000
[18] Pepinsky, T.B (2009). Economic Crises and the Breakdown of Authoritarian Regimes: Indonesia and Malaysia in Comparative Perspective. New York: Cambridge University Press.
[19] Phelps, Nigel A (2004). Triangular Diplomacy Writ Small: The Political Economy of The Indonesia-Malaysia-Singapore Growth Triangle. Pacific Review 17, no 3.
[20] Riyadi (2000). *Pengembangan Wilayah Teori dan konsep Dasar, dalam Pengembangan Wilayah dan Otonomi Daerah Kajian Konsep dan Pengembangan*. Penerbit Pusat Pengkajian Kebijakan Teknologi Pengembangan Wilayah Badan Pengkajian dan Penerapan Teknologi. Jakarta
[21] Rustiadi, Ernan (2011) *Perencanaan dan Pengembangan Wilayah*. Yayasan Pustaka Obor Indonesia
[22] Singarimbun M dan Efendi S (1995). Metodologi Penelitian Survey. LP3E. Jakarta.
[23] Streeten, P. (1990). *Gunnar Myrdal. World Development, 18(7), 1031–1037.* doi:10.1016/0305-750x(90)90084-b
[24] USDA Foreign Agricultural Services (2017), *Singapore Seafood Report 2017*. Gain Report Number SN7014.
[25] Van Grunsven, Leo (1998). The Sustainability of Urban Development in the SIJORI Growth Triangle: A Social Perspective. Thirld World Planning Review 20, No 2
[26] Wardono, Budi. Suryawati, Siti dan Erlina, D.M (2017) *Pengembangan Perikanan Tangkap Melalui Sentra Kelautan dan Perikanan Terpadu (SKPT) Kabupaten Natuna*. Prosiding Seminar Nasional Kelautan XII. Fakultas Teknik dan Ilmu Kelautan Universitas Hang Tuah, Surabaya 20 Juli 2017

Analysis of Flood Vulnerability Using GIS Method (Case Study : Beringin Watershed, West Semarang)

H Nugroho[1], Suharyanto[2], T Ariyanto[3] and R Arnadi[4]
[1234]Department of Civil Engineering, Faculty of Engineering, Diponegoro University
Jl. Prof. Soedarto, Tembalang, Semarang, 50239, Telp: (024) 7474770,
Fax: (024) 7460060

*Email : arnadirizal@gmail.com

Abstract. Land degradation in the Beringin Watershed causes an increase of runoff and have high potential for flood disaster. The fact is proven by the distribution of land use for settlements in 1995 which was 18.45%, increased to 22.65% in 2002, thus enlarging the peak discharge in the Beringin watershed. To overcome this problem, it is necessary to study the effects of land use conversion on flood discharge and formulate the concept of management of Beringin watershed. Maximum annual daily rainfall be calculated with Polygon Thiessen method for 3 rain stations: Mijen, Mangkang Waduk, and Tugu for 20 years data (1998 – 2017). For the land use analysis, we was compared 2015 year Beringin watershed land use data with RTRW Semarang (Spatial Plans) 2011-2031 data and identified by using ArcGIS software. Whereas for the discharge design, the results of the HEC-HMS program modeling with 3 small dam was constructed for 10-years return period is 178.9 m^3/dt. River normalization is design on the river downstream along ± 3.9 km, from the estuary (STA 0 + 6.2) to the Mangkang (STA 3 + 909.05) and from HEC-RAS simulation results shows that its can discharge safely.

Keywords: *Beringin Watershed, ArcGIS, Normalization.*

1. Introduction

Flood disaster is the most frequent natural disaster in Indonesia. Data shows that ± 40% of natural disasters are floods [1]. Rapid population growth and development have caused changes in land use in the city of Semarang. Many of the land that was originally open land / forest has turned into residential and industrial areas. This resulted in an increase in runoff and enlarged peak discharge in the Beringin watershed has resulting in an increase in the potential for flood disaster [2]. The Beringin watershed has a flood inundation area in Podorejo Village - 174.00 ha, Mangkang Wetan Village - 323.00 ha, Tambakaji Village - 4.00 ha, Wonosari Village - 28.00 ha, and Mangunharjo Village - 28.80 ha [3]. The distribution of land used for settlements in 1995 was 18.45%, an increase in 2002 to 22.65% of the total area in the Beringin Watershed. Uncontrolled land use has an effect on runoff, thus increasing the peak discharge in the Beringin watershed [4]. The occurrence of these floods, it shows that the quality of the Beringin River Basin (DAS) has a high degradation and needs to be reviewed for watershed management.

2. Methods

The methodology in this research is presented in the flow chart in Figure 1.

IOP Conf. Series: Earth and Environmental Science **328** (2019) 012041 doi:10.1088/1755-1315/328/1/012041

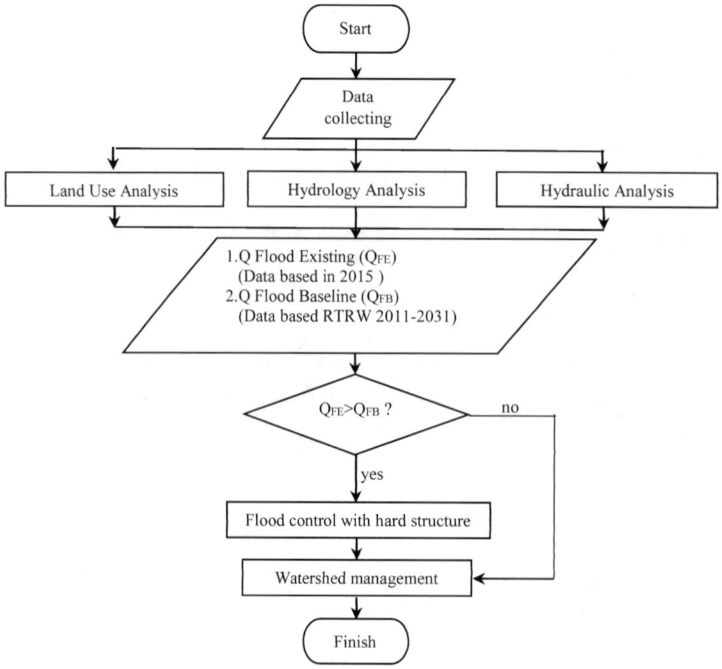

Figure 1. Research Flow Chart

3. Results and Discussions

3.1 The Result of Land Use Analysis

Land use analysis with GIS Method has obtained that Beringin watershed is divided into 31 sub-watersheds as Figure 2a and consisting of 3 types of soil, : Alluvial (C), Mediterranean (C), Latosol (D) as Figure 2b.

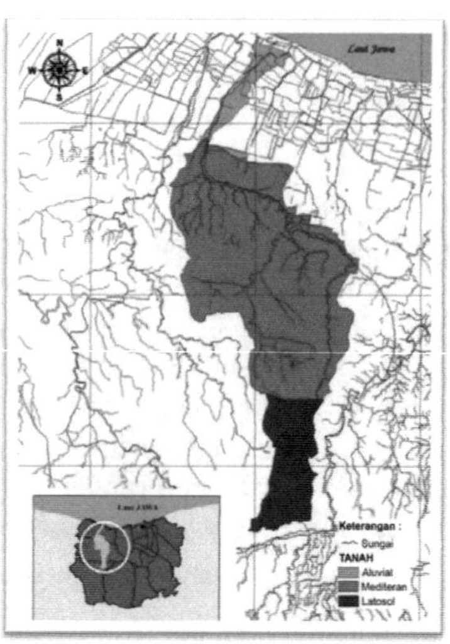

(2a).

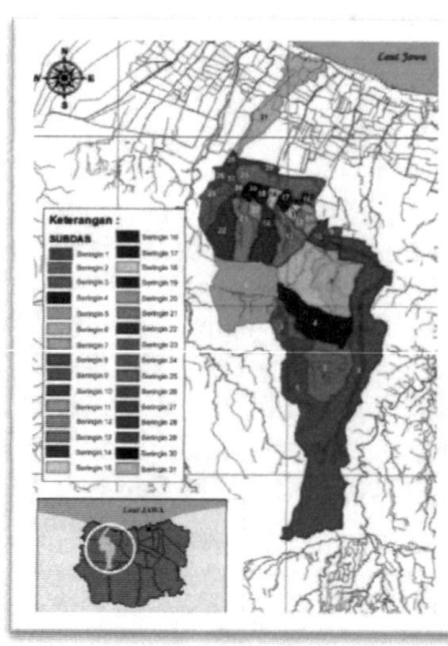

(2b).

Figure 2a. 31-Sub Watershed Beringin and Figure 2b. Three types of soil at Beringin Watershed

3.2 The Results of Hydrological Analysis
Hydrological analysis is needed to determine the design flood discharge which will be used to planning reservoir development.

3.2.1 Rainfall Calculation
The rain gauge stations are were used Tugu (Sta. 41), Mijen (Sta. 44), and Mangkang Waduk (Sta. 41c) with rainfall data during 1998-2017 (20 years). Then the Thiessen Method is used to calculate the maximum annual daily rainfall. Table 1. shows the effect of the rain station of the Thiessen coefficient. Table 2. shows the results of the calculation of maximum annual daily rainfall. And Figure 3. shows the location of the rain station and the division of area which is adhered to by Thiessen Polygon .

Table 1. Calculation of The Thiessen coefficient

No	Station	Area (km^2)	Thiessen Coeff
1.	Mijen (44)	9.57	0.28
2.	Tugu (41)	11.89	0.35
3.	Mangkang Waduk (41c)	12.38	0.37
	Total	33.84	1.00

Table 2. Maximum Annual Daily Rainfall (1998-2017)

No.	Year	Maximum daily rainfall (mm)	No.	Year	Maximum daily rainfall (mm)
1.	1998	72.09	11.	2008	100.21
2.	1999	108.36	12.	2009	170.26
3.	2000	164.09	13.	2010	68.96
4.	2001	82.67	14.	2011	63.43
5.	2002	59.07	15.	2012	96.97
6.	2003	51.40	16.	2013	120.54
7.	2004	104.74	17.	2014	167.33
8.	2005	48.79	18.	2015	111.78
9.	2006	148.11	19.	2016	83.58
10.	2007	83.34	20.	2017	127.37

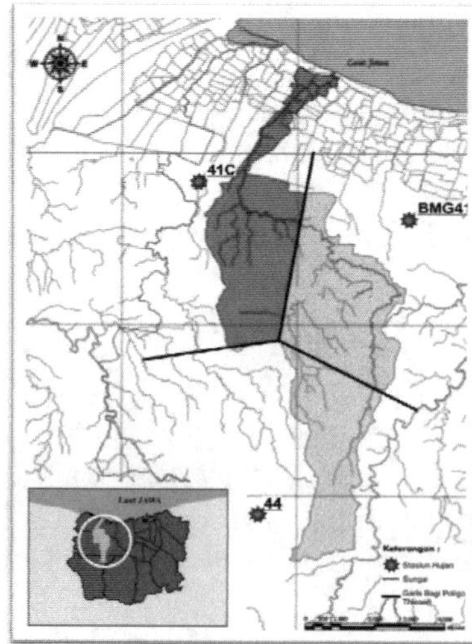

Figure 3. Location of the Beringin Watershed Rain Stations

Furthermore, the frequency analysis of rain data is carried out using 4 probability distributions: Gumbel, Normal, Log Normal, and Log Pearson Type III. After processing the data, we obtained values were skewness coefficient (Cs), kurtosis coefficient (Ck), and coefficient of variation (Cv) which are used as variables to select the type of rainfall data distribution.

Table 3. Result of The Frequency Analysis

Distribution	Requisite	Result	Information
Normal	Cs ≈ 0	0.508	rejected
	Ck ≈ 3	2.662	
Log Normal	Cs ≈ 0.258	-0.080	rejected
	Cv ≈ 0	0.084	
Gumbel	Cs < 1.1396	0.377	rejected
	Ck < 5.4002	2.662	
Log Pearson III	Cs ≠ 0	-0.080	accepted
	Ck ≈ 2.884	2.541	

Source : Triatmodjo, 2008.

Flood discharge was calculated using the HEC-HMS method. Before calculating the design flood discharge, this research needs to conduct the land use analysis to get the parameters of the HEC-HMS method. The results of the analysis are presented in Table 4

Table 4. Ratio of land use RTRW (*Spatial Plans*) 2011-2031 vs 2015

No.	Parameter	RTRW (Spatial Plans) 2011 - 2031		2015		Increase (+) / decrease (-)
		Area (km^2)	(%)	Area (km^2)	(%)	
1.	Land Use					
	- Forest	8.552	25.27	7.952	23.50	-
	- City (Industry, Mixed Business, Trade and Services, Education etc.)	6.933	20.49	8.193	24.21	+
	- Settlement	8.994	26.58	11.841	34.99	+

	- Agriculture	8.588	25.38	5.219	15.42	-
	- Ponds	0.774	2.29	0.636	1.88	-
2.	*Curve Number (CN)*	86.42		86.62		-
3.	*Impervious (I)*	22.78		23.56		+

Flood analysis using the HEC-HMS method with Q_{10th}, based on 2015 land use and the RTRW Semarang City (Spatial Plan) 2011-2031 has obtained hydrograph as shown in Figure 5.

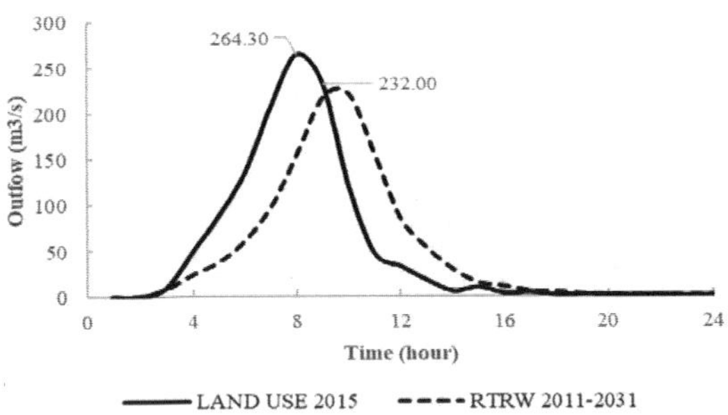

Figure 5. Estimated flood hydrograph for 10 years Return Period

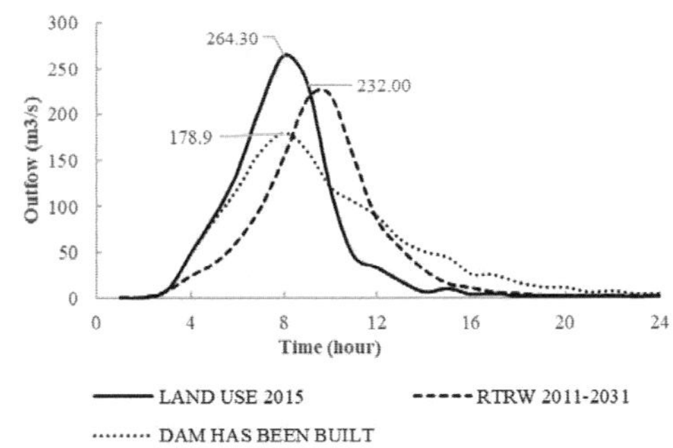

Figure 6. Flood hydrograph for 10 years Return Period if small dam has been built in Watershed

From Figure 5, it was concluded that 2015-flood discharge is exceeded than flood as designed 2011-2031 Spatial Plan, so it was necessary to proposed reservoir to reduce the existing flood discharge. Three potential reservoirs proposed in the Beringin watershed with total capacity of ± 7,504,548 m^3.
Flood simulation if 3 (three) new reservoirs in the watershed are constructed, can reduce the existing peak flood discharge from 264.3 m^3/sec to 178.9 m^3/sec or 32.31% below the peak discharge of the RTRW 2011-2031 for the period 10-years return, as presented in Figure 6.

3.3 The Results of Hydraulics Analysis

Hydraulic analysis aims to determine the ability of a river cross section to conduct the design discharge. As explained that one of the causes of flooding is because the river cross section cannot to drain the flood discharge. The HEC RAS simulation concluded that the cross section was unable to accommodate the flood discharge and the cross section needs to be expanded. In river cross-section analysis, the HEC-RAS program modeling method is used.

There are 5 (five) main steps in hydraulic simulation with HEC-RAS model:

1. Start a new project and giving the name of the project
2. Drawing a river flow scheme. The scheme is obtained by exporting data from Auto CAD Civil 3D. Elevations and river cross sections are included in the scheme.
3. Enter the geometry and hydrology data (river scheme, cross section, design discharge data, etc.).
4. Enter the boundary requisites; - Boundary Condition (upstream debit data) - Initial Condition (tidal data)
5. Perform hydraulic calculations. Simulation of river cross section with HEC-RAS is use design discharge i.g. 178.9 m^3/sec (Q_{10} years).

3.4 Cross Section Normalization Planning

HEC-RAS modeling was running in 2 conditions; 1). Running with the cross section existing, and 2). Running with the river cross section after normalized. Based on running the HEC-RAS modeling condition 1, it can be concluded that there are still several river cross sections are not able to accommodate flood discharge (*see Table 5, part : Cross Section Existing before normalization*). Flood is causes runoff of water and inundation in the downstream section of the river. The area of flood inundation can be seen in Figure 7.

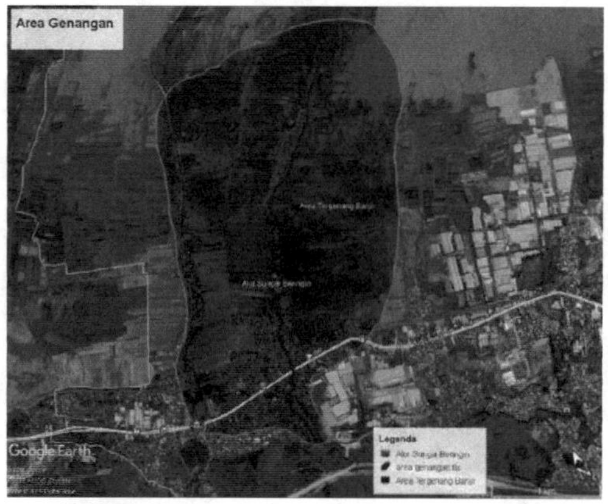

Figure 7. Flood Inundation Area

From Table 5. and Figure 7. it can be concluded that the construction of 3 small dams (embung) in the upstream area is still not enough to reduce the river water level in the downstream. Normalization is needed to accommodate the existing flood discharge. Normalization will be carried out in condition 2 by widening the cross-section of the river to 50 meters wide, with 6 meters height, 1:1 slope ratio and 1: 2 slope of embankment.

Table 5. Comparison of cross sections existing and after normalization (for example)

Point	Cross Section Existing (before normalization)	Cross Section (after normalization)
STA + 1438.89	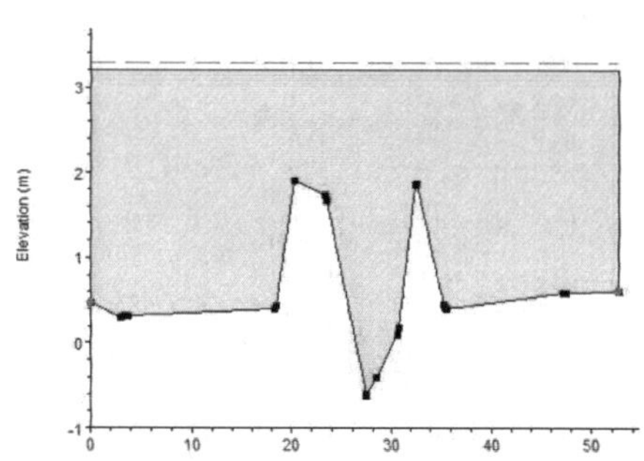	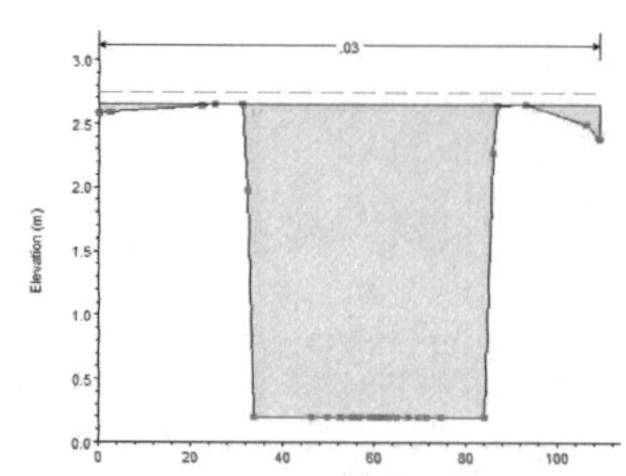

STA + 2426.96	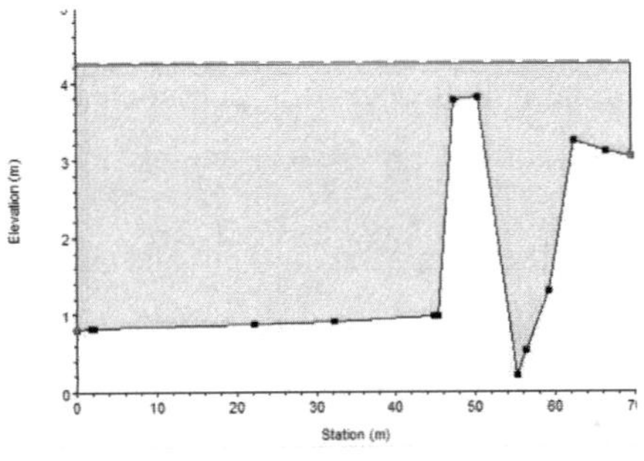	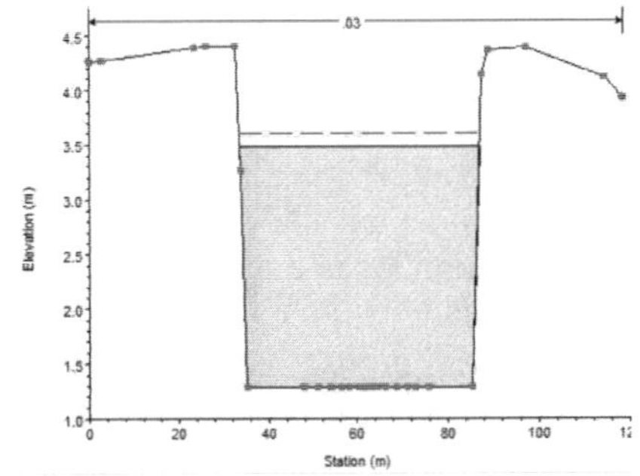
STA + 3711.44	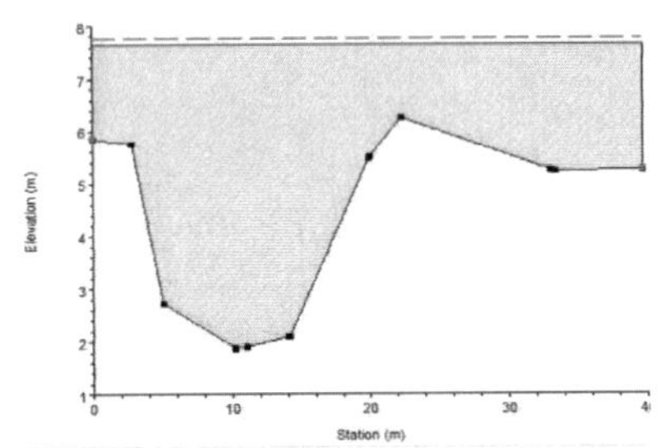	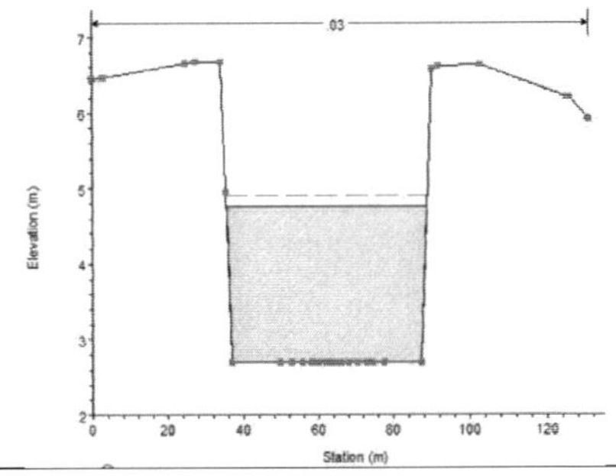
STA + 3909.05	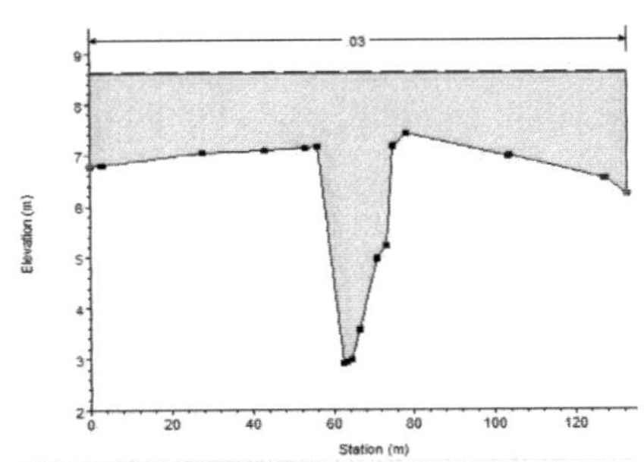	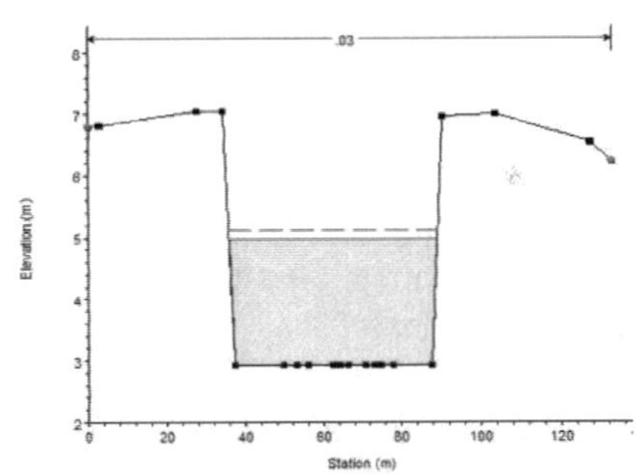

So that normalization in the downstream Beringin watershed by widening river cross section by 5 to 50 meters is a solution to reduce flood. According to Kodoatie and Sugiyanto (2002) there are several activities to handling floods with non-structural solution, for example:

a. Management of Watersheds
b. Land use arrangements
c. Utilization of the flood area
d. Forecasting and warning of flood hazards
e. Community Extension

4. Conclusion

Based on the land use, hydrology and hydraulic analysis, its can be concluded that:

1. Land Use in 2015 has exceeded the 2011-2031 RTRW, especially forest areas, agriculture, and ponds which have changed into residential and urban areas (industry, mixed business, trade and services, education etc.).
2. Land Use that deviates from the 2011-2020 RTRW causes an increase in peak flood discharge from 232.0 m^3/sec to 264.3 m^3/sec for Q_{10}.

3. Beringin Management river basin are divided into two parts: upstream and downstream. Upstream is construction of 3 reservoirs which has potential capacity of ± 7,504,548 m^3 can reduce peak flood discharge by 32.31% for Q_{10}. Downstream is needed by widening the river cross section existing to 50 meters wide to reduce the floods.
4. Increasing the effectiveness of Beringin Watershed management with non-structural solutions such as:

   a. Management of watersheds
   b. Control the development of inundation areas including land use regulations.
   c. Early warning flood system.
   d. Operational flood control.
   e. Society participation.
   f. Law - Enforcement.

Acknowledgements : This study was financially supported by Engineering Faculty Diponegoro University with Penelitian Strategis contract No : 2687/UN7.3.3/PG/2018 - 2 April 2018.

5. Reference

[1] Suherlan, 2001. *Bandung District Zoning for Flood Vulnerability Using Geographic Information System*. Bogor
[2] Prasetyo, W. 2016. Tesis, Management of the Beringin Watershed, Fakultas Teknik Undip
[3] BAPPEDA Semarang City, 2001. Flood Control in the Beringin River in Districts : Mijen, Tugu, and Ngaliyan.
[4] Kodoatie, R.J. dan Sjarief, R., 2009. Arrangement of Water Spaces (Tata Ruang Air). Yogyakarta : Andi Offset.
[5] ASEAN. 2016. ASEAN Homestay Standard.Jakarta: ASEAN Secretariat. www.asean.org/storage/2012/05/ASEAN-Homestay-Standard.pdf. Diakses 31 Maret 2018
[6] Hydrologic Engineering Center, 2010, *HEC-HMS Hydrology Modelling System, Hydrology Reference Manual*, Version 4.1, January 2010, U. S. Army Corps of Engineers, Davis, CA.
[7] Hydrologic Engineering Center, 2010, *HEC-RAS River Analysis System, User's Manual, Version 4.1*, January 2010, U. S. Army Corps of Engineers, Davis, CA.
[8] Triatmodjo, Bambang. 2008. *Applied Hydrology*. Yogyakarta : Beta Offset.

Architectural Concept for Homestay in Rural Area- A Case Study of Homestay Design in Sumbersari Village - Malang

H Sufianto[1], M Baskara[2], B Sugiarto[3], A Citraningrum[4] and M S Adhitama[5]

[1345]Faculty of Engineering, Brawijaya University
[2]Faculty of Agriculture, Brawijaya University

*Email: hsufianto@ub.ac.id

Abstract. Sumbersari, a village with wonderful natural views has the potential for ecotourism development in Malang City. As a part of UB Forest area, Sumbersari will be completed with homestay among its existing residential area. This study aims to develop a design guideline of homestay construction. The guideline is essential to ensure buildings and environment will be developed neatly and designed appropriately. A Number of data were collected through site observation and interviews with the local community. Data were examined to seek appropriate components in the guidelines acceptable by residents. This study proposed the size, rooms, layout, typical materials, structural system and typical furnishings for homestay.

1. Introduction

1.1 Background

Along with Indonesian tourism programs, the central government through local government of Malang induces village tourism visit program in the last ten years. One of the village areas around Malang potentially supports the program is Sumbersari. The village has a moderate contoured landscape, surrounded by amazing pine forests and coffee plantations. Located on the slopes of Mount Arjuna, this village has cool air and beautiful natural scenery. Populated by only 28 households, most of the dwellers are farmers of coffee and vegetables. (figure 1).

Over the last 5 years, the village was one of villages organized and developed by Brawijaya University, known as UB Forest. Small plantation industries have been created in this area along with research activities of students and lecturers of Brawijaya University. Several educational institutions in East Java are reported to conduct agricultural research activities in this village. Therefore, Sumbersari should be relevant as a stopover area of the visitors of UB Forest.

Related to the concept of ecotourism, on January 6, 2017 to coincide with the 54th Anniversary of Universitas Brawijaya Arjuno Fest, a music and arts festival in Sumbersari was held. The event was located in a field around the residential area and attended by more than 300 audiences. With a vast and relatively flat land and a beautiful backdrop of forest scenery, this village has the potential to become a tourist attraction for implementing similar events in the future.

Currently, regular visitors of Sumbersari are mainly researchers and students from Brawijaya University. In the future, Sumbersari village is expected to be developed as a potential ecotourism village in Malang. The implemented concept of ecotourism village in the future should be able to improve income for the dwellers and provide added value in the management of UB Forest area.

In contrast with potential natural resources around the village, the dwellers in majority are the low-level income society. The poor infrastructure, such as electricity, streets and roads of the village reduce economic improvement of the local society. Residents of Sumbersari complained about many pine plants planted near each other, which prevented the coffee plantations from fruiting, resulting in decreased coffee yields. In addition, the absence of primary school and health clinic requires the residents to travel to the nearest 3 km facilities.

This study aims to seek a solution to improve income for people who live in Sumbersari village. A concept of homestay was proposed for Sumbersari residents in order to invite more people, such as: tourists, researchers and students to visit and stay in the village for longer period. The concept of homestay, however, should be designed as simple as possible for the local society point of views and as familiar as possible to avoid rejection by residents.

1.2 Literature reviews

Homestay is defined as an alternative place to stay for commuters, living with the homeowner's family in the same residential building area and involving their daily lives with the host families and local communities [1] In this case, houses of residents (all or some parts) are rented for a certain period to visitors. It is a reasonable accommodation alternative, ideal for independent leisure travellers of all ages, interns, gap year students, students living abroad, researchers and anyone seeking a real and genuine travel experience. Homestay is chosen as a solution to improve economical prosperity among community as its advantages for both of hosts and guests. Homestay is chosen as an effort for empowering local communities to improve incomes. Living in a homestay for visitors will ave lodging costs. They will also benefit from quiet atmosphere, learning about local perspectives and the environment that are not easily found from a guidebook, a deeper understanding of culture and wisdom of locals as a new experience with a new atmosphere [3]. Meanwhile, the positive impacts for the hosts include: more job opportunities, cultural exchange, new knowledge, and maintain local traditions and wisdom.

Phisically, a homestay should follow certain technical requirement. It should have good and safe structures, designed and constructed with materials that reflect localities [1]. To some extent, homestay should have separate guest room to keep the privacy of the visitors. Electricity is highly required, while water supply and cleanliness of the room as well as comfortable furnishing are extremely important.

The planning of homestay in Sumbersari in fact, is in line with government programs, in which buildings should be developed from villages to suburbs. The growth of village homestay today indicates the trend of tourists who prefer to stay at home sharing compared to hotels. In 2020, visitor interested in home sharing in major cities of the world will increase by 15% from 10% (in 2016). Meanwhile, in Southeast Asia it will increase by 5% from 2% [5].

2. Methods

To find suitable design concept of homestay in Sumbersari, several data have been collected covering slope mapping around the village, environment, existing layout and site plan of houses, preference and opinion of Sumbersari residents. The existence of building utilities and infrastructures were considered during architectural design process. Physical data were collected through site observation while opinion and preferences of residents were sought through interviews. Collected data were used to analyse potential massing and positioning of homestay in each site, typical building materials used as well as rooms and furnishings essential in the unit.

3. Results and Discussions

3.1 Architectural design concept

Architectural design concept for homestay in Sumbersari should cover the principle of eco friendly architecture covering light and semi permanent building construction technology which is easy for local workmanship. The design should be simple and efficient in space. In addition, the concept should employ natural resources such as: carrying out natural ventilation system, optimizing natural lighting, capturing potential views and operating rainwater harvesting system.

3.2 Massing, slopes and distances subsection

The Sumbersari village consists of 28 families. Each house is commonly inhabited by one family. Houses are arranged linearly at the East and West sides of the road. All house faces toward main road. The buildings are located along the frontage of road while large farm is commonly located at the rear (figure. 1).

The following table shows the possibility of homestay location in each site. Considering the land slopes are lower at South area (figure. 1) and available distance from main buildings (Table 1), homestay was designed at the right side of the main building. Only several units are in the left and rear side.

Table 1. Building distances and sizes

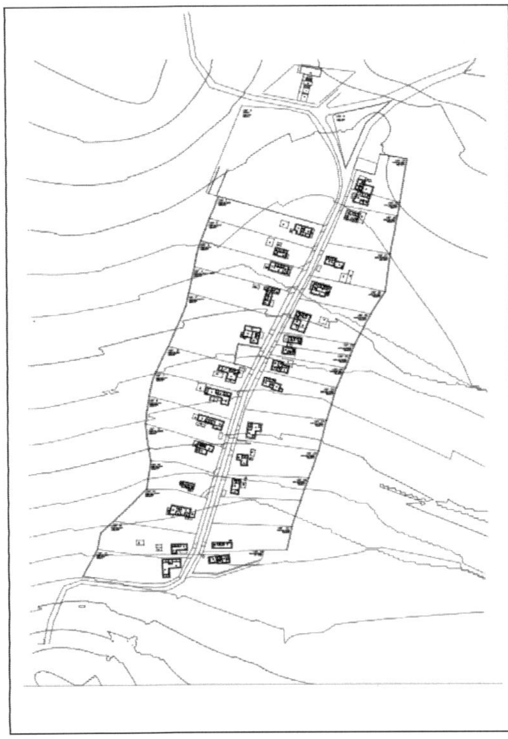

Site code	Front (m)	Right (m)	Left (m)	Possible homestay location(of main building)	Land size (m2)	Building size (m2)
UBF1	4	7.4	4	Right	1583	130
UBF 2	13	4.2	3	Right	1489	96
UBF 3	6.3	12.6	1.3	Right	1588	111
UBF 4	6.7	7.1	0.7	Right	1542	89
UBF 5	4.2	3.1	12.33	Left	2695	123
UBF 6	3.2	6.8	1.8	Right	1371	156
UBF 7	5.3	10.3	2.8	Right	1387	165
UBF 8	4.5	12.5	3.3	Right	1389	128
UBF 9	6.2	12.5	2	Right	1291	94
UBF 10	12	1.8	13	Left	1016	44
UBF 11	5	10.6	11.5	Right	1655	152
UBF 12	3	3.5	9.2	Left	1542	100
UBF 13	1.3	6	0	Right	1522	152
UBF 14	3.1	4.4	0	Rear	944	38
UBF 15	3.3	0.5	3.8	Rear	489	149
UBF 16	3.8	3	22.5	Left	1358	103
UBF 17	3.6	10.5	12.2	Right	1497	139
UBF 18	0	1.5	9.1	Left	1076	119
UBF 19	2.9	10	1	Right	1369	177
UBF 20	2.9	1.5	0	Rear	546	71
UBF 21	5.2	0.5	6	Left	647	40
UBF 22	4	3.7	2.6	Rear	866	119
UBF 23	3.2	5	18	Left	1763	83
UBF 24	5	10.5	2.9	Right	1537	88
UBF 25	3.5	11.9	8.2	Right	1544	66
UBF 26	5	3.3	3	Rear	1089	77
UBF 27	4.8	7.3	9.3	Right	1272	81
UBF 28	5	11.8	6.8	Right	1261	57

Figure 1. Existing slopes and building

Considering the existing conditions, a number of design concepts were proposed, including:

- The homestay unit is designed separately from the main building. The distance between two walls are as far as the height of the main building.
- The building is designed with a sleek shape along West to East, to avoid high solar radiation in tropical area.
- The design of the homestay should reflect the localities
- The homestay terrace is facing the road
- The homestay building can be seen as a whole from the road
- The parking area can be accessed directly and can be accessed easily from the site entrance
- Homestay should be completed with at least one carport heading toward terraces.

3.3 Building materials
Most buildings in Sumbersari use light weight wall construction either kalsiboard or woven bamboo, however, one house is made from brick wall. Tiles are not installed on the floors as in many houses' floors are made from compacted soil. Roof construction is made from a variety of materials, such as: asbestos sheet, burnt clay tiles, or a combination (Figure 2). Typical small opening for air ventilation are found in the facade of buildings to avoid cold breezes flowing inside the room, and consequently houses have low light indoors.

Figure 2. Typical building materials of existing buildings

In accordance with local weather conditions and affordable cost of construction, homestay materials should be made from cheap local materials which can easily be found around village and strong and safe. Stage construction is used to increase evaporation from the soil and reduce dampness. The selection of building materials should consider the ease of installation so that it can be carried out by the residents themselves. Therefore, materials which are chosen includes:

- Roof : Onduline tile completed with solar panel 245 Watt
- Bedroom walls : combination of pine wood walls and bamboo webbing with varnish coating
- Bathroom : bottom of the brick wall, top of the woven bamboo with varnish coating
- Column : Columns and beams of coconut wood structure
- Roof frame : Coconut wood roof frame
- Bedroom Floor : Varnish pine wood flooring
- Bathroom floor : Stucco brick with coarse ceramics
- Fondation : Concrete footplate
- Windows frame : Wood
- Bathroom : Fiber and squat toilet
- Pathways: Paving stone

3.4 Rooms
It was decided that one family would only have one single unit homestay with two-person capacity. The unit should be an individual building separated from the existing one. The unit should be able to accommodate guests who stay or forstaying or office working activities. Therefore, one unit would only be designed for at least 20 m2 comprising terraces, carport, bedroom and working space as well as individual bathroom. All rooms should be completely furnished, including: 2 working desks with chairs, small cupboard, bunk bed unit and electrical contact points.

Homestay unit should have rear terrace that is connected through pathways to kitchen room of the existing building. This access is important to maintain relationship between guests and host. In fact, most hosts traditionally receive guests in the kitchen instead of living room

IOP Conf. Series: Earth and Environmental Science **328** (2019) 012042 doi:10.1088/1755-1315/328/1/012042

3.5 Prototype of homestay design

From the design concept discussed above, initial homestay design is proposed in the following figures. The concept of eco friendly energy is applied through the use of natural materials, such as: coconut wood, pine wood, bamboo webbing, lightweight construction, pedestal foundations, ease of installation, simple construction technology, efficient use of space with compact furniture arrangement, as well as utilizing the potential of the site with the use of natural air and lighting, view optimization, and the application of rainwater harvesting systems. In addition, the concept of eco-friendly energy is implemented using solar cells on the roofs

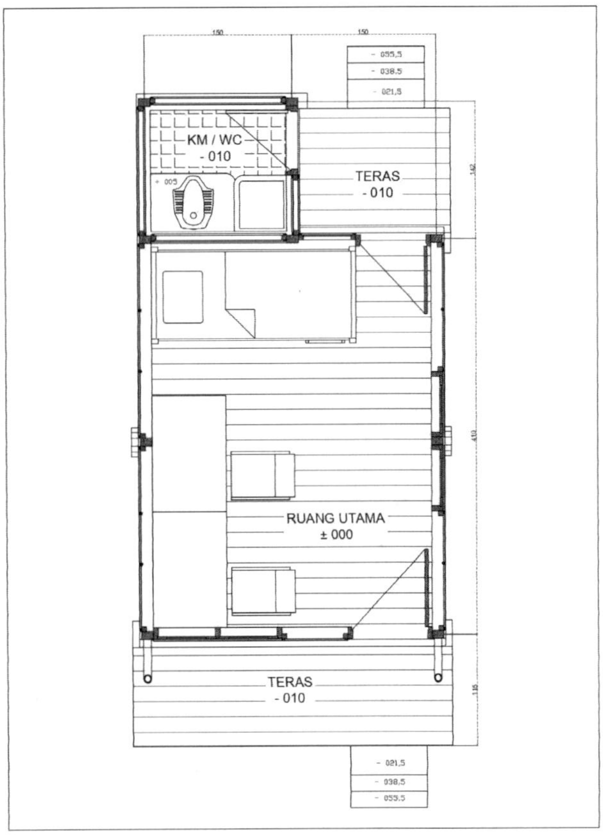

Figure 3. Proposed Plan for homestay

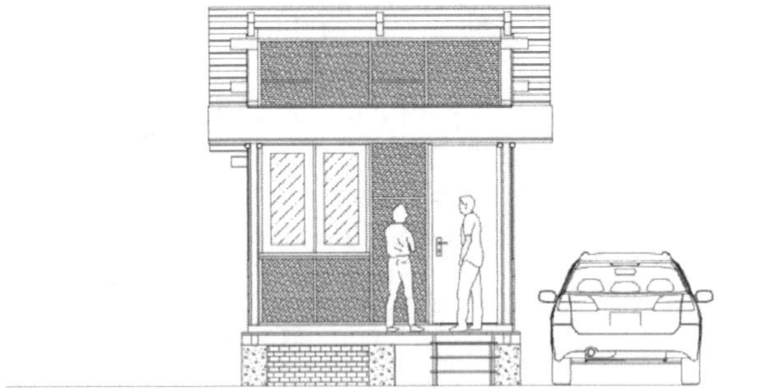

Figure 4. Front views of homestay

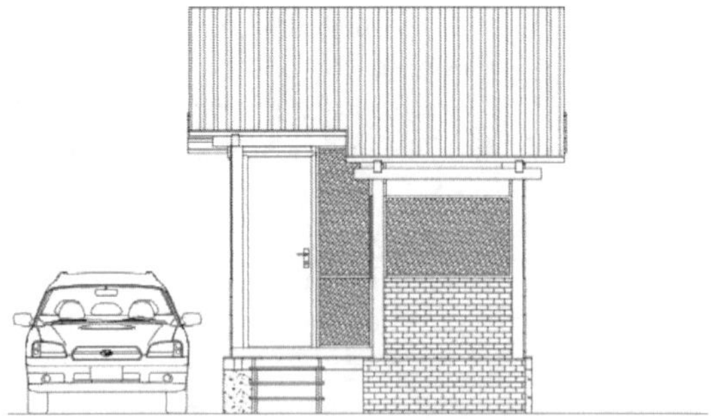

Figure 5. Rear views of homestay

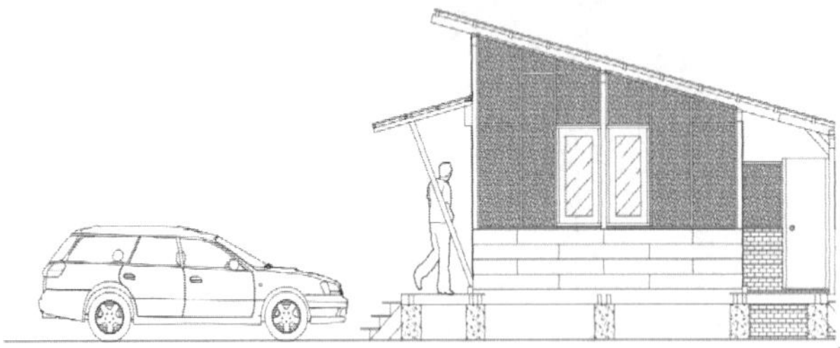

Figure 6. Side views of homestay

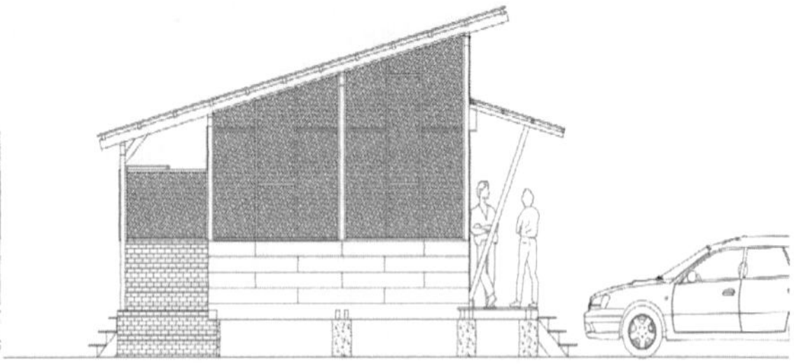

Figure 7. Side views of homestay

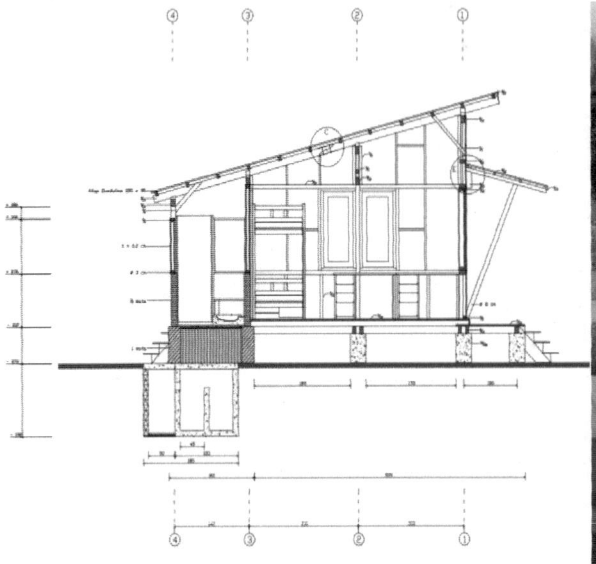

Figure 8. Section views of homestay

Figure 9. Image 3D of proposed homestay design

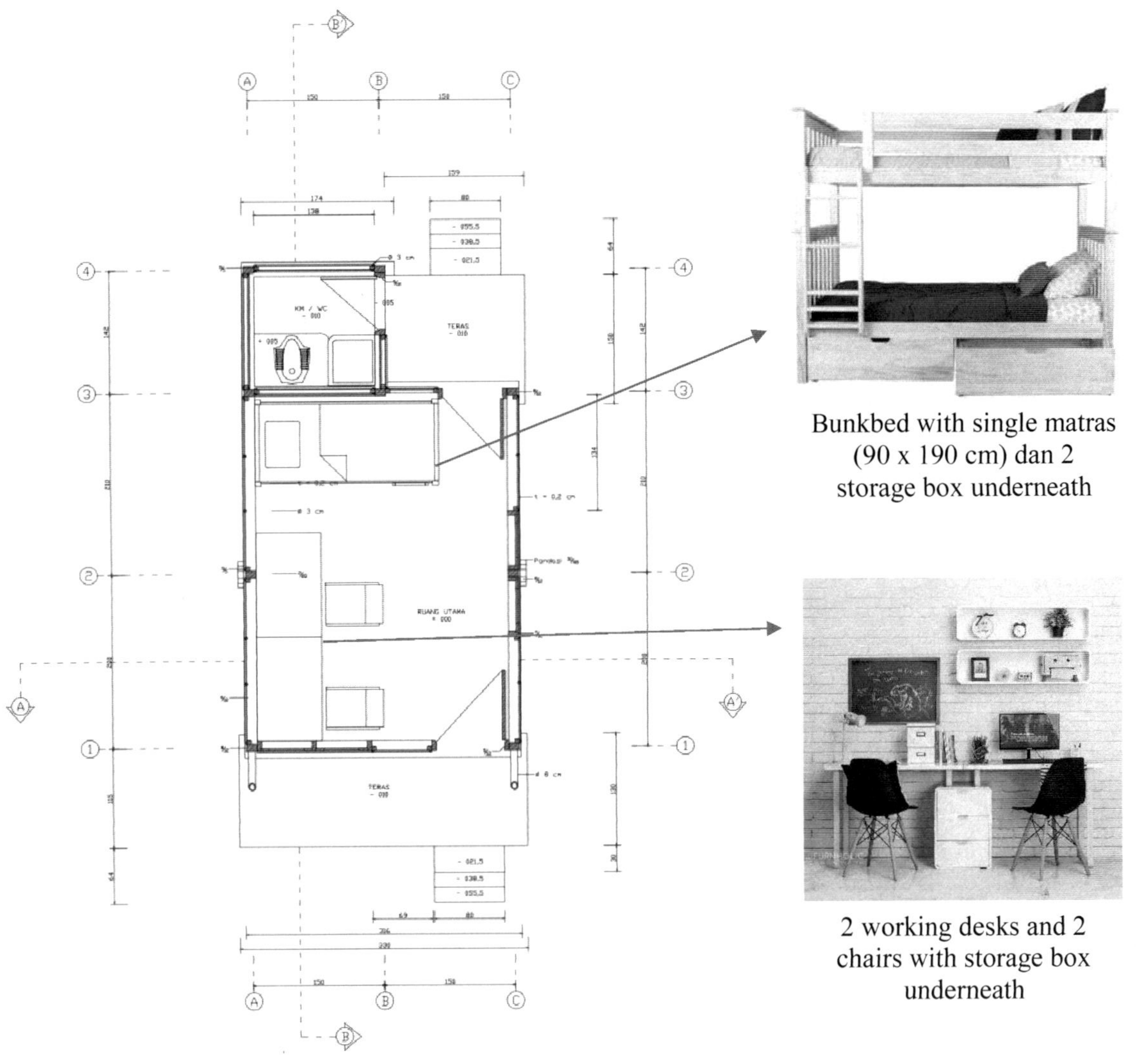

Figure 10. Images for room furniture

4 Conclusion

Sumbersari has the potential to be developed into an ecotourism village. Homestay is intended for both general tourists and researchers who will be active in UB Forest. The existence of homestay is intended by local residents to increase their income as well as to meet the needs of Universitas Brawijaya for researchers' accommodation facilities at UB Forest. Based on site observation and interviews with residents, design criteria for homestay are found as guidelines for design, building orientation, accessibility / parking, building area and capacity, room requirements, and connectivity to the main houses inhabited by residents.

References

[1] ASEAN. 2016. ASEAN Homestay Standard.Jakarta: ASEAN Secretariat. www.asean.org/storage/2012/05/ASEAN-Homestay-Standard.pdf. Diakses 31 Maret 2018

[2] Kementerian Desa, Pembangunan Daerah TertinggalDan Transmigrasi. 2015. Indeks Desa Membangun 2015. Jakarta: Kementerian Desa, Pembangunan Daerah TertinggalDan

Transmigrasi. http://kedesa.id/id_ID/repository/indeks-desa-membangun-indonesia/. Diakses 31 Maret 2018
[3] Sustainable Regional Economic Growth and Investment Programme (SREGIP). 2106. Pelatihan Homestay. www.sregip.or.id/wp.../Training-on-Homestay-Development.pdf.Diakses 31 Maret 2018
[4] Sumarlan. 2017. Sekeping Informasi dari "UB Forest" Malang. http://bp2sdm.menlhk.go.id/pusrenbang/index.php/component/content/article?id=24:sekeping-informasi-dari-ub-forest-malang. Diakses 31 Maret 2018
[5] Rachman, Vicky. 2017.Proyek Besar Kemenpar: Bangun 100 Ribu Homestay Desa Wisata. https://swa.co.id/swa/trends/proyek-besar-kemenpar-bangun-100-ribu-homestay-desa-wisata. Diakses 31 Maret 2018.
[6] Putrizahrah, Annisa. 2017. UB Forest Dusun Sumbersari. http://anisaputrizahrah015.blogspot.co.id/2017/05/ub-forest-dusun-sumbersari.html, 31 Maret 2017.

IOP Conf. Series: Earth and Environmental Science **328** (2019) 012043 doi:10.1088/1755-1315/328/1/012043

Neighbourhood Relationship among Villages in Gedangan District: Multidimensional Poverty Approach

I R D Ari[1], S Hariyani[2] and B S Waloejo[3]

[123]Department of Urban and Regional Planning, Brawijaya University, Jl. MT. Haryono 167, Malang City, East Java, Indonesia, 65145

Email: dwiari@ub.ac.id

Abstract. Poverty becomes one of the global problems especially in developing country. Poverty is condition when people have difficulty to access basic needs and unable access it. This research takes place at Gedangan District, Malang Regency. This research aims are to: i) measure multi-dimension poverty index and ii) find out neighbourhood relationship among villages in Gedangan District. There are two research methods which are are used: i) Multidimensional Poverty Index (MPI) – consists of three dimensions (health, education and living standard); and ii) Spatial Data Analysis. MPI value measurement indicates that there are 2 MPI levels: low level (4 villages) and medium level (4 villages). Therefore, based on the MPI measurement, there are 4 villages with medium level of MPI as the priority villages to deal with poverty. The result of LISA map by spatial data analysis indicates that cluster of Low-High is form by Sidodadi village. It means that there are inequality facilities services between low value area and high value area. Sidodadi gets better services than its neighbour because it gets better service from another district, and it is vice versa for the village that have high value but does not get good facility services from Gedangan District.

Keywords: Poverty Alleviation; Multidimensional Poverty Index; Spatial Analysis

1. Introduction

No Poverty in all forms and dimensions is one of SDG's important goals and is the greatest challenge faced by society and government. According to the most recent estimates, in 2015, 10 percent of the world's population lived on less than US$1.90 a day [1]. Poor people have difficulty in accessing food, clean water, and sanitation [2]. Moreover, poverty provides much effect, such as malnutrition, mental illness, infectious diseases, and dependence of drugs. Poverty problems are derived from lack of ability, unemployment, low income, low education's level, have a bad health, insecurity, and inconfidence [3]. The human development paradigm have a evolution that made theoretical base to measure multidimensional poverty [4].

Indonesia has 25, 95 million people or 9, 82% living below poverty line in March 2018. Most of them are living in Java Island spreading over six provinces that is 13, 34 million poor people or 8, 94% [5]. Indonesian government endorses strategies to alleviate poverty by 3 aspects, such as: implementing the comprehensive social protection, increasing the public services, and promoting sustainable livinghoods. Government's program in Indonesia in poverty alleviation consists of *Beras Miskin* (Rice for the poor), *Beras Sejahtera* (Rice for Prosperous), *Bantuan Pangan Non Tunai* (Food Aid Non Cash) and *Bedah Kemiskinan Rakyat Sejahtera* (Surgical Poverty People Prosperous) [6].

Malang Regency is the part of East Java Province having the highest number of poor peoples that is 293.740 poor people or 11, 4% from total population and one of the districts in Malang Regency is Gedangan District. Based on data from Gedangan District, the number of poor people in Gedangan District is 5.178 people. In general, poor people in Gedangan District also accept *Beras Miskin*. Gedangan District have 8 (eight) villages, such as: Tumpakrejo, Sindurejo, Gajahrejo, Sidodadi, Gedangan, Segaran, Sumberejo, and Girimulyo Village.

The understanding towards the poverty condition in Gedangan District plays a crucial role to recommend poverty policies based on priority village location. Therefore, an approach by figuring out the level of poverty index in Gedangan District is the important attempt. Poverty's level in Gedangan District was analyzed by using MPI (Multidimensional Poverty Index) focused on the 3 dimensions such as: health, education and living standard. To add, interrelationships among the variables are estimated by using spatial data analysis.

2. Methods

This research utilizes two methods of: Multidimensional Poverty Index and Spatial Data Analysis.

2.1. Multidimensional Poverty Index

Multidimensional Poverty Index (MPI) is one of poverty's level analyses and is popular due to being applied by Oxford Poverty and Human Development Initiative (OPHI) and United Nations Development Program (UNDP). However, the inventor of this method is Alkire and Foster in 2011. MPI method involves the three dimensions consisting of ten indicators at each. The indicators was divided into two indicators for health aspect (0,1667 per indicator), two indicators for education aspect (0,1667 per indicator) and six indicators for living standards (0,0556 per indicator) [7][8]. The MPI is a multiplication of the percentage of people who are poor or multidimensional headcount ratio (H) with intensity of poverty (A) [8]. The MPI formula can be expressed as follow:

$$MPI = H \ x \ A \qquad (1)$$

MPI = Multidimensional Poverty Index
H = Headcount
A = Intensity of Deprivation

Based on module of MPI Research Team in Indonesia [9], MPI or poverty level values can be classified as:

Very High : >0, 36
High : 0, 27-0, 36
Medium : 0, 18-0, 27
Low : 0, 09-0, 18
Very Low : <0, 09:

2.2. Spatial Data Analysis

Spatial data analysis is performed by utilizing Geoda Program (Geographic Data Analysis), which is free and open source software. Dr. Luc Anselin and his team introduced this software in 2003 [10], providing a user-friendly methods to analyze spatial data such as spatial data autocorrelation and basic spatial regression analysis [11]. Spatial analysis is an analysis engaging spatial data input by using arcgis tools to process spatial data that has been surveyed. Then, spatial data will be analyzed by using the Geoda tools. Spatial analysis have weights matrix to calculate the value of regression.

Weights matrix in Geoda Program can be operated with the space concept. This matrix contains about neighborhood relationship in a dataset. This study indicates that the neighborhood relationship among eight villages in Gedangan District with Queen Weight Matrix. Queen Weight Matrix defines a village's neighborhood as those with either a shared border or vertex. Connectivity histogram was displayed by the characteristics of weights matrix. Every bar in connectivity histogram was explained by the frequency from each number of neighbors in Gedangan District.

3. Result and Discussion

3.1Multidimensional Poverty Index

Table 1. Multidimensional Poverty Index for each Village in Gedangan District

No.	Villages	MPI Values	Classification	Contribution of deprivation in Dimension to overall poverty (%)		
				Education	Health	Living Standards
1	Tumpakrejo	0,12	Low	53	0	47
2	Sindurejo	0,20	Medium	54	0	46
3	Gajahrejo	0,25	Medium	56	0	44
4	Sidodadi	0,18	Medium	55	0	45
5	Gedangan	0,27	Medium	51	0	49
6	Segaran	0,15	Low	59	0	41
7	Sumberejo	0,14	Low	48	0	52
8	Girimulyo	0,16	Low	53	0	47

Based on Table 1, the result of MPI indicates that there are four villages classified as medium level and four villages classified as low level. The medium level of MPI occurs in Sindurejo, Gajahrejo, Sidodadi, and Gedangan. Meanwhile, low level of MPI occurs in Tumpakrejo, Segaran, Sumberejo and Girimulyo. The highest value of MPI is in Gedangan village with the value of 0, 27 MPI . MPI level map of each village is visualized in the following figure 1.

Figure 1. MPI Level of each villages in Gedangan District

IOP Conf. Series: Earth and Environmental Science **328** (2019) 012043 doi:10.1088/1755-1315/328/1/012043

3.2 Spatial Data Analysis

The first result of spatial autocorrelation is Morans Scatter Plot, providing a statistic (Moran's I) to determine the extent of linear association between the MPI values in a given village (x-axis) with the MPI values in neighbouring villages (y-axis). Each quadrant corresponds to a different type of spatial autocorrelations, such as: high-high and low-low for positive spatial autocorrelation; low-high and high-low for negative spatial autocorrelation. The type of spatial autocorrelation in a data distribution of MPI values in Gedangan District is visualized in the following figure 2.

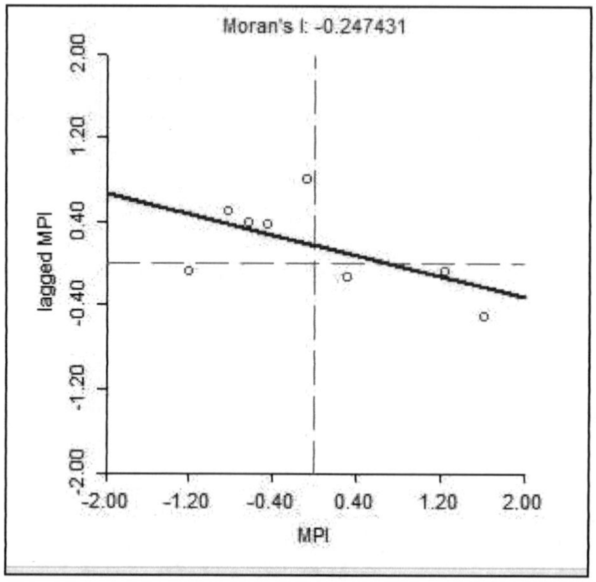

Figure 2. Morans Scatter Plot

Based on Figure 1, the values of MPI in Gedangan District are spread, indicating the value of Moran's I which is closer to the zero. The closer value of Moran's I to the zero means the absence of spatial autocorrelation. Contrastingly, the closer value of Moran's I to the one means the existence of spatial autocorrelation. The slope of the scatter plot indicates a little negative spatial autocorrelation in Gedangan District, as visualized in LISA Cluster Map below.

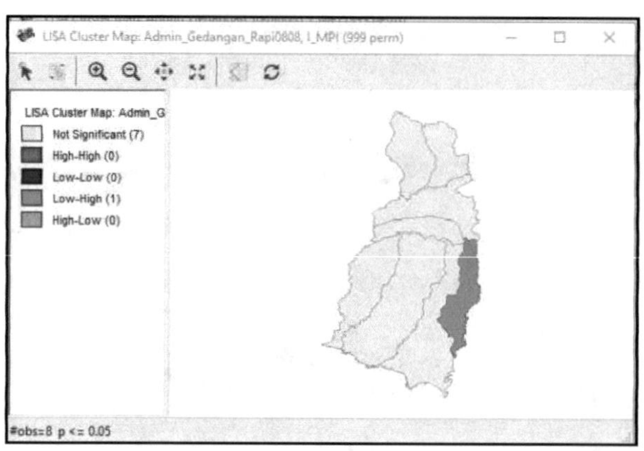

Figure 3. LISA Cluster Map

Based on figure 3, there is one type of cluster in Gedangan District, which is: the Low-High Cluster covers 1 village in Sidodadi Village, having Low MPI value correlated with high neighboring MPI values. Based on Morans Scatter Plot and LISA Cluster Map, it is apparent that the poverty in Gedangan District (which measured by MPI) tends not to form a cluster.

4. Conclusion

From the research, in order to lessen poverty level in the research area, there are two conclusions as recommended such as:

- Villages In Gedangan District only has two classification of MPI level, low level and medium level. The low level occurs in 4 villages, such as in Sindurejo, Gajahrejo, Sidodadi and Gedangan.
- Gedangan District, especially in Village belongs to Low-High classification in LISA cluster, having low value of poverty or MPI measure but surounded by village who has high value of poverty.

Acknowledgements

This research is funded by Indonesia Ministry of Research, Technology and Higher Education. Sincere gratitude is addressed for the solid team - Dwi, Rindang, Anes, Meri, Bayu, Nadine, Diah

References

[1] World Bank. 2019. Understanding Poverty. (online) https://www.worldbank.org/en/topic/poverty/overview

[2] United Nations Department of Economic and Social Affairs. 2018. Sustainable Development Goals. [Online]. Available: https://www.sustainabledevelopment.un.org/topics/sustainabledevelopmentgoals, 2018.

[3] E. T. Adhi. 2009. Pelayanan Sanitasi Buruk: Akar dari Kemiskinan. Bandung. Jurnal Analisis Sosial, vol. 14, pp. 76-87.

[4] Dehury, Bidyadhar and Sanjay K Mohanty. 2015. Regional Estimates of Multidimensional Poverty in India. Economic E-journal.

[5] Badan Pusat Statistik, Jumlah Penduduk dan Presentase Penduduk Miskin dan Garis Kemiskinan Menurut Kabupaten/Kota di Jawa Timur Tahun 2018, 2018.

[6] Badan Perencanaan Pembangunan Nasional. 2017. Strategi Nasional dalam Pencapaian Target Pengurangan Kemiskinan Tahun 2019.

[7] S. Alkire and J. E. Foster. 2011. Counting and Multidimensional Poverty Measurement. Journal of Public Economics, vol. 95, pp. 476-487.

[8] United Nations Development Programme (UNDP), Technical Notes Human Development Report, 2016.

[9] Research Team MPI Indonesia Perkumpulan Prakarsa. 2015. Modul Pelatihan Perhitungan Multidimensional Poverty Index Indonesia dan Penyusunan Pelaporan MPI Indonesia.

[10] L. Anselin, Spatial Econometric: Methods and Models. Netherlands: Kluwer Academic Publishers, 1988.

[11] X.Li. Term Used in GeoDa 1.12.1.129. [Online]. Available: https://geodacenter.github.io/glossary.html

Collaborative Governance in Providing Facilities of Sungai Bambu Child Friendly Integrated Public Space, North Jakarta City

I Rahmaningtyas[1] and A Y S Rahayu[2]

[12]Department of Public Administration, Faculty of Administrative Science, Universitas Indonesia, Depok, Indonesia

*Email: intan.rahmaningtyas@gmail.com

Abstract. The development of a country cannot be separated from development which includes the fulfillment of children's rights. Through the Regulation of the State Minister for Women's Empowerment and Child Protection of the Republic of Indonesia Number 11 of 2011, it was emphasized the importance of establishing a Child-Friendly City Development Policy. The regulation triggered the DKI Jakarta Provincial Government to remain committed to realizing the Eligible City of Children by creating Child-Friendly Integrated Public Spaces (RPTRA) in administrative cities. From 31 indicators set by the Ministry of Women's Empowerment (KPP) for Eligible Cities, 70 to 80 percent of them can be achieved through RPTRA. Due to the high budget needs, the government cannot operate alone but also collaborates with the private sector in its development. This study aims to find out collaborative governance conducted by the government, and the private sector in the provision of Sungai Bambu Integrated Child Friendly Public Space (Ruang Publik Terpadu Ramah Anak, abbreviated as RPTRA) facilities, North Jakarta City. This research is important because it can provide an overview and conclusions related to collaborative governance modelling in the construction of RPTRA. This study uses a post-positivist approach based on the Collaborative Governance theory by Emerson & Nabatchi (2015). Interviews and documentation studies were data collection techniques. The results of this study indicate there is a dynamic of responsive collaboration between one component and another. The Context System factors that influence the dynamics of collaboration are public resources and service conditions, policy and legal frameworks, socio-economic and cultural characteristics, political dynamics and power relations, and the history of conflict.

Keywords*: Collaboration, Collaborative Governance, Child Friendly Space*

1. Introduction

1.1 Background

Collaborative Governance can be defined as a regulation that regulates one or more public institutions directly involved with non-public stakeholders in a formal decision-making process, consensus-oriented, and deliberation aimed at making or implementing public policies or managing public programs or assets. Collaborative Governance can also describe the state of interdependence between factors. The desire to do collaborative governance arises because these actors need to express their willingness and willingness to establish closer relations with other actors [1].

One of the collaborations was carried out in the provision of Child Friendly Integrated Public Space (Ruang Publik Terpadu Ramah Anak, abbreviated as RPTRA) facilities in DKI Jakarta Province. Child Friendly Public Space is a public space that has characteristics as a public open park, a vehicle for children's play and growth, a part of the infrastructure and facilities of a child friendly city, green open space, and facilities for social activities equipped with various interesting games, CCTV surveillance, library space, PKK Mart, lactation room, and others [2]. Starting from the budgetary needs of the government for the fulfillment of these facilities, the government provides an opportunity for the private sector to participate in building RPTRA through the Corporate Social Responbility scheme.

The construction of RPTRA has been going massively since 2015. Within two years, the number of RPTRA, which amounted to 186 RPTRA at that time, continued to increase as 100 RPTRA were built using the APBD and 5 RPTRA from CSR. In 2018, the Head of the DKI Jakarta Housing and Settlement Area Office Agustino Darmawan said, the number of child-friendly integrated public spaces (RPTRA) was recorded at 291 RPTRA with details 68 RPTRA from CSR funds, and 223 from Regional Budget funds[3]. The latest data in March 2019 from the Department of Child Protection and Population Control (DPPAPP) stated that there were 308 RPTRA, of which 68 were from the government and private collaboration.

The first RPTRA which was built as a result of collaborative governance was RPTRA Sungai Bambu, Urban Village of Sungai Bambu, North Jakarta. The private company that played a role was PT Pembangunan Jaya with a CSR fund of Rp. 412,154,795 (four hundred twelve million one hundred fifty four thousand seven hundred ninety five rupiahs) [4]. Now the Sungai Bambu RPTRA is a pilot RPTRA and is visited by many other regions. Even so, there is no in-depth literature or information that can illustrate how this collaborative governance model can be a model for governments in other regions in realizing RPTRA.

Therefore, this paper aims to illustrate how collaborative governance in the provision of integrated public space facilities is friendly to Sungai Bambu children, and what kind of system context factors that influence Collaborative Governance in the Process of Providing the Facility of Child Friendly Public Space through Emerson and Nabatchi (2015) theory [5].

2. Methods

Using qualitative approach, case study research was conducted to answer some questions such as:

1. How is the collaborative governance in Providing the Facility of Child Friendly Public Space?
2. What kind of system context factors that influence Collaborative Governance in the Process of Providing the Facility of Child Friendly Public Space?

2.1. Area of the Study

North Jakarta is one of the five administrative cities which forms Special Region of Jakarta. The city has become one of the regions where child-friendly city program is implemented, and is followed by the construction of child friendly public space. This area has the highest number of RPTRA based on monthly data in November 2017.Massive developments resulted in 42 Child Friendly Public Spaces, of which 29 came from the Regional Budget (APBD), and 12 came from Corporate Social Responsibility funds.

2.2. Data collection

Data were collected through interview by collecting information from key persons who are deemed most aware of the communication and efforts regarding collaboration governance in providing the facility of child friendly public space.

2.3 Key respondents

Interviews were conducted with stakeholders who have contributed to the provision of facilities, such as the Manager of Sungai Bambu Child Friendly Public Space, Head of the Village, the Office of Women's Empowerment and Child Protection, and PT Pembangunan Jaya as a private sector.

2.4 Collaborative Governance Model

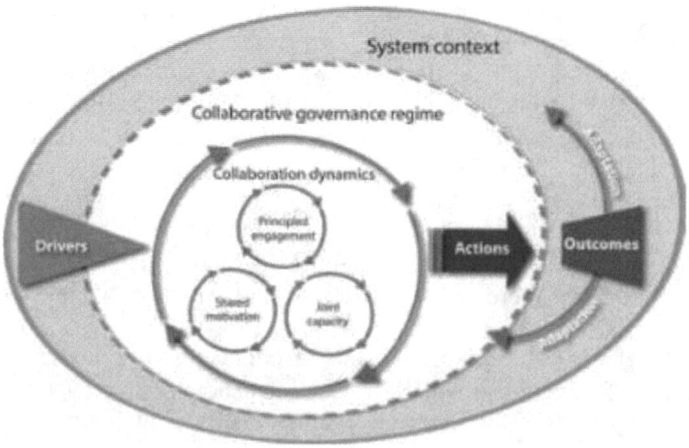

Figure 1. Collaborative Governance Regimes
Eemerson and Nabatchi(2015)

Emerson and Nabatchi sees the dynamics of the collaboration process as an orientative cycle of interactions. Emerson focuses on three main components of the interaction of the collaboration dynamics [5]. These components include: Principled engagement, namely things that occur continuously in collaboration; shared motivation is emphasis on reinforcement and improvement of the movement of shared principles; and the capacity for joint action, namely the existence of capacities, for example no differences and power imbalances. The system context can be interpreted as the dynamics that produce opportunities and challenges that directly impact collaboration, or included as a part of the initiation of collaborative governance. There are drivers on Emerson and Nabatchi's model in the form of Initiating Leadership, where these components play a role and can influence the other three components. At that time Jakarta was led by the Governor of the Jakarta Capital Special Region for the period 2014-2018, Basuki Tjahja Purnama who was the promoter of the development of Child Friendly Integrated Public Space.

3. Results and Discussions

With the support of data from in-depth interviews with several key figures, the result section informs about how dynamic collaboration is implemented. The form and initial direction of this regional and multisector collaborative governance initiative were shaped by drivers emerging from the system context.

3.1 Principled Engagement .

Principled engagement is the process of revealing the interests, values of actors, and efforts to construct common interests. Disclosures can be analyzed from why the actor joins the collaboration. There are several efforts made to realize the principle of engagement, such as Discovery, Determination and Deliberation.

3.1.1 Discovery.

The initial discussion about RPTRA was launched in September 2014 and was a collaboration between former Ms. Veronika Tan as chairman of the DKI Jakarta Family Welfare Development Program (PKK) and chairman of the BPMPKB (Community, Women's and Family Planning Empowerment Office) , Ms. Dien Emawati. In the discussion of the two leaders, a proposal for facilities and infrastructure for Child-Friendly Parks (TRA) was formulated. A meeting between Mrs. Veronika then emphasized the plan as Chair of the Family Welfare Empowerment Team (PKK) and Pak Sutopo Ristanto from PT Jaya Konstruksi and Pak Achmad Noerzaman from PT Arkonin and Ms. Eka Permanasari as a representative from Pembangunan Jaya University. In this initial discovery, it was concluded that there was a need to provide public facilities, which were not only to fulfill the needs of children but also to achieve all elements of the urban community. Ms. Eka Permanasari was developing the term of RPTRA with emphasis on space that can bridge communication between citizens and the government.

The private sector stated its findings in the form of social mapping in the field carried out by the social team (Yayasan Nurani Dunia and Pembangunan Jaya University) which aimed to identify the potential of the population as well as collect data on formal-informal organizations in the community around Sungai Bambu village. The social team also determines the types of activities of citizens and active figures in community activities without involving the relevant local government agencies and bureaucratic apparatus, while the government is based on urban and regional government data.

3.1.2 Determination.

The Letter of Agreement Number 78 of 2014 and number 178 / DIR-SK-12.14 as cooperation agreement between the government and private sector is the initial basis of the construction of RPTRA facilities. Here the government provides the most significant opportunity for the private companies to make RPTRA infrastructure facilities as a realization of their CSR programs. A memorandum of understanding was also carried out between PT Pembangunan Jaya and Pembangunan Jaya University about a community empowerment assistance program in the RPTRA environment, namely the Memorandum number 001A / PER-PRE / UPJ.01.05. A strong determination is continued by follow-up meetings to strengthen the next plan, such as the affirmation of the obligations that each party must fulfill, and the interchange they get. There is no regulation regarding RPTRA because it is this development that is the leading promoter and the basis for the provision of governor No. 196 of 2015 concerning the Guidelines for the Management of the Child-Friendly Integrated Public Spaces which were implemented after RPTRA Sungai Bambu construction was completed.

3.1.3 Deliberation

Deliberation was stated in a joint discussion either between the private sector and the central government, the private sector with the local government, and the Focus Group Discussion (FGD) involving all parties without exception, including the private sector, *kelurahan* or urban village, central government, surrounding communities, and non-governmental organizations. The FGD was conducted many times to represent the negotiated achievements and findings in the field to reach a mutual agreement on the form of infrastructure to be built.

This process was stated on 21 January 2015, in the establishment of a Focus Group Discussion on Information Dissemination & Plans for the proposed design of RPTRA facilities. The initial design of RPTRA development was carried out to the community of Sungai Bambu village by the CSR of PT Pembangunan Jaya Group, which included proposed land management, the purpose of use, and the form of a semi-permanent area. Then on 12 April 2015 Phase 2 of the FGD was conducted to disseminate the design revisions, community input was included in the design revision. The Focus Group Discussion stage also involves the village officer, local governments, as promoters to foster a sense of ownership, and community participation in the implementation of development. The village officer and the private sector agreed to coordinate the initial stages of development in cooperation by clearing the area of land to be built. For the physical facilities, a further FGD was conducted which was conducted four times, 15 March 2015, 21 March 2015, 28 March 2015 and 4 April 2015. The results of social mapping also strengthen the deliberation process. Focus group discussion is now the primary requirement and obligation for the development of RPTRA whether or not using a collaboration scheme between the private sector and the government.

3.2 Shared Motivation

According to Huxhamand Vangen in Emerson and Nabatchi (2015) shared motivation also strengthens and enhances the process of "mobilizing shared principles". Shared motivation is described as a cycle of self-reinforcement consisting of four mutually beneficial elements, namely mutual trust, shared understanding, internal legitimacy and commitment.

3.2.1 Mutual Understanding and Trust.

Relationships outside collaboration are also important while the relationship between the DKI Jakarta provincial government and PT Pembangunan Jaya has existed before collaboration in the construction of RPTRA. Their links are in other fields in the form of educational programs (Yayasan Pendidikan Jaya), infrastructure programs (Yayasan Marga Pembangunan Jaya) and sports programs (Pembangunan Jaya Raya). Good relations form a high level of trust between actors. The government understands that PT Pembangunan Jaya already has the expertise and experience in the infrastructure and social fields. Also, PT Pembangunan Jaya is part of the Regional Government-Owned Enterprise of the DKI Jakarta Province (BUMD) since 1961.

It was found that an increase in shared motivation could also be developed with informal interactions between the government and the private sector through social networks in personal and group forms - which would enhance mutual understanding. In the construction of the RPTRA, it is important to be as consistent as possible with the same figures related to RPTRA development planning stage to increase interpersonal trust in cross-border collaboration. From this, the private sector regularly provides information related to development in an informal way. Representatives of these actors can also give rise to the nature of leadership that supports the collective consensus. It should be remembered that in shared motivation each party has a different background, different views or conflicting desires. Unfortunately, due to a large number of parties involved, shared understanding does not occur evenly in collaboration.

3.2.2 Internal legitimacy.

Private companies communicate with various levels of government from the central government, local government and local communities in emphasizing that the business is undertaken are also socially based on the public interest. Likewise, the central government must be able to encourage the lower government to participate in direct meetings. This emphasizes that collaborative actors have a sense of dependency. There is a strengthening of the results of the principle of engagement in internal legitimacy, even though it is still in the process, but parties between levels have given motivation.

3.2.3 Commitment.

A commitment was formed as temporary good results from collaboration . Beside focus group discussion and social mapping, proof of commitment through the first "*gotong royong*" activities was also accompanied by the laying down of the RPTRA construction by the North Jakarta Mayor. PT Pembangunan Jaya also coordinated with the Regional Secretary, BPMPKB, Public Welfare Bureau, Parks Service, Education Agency, Sports Agency. After that, it was followed by the signing of the pact of the integrity of the citizens to commit to securing and succeeding the process of building Sungai Bambu RPTRA. For *gotong royong* activities to be effective and directed, for the technical implementation of mutual cooperation and the type of work to be carried out at the time of mutual cooperation coordinated with the CSR who carried out the construction of the RPTRA.

The nature of this commitment develops over time, like trust, mutual understanding, and internal legitimacy develop together and in turn, are reinforced by principled involvement. Along with the construction of facilities and infrastructure, the government is not only a party that participates in deliberative practices, but it is also a party that carries out direct supervision 3-4 times in the field, for three months of construction. The role was carried out directly by the village officer, especially the chief of the office at that time namely Mr. Suwanto and Secretary of the Village, Ms. Trinanda. The private sector is also responsible for repairing several damaged facilities within a period of 6 months after the inauguration before the handover minutes (BAST) are held, one of which is a swing which is then repaired directly by PT CMNP and PT TMMIN. However, there is one facility that is not built until the handover from the private sector to the government (BAST), that is the construction of a soccer field.

3.3 Capacity for Joint Action

Capacity for joint action is a variety of results from cross-functional elements to produce effective action, due to the adequate capacity of the factors.

3.3.1 Institutional procedures and agreements

Institutional arrangements are carried out to maintain cross-sector interactions over time. When the collaboration process begins to involve other parties, there must be rules to regulate interactions between parties in detail. The agreements made by private companies after being carried out include the submission of files regarding regional needs to local governments, such as urban villages, city and central regions. Reporting the amount of budget used until details of the Handover Minutes from the private sector to the provincial government is made into consideration. Given that Sungai Bambu RPTRA was the first RPTRA born from the collaboration between private company and goverment, it became a lesson for both parties. From the above processes, the resulting structure is innovative and "quite flexible" where the government provides flexibility for the organization to carry out deliberation process, as well as development based on problems and needs in the field. The private government is given the authority to develop broader sectoral networks and relationships. Private companies that have previously carried out the construction of RPTRA have full support if they want to carry out development elsewhere, as well as carry out activities and eventually funding RPTRA for activities and social activities.

3.3.2 Leadership
There is a leadership role in this form of collaboration; and the roles are (1) Gaining support from North Jakarta Mayor to lay the first stone (2) Leaders initiate meetings at the central and regional levels such as socialization and group discussion forums (3) The leader also plays a role as a facilitator in some programs, such as DKI Jakarta Province Family Welfare Development (PKK) representative as the actor who leads the entire process of the Focus Group Discussion, namely Mr. Andi M. Jufri (4) Leaders represent the overall collaboration carried out in the construction of facilities and infrastructure (5) leaders try to distribute knowledge by making periodic reports (6) encouraging the use of technology in informal communication between actors, and (7) leaders advocating for the draft of regional regulations of RPTRA which will arrange further about the Handover Minutes and the authority of the private sector in development.

3.3.3 Knowledge
Knowledge is information needed by each party in collaborating. Although some have been disclosed at the meeting, but because of the many actors and layers of government involved sometimes the distribution of knowledge does not occur smoothly. As for the source of knowledge in the form of social mapping that has been done before as well as the results of the forum group discussion, they later became the basis of the process of building infrastructure. There are still individuals or actors who do not get the results of the knowledge evenly. The actors who serve as the source of knowledge in the construction of facilities (include activities and physical) is the Provincial Government of DKI Jakarta, Child Protection Empowerment Office and Population Control of DKI Jakarta Province, DKI Jakarta Province Family Welfare Development (PKK). Also PT Pembangunan Jaya Group and PT Arkonin as a source of physical design.

3.3.4 Resources
Financial funding for development, administrative implementation of development such as requests for permission to establish semipermanent buildings, academics and sociologists as leaders of community mapping and contractors have been fully borne and fulfilled by the private sector. As for the government, it provides support in the form of direct and indirect monitoring, facilitating and accompanying meetings of group discussion forums, gathering mass, providing new links to relevant parties needed by the private sector in development.

3.4 System Context
The system context can be interpreted as the dynamics that produce opportunities and challenges that directly impact collaboration. Following are the prominent conditions of collaboration in the process of building child friendly space facilities. There are six prominent conditions of a system context:

3.4.1 Public resource or service conditions. Public spaces such as city parks, youth centers or other public places that already exist in Jakarta have not been able to accommodate the needs for fulfilling children's rights.

3.4.2 Policy and legal frameworks. Based on the Regulation of the Minister of Women Empowerment and Child Protection of the Republic of Indonesia Number 11 of 2011 concerning Child Friendly City

Development Policy. Government affairs in the field of child protection in the form of policies, programs and activities for guaranteeing the fulfillment of children's rights so that children can live, grow, develop, and participate optimally in accordance with the dignity and dignity of humanity, and receive protection from violence and discrimination, is a responsibility of the Regency / city administration. One indicator of a Child-Friendly City is the availability of facilities for creative activities and child-friendly recreation outside of school which is accessible to all children. One of the manifestations is the massive development of the Child Friendly Integrated Public Space (RPTRA) by the DKI Jakarta Provincial Government. Also, the other basic rules underlying the Governor of DKI Jakarta Regulation No. 112 of 2013 concerning Social Responsibility and Business Environment as an "umbrella" of the Corporate Social Responsibility program.

3.4.3 Socioeconomic and Cultural Characteristics are expressed, such as income and education level, health level of community, ethnic differences, etc.

- Population growth in DKI Jakarta is increasing every year. Based on data from the Central Statistics Agency, the population of DKI Jakarta in 2015 reached 10.18 million. Then it increased to 10.28 million in 2016, and increased to 10.37 million in 2017. This means that for the last two years the population in the Capital City has increased by 269 people per day or 11 people per hour [6].
- Looking at the density of the population of DKI Jakarta Based on the Age group, the population aged 0-19 is 3,230,265 or 32% of the total population. The population in that categorization refers to the definition of "child" as written in Law No. 39 of 1999 concerning Human Rights [6].
- The fulfillment of children's rights has not been achieved in DKI Jakarta. As with cases affecting children there was a significant increase from 2011 to 2014 [7]'
- Regarding the population density of Sungai Bambu Village, with an area of 2.3640 km2 the village has 35,869 residents resulting in a density of 15,173 / km2 [8]. This figure shows the importance of an empowerment forum not only for children but also for all levels of society.

3.4.4 Political Dynamics and the Power Relations .First, there is a form of agreement that the idea of fulfilling children's proper rights will be initiated into a superior program of the DKI provincial government. Second, As the need to fulfill people's welfare, the government as a policymaker never closes the possibility of relations and other sources from cross-sectoral parties as well as compliance with local government for central government instructions.

3.4.5 History of Conflict, a positive conflict occurred where the DKI Jakarta provincial government had previously carried out various collaborations and collaborations with various private parties for the sake of fulfilling people's welfare, for example: giving free bus services and garbage truck, revamping the park like a city park, and planting productive trees on vacant land.

4. Conclusion

Based on the results and discussion, it can be concluded inside the Collaborative Governance in Providing Facilities of Child Friendly Integrated Public Space, North Jakarta City; there is a dynamic of collaboration that mutually reinforcing and have a reciprocal relationship. This can be seen from the indicators of collaboration dynamics, namely principled engagement, shared motivation, and capacity for join action that related to each other and influence each other. Among the components that have not yet been achieved perfectly , one of which is in the "Shared Motivation". Because of the large numbers of parties involved, there is an unequitable understanding among collaborative actors. Also in "Capacity for Joint Action" section, knowledge distribution is a little difficult to do because of the large number of government levels and the widespread parties in collaboration. The sustainability of the collaboration dynamics is inseparable from the system context that initiated it. Emerson &Nabatchi stated that the more system context factors that arise, they can influence and can strengthen the success of the implementation of the dynamics of collaboration itself. In this collaboration, five of the six factors are included. It also does not rule out the possibility that the forms of collaboration dynamics can also affect the system context.

Based on the results of conclusions, the author can provide suggestions: In the collaboration forum the formation of joint routines must be formed more intensely, especially, for example, meetings at the *kelurahan* or urban village. It should involve all the apparatus, not just the head of the village. There is management about the role of each government actor including all private parties in a more complex written form. This is especially important in the provision that does not only involve a private company. Collaborative leaders should minimize the replacement of actors in each collaboration meeting or forum to prevent disinformation and unequitable understanding. To all collaborative actors, there is a need to create a culture of opinion sharing more actively to monitor and evaluate each other. The need for a deeper understanding of system context is important so that collaborative members do not understand that the conditions in the context system are only existing things, which cannot be changed, but are conditions that can be monitored and added over time.

References

[1] Ansell, Chris and Alison Gash. (2007). *Collaborative Governance in Theory and Practice. Journalof Public Administration Research and Theory.*

[2] Hernowo, Endy and Navastara Ardy.(2017). *KarakteristikRuangPublikTerpadu Ramah Anak(RPTRA) Bahari di Kecamatan Cilandak Jakarta Selatan.* InstitutTeknologiSepuluh November. Surabaya

[3] Carina Jessi. (2018). *Pemprov DKI Stop Bangun RPTRA tetapi Swasta Tetap Boleh Lakukan.* https://megapolitan.kompas.com/read/2018/03/05/16495211/pemprov-dki-stop-bangun-rptra-tetapi-swasta-tetap-boleh-lakukan

[4] Faedlulloh Dodi,Prasetyanti Retnayu, and Indrawati. (2017). *Initiating Deliberative Democracy-Based Public Space: A Study on Dynamic Management of Child-Friendly Integrated Public Space (RPTRA) in North Jakarta.* 17 Agustus 1945 Universitas, Jakarta.

[5] Emerson, K., and Nabatchi, T. (2015). *Collaborative Governance Regimes.* Washington, DC: Georgetown University Press.

[6] Badan Pusat Statistik. (2017). *Jumlah Penduduk DKI Jakarta berdasarkan Usia.*

[7] P2TP2A DKI Jakarta. (2017). *Data Penanganan Kekerasan pada Anak Tahun 2017.*

[8] Badan Pusat Statistik Kota Jakarta Utara. Jakarta Utara Dalam Angka.(2017). *Jakarta Utara Dalam Angka.*

Waste Bank System as a Tool to Increase Regional Income and to Manage the Environment

N L Widyaningsih

Regional and Urban Planning, Esa Unggul University

*Email: nhwidya@yahoo.com

Abstract. The increasing number of people in Indonesia and modern economic activities create mobile lifestyle. The demand for goods creates more supply into the market and it needs more economic resources. Production and consumption create waste and pollution. The waste from the household causes tremendous negative impacts on the social, economic, and environmental aspects of life. To solve the unmanaged waste, people build a waste bank where people can collect and give their waste to the waste bank in which it is a unique independent system. Based on the qualitative research at the Central Waste Bank at Makassar City (2019) which has almost one thousand waste bank units and the previous research at Pulo Kambing Waste Bank in East Jakarta (2018) which is already advanced level organization, there are two main goals that it can be seen through the waste bank system. The Central Waste Bank at Makassar City has a top-down approach based on the regional waste management policy. Pulo Kambing Waste Bank in East Jakarta has a bottom-up approach to build their organization capacity. The two main goals that can be achieved through the waste bank system are to increase household income and to manage the environment, focusing on the 3R concept (reduce, reuse, and recycle).

Keywords: waste bank, household income, environment

1. Introduction

Indonesia has more than 250 million people in 2018 and has waste management issue in urban and rural areas. Jakarta as the capital city has the most unmanaged waste number and we can see it in every corner and in the gutter/river, especially on the empty land. The phenomena has been accumulated and causes many problems on the social, economic, and environmental aspects.

Until now, the waste management is still used the old paradigm, where the household collects the solid waste in the trash bin (*tong sampah*) and the sanitary officer carried it into the temporary waste place (*tempat penampungan sementara* or TPS) before they bring it to the open-dumping landfill (*tempat pembuangan akhir* or TPA). The number of household waste is increasing because of the population number and the variety of their consumption. On the other hand, there are huge supplies of the new products in the market. This household consumption type is also different between urban and rural areas.

The household consumption can be divided into nonfood products (such as clothes, shoes, and others) and food products (such as vegetables, meat, and fruits). After people consume the products, it become waste. We can categorize the solid waste into inorganic waste (not biodegradable) and organic waste (biodegradable). The inorganic waste can be plastic, metal, and glass materials. The organic waste can be vegetables, meat/fish, and fruits.

The household solid waste must be separated from the lowest level, which are the households, but in Indonesia, they do less separate their own waste. Based on my doctoral research in 2014 entitled e "The Linkage Model between Sustainable Consumption and Household Waste Management System", there are many factors that can influence people to do the 3R concept (reduce, reuse, and recycle) on the waste management system. Primarily, we can see from the outside and inside decision-making factors of people to do their household consumption, such as: (a) people's motivation (inside factor) to consume products based on their needs and the utility; and (b) people's lifestyle (outside factor) to consume products based on other people influence and the media advertisement. I believe that people can act as homo-economicus and also can be act as homo-ecologicus. It means that people can do the economic activities (consumption and production) as well as be responsible for the environmental condition (management and protection).

There are many bad impacts on the social, economic, and environment if we do not manage the household solid waste. The new paradigm on the waste management system starts from the source of the waste problems itself. So, it will reduce the number of waste or reduce the percentage of the waste before getting into the open-dumping landfill. We cannot depend on the open-dumping method anymore because it causes many health issues in the future to the land/air/water and to the local people. There is less space for the waste because the housing demand is always high every year.

To solve the waste management issue, local people create a waste bank to reduce the unmanaged waste and to increase their household income by selling the waste. In this paper, I will focus on the Central Waste Bank at Makassar City and Pulo Kambing Waste Bank in East Jakarta. They have different approaches, such as: (a) top-down regional policy on waste management system, and (b) bottom-up from local people aspiration. The waste management system through a waste bank system will give impacts to the regional planning and policy.

2. Methods

This research paper used qualitative method to gather information more detailed on the lowest level of administration, such as Rukun Tetangga (RT). Not all data are available on the statistical office or online system in Indonesia, so we need to come closer to the field area of study. I conducted unstructured questionnaires and Focus Group Discussion (FGD) with the informants in the Central Waste Bank at Makassar City on January 2019 and Pulo Kambing Waste Bank in East Jakarta in 2018. I followed their development of their waste bank system, especially on their program or activity; reducing the waste; capacity building for the management; and the number of waste bank clients. The locations for both waste banks are very strategic in the central city, which is close to the housing area, market or store, and school.

3. Results and Discussions

3.1 The Central Waste Bank at Makassar City

The Central Waste Bank at Makassar City has almost one thousand waste bank units (WBUs) but only 700 WBUs are still active. I visited the place on January 2019 and interviewed the manager of the central waste bank (Mr. Nasrun). After the interviewe, I visited some of their WBUs, such as: (a) WBU under Mr. Faisal that applied the gold savings system under Corporate Social Responsibility (CSR) program from PT Pegadaian; and (b) WBU Sukses Abadi that has already used the biogas system since 2015 where people can exchange their waste into daily needs, such as rice, eggs, and cooking-oil. The WBUs is located in the housing area that is packed of small-medium size housing.

In Makassar, the local government used every aisle wall to draw traditional slogan to strengthen socio-cultural engagement among all leves of society. Based on my observation, each aisle has different picture and it is colorful. The distance between WBUs is not too far because local government ignites the local people to build the WBU in each Rukun Warga (RW). It is about 10 RTs and each RT consisting around 50 households. The waste bank system in Makassar city achieved a national award, such as the Best System for Reducing The Waste. The address for the waste bank is on Jl. Toddopuli Raya, Panakkukang, Makassar City - South Sulawesi. This central bank can also pick up the waste from all their 700 WBUs without any burden of pick up schedule daily. They use the WhatApps (WA) application to arrange the schedule among the WBUs. Below is the administrative map of Makassar City:

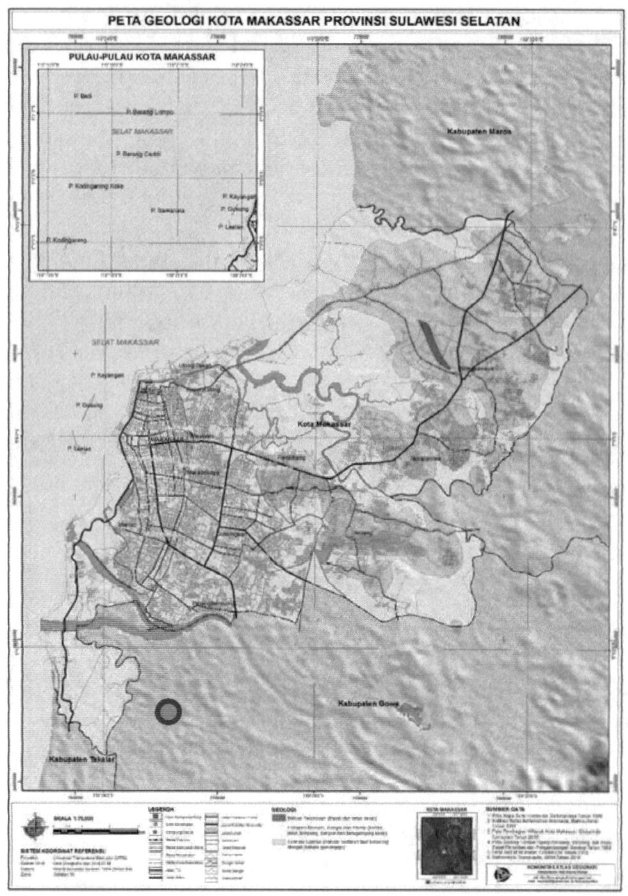

Figure 1. The Administrative Map of Makassar City

The Central Waste Bank at Makassar City was built in 2011 under the regional waste management system policy. The central waste bank can reduce the percentage of the waste that goes to the TPA around 30% annually and the local government sees the local people's need. The local government has used the cultural system approach that really touched the local people's heart, for example: they labeled the dump truck with the name "Tangkasarong-Tabungan Bank Sampah Anak Lorong". It means the waste bank is from aisle people and to aisle people.

The waste management system in Makassar city puts its concern on the cultural and social level first rather than on the economic side. It is because they believe that waste management system through a waste bank will be sustained if they put attention on their cultural and social aspects. The WBUs receive the sorted inorganic waste from the households. There are around 49 items that can be accepted at the Central Waste Bank at Makassar City. The households as a member of the WBU is already do the collecting and sorting of their waste. Usually, they separated into plastic, metal, and glass materials (there are more sub-categories under each item). The price system is based on the price from the waste collector daily. Local government has special budget for the Central Waste Bank at Makassar City (Rp. 2.8 billion on the following year).

There is Regional Act No. 4/2011 about Waste Management in Makassar City where the local government will provide the financial to support the waste management system and technology to apply the 3R concept. On this regulation, the local government coordinates all the stakeholders that involve on the waste management system from the lower level administration, such as RT and RW. Besides that, the local government will give some awards to: (a) the best innovation on the waste management system; (b) the best report for anyone who brakes the law; (c) the best reducing waste method; and (d) the best management on the system overall.

The interviewe that I conducted with the non-government organization (NGO) a.k.a. Yayasan Peduli Negeri (YPN), showed that the central waste bank has consistent supports for the socialization to the local people regarding the household waste management and the waste bank system. This NGO is oriented on society empowerment, improving th environmental condition, improving health quality of the people, and increasing education quality too. The NGO gets funding from Unilever Company as their CSR program through the WBUs. They give incentive financially for the active WBUs. They conduct training regularly to all the WBUs in Makassar city by hiring undergraduate students.

The simple mechanism of the household waste management system in Makassar City is: (a) household solid waste is being sorted into plastic, paper, glass, and metal materials; (b) people carry the waste or ask the WBU's officer to pick up the waste; (c) the WBUs will weigh the waste and put the waste measurement note on the waste bank saving book; (d) the WBUs will transport the waste to the Central Waste Bank at Makassar City or they can call the officer to pick up the waste; (e) the Central Waste Bank officer will weighted the waste and report to each WBU about the daily price from the waste collector; and (f) there is a unique waste bank system that transforms the waste bank saving book into gold saving book. During the pickup time, they do not have any significant issue based on the distance from the central waste bank and WBUs locations.

The transformation from waste into gold is the most lucrative method for low-income level family to manage their household waste. One of the owned-state enterprises that do their CSR program on this system is Pegadaian Company. This company builds more than 25 waste banks under the GADE Clean and Gold action program in Indonesia. Not only their CSR program through the waste bank but also they have helped the WBUs by providing them the waste transportation vehicle/stuff. One kind of WBU with the gold saving system is under Mr. Faisal that I visited last January 2019. Here is the interview note:

> "The waste bank unit in Makassar city helps the local people to clean their dirty environment and after they bring their waste into the WBU, they realize that their waste can become money or gold saving. Some of the WBUs can accept waste to buy people daily needs, such as rice, cooking-oil, and eggs. The local government is very supportive to build a WBU on each RW. The transportation problem to pick up the waste in the narrow aisle can be handled by using smaller trash vehicle (*gerobak*)."

The WBU Sukses Abadi has a different story where they used the organic waste into biogas energy to cook everyday since 2015. The biogas machine is very simple in which we just put our daily organic waste into the big bin machine and wait for it about 5 hours and we can use the gas to cook for 1.5 hour maximum. The WBU received the biogas technology from one of the university in Bandung. After 3 years using the biogas machine without any damaged/issue, they will receive more than 100 biogas machines in the future. The utility scope of the usage of the biogas machine will be bigger than only for 1 household unit.

The mechanism of the WBUs and the central waste bank are:

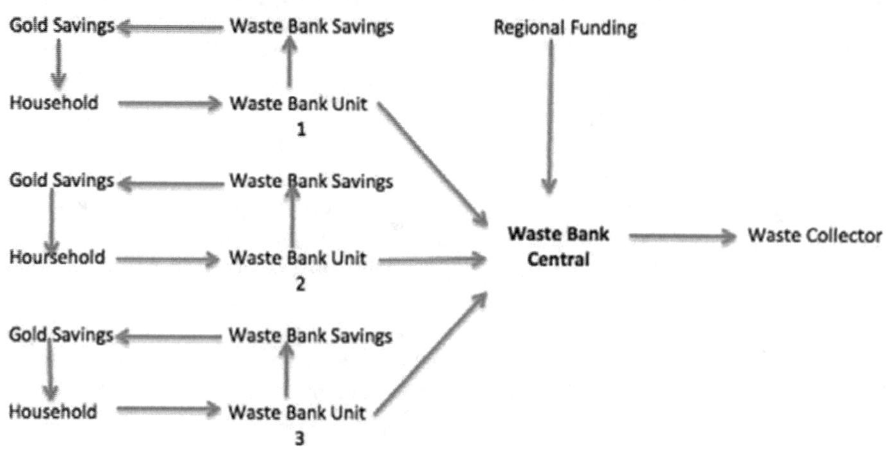

Figure 2. The Mechanism of the Central Waste Bank at Makassar City

3.2 Pulo Kambing Waste Bank

Pulo Kambing Waste Bank is located on Jl. Swadaya No. 1, Komplek PLN, Gas Klender - East Jakarta. It was built in 2014 with the main goals to help low-income family and reduce the unmanaged solid waste around the area. The innovation comes from the housewives and they have strong motivation to manage the environment. They believe by sorting the waste in which the waste can have more economic value and utility. The waste bank is a mediator to build social relationship and to create independent activity based on the local people capability and capacity.

The leader of this waste bank is Mrs Vera and she is a motivator for her team as well as for the neighborhoods in every RT/RW. The first waste bank place was in her home garage (around 2 meter by 3 meter). She taught the other housewives how to sort the waste into inorganic and organic waste. She shared the work among the team member to: (a) look for the waste collector; (b) motivate other housewives to join the waste bank; and (c) indirectly to create economic activity based on each member location.

In 2018, the waste bank was not only for the waste management system but it also expands the program into environmental education program and do the training about waste management in school (from elementary to high school level). Besides that, the activity grows to the gold savings from waste savings book. There is also a support from the private sectors, such as PT Antam Logam Mulia and Kalbe Farma to legalize the organization (a.k.a *Koperasi*). The more variety of their programs makes them received the highest award on environmental, such as Kalpataru on Environmental Pioneer and the Best Waste Bank in East Jakarta.

From the first Pulo Kambing Waste Bank location (2014) to the location now (2016) is not far. Their location is in the housing area and East Jakarta area is the most densely populated in Jakarta province. I conducted many observations (2017 and 2018) and interviewed with the main team member of the organization of Pulo Kambing Waste Bank. One of the most interesting notes from my discussion with the team member is:

> "At the beginning, we did not know what we should do to reduce the unmanaged waste that made our environment dirty which increased the disease occurrence. We just trusted on ourselves to build our teamwork together despite no appropriate place for a waste bank and knowledge to manage a waste bank. Now, we have more representative waste bank place and office to do more than just waste management system."

Pulo Kambing Waste Bank is an example of many waste banks (5.427 waste bank in 2017) in Indonesia in rural and urban area, where they do not receive any assistant (training and funding) from the local government and the private sector. Below is the East Jakarta Province administrative map:

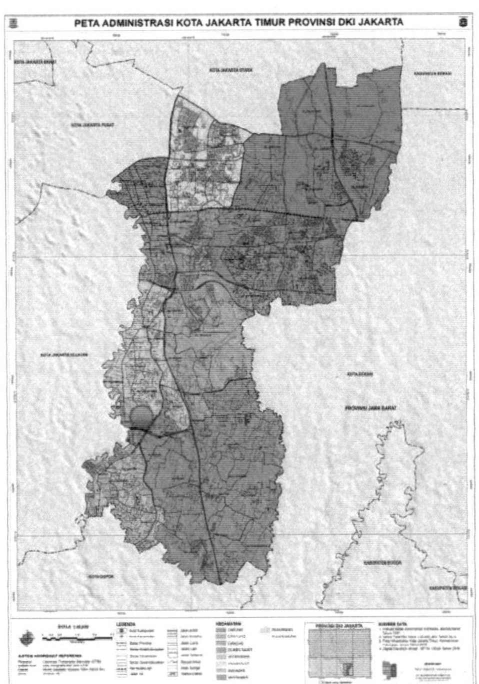

Figure 3. The Administrative Map of East Jakarta

The Central Waste Bank at Makassar City and Pulo Kambing Waste Bank in East Jakarta showed that the waste bank mechanism in the waste management system could reduce the unmanaged household solid waste but no significant statistical data available. Based on my interviewe, they could reduce the percentage of the waste from TPS to TPA. The quality of the environment affects the health condition.

The waste bank mechanism can increase the household income. Usually, after they sort and collect the waste, they must clean it before they bring it to the waste bank or WBU. People will get higher price for their cleaned inorganic waste.

Based on the Ministerial Regulation No. 13/2012 for Reduce, Reuse, and Recycle Implementation trough Waste Bank, there are two mandatory requirements for each waste bank: (a) building construction; and (b) management system of the waste bank. I reviewed briefly each of the waste banks under this law:

Table 1. Significant Aspects on Waste Bank

No	Component	Central Waste Bank at Makassar City	Pulo Kambing Waste Bank in East Jakarta
1	Building Construction		
	1.1. Floor	Good	Good
	1.2. Wall	Good	Good
	1.3. Door/Window/etc	Good	Good
	1.4. Fence/Park	Good	Good
	1.5. Drainage	Good	Good
	1.6. Storage	Good	Good
2	Management System		
	2.1. Clients	Waste Bank Units	Local People
	2.2. Staff/Officer	Government Officer	Housewives
	2.3. Waste Collector	Good	Good
	2.4. System	Using WhatApps	Clients come in
3	Economic Aspect		
	3.1. Increase Income	Good	Good
	3.2. Empowerment	700 Active WBUs	Create Home Industry
	3.3. Development	Biogas System	EcoEducation
4	Social Aspect		
	4.1. Increase Skills	YPN	Learning by Doing
	4.2. Health Condition	Good	Good
5	Environment		
	5.1. Reduce Waste	Good	Good
	5.2. Reduce Pollution of Air/Water/Land	Good	Good
6	Cultural Aspect	Homogen	Heterogen
7	Political Aspect	Top-down Approach	Bottom-up Approach

Source: Research Result, 2019.

4. Conclusion

The Central Waste Bank at Makassar City and Pulo Kambing Waste Bank in East Jakarta is able to reduce the unmanaged household solid waste in the urban area; increase household income through sorting and recycle the inorganic waste; and build society empowerment based on each local capacity and capability.

References

[1] Ministerial Regulation No. 13/2012 for Reduce, Reuse, and Recycle Implementation trough Waste Bank
[2] Bank Sampah Pusat Kota Makassar. Tabungan Bank Sampah Anak Lorong. www.komunita.id
[3] Bank Sampah Pulo Kambing. www.pulokambing.com. Teknologi Pengolahan Daur Ulang Sampah. Cecep Dani Sucipto. Gosyen Publishing. 2012.
[4] Linkage Model between Sustainable Consumption and Waste Management System. NiLuh Widyaningsih. University of Indonesia. 2014. Dissertation.
[5] Integrated Solid Waste Management: a Life Cycle Inventory. Forbes McDoughgall, Peter White, Marina Franke, Peter Hindle. Blackwell Publishing. 2001.
[6] Ekonomi Wilayah dan Perkotaan. Sjafrizal. Rajawali Press. 2014.
[7] Menyulap Sampah Jadi Rupiah. Mumtaz Media. A Guruh Permadi. 2015.
[8] Pengelolaan Sampah. Hilarion Widyatmoko. Penerbit Universitas Trisakti. 2011.
[9] Contextual Factors Influencing Household Recycling Behaviors: A Case of Waste Bank project in Mahasarakham Municipality. Wichitra Singhirunnusorn, Kidanun Donlakorn, Warapon Kaewhanin. SciVerse ScienceDirect. 2011.
[10] Waste Bank as Community-based Environmental Governance: A Lesson Learned from Surabaya/ Dyah Retno Wjayanti, Sri Suryani. ScienceDirect. 2014.
[11] Waste Management Scenario through Community Based Waste Bank: A Case Study of Kepanjen District, Malang regency, Indonesia. International Journal of Environmental Science and Development, Vol. 5, No. 2, 2014. Hasfarm D. Purba, Christia Meidiana, and Dimas W. Adrianto.

Accelerating Economic Development through Technopark: The staging of National Science-Technopark Formation Process in Indonesia

N Maninggar[1]

[1]The Agency of Assessment and Application of Technology (BPPT), South Tangerang Banten

*Email: nimas.maninggar@gmail.com

Abstract. Technopark has become a widely cited term when discussing about innovation and technology. The world economy concept has been shifting from traditional based economy to knowledge based economy. This concept fosters an understanding of technopark, functioned as a seedbed of innovation to gain regional development, which is later popularly implemented. There are phases in technopark development in achieving its entire function. In the developing countries, technopark is assumed to be still in its infant phase. Indonesia is one of the developing countries that have declared the policy of technopark establishment in its national mid-term development plan document of 2015-2919. However, a number of national governmental agencies has already taking the role of technopark, even before the policy was issued. The objective of this paper is to explain the technopark stage level in Indonesia as well as to evaluate the assumption on technopark stage in developing country. The method used is case study on two Indonesian national technoparks which will be assessed through two characteristics that forms a technopark, such as physical and functional dimension. The result indicates that the assumption is not entirely correct. Indonesia's national technopark, despite in its early phase organizationally, has already shifted from its early phase towards development phase functionally.
Keywords: technopark, economic development, staging of development

1. Introduction

Technopark development as a means to drive regional development has become policy that was being implemented in many countries [8, 10, 15, 16]. Technopark is defined as an area development with the concept of economic and social development which consists of a centre of science activities, production activities (industry and businesses) which are linked to accelerate the development of innovation, diffusion and learning [3, 4, 7, 14]. The increase of its application in many countries is in line with the increase of necessity for innovation and technology as a result of the collaboration among innovation creator, such as university and research institutions, and the industry. The successful phenomena of technopark such as Silicon Valley, route 128 and Cambridge in driving regional economic growth by attracting industrial investment, scientist and entrepreneur, has triggered and inspired the implementation of such technopark in many other countries [7, 12, 19].

Regional authorities and technopark management will adjust the mission and vision of its development [2, 19]. Thus, it addresses the reason of variations in area development, organizational structure, research focus and industrial partner to the area. However, there are general pattern in technopark development stages. In high income and developed country, the development of technopark does no longer focus on boosting local and regional economy, but on diffusing the innovation that will globally influence social-cultural. Silicon Valley's major products such as computer, digital watch, laser and cordless telephone have become the common lifestyle in the current era [19]. This advanced stage development was suspected differently with the development stage in the developing country due to the economic development phase, primarily in Southeast Asia which is still in the transitional phase between institution and infrastructure necessity to innovation [17].

Indonesia becomes one of the developing countries in South East Asia that has recently includes technopark development policy to its National Mid-Term Development Plan (RPJMN) of 2015-2019. The central government is committed to develop and revitalize technopark to drive the development of innovation. Even though the policy for developing new technopark has only been issued in 2015, a number of national agency such as Indonesian Institutes of Science (LIPI) and The Agency of Assessment and Application of Technology (BPPT) has been already been functioned as technopark. There was a presumption that there is a difference in the development staging between the organization and the function. Therefore, it is necessary to analyze the technopark development stage from organizational and functional perspective to improve and also explain stage development achievement especially in developing country in moving towards innovation driven economy.

Revealing the technopark development stage will be beneficial for both the developer and also the local government to maximize investment and other basic necessity for supporting innovation creation in the area [5, 19]. This paper will try to explore technopark developments from the physical development and organizational-function perspective. The qualitative study method will be utilized on two national technoparks in Indonesia, which are the NTSP-BPPT and CSTP. This paper is divided into a couple of sections for complete exploration result, such as: preliminary section, literature review which emphasizes on technopark evolution process from both physical and function perspective, research method, general view of Indonesia's policy on technopark, followed by result's review. The result's review reveals that there is indeed a different staging between technopark physical and functional development. Even though the physical development is still in infant phase, the function development has already reached development phase. Supports in the form of acceleration facility and easiness in establishing independent body, physical development and operational cost funding, have been able to drive the national level of technopark.

2. Methods

The method used in this study is qualitative study case on multiple case. Multiple case studies are beneficial to explore more broadly the studied cases [18]. The case in this study is National Technopark in Indonesia. National technoparks are central government initiated technoparks as a location for research result incubation and its development as commercial products [7]. National Science Technopark-BPPT (NSTP-BPPT) and Cibinong Science Technopark-LIPI (CSTP-LIPI) are the two case studies selected for this paper. In the 2015-2019 RPJMN, it was stated that both bodies are the central government priority in the effort for innovation development. CSTP-LIPI even becomes one of pilot projects for Indonesia's Technopark [7].

The exploration of information was conducted through in depth interview with NSTP's management, exploring the kind of support was granted by the government, by private sectors and the public; providing benefits for the local and regional public towards the technopark itself.

Stage analysis is being conducted by evaluating the development stage of both NTSP by physical and functional criteria (presented in table 1). There are three stages in each criteria which are initial, developmental and advanced stage. In the initial stage, the technopark is still in the physical and area mission development. The development stage means that technopark is already has mission performing its function as innovation creator. Advance stage means that the area is able to do spin-off to other area.

Table 1. Development Staging of Technopark

Physical Development Staging				
	Initial	Developmental	Advanced	Source
Organizational Structure	The organization structure is still a sub to the main body	Independent organization structure	The organization already expand and it has other network structure	Martínez-Cañas & Ruíz-Palomino, 2011
Physical Construction	On progress of building construction	The building has already been build	expanding building construction in other region	Martínez-Cañas & Ruíz-Palomino, 2011
Clustering	Organizational Proximity	Specific area	Innovative Milieu-spin off technology	Capello, 1999; Nahm, 2000
Location	Remote area	Urban corridor	Metropolitan area	Nahm 2000
Function Development Staging				
	Initial	Developmental	Advanced	Initial
The variety of industry	SME local product	SME local product and High-tech industry	High-tech Industry with global market	Amirahmadi & Saff, 1993; Goldstein & Luger, 1990; Rhee dkk., 2010; Zhang, 2005.
Mission	Reducing poverty	Local and regional economic development	Branding and creating international market and also property development	Rhee et al., 2010; Zhang, 2005
Activities	Business incubator	Incubator, R&D (seedbed innovation)	Incubator, R&D, Technology transfer, land leasing	Zhang, 2005
Networking	Local industries	Regional and national industries and also universities	Global industries and universities-collaboration between STP in other countries.	Martínez-Cañas & Ruíz-Palomino, 2011

3. Results and Discussions

The development of CSTP and NSTP-BPPT become the top priority agenda in the national mid-term development plan of 2015-2019. To support the program, the assessment of physical factor is required to discover the development staging of national technopark organization and building construction. In general, the CSTP and the NSTP-BPPT are not yet an independent organization. The organizational structure is still a part of its main organization. The CSTP is a sub organization centre of innovation while NSTP-BPPT is still attached to the Centre of Technology Policy Assessment. The construction of the building is still in progress with varying of mansion type and the target due date. The CSTP is planning to build pilot plant for biotech, building tenant, guest house, community and business centre and also for Cibinong Innovation Convention Centre until 2021. The NSTP-BPPT is still focusing on construction of tenant building until 2019. The location of its building is on remote area (far from their main organization/ headquarter). The Indonesia Institute of Science (LIPI) is the main organization for CSTP located in South Jakarta. The Agency of Assessment and Application of Technology becomes the main organization for NSTP-BPPT located on Central Jakarta.

Functionally, Technopark assessment is useful for revealing the benefit of establishing national technopark for innovation and economic development. CSTP and NSTP BPPT are under coordination of the National Research Institute and become reference for other organization in R&D development in Indonesia. The function of both technopark is focused on optimizing collaboration between the research centers inside the technopark area with other organization outside technopark area (depicted in table 2). The mission is not on poverty reduction, but on regional economic development by applying incubator processes to create new entrepreneur. Networking and cooperation among actors has reached national scale. The type of business cooperation has been penetrating to high-tech industry such as biotech, biopharmaceutical and ICT.

Table 2. Indonesian National Technopark Development Stage

Physical Criteria	CSTP	NSTP-BPPT
Organization structure	Organization structure is still attached to The Centre of Innovation LIPI	Organization structure is still on planning phase. Currently, NSTP is still just an activity program in the Centre of Technology Policy Assessment-BPPT.
Construction building	Construction of building such as pilot plant for biotech, building tenant, guesthouse, community & business centre and also Cibinong Innovation Convention Centre targeted to be finish in 2021	Focused on construction of tenant building, targeted to be finished in late of 2019.
Clustering	The building complex in the Technopark area will benefit from the transportation and collaboration efficiency,	The building complex in the Technopark area will benefit from transportation and collaboration efficiency,
Location	Located on Cibinong. The distance of Cibinong and its main organization (LIPI) is 46 km.	Located in South Tangerang, 33 km from its main organization (BPPT) in Centre Jakarta
Function Criteria	CSTP	NSTP-BPPT
Variety of Industry	Focusing on increasing the sustainability of traditional SME and also industrial technology services, it also supplies the technology for high tech industry such as biotech and biopharmaceutical.	Focusing on producing new entrepreneur and industrial technology services for agro industry, manufacture and ICT sector.
Mission	Increasing the number of new entrepreneurs who use technology produced by LIPI and to increase knowledge and	Commercialization hub to push the usage of technology produced by BPPT by the industry and to produce new entrepreneur.

Physical Criteria	CSTP	NSTP-BPPT
	community's productivity	
Activities	Business incubator activated by providing co workspace for new entrepreneur, technology services to the SME connecting them to the technology creator in LIPI, accommodate training for SME as well as for new entrepreneur, intellectual property and business licence consultation.	Tenant Incubator, providing training for SME and new entrepreneur, giving access to financial institution to fund their business.
Networking	Universities, industry, government and community organization such as farmer community both local and national	Universities, industry, government

4. Conclusion

The main purpose from a region for developing technopark is for driving the growth on innovation and commercialization of technology by connecting and achieving its creator to its user, having different development stage between each other. The condition will depend on the mission and the development scenario of technopark management [18]. In the developed country, technopark development has already reached advanced stage focusing on high-tech industry. Its product diffusion globally influences socio-cultural development. According to Rhee (2010) technopark stage in developing country is still in its infant phase, but this opinion is not entirely true [15].

Indonesia as a developing country has national technopark that is still in the infant phase of physical development stage. However, its functional development stage has reached development phase. The functional development achievement is more advance compared to its physical development, as the technopark has already performed its function in spite of being a sub agency of a main institution. The main institution as a national research institution has been able to give some leverage on some of the technopark functions such as hub service between technological innovations to industry, or incubator for SME, though some of its building and laboratories are still under construction.

Based on those the identification of Indonesia's National Technopark development staging, then agency reformation and building construction should become the first development strategy executed by the management. Independent organization will spur the innovation of creation by establishing a conducive environment for the employees [1]. Central government's support shall assist in forming NSTP as an independent structure under a main body, along with providing funding for physical development and operational cost to drive the development of technopark national level.

Reference

[1] Altunoğlu, A. E., & Gürel, E. B. B. (2015). Effects of leader–member exchange and perceived organizational support on organizational innovation: The case of Denizli Technopark. *Procedia-Social and Behavioral Sciences*, *207*, 175–181.

[2] Amirahmadi, H., & Saff, G. (1993). Science parks: a critical assessment. *Journal of Planning Literature*, *8*(2), 107–123.

[3] Bakourossis of the science parks of Greece. *Technovation*, *22*(2), 123–128.

[4] Benko, G. (2000). Technopoles, high-tech industries and regional development: A critical review. *GeoJournal*, *51*(3), 157–167.

[5] Cabral, R. (1998). Refining the Cabral-Dahab science park management Paradigm. *International Journal of Technology Management*, *16*(8), 813–818.
[6] Capello, R. (1999). Spatial transfer of knowledge in high technology milieux: learning versus collective learning processes. *Regional studies*, *33*(4), 353–365.
[7] Castells, M., & Hall, P. (1994). *Technopoles of the world: The making of 21st century industrial complexes*. New York: Routledge.
[8] Goldstein, H. A., & Luger, M. I. (1990). Science/technology parks and regional development theory. *Econ*, Y. L., Mardas, D. C., & Varsakelis, N. C. (2002). Science park, a high tech fantasy?: an analy *omic Development Quarterly*, *4*(1), 64–78.
[9] Indonesian Govermenet WorkPlan, 2017
[10] Martínez-Cañas, R., & Ruíz-Palomino, P. (2011). 25 Years Of Science Parks In Spain: Towards A New Model Of Development.
[11] Nahm, K.-B. (2000). The evolution of science parks and metropolitan development. *International Journal of Urban Sciences*, *4*(1), 81–95.
[12] National Research Council, & others. (2013). *Best practices in state and regional innovation initiatives: competing in the 21st century*. National Academies Press.
[13] National Mid-Term Development Plan (RPJMN) 2015-2019
[14] Peck, F., Stone, I., & Esteban, M. (1996). Technology parks and regional development in the southern European periphery: the Andalucia case. *European Urban and Regional Studies*, *3*(1), 53–65.
[15] Pike, A., Rodríguez-Pose, A., & Tomaney, J. (2006). *Local and Regional Development*. New York: Routledge.
[16] Rhee, J., Hassan, A. E., & Saitova, R. (2010). Evolution of Technoparks: an instance towards Regional Boost for Developing Countries: Experience from Korean Technoparks. *Acadaemia. edu*, *12*(23), 2010.
[17] World Economic Forum. (2017). *The Global Competitiveness Report 2017*. Geneva.
[18] Yin, R. K. (2009). Case study research: Design and methods fourth edition. *Los Angeles and London: SAGE*.
[19] Zhang, Y. (2005). The science park phenomenon: development, evolution and typology. *International Journal of Entrepreneurship and Innovation Management*, *5*(1/2), 138. https://doi.org/10.1504/IJEIM.2005.006341

IOP Conf. Series: Earth and Environmental Science **328** (2019) 012048 doi:10.1088/1755-1315/328/1/012048

Risk Management of Flood Disaster (A Case Study of Kendal River)

P H Kuswanto[1] , M A Wibowo[2], and F Hermawan[2]

[1]College Student MMRI Study Program, Diponegoro University, Semarang, Indonesia

[2]Civil Engineering Masters Program, Diponegoro University, Semarang, Indonesia

*Email: prisanhery@yahoo.com

Abstract. Kendal River, which runs through the city of Kendal, often overflows causing flood for the immediate areas. This study uses the concept of risk management aimed at relevant stakeholders. Data analysis of this study applied the Risk Breakdown Structure (RBS) method to describe each classification of risk sources to risk sub-classification. The results depicted that there were 20 (twenty) risks of causing flooding in Kendal City, with 5 (five) risks being the main cause. According to the Central Java Provincial Government, there was a 35% risk at the extreme level, 60% at the high level, and 5% at the middle level. On the scale of risk acceptance, there were 35% classified as unacceptable, and 65% classified as unexpected. According to the Kendal District Government, there was a 60% risk at extreme levels and 40% at high levels. On the risk acceptance scale, there were 60% classified as unacceptable and 40% classified as unexpected. The results of this study expected to provide a guideline for stakeholders in risk response according to their role both at the Provincial Government and District Government levels based on priority scales.

Keywords : Flood, Risk Management, Stakeholders, RBS

1. Introduction

1.1 Background

Kendal River is one of the rivers in Bodri Kuto River Area (RA) in Central Java province divided into 12 (twelve) watersheds, one of which is Kali Kendal watershed (Dinas PSDA Provinsi Jawa Tengah, 2012). Kendal River flows through 10 (ten) sub-rurals in Kendal City Sub-District, including: Sukodono, Trompo, Kebondalem, Kalibuntu Wetan, Pegulon, Pekauman, Patukangan, Ngilir, Balok, and Bandengan [1]. Indonesian President Regulation No. 12 of 2012 concerning Establishment of River Areas related to List of Watersheds (DAS), establishing the pattern of management of Water Resources in the cross-district river area (Bodri Kuto RA) declares the authority and responsibility of the Provincial Government, specifically the Public Works Water Resources and Spatial Planning Office (*Pusdataru*) of Central Java Province [2].

Based on Indonesian Law Number 24 of 2007 concerning Disaster Management, the disaster risk management approach in the preparation of disaster management plans is initiated from Government initiatives and commitments, by identifying disaster risks, choosing disaster risk reduction actions, regulating actors, and allocating tasks and authorities, as well as utilizing the available resources [3]. This research focuses on the disaster mitigation phase, during which no disaster occurs. The results of the study are expected to provide input and description to stakeholders related to the management of

Kendal River and also in flood management, Therefore, the flood disaster management programs is expected to be effectively, synergistically, and efficiently performed.

1.2 Research Inquiries
1. What are the main causes of Kendal River's frequent overflow, causing flooding in Kendal City?
2. How are the risks which become the main targets for the management of flood disaster in Kendal City from the perspective of the relevant stakeholders?
3. How much impact does this risk have on the main targets for the management of flood disaster in Kendal City from the perspective of the relevant stakeholders?
4. What is the response to the risks which become the main targets for the management of flood disaster in Kendal City from the perspective of the relevant stakeholders?

1.3 Purpose of the Study
1. To identify the main causes of flooding in Kendal City as a result of Kendal River overflow.
2. To analyze the risks which become the main targets in the management of flood disaster in Kendal City from the perspective of the relevant stakeholders.
3. To analyze the extend of how the risk that are the main targets impacts the management of flood disaster in the Kendal City from the perspective of the relevant stakeholders.
4. To analyze the response in order to prevent and deal with risks which become the main targets in handling floods in Kendal City from the perspective of the relevant stakeholders.

2. Methods

2.1. Research Methodology
This study applies descriptive methods or case studies with a combination of approaches (mixed methods), between qualitative and quantitative. According to Soerjono, research with descriptive methods or case studies aims to learn in depth about one of the real symptoms that exist in people's lives that can be used to examine a situation, groups, local communities, institutions or individuals [4]. Descriptive method will explain whether program implementation is in accordance to the technical/implementation guidelines.

2.2 Samples and Respondent Selection
The respondents were chosen by using purposive sampling method. Purposing sampling method is a technique in which a set of consideration is applied in choosing research sample [6]. According to Bouma, by using the Purposive Sampling method, the researcher believes that they can use their discretion or intuition in selecting the best people or groups to study or to provide the most accurate information [7]. Table 1 and Table 2 indicate the description of respondents for Qualitative and Quantitative methods.

Table 1. Respondents for In-depth Interviews using Qualitative Methods

No.	Institution	Number	Experience
1	Public Works Water Resources and Spatial Planning Office of Central Java	1 person	Managing Flood Disaster in Kendal City ≥ 10 Years
2	Public Works Water Resources and Spatial Planning Office of Bodri Kuto Semarang	2 persons	Managing Flood Disaster in Kendal City ≥ 10 Years
3	Planning, Research and Development Office of Kendal District	1 person	Managing Flood Disaster in Kendal City ≥ 10 Years
4	Public Works and Public Housing Office of Kendal District	1 person	Managing Flood Disaster in Kendal City ≥ 10 Years

No.	Institution	Number	Experience
5	Regional Disaster Management Agency of Kendal District	1 person	Managing Flood Disaster in Kendal City ≥ 10 Years
6	Women and Children Empowerment Office of Kendal City	1 person	Managing Flood Disaster in Kendal City ≥ 10 Years
	Total Respondents	**7 persons**	

Table 2. Respondents for Research Interviews and Questionnaires using Quantitative Methods

No.	Institution	Number	Experience
1	Public Works Water Resources and Spatial Planning Office of Central Java	2 persons	Managing Flood Disaster in Kendal City ≥ 8 Years
2	Public Works Water Resources and Spatial Planning Office of Bodri Kuto Semarang	2 persons	Managing Flood Disaster in Kendal City ≥ 8 Years
3	Public Works and Public Housing Office of Kendal District	4 persons	Managing Flood Disaster in Kendal City ≥ 8 Years
	Total Respondents	**8 persons**	

2.3 Data Analysis

Data analysis in this study utilizes the process according to the stages of risk management, especially for floods in the City of Kendal. The stages of risk management carried out are depicted in Figure 1 below:

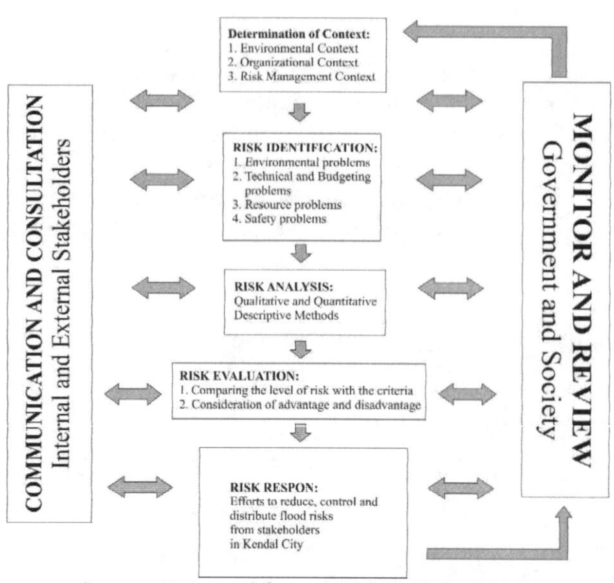

Source: Extracted from AS/NZS 4360 (2004)

Figure 1. the Flood Disaster Risk Management Process For Kendal River

1. Risk Identification

IOP Conf. Series: Earth and Environmental Science **328** (2019) 012048 doi:10.1088/1755-1315/328/1/012048

Based on risk classification in Godfrey, *et al.* [9], the author limited to only four main risk sources to be more detailed and in-depth, such as: environmental issues, technical and budgeting matters, resource problems, and security/safety issues.

2. Risk Analysis and Evaluation

The risk analysis in this study was carried out by using the Risk Breakdown Structure (RBS) method. According to the Project Management Institute, RBS provides a list of categories and subcategories that might arise in a project [10]. RBS is advantageous as a reminder for at-risk population of the risk sources that might arise on the project. The amount of risk according to Godfrey, *et al* is determined from the value of multiplication between the probability/frequency value and the impact/consequence value [9]. Risk probability is often interpreted by the term risk likelihood, while the consequence if the risk occurs is known as risk impact, and the significance of risk defines risk exposure calculated by using the formula of:

$$Risk\ Exposure = risk\ likelihood\ x\ risk\ impact \qquad (1)$$

The basis for determining the frequency scale/value and impact scale/value is presented in following Table 3 and Table 4:

Table 3. Level and Scale of Likelihood

No.	Frequency Level	Likelihood (%)	Scale
1	Very seldom	< 20	1
2	Seldom	$20 \leq x < 40$	2
3	Sometimes	$40 \leq x < 60$	3
4	Often	$60 \leq x < 80$	4
5	Very often	≥ 80	5

Source *Extracted from Godfrey, et al.* (1996)

Table 4. Level and Scale of Consequence

No.	Frequency Level	Likelihood (%)	Scale
1	Very Low	< 5	1
2	Low	$5 \leq X < 15$	2
3	Moderate	$15 \leq X < 45$	3
4	High	$45 \leq X < 80$	4
5	Extreme	≥ 80	5

Source: *Extracted from Godfrey, et al.* (1996)

Risk Evaluation with the RBS method is conducted by plotting the value of frequency and impact into the risk level matrix map to obtain the risk level of each risk that occurs. The risk matrix map is presented in following Table 5:

Table 5. Risk Level Matrix

			CONSEQUENCE				
			Insignificant	*Minor*	*Moderate*	*Major*	*Catastrophic*
			1	2	3	4	5
LIKELIHOOD	*Almost Certain*	E	Moderate	High	Extreme	Extreme	Extreme
	Likely	D	Moderate	High	High	Extreme	Extreme
	Possible	C	Low	Moderate	High	High	Extreme
	Unlikely	B	Low	Low	Moderate	Moderate	High
	Rare	A	Low	Low	Low	Low	Moderate

Source: *Extracted from Godfrey, et al.* (1996)

The risk value is obtained from the multiplication of frequency / probability with the impact determines the Risk Acceptance Scale for sources of risk causes of flood disasters. Risk acceptance indicators are presented in the following Table 6:

Table 6. Risk Acceptance Indicator

No.	Risk Acceptance Indicator	Risk Value
1	Negligible	$X < 3$
2	Acceptable	$3 \leq X < 8$
3	Undesirable	$8 \leq X < 15$
4	Unacceptable	$X \geq 15$

Source: *Extracted from Godfrey, et al.* (1996)

3. Treatment/Risk Response

This stage depicts the result of in-depth interviews conducted by the author to the respondent. The results of this in-depth interview are presented in the form of tables and risk responses, risk response strategies, and which leading sector of each identified risk are then determined.

4. Risk Monitoring and Review

This stage presents the result of in-depth interviews and brainstorming (combining ideas/views) of the author with the respondents relating to monitoring activities and reviews coupled with secondary data collection. This stage indicated the monitoring and review activities that were carried out and the mechanism by the Central Java Provincial Government represented by the Public Work in Water Resources and Spatial Planning Office of Central Java. In addition, the Public Work in Water Resources and Spatial Planning Office of Bodri Kuto with the Kendal Regency Government are represented by the Public Works and Public Housing Office.

3. Results and Discussions

The concept of risk management for natural disasters, especially floods that occur in the city of Kendal due to the overflowing of Kali Kendal, in this section is limited to 3 (three) main contexts according to the author's limits, including: the context of the environment, the organizational context (stakeholders) and the context of risk management that has been carried out so far.

3.1 Risk Identification

Risk identification in the flood disaster risk management process in Kendal City was conducted by in-depth interviews with each respondent. The results of risk identification are presented in in the following Table 7:

Table 7. Risk Identification

No	Source of Risk	Risk
1	Environmental Issues	a. High level of sedimentation. b. The wet cross section of the river does not hold sufficient water at the peak of wet season. c. Shallow river slope. d. Accumulation of garbage in the river. e. The merging of the Kendal River and Buntu River mouths. f. Shift in land utilizations and the number of buildings on the river banks. g. Fishing boats hindering the flow of river water. h. Overflow of water from Blorong River.
2	Technical and Budgeting Issues	a. Poor coordination between stakeholders. b. Lack of commitment of stakeholders during land acquisition phase. c. Kendal River is situated in a higher elevation than the drainage system of its surrounding settlements. d. Provision of OP and rehabilitation budgets is very dependent on conditions in the field.

No	Source of Risk	Risk
		e. Lack of community involvement. f. Poor city drainage system. g. Lack of accessibility to early disaster information systems.
3	Resources Issues	a. OP River in Kendal River has just been initiated in the last three years. b. Equipment limitations, especially when there are simultaneous floods in other locations.
4	Security Issues Safety	a. The location is in the center of transportation, economic and public service activities. b. Damage to commodities in the agricultural area. c. The existence of densely populated settlements in the immediate vicinity of Kendal River.

3.2 Risk Analysis and Evaluation

The followings are the results of the Risk Analysis between the Provincial and District Governments:

a. Provincial Government

- 7 (seven) highest risk level, which are: 35% at the extreme level with a risk value of 20 (twenty), 1 (one) levels of risk namely 5% at the medium level with a risk value of 12 (nine) remaining, 12 (twelve) levels of risk that is 60% at high level with risk values ranging from 12 (twelve) to 16 (sixteen).
- The greatest impact risk based on the four main risk sources according to the Provincial Government is the risk originating from the shifting of land functions and the density of buildings on river borders, while the highest risk based on frequency and impact is the existence of densely populated settlements on both sides of Kendal River.

b. District Government

- 12 (twelve) highest risk level, which are: 60% at the extreme level with a risk value of 25 (twenty), the remaining 8 (eight) levels of risk are 40% at the high level with a risk value from 12 (twelve) to 16 (sixteen) .
- The risk that has the greatest impact is based on four main risk sources from the District Government's point of view lacking the capacity of the river cross section at the peak of the rainy season, while the highest risk based on the frequency and impact value becomes the high level of sedimentation on the river body.

Risk Evaluation Results based on the impact of the risks on the acceptance scale indicate 7 (seven) risks or 35% which are classified as unacceptable, and 13 (thirteen) risks or 65% which are classified as undesirable according to the Provincial Government. Whereas, according to the Regency Government there are 12 (twelve) risks or 60% classified as unacceptable, and 8 (eight) risks or 40% classified as undesirable (not expected). The study revealed that the Provincial Government and District Government had different views on 9 (nine) items of risk or 45% of all items of risk; in other words, almost half of all items of risk are in debate.

3.3 Treatment/Risk Response

In every risk that occurs in the field, every stakeholder, both at the Central Java Provincial Government and Kendal District Government have their respective roles in responding to these risks, in addition to their obligation to cooperate in synergy. Every response to these risks must be delivered by all relevant parties with an equal role. Table 8 illustrates an example of a risk response, specifically responses to extreme levels of risk.

Table 8. Examples of Response to Risk with Extreme Levels

Risk	Risk Response Strategy
High levels of sedimentation.	a. Holding as much water as possible or build water catchment areas in the upper (upstream) area in two ways, by vegetative and structural methods.

Risk	Risk Response Strategy
	Vegetative: planting trees (conservation) by means of terracing, planting trees with strong roots and long-term economic values such as fruit trees. In principle, the soil surface is covered with a canopy, preventing from direct raindrops, and then the remaining leaves will formpeat, which in turn will retain water. Structural: The most impactful structures are check dam, water reservoir, and sediment and reservoir control structures. At the household level, roraks (holes in the ground covered with plastic) can be made in case of sediment buildup, to collect and return the sediments to its original place. Dams can also be built, noting that the impact is small because the dam's main function is to raise the water level and control water, not to reduce flooding. b. Performing normalization of river areas, if conditions in the field are conducive (still possible on certain spots). c. Improving the bridge design along the Kendal River to prevent the girders to be too low and removing the pillar structure in the middle that can cause sedimentation. d. Controlling the street vendors and illegal buildings along the river banks by collaborating with the Civil Service Police Unit at the Provincial and District levels, related to the enforcement of regional regulation.
Lack of water storage capacity in the wet section of the river during the peak of the rainy season	a. Performing normalization of river bodies must be carried out, with priority areas of rehabilitation starting from the Trompo dam to the Bandengan village. b. Ensuring that river boundaries in conservation areas must be free from buildings and services for normalization activities must be carried out. There must be an inspection road to transport the normalized sediment in the border area. c. Redesigning the river body to add water storage capacity, because the wet filter that should be able to accommodate ± 25 m3 in fact has now run out when it is filled with ± 10 m3. Widening of the river can be widened if possible, or dikes can be made to raise both banks of the river. Widening the river requires land, and making dikes means increasing safety and security risks. d. Separating theKendal River with Buntu River estuaries.
Accumulation of garbage in the river.	a. Disseminating and guiding the community around the river to increase awareness so as not to carelessly dispose of garbage. Currently the KSI (Indonesian River Congress), River School, or Asia River Conference have often been held in relation to the concerns of river conditions in Indonesia in general, especially in Central Java. b. Empowering of the community regarding waste, especially green waste, for example counseling in waste sorting and composting and working with farmers to use the resulting product as organic fertilizer. c. Pursuing a synergy between the Provincial Government and the District Government; for example, the Provincial Government is tasked with taking garbage from irrigation canals and river bodies, then the District Government takes the garbage to be transported to the Final Disposal Site (TPA).
Shift in the land utilization and the number of buildings on the river border	a. Controlling and utilizing the impacted area immediately, for example, to be made as a Green Open Space (RTH). b. Pursuing synergy between the Provincial Government and District Government, not only with the technical service but also with the BPN/ATR (State Land/Agrarian Agency and Spatial Planning Agency), to track residents in possession of land certificate in that location. c. Promoting law enforcement / regional regulations concerning River boundaries for existing buildings, and spatial control for illegal buildings, required for structuring the Kali Kendal Watershed upstream.
There are densely populated settlements in the immediate area of Kali Kendal	a. Creating disaster mitigation maps, for example maps of shelter/evacuation locations, evacuation routes, locations designated as public kitchen places, and so on. b. Conducting spatial planning by considering aspects of flood disaster mitigation. c. Providing immediate solutions, for example by providing water pumps to accelerate the reduction of inundation. d. Promoting the importance of conservation of river border areas to the local community.

3.4 Risk Monitoring and Review

The monitoring and review activities are carried out by the Government through related agencies, both by the Central Java Provincial Government and the Kendal District Government, in relation to water resources management, particularly in the management of Kendal river starting from upstream to downstream, which are conducted annually.

4. Conclusions and Recommendations

4.1 Conclusions

1. From the 20 (twenty) risk factors identified as the main causes of the flood disaster in Kendal City due to Kali Kendal overflowing, there are 5 (five) risk factors with extreme risk values, including: high sedimentation, lack of water storage capacity for river cross sections during rainy peaks, poor city drainage system, land conversion and the number of buildings on the river border, and the existence of densely populated settlements in the immediate area of Kendal River.
2. Risk analysis indicates that at least 60% of the factors identified as the cause of flooding in Kendal City have high risk values, requiring a better handling and focus on risks with extreme values as a priority scale.
3. Every stakeholder, both the Central Java Provincial Government and the Kendal Regency Government, have their respective roles in the treatment / response to these risks, and both need to work in synergy. The difference that can be underlined is that the Provincial Government is more concerned with the technical planning and management section in the physical or infrastructure section, while the Kendal District Government is more concerned with its social aspects.

4.2 Recommendations

1. The results of risk identification in accordance with the risk source are expected to function as a guide for stakeholders in conducting risk mitigation based on the main tasks and functions of each Regional Apparatus Organization (OPD), both the Central Java Provincial Government and Kendal District Government.
2. The risk treatment / response that have been carried out should be improved by prioritizing treatment according to the frequency of the incident and its impact scale, and by focusing on extreme levels of risk both from the perspective of the Provincial Government and District Government, without diminishing focus to other risks.

References

[1] Central Bureau of Statistics. (2016), *Kecamatan Kota Kendal Dalam Angka 2016*, Kendal.
[2] Republic of Indonesia Presidential Regulation Number 12 of 2012, concerning *The Establishment of River Watershed (DAS) Areas per River Region (RA) in Central Java*, Jakarta.
[3] Republic of Indonesia Law Number 24 of 2007, *concerning Disaster Management*, Jakarta.
[4] Soerjono, S. (1990), *Sosiologi Suatu Pengantar*, Jakarta: Raja GrafindoPersada, PT.
[5] Sugiyono. (2012), Metode Penelitian Kuantitatif Kualitatif dan R&D, Bandung: Alfabeta, CV.
[6] Ismiyati. (2011), *Statistik dan Probabilitas Untuk Teknik Bagi Peneliti Pemula*, Semarang: Magister Teknik Sipil Program Pasca Sarjana Universitas Diponegoro.
[7] Bouma, G. (1993), *The Research Process*, Revised Ed., Oxford: Oxford University Press.
[8] AS/NZS 4360. (2004), *Australian/New Zealand Standard Risk Management*, Joint Technical Committee OB-007 Risk Management.
[9] Godfrey, Patrick S., Sir Halcrow, W., and Partners Ltd. (1996), *Control of Risk: A Guide to the Systematic Management of Risk from Construction*, Westminster London, Construction Industry Research and Information Association (CIRIA).
[10] Project Management Institute (PMI). (2000), *A Guide to The Project Management Body of Knowledge, PMBOK Guide*, Newtown Square, Pennsylvania, USA.

Improving Community Services and Infrastructure in Remote Areas by utilizing Community Need Assessment (CNA) Approach. (Case Study: Kampung in Mahakam Ulu Regency)

R A Vidianti[1], R N Istiqomah[2], and A R T Hidayat[3]*
[1]Graduate student of Urban and Regional Planning Department, Brawijaya University
[2]Undergraduate student of Urban and Regional Planning Department, Brawijaya University
[3]Urban and Regional Planning Department, Brawijaya University

*Email: a.r.taufiq.h@ub.ac.id

Abstract. The remote area is defined as a remote (rural / settlement) location within an area One of the remote area locations is Mahakam Ulu Regency. One of the problems faced in remote areas locations selected by researchers is the availability of commnunity services and infrastructure. This research explores the condition of commnunity services and infrastructure in the Kampung Lutan, Long Hurai and Muara Ratah which are being the priority villages in the Mahakam Ulu Regency. Exploration results are expected to provide input for determining the needs and hopes of village commnunity services and infrastructure by involving community participation. Community Need assessment (CNA) approach was applied in this study involving process of collecting, analyzing and reporting information about community needs and the amount of capacity or strength that exists in the community to fulfill those needs. By understanding the needs and hopes of village community service and infrastructure, it can facilitate the government in establishing the priorities of village development to observe the social aspects of the community. CNA analysis addressing a number of facilities and infrastructures in the three villages tended not to require additional units, but to improve the quality of the commnunity services and infrastructure.

Keywords: *village, remote area, community services-infastructure, community-needs-assessment*

1. Introduction

Infrastructure development becomes one of the important aspects in regional development, with the increasing number of infrastructure that can support the community's economy. The construction of physical infrastructure especially in villages has a vital function, in supporting community in daily life and accelerating development in order to improve the socio-economic life of the community [1]. Republic of Indonesia Law number 6 of 2014 [2] concerning Villages emphasizes the importance of infrastructure development and empowerment of village communities to increase access for services and economic activities. One factor behind the village's lagging was also influenced by the lack of availability of community service and infrastructure. As result, the government authorizes and manages the development directly to the village [3] the minimum availability of community services and infrastructure, one of which is in the remote area. Remote area is defined as a remote (rural / settlement)

IOP Conf. Series: Earth and Environmental Science **328** (2019) 012049 doi:10.1088/1755-1315/328/1/012049

location within an area [4]. One of the remote area locations is Mahakam Ulu Regency, which is a new autonomous region ratified on December 14, 2012 and also included in the border area. Mahakam Ulu Regency is a proliferation of a district of West Kutai Regency. Unlike in Java, accessibility between villages tends to use water transportation requiring longer time.

In order to achieve equitable development, it is not possible to do it simultaneously at the same scale. Central and regional governments have limited human and financial resources. Therefore, development always has a priority scale. Mahakam Ulu Regency has a priority development village. The village is a growth pole for other villages in the district. This study was intended to determine the needs of community services and infrastructure in the priority villages in Mahakam ulu Regency with a community participation approach, such as Lutan (Long Hubung District), Long Hurai (Long Bagun District), and Laham (Muara Ratah District).

1.1 Literature Review

Village in general means that a community consists of a variety of diverse ethnic or ethnic communities who live in one area and in groups with a simple lifestyle that has wisdom and its rules and are being practiced in daily life. Village is a subsystem of a broader whole area called a country. Village as a system has components in physical, human, and social institutions. Muhammad [5] detailed the components of the village as follows;

a. Agricultural and environmental resources
b. Economy of rural areas
c. Social institutions
d. Human Resources
e. Physical facilities and infrastructure

Figure 1. Goals of infrastructure development

In this study, in order to determine the needs and hopes of the community, the Community Need Assessment approach was applied. Based on the Participant Workbook-,CNA or Community Need Assessment [6] defined as identifying community needs in providing services to fulfill the needs. Other sources define community assessment as a process of collecting, analyzing and reporting information about community needs and the amount of capacity or strength that exists in the community to fulfill those needs. Community Need Assessments has objectives to give community leaders an overview of current local policies, systems and strategies for environmental change and help identify areas for improvement. With this data, communities can map to create strategies to make positive and sustainable changes in their communities. The application of CNA is a form of accommodation for community participation in development. Thus, the objectives of rural development, called bottom up planning can be achieved. The results from CNA expected to deepen the government programs and policies to achieve target.

2. Methods

In this study, the scope of the area studied covers 3 villages in Mahakam Ulu Regency, such as Lutan (Long Hubung District), Long Hurai (Long Bagun District), Muara Ratah (Laham District). These three villages are priority development villages in Mahakam Ulu Regency. As for the flow of this research, it begins by identifying the conditions of the research area and identifying problems. After the problem is identified, the next step is conducted through data collection. Data collection is gathered by primary and secondary surveys. Forum Group Discussion (FGD) with local communities are extracted from the information related to the direction of availability of village community services and infrastructure based on community perceptions. After the existing condition data and direction were obtained, an analysis of the needs and hopes of village facilities and infrastructure were carried out. In accordance with the research objectives, the needs of community services and infrastructure are prioritized in Mahakam Ulu Regency with a community participation approach,-. To obtain the required data, a primary survey was conducted by semi-structured interviews, field observations, and FGDs. The researchers also observed the system or customary norms as the social side of society applied in a community. The analysis applied is an analysis of needs and hopes. In this analysis, a comparison was made among the existing conditions of village infrastructure, direction of the availability of village community service, infrastructure based on community perceptions, and consideration of important village leaders (key person).

3. Results and Discussion

3.1 Overview of the Village in Mahakam Ulu

Mahakam Ulu Regency is a part of Mahakam Ulu Regency which established based on Law Number 2 in 2013. Mahakam Ulu Regency consising of 5 districts, such as Long Apari District, Long Pahangai, Long Bagun, Laham, and Long Hubung. Mahakam Ulu Regency has its capital in Ujoh Bilang District. As a new autonomous district in the border areas, it is a challenge in development and infrastructure needs for settlements. In this study, the location of the studio was divided into three villages located in different districts which are Lutan, Long Hurai and Muara Ratah Lutan located in the Long Hubung District and divided into 4 Neighborhood Units (RT). Long Hurai is divided into 2 Neighborhood Units (RT). Muara Ratah is located in the Laham District area and divided into 2 Neighborhood Units (RT). Main referred access to the village is mainly by air transportation. Basically the village is a Dayak tribe community in Long Hurai and Lutan,: whereas, Dayak tribe in Muara Ratah mainly consists of the Bahau Dayak tribe.

3.2 Conditions for Community Services and Infrastructure

3.2.1 Condition of Community Services

Community services are the most important supporting tools in social activities or economic activities. Therefore the existence of community services in an urban or rural area is required to build the social and economic structure of a settlement [7]. Meanwhile, infrastructure is defined as the basic necessity to fulfill the principles of a healthy and comfortable environment [7]. The condition of community services and infrastructure in the research area coverage can be explained by the following table.

Table 1. Conditions for the Existing Community services

Community services	**Kampung**		
	Lutan	**Long Hurai**	**Muara Ratah**
Education facilities	4 educational facilities consist of 2 Kindergaerten units, 1 elementary unit, and 1 junior high school unit.	2 educational facilities, consist of 1 kindergarten unit and 1 elementary school unit.	3 educational facilities consist of 1 kindergarten unit, 1 elementary school unit and 1 junior high school unit.
Health care facilities	1 unit of auxiliary local health center and 1 unit of ***Posyandu***	1 unit of auxiliary local health center and 1 unit of ***Posyandu***	1 village polyclinic unit
Government Community Service and Public Services	1 unit of Village hall office, 1 "lamin adat" and 3 security post units	1 unit of Village hall office, 1 "lamin adat" and 2 security post units	1 office unit of the Village Head and 1 unit of the General Assembly Hall. and 1 unit of security post
Worship Community Service	3 units of Worship Community Service consist of 1 unit of mosque, 1 unit of Catholic Church, and 1 unit of Christian church	2 units of church	1 unit of mosque

Community services	**Kampung**		
	Lutan	**Long Hurai**	**Muara Ratah**
Trade and Service of Community Service	11 units of waroong	2 units of waroong	2 units of waroong

Table 2. Conditions for the Existing Infrastructure

Infrastructure	**Kampung**		
	Lutan	**Long Hurai**	**Muara Ratah**
Road infrastructure	Road Pavement in Kampung Lutan in the form of cement roads and dirt roads.	64% of Road pavement is in the form of concrete	Road Pavement in the form of concrete roads and dirt roads in good condition and lightly damaged.
Dock	there is a Dock located on RT 02	there are 5 docks located on RT 01 and 02	2 unit of docks
Sanitation and Waste Disposal Infrastructure	33 houses from 164 houses that do not have private sanitary installation.	There is no infrastructure for waste water treatment or communal WWTP	1 communal sanitary installation in Kampung Muara Ratah and there are no retributon and waste water treatment.
water supply	Each house in the kampung Lutan has water reservoirs and piping facilities which had been built in 2017.	The need for clean water in Kampung Long Hurai use pumping system.	Fulfilling the necessity of clean water from wells
Telecommunication	BTS towers that emit signals are not found in Kampung Lutan.	Based on the survey results, 80% of the people in Kampung Long Hurai use mobile phones as telecommunication tools.	The condition of telecommunication distribution is served only by telephone and SMS.
Electricity	The utilization of communal and private generators.	The utilization of a personal generator set that is used between 18.00 - 00.00 WITA.	In Muara Ratah electricity is only availible at 18.00-24.00.

3.2.2 Community Needs Assessment (CNA) Analysis

The necessity of village community service and infrastructure was carried out through an analysis of Community Need Assessment involving the opinions of the community of Kampung Lutan, Long Hurai and Lutan through the Focus Group Disscussion. The results obtained are as follows

Table 3. community assessment

Community Services and Infrastructure	**Assessment**		
	Kampung Lutan	**Kampung Long Hurai**	**Kampung Muara Ratah**
Education facilities	• No additional unit needed. • Renovation for kindergarten	• No additional units are needed. • Need to build a kindergarten building	• No additional units are needed. • Renovation and widening educational buildings are required

IOP Conf. Series: Earth and Environmental Science **328** (2019) 012049 doi:10.1088/1755-1315/328/1/012049

Community Services and Infrastructure	**Assessment**		
	Kampung Lutan	**Kampung Long Hurai**	**Kampung Muara Ratah**
		• For elementary schools, the building should be renovated.	
Health care Facilities	• No additional unit is needed. • Renovation of auxiliary local health center and adding some rooms • Adding posyandu facilities	• No unit is needed • Rooms renovation are needed as a supporting facility of posyandu • Adding posyandu facilities	• No additional unit is needed. • Renovation and adding facilities of auxiliary local health center and adding some rooms
Government Community Service and Public Services	• Repair / rebuild high-ranking offices • widening of lamin adat buildings • addition of 1 unit of security post	• Repair / rebuild high-ranking offices • Widening of Lamin Adat buildings • Addition of 1 unit of security post (at RT 1)	• Construction of a one-roof government office • Additional security post in each RT
Worship Community Service	• Renovation of mosques and churches	• Additional mosques are needed • Adding complementary facilities and infrastructure • Renovation of church buildings (RT 2)	• Renovating mosque
Trade and Service of Community Service	• Additional units are needed. • Adding 7 store units and build 1 market and Koperasi	• Additional units are needed. • adding 5 store units and construction of 1 BUMDES / Koperasi	• Additional units are needed. • adding 4 shop units and construction of 1 market unit and Koperasi
Road infrastructure	• Building the main road from soil to concrete / asphalt. • Adding the amount of street lighting lights on the main road	• Building road RT 2 pavement from soil to concrete / asphalt. • Adding the amount of street lighting lights on the main road	• Building the main road from soil to concrete / asphalt. • Adding the amount of street lighting lights on the main road
Dock	• Improving the facilities quality of dock	• Improving the quality of dock infrastructure • Improving road access from the dock to the village	• Improving the quality of dock infrastructure • Improving road access from the dock to the village
Sanitation and Waste Disposal Infrastructure	• Construction of communal sanitary installation in public service facilities, • Private sanitary installation	• Construction of communal WWTP of the dock infrastructure.	• Construction of Waste Water Management Installation (WWTP) • Construction of new private sanitary installation for new settlements
water supply	• Provision of communal clean water managed by the	• Provision of communal clean water managed by the Long Hurai BUMDES / BUMKAM	• Provision of communal clean water managed by the Muara Ratah BUMDES / BUMKAM

Community Services and Infrastructure	Assessment		
	Kampung Lutan	**Kampung Long Hurai**	**Kampung Muara Ratah**
	Lutan BUMDES / BUMKAM.		
Telecommunication	• Build addtitional BTS in Kampung Lutan	• Need extra telecommunication signal capacity in Kampung Memahak Besar (BTS location)	• Build addtitional BTS in Kampung Muara Ratah
Electricity	• Development of BUMDES / BUMKam in PLTD Management in Kampung Lutan	• Development of BUMDES / BUMKam in PLTD Management in Kampung Long Hurai	• Development of BUMDES / BUMKam in PLTD Management in Kampung Muara Ratah

The results of the two tables above are then carried out to determine the needs and hopes of the community service and infrastructure in each village. It is carried out to find the most required aspects of the development

Table 4. The analyzes of needs and hopes

Community Services and Infrastructure	Necessity	Hopes
Education Facilities	• In the three villages there is no need to add educational units. • improvement of the conditions for the construction of Kindergarten (Kampung Lutan and Muara Ratah) • construction of its own building for Mattandaurp Middle School (Kampung Lutan) • improving the quality of elementary school buildings (Long Hurai Village)	• need additional Elementary School teachers (Kampung Lutan) • construction of new buildings for kindergarten (Kampung Long Hurai)
Health care facilities	• In the three villages there is no need for additional health facilities • addition of facilities and auxiliary local health center rooms (Lutan, Long Hurai and Muara Ratah)	• -
Government Community Service and Public Services	• Renovating / rebuilding high-ranking offices (Lutan and Long Hurai) • Widening of "lamin adat" buildings (Lutan and Long Hurai)	• addition of a security post (kampung Muara Ratah and kampung Long Hurai) • construction of a one-roof government office (Kampung Muara Ratah)
Worship facilities	• Renovating of churches and mosques (Kampung Lutan) • adding of mosques (Kampung Long Hurai) • improving the quality of church buildings (RT 2) (Kampung Long Hurai) • Adding of complementary facilities and infrastructure (Kampung Long Hurai)	• improving the quality of mosques (Kampung Muara Ratah)
Trade and Service of Community Service	• the addition of 7 stores and the construction of 1 market and a Koperasi (Kampung Lutan) • addition of 5 store units and construction of 1 BUMDES / Koperasi unit (Kampung Long Hurai)	• -

Community Services and Infrastructure	Necessity	Hopes
	• the addition of 7 stores and the construction of 1 market and a Koperasi (kampung Muara Ratah)	
Road infrastructure	• Road repairing especially on RT 2 in the form of improving types of pavement from soil to concrete / asphalt (kampung Long Hurai) • Adding the amount of treet lighting lights on the main road (kampung Long Hurai, Kampung Lutan, kampung Muara Ratah)	• -
Dock	• Repairing the access road from the dock to the village (kampung Long Hurai and kampung Muara Ratah)	• Improving the quality of the dock facilities
Sanitation and Waste Disposal Infrastructure	• Construction of Waste Water Management Installation (WWTP) • Private sanitary installation assistance for houses which do not have private sanitary installation. (kampung Lutan)	• construction of Communal private sanitary installation in public service facilities, (kampung Lutan)
Clean water	• Provision of communal clean water supply managed by BUMDES / BUMKAM (Kampung Lutan, Long Hurai and Muara Ratah)	• -
Telecommunition	• addition of telecommunication signal capacity in Kampung Memahak Besar (BTS location) (kampung Long Hurai)	• Adding BTS (kampung Lutan and Muara Ratah)
Electricity	• Developing BUMDES / BUMKam in PLTD Management (kampung Lutan, Long Hurai and Muara Ratah)	• -

Based on the results of the analysis, it could be concluded that for education and health facilities, the availability is sufficient; thus, there are no addition required. The direction of the two facilities is to improve the quality of buildings or supporting facilities. Facilities for trade services, government and public services, and worship need additional units because the existing conditions are still not available or still insufficient. The village community is bounded closeness to relationships and social activities. Government and public services, and worship facilities are located in which they interact with the general public, the government, and religious leaders and worship. Besides, trade and services are absolutely necessary for them. At present the community is not too concerned about the price of goods. They need the availability of goods at the nearest stores.

The community needs to improve the quality of roads, the construction of WWTPs, the access from the dock to the village and BUMDES / BUMKam in the management of clean water and electricity. Moreover, the development of *Koperasi* (BUMDesa) will support and encourage the emergence of new economic activities. In line with the government's program on sanitation and waste disposal, it is important to protect the environment (because the villagers depend on nature). Communities are aware that bad sanitation and waste management can pollute the environment. Electricity is demanded to support social and economic activities. Even though there is electricity, its availability is still limited. By understanding the needs and hopes of village community service and infrastructure, it can facilitate the government in establishing the priorities of village development to observe the social aspects of the community. By the formation of community participation in the village planning process, communication was also built between the people as parties who felt the direct impact of the development and the government as policy makers [8].

4. Conclusions

From the identification results, the researchers concluded that there are still problems with the availability of community service and infrastructure in Kampung Long Hurai, Lutan and Muara Ratah, mainly due to located in the remote area in East Kalimantan. When surveying the existing conditions, it was apparent that residential facilities tended to be comfpleted but some facilities had unfeasible conditions.Besides, the infrastructure conditions present the main problems, such as telecommunications, electricity and roads infrastructure. The electricity in the three villages is still not available for 24 hours. The results of the identification were then delivered as the input into Forum Group Discussion (FGD) activities in each village and for the community participated in providing advices for village infrastructure in addressing the necessity and hopes. Based on the analysis, it can be summarized that the community of those villages needs the additional community service and infrastructure. They realize that the need for development is absolutely high compared to the conditions of village and the government financial capability. Thus, it is sensibly reasonable if the community proposes the improvements and completions of existing facilities.

Acknowledgment
This research was funded by the government of Mahakam Ulu Regency.

References
[1] Agustina 2017 *studi tentang pembangunan infrastruktur fisik di desa long-nah kecamatan muara ancalong kabupaten kutai timur* eJournal Administrasi Negara Volume 5 , (Nomor 3) 2017: 6211-6224
[2] Undang Undang Republik Indonesia nomor 6 tahun 2014 tentang Desa
[3] Kementerian Desa, pembangunan Daerah teritinggal dan Transmigrasi *2016 Buku Teknis Sarana dan Prasarana permukiman desa*
[4] Macintyre R and Macdonald J 2011 *Remote from What? Perspectives of Distance Learning Students in Remote Rural Areas of Scotland* International Review of Research in Open and Distance Learning Vol. 12.4
[5] Muhammad 1995 *Komponen desa* Salemba empat
[6] Centers for Disease Control and Prevention (CDC) *2013 Community Needs Assessment. Atlanta, GA*
[7] Sutikno R 2011*Perencanaan infrastruktur permukiman perkotaan* Malang : Universitas Brawijaya
[8] Laily Elida I N 2015 *Partisipasi Masyarakat Dalam Perencanaan Pembangunan Partisipatif* Surabaya : Universitas Airlangga *(jurnal Kebijakan dan Manajemen Publik Volume 3, Nomor 2, Mei-Agustus 2015)*

The Influence of Social Capital and Water Infrastructure on The Level of Happiness Index in Rural Area

R Alfiah[1], R N Listyawati[1] and D J Koesoemawati[1]
[1]Urban and Regional Planning Department, Faculty of Engineering, University of Jember, Indonesia

*Email : rindangalfiah@unej.ac.id

Abstract. Nowadays, access to drinking water services in Indonesia is below the SDG's target where 100% of the people must have access to clean water, but based on data from the Ministry of Public Works and Public Housing (2018), there are only 72 percent of Indonesians having access to piped water through The Indonesian Water Supply Company (PDAM) and only 7,474 villages (out of 72,944 villages) have access to piped clean water. To overcome the limited access to drinking water in rural areas, communities are supported to establish community-based drinking water organizations (HIPPAM) to manage local water supply services. Increasing access to adequate drinking water is expected to influence the level of population welfare measured in the happiness index. Sumbermujur Village is one of the villages that can manage water by establishing HIPPAM. From the SNA calculation, it can be seen that the value of social capital in Sumbermujur village are classified as good. The ties formed from the HIPPAM community are very strong. Based on community characteristics and values on the 10 happiness index variables, it was found that all hamlets in Sumbermujur Village belong to the "happy" group. The social relations variable in the happiness index has a similar characteristic with environmental conditions and family harmony.

1. Introduction

The Indonesian Government is currently starting to implement the concept of sustainable development through the Sustainable Development Goals (SDGs) program that integrates economic development with socio-cultural and environmental aspects. Strengthening social capital is very appropriate in empowering rural communities. Social Capital is needed to accelerate regional economic development and community welfare because social capital has an important role in linking community groups [1]. The higher the social capital owned by a family / community group, the higher the level of their welfare [2]. The important variables in social capital are local associations and community characteristics. Social capital that influences the level of well-being is also related to the level of happiness of the community [3]. The relationship between social capital and the level of happiness of the community can be seen from trusts that are built between individuals or communities. Determinants of happiness in Indonesia are income, education, health and social capital [3], which are in line with sustainable rural development program (P2KPB) with one of its attributes being the availability of clean water infrastructure through inclusive institutions and rural communities. Sumbermujur Village, which is located in Candipuro Subdistrict, is one example of a village that is implementing sustainable principles by utilizing the natural potential around it. There are water utilization activities that are inseparable from community participation and community social capital through HIPPAM Tirta Lestari organization which assists the village in clean water management in Sumbermujur Village. This research is needed regarding Social

Capital to produce a description of the relationship characteristics of Sumbermujur Village community in water resources management participation activities. The characteristics of community participation in bamboo forest management to maintain the quality of these springs will then be evaluated to determine the level of sustainability in social, economic and environmental aspects. Calculation of happiness index is used as one of the ingredients of public policy making (BPS, 2014) with a component of life satisfaction. Later it will be known that the achievement of welfare level will help in making the right policies in developing Sumbermujur Village based on sustainable principles.

2. Methods

Input data for this research was collected through depth interview with the key person in Sumberrejo Village such as head of Sumbermujur Village, Head of HIPPAM, or Head of Ecah Hamlets. In order to measure the social capital, affiliation data is needed through questionnaire survey with main question about community membership in some community groups in Sumbermujur village. Total respondent for villagers in Sumbermujur Village are 330 respondents. The respondents distributed proportionally based of the number of HIPPAM member and the number of all villagers in each hamlets. The second aim, to measure the happinex index measured from nine aspects of national happiness, namely: psychological calmness, health, education, time use, cultural resilience and diversity, governance, community vitality, environmental resilience and diversity, and living standards.

3. Results and Discussions

3.1. Social Capital of Community

Calculation of the level of community participation is obtained from the calculation of the Rate of participation with the formula Rate of Participation = Sum of Diagonal Matrix / Number of Respondents with data from interviews with community participation in institutions in Sumbermujur Village. Based on the calculation on community participation in institutions in Sumbermujur Village, it can be seen that the value of the level of community participation/respondents is the highest in Krajan hamlet with a value of 2.49, which means that the community joins 3 out of 7 institutions in Sumbermujur Hamlet. Meanwhile, the lowest value is in Umbulrejo Hamlet where the value of the participation level is 1.07, which means that the average community only follows one institution, namely recitation of Quran other than HIPPAM. The value of the participation rate that is classified as moderate and low in this calculation is because the participation rate is calculated based on the amount of community participation in the institution in Sumbermujur Village.

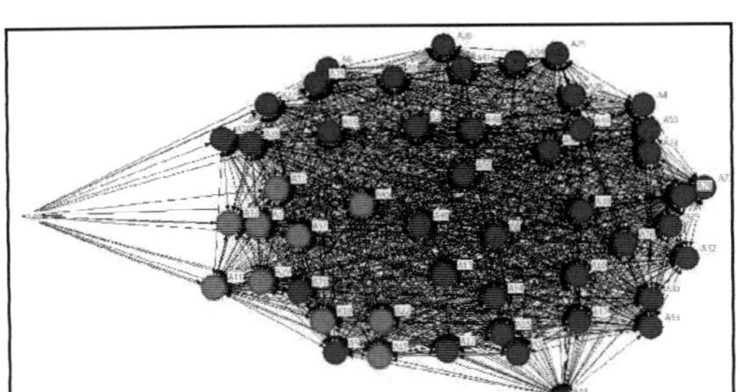

Figure 1 Netdraw of Krajan Hamlet

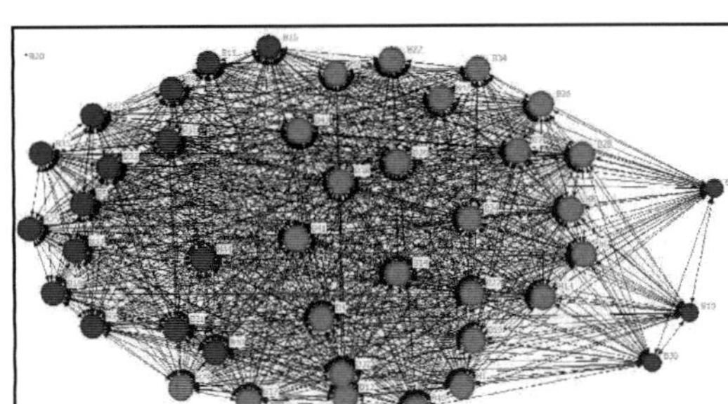

Figure 2 Netdraw of Umbulsari Hamlet

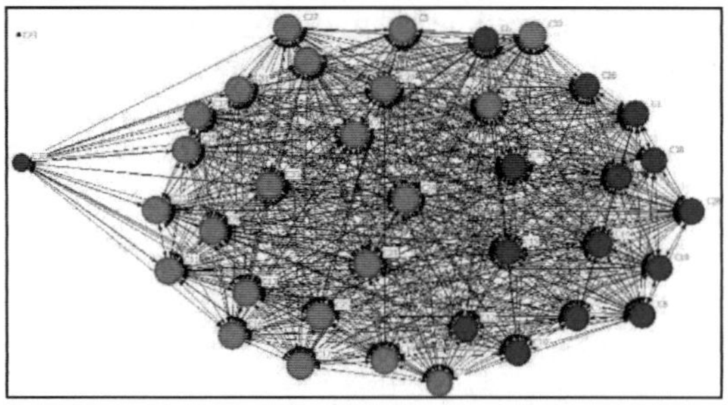

Figure 3 Netdraw of Kebonseket Hamlet

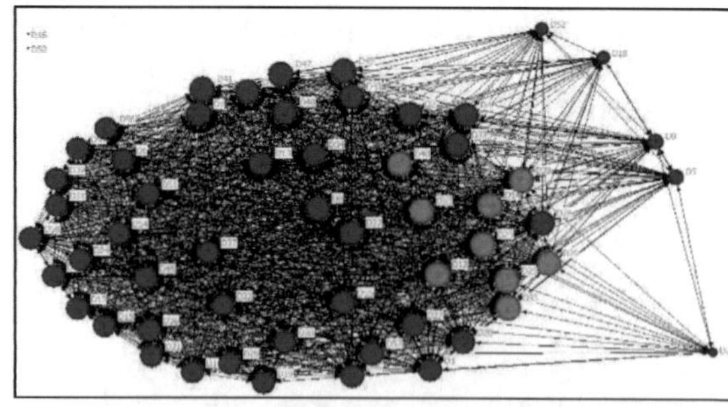

Figure 4 Netdraw of Umbulrejo Hamlet

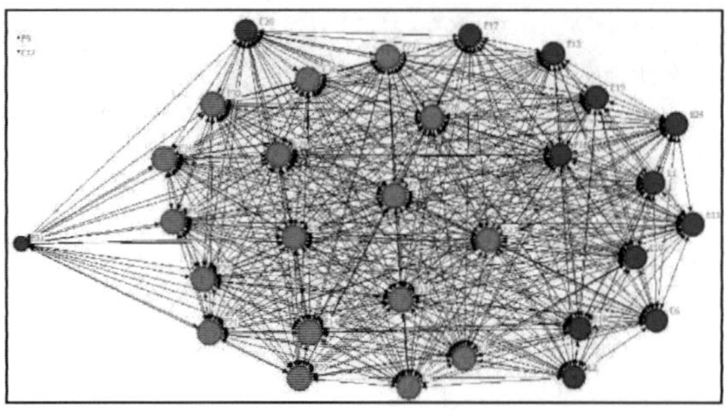

Figure 5 Netdraw of Wonorenggo Hamlet

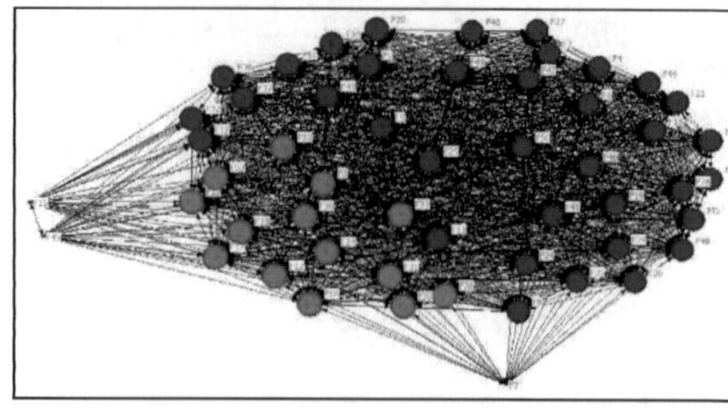

Figure 6 Netdraw of Banjarejo Hamlet

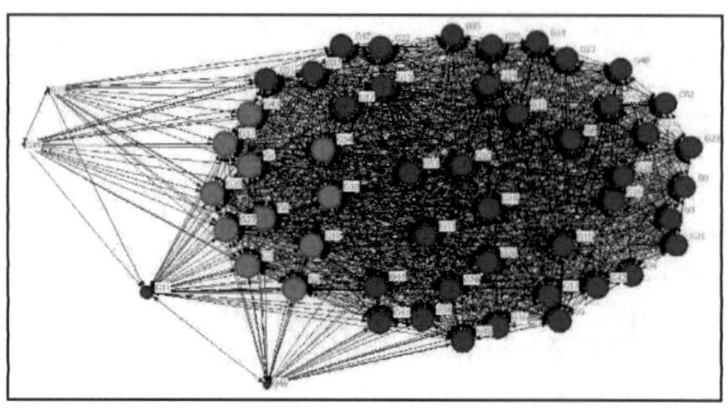

Figure 7 Netdraw of Sidorejo Hamlet

3.2. Happiness Index

Happiness index is one indicator to measure people's welfare. The higher the index, the higher the level of people's welfare, meanwhile the lower the level of happiness of the population,the less welfare of the community. In this study, happiness index includes several variables found in the aspects of people's lives. Ten aspects of happiness in calculating the happiness index in Sumbermujur Village are health, education, occupation, income, social relations, availability of leisure time, condition of the house and assets, environmental conditions and security conditions.

Table 1. Happiness Index

Variable	*Krajan*	*Umbul sari*	*Kebon seket*	*Umbul rejo*	*Wono renggo*	*Banjarejo*	*Sidorejo*
Health	63,44	51,37	61,46	47,74	60,92	71,19	55,95
Education	62,19	38,84	58,14	61,84	64,91	62,02	61,34
Occupation	68,83	71,94	66,47	73,04	57,10	70,02	69,80

Variable	*Krajan*	*Umbul sari*	*Kebon seket*	*Umbul rejo*	*Wono renggo*	*Banjarejo*	*Sidorejo*
Income	66,09	59,46	73,27	74,94	72,12	74,10	75,89
Security	60,00	72,24	59,52	59,15	80,81	74,67	46,40
Social relations	70,32	71,50	61,78	57,37	75,02	79,79	50,83
Time availability	70,08	70,52	60,50	60,74	68,71	65,84	73,09
Assets and house condition	58,74	67,72	63,58	50,49	46,85	66,32	68,78
Enviromental condition	68,40	62,09	63,38	64,46	75,49	44,71	62,63
Family	40,94	56,29	62,81	72,31	70,43	60,34	67,09

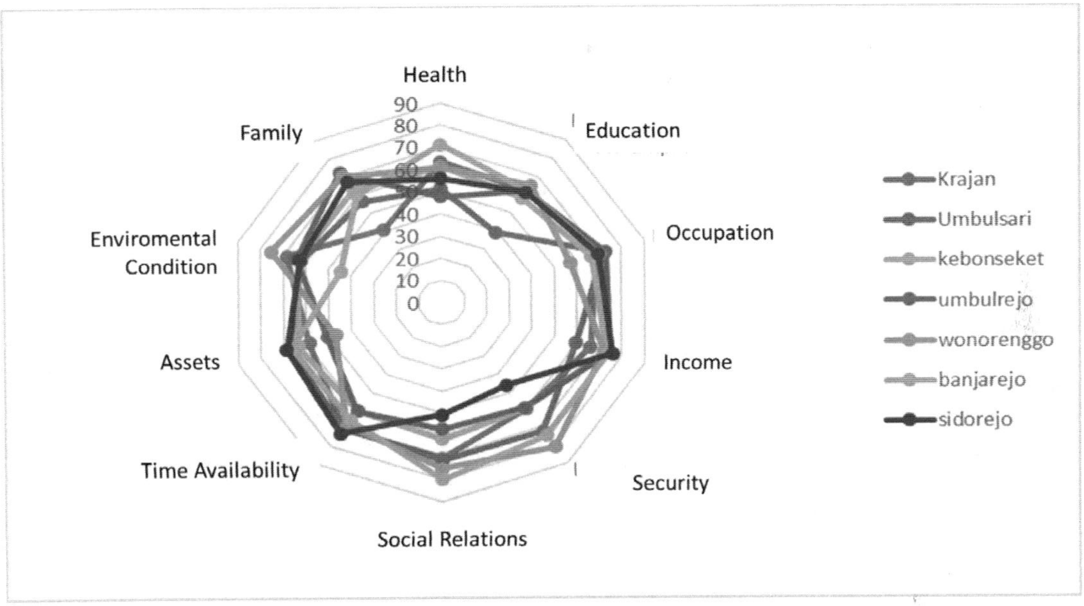

Figure 8. Value of Happiness Index

The calculation of happiness index shows that all hamlets in Sumbermujur Village are classified as happy groups. The highest aspect of happiness in each hamlet is different.

Table 2. Highest aspect of Happiness Index in each Hamlet

No	Hamlet	Aspect
1	Krajan	Social relations
2	Umbulsari	Security
3	Kebonseket	Income
4	Umbulrejo	Income
5	Wonorenggo	Security
6	Banjarejo	Social relations
7	Sidorejo	Income

The aspect of social relations is one aspect of happiness with the highest value in Sumbermujur Village proving that social relations are very influential for community life in Sumbermujur Village. Social relations are described through the number of organizations that are followed by the community, the frequency or activeness of the community in following the organization which means community participation in various activities and institutions in Sumbermujur Village. This high value indicates that community participation as one of the variables measuring social capital can be considered quite good in Sumbermujur Village. The people's livelihood as farmers in Sumbermujur Village cannot be separated from the need for water. A good HIPPAM institutional condition, which is capable of managing springs in Sumbermujur Village plays an important role in increasing agricultural yields and household income in Sumbermujur Village. Therefore, good utilization of water by HIPPAM which increases crop yields and income of people who depend their lives on agricultural land is one of the aspects of happiness with high value in Sumbermujur Village.

Social relations described through community participation is one of the variables used to measure social capital. Social capital itself cannot be separated from the context of sustainable development because good social capital will facilitate the occurrence of activities between individuals, households and efficient groups through the availability of information for the development of the population at low, easy, and beneficial to the community. Social capital plays a role in increasing regional growth and development through increasing the provision of public access to the availability of capital, education, health and security. Social capital and social interaction will build. social integration which plays an important role in efforts to achieve economic prosperity in the community and efforts to achieve sustainable development.

4. Conclusion

In general, it can be said that Sumbermujur Village community is a relatively happy community. From the results of the study, there are two types of community groups in Sumbermujur Village based on SNA calculations and happiness index calculations. The two groups of people are:

1. Community groups with low Rate of Participation values and high density values that have a happiness index are classified as "happy".
2. Community groups with low Rate of Participation values and moderate density values that have happiness indexes classified as "happy".

5. References

[1] Ari, Ismu R D. 2013. *Social Network of Memberships in Community Groups*

[2] Suandi. 2014. *Hubungan Modal Sosial dengan Kesejahteraan Masyarakat*. Jurnal Komunitas (Research & Learning in Sociology and Anthropology Volume 6)

[3] Rahayu, T. P. 2016. *Determinan Kebahagiaan di Indonesia*. Jurnal Ekonomi dan Bisnis Volume XIX No. 1 (ISSN 1979-6471), 149-170.

Environmental Scanning: In Creating Strategic Planning for the Education of Persons With Different Ability

R D Mustika[1] and K P Rahmayanti[2]

[1] Department of Public Administration, Faculty of Administrative Science, Indonesia University, [2] Department of Public Administration, Faculty of Administrative Science, Indonesia University

*Email: [1]rizkydefiramustika@gmail.com, [2]krisnarahmayanti@ui.ac.id

Abstract. This study aims to do environmental scanning in order to identify factors that influence the education policy for difable group. The current condition indicates that Indonesia's Human Development Index (HDI) towards 2017 is ranked in 116th with a final score of 0.694 which is still relatively low. As an effort to improve HDI, Indonesia conducts human development through education as basic rights of every human being and as one of the public service products that must be obtained by the community including all Indonesian people, particularly people with different abilities and children with special needs (ABK). According to the Ministry of Education and Culture (2018) out of as many as 1.5 million people with different ability, only 300,000 crews who have become students in 2017. Thus, this condition is disadvantageous for 1.2 million ABK who cannot get proper education in Indonesia. The research utilizes the qualitative approach and data collection techniques through secondary data analysis. This study found out that there are several factors which impact the ability of all levels of government to build strategic planning in the education sector for difable group such as limitations and unequal commitments from local and national governments, limitations in implementing regulations, unequal opportunity to obtain sufficient livelihoods that affect the income they earned, exacerbated by unequal perspective regarding the need for quality education for people with different abilities, by which the people with different abilities has not been prioritized for education. In addition, the use of assistive technology in the field of education for persons with different abilities is very minimal in Indonesia, as well as geographical factors as the main determinant factors that influence the issue of education for persons with different abilities

Keywords: Basic Public Service, Inclusive Education, People with Different Abilities, Children with Special Needs

1. Introduction

Development in the educational field in Indonesia has been one of the top priorities, especially in recent years due to its being a fundamental element in the human aspects for the development. In 2017, the Human Development Index (HDI) in Indonesia was still at 0,694 categorized as the medium level, despite the fact that Indonesia is still ranked at 116 from about 180 countries [1]. The main legal framework for the educational field in Indonesia is regulated in the Law no 20 the year 2003 stating about the national education system. Education becomes basic rights of every human being and as one of the public service products that must be obtained by the community. Therefore, both fair participation and qualified education must be the main indicator of the implementation of SDGs in the education sector [2]. Referring to these arguments, education in Indonesia should be implemented inclusively by involving every element of the citizen without distinction (including the marginalized and those having special needs).

On one of the EFA's Global monitoring report entitled "Reaching the Marginalized", those who have special needs or people with different abilities/difable are recognized as one of the most implicit factors with a big impact in the marginalization of education [3]. Marginalization means an imbalance in the acquisition of opportunities in the economic, social and educational fields by a group of people [4]. In the context of this study, difable groups often do not have equal opportunities in the field of education, such as access to education and quality of education because difable groups have not become a priority group by the government. This condition is proven through statistical data which explains that as many as 6 million Indonesians who are disabled are 1.5 million people included as school-age children with only 300,000 children who have become students in 2017 [5]. In dealing with this issue, inclusive education can be one of the solutions applied in Indonesia for the inequality that occurs for all this time towards the people with different abilities.

Through inclusive education, the school can have an active role in creating and establishing an inclusive society. Inclusive education emphasizes the perspective of acceptance of the diversity that exists and embraces it as an important value of civil society whether in school, professional, cultural or in the community [6]. Inclusive education is not only the answer to problems in the education sector, but the impact will also cause ripple-effects to other sectors as well as social societies, economics, culture, and politics. However, in practice, inclusive education is still not optimal; this is proven by the imbalances that occur in the implementation of inclusive education that still does not meet the needs of Indonesian difable groups, especially children with different abilities. As the statistics revealed by the [5], there are still 1.2 million ABK/ children with different abilities who have not been able to get an education in Indonesia, even though education has been implemented in 23 provinces and 102 districts/cities in Indonesia with a total of 31,724 inclusive schools with 128,192 students [5]. Based on that condition, an evaluation must be carried through efforts to do environmental scanning of inclusive education. Therefore, this research question of this study is "what factors that influence the education policy for difable group?". The purpose of this study is to do environmental scanning in order to identify factors that influence the education policy for difable group.

Environmental scanning (ES) is a process of identifying information about the external environment of organization. ES can be conducted through a PEST analysis, which is an acronym for Political, Economic, Social and Technology analysis as one of the tools for helping to illustrate the framework of macro-environmental factors in strategic management [7]. The paradigm of PEST analysis emphasizes the ability to adapt to the change of the macro environment to develop a strategy that can balance between the capabilities that already possessed and the existing conditions of the macro environment. PEST analysis is the media that becomes the evaluation stage of the strategic management process [8]. The development of the PEST analysis concept, which is PESTEL analysis, incorporates Environmental and Legal elements into the analysis of the macro environment. The focus of PESTEL analysis is to identify trends that occur in the macro environment. Here are some following explanations about the factors in the PESTEL analysis concept [9]:

1. Political factors can be categorized into three levels that are: 1. Supranational which covers the supervision of political aspects of another country. These aspects are necessary for the development of an appropriate strategy in adapting to the change in the political environment that might happen in the future time; 2. National, which includes the process of evaluating of a political condition inside the country itself that covers several policies, programs, activities and regulations that applied; and 3. Subnational that is supervision of political aspects of a districts or special region [8]. Identification of political factors includes government policies, resource allocation policies, needs, desires of stakeholders, support from interest groups and the influence or pressure of other

actors on an international scale, changes in power and relationships between actors or groups, and the direction of future political movements.

2. Economic factors can affect the multiple elements of weather in the region, national, or even global scope, including the economic situation of a societal group such as jobs, taxes, mobility, etc. Then, the economic situations and prospects of certain sectors that are discussed include conditions of infrastructure, availability of resources, the direction of economic trends, intervention from the government are also included in this factor.
3. Social factors cover the existing condition and situation of a society such as, general public health, education level, access to public services, and the public perception of a phenomenon that occurs. Besides, culture also takes an active role in this social factor, so is the historical aspect of a society or a nation, social change dynamics, knowledge and the attitude towards a certain issue, and the credibility of a mass and informational media such as newspaper.
4. Technological factor, which can map the change in the science and technological fields to understand the effect that may occur in a certain sector [10]. This includes access to technological marketing facilities, trends in the use of existing technologies, a potential for innovation, issues related to technology itself such as issues of patent ownership.
5. Environment factors, contextually relate to the issues of the natural environment. Such issues include the regulation of the natural environment, the effect of human activities to the environmental trends or forecasted futures around the developments in the realm of the environment, including geographical conditions.
6. Legal factors, emphasizes on the existing or future regulations related to the human rights, the impact of existing regulations on other relevant factors and on society, international agreements, regulations being made or the possibility of changes in regulations and ethical issues in society.

An inclusive education policy urges suitable governance in the policy-making process. This paper aims to analyze current factors that may hinder the existence of collaborative governance of the education policy particularly for people with different ability.

2. Methods

The methodology in this research utilizes the qualitative approach allowing the researchers to explore the meaning of a social phenomenon and make interpretations of the data collected [11]. Based on its purpose, this research is considered as a descriptive study aiming to provide an in-depth description of a situation, activity, social condition or relationship with contextual understanding [12]. In this context, the research focuses on analyzing the factors that influence the strategic planning approach in the education sector for people with different abiliy in Indonesia. Data collection techniques are conducted through secondary data by literature studies including books, newspapers, scientific journals, previous research results, various documents of agencies and institutions such as United Nations, UNICEF, UKAid Department for International Development, and Ministry of Education and Culture relevant to the PESTEL analysis theory. Data analysis is conducted by applying descriptive analysis to exploit and describe the data that has been collected from the mentioned sources regarding a social phenomenon which is education for Persons with Different Ability in Indonesia. Hence, this study is done by breaking down each factor of the PESTEL analysis and looking at the reality of education for Persons with Different Ability in Indonesia regarding to each and every factor determined in the theory to then describe the information comprehensively.

3. Results and Discussions

Education for those who have different abilities had already become a national agenda in many countries, even globally. In Indonesia, inclusive education is the national commitment to the international agreement declared in The Salamanca Statement and Framework for Action on Special Needs Education by the year 1994 carried out by UNESCO. Inclusive education is one of three approaches on educating children with special needs [3], the differences between the three approaches are as follows: 1) Segregation, is education for children with special needs (*Anak Berkebutuhan Khusus* in Bahasa /ABK) which is carried out through the provision of special schools such as "*Sekolah Luar Biasa*" (SLB) or home -schooling; 2) Integrated education, is an education process carried out in public schools, but there is a class or special unit for ABK; 3) Inclusive education, is an ABK education effort carried out effectively in public schools whose systems have been adapted to meet the needs of all Persons with Different Ability (PWDs) and Persons without Different Ability students. The preferred system for ABK includes providing facilities such as assistive technology that is needed in sign language and other methods to be ascertained that both students have the same educational accessibility [13]. However, in fact, all-inclusive education for Persons with Different Ability in Indonesia has not obtained the same opportunity to obtain an education. The following is the result of the analysis of factors influencing strategic planning approach in education for Persons with Different Ability based on the theory used in this research, such as:

1. *Political factors*. The political factors that can affect the education sector in Indonesia are government policies, support or pressure from certain societal or interest groups, and existence of actors who involved in the development of educational sector for people with different abilities in Indonesia. In reality, several obstacles are portrayed in the field of inclusive education, including the distribution of the Indonesia Smart Card (KIP) as a tool from the Smart Indonesia Program (PIP) that has not been well targeted and has not yet been distributed specifically to community groups with middle to lower economic conditions and Children with Special Needs (ABK). However, the two groups are the highest school dropout group in Indonesia. ABK also experiences obstacles to accessing education because of the uneven and minimal existence of inclusive schools compared to regular schools in Indonesia [14]. In addition, the constraints are also experienced in the provision of human resources which are the limited number of Special Facilitator Teachers (GPK) in Indonesia compared to the number of ABK itself. At the level of junior high school education (SMP), in Indonesia, there are as many as 24,985 ABK with only 1,101 GPK [15]. This condition triggers the majority of ABK to not get GPK for their education appropriately. Then the number of schools available is fairly minimal compared to the number of ABK in Indonesia. In implementing special education, teachers who can guide ABK are obviously needed, considering several special competencies that must be owned by a GPK in guiding children with special needs.
2. *Economic Factors*. The economic factor becomes one of the most important aspects and has a huge impact on education or any field. From the government point of view, the education sector has been one of the priority in the national budget of Indonesia (APBN) which the portion of the allocation is one the fifth or 20% of all the budget owned by the Indonesian government, and carried out since 2009. In 2017, President Joko Widodo gave instructions to the Indonesian government to form sovereign wealth fund (*dana abadi*) to ensure the effectiveness and rightfulness of the budget allocation [16]. Contradictory to the big budget, the development of Indonesian education is still not progressive enough to help improve the quality of human resources in Indonesia. Poverty becomes the cause of the consequences as experienced by Persons with Different Ability [3]. Based on these arguments, the conditions experienced by Persons with Different are often associated with long-term poverty because Children with Special Needs (ABK) often do not have the opportunity to obtain sufficient education that affecting their livelihoods such as low income caused by the lack of skills and the job market for people with these conditions is practically non-available. ABK with an adequate educational background tend to get a job easily as they have the ability and knowledge that can support when doing their jobs. Persons with Different Ability in Indonesia are still considered to have limitations. Thus, this group of people with disabilities often considered belongs to the lower-class society. The economic conditions possessed by Indonesian difable groups can be seen through the 2013 Basic Health Research (*Riskesdas*) statistical data which states that the gap that occurs is illustrated by the Difable Group Ownership Index in the lowest group at 15,2 percent and only in the top 8,3 percent [17]. Therefore, the data implies that the prevalence of difable is higher on the lower ownership index.
3. *Social factors*. Demographic conditions in Indonesia continue to change. Regarding the difable group, according to the results of the 2018 Basic Health Research (*Riskesdas*), the proportions of

persons with different abilities based on their age distribution are as follows: 1) Ages 5-17 is 3.3 percent; 2) Age 18-59 years is 57.6 percent; 3) Age ≥60 is 2.6 percent [18]. It can be seen from these data which the majority of difable groups are at 18- 59 years of age who are classified as productive age. Under these conditions, there are a variety of perceptions regarding individuals with different abilities among Indonesian society. Indonesian people tend to judge that Persons with Different Ability is a burden for a family, the condition of the disabled is also a burden for individuals who have it. If viewed from the education acquisition figures, the average education obtained by Persons without Different Ability is 6.5 years while for Persons with Different Ability (moderate level) is only 4.4 years while for those with a severe condition is only 2.8 years [19]. This inequality is due to perceptions that have developed in the community where Persons with Different Ability is considered for not having the opportunity to have a job with high income. Besides, access to education for people with disabilities is still difficult encouraging families from difable groups to be reluctant to send their family members to school [19]. Thus, there is a duty to change perceptions held by the community towards the difable group. This is not easy task due to the stigma about difable groups having various limitations since the past attached to almost every single individual in the community.

4. *Technological factors*. In implementing inclusive education, various support systems such as technology that are used as learning media should also be adapted to the needs of all students. In optimizing the teaching and learning system for Persons with Different Ability, there is a technology called 'assistive technology' as an adaptive and rehabilitative tool for Persons with Different Ability (including technology to support the difficulties experienced by groups of people with disabilities) [20]. The use of assistive technology in inclusive education emphasizes is aimed to help the learning process; thus, ABK can get the same knowledge as non-ABK. In Indonesia alone, the use of appropriate facilities and infrastructure in the implementation of education in Indonesia has never been on the side of people with disabilities. The Chairperson of the Indonesian Disability Association (PPDI) argues that access to facilities and infrastructure facilities to education is still minimal, especially in the municipalities [21]. The technology used is not yet innovative enough, only maintaining supporting facilities that have been previously owned.
5. *Environmental factors*. The dominating environmental factors in relation to the issues discussed in this paper are geographical conditions. Indonesia is an archipelagic country where the infrastructure development has not yet been completed in all corners of the region; therefore, there are still many locations in this country that have not been reached by development itself, including in the education sector. People with different abilities in many rural communities do not have access to inclusive schools because their existence is still practically non-existence in various regions [21]. This access can be caused by various things, such as the absence of transportation and access to the location, lack of road infrastructure to schools, geographical conditions that include the land, the shape of the region and geographical factors that do not support the government to hold inclusive schools in various parts of Indonesia.
6. *Legal factor*. Equitable education becomes part of the mandate of the constitution of 1945 in the Indonesian Constitution, especially in Article 28C, paragraph (1) stating that "*everyone has the right to develop themselves through fulfilling their basic needs, the right to receive education and to benefit from science and technology, art and culture in order to improve their quality of life and for the welfare of humanity*". From these regulations, the even distribution of education must reach the entire community regardless of the difference in the physical and mental state of a person, economic, social or geographical conditions. Continuing the basic regulation, in order to support equal opportunity to have education, the Indonesian government issued Ministerial Regulation No. 70 of 2009 concerning Inclusive Education for Persons with Disabilities which was later supported by Law Number 8 of 2016 about Persons with Disabilities. The regulation is expected to support the implementation of inclusive education in Indonesia in order to improve the quality and accessibility of the entire community towards education so that Indonesia has quality human resources. Even so, the basic regulations in the process of implementation are still met with various

IOP Conf. Series: Earth and Environmental Science **328** (2019) 012051 doi:10.1088/1755-1315/328/1/012051

limitations. The limitation concerning the regulation still does not reach all regions in Indonesia. The central government has not socialized inclusive education comprehensively, and there has not been a unified understanding about inclusive education itself between the central and local government. Therefore, there is a risk that local governments do not understand or have different perceptions about inclusive education leading to the lack of regional regulation, regulation of governor, regulations of regents / mayors [22]. This also encourages regional governments to neglect inclusive education as one of their main agendas or priorities. Consequently, local government budget/APBD do not allocate such program.

4. Conclusion

The result of this study based on the theory of PESTEL Analysis concludes that several factors influence the government in the preparation of strategic planning in the education sector for the people with different abilities. In the political and legal fields, there are limitations and unequal commitments from local and national governments, specifically in the political sphere, tools from the Smart Indonesia Program (PIP), which is the Smart Indonesia Card (KIP has not been right on target and has not yet been distributed specifically to community groups with middle to lower economic conditions and Children with Special Needs/ABK) .In the economic field, it was found that Children with Special Needs (ABK) often did not have the opportunity to obtain sufficient livelihoods; thus, the income they earned was also low. Furthermore, in the social sector, it was found that there was an unequal perspective regarding the need for quality education for people with different abilities. In general society, people with different abilities has not been prioritized for education. Then, in the field of technology, the use of assistive technology in the field of education for persons with different abilities is very minimal in Indonesia. In the field of environment, geographical factors become the main determinant factors that influence the issue of education for persons with different abilities.

References

[1] United Nations Development Proramme, "Human Development Indicators," 15 07 2018. [Online]. Available: http://hdr.undp.org/en/countries/profiles/IDN#. [Accessed 24 02 2019].

[2] Education International, "Equitable Quality Education: a Precondition for Sustainable Development," Retrieved: 27 1 2019. [Online]. Available: https://sustainabledevelopment.un.org/content/documents/3721education2.pdf

[3] Department for International Development UKAid, "Educating Children with Disabilities," Department for International Development UKAid, United Kingdom, 2011.

[4] Alfitri, Pembangunan Masyarakat Marginal, Palembang: UNSRI Press, 2016.

[5] Ministry of Education and Culture , "Tantangan dan Upaya Meningkatkan Pendidikan Inklusi," 24 11 2018. [Online]. Available: http://pklk.kemdikbud.go.id/portal-utama/read-news/Tantangan-dan-Upaya-Meningkatkan-Pendidikan-Inklusi. [Accessed 26 01 2019].

[6] Directorate General for Development Cooperation, "Inclusive Education for Persons With Disabilities and Development Cooperation," Directorate General for Development Cooperation, Rome, 2015.

[7] D. A. Gupta, " Environmental and pest analysis: An approach to external business environment," *Merit Research Journal of Art, Social Science and Humanities,* vol. 1, no. 2, pp. 013-017, 2013.

[8] T. S.-B. a. D. Galea, "PEST Analysis," in *Wiley Encyclopedia of Management*, Chichester, John Wiley & Sons, Ltd, 2014, p. _.

[9] UNICEF, "SWOT and PESTEL: Understanding Your External and Internal Context For Better Planning and Decision-making," 7 2 2019. [Online]. Available: https://www.unicef.org/knowledgeexchange/files/SWOT_and_PESTEL_production.pdf.

[10] J. Tribe, "Chapter 4: The External Environment: PEST," in *Strategy for Tourism*, Oxford, Goodfellow Publishers, 2008, p. 69.

[11] J. W. Creswell, Research Design: Qualitative, Quantitative, and Mixed Methods Approaches, United States of America: SAGE Publications, Inc, 2014.
[12] W. L. Neuman, Basics of Social Research: Qualitative & Quantitative Approaches, Harlow: Pearson Education Limited, 2014.
[13] S. D. Kamga, "Inclusion of learners with severe intellectual disabilities in basic education under a transformative constitution: a critical analysis," *The Comparative and International Law Journal of Southern Africa,* vol. 49, no. 1, pp. 24-52, March 2016.
[14] U. N. Fadhilah, "Ini Tujuh Masalah Pendidikan di Indonesia Menurut JPPI," 03 05 2017. [Online]. Available: https://republika.co.id/berita/pendidikan/eduaction/17/05/03/opchjr354-ini-tujuh-masalah-pendidikan-di-indonesia-menurut-jppi. [Accessed 24 01 2019].
[15] Ministry of Education and Culture, "Gambaran Sekolah Inklusif di Indonesia: Tinjauan Sekolah Menengah Pertama," Kementerian Pendidikan dan Kebudayaan, Jakarta, 2016.
[16] B. C. A. Nursyifani, "FOKUS APBN 2018: Dari Infrastruktur Prioritas Hingga Pengentasan Kemiskinan," 05 04 2017. [Online]. Available: https://ekonomi.bisnis.com/read/20170405/9/642786/fokus-apbn-2018-dari- infrastruktur-prioritas-hingga-pengentasan-kemiskinan.
[17] Ministry of Health Republic of Indonesia, "Buletin Jendela Data & Informasi Kesehatan, Semester 2," Kementerian Kesehatan RI, Jakarta, 2014.
[18] Bureau of Communication and Community Services, Ministry of Health of the Republic of Indonesia, "Depkes.go.id," 2 11 2018. [Online]. Available: http://www.depkes.go.id/article/view/18110200003/potret-sehat-indonesia-dari-riskesdas-2018.html.
[19] L. C. Suarez and D. Contreras, "Disability in Indonesia: What can we learn from the data?," The Australia Indonesia Partnership for Economic Governance, Melbourne, 2017.
[20] F. K. Ahmad, "Use of Assistive Technology in Inclusive Education: Making Room for Diverse Learning Needs," *Transcience,* vol. 6, no. 2, pp. 62-77, 2015.
[21] R. L. Sulistyawati, "Sarana Prasarana Pendidikan Penyandang Disabilitas Minim," 3 12 2018. [Online]. Available: https://www.republika.co.id/berita/nasional/umum/18/12/03/pj5xxk384-sarana-prasarana-pendidikan-penyandang-disabilitas-minim.
[22] Ministry of Education and Culture, "Tantangan dan Upaya Meningkatkan Pendidikan Inklusi," 24 11 2018. [Online]. Available: http://pklk.kemdikbud.go.id/portal-utama/read-news/Tantangan-dan-Upaya-Meningkatkan-Pendidikan-Inklusi. [Accessed 26 01 2019]

IOP Conf. Series: Earth and Environmental Science **328** (2019) 012052 doi:10.1088/1755-1315/328/1/012052

Local Wisdom as a Foundation for Regional Policy in Water Conservation (Case Study: Pondok Pucung Betawi Community)

R E Oppusunggu[1, 2] and C Lusiana[1, 2] and M Katoppo[1, 3]

[1] Design as Generator, South Tangerang, Indonesia
[2] Lecturer, School of Design, Pelita Harapan University, Tangsel, Indonesia
[3] Acting Dean, School of Design, Pelita Harapan University, Tangsel, Indonesia

*Email: 200rumahbesi@gmail.com

1. Introduction

1.1 Research Background

In 30 years, since the 1980s, Jakarta population has been growing into 10 million, while it sums up in total of 30 million people living in the area of *Jabodetabek* (Jakarta-Bogor-Depok-Tangerang-Bekasi cities). This massive changing becomes a driven power that also turns Pondok Pucung (PP), an area at Southern Tangerang turning from rural to urban district. Once, this vast land was inhabited by local people, Betawi indigenous community, who lived in harmonious ways with the nature as farmers. Now PP, in a mere three decades, is changing mostly into mega modern real estates and the remaining PP has become "islands of *kampongs*" surrounded and divided by these real estates. Thus, consequently the area has been facing challenges of losing its green open space and clean water availability. To the worse, PP also has to deal with cultural changes especially of losing local knowledge and wisdom on how to deal with the sustainability of their environment including on how to conserve water.

1.2 Research Statement

This paper aims to record the local wisdom of PP Betawi community in conserving water before massively changed in the span of only 30 years. The findings and outcomes especially the acknowledgement of the local wisdom of the previous indigenous generations will be a fundamental recommendation for culture-based regional planning on water conservation.

1.3 Theory Discussion

According to Nirwono and Ismaun (2011), a green open space ensures the balance of city ecology including hydrology and clean water availability, micro climate and clean air, and the aesthetic of the city [1]. Similarly, in his UN-HABITAT Lecture Award Series, Friedmann (2007) stressed that there are seven tangible assets for wealthy cities, including: (1) human assets: people and the quality of their lives and livelihood, (2) organized civil society: multiple self-organizing activities of local citizens, (3) the region's heritage of its built environment and the distinctiveness and vibrancy of its cultural life, (4) intellectual and creative assets, (5) natural assets: farm, watersheds, lakeside, etc. (6) environmental assets: air, water, land, (7) the quality of (urban) infrastructure [2].

In different perspective but in parallel notes, Oliver (2000) argues that vernacular traditions, cultural thinking and doings, have evolved to meet socio-cultural needs helping local people to live in harmony not only with their fundamental aspects of community life but also with their nature [3]. Melenchon (2012) in more specific notes on water points out that "if we want to solve water problems, then we should see water as a cycle", revealing that water is not only as ground water that we pump up and drink, but water as it comes down in raindrops, runs on the ground and rivers, absorbs into the grounds, and evaporates back into the air. A cycle portrays the origin of clean water to distribute, use, and conserve [4].

Julian Steward, an American anthropologist, (in Pillai, 2013) concludes that cultural change is cultural ecology. This process is a creative adaptation to circumstances or physical/natural environment and resources such as terrain, materials, weather, and is a creative adjustment to other groups. This cultural change is reflected through the use of technology, social grouping and economic organization. Moreover, Pillai (2013) states by studying cultural ecology, the interrelationship between people, place and use, reveals how the people have been subjected to or have reacted to certain historical and environmental circumstances for their sustainability. Thus, one way to study this is through cultural mapping [5].

2. Methods

Pillai (2013) explains that cultural mapping is a way to identify and record the occurrence of cultural tangible and intangible elements of a site. This involves synthesizing and interpreting data to discover significance, pattern or character of a culture in response to people, place and use. Cultural mapping consists of framing (preparations to conduct a mapping), mapping (collecting and recording data on the site), and evaluation [5].

In framing stage, the researchers formulate list of questions distributed evenly into 4 focusing on: profile, people, place and use (presented in Table 1). The questions are formulated to compare the conditions of today and previous generations. To recollect data of previous generations, the researchers interviewed the informants whom are leaders and elders at the age of 40s and above. To recollect today data, the existing conditions were observed.

In mapping stage, there are four key informants: (1) Mr. Syarifudin (A), male, age 46-year-old, lives at Karya Usaha Street – West of PP, works at PP District Office. His grandfather was a farmer and his father was a builder. His grandfather and father had paddy fields and *kebons* –backyards that well grown by fruitful and big trees. Now his house is wall-to-wall to his siblings' houses and their rented houses. (2) Mr. Rosib Sulaeman (R), male, age 53-year-old, lives at Kampung Rawa Street – South of PP, works at PP District Office. His grandfather and father were farmers. He inherited a piece of land from his father that is his home now. His grandfather and father had paddy field, *kebon* and *bulak* –a field to growth vegetable and other starch plants. Now he lived in quite crowded neighborhood. (3) Mr. Moerdan (M), male age 63-year-old, lives at Mushola Street – West of PP, works as chicken farmer. His grandfather was district officer and his father worked in a shop selling log wood for cooking in Tanah Abang, Jakarta. He is the elder in the community and once a neighborhood leader. He lives on his inherited piece of land from his mother. His house is close to his daughter's, his sibling families', his rented houses and the outsider settlers' houses, who had bought some lands from his family. (4) Mrs. Hj. Sati (S), female, age 66-year-old, lives at Mushola Street – North West of PP, works as housewife. Her grandfather and father were farmers. Before married, she lived near to one of the rivers passing through PP. Now she lives surrounded by her rented houses and her children's houses and their gardens.

In evaluation stage, the researchers provide the assessment of cultural elements, site, character, significance of assets and resources on the site.

3. Results and Discussions

Pondok Pucung area, in Joniansyah (2010), is a 400 acres district, situated in south west of Jakarta, and a part of Pondok Aren sub district, in Southern Tangerang regency of Banten province. Most of

IOP Conf. Series: Earth and Environmental Science **328** (2019) 012052 doi:10.1088/1755-1315/328/1/012052

PP area is the house for 25,000 inhabitants with Betawi people as the remaining local people [6]. Joniansyah (2010), and Jurnal Tangsel (2014) pointed out the comparison to other areas of Southern Tangerang regency, in which PP is losing 20% of its rainwater conservation area [6], [7]. Djarot (2014) also argues that PP suffers flood due to the turning of rainwater conservation areas into housings [8].

(a) (b) (c)

Figure 1. (a) The Area of Pondok Pucung, (b) The water catchment area,(c) The flooded area [8]

Mapping Stage: People, Place, Use

In the mapping stage, the data cover the 4 previous generations of Betawi people who chose to expand their families (people), where they built the next dwellings (place), and the rituals for the next generation houses concerning weather, water, materials and location (use) (Table 1).

Table 1. Mapping of People, Place and Use of Pondok Pucung Area in Its Relations to Water

A	PEOPLE	
(1)	*How many of your family generations have been living in PP?*	As far as all the interviewees remember, their families have been living in PP from 3 previous generations. Furthermore, S remembers that her both grand families were from PP. R adds the data that everyone who lived in PP were relatives and families.
(2)	*Was there any habit to welcome rainy season in your family / neighborhood?* *How about now?*	According to A, comparing to today's program, there was no such program like saving rainwater in the old days, when most of the lands were well grown with paddy, vegetable and big fruit trees. There were enough rainwater infiltration areas. The water kept on coming out from the wells, as to why the people had never lacked of water. R says, when rain came, people worked in dry land called *bulaks* served as fields to grow vegetables and other starch plants. S adds, people also started sewing paddy seeds (*nandu*) in paddy field and swamps, keeping the fields and swamps clean from grass (*matun*), and then harvested brown paddy (*motong padi*) in 3 months. People also went to the paddy fields, the rivers and the *kalenans* – the opened dug channels on the ground for runoffs to end into the paddy fields- to catch fish. Now the environment is different, consequently they do not do their old habits and they do not do anything special.
(3)	*Was there any habit to prepare the coming of draught season in your family / neighborhood?* *How about now?*	The dwelling areas of PP had always enough water. Nevertheless, if the wells dried, says A, people would call well diggers. According to R, the crops were bad in draught seasons so people made the famous PP's *kubils* -wooden shoes heels- and also *pandanus* plaited mats and hats. A added that some went to Jakarta to work. No special preparation that they do. But if the water pump cannot draw any water, people will call plumbing service to dig into 15-20 meters deep.
(4)	*Was or is there any folk story about rainy season,*	At the end of draught seasons or to end the long draught seasons, people would put some cooked rice on the streets, says M, and in the corners of the *bulaks,* says R,

IOP Conf. Series: Earth and Environmental Science **328** (2019) 012052 doi:10.1088/1755-1315/328/1/012052

	dry season or water?	and prayed for rain to come.
B.	**PLACE**	
(5)	*What and where were your family's clean water resources, grey and black water treatments?*	According R, back then people threw a stone from their house, thus the location of the stone landing would become the place to set their wells. A says, generally the 8-meter-deep wells were situated at the back of the house, at the west or the south. Some families put their wells in front of the houses so they could share the clean water with the surrounding neighbors. S adds, the wells were usually situated close to the kitchen doors. A says, some people also went to the *galurs* -a pond was located near to paddy field and got its water from the higher paddy fields- and the *empangs* in paddy field to wash their clothes. S adds, for a few months after delivering baby, she went to the river for washing. A says that not far from the well, there was an *empang* –an opened grey water disposal pond with catfish living in it. Since back then, there was no channel for clean or disposal water. M says that around 5 meters from the well, there was a *cubluk* –a roofless toilet on a pit in the ground. R adds that a *cubluk* could be as far as 30 meters away from the well.
	How about now?	Now each house has each septic tank with infiltration well for its grey and black disposal water. The septic tank is around 5 meters away from clean water resource. Interviewees' neighborhoods have unmaintained and shallow sewers that mostly active only for rainwater runoffs. Meanwhile, the neighborhoods of housing estates dump their grey water to the sewer that some end into the rivers and big water disposal ponds made by developers.
(6)	*Then, how did your family decide new house locations related to clean water resources and grey and black water treatments?*	A says that a new house location for a new family was decided according to his/her land inheritance from the parents. The facade was accidental decided by parents so not to face or align to the other facades. S adds, her late husband arranged their children's houses surrounding their house. R adds, the previous generations set a new house for a new married son/daughter as far as 50 meter from parent's house. Then the new family would apply the same conditions for clean, grey and black water.
	How about now?	Their newly married son/daughter will live in his/her inherited piece of land and apply the nowadays water condition. If there is no more land to share, the new family will live with their parents. M adds, the parents can buy a piece of land in other place and apply the nowadays water condition.
(7)	*Were there any other clean water resources and where were they?*	M says, the clean water resource for domestic needs were only from the wells. Other than that, people got clean water from the *galurs,* says A, *empangs,* says R and rivers, says S.
(8)	*How is the condition of the old clean water resources now?*	M says, the wells were closed mostly because of safety consideration. Now they use electric pump to pull ground water. A says, since 1980, the *galurs* and the *empangs* were gone at the same time the developers started to build the housing estates. S says the rivers were wider with ground & grassy riverside for the rivers to overflow. A adds, adjacent to the rivers were paddy fields. Now the rivers are still there but the riversides are less and the paddy fields had turned into housing estates. Back then, all the water resources were clean. People drank water without boiled it; R used to drank water from the *empang* in paddy field when he got tired from shepherding his father's flocks. M used to have a terracotta jar & wooden coconut dipper for passerby quenched their thirst. S only boiled water for tea or coffee. Now the rivers still have some fish but the water is brown and even black muddy, contains garbage and produces bad odor. In heavy rain, the rivers overflow and paralyze the closest streets.
C.	**USE**	
(9)	*How to decide the locations for agriculture fields, plantations/gardens, farmings, cementaries, factories and housings built by developers?*	According to R, there were dry land and wet land with its own never ending spring or, adds A, with its runoffs from other *kampongs* (villages). R says, people set their paddy fields in wet lands, *bulaks* in dry lands, and, adds A, let fruitful & big trees grew in *kebons* that surrounding people houses. A says, the highest land were the dwelling areas and the lower lands, in order, were the *kebons,* the *bulaks,*and the paddy fields and the swamps as the lowest grounds. Now the dwelling areas remain but get crowded. The *kebons* are rare and small. The *bulaks* and paddy fields had turned into the housing estates and their infrastructures & facilities. In 1975, M says, a businessman bought a piece of *kebon* to start a chicken

		farming. The man also rented other higher lands and expanded his farming. Afterwards he financed some members of the community to start their own home-chicken-farming in any possible space in their lands. According to R, there was also small cow farming, belonged to the community as a part of government farming program. The farm was located close to the paddy fields. Now, M is the only chicken farmer survives. The rest has gone or got financial lost forcing them to sell their lands. The cow farming is also gone when the paddy fields were sold and the government program stopped. To date, the positions of cemeteries have been in the lands of *wakaf* – property donated for religious or community uses in Islam, which helps the community to zone together their family graves that previously scattered in each private land. In PP community cemeteries, there were sacred graves of each neighborhood religious elders. Since 1980 PP had changed into housing estates. According M, the chicken farming changed into housing estates since the businessman sold his land to the developer. A says, there were local mediators who worked for the developers that bought piece by piece of land from his neighbors. When the sold lands –mostly paddy fields and *kebons*, and some parts of his neighbors' houses that were in the developers' master plan- were accumulated together and big enough, the developer started the housing development. A says, the first housing estate (Pondok Pucung Indah 1) were rubber plantations which were higher than the second one (Pondok Pucung Indah 2) that was paddy fields. The developer had leveled both areas with cut and field. However, both are still lower than the dwelling areas. A adds that his father generation was not keen on farming anymore. Therefore initially they sold the paddy fields than the *kebons*. Moreover, the developer needed big open areas and those areas were ideal.
(10)	*What was the use of clean water?* *What is the use of clean water now?*	S says, she mostly used water for drinking and domestic needs. Meanwhile, A says that for livelihood, people went to the *galur* to catch fish; to soak logs of wood and bamboos for building materials; and to soak *pandanus* leaves for woven mat and hat materials. People also went to the *galur* to wash clothes. R adds that people use water from the *empang,* in paddy fields, for watering the *bulaks,* cow farming needs, and water for private flocks. M says, he used water from the well for his farming. M still uses clean water for his chicken farming. The rest of them use clean water for drinking and domestic needs.
(11)	*How did your family and neighbors take care of the clean water resources so the resources stayed clean and available?* *How do your family and neighbors do now?*	According S, people built a wall around the well to prevent soil or mud to contaminate the water. R adds that people would dig the mud out of the well bottom to ensure the clearness of the water. They do nothing special because the ground water in dwelling areas is still available and as clear as before. But R says, it is not as clean as before when the dwellings get denser and crowded and leave no ideal distance between a septic tank and infiltration well to a clean water resource. S understands if some of her neighbors are lack of water in draught seasons as the result of more and more people live in her neighborhood. According to R, people need to buy clean water especially the ones who live in the previous swamps or paddy fields.
(12)	*How did your family and neighbors manage the leftover, grey and black water? How about now? Did your family and neighbors reuse grey and black water after you purified the water?* *How about now?*	S says if the *empangs* were too full, people dug out the deposit to the brinks of the *empangs*. Moreover people always kept catfish in the *empangs.* A adds, there were always the *kalenans* to make sure the overflow of the *empangs* passed the *kebons* and ended into the paddy fields. If a *cubluks* were full, R says, people would cover them and make a new ones next to the old ones. Now people use septic tanks and infiltration wells. They do not purify and reuse grey or black water.

Evaluation Stage: People, Place, Use

IOP Conf. Series: Earth and Environmental Science **328** (2019) 012052 doi:10.1088/1755-1315/328/1/012052

The researchers evaluate the findings from cultural mapping which consists of assessing cultural characters and significances, and analysis to come into outcomes or conclusions. The discussions are divided into three main focuses of: People, Place and Use as presented in Table 2.

Table 2. Evaluation of People, Place and Use of Pondok Pucung Area in Its Relations to Water

A	**PEOPLE**	
(1)	*The people of PP once lived harmoniously and in accordance with their natural environment, as they only used what they needed.*	The indigenous people of PP have been living in clustered family lands until now. In the past when the families expanded because of marriage, the new ones would move and live 50 meters away from the established family settlements. The distance was a calculation of environmentally and socially impact considerations. They lived mostly as paddy field farmers until the 1980s and chicken farmers until the 1990s. At that time, they lived harmoniously with nature and only took whatever necessary, including in terms of water usage and distribution. The condition tremendously was altered by the coming of developers in the mid the 1990s.
(2)	*To PP, modernity changed the environment, thus changed the way of living and the culture of the people. The changes expenses are the disappearing of natural environment (physically) and local knowledge/ wisdom on how to live environmentally friendly.*	The community has been experiencing the changing types of work over the years. Whereas once they were farmers with knowledge associating to the nature, then they were builders and workers with crafting skills, and now they tend to be officers with managerial knowledge, as their lands are subsided and turning to be other properties and functions. However, the community has been enjoying access to urban modernity since 30 years ago through a considerable distance and railways to cities and towns such Jakarta. Nevertheless, when the modern development in the 1990s started, it excluded the local knowledge and wisdom of PP people regarding the natural environment.
B.	**PLACE & USE**	
(3)	*The people of PP once had a clear, sustainable and environmentally friendly system and knowledge on how to manage their settlements area.*	The people of PP was used to arrange their settlements as follows: - Currently, the dwelling areas have been occupying the highest topography where the community lives in a family clusters. The expansion of a new settlement for a new family took a distance of 50 m away. The 8-meter deep clean water resource (well) was in 5-meter radius from the house. The opened grey water disposal (*empang*) with it organic treatment (catfish living in it) was 5-meter further away from the clean water resource. The opened black water disposal (*cubluk*) was situated 30-meter away from the house. They never dumped their black water disposal to any body of water. The community made *kalenans* to pass overflow, especially in rainy seasons, from the *empangs* (close to houses) to the *kebons* and to end to the paddy fields. - The community grew organically (without any chemical) plants for their food in their areas except their dwelling areas. These areas were productive and they divided them into 2: The wet land (paddy field) having never ending spring or runoffs and dry land (*bulaks, swamps* and *kebons*) only having water in rainy seasons. The community only farmed in the paddy fields, swamps and the *bulaks*, meanwhile they - They also build water controlling systems, in a form of *galur* ponds serving as the meeting point of all surrounding runoff from other kampongs and *empang* ponds (with their never ending springs) in paddy fields serving as communal clean water resource. - The bodies of water (rivers, *empangs, galurs,* and *kalenans*) gave them fish without they cultivated the crops. - It is hard effort to maintain the *kebons.* Yet, all never failed to give them prolific crops. - They had riversides as the lowest ground separating their paddy fields from the rivers, thus functioning also as absorption areas when the rivers overflew in the rainy seasons.

(4)	*The people of PP once saw that water was embedded in their daily lives and as inseparable parts of the natural environments they lived harmoniously within.*	The people of PP saw water as follows: - As for the needs of clean water, in the past they never had a problem, because the clean ground water was never polluted and located in the vast highlands. The availability was relatively in insubstantial depth. However, today because of the environmental changes and how they were pushed to live in denser neighborhoods, the community starts having water scarcity in draught seasons and digging deeper for clean ground water resource. - As for the use of water, in the past they used it for many things besides the primary needs of daily activities (to drink, cook, washed, etc.), such as for ponds, irrigating paddy fields and others. Since the lands subsided and the farms were long gone, the local only used water for primary activities. - The quality of water in the past was also good. The ground water was fresh and clean, the water that ran into the ponds, paddy fields and even in the river was clean. They drank uncooked ground water and claimed that they were healthy. They even had a tradition of sharing water; whey put water in ceramics jar outside houses for strangers/passerby consumed it freely. Now the quality of ground water is still good, but scarce and coming out from deeper ground. - Because of the respect of nature in their settlements (always located in the highland surrounded by *kebons*) and water system -designated channels for runoffs (*kalenan)* from dwelling areas to rivers and lands for river overflows (riversides) and runoff junctions (*galur*)- back then the people of PP never experienced flood. But now, because of the rampant development and the loss of open and green space, some of the housing estates experience flood. - Because of the water management system and vast open spaces, the people of PP in the past never had a problem with polluted ground water resources. At the present, in the dense dwellings and the modern knowledge of grey and black water disposal, many houses cannot apply the standard requirement distance of 5-10 meters between clean water resource to septic tank and its infiltration well thus fail to provide adequate water absorption.
(5)	*Modernity separates the people of PP from the environments through modernized system on how they interact with its environment, emphasizing on its easiness and practicality (i.e.: concrete roads, sewer, septic tanks, jet pumps, etc.) not the wisdom that usually a product of long thoughts and experiences.*	

4. Conclusion

4.1 The Recommendation: Culture-based Regional Planning on Water Conservation

Upon referring at Friedmann's 7 assets (2007) [2], the previous generations of Pondok Pucung indigenous Betawi community in Southern Tangerang, Banten, West Java had at least 4 assets, which were: (1) human assets (people and the quality of their lives and livelihood), (2) organized civil society (multiple self-organizing activities of local citizens), (3) natural assets (farm, watersheds, lakeside), and (7) environmental assets (air, water, land). The irony is due to the development of urban housing and infrastructure in which all of these assets are dissipating. The previous generations not only lived in harmony with their fundamental aspects of community's life but also with their nature. Oliver (2000) states [3] and mentions that the water as a cycle of equal activities of distribution utilizes conservation. Melenchon (2012) also suggests the similar idea [3]. Consequently, all of this wisdom was disappearing in alarming rate due to the modern developments and exclusive practices of knowledge.

By such presented phenomenon, the researchers offer several recommendations regarding water management and conservation policy by:

1. **Physical developments** especially the expanding of human settlements should respect to the existing natural water conditions and places as to make sure the water cycle can happen naturally, where the cycle needs places for raindrops to permeate, plants to hold ground water, springs to

leap out, runoffs to be channeled without deluge, overflows pour out without flood. In addition to this, ndevelopments and expansions should respect the natural topography as it usually was zoned by previous indigenous generations accordingly to water availability and cleanness.

2. **Modernity and development** which are always underway. However, the process should not exclude the knowledge of indigenous people, that seldom was considered traditional and old fashioned, and replace it merely because it is has not yet been proven and recorded scientifically. There should be synergies between local wisdom and modern scientific approach in building up policies including in water management and conservation policy aiming in a socio environmentally fairness area [9].
3. **The development of lapse** causing water management and conservation problems in PP, which should be valuable inputs for new design policies and evolving the user/settler needs and intentions. This can catalyze into a revelation, to become more adaptable, autonomous and resilient [10].
4. **The community involvement,** in order to become resilient, according to Lee (2016) the community needs to be actively included and engaged in the development process in generating district policies. They also need to be involved in decision making, implementations and evaluations. In the case of PP and any other areas that have changed because of modern developments, the approach should be the people-centered policy reconstruction. The policy should neither a top-down nor a bottom-up approach, but rather providing the community with the means to become the authority of their own decisions, supporting them to make appropriate decisions, and building their capacity to carry on the necessary action and work in a long run [10].
5. **Ecological democracy promotion** as a base for generating resilient water management and conservation policies, actions should respond to understand natural processes and social relationships within the indigenous own locality and the larger environmental context. According to Hester (2006) this will cause us to creatively reassess individual needs, happiness and long term community goods in the places we inhabit [11].
6. **The three approaches** that Hester (2006) suggested in forming the policy of water management and conservation, through: [11]
   a. *Enabling Form:* to reform our approach and act as communities, where we know each of our neighbors, their needs and abilities in terms of water management and conservation, thus to facilitate co-working communities that share specialties, tools, knowledge and local wisdom to solve a problem of water management and conservation collaboratively. Enabling form will provide the sense of centeredness of place and how we as people interconnected to each other and to our landscape as well, thus instill the responsibility to others and the environment. This in turn will develop rootedness and a collective destiny that is tied to place and inspired a shared higher civic purpose.
   b. *Resilient Form:* to reform our cities to become ecologically resilient. Resilient cities derive from the particular character of the surrounding ecology, including hydrology, and not reliant on merely technological advancement.
   c. *Impelling Form:* to reform our cities to impel us by joy. Impelling form invites us to be our natural selves awakening our consciousness of our oneness with distinctiveness within the ecosystem, which results in a sense of identity with the place we live and produces multiple avenues for the stewardships for healthier and sustainable earth. In this sense, it will be a celebrating live, as an evident of actual past relationship in Pondok Pucung (PP) local Betawi community in Southern Tangerang, Banten, West Java with their environment, where water was considered as a full cycle and integrated elements of their lives.

References

[1] Nirwono J and Ismaun I 2011 *RTH 30%! Resolusi (Kota) Hijau* (Jakarta: Gramedia) p 35-41
[2] Friedmann J 2007 *The wealth of cities: Towards an asset-based development of newly urbanizing regions* (UN-HABITAT Lecture Award Series No.1, United Nations Human

IOP Conf. Series: Earth and Environmental Science **328** (2019) 012052 doi:10.1088/1755-1315/328/1/012052

Settlements Programme (UN-HABITAT), Nairobi) https://mkusumawijaya.wordpress.com/2009/06/03/kekayaan-kota-menuju-pembangunan-berdasarkan-aset-di-wilayah-wilayah-yang-baru-mengalami-urbanisasi1/

[3] Oliver P 2000 *Ethic and Vernacular Architecture* (Ethics and the Built Environment) ed Fox W (London: Routledge) p 115-126

[4] Melenchon J 2012 *Aturan Hijau untuk Eko-Sosialisme* (Bandung: Ultimus) p 64-70

[5] Pillai J 2013 *Cultural Mapping, A Guide to Understanding Place, Community and Continuity* (Petaling Jaya: Strategic Information and Research Development Center) p 6-10

[6] Joniansyah 2010 *Resapan Air di Tangerang Selatan Menyusut 20 Persen* (Tempo.co) https://metro.tempo.co/read/253744/resapan-air-di-tangerang-selatan-menyusut-20-persen

[7] Jurnal Tangsel 2014 *Mengenal 9 Situ di Tangerang Selatan* (Jurnal Tangsel) http://jurnaltangsel.blogspot.com/2014/09/mengenal-9-situ-di-tangerang-selatan.html?m=1

[8] Djarot B 2014 *Ulasan Usulan Awal Kawasan Pondok Pucung* https://www.slideshare.net/mobile/baskorodjarot/ulasan-usulan-awal-kawasan-pondok-pucung

[9] Astuti, W K 2017 *"Air dan Dialektika Kota Pariwisata: Merebut Hak atas Kota dan Air di Yogyakarta* (Kaliurang: Wacana Jurnal Transformasi Sosial No.35/tahun XIX/2017) p 91-116

[10] Lee, A J 2016 *Resilience by Design* (Switzerland: Springer Publishing) p 3, 24-30, 34

[11] Hester R T 2006 *Design for Ecological Democracy* (Cambridge, Massachusetts- London, England: The MIT Press) p 3, 8-9

IOP Conf. Series: Earth and Environmental Science **328** (2019) 012053 doi:10.1088/1755-1315/328/1/012053

Spatial Model of Coastal Community Vulnerability of Puger District to Tsunami Disaster Hazard

S Hariyani[1], A Susilo[2], E B Kurniawan[3], and F Shoimah[4]

[134]Departement of Urban and Regional Planning, Faculty of Engineering, Brawijaya University, Malang, Indonesia
[2]Departement of Physics, Faculty of Mathematics and Science, Brawijaya University, Malang, Indonesia

*Email : septianahariyani@ub.ac.id

Abstract. Puger coastal region is prone to tsunami and earthquake disasters. Puger District lies in the coastal zone which is directly adjacent to the Indian Ocean. Village areas included in the Coastal District of Puger are villages of Puger Kulon, Puger Wetan, Mojomulyo, and Mojosari. The potential for tsunami disaster on Puger Beach can affect the safety of the coastal communities of Puger. The location of community settlements is close to the beach, with relatively poor understanding of the danger of tsunami. Moreover, the negligence of an established disaster mitigation plans for tsunamis in Puger coastal region causes high vulnerability in case of tsunami; hence, casualties and property loss can affect the economy of the surrounding community. Based on these conditions, the study aims to assess the level of social, economic, physical and environmental vulnerabilities modelled spatially to determine the level of vulnerability of the community with the aid of Arc GIS 10.2 software. The analysis method employs scoring analysis and spatial analysis. The research output is a spatial model in the form of a community vulnerability map to the threat of a tsunami disaster.
Keywords: Spatial Model, Tsunami, Community Vulnerability, Puger

1. Introduction

Indonesia as an archipelago has a total of approximately 18,000 islands with coastlines reaching 92,000 km, making Indonesia one of the largest archipelagic countries in the world [1]. Indonesia's territory is largely a coastal area having a lot of potentials. The location of Indonesia is in the very active four-plate boundary zone leading to very high tectonic and volcanic activity; hence it can potentially cause tsunamis [2].

The Indo-Australian Plate collides with the Eurasian Plate off the coast of the Indian Ocean, Java Island and Nusa Tenggara whereas the Pacific Plate is located in the north of Irian and North Maluku. Around the location of the plate meeting, there is an accumulation of collision energy collected until a point until the earth's layer is no longer able to hold back a pile of energy which then dislodges a tectonic earthquake [3]. The quake is partly centered on the Indian Ocean floor, and some could trigger large ocean waves called tsunamis. The Ministry of Public Works stated that the south of Java is included in Region 4, which means that the southern part of Java Island has high potential to experience earthquakes triggering tsunamis because the point of the earthquake is in the Indian Ocean.

On the other hand, the protection of coastal areas related to mitigation efforts against tsunami wave hazards is very rare. This is due to the fact that the Indonesia has great width of coastline directly opposite to the plates and faults of the earth which has the potential to cause earthquakes and tsunamis.

IOP Conf. Series: Earth and Environmental Science **328** (2019) 012053 doi:10.1088/1755-1315/328/1/012053

The Indonesian coast which includes the tsunami-prone criteria stretches from West Indonesia, covering Aceh, Nias, Padang, Bengkulu, Lampung, then Pangandaran, Pacitan, Banyuwangi, Jember to East Nusa Tenggara.

Jember Regency is one of the South Coast coastal areas in East Java which is also prone to tsunamis. Puger coastal region is prone to tsunami and earthquake disasters. Puger District is included in the South Coast coastal zone which is directly adjacent to the Indian Ocean. Village areas included in the Coastal District of Puger are Puger Kulon, Puger Wetan, Mojomulyo, and Mojosari. The potential for a tsunami disaster on Puger Beach can threaten the safety of the coastal communities of Puger. The location of community settlements is close to the coast with poor knowledge and understanding of the threat of tsunami hazards. To make it worse, there is no tsunami disaster mitigation plan in the Puger coastal area, causing high vulnerability in the event of a tsunami disaster. It can greatly cost lives and property loss, affecting the economy of the surrounding community.

Based on these conditions, the study aims to assess the level of social, economic, physical and environmental vulnerabilities modelled spatially to determine the level of vulnerability of the community with the aid of Arc GIS 10.2 software.

2. Methods

The approach taken in this study is a spatial modelling approach which describes the real conditions in the field. The research method of data analysis is to support the process of data modelling and analysis. The analytical method used in the research is scoring analysis and spatial analysis. Variables, sub-variables, and parameters used in the study are taken from the previous research, literature review, and government regulations (as presented in Table 1).

Table 1. Research Variable

Variable	Sub Variables	Parameter	Source
Hazard	Estimation of Wave Height	✓ Water depth (bathymetry) ✓ Elevation (topography) ✓ Wave height ✓ Flow depth ✓ Inundation length (inundation distance) ✓ Height of run-up ✓ Tsunami hazard scenario	✓ [4] ✓ [5]
Vulnerability	Social Vulnerability	✓ Population density ✓ Population growth rate ✓ Female population ✓ Level of education	✓ [6] ✓ [7] ✓ [8]
	Environmental Vulnerability	✓ Environmental protection zone (protected forest, green open space, water catchment)	
	Physical Vulnerability	✓ Built area ✓ Building density ✓ Road damage	
	Economic Vulnerability	✓ Poverty level ✓ Area of productive land ✓ Vulnerable labor sector	

The analysis used is:

2.1. *Scoring Analysis*

Scoring analysis is conducted by giving a score on each variable and parameters used in the study. The first step is scoring the parameters of each sub-variable; afterwards, the second scoring will be completed on the sub-variables of each variable. Hence, the final results are obtained in the form of the level of vulnerability of the community to the tsunami disaster. The scoring analysis does not utilize standardized classification or theoretical review because the resulting level is not well distributed. In addition, there are several indicators which do not have standards. Therefore, a classification calculation is adjusted according to the indicators used in the study as well as the condition of the study area using Equation 1:

$$\text{Class Length} = \frac{(Max.Score - Min.Score)}{3} \quad \text{.........(1)}$$

2.2. Spatial Analysis

Spatial analysis is carried out by overlaying threat maps and vulnerability maps based on matrix presented in Table 2 [9]. The spatial analysis is carried out using Geographic Information System tools in the form of ArcGIS 10.2 software.

Table 2. Hazard & Vulnerability Maps Overlay Matrix

Hazard / Vulnerability	High	Moderate	Low
Low	Moderate	Low	Low
Moderate	High	Moderate	Low
High	High	High	Moderate

Source: Sudibyakto & Priatmodjo (2016)

3. Results and Discussions

3.1. Overview of Puger District

Puger District consists of 12 villages, Mojomulyo, Mojosari, Puger Kulon, Puger Wetan, Grenden, Mlokorejo, Kasiyan, East Kasiyan, Wonosari, Jambearum, Bagon, and Wringin Telu. The villages directly facing the Indian Ocean are Mojomulyo, Mojosari, Puger Kulon, and Puger Wetan Villages. Puger District consists of 37 hamlets, 225 community groups and 657 neighborhood groups. Puger District has an area of 73.53 km2, with land use classification in 2016 consisting of 4,039 ha of rice fields, 933.4 ha of moorlands, 67.2 ha of ponds, 1,895 ha of buildings and courtyards and 7,357 ha of other uses.

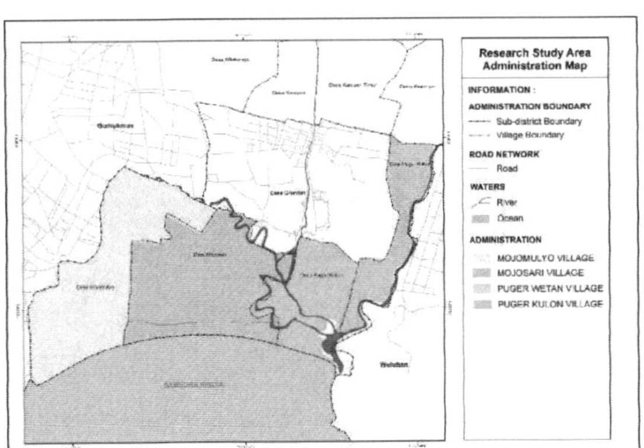

Figure 1. Administration Map of Research Study Area

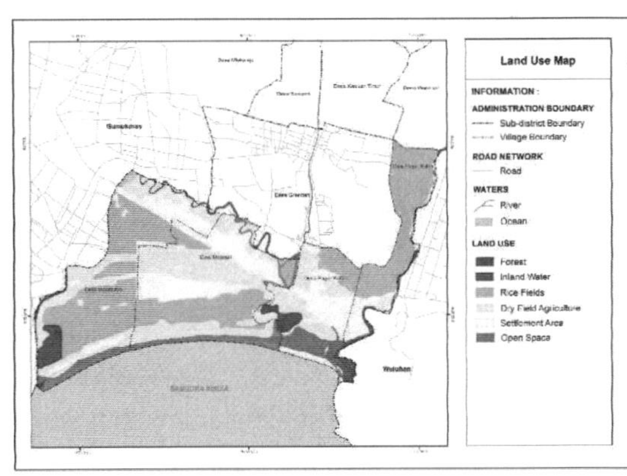

Figure 2. Map of Land Use

3.2. Hazard Analysis

A disaster threat is an event which can cause a disaster [10]. Tsunamis are ocean waves caused by undersea geological processes in the form of earthquakes, volcanic eruptions, landslides, and the fall of meteors in the sea [11]. Potential areas for tsunami hazard are categorized to high, moderate and low hazard potential. Tsunami hazard map provides information about the possibility of a tsunami and the tsunami hazard category classification by the tsunami wave height range (Table 3.) [12]. Table 4 presents an extensive exposure to rural areas by the tsunami hazard level.

Table 3. Classification of Tsunami Hazard Categories

Wave Height Range (meters)	Tsunami Category	The Color of Affected Area
0.1 - <0.5 meters	Small Tsunami	Green
0.5 - <3.0 meters	Moderate Tsunami	Yellow
≥ 3.0 meters	Large Tsunami	Red

Source: InaTEWS (BMKG, 2008)

Table 4. Area of Each Village Based on the Level of Potential Disaster Hazards

Village	Area Size Based on Potential Danger Level (Ha)		
	High	Moderate	Low
Mojomulyo	305.15	502.27	162.85
Mojosari	339.55	519.10	133.90
Puger Kulon	147.56	294.68	3.36
Puger Wetan	-	361.30	170.65

Source: GITEWS, 2007

The widest area of the village included in the high, moderate and low tsunami hazard level is Mojosari Village, with a high tsunami hazard area of 339.55 Ha, moderate tsunami hazard of 519.1 Ha, and low tsunami hazard of 133.9 Ha. Puger Kulon village has a high tsunami hazard level of 147. 56 Ha, a moderate tsunami hazard level of 294.68 Ha, and a low hazard level of 3.36 Ha. Mojomulyo Village has a high tsunami hazard area of 305.15 Ha, moderate tsunami hazard level of 502.27 Ha, and a low tsunami hazard level of 162.85 Ha. Puger Wetan village has a moderate tsunami hazard level of 361.3 Ha, and a low hazard level of 170.65 Ha. Mojosari Village has a high tsunami hazard area of 339.55 Ha, moderate tsunami hazard level of 519.1 Ha, and a low tsunami hazard level of 133.9 Ha. The widest area of the village with the lowest tsunami hazard level is Puger Wetan Village with an area of 170.65 Ha and a moderate hazard level of 361.3 Ha.

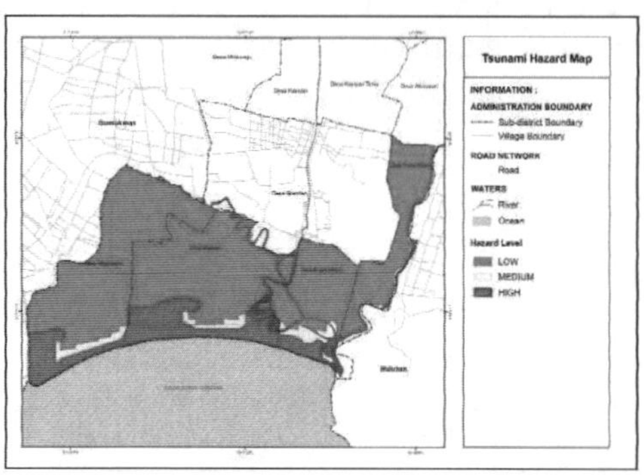

Figure 3. Map of Tsunami Hazard

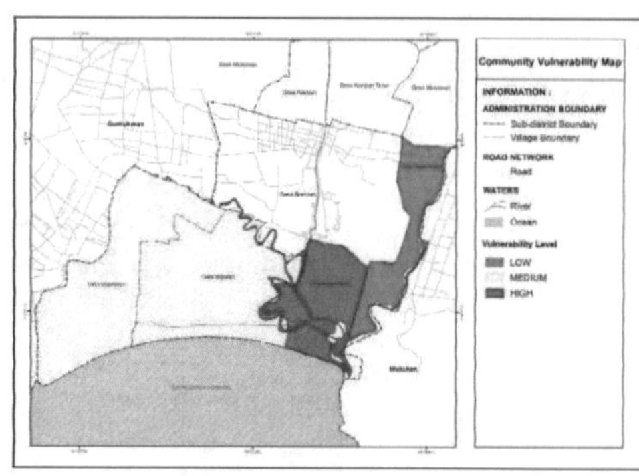

Figure 4. Map of Community Vulnerability

3.3. Vulnerability Analysis

The vulnerability is a condition of a society which causes an inability to deal with disaster threats [5].

3.3.1 Economic Vulnerability

The indicators used in the analysis of economic vulnerability are poverty levels, the area of productive land and vulnerable sector labor as presented in Table 5. Scoring results for the economic vulnerability in each village in the coastal area of Puger Beach show that Mojomulyo Village and Puger Kulon Village have a classification of low economic vulnerability. Mojosari Village has a classification of moderate economic vulnerability to disasters; meanwhile, Wetan Puger Village has a classification of high economic vulnerability.

The high economic vulnerability of Puger Wetan Village comes from high percentage of productive land area, which accounted for 75.73% of the total land area of Puger Wetan Village, including the land area of the agricultural sector, plantation sector, and ponds. The productive land is used by the community as a source of income; hence, if the area of productive land belonging to the community is affected by a disaster, the community will suffer from economic resource loss. In addition, vulnerability also comes from high percentage of vulnerable sector labor, which accounted for 46.07% of the total population of Puger Wetan Village, relying on sea fishing. The vulnerable sector affected in the event of a tsunami disaster is the fishery sector in which the residents making their living as fishermen will lose their livelihood for a specific duration.

Table 5. Economic Vulnerability of Villages in Puger District

Indicators	Village			
	Mojomulyo	Mojosari	Puger Kulon	Puger Wetan
Poverty Level (%)	10.33	10.90	5.06	4.86
Area of Productive Land (%)	40.16	49.27	34.81	75.73
Vulnerable Labor Sector (%)	6.31	0.00	31.18	46.07
Poverty Level Score	3	3	1	1
Area of Productive Land Score	1	2	1	3
Vulnerable Labor Sector Score	1	1	3	3
Economic Vulnerability Classification	Low	Moderate	Low	High

3.3.2 *Environmental Vulnerability*

The environmental vulnerability indicators used are environmental protection zones (Table 6). The wider the scope of the environmental protection zone, such as protected forest, green open space, water absorption, the higher the rain catchment area or water absorption to absorb tsunami water. Therefore, the environmental vulnerability is lower. The width of the environmental protection zone used is the area of green open space in the form of moorings and shrubs because there is no use of forest land in the coastal villages of Puger District. The results of the assessment of the level of environmental vulnerability in the coastal villages of Puger District obtain two levels of vulnerability classification, for villages included in high environmental vulnerability, Puger Wetan Village. However, Mojomulyo Village, Mojosari Village and Puger Wetan Village are included in the level of low environmental vulnerability classification.

Table 6. Environmental Vulnerability of Villages in Puger District

Indicators	Village			
	Mojomulyo	Mojosari	Puger Kulon	Puger Wetan
Environmental Protection Zone (%)	20.81	18.34	22.67	3.76
Environmental Protection Zone Score	1	1	1	3
Environmental Vulnerability Classification	Low	Low	Low	High

3.3.3 *Physical Vulnerability*

Indicators used in physical vulnerability include built areas, building density, and road damage. Scoring results for physical vulnerability indicate that Mojomulyo and Mojosari villages have a moderate physical vulnerability classification, Puger Wetan Village has a low classification of physical vulnerability to disasters while Puger Kulon Village has a high vulnerability classification (Table 7).

The high physical vulnerability of Puger Kulon Village comes from the high percentage of the built area of 36.40% from the total land area of Puger Kulon Village. Areas with high built areas are more vulnerable when tsunami disaster occurs. In addition, vulnerability also comes from the high percentage of road damage, which is 36.7% of the total road length in the Puger Kulon village. The road network is one of the supporting infrastructures for disaster, which is used as a disaster evacuation infrastructure. Villages which have a high percentage of damaged roads include areas that are classified as vulnerable because they can inhibit the evacuation process and the process of rescuing themselves when disaster occurs.

IOP Conf. Series: Earth and Environmental Science **328** (2019) 012053 doi:10.1088/1755-1315/328/1/012053

Table 7. Physical Vulnerability of Villages in Puger District

Indicators	Village			
	Mojomulyo	Mojosari	Puger Kulon	Puger Wetan
Built Area (%)	37.1	30.1	36.4	11.9
Building Density (unit/ha)	13	5	28	53
Road Damage (%)	22.9	22.6	36.7	10.3
Built Area Score	3	3	3	1
Building Density Score	1	1	2	3
Road Damage Score	2	2	3	1
Physical Vulnerability Classification	Moderate	Moderate	High	Low

3.3.4 *Social Vulnerability*

Indicators of social vulnerability include population density, population growth rate, the percentage of the female population and the percentage of people with low education in the coastal village of Puger District. Puger Kulon Village has a low social vulnerability classification. Puger Wetan Village and Mojosari Village have a classification of moderate social vulnerability to disasters. Meanwhile, Mojomulyo Village has a high classification of social vulnerabilities (Table 8).

The high social vulnerability of Mojomulyo Village comes from high percentage of the female population, which is 50.22% of the total population of Mojomulyo Village. The population of women is more vulnerable to the tsunami because women tend to need assistance and are physically weak in the evacuation process. Thus, they can increase vulnerability when tsunami disasters occur. In addition, vulnerability also comes from the percentage of non-school residents and the latest education in Elementary School and those who are still in elementary school which is 69.20%. In general, knowledge about disasters is still relatively poor, and the residents are not ready to anticipate disasters, categorized in the population vulnerable to disasters.

Table 8. Social Vulnerability of Villages in Puger District

Indicators	Village			
	Mojomulyo	Mojosari	Puger Kulon	Puger Wetan
Population Density (people/ha)	12	12	39	25
Population Growth Rate	0.4567×10^{-2}	0.4554×10^{-2}	0.4558×10^{-2}	0.4573×10^{-2}
Female Population (%)	50.22	50.04	49.48	49.25
Level of Education (%)	69.20	71.80	57.88	63.85
Population Density Score	1	1	3	2
Population Growth Rate Score	2	1	1	3
Female Population Score	3	3	1	1
Level of Education Score	3	3	1	2
Social Vulnerability Classification	High	Moderate	Low	Moderate

3.3.5 *Vulnerability Scoring Analysis*

Based on the Head of BNPB Regulation No. 2 of 2012 about General Guidelines for Disaster Risk Assessment, the tsunami vulnerability index has the following weighting concepts.

Vulnerability = (0.25 x physical vulnerability score) + (0.25 x economic vulnerability score) + (0.4 x social vulnerability score) + (0.1 x environmental vulnerability score)

Overall, disaster vulnerability assessment in coastal villages in Puger is a result of the level of vulnerability scoring including aspects of physical vulnerability, social vulnerability, economic vulnerability, and environmental vulnerability. Therefore, the overall vulnerability level of the disaster in the coastal villages in Puger District of Jember Regency is presented in Table 9.

Table 9. Community Vulnerability of Villages in Puger District

Village	Physical Vulnerability (0.25)		Social Vulnerability (0.25)		Economic Vulnerability (0.4)		Environmental Vulnerability (0.1)		Total	Classification
	Value	Value x Weight	Value	Value x Weight	Value	Value x Weight	Value	Value x Weight		
Mojomulyo	6	1.50	9	2.25	5	2.00	1	0.10	5.85	Moderate
Mojosari	6	1.50	8	2.00	6	2.40	1	0.10	6.00	Moderate
Puger Kulon	8	2.00	6	1.50	5	2.00	1	0.10	5.60	Low
Puger Wetan	5	1.25	8	2.00	7	2.80	3	0.30	6.35	High

3.4. *Community Vulnerability of Puger District to the Tsunami Disasters Hazard*

The analysis of the vulnerability of the community to the tsunami disaster was carried out by overlaying a threat map and vulnerability based on Table 2; hence, a spatial model was obtained in the form of a community vulnerability map against the threat of a tsunami disaster. The analysis shows that villages with high vulnerability to the tsunami disaster are part of Mojosari Village and part of Puger Wetan Village. A moderate vulnerability village are part of Mojomulyo Village, part of Mojosari Village and part of Puger Wetan Village. Meanwhile, Low vulnerability Village is Puger Kulon Village.

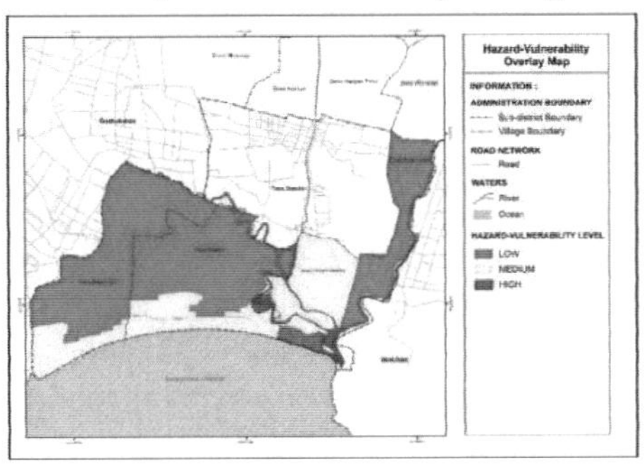

Figure 5. Map of Hazard-Vulnerability Overlay

4. Conclusion

Based on the results of the analysis which have been carried out, it can be concluded that the vulnerability of the Puger District community to the tsunami disaster is as follows:

1. The village in Puger District which is directly adjacent to the Indian Ocean is Mojomulyo, Mojosari, Puger Kulon, and Puger Wetan Villages. Puger Kulon village has a high tsunami hazard level of 147.56 Ha, a moderate tsunami hazard level of 294.68 Ha, and a low hazard level of 3.36 Ha. Mojomulyo Village has a high tsunami hazard area of 305.15 Ha, moderate tsunami hazard level of 502.27 Ha, and a low tsunami hazard level of 162.85 Ha. Puger Wetan village has a moderate tsunami hazard level of 361.3 Ha, and a low hazard level of 170.65 Ha. Mojosari Village has a high tsunami hazard area of 339.55 Ha, moderate tsunami hazard level of 519.1 Ha, and a low tsunami hazard level of 133.9 Ha.
2. Vulnerability in the study was reviewed from four sub-variables: physical vulnerability (built land, building density, and road damage), social vulnerability (population density, population growth rate, female population, and education level), economic vulnerability (poverty level, productive land area, and workers in vulnerable sectors) as well as the vulnerability of the environment (environmental protection zone). Based on the social vulnerability analysis, of the 4 coastal villages in Puger District, 1 of which has high social vulnerabilities with 2 villages have moderate social vulnerability and 1 village has low social vulnerability. Based on the economic vulnerability analysis, of the 4 coastal villages in Puger District, 1 of which has high economic vulnerabilities with 1 village has moderate

economic vulnerability and 2 villages have low economic vulnerability. Based on physical vulnerability analysis, of the 4 coastal villages in Puger District. 1 of them has high physical vulnerabilities, with 2 villages have moderate physical vulnerability and 1 village has low vulnerability. Based on environmental vulnerability analysis, of the 4 coastal villages in Puger District, 1 of which has high environmental vulnerabilities, with 3 villages have low environmental vulnerability. Village with a high level of vulnerability is Puger Wetan Village. Villages with moderate vulnerability are Mojomulyo and Mojosari Villages, and villages with a low level of vulnerability are Puger Kulon Village.

3. Based on the results of the overlay analysis of the hazard map and vulnerability map, villages with high vulnerability to the tsunami disaster are part of Mojosari Village and part of Puger Wetan Village. Moderate vulnerability villages are identified in part of Mojomulyo Village, part of Mojosari Village and part of Puger Wetan Village. Village with low vulnerability is only Puger Kulon Village. Mojosari Village and Puger Wetan Village have a high vulnerability to the tsunami disaster. In terms of threats, they have a higher threat level than other villages because the 1994 tsunami was moderately affected and was located closest to the earthquake epicenter, causing the tsunami in Banyuwangi Regency. Mojosari village also has a high vulnerability to disasters, especially in terms of economic, physical, and social aspects. Therefore, efforts in tsunami disaster risk reduction in Mojosari Village are carried out through improving economic conditions and improving social conditions by prioritizing the education sector including the provision of trainings specifically for women. Puger Wetan Village also has a high vulnerability to disasters from all four aspects of vulnerability. Therefore, tsunami disaster risk reduction efforts in Puger Wetan Village are carried out through improving economic, social and environmental conditions through good environmental management measures.

References

[1] Maryani, Elok. 2008. Model of Mitigation Socialization in Disaster-Prone Local Communities in West Java. Bandung: Research on Higher Education Grants.

[2] Meteorology, Climatology, and Geophysics Agency. 2014.

[3] Fauzi and Wandono. 2005. Determination of Epicenter for Analysis of Tsunami Runoff Risk. *Proceedings of the Tsunami Seminar in the Framework for the Application of Research Results for Tsunami Disaster Management in Indonesia* (2002-2003-2004). pp 163-169. Jakarta: BPPT Press.

[4] German-Indonesian Cooperation for a Tsunami Early Warning System. 2007

[5] Head Regulation of the National Disaster Management Agency Number 02 of 2012 concerning General Guidelines for Disaster Risk Assessment.

[6] Sumekto, Didik Rinan. 2011. Disaster Risk Reduction Through Analysis of Community Vulnerability and Capacity in Facing Disasters. Yogyakarta.

[7] Habibi, Marbruno & Buchori, Imam. 2013. Spatial Model of Socio-Economic Vulnerability and Institutional Against the Disaster of Mount Merapi. *Regional and Urban Planning Engineering Journal*. II (1): 1-10. Semarang: Diponegoro University.

[8] Miladan, N. 2009. Study of Semarang's Coastal Area Vulnerability to Climate Change. *Thesis*. Semarang: Diponegoro University.

[9] Sudibyakto & Anggit Priatmodjo. 2016. Disaster Risk Management In Gunung Padang Cultural Heritage, Ciamis, West Java. *Indonesian Disaster Research Journal*. II (1): 50-58. Yogyakarta: Gajah Mada University.

[10] Republic of Indonesia Law Number 24 Year 2007 concerning Disaster Management.

[11] Minister of Energy and Mineral Resources Regulation Number 15 of 2011 concerning Guidelines for Volcanic Disaster Mitigation, Soil Movement, Earthquakes and Tsunamis.

[12] Indonesia Tsunami Early Warning System. 2008. Jakarta: Meteorology, Climatology, and Geophysics Agency.

Problems and Challenges in Sustaining Rural Land Resources

T D Insani[1], I Rudiarto[2] and H B Wijaya[3]

[123]Department of Urban and Regional Planning, Diponegoro University, Semarang, Indonesia

tia.insani@yahoo.com

Abstract. The discussion in this paper brings up issues on how the land resources utilization is done mainly as a respond to socioeconomic development instead of achieving sustainability. Delanggu sub-district of Klaten Regency and Kledung sub-district of Temanggung Regency were chosen as the study areas. We collected data and information by conducting Focus Group Discussion in each study area with relevant stakeholders and analysed the issues further using SWOT and problem analysis approach. The result showed that even though land use change is a common issue shared by both study areas, the drivers as well as the approaches taken by the community and stakeholders from each sub-district are different. Kledung sub-district as a rural mountain area appears more sustainable in term of land resource utilization than urbanized Delanggu sub-district which is a plain area. The findings call more challenges for future adaptive approaches to answer the notion of sustainable agriculture in rural area.

Keywords*: rural resources, land use, land resource utilization*

1. Introduction

Rural areas have a great diversification in terms of physical environment, ecology, and pattern of land use as well as their socioeconomic characteristics [1],[1], although it is common to find that agricultural activities are where the big part of its community's income share is generated from. However, the increasing agricultural activities in rural areas has caused direct and indirect impacts on the decreasing rural resources. Several key issues at regional level that threaten the existence of natural resources include land degradation, water availability, and loss of biodiversity [3]. At the local level, the problems that arise around rural area's agricultural activities are shown through the socioeconomic impacts on its community. From time to time farmers' families are constantly adapting to the stresses on environment and socioeconomics conditions that influence the existence and availability of natural resources and biodiversity [4].

Many rural areas in Indonesia, especially on Java, have experienced a transition process in a long period of time that was determined and influenced by the activities of their rural communities [5]. This long transition process has helped shape the pattern of rural resource utilization. Resources in rural contexts are mostly used as inputs for agricultural production and closely related to environmental conditions. With such context, agricultural activity is considered an effort in managing interactions between resources and environment for output optimization [6]. Resource utilization in the rural area indicates the relationship between economic and environmental aspect. Income gain by the farming family represents the economic aspect while land resource utilization represents the environmental aspect. Farming families are the main player between those two aspects where the decision on how to utilize the resource very much depends on the family [7], [8]. While resources like land, water, and human resources are vital for the agricultural activities, the practice of the activities by rural communities are also resulted in the unsustainability of both resources and agricultural activities. This dynamic can then be seen through the shift of land use practice as well as the socioeconomic conditions of the rural communities.

1.1 Study Area

Two different rural areas were selected for this study to show how the community characteristics in each rural area affect the land use and resource utilization practices. Delanggu sub-district (100-200 meter above sea level) in Klaten Regency will be discussed to represent plain rural area and Kledung sub-district (1100-1500 meter above sea level) in Temanggung Regency for mountainous rural area. Based on the community socioeconomic characteristic and agricultural activities in both areas, we seek a deeper understanding about issues faced by the community through focus group discussion (FGD) that involves all the village head and sub-district head. The discussions bring up issues on why most spatial planning in rural areas happens as a respond of socioeconomic development instead of achieving sustainability of rural resources. In this paper we analyze the issues further by implementing problem tree and SWOT approaches which aim to get more rounded understanding on how land use and resources utilization practice in rural areas are both affecting to and affected by the socioeconomic characteristics of the rural community.

2. Methods

In achieving the objectives of the study, the research team has conducted focus group discussions (FGDs). The FGDs are meant to collect information and perceptions of the stakeholders regarding the sustainability of resources and biodiversity in village spatial planning. The discussion also provides how the stakeholders handle problems around land use and resources utilization and what their strategic decisions on the matter implicates the overall practice of rural spatial planning in the study areas. The analysis tools used are problem trees and the SWOT analysis. Problem tree analysis helps to map issues and its causes as well as effects so that it is easier to form solutions for the problem [9]. Developed by Humprey [10], SWOT (Strength-Weakness-Opportunity-Threat) is an analytical tool to find the best strategy in decision making based on internal strengths and weaknesses as well as opportunities and challenges which comes from external factors. By identifying internal strength and weaknesses as well as external opportunities and threats, this instrument provides a simple approach to assess the best way of implementing a strategy [9]. By using these two analysis tools, the writer needs to breakdown the situations faced by rural community that mostly still has to depend on agricultural activities despite land use issues and decreasing quality of supporting resources.

3. Results and Discussions

3.1 Socioeconomic Characteristic and Land Use Change

Rural areas with dominant agricultural activities as its community livelihood generally have agricultural land owned by the farming families that are located nearby their settlement areas. It is common to find most of their houses located next to their farming or paddy fields. Even though both Delanggu sub-district and Kledung sub-district are rural areas, family income composition shows a different trend that may lead to the difference on each socioeconomic characteristic. As seen in Table 1, family income in Kledung sub-district dominantly makes up from 96% farm income and only 4% from off-farm income. Meanwhile, Delanggu sub-district has its community more dependent on off-farm income as it makes up 61% of the family income in the sub-district whereas farm income has 39% share. Such result can come from the condition that Delanggu sub-district which is a plain area located near the urban area of Solo City. On the other hand, Kledung sub-district is far from any city given its location in mountainous area hence more dependent on farming. This indicates that based on their economic activities, the community in Delanggu sub-district is more likely to have an urbanized characteristic than the community in Kledung sub-district.

Table 1. Composition of Family Income[a].

Sub-District	Area Type	Composition	
		Farm income	Off-farm income
Delanggu	Urbanized Rural Area	39%	61%
Kledung	Mountainous Rural Area	96%	4%

[a] Rudiarto et. al., 2018.

The socioeconomic condition is also supported by the land use change in both study areas over the last 15 years from 2000 to 2015. As shown on Figure 1, yellow area that represents settlement and built areas area growing in both Delanggu sub-district and Kledung sub-district. However, the change is apparent in Delanggu sub-district where settlement and built areas are growing vastly to most of the sub-district. The maps show how these yellow areas come in substitute of green areas that represent agricultural lands such as farm fields and paddy fields. Whereas in Kledung sub-district the growth of yellow areas are not significant and only concentrated along the main road. As it is geographically located between two mountains, Mount Sumbing and Mount Sindoro, the major parts of Kledung sub-district are sloping areas that is not only relatively harder to develop but also is part of protected forest area. It can be seen on the Figure 1 that farm fields and irrigated fields are still dominant as it grows nearing the part of shrubs area on the mountainsides of Mount Sumbing and Mount Sindoro.

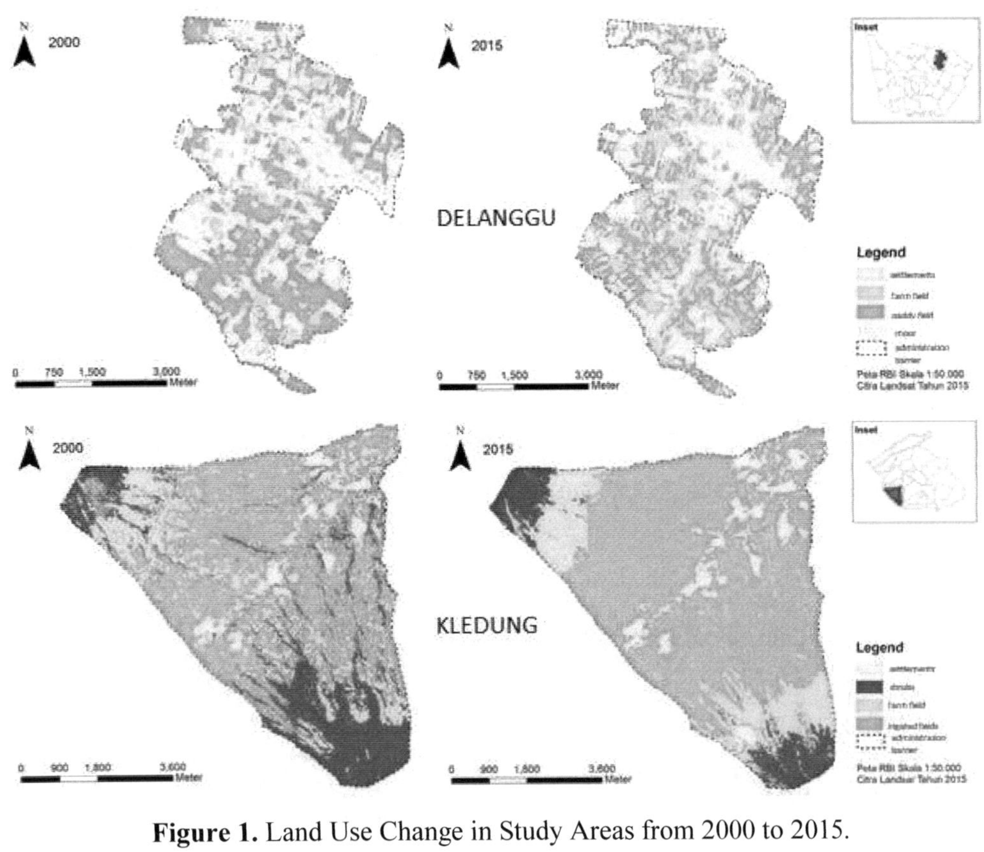

Figure 1. Land Use Change in Study Areas from 2000 to 2015.

The existence of settlements in rural area is central as it may reveal the relationships between land and the community [11] [12]. Generally, rural settlements are in high pressure of the urbanization and have been significantly transformed into a more urbanized area [13]. According to the measurement of land use cover change from landsat maps of year 2000 and 2015, Delanggu sub-district has a surging of settlement and built areas up to 61,11 ha while agricultural lands and grass have decreased as much as 60,17 ha and 0,94 ha respectively. The number shows that the decreasing green areas happened to accommodate more development of settlement and built areas. The same thing also happens in Kledung sub-district where the agricultural lands have decreased by 24.39 ha. Unlike in Delanggu sub-district, the decreasing agricultural lands in Kledung sub-district is not only a shift to settlement and built areas (20.88 ha) but also to grass (3.51 ha). It indicates that some agricultural lands are gone neglected or unproductive hence the size of grass areas is increased. The land use portion of settlement and built areas in Delanggu sub-district is 27% of the total area while in Kledung sub-district it is only 6%. The difference on the settlement and built areas size once again shows that Delanggu sub-district is becoming an urbanized area while Kledung sub-district still preserves its rural characteristic.

3.2 Problems of Land Resource Utilization

As has been discussed on the previous section, land conversion is one common issue that is present in both sub-districts. On the one hand, land use shift to more built areas can indicate a positive development in rural areas where the community is benefited socially and economically. On the other hand, if the land use changes disrupt the previously established way of life of the farming community, the adjustment made could detrimentally affect how rural resources such as land, water, and human resource are utilized. Here, through SWOT analysis as presented in Table 2, we look closer to how the changes in land use are at the centre of the problems surrounding utilization of land as an integral resource for agricultural activities in Delanggu sub-district and Kledung sub-district.

3.2.1 Delanggu sub-district

According to Bappeda of Klaten Regency [14], Delanggu sub-district is planned to be one of the urban activity centres in the region. It is a local activity centre that serves regional scale activities that include trade and service, urban settlements, agriculture, and industry among others. Delanggu sub-district is also allotted for agricultural food crops and horticultural production centre, especially paddy. The average land area managed by the community is 0.54 ha per household with an average yield of 11.2 tons per ha per year [8]. Although most of the land in Delanggu sub-district is utilized as paddy fields, the land along its main road are now converted to built areas for settlements and business.

Based on the results of focus group discussion (FGD) with the heads of 16 villages in Delanggu sub-district, the conversion of paddy fields is increasingly prevalent over the past decade. Many land conversion cases are done without legal permit as there is also incongruity between practices and the spatial planning regulation. Land that is supposed to be green areas could be made available for development or are re-planned to accommodate market demand over land. At the same time, the village administration also does not hold any authority to hinder land owners that choose to sell their land to developers or build it into another house.

Aside from the regulation issue on spatial planning, economic motives are in the heart of the problems surrounding land conversion in Delanggu sub-district. Over the years, the community finds that productivity of their paddy fields are decreasing significantly with the production cost greater than their earnings from it. Some factors such as planthopper pests, decreasing irrigation water - mainly from Cokro springs, and low quality of the seeds come into play in this issue. As the farming community no longer can rely on their livelihood in the agricultural activities alone, they end up looking for more reliable income options outside of farming.

Adding urbanization, which is already present in the community, to the situation, it is then inevitable to see a declining trend of the community participation in agricultural activities. Younger generation prefer to work in service or industry sectors while the older ones can no longer handle the works needed in farming. The results are the ever decreasing productivity and inclination to give up agricultural activities. Both effects perpetuate more land conversion as selling or developing their land to meet their financial needs has more cost effective than keeping it as farming field. However, in 2018 national government through Ministry of Agriculture has already initiated a development program in which the farming community in Delanggu sub-district is given seeds of new paddy variety that can be harvested four times a year. It aims not only to boost the productivity of the community's agricultural activities but also as an effort to slow down the land conversion.

Table 2. SWOT Analysis of the Land Resources in Delanggu and Kledung.

	Strength	Weakness	Opportunity	Threat
Delanggu	1. one of the centers of urban activities in the region	1. high conversion of irrigated land/paddy fields to settlements and business	1. provision of new paddy variety that can be harvested four times a year is already on going	1. incongruity between practice and spatial planning regulation
	2. one of the center of food crop agriculture in the region	2. declining community's participation in agricultural activities		2. decreasing availability of irrigated water from Cokro springs due to bottled water industry
	3. 72% of its total area are agricultural land	3. low work force in agricultural works		3. high demand of land from developers
		4. declining productivity of agricultural sector		4. urbanization
Kledung	1. highland with potential to be developed as agropolitan	1. sloping areas prone to erosion and landslide	1. high demand of tobacco	1. uncertain climate and season
	2. one of the centers of tobacco production in the region	2. agricultural land conversion	2. local government's plan to develop ecotourism in the sloping areas of Mount Sumbing and Mount Sindoro	2. low price of tobacco
	3. abundant availability of water	3. decreasing soil fertility		
		4. declining production of tobacco		
		5. poor access infrastructure		

3.2.2 Kledung sub-district Situated on a highland, Kledung sub-district is potential for agropolitan and ecotourism. Kledung valley is one of the main catchment areas around Mount Sumbing and Mount Sindoro, hence most of its land use is green area. There are three springs: Si Gandul springs, Sidandang springs, and Tuk Sewu springs, for their main source of clean water and irrigation. This mountainous land has a moderate to high potential to erosion and is prone to landslide. Conservation is done to reduce the risk of erosion and landslide by providing hedgerows and the application of terrace systems on the community's farm land. Most irrigation systems used in agricultural activities are non-technical and rain-fed. Rainy season becomes a big challenge for the agricultural activities in Kledung sub-district as water inundating the farm land while the commodity are the types that does not need a lot of water.

The farming community in Kledung sub-district manage an average land of 1.15 ha per household with an average yield of 9.8 tons per ha per year [8]. The main agricultural commodities are tobacco, garlic, onions red, chili, coffee, paddy, corn, and cabbage. In general, there was a decrease in agricultural production in Kledung sub-district for the last 15 years with tobacco being the commodity with the highest declining number of production. This declining production is caused by several issues but mainly because the uncertain climate conditions make it difficult to maintain the ideal environment for tobacco cultivation. Failure in harvesting could reach up to 50% in a year and this drives some farmers to stop farming all together and prefer to sell their land as a way to save them from financial crisis.

The decline in tobacco production and the inability to maintain the quality of crop yields make the selling price of tobacco become very low and uncertain. Tobacco farmers in Kledung Sub-district are then looking for a solution by implementing intercropping and crop rotation with other commodities such as coffee and vegetables. Some others even switch to other commodities with a lower risk of loss than tobacco. Nevertheless, there are also many farmers who have not made any efforts to increase their land productivity as the big part of the community by considering their farm production is already enough to help meet their family needs.

Another challenge related to land resources in Kledung sub-district is high erosion that causes decreasing land fertility. A lot of farm lands are on nearly 90 degrees slope area so its layers of soil are easily carried away by the flow of water when raining. Many farmers then use chemical herbicide to prepare their land for every planting cycle so that it reaches certain fertile conditions. Even though it is done to increase soil fertility, continuous use of chemicals actually further reduces soil fertility on the long run. The process of preparing their land without using herbicides is actually possible, however it requires them more cost as well as to do extensive plowing while nowadays it is increasingly difficult to find human resource for the work. It is also not to mention that the process of land plowing is already hard enough with the poor access infrastructure for the topography condition of most Kledung sub-district areas.

3.3 Formulating Problems and Challenges

Discussion on the problems of land resources in Delanggu sub-district and Kledung sub-district helps to see how even though land conversion issue is present in both areas, the drivers of land use change as well as the approaches taken by the community and stakeholders from each sub-district are different. Figure 2 sums up the main threats for the land resources in both sub-district and the current approach each has in facing the problems.

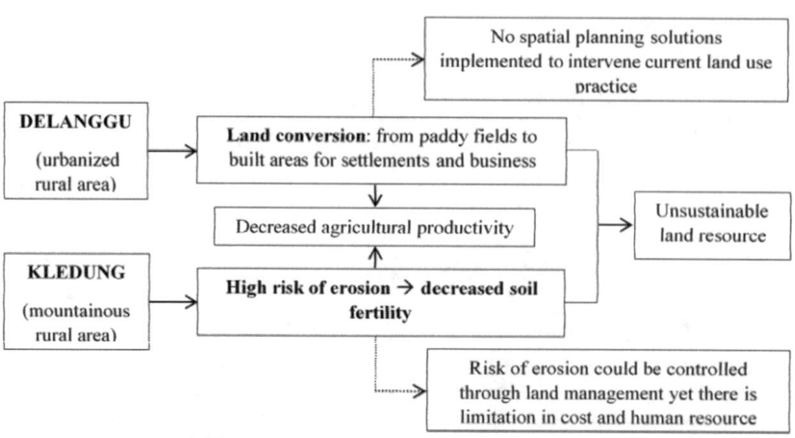

Figure 2. Sustainability Problems of Land Resources in Delanggu and Kledung.

Kledung sub-district with the lower level of land conversion and more manageable problems has a higher chance to keep its land resource sustainability. On the other hand, Delanggu sub-district problems are inevitable as the result of urbanization that seem to drive its local development. The community alone cannot do much change when the needed intervention on spatial planning is neglected by the authority, which in this case is regional government.

4. Conclusion

The study shows that even though land use change is a common issue shared by both study areas, the drivers as well as the approaches taken by the community and stakeholders from each sub-district are different. The sustainability level of rural land resource decrease in accordance to the development of its surrounding area when the needs of land for non-agricultural activities grow rapidly. Land use shift to more built areas is benefitted socially and economically even though at the same time such disruption to an established way of life of the farming community can detrimentally affect how rural resources, especially land resource, are utilized. Kledung sub-district as a rural mountain area appears more sustainable in term of land resource utilization than urbanized Delanggu sub-district which is a plain area. The two findings call more challenges for future adaptive approaches to answer the notion of sustainable agriculture in rural area.

Acknowledgement
This paper is part of the research on "Spatial Planning Modelling for Sustainable Rural based on Resources and Biodiversity in Central Java - Indonesia" funded by Diponegoro University.

References

[1] Huby, M., Cinderby, S., Crowe, A. M., Gillings, S., McClean, C. J., Moran, D., ... & White, P. C. L. (2006). The association of natural, social and economic factors with bird species richness in rural England. *Journal of Agricultural Economics*, *57*(2), 295-312.

[2] Wang, J. F., Li, X. H., Christakos, G., Liao, Y. L., Zhang, T., Gu, X., & Zheng, X. Y. (2010). Geographical Detectors-Based Health Risk Assessment and its Application in the Neural Tube Defects Study of the Heshun Region, China. *International Journal of Geographical Information Science*, *24*(1), 107-127.

[3] Hurni, H. (1997): Concepts of sustainable land management. Proceeding of the International Conference on Geo-Information for Sustainable Land Management. ITC, Enschede.

[4] Gregory, P. J., Ingram, J. S., & Brklacich, M. (2005). Climate change and food security. *Philosophical Transactions of the Royal Society of London B: Biological Sciences*, *360*(1463), 2139-2148.

[5] Rudiarto, I. (2009), Sustainable land use concept in mountain area of Java, Indonesia: In Farming and Rural Systems Economics and Biodiversity in the Tropics - Proceedings International DAAD Alumni Summer School, Margraf Publisher – Germany.

[6] Almekinders, C., andJ. Hardon. (2001). The role of genetic resources inrural livelihood systems. UPWARD working paper series No. 6. WUUPWARD.Wageningen.

[7] Doppler, W. (2006): Resources and livelihood in mountain areas of South East Asia: Farming and rural systems in a changing environment. Margraf Verlag. Wekersheim.

[8] Rudiarto, I., Handayani, W., Wijaya, H. B., & Insani, T. D. (2018). *Spatial Distribution of Socioeconomic Characteristics in Rural's Java: A Case from Three Different Rural Areas in Central Java. IOP Conference Series: Earth and Environmental Science, 123, 012024.*

[9] Start, Daniel & Hovland, Ingie. 2004. *Tools For Policy Impact. A Handbook For Researcher.* RAPID. Westminster London.

[10] Humphrey, Albert (December 2005). *"SWOT Analysis for Management Consulting". SRI Alumni Newsletter* (SRI International).

[11] Jones, E. E. (2010). An analysis of factors influencing sixteenth and seventeenth century Haudenosaunee (Iroquois) settlement locations. *Journal of Anthropological Archaeology*, *29*(1), 1-14.

[12] Palmisano, G. O., Govindan, K., Loisi, R. V., Dal Sasso, P., & Roma, R. (2016). Greenways for rural sustainable development: An integration between geographic information systems and group analytic hierarchy process. *Land Use Policy*, *50*, 429-440.

[13] Dumreicher, H. (2008). Chinese villages and their sustainable future: the European Union-China-Research Project "SUCCESS". *Journal of environmental management*, *87*(2), 204-215.

IOP Conf. Series: Earth and Environmental Science **328** (2019) 012054 doi:10.1088/1755-1315/328/1/012054

[14] Bappeda of Klaten Regency. (2011). Regional Spatial Planning Document of Klaten Regency year 2011-2031 *(Rencana Tata Ruang Wilayah Kabupaten Klaten Tahun 2011 – 2031).* Klaten.

The Impact of Spatial Patterns of Bunaken Subdistrict Settlements on Mangrove Ecosystems in Manado Coastal City

V A Kumurur

Architect Department, Sam Ratulangi University, Manado

*Email : veronicakumurur@unsrat.ac.id

Abstract. Based on the results of the studies by Sasauw et al (2016), the ecological conditions in the area of mangroves in Manado city is unstable, evidenced by the low diversity of species *R. apiculata*, although density and frequency of species have high value. Bunaken is one sub-district in the coastal city of Manado, consisting of five sub-rurals, where the 3 villages (Molas, Meras and Tongkaina Neighborhood) are located in the coastal area and 2 sub-districts, such as Pandu and Bailang neighborhood are in mountainous regions. Residential area bordering the Bahowo mangrove area (84 ha) is green open space as Manado city protected area. To obtain and analyse data on settlement patterns, the ArcGis software with satellite map is utilized and observations of the infrastructure condition are conducted in the three areas of urban villages located on the shore (adjacent to the area of mangrove forests). The results indicated that the housing pattern of three villages is the combination form (linear and cluster), but linear form is more dominant. In general, the pattern of residential infrastructure networks, especially roads and drainage channels (grey water) is dominant in the direction of the road network and ends in the mangrove forest area.

1. Introduction

The growth and development of the city become very significant centre of consumption of resources, where in the last decade, more people live in cities than outside them [1, 2] and this trend will continuously increase to 90% by 2100 (Fragkias, Lobo, Strumsky, & Seto, 2013) [2]. The rapid development of the city increases the demand for urban housing and Manado as one of the developing cities in Indonesia. In the past, several rural areas become now urban areas, including the Bunaken sub-district. Based on the City Regulation of Manado Number 1 in 2014, the city of Manado is divided into 6 sub-city service centres, where the Bunaken Subdistrict is located in the sub-service centre city V with three service functions, including: 1) settlement, 2) trade & services, and 3) tourism. Convergence of economic activities in coastal areas of the city of Manado can lead to land use conflicts that are very difficult to resolve, reducing the quality of coastal resources without management and protection (Sjafi'i 2001).

The mangrove forest area covers the Bunaken and Bunaken sub-districts, as one of the wetland resources in the coastal area and as a buffer for life and natural wealth that is important to environmental services for the city of Manado. According to the Presidential Regulation of the Republic of Indonesia Number 73 of 2012 concerning the National Strategy for Mangrove Ecosystem Management, it is necessary to protect, preserve and sustainably use for the welfare of the community.

IOP Conf. Series: Earth and Environmental Science **328** (2019) 012055 doi:10.1088/1755-1315/328/1/012055

Sasauw et. al (2016) found that mangrove forest diversity was in an unstable condition, evidenced by the low diversity of species of *R.apiculata*, although the density and frequency of species had high values. Whereas, mangrove forests function as a link between marine and terrestrial ecosystems important for the stability and maintenance of various adjoining ecosystems such as seagrass, coral reef, and other marine ecosystems, providing habitats for juvenile crustaceans and fish [3]. Mangroves also contribute significant quantities of organic matter (OM) through litter fall that can be transported offshore (Odum and Heald, 1972; Alongi, 1990) [3]. Prasad et al discovered that amount of litter produced in a mangrove forest directly affects the particulate organic matter (POM) between mangrove forests and adjacent marine habitats [3]. Nowadays, many of the most productive mangrove forests in the world lose a significant fraction of their net primary production to coastal waters (Robertson et al., 1992; Jennerjahn and Ittekkot, 2002) [3].

As a result, population growth has resulted in increased demand for residential land, while the community settlement pattern in the coastal area of Bunaken Subdistrict is on a sloping topography with the main drainage network in the gravitational area leading to the protected area of the Bahowo mangrove forest. Alberti et al and Marzluff (2001) expressed that the urbanization has an impact on change natural habitats and species composition (Blair, 1996), disrupting hydrological systems (Arnold and Gibbons, 1996; Booth and Jackson, 1997), and modifying energy flow and nutrient cycles (McDonnell and Pickett, 1990; Vitousek et al., 1997; Grimm et al., 2000) [4]. Due to the settlement pattern in the Bunaken sub-district, there was a change in the quality of mangrove natural forests.

Tongkagie (2003) noted that Bunaken National Park (BNP) encompasses about 89,056 hectares of land and sea area, divided into a southern mainland section (the Arakan-Wowontulap coast, protected primarily for its old-growth mangrove forests and dugong population) and a northern island section (including five islands with its fringing coral reefs and a mainland section). The area of mangrove forest in Bunaken sub-district included mainland section of BNP located in the urban area of Manado, as a green open space for the urban neighborhoods area of Molas, Tongkaina and Meras.

Today, the city of Manado has currently been planning the development of its territory, including at the sub-district Bunaken. The development plans refer to the guidelines of coastal area and small islands zoning plan (RZWP3K), whereas the designation of coastal areas is the area with general use, conservation areas, strategic areas and grooves. However, according to Law Number 1 of 2014, the classification of areas in coastal areas and small islands, includes: conservation areas, public use areas, sea lanes, certain national strategic areas. The coastline length of the three villages covers: Molas 3.77 km, Tongkaina 3.77 km and Molas 4.13 km used as a public utilization area and conservation area

Table 1. Distribution of Allocation of Coastal Areas

Zone	Utilization
Public utilization area	Tourism
	Settlements
	Port
	Agriculture
	Mining
	Fishery
	Aquaculture
	Industry
	Public Facilities, and/or
	Other uses are in accordance with bio geophysical characteristics of the environment.
Conservation area	Coastal and island conservation areas small
	Maritime Conservation Area
	Marine Protected Areas
	Shoreline

IOP Conf. Series: Earth and Environmental Science **328** (2019) 012055 doi:10.1088/1755-1315/328/1/012055

Zone	Utilization
Particular national strategic areas	Management of maritime boundaries of sovereignty country
	State Defence And Security
	Management Of World Heritage Sites
	Community Welfare; And / Or
	Environmental Conservation
Sea lanes	Cruise flow
	Underwater Pipes / Cables; And
	Migration Of Marine Biota

Source: the guidelines of coastal area and small islands zoning plan (RZWP3K), 2013

Urban patterns influence the feasibility of using alternative systems to supply resources and services to the urban population; thus, indirectly affecting their environmental impact [4]. Kostof (1983 and Kardi et al (2014) expressed settlement patterns are based on the nature of the community consisting of two types, such as: Community Groups, face to face and combination types. Sub Community Groups are clustered settlement patterns, consisting of several units or groups of residential units, focusing on important spaces, such as drying, public open spaces, mosques and so on (Figure 1). Whereas, face to face type is the settlement pattern of linear shape, between units of residential housing and linearly along the bearing hub of activity that contained a boat mooring or dock, drying room, markets and so on (Figure 1).

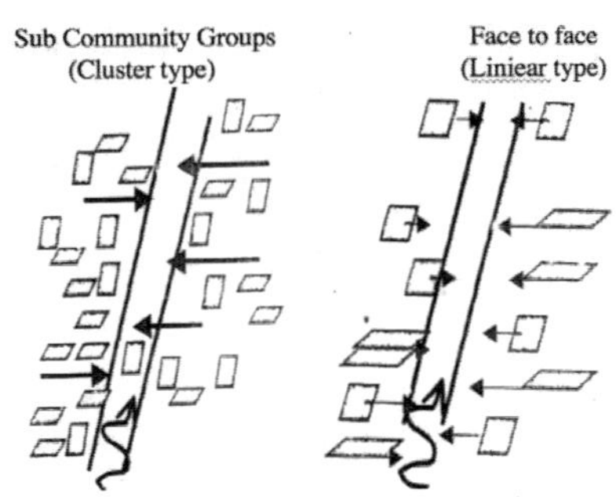

Figure 1. Distribution patterns of settlements based on the community [6]

2. Methods

The study was conducted in three urban villages on the coast of Bunaken Sub-district, Manado City, including Kelurahan Tongkaina, Molas, and Meras (Figure 2). The map of Bunaken was obtained from Google Earth based on the Google Earth imagery projecting a map of the settlement. Primary data was obtained from field surveys for actual data validation of object settlements.

IOP Conf. Series: Earth and Environmental Science **328** (2019) 012055 doi:10.1088/1755-1315/328/1/012055

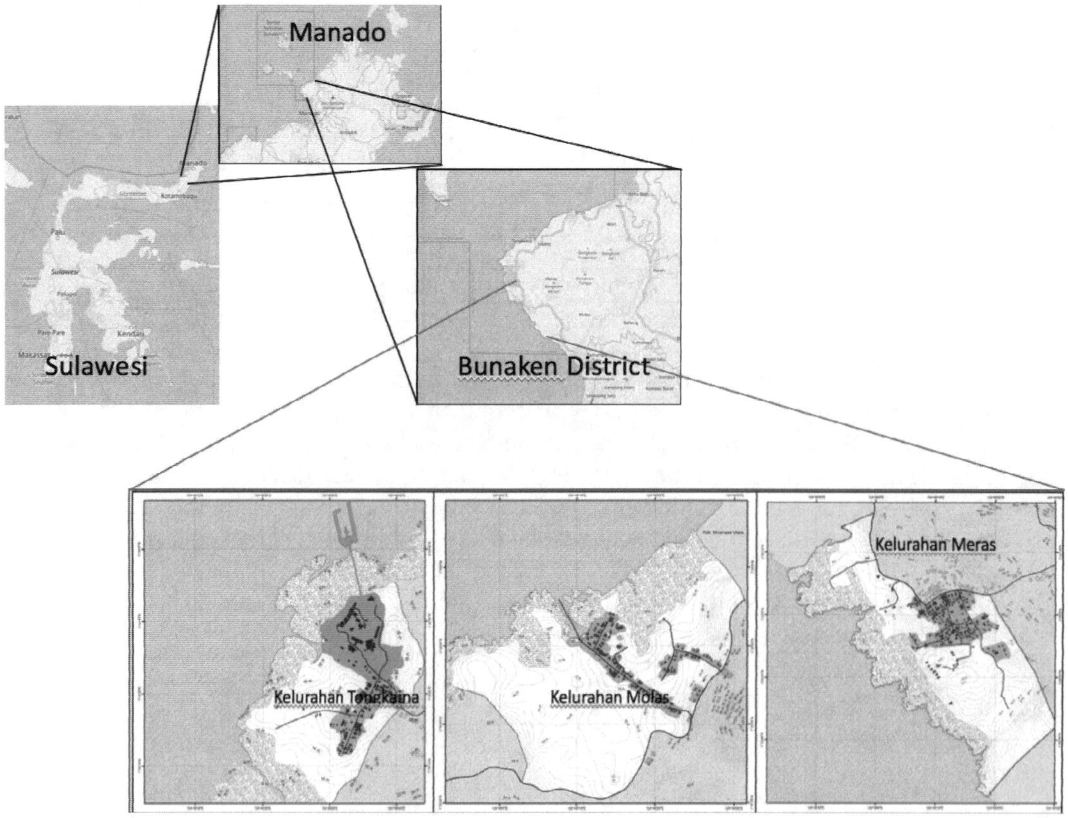

Figure 2. Map of Research Location

.

Visual analysis was conducted by the Google Earth imagery interpretation, to identify and interpret the image based on the appearance as seen in the image. Appearance seen in the image will be similar to the appearance of an existing field. The images obtained from Google Earth are digitized to obtain the settlement pattern of the Tongkaina village, Molas village and Meras village. Analysis of the distribution pattern by visual interpretation method is conducted by observing the settlement pattern, either cluster-type or linear type.

3. Results and Discussions

From the results of the field survey, it was found that the settlement pattern of the Molas Village was generally a linear pattern following the road pattern. Settlement land is at an altitude of 8-18 meters above sea level (Figure 4). Building orientation in this research area faces from north-south with its orientation facing the road (face to face) (Figure 5)

Figure 3. Contour Map of residential land in Molas Village

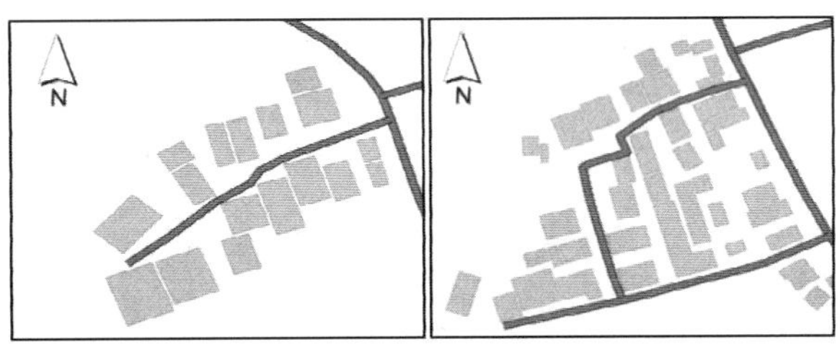

Figure 4. Settlement pattern of Molas Village

From the survey, it was found that the general settlement patterns in the Meras Village area were cluster patterns following the path and clustering. The residential land area is at an altitude of 22-54 meters above sea level (Figure 6).

Figure 5. Contour Map of residential land in Meras Village

The orientation of the building faces from north-south and west-east with its orientation facing the road (Figure 7)

IOP Conf. Series: Earth and Environmental Science **328** (2019) 012055 doi:10.1088/1755-1315/328/1/012055

Figure 6. Settlement pattern of Meras Village

Based on the results of the field survey, it was found that the settlement pattern in Tongkaina Village was a longitudinal linear pattern following the main road. The residential area is at an altitude of 12-28 meters above sea level (Figure 8) and the orientation of the building faces from north-south and west-east with its orientation facing the road (face to face) (Figure 9).

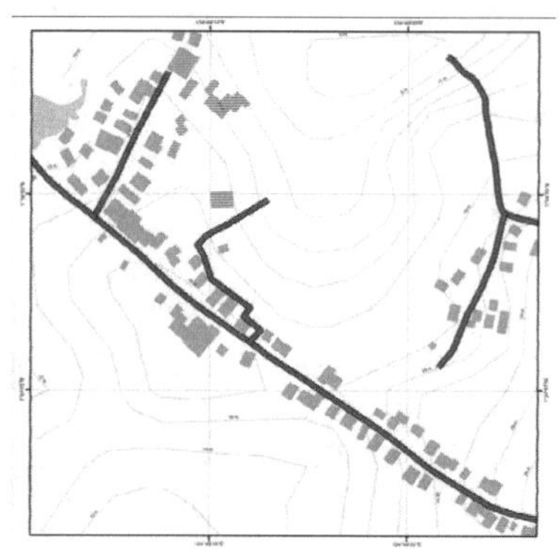

Figure 7. Contour Map of residential land in Tongkaina Village

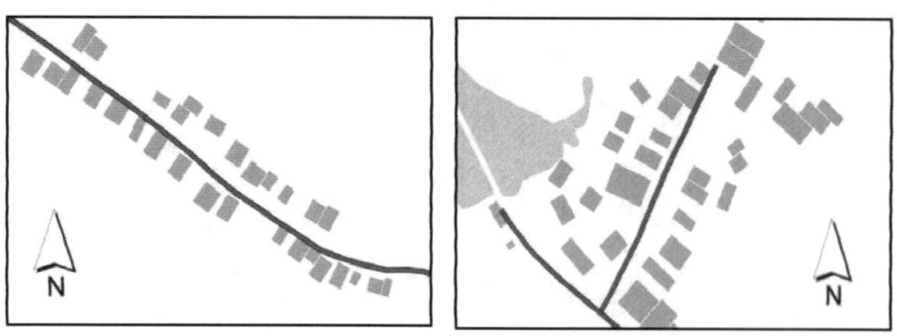

Figure 8. Settlement pattern of Tongkaina Village

4. Conclusion

Settlement patterns in the villages of Molas, Tongkaina and Meras are linear patterns with orientation facing the main road and at an altitude of 8-54 meters above sea level. Waste water flows from every house flowed into urban drainage and ends in sea waters or mangrove habitat.

Acknowledgment

This research was funded by the University of Sam Ratulangi through Unsrat Research Advancement schemes in fiscal year 2018.

References

[1] Swilling, M., & Annecke, E. (2012). Just Transitions: Explorations of Sustainability in an Unfair World. Claremont, South Africa : Tokyo, Japan: United Nations University Press

[2] Paola Pulido B. 2015. The Behaviour of Cities: Theoretical Framework for Urban Metabolism from an Ecological Perspective. Thesis. Univeriteit. Utrecht

[3] Prasad B.K, Dittmar T and AL Ramanathan. 2010. Organic Matter and Mangrove Productivity in Management and Sustainable Development of Coastal Zone Environments. Springer. UNESCO. New Delhi India.

[4] Alberti. M et al. 2007. The impact of urban patterns on aquatic ecosystems: An empirical analysis in Puget lowland sub-basins. Landscape and Urban Planning 80 (2007) 345–361

[5] Toengkagie, M.A. 2003 . Bunaken National Park, North Sulawesi (Indonesia) Charting a new course for the co-management of the Bunaken National Park North Sulawesi, Indonesia. http://agris.fao.org/agris-search/search.do?recordID=PH2004000965 diakses 20 Januari 2019

[6] Wardi et al. 2014. Characteristics And Changes In Patterns Of Settlement Fishermen Karang Panas Environment, South Ampenan Mataram City. Jurnal Penelitian UNRAM, Agustus 2014. Vol 18 No. 2

IOP Conf. Series: Earth and Environmental Science **328** (2019) 012056 doi:10.1088/1755-1315/328/1/012056

Village Development Index of Probolinggo Coastal Villages Case study: *Bhinor Village, Paiton District*

Y E Prasetya[1], A R T Hidayat[2]* and D Dinanti[2]

[1] Student of Urban and Regional Planning Department, Brawijaya University
[2] Urban and Regional Planning department, Brawijaya University

*Email: a.r.taufiq.h@ub.ac.id

Abstract. One of the factors affecting the development of villages through Village Development Index (IDM) is the access to information and communication technology. Paiton District has the highest level of tele-density and serves as strategic areas of Probolinggo Regency. The activities of the Paiton Steam Power Plant affect the socio-economic conditions of Bhinor community as well as an expansion of the coastline. This research aimed to find out IDM in Bhinor Village related to the existence of the Paiton Steam Power Plant. This study employed IDM composite index and literature review. The results indicated that characteristics of ICT infrastructure services in Bhinor Village tended to BTS services as well as Radio and Telephone. Paiton Steam Power Plant contributes positively to the social and economic activities of the residents in Bhinor Village. In environmental aspects, it produces pollution and changes in the coastline due to abrasion. This is in accordance with the results of the IDM calculation, in which Bhinor Village has a high value on the Economic and Social Security Index. It has a low value on the Environmental Resilience Index. Villagers can use ICT availability to deal with environmental issues and improve Environmental Resilience.
***Keywords:** village development index, ICT, village, Paiton Steam Power Plant*

1. Introduction

1.1 Background

Opportunities to create prosperity, justice, and regulate the village in their own way have been provided by the government with the existence of the Village Law [1]. The Law No. 6 issued in 2014 concerning Villages has developed a new paradigm and concept of village management policies. The village will be treated differently than before. The position of the village is no longer subnational, but is located in the regency/city area [2]. In the 2015-2019 National Mid Term Planning (*Rencana Jangka Menengah Nasional* or RPJMN), one of the targets of village development is to reduce the number of disadvantages village to 5,000 villages and to increase the number of independent villages in at least 2,000 villages in 2019 [3].

Indeks Desa Membangun (IDM) or Developing Villages Index was developed to strengthen efforts to achieve the development goals of the Village and Rural Areas. The development goals require locus (Village) clarity and development status. The IDM not only can provide benefits to investigate the development status of each village that is closely related to its characteristics, but also can provide development as an instrument of target in achieving RPJMN 2015-2019 and the coordination in village development [4]. IDM calculations at 73,709 villages are based on the 2014 village potential data (data potensi desa or PODES) with an average number of 0.566 indicating the following data [4]:

IOP Conf. Series: Earth and Environmental Science **328** (2019) 012056 doi:10.1088/1755-1315/328/1/012056

- Very Disadvantaged Villages: 13,453 Villages or 18.25%
- Disadvantaged Villages: 33,592 villages or 45.57%
- Developing Villages: 22,882 villages or 31.04%
- Developed Village: 3,608 villages or 4.89%
- Independent Village: 174 villages or 0.24%

The District that has the IDM lower than the average IDM Province Index is Probolinggo Regency with a value of 0.6303 (Ministry of Village, Development of Disadvantaged Regions and Transmigration, 2015). It is advised that some villages in Probolinggo District should be improved and developed in order to form independent villages.

One of the factors that influence Village Development based on the IDM is access to information and communication technology. Rural telecommunications development has a high and significant benefit to rural development. Besides, socio-economic provides significant advantages in social psychological and socio-cultural aspects [5].

The real conditions in Probolinggo Regency indicate a number of regions where there is no signal of telecommunication. However, about 80% of rural communities have used media both Information and communication technology in the form of wired and wireless [6]. The gap occurs between the northern and southern regions of Probolinggo Regency, where the northern region has a high level of teledensity, in contrast to the southern region. One of the regions with the highest level of tele-density is in the Paiton District area.

Being a region with a high level of teledensity, Paiton District becomes one of the strategic areas situated in the Probolinggo District Spatial Plan [6]. Paiton District serves as the North Coast Strategic Area with the Paiton Steam Power Plant as the Power Plant Centre for Java and Bali. Paiton District is located in the coastal area and traversed by the North Coast of Java. The coastal area of Paiton District is one of the coastal waters with quite high frequencies of people activities. Community activities ranging from aquaculture and fisheries, industries, and power plants where they are located at coastal area of Paiton.

The activities of the Paiton Steam Power Plant have affected the socio-economic conditions at Bhinor village as the location of the power plant due to industrial activities of Paiton Steam power Plant [7]. According to some of these issues, the IDM of Bhinor Village will be investigated. The study is carried out related to the characteristics of the implementation of ICT infrastructure in Bhinor Village and its relationship to the strategic issues of the Region at Paiton Steam Power Plant area.

1.2 Literature review

A. Developing Villages Index.

IDM classifies villages in five (5) statuses as follows [4]: (i) Very Disadvantaged Villages; (ii) Disadvantaged villages; (iii) Developing Villages; (iv) Developed Village; and (v) Independent Villages. The Village Classification is applied to reveal the diversity of characters in each village in a range of scores from 0.27 to 0.92 of the IDM. The classification in the 5 Village statuses is also to sharpen the determination of the status of village development and at the same time to recommend the necessary policy interventions. The status of Disadvantaged Villages explained in two statuses of Disadvantaged Villages and Very Disadvantaged Villages where the situation and condition of each village requires different policy approaches and interventions. Dealing with Very Disadvantaged Villages, the researchers will divide the level of affirmation of their policies compared to Disadvantaged Villages.

Developing Village is commonly related to the situation and conditions in the status of Disadvantaged Villages. Meanwhile, Very Disadvantaged Villages can be explained by vulnerability factors. If there is a pressure of vulnerability factors such as the occurrence of economic shocks, natural disasters, or social conflicts, it will affect the status of a developing village to drop down to a disadvantaged village. If a natural disaster factor is handled without rapid and appropriate responses, or the occurrence of a social conflict continues to occur, it is very risky that the status falls from Disadvantaged Village to a Very Disadvantaged Village.

In addition, the ability of the Developing Village to manage power, especially related to potencies, information value, innovation / initiative, and entrepreneurship will support the progress of the Developing Village to a Developed Village. The classification of village status based on the IDM is also directed at strengthening efforts to facilitate the promotion of the Village to achieve Independent Village status.

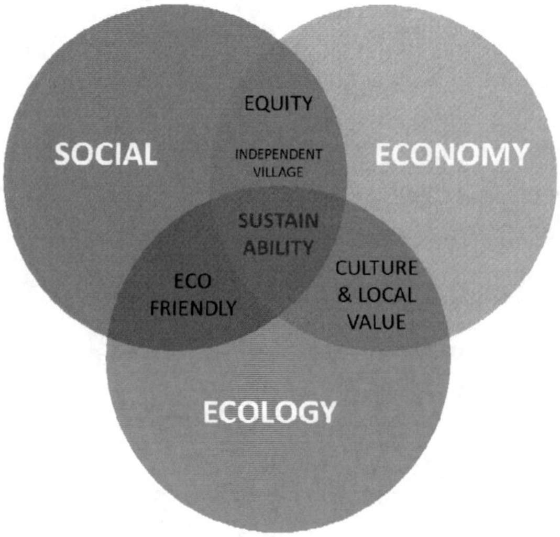

Figure 1. Three dimension of IDM [4]

IDM is a composite of social, economic and ecological resilience. The IDM is categorized according to 3 (three) dimensions and is further developed in 22 variables and 52 indicators.

B. ICT Infrastructure

ITU [8] develop ICT development index that represent how ICT is being used and its development. It consist of 3 main aspect, ICT readiness (i.e. infrastructure access); ICT use (i.e. intensity); and ICT capability (i.e. skill). ICT readiness holds important role because it allow ICT user utilize the technology. ICT is generally understood by the public as matters relating to computers and the internet. However, some people consider it as a very narrow view because traditional electronic equipment such as television, radio, telephone, and newspapers also carry information. ICTs are generally grouped based on the duration of the ICT used. The categorization is divided as follows [9]:

1. New ICTs: Computers, satellites, wireless communications (including mobile phones), internet, e-mail, and multimedia which are included in the category of new technology. The concept behind this new technology is a digital device.
2. Old ICT: Radio, television, telegraph, telephone. In general, these electronic gadgets are required to use analogue transmission techniques.
3. Very old ICTs: Newspapers, books and libraries. These media have been the most popular communication tool for years.

C. Relationship between ICT Infrastructure and Village Development

Information and Communication Technology provides benefits to ICT users in socio-economic, social psychology, and socio-cultural aspects [5]. The benefits of socio-economic are several indicators which has been showed as the benefits of Information and Communication Technology in rural communities. In the socio-economic aspects, Information and Communication Technology delivers benefits in the welfare aspects for the village community. For community leaders, Information and Communication Technology is very important in dealing with problems / affairs that occur in community institutions. In addition, the social-psychological aspects, Information and Communication Technology can increase prestige (pride) for the region. The increasingly close relationship between family and neighbours can be beneficial for Information and Communication Technology.

In the socio-cultural aspects, activities supported by Information and Communication Technology can accommodate the information distribution. By the existence of Information Technology and Communication, population mobility to access the region is easier to sustain the wellbeing of countryside community. According to research conducted by [5], it indicated that ICT affects 3 aspects such as social, economic and cultural aspect, including the IDM.

D. Steam Power Plant

Steam power plant is a power plant that employs steam power as the main driver of the turbine to produce electricity. Steam power plants runs in the form of hot water which is higher in temperature than the water temperature before being used for cooling down. The amount of cooling water needed depends on the maximum capacity of the units at the steam power plant. In general, cooling water at full load for each cooler with a relatively high temperature requires a lot of volume, and is continuously discharged into local waters. The cooling water receiver will gradually be affected by rising temperatures due to the discharge of cooling water [10]. Any prominent business in the rural area will affect to socio-economic of the local community as well as land use.

2. Methods

This study employed the calculation of the IDM and literature study which covers data collection from related agencies as well as reviews theories from the existing journals to obtain descriptive analysis. Calculation of the IDM is prepared by ensuring the availability of data sourced from the Village Potency, published by the Central Bureau of Statistics. IDM is a composite index that is built from social, economic and cultural dimensions. The three dimensions consist of variables, and each variable is revealed to be an operational indicator. The procedure for obtaining the IDM is as follows[4]:

1. Each indicator has a score of 0.5; the higher the score reflects the higher level of significance.
2. Each indicator score is grouped into variables, resulting in a variable score. The total variable score is then formulated into a following index:

$$Variabel\ Index = \frac{\sum \text{Indicator X}}{\text{Maximum Score (X)}} \qquad (1)$$

3. The index of each variable becomes a Composite Index called the Village Build Index (IDM).
   $$IDM = 1/3\ (IKS + IKE + IKL)$$
   IDM: *Developing Villages Index*
   IKS: Social Security Index
   IKE: Economic Resilience Index
   IKL: Environmental Resilience Index (Ecology)
4. The status of each village classification is carried out by calculating the range obtained from the maximum and minimum values. The range value obtained is the limitation of the status of each village, so that five village status classifications are established.

The study investigated the IDM in the strategic area of North Coastal affected by the existence of the Paiton Steam Power Plant to assess its impacts and potencies and the existing problems related to the IDM. In addition, it will also be discussed descriptively about Information and Communication Technology related to the number of telephone connections, the number of households with Internet access, the number of BTS, the number of households equipped with radio and the number of households equipped with TV; the relation to the IDM will be discussed as well.

3. Results and Discussions

3.1 General view of study location

Bhinor village is the largest village in the Paiton District. With a total population of 2,445 people who inhabit 13.97 ha. Most of the population is less than 40 years old. Many villagers work in the agricultural sector as farmers and farm labourers (444 people). The population working for the company is very small although its location is near to Paiton Steam Power Plant. They actually are benefited by selling and offering services to Paiton Steam Power Plant's workers. The village land is mostly managed for settlements and agriculture. Agricultural land dominates up to 90%. The location of the village is in the northern coastal region of the island of Java, but the village of Bhinor has no fish ponds. Bhinor village community has gained access to community services and infrastructure, including telecommunication. This is indicated by the high social and economic dimensions of the IDM. The following are the results of IDM calculations for Paiton District. Telecommunication infrastructure in the village of Bhinor is very complete. This village is passed through the main fibre optic line. There are at least 3 BTS including special BTS at the Paiton Steam Power Plant. It is quite reasonable that tele-density at Bhinor village is in very good state.

3.2 IDM of Bhinor and Paiton District

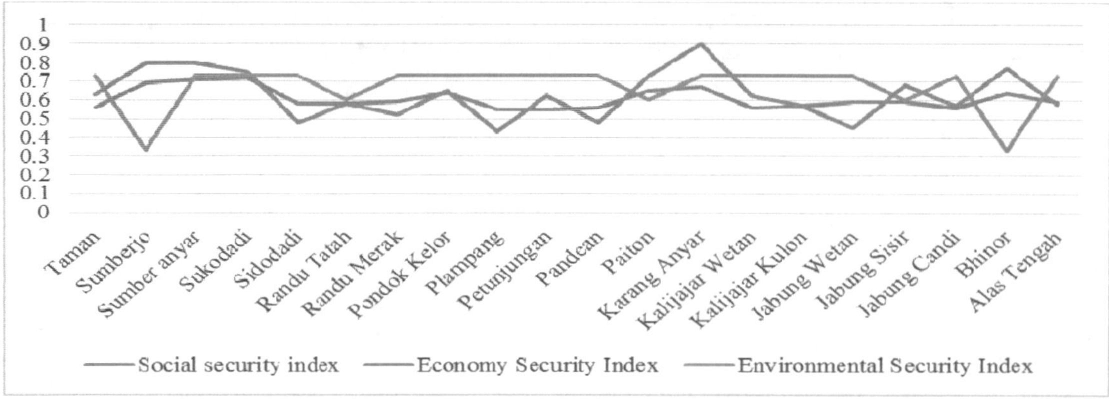

Figure 2. Social, economy, and environmental index of Paiton Distric

Figure 2 indicated that the average Social Security Index at Paiton District is 0.61. There are 13 villages that have a Social Security Index below the average (i.e. 0.61). The lowest Social Security Index is Plampang and Petunjungan because there are no health and education facilities in these two villages. The highest Social Security Index in Paiton District is in Sukodadi Village because there are junior and senior high school education facilities, health facilities are in the form of local health care centre and post office facilities.

According to Economic Resilience Index, Paiton District has an average Economic Resilience Index of 0.63 which is included in the developing category. There are 11 villages that have a Social Security Index below the average. The lowest Social Security Index is Plampang because there are no economic facilities in this villages. The highest Social Security Index in Paiton District is Karanganyar Village because there are economic facilities in the form of markets, shops and bank offices.

Paiton District has an average Environmental Resilience Index of 0.67 which is included in the developing category. There are 11 villages that have a Social Security Index below the average. The lowest Social Security Index is in Bhinor and Sumberejo villages because the two villages are areas where Paiton Steam Power Plant is located. The highest Environmental Resilience Index can be found in other villages except Randu Tatah, Bhinor, Sumberejo, and Jabung Sisir. These conditions can be seen in the following picture:

According to the research that has been done, it was found that total IDM Value at Bhinor village is 0.58 with a Social Security Index is 0.64, an Economic Resilience Index is 0.77 and an Environmental Resilience Index is 0.33. Based on these values, it can be concluded that Bhinor Village has a good development in the economic and environmental sectors. This is related to its location on the northern coast line with facilities and infrastructure to support existing social activities. However, the environmental aspect has a low value due to the large amount of pollution in the area, especially from the activities of the Paiton Steam Power Plant.

IOP Conf. Series: Earth and Environmental Science **328** (2019) 012056 doi:10.1088/1755-1315/328/1/012056

3.3 The Characteristics of ICT Infrastructure Services in Bhinor Village

Fixed line service

According to the results of the survey, it was found that the Telephone Network Service Level in Paiton District was 1%, indicating that telephone was very rarely used. The telephone network service in Paiton District is only in urban areas, such as at Sukodadi, Sumberanyar, Karanganyar and Paiton Villages. Bhinor village is a village outside Paiton, with no telephone connection. Although it is not as an urban area, Bhinor Village is an industrial area so that industrial development should require infrastructure development to support the existing activities [11], one of which is telephone network as a supporter in industrial activities.

Internet and BTS services

Similar to telephone networks, the distribution of internet networks is dominated by Paiton District urban areas. The level of internet service in Paiton District is 2.7% with the highest level of Internet service in Paiton Village. The internet today does not act as a solution to overcome geographical boundaries as predicted by many experts. Instead the Internet is currently only a supporter of activities and business in urban areas [12]. This is applied at Paiton District with the internet network employed as a supporter of activities, especially education and business activities in urban centre areas. In addition, businesses in the form of internet cafes are also located in these urban areas. Although Bhinor Village currently has no internet network, the level of tele-density is relatively high with the presence of BTS. The existence of BTS should be utilized, especially related to access of information through mobile phones.

Television and Radio Services

ICT Infrastructure Services in the form of television in Paiton District are 35% in a fair distribution. Unlike telephone, internet and BTS services, almost every home has a television. In addition, television in rural areas plays as substitution for other ICT services that do not reach rural areas in Paiton District. Television is recognized as the most important media for communication of rural populations in developing countries [13]. This applies to Paiton District for rural areas that have not been served. Other ICT infrastructure has been covered by television services. The level of radio service in Paiton District is 5% with a fair distribution in each village.

Similar to television, radio utilizes as a substitution to obtain information that cannot be accessed by using the internet. Radio is currently still considered important for the information necessity in rural communities. The existence of radio has made tremendous progress in the socio-cultural life [14]. Television service that can be reached by Bhinor Village is only 1.81% while radio is 0.03%, indicating a very low ICT infrastructure services that can be accessed by people at Bhinor Village.

3.4. Impact of PLTU on Social Economy

The construction of the Steam Power Plant (PLTU) in 1983 in Paiton District became a power plant centre which was established by the government with the assistance of foreign capital or private companies. It was aimed to fulfil electricity necessity in Java and Bali as the beginning of a change in balance in a system (community) at Bhinor village. Since the 1980s the government has turned the function of settlements into industrial estates by paying compensation causing many residents to leave work as farmers and to become labours at the Paiton Steam Power Plant (PLTU).

Job Opportunity Aspects

In the aspect of employment opportunities, the changing of working types of Paiton Steam Power Plant has been quite significant. When the Paiton Steam Power Plant was established, several new job opportunities for the service business began to develop, as seen from the trading business provided for the workers. Almost along the main road, there are stalls providing the workers necessity. With the ease of transportation routes to the village, it was beneficial for some residents who offer services such as motorbike or taxi drivers to empower motorized vehicles to get income. The broaden variety of occupations at the Bhinor villagers after Paiton Steam Power Plant established, it turned out to be very instrumental in changing socio-economic conditions. The majority of Bhinor village people has multiple jobs improving the income of each individual along with enhancing the quality of life.

Aspects of Social Interaction

Changes in social interaction were also very visible at Bhinor village community. According to the results of the study, it was found that there were changes in the socio-economic aspects of the Bhinor Village community related to the construction of the Paiton Steam Power Plant. Many Bhinor villagers at that time preferred working as PLTU employees, employment opportunities were increasing, thus affecting the level of income of the community. The latest changes in social interactions affect the concessions in physical and economic dimension, as well as the social community. The survey results revealed that the Paiton Steam Power Plant had a positive impact on the social and economic activities of the villagers of Bhinor, starting from increasing income to creating new employment opportunities.

3.5 Significances of Paiton Steam Power Plant on the Environment

According to research conducted by Saptarini [15], it was investigated that coastline changed from 1993-1996 and from 1996-2009. The shoreline had been changed in the coastal area of Paiton District in 1993-1996 predominantly caused by a coastal abrasion process for 56.1 ha area. The abrasion caused environmental changing inrice fields (658,158 ha), forests (193,744 ha) and industries (15.125 ha). The coastline change in 1996 - 2009 was related to the coastal abrasion process which caused a change in land area of 12.4927 ha. The abrasion causes limitation of land potency to be used. The type of land affected by abrasion consists of farm, industry and ponds, each of which is as wide as 870,903 ha, 73,667 ha and 85, 9307 Ha. In the period of 1996-2009, there was an addition in the coastline length for 1,193,489 m. The changing of land area due to the accretion process from 1993 to 1996 was 38.5345 ha. Besides, the changing of land area due to accretion from 1996 to 2009 was 11,863 ha. In addition, other environmental impacts that occured around the Paiton Steam Power Plant were related to the sustainability of biodiversity, especially marine biota around the PLTU area. In 2009-2010 indications of cases of coral bleaching in coastal locations in East Java were found around the Paiton Steam Power Plant [16]. From the previous study it was investigated that the problem with the Paiton PLTU was related to environmental aspects, especially in terms of pollution and shoreline changes due to abrasion. This is directly proportional to the calculation results of the Environmental Resilience Index in Bhinor village which has the lowest yield.

E. The Role of ICT Infrastructure in overcoming the problems of village development

From the results of research and calculations that have been carried out, Bhinor Village is included in the category of underdeveloped villages with the IDM value of 0.58. There are problems in the environmental aspects due to the large amount of pollution in the area mainly from the activities of Paiton Steam Power Plant. The existence of the Paiton Steam Power Plant and its waste disposal are inescapable as one of the consequences of industrial development. In order to minimize the negative impact, community response and education to the community are vital in addressing these problems so that the presence of waste disposal does not interfere the community activities.

In addition, information technology has been greatly affected by economic, educational, environmental, consumer behavior related to the environment, and personal life. With the existence of information and communication technology, people have experienced opportunities to share knowledge, thoughts, and to access valid information easily and quickly [17]. In Bhinor Village, the ICT infrastructure available consists of BTS, Radio and Television. BTS plays an important role in the spreading information through mobile media, which is more accessible and available while television and radio can provide information in audio-visual forms related to knowledge in managing waste disposal problems. Since rural areas in Indonesia are facing digital divide challenges [18], the usage of internet for obtaining information to manage waste is doubtable.

4. Conclusions

The results of the study indicated that the Paiton Steam Power Plant had positive significances on the social and economic activities, starting from increasing income to creating new employment opportunities. The problem arose from Paiton Steam Power Plant is related to environmental aspects, especially in terms of pollution and shoreline changes due to abrasion. This is directly proportional to the calculation results of the Environmental Resilience Index in Bhinor village which has the lowest yield. This is in accordance with the results of the IDM calculation, where Bhinor Village has high values on the Economic Resilience index and the Social Security Index, with a low value on the Environmental Resilience Index. The availability of Information and Communication Technology, should be utilized in tackling the existing environmental problems through environmental education so that information can be disseminated to the village community.

References

[1] Palupi, Sri et all. (2016). Guidebook for the Implementation of Rights-Based Village Laws. Jakarta: Lakpesdam PBNU
[2] Kurniawan, B. (2014). Book 5 : Independent Village, Village Development. Jakarta: Ministry of Village, Development of Disadvantaged Regions and Transmigration
[3] National Development Agency. (2014). National Medium Term Development Plan 2015-2019. Jakarta: Beppenas
[4] Ministry of Village, Development of Disadvantaged Regions and Transmigration. (2015). Indeks Desa Membangun (Developing Villages Index). Jakarta: Kemendes PDT.
[5] Supriadi, Dedi. (1992). Study of Customer Profiles and the Benefits of Rural Telecommunications on Social-Economics. Jakarta: PT. Telekomunikasi Indonesia
[6] Bappeda Probolinggo Regency. (2012). Probolinggo Regency Spasial City Planning. Probolinggo Regency: Bappeda
[7] Istiono, Feri. (2011). Evaluation of Coastal Zone and Land Cover Changes with Remote Sensing Data: A Case Study of Pasuruan, Probolinggo, and Situbondo Coastal Areas. Journal of Civil Engineering and Planning. Institut Teknologi Sepuluh November. Surabaya
[8] International Telecommunication Union. (2009). Measuring the InformationSociety. Geneva: International Telecommunication Union
[9] Greenberg, A. (2005). ICTs for Poverty Alleviation: Basic Tool and Enabling Sector. Stockholm : Sida
[10] Hutomo M., Arinardi O.H. 1992. Impact of Power Plants (Especially Thermal Waste) on Aquatic Ecosystems, Oseana, Volume XVII, Nomor 4: 135-158
[11] Johnson L Leland. (1991). Advances in Telecommunications technologies That May Affect The Location of Business Activity. Santa Monica : RAND
[12] Gorman, S. P. And Malecki, E. J. (2000). The networks of the Internet: an analysis of provider networks in the USA, Telecommunications Policy, 24, pp. 113–134
[13] FAO. (2001). International workshop on farm Radio broadcasting; Rome, 19–22 February 2001. Information and communication technologies Servicing farm radio: New contents, new partnerships. Final Report. Rome: FAO
[14] Amadu, FM and Alhasan A. 2018. Community radio in rural development in Northeastern Ghana: experiences of radio Gaakii in the Saboba District. International Research Journal of Social Sciences ISSN 2319–3565 Vol. 7(5), 1-12, May
[15] Saptarini, D., Muzaki F.K. (2010). Study on Coral Lifeform and Species that Susceptible of Bleaching in PLTU Paiton Water, JIWECC
[16] Prameswari, S., Agus, A., & Azis, R., (2014). Kajian Dampak Perubahan Garis Pantai Terhadap Penggunaan Lahan Berdasarkan Analisis Penginderaan Jauh Satelit di Kecamatan Paiton, Kabupaten Probolinggo Jawa Timur. Jurnal Oseanografi 3(2), 267-276.
[17] Seçken, N. (2005). Sthe Relations Between Global Environmental Awareness And Technology. The Turkish Online Journal of Educational Technology – TOJET January 2005 ISSN: 1303-6521 volume 4 Issue 1 Article 8
[18] Onitsuka, K., Hidayat, A. R. R. T., & Huang, W. (2018). Challenges for the next level of digital divide in rural Indonesian communities. Electronic Journal of Information Systems in Developing Countries, 84(2). https://doi.org/10.1002/isd2.12021

Spatial and Temporal Patterns of the Decreasing Rice Field in Sidorejo Sub-district, Salatiga City, Central Java Province, Indonesia

A Damayanti[1], A Setiadi[2], A Febrianti[3], B A Setyawan[4], P Angelika[5], and R P Setiadi[6]

[1]Department of Geography, Indonesia University, Pondok Cina - Depok

*Email: astrid.damayanti@sci.ui.ac.id

Abstract. The agricultural land conversion and all its utilization happens as time goes by. This conversion commonly happens in Indonesia, including in Sidorejo District, Salatiga City. Sidorejo is one of the potential areas of agriculture, especially for agriculture area in Central Java. The agriculture area in Sidorejo has decreased by 3.12 km^2 in the past 10 years (2007 - 2017). The purpose of this research is to map changes of rice field area in spatial and temporal manner. The research variables include topography, river network, farmer education level, farmers' income, and population. This study utilizes primary data obtained from interviews and questionnaires, while the secondary data is obtained from Centre of Statistical Bureau, Geospatial Information Bureau, and USGS DEM. The analytical method used in this study is overlay and transects in a systematically random sample. The results indicated that rice fields area changes in the western and eastern parts of Sidorejo for 10 years (2007-2017) which were generally triggered by population. Rice field area changing in west correlates with topography, river networks, education levels and low farmer incomes. Meanwhile, rice field area changing in east correlates with the level of education and high income of farmers.

Keywords: *Agrilucture, Land use, Salatiga.*

1. Introduction

Land is the earth surface with physical characteristics such as: provision of water and cover plants which are found as a human activity space [1, 2]. Land use becomes an aspect of human mastery as a description of human behavior towards land to achieve the desired objectives of the land use [3, 4]. Conversion of the use of agricultural land and its type occurs all the time. Conversion of agricultural land to non-agricultural land becomes one of the central issues of agricultural development due to its significantly negative impact on food production and environmental problems [5, 6, 7].

Land use conversion commonly occurs in Indonesia, including the District of Sidorejo, in Salatiga City [5]. Every development program regarding infrastructure requires suitable land. On the other hand, the need for human habitation increases along with the increase in population. The conversion of agricultural land use in Sidorejo District generally occurs into paddy fields. Sidorejo is one of the agriculture areas, especially rice potentially. Although it has become a city, Sidorejo still has extensive

IOP Conf. Series: Earth and Environmental Science **328** (2019) 012057 doi:10.1088/1755-1315/328/1/012057

agricultural land. According to BPS data in 2017, the total agricultural land area in Sidorejo District is around 3.78 km^2. The decrease in agricultural land area in Sidorejo Subdistrict from 2007 to 2017 reached 3.12 km^2. The implication of conversion of agricultural land to farmers is the change of agrarian structure in the farming community [5]. These implications are obvious after land conversion happens for a long period of time. Conversion of agricultural land to non-agricultural land becomes one of the central issues of agricultural development due to significantly negative impact on food production in addition to other socio-economic aspects and environmental problems [6].

The purpose of this research is to spatially and temporally map agricultural land area changes in Sidorejo District. This research is very important to determine the condition of agricultural land use with the assumption of a comparison of multitemporal population number (2007 – 2017) [7]. This research is employed as a reference in developing appropriate policies for the future land use management

2. Methodology

Sidorejo District is one of the Districts in Salatiga City. Geographically, Sidorejo is at 110^0 29 '38.4 "E and 7^0 18' 36" S. Sidorejo District is located in the north of Salatiga City. Sidorejo has an area of approximately 15.84 km^2 divided into six sub-districts of: Salatiga, Sidorejo Lor, Blotongan, Pulutan, Bugel, and Kauman Kidul.

The variables used in this study are topography, river, farmer's education level, farmer's income and population. Topography and river are obtained from the results of processing map data that has been verified in the field. Farmer's income and farmer's education level are obtained from interviews with farmers directly. The population is obtained from Statistical Centre Bureau data. All of these variables were correlated with changes in rice land use in Sidorejo District. Maps of multitemporal land use (2007 & 2017), river data in 2017, road data in 2017, and irrigation area data in 2017 were obtained from the Geospatial Information Agency. In addition, Digital Elevation Model data was obtained from USGS in 2017. The sampling technique used in this research is the stratified random sampling technique. The point of the sample was based on the distribution pattern of paddy fields in Sidorejo District. The pattern of the distribution of rice fields depicted in Sidorejo is rice fields in the west and east.

The analytical method used in this study is overlay and extraction of transect lines. The GIS Overlay implemented in this research is the overlay of several thematic maps arranged in layers with the aim at identifying correlations between them [8, 9, 10, 11]. Thematic maps data that are stacked in layers include: river, irrigation area, land use, and administrative boundaries. The technique of extracting the transect is obtained by cross-sectioning on a map with a description in general on the appearance of the transect line and is interpreted through legend [12]. Transect line drawings were implemented on seven research samples focused on the use of rice fields.

3. Result and Discussion

3.1. Sidorejo Land use in 2007

Sidorejo land use in 2007 had a total area of 15.48 km^2 (Figure 1). The land use covers rice fields land and non-rice land. The use of rice fields includes irrigated and non-irrigated rice fields. The use of rice fields has a total area of 4.46 km^2 (28%) which is concentrated mostly in Blotongan, Pulutan and Kauman Kidul. The use of non-rice land is for settlement, office affairs, plantations, grassland, and yard with a total area of 11.02 km^2 (72%). The use of non-rice land is mostly concentrated in the Salatiga, Sidorejo Lor, Bugel and Blotongan.

3.2. Sidorejo Land use in 2017

Land use area in Sidorejo District in 2017 has the same total area as in 2007, which is 15.48 km^2 (Fiqure 2). The use of rice fields land has a total area of 3.20 km^2 (20%). The use of rice fields is still concentrated mostly in Blotongan, Pulutan, Salatiga and Kauman Kidul. Whereas, the non-rice land use is developed into several categories, such as: trading, settlements, education, graveyard, yard, medical facility, industry, social facilities, plantation, and office affairs. The total area of non-rice land is 12.28 km^2 (80%). The use of non-rice land is still concentrated in the Salatiga, Sidorejo Lor, Bugel and Blotongan.

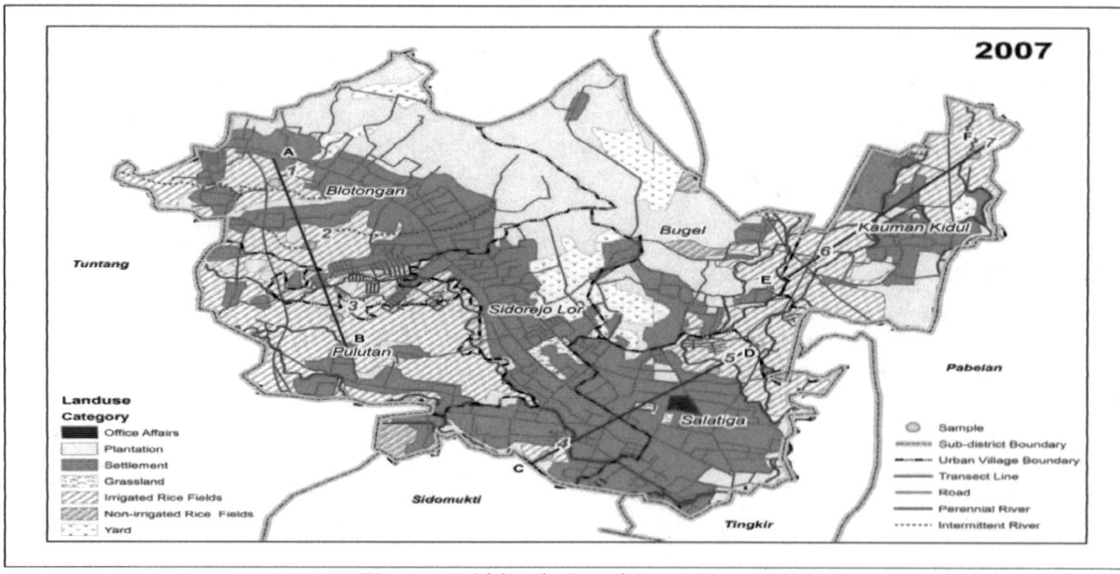

Figure 1. Sidorejo Land Use Map (2007)

Figure 2. Sidorejo Land Use Map (2017)

3.3. Multitemporal Rice Field Area Dynamicin Sidorejo (2007 – 2017)

The land use of irrigated and non-irrigated rice fields in Sidorejo District has a total decrease of 1.26 km^2 (29%) (Figure 3). The biggest decrease in rice fields occurred in Blotongan with a total decrease of 0.42 km^2 (33%). In addition, the type of rice fields located in Blotongan is transformed into non-irrigated rice fields. On the other hand, rice fields from irrigation type to non-irrigated type occurred in Pulutan and Sidorejo Lor, although each of their total decrease was very small with 0.02 km^2 (1%) and 0.06 km^2 (4%). Rice fields with unchanged type are in the Kauman Kidul, Bugel, and Salatiga with each of their total decrease in rice fields are 0.32 km^2 (25%), 0.20 km^2 (15%), and 0.24 km^2(19%). Besides, the west Sidorejo has changed significantly its type of rice fields in 10 years (2007-2017).

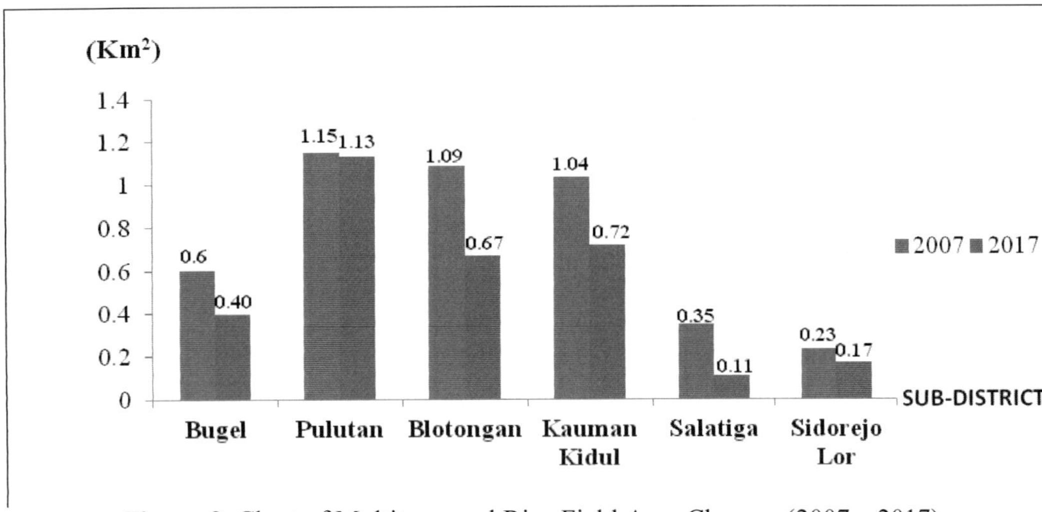

Figure 3. Chart of Multitemporal Rice Field Area Changes (2007 – 2017)

3.4. Correlation Between Physical Aspect and Rice Fields

Altitude is the physical aspect related to the land use [1]. Land use for rice plants tends to be placed in lower places. Rice plants require warm temperature to grow optimally at low altitudes. Based on the theory of the land evolution scheme, the area of residential settlements is above the height of the wetlands [1].

The AB transect line presents land use condition in 2007 from Blotongan to Pulutan (Figure 4). The first sample point on the AB transect line (26m ASL) is irrigated rice land use. The second sample point on the AB transect line (27.9m ASL) is irrigated rice land use. Sample points 3 on the AB transect line (29.5m ASL) is non-irrigated land use. The second sample point on the A'B' transect line (27.9m ASL) is non-irrigated rice land use. The sample point 3 on the A'B' transect line (29.5m ASL) is the land use of the education area.

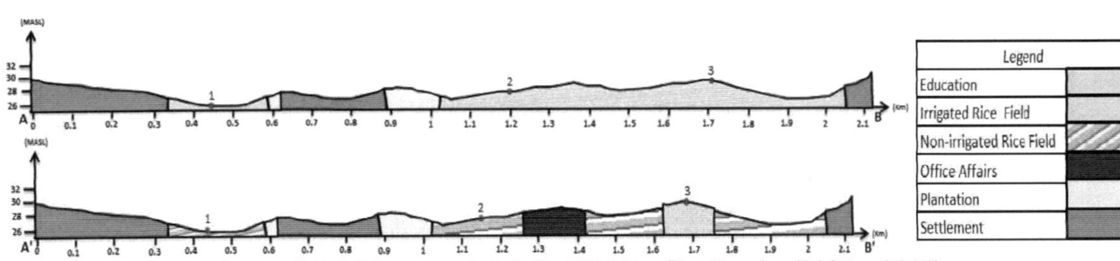

Figure 4. Multitemporal AB & A'B' Profile Graphs (2007 – 2017)

The CD transect line is a transect line of land use condition in 2007 from Sidorejo Lor to Salatiga (Figure 5). Sample point 4 on the CD transect line (26.7m ASL) is irrigated rice land use. Sample point 5 on the CD transect line (28.4m ASL) is irrigated rice land use. The C'D' transect line is a transect line for 2017 land use conditions from Sidorejo Lor to Salatiga (Figure 5). Sample point 4 on the transect line C'D' (26.7m ASL) is irrigated rice land use with reduced area than previously known (2007). Sample point 5 on the transect line C'D' (28.4m ASL) is settlement land use.

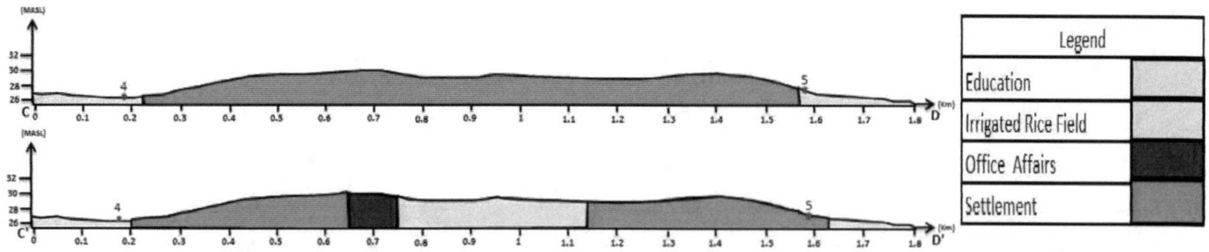

Figure 5. Multitemporal CD & C'D'Profile Graphs (2007 – 2017)

The EF transect line is a transect line of land use condition in 2007 from Bugel to Kauman Kidul (Figure 6). Sample point 6 on the EF transect line (30m ASL) is irrigated rice land use. Sample point 7 on the EF transect line (28m ASL) is irrigated rice land use. The E'F' transect line is a transect of land use condition in 2017 from Bugel to Kauman Kidul (Figure 6). Sample point 6 on the E'F' (28m ASL) transect line is settlement land use. The sample point 7 on the E'F' transect line (28m ASL) is trading land use area.

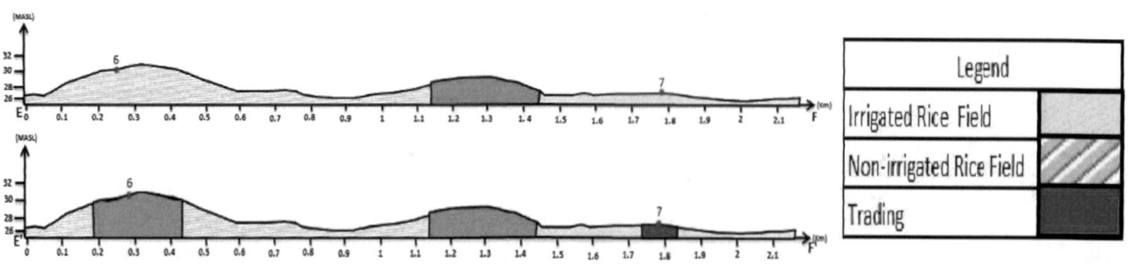

Figure 6. Multitemporal EF & E'F' Profile Graphs (2007 – 2017)

The low topography in Sidorejo is utilized more by the residents as rice fields. The high topography is used by the population as an area of built up land such as settlements and other built-up land facilities [14]. The topography in the west of Sidorejo is lower when compared to east Sidorejo. This situation allows the river network in the lower to reach Sidorejo blockage due to sedimentation transported from the upstream of Sidorejo [6]. The type of river that is in the Blotongan is turned into a perennial river drying the irrigation area around the Blotongan paddy field. The change in the type of river and the irrigation area causes the type of irrigated rice fields in Blotongan turned into non-irrigated rice fields. Although the irrigation area in Pulutan is not dry, the irrigation area in Pulutan has been distributed to people's homes causing the dried up and changing their rice fields type into non-irrigated rice fields. The types of rice fields in the east of Sidorejo survive as irrigated rice fields because the river network and irrigation areas are not clogged and dried up.

3.5. Correlation Between Human Aspect and Rice Field

The population in Sidorejo has increased multi-temporally (2007 - 2017). The population in all Sidorejo in 2017 was 21738, which was more than the total population in 2007 (Figure 7). The increasing population of Sidorejo Sub-district resulted in increasing housing needs [7]. The use of vacant land in Sidorejo Sub-district is so limited that most residents of Sidorejo Sub-district use rice fields to be converted into settlement areas and other built facilities.

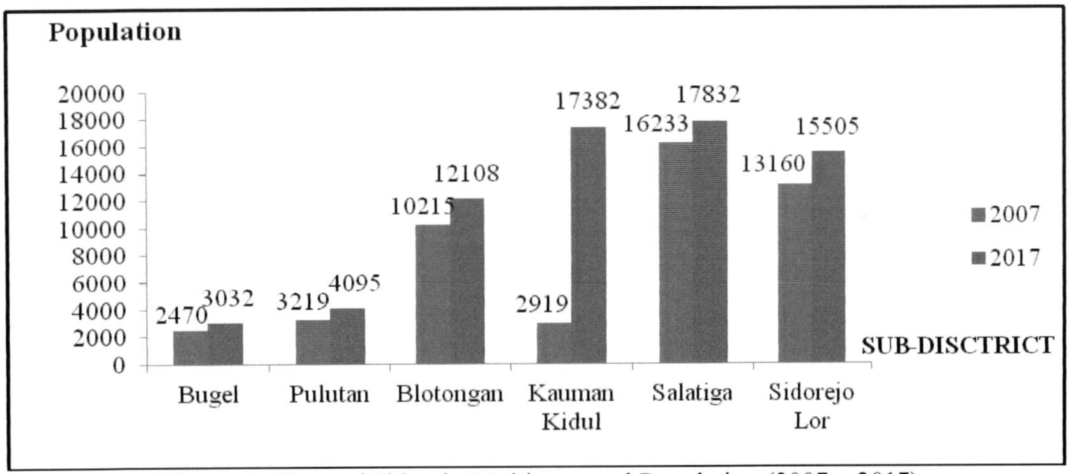

Figure 7. Chart of Sidorejo Multitemporal Population (2007 – 2017)

Based on table 1 column (b), the average level of education of farmers in west Sidorejo is completed only at the elementary level. While the education level of farmers in east Sidorejo is completed on average to high school level. This has an effect on the readiness of urbanization that occurs in Sidorejo Sub-District which has an impact on their rice fields [15]. Farmers in the eastern part of Sidorejo, especially the Kauman Kidul, tend to make their fields as part of catfish farms to earn more income. While some rice farmers in the Salatiga and Bugel use their paddy field compensation money which has been partially converted into infrastructure and local residents' settlements for business or work in fields other than their rice fields. Most farmers in the western part of Sidorejo only depend on their paddy fields which are reduced and changed their type without any innovation and creativity.

Based on table 1 column (c) and (d) shows the income of farmers is influenced by the number of harvests. Rice fields that have one to two harvest frequencies are non-irrigated rice fields. Rice fields that have a harvest frequency of two to three times are irrigated rice fields. The income of farmers in the west is less when compared to farmers in the east. Farmers in the east get the benefit from rice fields with a harvesting frequency of two to three times a year plus the efforts of farmers in other job or business. Farmers in the west can only harvest in one to two times a year because they are waiting for the rainy season to care for their rice having no business in other job.

Table 1. Farmer's Income, Education, and Rice Harvest Frequency

Farmer (a)	Education (b)	Rice Harvest Frequency (c)	Income per harvest (d)
1	Elementary	twice	IDR 1.500.000
2	Elementary	once	IDR 1.000.000
3	Elementary	twice	IDR 2.000.000
4	Elementary	twice	IDR 3.000.000
5	Senior High School	twice	IDR 3.000.000
6	Senior High School	three times	IDR 4.000.000
7	Senior High School	three times	IDR 4.000.000

4. Conclusion

Changes in irrigated and non-irrigated rice fields from 2007 to 2017 decreased by 1.26 Km^2. The change in non-rice land from 2007 to 2017 increased by 1.26 Km^2. The most reduced rice fields occurred in Blotongan with a decrease of 0.42 Km2. On the other hand, the type of rice fields in the

Blotongan has changed significantly from irrigated rice fields to non-irrigated rice fields. The least reduced paddy fields occurred in the Pulutan with a decrease of 0.02 Km^2. Even though the paddy fields decreased the least, the types of rice fields in the Pulutan Village changed significantly also from irrigated rice fields to non-irrigated rice fields. Changes in area and type of paddy fields are influenced by altitude, river network, population, farmer education level, and farmer's income.

Acknowledgments
The authors would like to thank the Department of Geography, Faculty of Mathematics and Natural Science, Indonesia University, which has facilitated the second field study class for the research activities of undergraduate students and the lecturers in Salatiga City.

References

[1] Sandy M I 1999 *Tanah Muka Bumi* vol 1 (Jakarta: Geography Department, Universitas Indonesia) p 1
[2] Koji T, Mattulada, and Narifumi M 1986 *Environment, Landuse, and Society InWallacea* vol 5 (Japan: Kyoto University) p 105
[3] Frederick S 1999 *The Living Landscape* vol 2 (United States: Arizona State University) p 28
[4] Benjamin C and Peter K 2010 *Global Environmental Forest Policies* (London & Washington: Earthscan) p 159
[5] Marko K, Zulkarnain F, and Kusratmoko E 2016 *Coupling of Markov Chains and Cellular Automata Spatial Models to Predict Land Cover Changes* (Jakarta: Geography Department, Universitas Indonesia) **47** 012032
[6] Alistair F P 1985 *Geomorphology and Trends* vol 1 (London) p 56
[7] Horstmann K and W Rutz 1971 *The Population Distribution on Java* (Tokyo: Institute of Developing Economies) p 29
[8] Gautam N C and Chennaiah G C 2007 *Landuse and Landcover Mapping and Change Detection in Tripura Using Satellite LANDSAT Data* **6** p 517-528
[9] Tunrayo A, Xiangming X, Kai S, ad Chandresahekar B 2008 *Mapping and Monitoring Agricultural Landuse in West Africa* (Africa: Association of Remote Sensing of the Environment) PMB 5320
[10] Supriatna 2001 *Dasar-Dasar Sistem Informasi Geografis* vol 1 (Jakarta: Geography Department, Universitas Indonesia) p 1
[11] Nicholas C 1997 *Exploring Geographic Information Systems* vol 1 (Washington DC: University of Washington) p 105
[12] Kang T C 2006 *Introduction to Geographic Information Systems* vol 5 (New York: McGraw-Hill Education) p 281
[13] Dwyer D J 1964 *Irrigation and Land Problems in the Central Plain of Luzon* vol XXTH International Geographical Congress pp 236 – 246
[14] Sandy M I 1996 *Republik Indonesia Geografi Regional* vol 3 (Jakarta: Geography Department, Universitas Indonesia) p 24
[15] Sayogyo 1985 *Pertanian dan Kemiskinan di Jawa* vol 1 (Jakarta: Gramedia Indonesia) p 16

Space Utilization and Transformable Architecture of Peri-Urban Co-Living Concept in Rancaekek, Bandung

A N Saputra [1], D G Lineker[1], H E Hibaturrahim[1], D K Nilla [1], RSobandi[2], AS Ekomadyo [3]

[1]Architecture Undergraduate Study Program; School of Architecture, Planning, and Policy Development;Bandung Institute of Technology; Jalan Ganesha no 10, Bandung; Indonesia
[2] Tunas Nusa Educational Foundation, Bandung, Indonesia
[3]Architectural Design Research Group, School of Architecture, Planning, and Policy Development; Bandung Institute of Technology; Jalan Ganesha no 10, Bandung; Indonesia

*Email: aliefianto03@gmail.com

Abstract. Peri-urban issues are generated by the problem of uncontrolled increase of settlement land-use that converts and decreases the agriculture and water absorption land-use. Development improvement is needed to support food protection and environment sustainability from urbanization pressure.Co-living Rancaekek in Bandung's peri-urban area is a concept which aims to improve the effectiveness of land use by providing multifunctional living space and household food production through hybrid infrastructure. This residence consists of five families living together in a modular housing unit with an area of 21 m^2. Each unit supports families with different responsibilities: farming; food processing; capacity building; building and infrastructure management; and special units for external researchers. The analysis in this paper provides evidence that each responsibility requires specific spatial planning and multifunctional furniture that can support various activities, tools, and processes to enable adaptation in limited space. The solution to space utilization is determined through the proposed movement of furniture based on a review of transformable architecture, namely flexibility and functionality of the space being utilized, to maximize functions in a minimum size space. This research requires deeper research and/or experiments on materials, community acceptance, and a more in-depth review of furniture design in accordance with the concept of transformable architecture.

Keywords: peri-urban, co-living concept, transformable architecture

1. Introduction

Peri-urban areas can be described as the land between urban and rural areas that have mixed characteristics of both city and countryside. Therefore, urban and rural uses often clash and encourage conversion of many things such as land, profession, and infrastructure. Land conversion cannot be avoided while agricultural and water absorption land transformed into industrial and residential areas. It can be a threat for food production and groundwater availability. One of peri-urban concerns is how to keep the agricultural land as a resource and how to keep environmental stability against urbanization pressures.

Peri-urban Co-living concept emerges as a concern to mediate agricultural rural living and industrial and service urban living. In Rancaekek, a peri-urban area of Bandung, co-living is designed to address water issues and food security by creating integrated housing with combined agricultural and service activities. This concept is aimed to reduce carbon footprint and contribute to environmental development by generating residents' environmental awareness lifestyle. In this place, four families participate in co-living, work together to meet their needs, and share each duty. These four families inhabit identical houses which are built integrated in one common land.

However, the distinction in work roles among families poses an issue in similar occupancy, including the absence of adjustment for supporting needs that follows the duties of each family within the identical houses. The co-living design is considered flexible, where the spaces are designed to accommodate a variety of functions and are not fixated for one main function [7]. Based on these problems, the objectives in this paper is to find adjustments towards the occupancy of each family based on transformable architecture to support their lives to the fullest.

1.1 Study of Transformable Architecture

The movements in architecture can be interpreted in several perspectives to meet their purposes. The moving architecture defined in several definitions, including kinetic, responsive, transformable, and adaptable [6]. To address the limitation in space, one of the defined categories that can be utilized is transformable architecture. Transformable can be interpreted as the ability of an object to make changes to adjust to environmental conditions. So, in more detail, transformable architecture is a design that can vary according to different functions [7]. Referring to the classification made by [7], these transformations are regrouped into four objectives to address climate problems, space openness, flexibility, and functional transformability.

In the context of co-living, the concept of flexibility and functionality is chosen as an approach to solve space limitations and save extensive use of space. Flexibility is a concept of open space with the minimal partition for various functions. Meanwhile, the functional concept of transformability is a strategy to utilize space for more than one function, used at different times to save space. The difference between flexibility and functional transformability is that in a flexible space, users can use the free space for extensive functions, while in space with functional concepts of transformability, its function has been determined. However, the space can be replaced repeatedly or can be returned to their initial conditions.

1.2 Co-living Design Concept in Rancaekek Bandung

Co-living in Rancaekek is designed to attain resilience in Bandung peri-urban area. It accommodates the need for supported value, market, and ecological connectivity within the region to reach an integrated network of the social, economic, and physical for a peri-urban area. [1] The house used for this concept is located in an area of 165 m^2 which can accommodate four housing units measuring 21 m^2 with minimum border and still produce at least 66 kg of carbohydrates. The co-living house is designed to accommodate four families with shared rooms and private spaces for each family. The units of each family are arranged stacked with one another to form circulation and privacy for each unit, added with farming space to fulfill the minimum amount of carbohydrates.

The co-living experiment will be carried out by four families in four units under these hypotheses:

- Inhabiting the first floor with two units are farmer families and the family of the building management and infrastructure team.
- The second floor has two residential units which are occupied by the family who processes food and the family of capacity builder.
- The third floor has one residential unit which will be occupied by researchers.

- Each family consists of four members; two adults (parents) and two children under 14 years of age

These four families have different responsibilities to maintain the sustainability of the co-living. The difference between co-living and co-housing lies in the responsibility of the family members. Families in co-housing only live together without sharing responsibilities, while families living in co-living share the responsibility to maintain the sustainability of co-living to be independent and self-sufficient. A brief description of the residents includes:

- The family of farmer has the role to plant and harvest food for consumption of the co-living families, through conventional farming, vertical farming, and roof gardening.
- The food processor family processing food from raw ingredients into consumable food to supply the five families' nutrition.
- The family of building and infrastructure managers has the role to manage and provide the necessity needed in the infrastructure and ensure the functions running in the building and infrastructure
- The capacity builder family has the role to improve the capacity level of soft skill and hard skill in the form of education for the families living in co-living

The designs in this paper consist of four typical house blueprints that share similar specifications. To support the needs of its residents, on the first floor there are communal spaces, conventional gardens, car parking space and parking space for three-wheeled motorcycles. At the split level towards the second floor, there is a drying room for units on the first floor. On the second floor there are two communal spaces for joint activities while on the top floor there is a rooftop garden (fig.1).

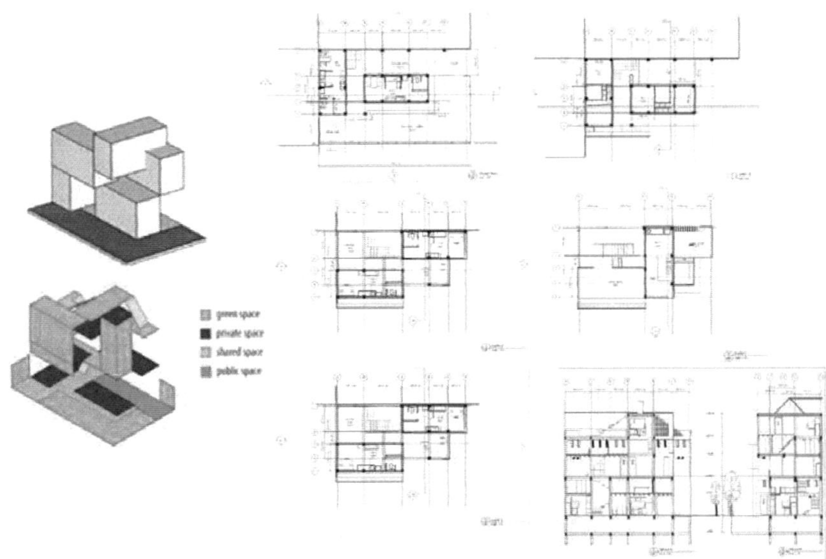

Figure 1. Co-living building concept
(Source: Sobandi, 2017)

The similarity of each unit floor plan indicates flexibility of use and unspecified activities [9]. However, Rancaekek co-living concept has a specific errand for each family that promotes the needs of supporting spaces with different specifications and purposes. The complexity of requirements is the main reason for the need of variations and specification in design [13]. The specification-based design approach is often found in limited area housing design [13]. Limited area requires specific functions and reduces the possibility of space to the level that it is only provided by the designer.

Therefore, Rancaekek co-living concept design requires adjustments from a typical residential unit to a specific unit for each family, including the adjustment of furniture and layout.

2. Method

This research employs descriptive and qualitative research. Descriptive research is done by conducting simulations to obtain the idea of the setting and the activities that occur within the object of research. The description presented is a variable related to the events that occur within the object of research whereas qualitative research is concerned towards assessing and carrying out analysis based on perceptions and cognitive constructs on the object of research.

The data collection in this study is obtained through studies of archives and simulations. Literature studies are also used to understand the basic idea of the concept, starting from the background of co-living in peri-urban and its relation in a broader context, also the concept of multifunctional space. In this study, multifunctional space theory is elaborated in a discussion of the theory of transformation in architecture that examines the flexibility of space and its completeness.

Transformable architecture directs to design flexible space when users want a space not only for one function but can be easily used for other functions as well [7]. Simulation in the form of role-playing is conducted to identify the families' daily routines in one week. Each role will further be processed into space needs that are related to flexible space theories. Functional transformability is classified again into four criteria for space saving solutions including movable functional walls, where functional components are placed in a field which can be moved indoors; function on the outside, functional components are placed in fields in the room (walls, roof or floor) to be moved manually for use; functions in the center, functional components that are stored centrally to be used by transferring to surrounding spaces; and movable walls, which divide the spaces for two functions carried out simultaneously. The four solutions are defined as ways to access various needs provided by furniture as complimentary in a space.

3. Results and Discussion

A role-playing simulation has been done as a method to collect data of all the families' activities in a week based on their specific role and responsibility. This simulation is aimed to have an objective result from the setting and activities that occur in co-living. The result of the simulation is qualitative data that could describe the possible and suitable activity related to the families' work function. These data are then used for the next analysis to differentiate particular spatial utilization from each family' to response to their role in co-living society.

Table 1. The schedule of daily activities for each family in one week, related to their respective responsibilities.

Time	Monday	Tuesday	Wednesday	Thursday	Friday	Saturday	Sunday
05.00 - 06.00	Watering the plants and giving fertilizer						
06.00 - 07.00	Checking the mushroom's bag log, feeding the chickens and fishes						
07.00 - 08.00						Fruits and vegetables harvest	Morning Workout Periodical infrastructure checking
15.00 - 16.00		Making snacks for a week	Education on gardening for children and parents	Making mushroom bag log, Breadfuit (*sukun*) or sweet potato harvest	Education on gardening for children and parents	Corrective maintenance infrastructure	Storing harvests
16.00 - 17.00					Periodical infrastructure checking		Health check-up, assisting *karang taruna*
17.00 - 18.00	Processing harvest crops				Making compost	Cooking meal together	
18.00 - 19.00						Life skill workshop barbeque/ Watching movie together	
19.00 - 20.00					Processing food waste into compost		

Farmer family

Food processor family

Capacity builder family,

Building manager and infrastructure family

3.1 Farmer Family
The farmer family fulfils the needs of food for all residents by utilizing the occupied land owned in the building. From what has been provided, the farmer family can manage and fulfil the number of nutritional needs for all residents. To maintain this, the farmer family is responsible for the process of planting up to harvesting food. The process of managing the land is completed through conventional agriculture, vertical farming, and roof farming. To carry out this function, the farmer family is expected to have knowledge about types of food production. They must be able to plan and manage crop-planting time, and have insights on the types of food combinations.

Table 2. Farmer family activities in a week.

Family	Activity	Time	Room Needed
Farmer	Watering plants and giving compost	Every morning, one hour	Gardening equipment storage
	Checking the mushroom's *bag log*, feeding the chickens and fishes	morning, one hour	Mushroom bag log storage, Food storage
	Making mushroom *bag log*, *Breadfuit* or sweet potato harvest	Thursday afternoon, two hours	Gardening equipment storage
	Making compost, collecting wastes of dried leaves	Friday afternoon, one hour	Gardening equipment storage, waste and compost shelter
	Fruits and vegetables harvest	Saturday morning, one hour	Harvesting equipment storage, harvest storage
	Storing harvests	Sunday afternoon, one hour	Harvest storage

3.2 Food Processor Family
The food processor family is fulfilling the nutritional needs of all residents by adding value to the crop. To do so, the farming family is responsible for the number of processed foods for a year. To carry out this function, the food processor family is expected to have the ability to process food with the skill they have, they can plan food preparation schedule and easily manage the process. The ability to process food includes knowledge about simple recipes, the ability to use cooking utensils well, the ability to store good food, as well as processing leftovers. Before the food processing process, food processor families need to have insights regarding the nutrients needed in food. During this process, this family also needs to be innovative by developing a wide range of menu for processing foods.

Table 3. Food Processor Family activity in a week

Family	Activity	Time	Room Needed
Food Processor	Processing and storing harvests	Monday afternoon, one hour	Food storage
	Making snacks for one week	Tuesday afternoon, two hours	Kitchen
	Processing food waste (Compost)	Friday night, one hour	Compost storage
	Cooking meal for everyone	Saturday afternoon, one hour	Communal room
	Storing harvest	Sunday afternoon, one hour	Harvest storage

3.3 Building and Infrastructure Manager Family
The building and infrastructure manager family has the role to provide the infrastructure needed by residents and ensure that the infrastructure and building system run well. This family has the

responsibility to control the building maintenance, including piping, electricity, building construction, and the use of hybrid technology needed for efficient and sustainable water and electricity. The technology used includes solar panels, drip irrigation systems, and rainwater harvesting. There are two types of control that must be done including preventive maintenance as a routine and preventive maintenance control; cleaning, inspecting, adjusting, updating, etc., and corrective maintenance, which is a repair, removal, replacement, or modification of damaged items or systems. This routine is done twice a week (corrective maintenance). Besides, this family has the chance to seek extra income as a handyman available for the community as well as construction and building consultant.

Table 4. Building and Infrastructure Manager Family's Activity in a week

Family	Activity	Time	Room Needed
Building and Infrastructure Manager Family	Preventive Maintenance #1	Monday morning, one hour	Tools storage
	Preventive Maintenance #2	Friday afternoon, one hour	Tools storage
	Corrective Maintenance #1 and #2	Saturday afternoon, one hour	Tools storage
		Saturday afternoon, one hour	Tools storage
	Watering plants by drip irrigation system	Everyday (automatically)	Tools storage
	Processing grey water and rain water harvesting	Everyday (automatically)	*Grey Water* storage tank Rainwater storage tank

3.4 The Capacity Builder Family

This family is the initiator of the residents' communal activities to improve the quality of social relations between each family, children's education related to the importance of food security, and hard skill training to carry out necessary practices such as gardening, food production, treatment, and life skills that can be used daily. In addition, this family also has the role of conducting quality improvement education on health, soft skills, intrapersonal communication, and interpersonal communication. The form of activities that can be carried out for capacity building is in the form of training, courses and workshops and activities that actively involve all residents of co-living, such as exercising, watching movies together, and cooking. A health control system can be done by working together with local health clinic (*posyandu).*

Table 5. The Capacity Builder Family's Activity in a week

Family	Activity	Time	Room Needed
Capacity Builder	Morning Workout	Sunday morning	Communal room
	Education on gardening for children and parents	Wednesday and Friday afternoon	Communal room
			Books Storage
	Health Check	Sunday afternoon	Health equipment storage
	Assisting *Karang Taruna*	Sunday afternoon	Communal room
	Life-skill workshop	Saturday night	Workshop equipment storage
	Barbeque	Saturday night	Communal room
	Watching movies together	Saturday night	Projector storage

All results of the analysis of space requirements in each family are combined to obtain a space utilization solution that refers to the Reuter classification. Adapted to the needs of the activity, this classification provides furniture solutions with suggestions for the space needed and possibilities of furniture's movement to save space. The wide range of the movement possibilities from one to another is needed to fulfil the space-saving requirements of furniture suitable for each family to do their activities.

Table 6. Analysis results of spatial utilization solutions referring to [7] classification.

Family	Room needed	Suggested area for storage	Space-saving solution	Possible movement
Farmer	Gardening equipment storage	Inside or outside the housing unit	Movable functional walls	Slide
	Mushroom bag log storage	Yard or communal room	Functions on the outside	Swivel
	Food storage			
	Compost waste storage	Yard	Functions in the center	Slide
	Harvest equipment storage, harvest storage	garage/ yard/ communal room	Movable functional walls	Slide
	Harvest storage	*Rooftop garden*, garden	Functions on the outside	Flap
Food Processor	Food storage	kitchen	Movable functional walls	Fold
	Cooking equipment storage	Inside housing unit (kitchen or family room)	Functions on the outside	Fold
	Food Processing	Kitchen	Functions in the center	Slide
	Compost Storage	Conventional garden	Functions in the center	Slide
	Cooking meals together	Communal room and garden	Movable walls	Fold
	Harvest room storage	Communal room	Moveable functional walls	Slide
Building and Infrastructure Manager	Tools equipment storage	Communal room	Movable functional walls	Rotate
	Grey Water storage tank	Outside housing unit, outdoor, children are unable to access	Movable walls	Fold
	Rainwater storage tank			
Capacity builder	Communal room	Outside housing unit, outdoor	Movable walls	Fold
	Books storage	Indoor communal room	Movable functional walls	Slide
	Health equipment storage	Indoor communal room	Functions on the outside	Slide
	Workshop equipment storage	Communal room	Movable functional walls	Slide
	Projector storage	Indoor housing unit	Functions on the outside	Slide

4. Conclusion

Based on the simulation that has been done, the design of a typical flexible space has not been able to accommodate the activities of Co-living occupants. The adjustment of Co-living's residential in Rancaekek can be done with the concept of functional transformability, which is a concept of adapting the use of space for various functions. This concept requires a transformation of functional space into another required space that is further continued by analyzing its furniture needs. In the adaptation, multifunctional space requires furniture with the criteria that suit best with transformable architecture solutions and the proposed transformation. The results of this study indicate that a typical unit plan on the Rancaekek co-living concept should be supported by furniture designs that fit the criteria of functional transformability. The design of furniture is made specifically for each type of family based on their work. Therefore, we recommend further research on furniture design that is suitable to the criteria required for Rancaekek's co-living stakeholders. In addition, experiments are also needed regarding furniture materials and community acceptance towards this concept.

Solution of transformable architecture's flexibility and functionality for co-living design in Rancaekek Bandung is a part of social concern exercise into built environment design. The social puts-built environment design not only concern about form and technical matters, but also about how design becomes an effort to provide human right through involving and empowering the societies [10] [8], [4]. Social space produced by architectural design represents the human complex social desire for a better future (2010:29613). Here, architecture becomes relational, as mediator for social action programs by shaping, conditioning and facilitating the possible sociality [12].

Acknowledgements
We would like to thank. R Sobandi, Mrs. from Tunas Nusa Foundation, who provided insight to Peri-urban and Co-living concept from the actual social and environmental activity at Rancaekek, Bandung that inspired this research. We would also like to show our gratitude to A S Ekomadyo, Drs. For sharing his pearls of wisdom with us in the form of assistance, comments, and reference during the course of this research that greatly improved the manuscript of this research.

References

[1] Chang D 2017 *Co-living in Bandung Basin Peri-Urban: An Interdisciplinary Model towards Sustainable Urbanization* HABITechno International Seminar – Ecoregion as a Verb of Settlement Technology and Development.
[2] De Paris S R and Lopes C N L 2018*Housing flexibility problem: Review of recent limitations and solutions* Frontiers of Architectural Research, **7**(1), p 80-91. doi:https://doi.org/10.1016/j.foar.2017.11.00
[3] Dovey K 2010 *Becoming Places: Urbanism/ Architecture/ Identity/ Power* London: Routledge
[4] EkomadyoAS and YuliarS2014*Social Reassembling as Design Strategies'. 5th Arte Polis International Conference and Workshop Reflections on Creativity: Public Engagement and The Making of Place Arte-Polis 5, 8-9 August 2014, Bandung, Indonesia*. Science Direct - Procedia - Social and Behavioral Sciences 184(2015) 152–160, 1877-0428 © 2015, doi: 10.1016/j.sbspro.2015.05.075
[5] Kronenburg R 2007 *Flexible:Architecture that Responds to Change* London: Laurence King.
[6] Lee J D 2012 *Adaptable, kinetic, responsive, and transformable architecture: an alternative approach to sustainable design* (Doctoral dissertation). University of Texas.
[7] Reuter R 2017 *Space-saving Techniques by the Use of Transformable Architecture.*
[8] Sanoff H 2000 *Community participation methods in design and planning*New York: John Wiley and Sons.
[9] Schneider T and Till J 2005 *Flexible housing: opportunities and limits* ARQ: Architectural Research Quarterly, **9**(2), p 157-166.
[10] Sommer R 1983 *Social design: creating buildings with people in mind* New Jersey: Prentice Hall Inc.
[11] Thogersen N2017 *Small spaces needs smart solutions: Designing furnitures for small spaces, in connection with human wellbeing*
[12] Yeneva A 2009 *Making the Social Hold: Towards an Actor-Network Theory of Design. Design and Culture***1**(3), p273-288, doi: 10.1080/17547075.2009.11643291
[13] Živković M, Keković A, Kondić S 2014 *The Motives for Application of The Flexible Elements in the Housing Interior* Facta Universitatis, **12**(1), Architecture and civil engineering, p 41-51. doi 10.2298/FUACE1401041Z

Spatial Pattern of Land Cover Change in The Coastal Area of Gresik Regency, Indonesia Using Land Change Modeler

A W Hasyim[1], **R E P Sianturi**[2], **Ar R T Hidayat**[1]

[1] Lecturer at Urban and Regional Planning Department, Faculty of Engineering, Brawijaya University, Malang, East Java, Indonesia

[2] Undergraduate student at Urban and Regional Planning Department, Faculty of Engineering, Brawijaya University, Malang, East Java, Indonesia

*Email: awahidhs@ub.ac.id

Abstract. Industrial activities could cause multiplier effects; one of them is increasing other supporting activities such as housing, trade and services, and transportation. The rapid growth of the industrial sector in Gresik Regency and the existence of the Java Integrated Industrial Port Estate (JIIPE) can trigger new activities, that could led the change in land use and land cover. This study aims to observe the patterns of land cover change before and after the implementation of the regional spatial plans (RTRW) from 2002 to 2012. The results indicated that, in 2002 to 2007 there was an increase in built-up area by 6016 ha while green open space and empty land decreased by 6416 ha. The implementation of the shifts towards the border areas has no longer occurred in urban areas.

Keywords: *land use, land cover, land change modeler*

1. Introduction

The industrial sector in Gresik regency becomes the dominant sector with value-added created by 2015 was almost IDR 49 trillion ($3, 46 billion). Processing industry contribute about 48,2 % in 2015 gross domestic product (GDP) of Gresik regency [1]. In 2015, the number of new industrial permits are 55 companies. It has increased 12% compared to 2014 with only 49 companies [2]. Due to the increased industrial growth, the Indonesian government in 2013 built the Java Integrated Industrial Port Estate (JIIPE) with the aim of facilitating the industries in Gresik Regency.

The construction of JIIPE itself could cause multiplier effect in coastal area. New industries tend to build its factory near JIIPE to save the transportation cost. Increasing industrial activities in coastal area could led to increasing supporting activities such as housing, trade and services. Increased activities could result in land use and land cover change with increasing built-up areas and decreasing non-built-up areas. The construction of JIIPE itself is in accordance with the regional spatial plans (RTRW), but changes in land cover resulting from the multiplier effects of JIIPE and other industries could led to the uncontrolled land cover changes. The regional spatial plans (RTRW) was made with the main goal of community welfare by controlling land use to maintain environmental sustainability, thus enabling optimal and sustainable use of space. This study examines the patterns of land cover changes before and after the implementation of regional spatial plans (RTRW), to observe the differences on land cover changes that occur in the study area before and after the regional spatial plans (RTRW).

2. Materials and methods

2.1. Study area

The coastal area of Gresik Regency (7°01'19.14"S and 112°39'01.16"E) with 833, 59 Km2, is located in Indonesia's East Java province. The study area consists of ten districts, with six districts are located along coastline (seen in figure 1). About 50% of the study areas are dominated by fish farm areas. The area is not considered fertile due to the lack of agricultural area and productive fish farm area. Sedimentations occur in the estuary of Bengawan Solo River, especially in the Northern area where Ujungpangkah district is located. Reclamation ground or landfill could be found along coastline in Manyar, Gresik, and Kebomas districts. Most of industrial estates in coastal area were formed from land reclamation process.

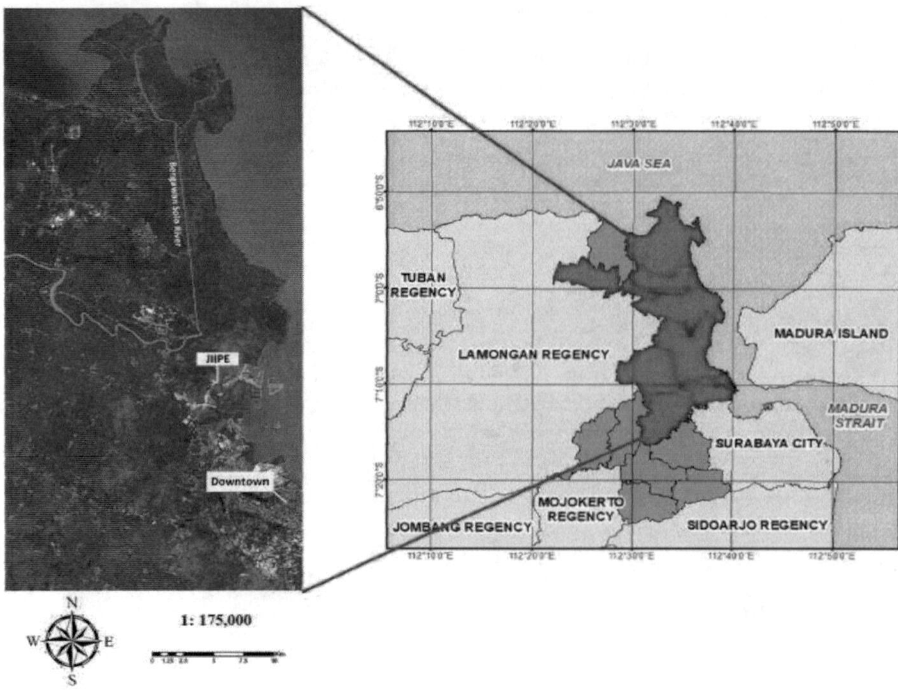

Figure 1. Illustration of study areas

2.2. Land cover mapping and accuracy assessment

Three Landsat 7 images from the years 2002, 2007, and 2012 with TM sensors (path/row: 118/65), were utilized to produce historical land cover maps. The radiometric correction was performed by using top of atmosphere correction [3] and dark-object subtractions [4] methods. The geometric calibration is less required due to the root mean square image (RMSE) of all images of < 0.5 pixel [4]. Only images with less than 20% were selected [5]. The failure of Scan Line Corrector (SLC) in Landsat 7 caused data gaps. Neighbourhood Similar Pixel Interpolator (NSPI) algorithm was employed for filling the data gaps in 2007 and 2012 images. The algorithm assumes that neighbourhood pixels close to the un-scanned (gap) pixel share have similar spectral characteristics and temporal patterns of changes to the un-scanned pixel(s). By combining the information provided by the neighbourhood pixels, the value of the missing pixels can be determined [6]. Contrast stretching and colour composite images from three bands (543) were created.

Training site were selected for each year by on-screen interpretation of the colour composite image and by reference data obtained from National Mapping Agency of Indonesia (1:25,000 topographic map) and Development Planning Agency of Gresik regency (1:100,000 land use and land cover maps). A supervised classification with maximum likelihood algorithm (MLA) was used to classify colour composite image [4]. Image classification was performed with ENVI 5.1 [7]. Accuracy assessment of land cover maps was carried out by confusion matrix or error matrix [8], with a stratified random sampling was employed to select a total 360 points for all the land cover maps. These points then were compared with the ground control points obtained from GPS [4]. Overall accuracy and Kappa coefficients were calculated, and images with > 80% value were considered to be accurate [9].

2.3. Change Analysis
Change analysis tools in Terrset Land Change Modeller (LCM) were used. Land cover maps for 2002 with 2007 and 2007 with 2012 were utilized for changes detection [4]. Change detection was performed by cross-tabulation method for each land cover classes [10]. Gains and losses of land cover from each classes and trend maps were evaluated [4], [10].

3. Results and Discussion

3.1. Accuracy assessment of land cover maps
Multi temporal land cover maps from years 2002, 2007, and 2012 were produced from maximum classification method and were validated with controlled ground control points. The overall Kappa and overall accuracy of each land cover maps were calculated by confusion matrix. The overall Kappa of 2002, 2007 and 2012 land cover maps was 80%, 89% and 93%, respectively (table 1). The commission and omission error of each land cover classes were also obtained from confusion matrix (**table 2**).

Table 1. Overall and kappa accuracy of each LULC maps

LULC Maps	Overall accuracy (%)	Overall kappa accuracy (%)
2002 (obtained August 23rd)	83	80
2007 (obtained August 5th)	91	89
2012 (obtained March 27th)	94	93

Table 2. Commission errors and omission errors of each land cover classes maps

LULC classes	2002 CE[a] and OE[b]	2007 CE[a] and OE[b]	2012 CE[a] and OE[b]
Sea	3.3, 3.3	1.8, 6.7	0, 1.7
Built-up areas	21.1, 6.8	7.9, 3.3	7.3, 8.3
Agriculture (rice fields)	15.3, 0	9, 0	0, 0
River or water beds	0, 48.3	4.4, 26.7	0, 16.7
Fish farm areas	28.6, 11.3	17.8, 7.5	16.5, 1.3
Green open space and/or empty land	12.5, 30	6.3, 10	4.3, 6

[a] Commission errors (%)
[b] Omission errors (%)

3.2. Change Analysis
Analysis of land cover changes in this study is divided into two parts to facilitate the understanding of changes that occur before and after the implementation of regional spatial plans.

3.2.1 Before spatial plans (2002-2007)
Green open space and empty land experienced the highest decline compared to other land cover classes (6416 ha). Built-up areas becomes the highest land cover (6016 ha) and has not decreased. Industrial areas, warehouses and settlements began to grow a lot in this period. Rice fields and fish farm areas also experienced more increase compared to the decline. In this period, the majority of the community still worked in the agriculture and fisheries sector because the industrial sector had not developed as rapidly as it is today. The community extends agricultural and fishery areas by utilizing open green space and

existing empty land. This causes the highest decrease in green open space and empty land as the conversion expands into rice fields and fish farm areas '**figure 2**'.

The highest trend of land cover change from non-built-up areas into built-up areas occurred in urban areas, including Gresik, Kebomas and Manyar Districts. The majority of land use changes in urban areas are for housing and industrial areas development. The location is considered strategic for industrial investors and housing developers because there are already toll roads that connect Gresik Regency and Surabaya City. Other locations with higher trends are in the border area with Lamongan Regency, such as Ujungpangkah, Dukun, and Cerme Districts. In this period, limestone-mining activities began to develop in Ujungpangkah District, causing a high trend of land cover change due to the existence of new residential areas that grew around the mining area. The majority of land use change in Dukun and Cerme District for the new residential areas is presented in'**figure 4**'.

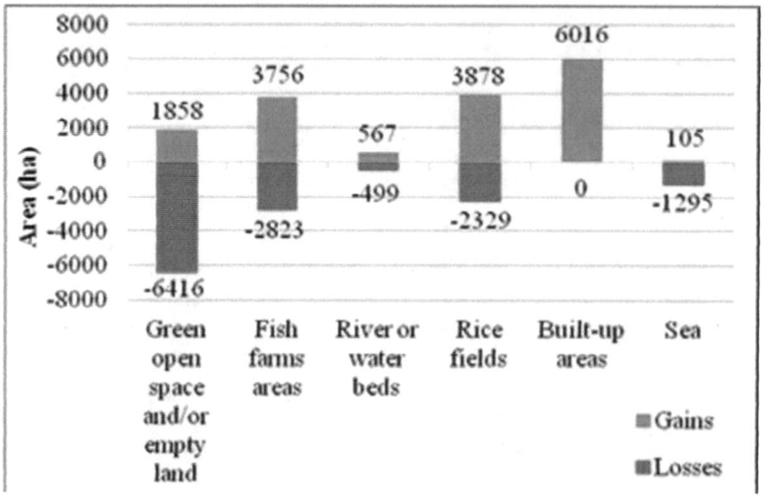

Figure 2. Gains and losses of each land cover classes in 2002-2007

3.2.2 After spatial plans (2007-2012)

Regional spatial plans of Gresik Regency (RTRW) has been ratified in 2010 and has a validity period up to 2030. In 2007 to 2012, built-up areas are the highest increase among all land cover classes (4246 ha) not as much as before the spatial plans implementation. The highest decline was experienced by rice fields (3839 ha). Open green space, empty land and fish farms areas have also increased and decreased with the same ratio. Open green space and empty land have increased because the rice fields and fish farms areas have been processed into empty land before turning into built-up areas, causing green space and empty land to decline and increase in the same time with almost balanced ratio. The community into fish farms areas processes the amount of sedimentation in the river estuary. The fisheries sector is still one of the main livelihoods of coastal communities even though its productivity is not as high as before. Sedimentation is also processed into fish farms areas that will be processed by industrial investors into empty land. This causes the fish farms areas also have a balanced decline and increase. The enactment of the spatial plans reduces the amount of reclamation in the coastal area, where the decline in sea is not as much as the previous period. However, the decline in river or waterbeds that changed the river boundary into built-up areas indicates the weak supervision and implementation of the spatial plans regarding river-border conservation areas as presented in **figure 3**.

Locations with high trends after the spatial plans implementation have shifted towards the border of Lamongan Regency and no longer in urban areas. Districts that have high trends of change are Sidayu, Dukun, Ujungpangkah and Dukun. The majority of changes are for the new settlements. Industrial workers who cannot afford to buy houses in urban areas choose to live in the border area causing a high trend of land cover change. National roads with good asphalt pavement roads in '**figure 5**' have completed the border area.

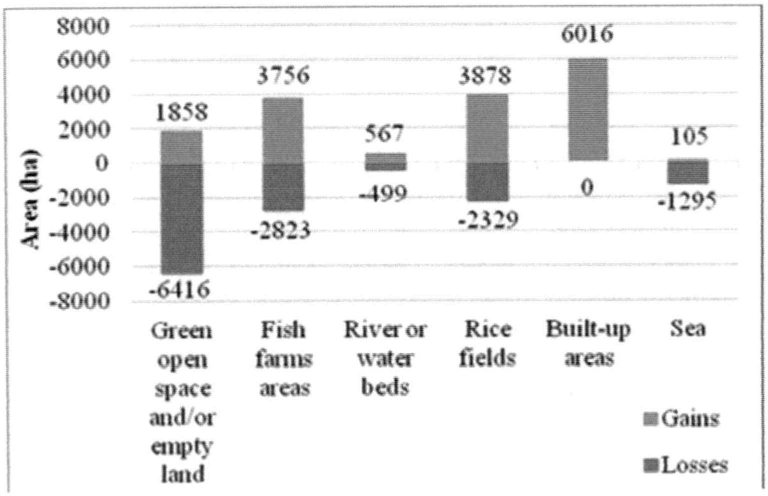

Figure 3. Gains and losses in 2007-2012

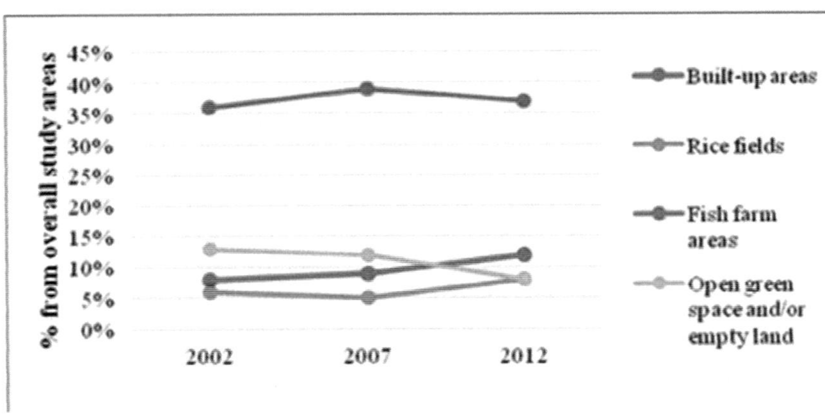

Figure 4. Land use and land cover trends in amount of area changes

IOP Conf. Series: Earth and Environmental Science **328** (2019) 012059 doi:10.1088/1755-1315/328/1/012059

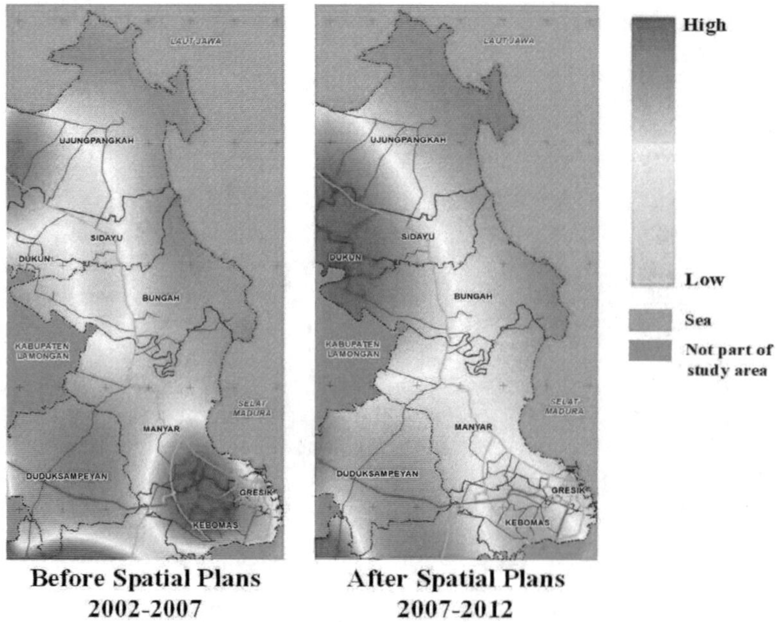

Figure 5. Land use and land change location trends

4. Conclusions

The amount of changes in land cover before the implementation of regional spatial plans (RTRW) is more extent than after the implementations. Built-up areas have not decreased either before or after the implementation of regional spatial plans (RTRW). Built-up areas increased by 6016 ha prior to the enactment of the spatial plans, but decreased after the enactment of the spatial plans with only 4264 ha. Thus, it indicates that the enactment of the spatial plans could control changes in land cover from non-built-up areas into built-up areas. Open green space and empty land did not decrease as much as before the implementation of spatial plans. However, the river border areas has been more decreased than before the enactment of spatial plans, which indicates that Gresik District Government does not carry out a policy of conserving river border areas that are part of the regional spatial plans (RTRW).

Built-up areas always increase both before and after the implementation of spatial plans, while open green space and empty land always decline. The fish farms areas had experienced an increase in 2007 because the industrial sector was still not developing as rapidly as it is today. Thus, the majority of coastal communities work as fish farmers. In addition, 2007 was the time of increased popularity Vaname shrimp cultivation; so that the community intensified fish farm, areas to meet the demand for Vaname shrimp. However, at the time of spatial plans implementation in 2012, the fish farms areas were decreased. Despite experiencing the decline after the enactment of the spatial plans in 2012 (3839 ha), rice fields have increased after experiencing a decline in 2007. The increase in rice fields occurred due to policies related to sustainable food agriculture (LP2B) which was a valid start in 2013, resulting an increase in rice fields in 2012 as a form of fulfilment the policy allocation '**figure 5**'.

Location of the highest trends in land cover changes before the implementations of regional spatial plans (RTRW) occurred in urban areas and border areas in Ujungpangkah District. After the implementation of the spatial plans (RTRW), the highest trend location occurred in the border areas. The implementations of spatial plans determining land cover change in urban areas so that the trend shifts towards the border region. In addition, expensive land prices in urban areas have caused border areas to grow into new settlements because of cheaper land prices (as presented in **figure 5**).

Besides being cheaper, the availability of empty land in urban areas is limited before the implementation of spatial plans, decreasing the trend of land cover change in urban areas. The implementation of the spatial plans causes the location of the trend to shift, which is no longer near the city center, but rather in the border area between Gresik Regency and Lamongan Regency.

References

[1] Badan Pusat Statistik Kabupaten Gresik 2015 *PDRB Kabupaten Gresik Tahun 2015* (Gresik-Badan Pusat Statistik)

[2] Badan Pusat Statistik Kabupaten Gresik 2015 *Kabupaten dalam Angka Tahun 2015* (Gresik-Badan Pusat Statistik)

[3] Barsi J A, Barker J L, Schott J R 2005 *Proc. Symp. IEEE International Geoscience and Remote Sensing (Toulouse)* vol 7 (New York: IEEE) http://dx.doi.org/10.1109/IGARSS.2003.1294665

[4] Joorabian S, Gholamalifard M 2015 Scenario-based land cover change modelling and its implications for landscape pattern analysis in the neka watershed, iran *J. Remote Sensing Appl.: Society and Environment* 1 1-19 http://dx.doi.org/10.1016/j.rsase.2015.05.001

[5] Scharsich V, Mtata K, Hauhs M, Bogner C 2017 Analysing land cover and land use change in the matobo national park and surroundings in zimbabwe *J. Remote Sensing of Environment* 194 278-286 http://dx.doi.org/10.1016/j.rse.2017.03.037

[6] Gaohong Y, Gregoire M, McCabe F M 2016 Gap-filling of landsat 7 imagery using the direct sampling method *J. Remote Sensing* http://dx.doi.org/10.3990/rs9010012

[7] Canty J M 2014 *Image Analysis, Classification and Change Detection in Remote Sensing: With Algorithms for ENVI/IDL an Phyton* (Florida-CRC Press)

[8] Rwanga S S and Ndambuki J M 2017 Accuracy assessment of land use/land cover classification using remote sensing and gis *Int. J. of Geosciences* 8 611-622 http://dx.doi.org/10.4236/ijg.2017.84033

[9] Pontius R G and Milones M 2011 Death to kappa: birth of quantity disagreement and allocation disagreement for accuracy assessment *Int. J. of Remote Sensing* 32 4407-4429 http://doi.org/10.1080/01431161.2011.552923

[10] Olmedo M T C, Pontius R G, Paegelow M 2015 Comparison of simulation models in terms of quantity and allocation of land change *J. of Environmental Modelling and Software* 69 214-221 http://doi.org/10.1016/j.envsoft.2015.03.003

Literature Review on Spatial Order of Ancient Javanese Cultural Landscape: It May Contains Spatial Continuum Idea

C Maulidi[1], W I Rukmi[1]

[1] Department of Urban and Regional Planning, Brawijaya University, Malang 65145, Indonesia

Email: c.maulidi@ub.ac.id

Abstract. Popular paradigm defines urban area as a Growth-Pole which has been expected to give trickle-down effect and spread welfare to its hinterlands and rural areas. In fact, urban growth tends to generate resource exploitations from its hinterland to comply urban necessity. Those, sharpen the dichotomy between urban and rural welfare and generate Desakota phenomenon. New paradigm sought as an alternative strategy seeks balance in the role of rural-urban linkages. This study aimed to bring out a model of spatial order that had existed in ancient Javanese civilization which followed a philosophy of maintaining the balance between human and nature in utilizing environment. Qualitative research was conducted through content exploration on ancient Javanese civilization references. Those contents were cross-diachronic to unravel the connection between politic and economic situation when that spatial order model had existed. *Wenua* (village) was located in a configuration that followed four geographic directions: at north, south east, west, and one in the middle of the fourth villages. Market was held rotationally among five villages through rotation *Pekan Mechanism.* This rotation system gave each village opportunity to become central pole for the economic activity in the region. Therefore, economic vitality was able to spread equitably over the villages.
Keyword: spatial order, javanese, cultural landacape

1. Introduction

Cultural landscapes is a product of natural processes of interactions between humans and nature, over a long period of time [1]. Cultural landscape is formed from continuous landscape reorganization by indigenous people in order to adapt their land use and spatial structures to achieve the ever-changing needs of humans from time to time. It is recognized as a multifunctional landscape that provides diverse benefits for people, produces goods and products, supports and delimits local resource management, improves culture, and so on. Cultural landscapes, therefore, presumably contain principles in building modern societies, and as a more harmonious model of interaction between humans and nature [2]. In addition, there has been an awareness of the need to explore local content to address some issues in the current spatial planning [3].

Urban and regional planning discourses are always evolving concomitant with the growing dynamical issues. Spatial planning theory in regional and urban planning studies was built on the foundation of location theory. This theory investigated spatial order with emphasis on the economic view point. Location theory has been formulated on the basis of the geographical location of economic resources and their relation to the location of various types of activities, both economic and social activities [5]. One of the spatial elements is ‘distance’ which is considered to be a 'disturbance' when humans relate to or move from one place to another [6]. Although this theory has not been much discussed, it has been formulated since at the beginning of nineteenth century. Empirically, it can be observed that service centers are generally urban in nature. As a location with the highest vitality than its surroundings, urban areas pretend to be an economic polar, or as central place. Such circumstances are universal and are tried to be explained by economists or geographers pioneered, Walter Christaller. Economist Von Thunen saw the differences in land uses from the point of difference of distance to the market which was reflected in land value. Weber specifically analyzed the location of the industry. Above three figures are considered pioneers or foundation creators in terms of location theory.

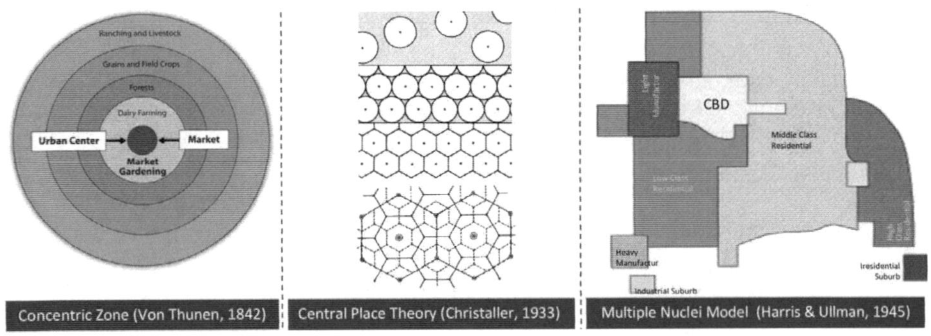

Figure 1. Classic Theories of Spatial Order

Refer to Christaller, service centers tend to be scattered within the territory according to a hexagon-shaped pattern. A region has a functional linkage between one center and its surroundings. The development of central places depends on the consumption of central goods influenced by population, demand and supply, and price, as well as regional and transportation conditions, known as "Central Place Theory." A region has dependence on other regions. Each region has advantages over the others, so the area has facilities that can serve the needs of the population within a broader radius and other residents will come to the region to meet their needs. Popular paradigm defines urban area as a Central-Place that is expected to give trickle-down effect and spread effect to its hinterlands and rural areas [5]. In fact, urban growth generates backwash effect, a form of resources exploitation on its hinterland to comply urban necessity [7]. Those sharpening dichotomy between urban and rural welfare generates *Desakota* phenomenon, which has a negative impact which gives stress to ecosystems and natural resources.

The Desakota phenomenon illustrates the impact of rural or suburban dependence to urban center. Desakota is a term used in urban geography that is used to describe areas in the extended surroundings of large cities, in which urban and agricultural forms of land use and settlement coexist and are intensively intermingled. The term was pointed by the urban researcher Terry McGee of the University of British Columbia around 1990. It comes from Bahasa Indonesia words: *desa* "village" and *kota* "city" [6]. Desakota areas typically occur in Asia, especially South-East Asia. Examples can be found in the urbanised regions of Java, such as the densely populated, delta-shaped areas on the peripheries of the Jakarta agglomeration (Jabodetabek), and also the extended metropolitan regions of Bangkok or Manila [6]. Outside South-East Asia, areas with comparable features have been described in China, India, Japan, Taiwan, and South Korea [8].

Desakota regions are characterised by high mobility of goods and services and rapid change in patterns of settlement. They are characterised by high population density and intensive agricultural use (especially wet-rice cultivation), but differ from the densely populated rural areas by more urban-like characteristics [9]. These are developed transport networks, high population mobility, increasing activity outside the agricultural sector, the coexistence of many different forms of land use, more female participation in paid labour, and unregulated land use [10]. The environment is being stressed differentially across varied geographies by the processes of technology-led economic globalization, which in turn have brought about rapid social responses of livelihood transformations in the form of migrations and more market-dependent activities. Environmental services that were rarely a concern in rural livelihood systems now are becoming of critical importance – air and water quality, for example, have implications for crop production as well as health. These may also be linked to poverty in very different ways than has been traditionally conceived of in rural areas. In rural regions, the link between poverty and ecosystems services has generally been conceptualized in terms of access to resources [11], [12], [13].

Those phenomenon leads to the development of spatial equilibrium theory in urban and regional planning. New paradigm sought as an alternative strategy seeks balance in the role of rural-urban linkages (rural urban linkages). This study aimed to find out a model of spatial balance that had existed in ancient Javanese civilization. Javanese civilization examined urbanism since the 10th century. Javanese cultural landscape order followed the *Vastusastra* principles, steeped in philosophy to maintain the balance between human and nature.

2. Methods

This is research implemented the documentary research method. It is refers to the analysis of documents that contain infor-mation about the phenomenon we wish to study [14], that is ancient Javananese civilization. Payne and Payne [15] stated the documentary method icludes categorising, investigating, interpreting and identifying focused information, which are most commonly written at documents. Documentary sources selected by quality control criteria, these are: authenticity, credibility, representativeness and meaning [16]. *Authenticity* refers to whether the evidence is genuine and from impeccable sources, such as the results of translations of various ancient manuscripts and inscriptions from ancient Java; *credibility* refers to whether the document is written by historians and archeologists who are concerned with ancient Javanese studies; *representation* refers to whether the document being consulted contains a narative picture about social, politics and culture in ancient Java periode; and *meaning* refers to whether the content of the document is clear and not multiple interpretations. Content analysis applied by grouping information: *Philosophy and beliefs* that underlie the ancient Javanese people's thinking; *livelihood* and *economic systems* that has been existed in Ancient Java. These are spatially mapped and how they are intertwined in the spatial order of landscapes is discussed further.

3. Results

3.1. Philosophy

Ancient Javanese civilization in a structured space, is at least recorded to have existed since the 9th century AD. Michael D. Coe [17] argued there are two types of civilizations that can grow in tropical forest areas as the original landscape of Java Island, namely unilateral and organic. Characteristic of civilization is the pattern of relationships among members of society that is mechanical solidarity. This pattern is characterized by relationships that emphasize the values of uniformity, the system of sanctions that are coercive and tend to be faithful to tradition. Religion used as a moral source and the basis of social control is applied as a law that tends to be coercive.

The ancient Javan had been recognized as urban civilization since 9th century. Hindu teaching background is still prominent and coloring ancient Javanese society. Spatial structure of the landscape in its existence has arranged correspond to a Hindu cultural value, i.e. *Vastu Shastra* philosophy. *'Vastu'* means places where immortals and mortals dwell. It classifies cosmos into earth (*bhumi*), house (*harmya*), vehicle (*yana*) and furniture / seating (*sayana*), defined to achieve harmony in living environment [18]. These conceptions are applied in a wide range of structures, such as: in utilizating neighborhood, temple architecture, painting and sculpturing. *Vastu Shastra* for every site and country differs as per the various regional factors of that place. Van Ossenbruggen [19] mentioned there was a local conception that organize relationship among javenese villages, it was known as *mancapat-mancalima*. It is a five-elements principle which is one of the symbols of cosmos harmonization according to Javanese philosophy [20]. Villages as living place for human are located in a 5-pole configuration, one village is located at the middle and the other four villages are located at four directions of the village: north, west, east and south. However, the location of these villages may not always be precise in the direction of the four or eight corners of the wind, because in the direction of the winds there are rivers, seas, ravines, lakes, or forests that are impossible to live in [21]. Besides that, Javanese also implement the 5-elements principle in calender, that is *pancawara* or 5 days in a week: *Manis, Pahing, Pon, Wage* and *Kliwon* [22].

3.2. Livelihood

In the old Javanese society, the king lived in a palace located in the imperial capital surrounded by a brick wall. Inside the palace also lived royal family and servants. Outside the palace wall (but its still inside of city walls) lived the crown prince and princess with his family, palace officials. The crown prince, princess, and nobleman also had an area outside the capital city or *watak.* They acted as local rulers.

Common people lived in clustered villages or *wenua*. They are generally farming, trading, working on home industries producing food such as sugar and oil, daily necessities such as charcoal and firewood, lime, pottery, wicker, snare trap and snares. A village may accept a gift as an autonomous village or *sima*, if it had services to the kingdom during war and/or becomes a location for sacred religious ceremonies. Although most villages was agricultural, there are also types of villages inhabited by a community of craftsmen who specialize in certain goods production such as blacksmiths or pottery, its called *Staphaka* [22]. A person who taught the expertise to community, appointed as a village ruler.

3.3. Economic and Market

The river-ways had an important role for mobility in the ancient Javanese civilization, including for logistics trasportation. Two Rivers as major for ancient Javanese civilization is Kali Brantras and Bengawan Solo River. Both linked Java's inland to coastal areas, and vice versa. Various types of commodities traded on the market are mainly agriculture commodities, such as: rice, fruits, coconut, cucumber, flower, livestock (cow, goat, chicken and eggs), fresh fish and dried fish. Home industry products such as baskets, salt, whiting, home furnishings from pottery, and iron and copper [22].

The *Garaman* inscription dated back to 975 *Caka* (1053 AD) shows the linkage between the location of the villages and the market location. It is mentioned the existence of traders who take merchandise from other villages and sell it in the villages around *Garaman* Village without being reprimanded by officials of the five villages. These mentioned that there was an organization among 5 villages in trading management, and suggested that every village had a market. *Pangumulan*, an inscription dated back to 824 *Caka* (902 AD) mentioned traders' behavior in ancient Javanese who used to move from village to another where was a market open alternately among 5 villages. Based on the information it can be imagined that traders brought merchandises to move to the surrounding villages to trade. The paths that traders used are land routes and river routes [22].

3.4. Spatial Order of the Ancient Javanese Market System

In his explanation of the types of primitive economic systems, Jacob and Stern [23] mentioned that simple exchanges began to exist in the advanced food gathering economies. At that time, the public began to recognize surplus, so that its excess production was channeled in exchange, either by barter, the agreed exchange (gold / silver), or exchanged as inter-community gifts. The objects that were exchanged as gifts are usually objects that symbolize the wealth of society. While the objects that were traded are two kinds. First are items that met the primary needs such as clothing and shelving, and second is secondary goods that have social meanings and functions, made by artisan communities such as ornaments, weapons, beautiful clothes, silk umbrellas.

The need for the distribution of products in the emergence of the market as a place to exchange the results of production and thus became an indicator of economic progress of the community. The market also served as a political meaning to the ruler, which was the indication of the welfare of the people and guaranteed security to transact. The Old Javanese market had various roles, and could be seen as a complex system. The system in question was an interrelated organization forming a complex system unity [24].

IOP Conf. Series: Earth and Environmental Science **328** (2019) 012060 doi:10.1088/1755-1315/328/1/012060

The Old Javanese Market System was not only a system of exchange of goods and services between communities that had specific products, but also as a form of social norms that were institutionalized in cultural institutions [25]. This was evident from integrating the application of the concept of *mancapat* and *mancalima* with the concept of *pancawara* in the spatial order of villages and the rotation of market operations in five villages. Each village had a market, and was annulled every day. On the day *Kliwon,* the market was held in the center village, on the day of *Legi* or *Manis,* the market was held in the eastern village, on the day *Pahing* the market was held in the southern village, on the day *Pon,* market was held in the western village, and on the day *Wage,* market was held in the northern village [19]. Among the five markets, the *Kliwon* market was considered a center, and of course, it had greater economic vitality than other villages.

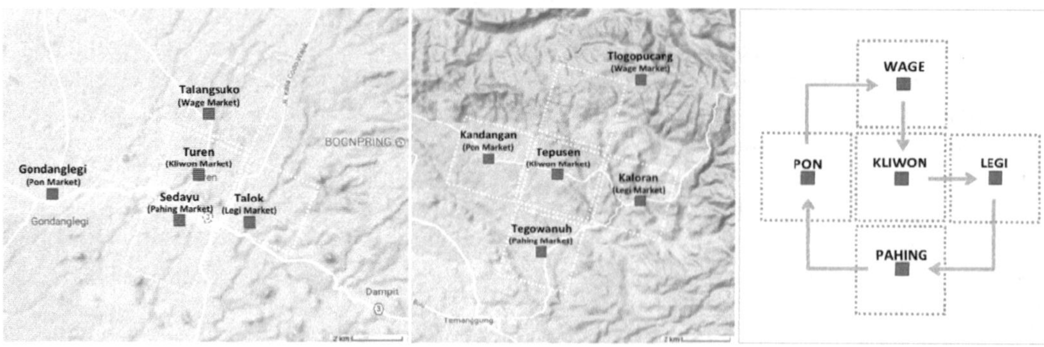

Figure 2. Spatial Order of Old Villages Market in Malang (left) and Temanggung (middle); Market Rotation Model (right)

Ethnoarchaeology research results in Malang Regency and Temangggung also found that in general, a village market day revolves in one rotation or *sepasar* [25]. For example, in Kabupaten Temanggung, the market in Tepusen village (which is considered to be the central market) recognizes *Kliwon's* day as market day; The village of Kaloran, located in the east of Desa Tepusen, recognizes *Legi* day as market day; The village of Kandangan located west of Desa Tepusen recognizes *Pon* day as market day; Tegowanuh village located in the southern village of Tepusen recognizes *Pahing* as market day; But there was no market that really lies to the north of Tepusen Village, but there is Tlogopucang Village which is located in the northeast which recognizes *Wage* day as market day. Thus, one village market can enter the rotational unit of economic vitality. Similarly, the market in Malang Regency, generally open every day to keep the market day twice in the market, for example Turen market recognizes *Kliwon* as market day. The other three markets, namely Talangsuko market located north of Turen, Talok Market in the east, and Sedayu Market located in the south are simple markets.

4. Discussion

The application of the concept of "*panasta* village or *mancapat*" and the concept of "*pancawara* or *sepasaran*" created an integrated spatial suckler villages on Java landscape and composed the economic vitality throughout the region. The integration of these two applied concepts are in line with the backdrop of social and cultural conditions of unilateral character from ancient Javanese history with a strong collective attachment to the values of the Hindu religion and philosophy *Vastusashtra*. Yunus [26] saw this as an ideal model for the survival of living environment where it should also be a place for working, opportunities, circulation, housing, recreation, and other living facilities.

Market rotation preceded by the "*Pon*" held at the central market or "*watek*" markets, continued by the "*Umanis / Legi*" market to the east, the "*Pahing*" in the south market, the "*Pon*" in the west market west, and the rotation ends on "*Wage*" day in the northern market. Continuous cycles of every "target"

IOP Conf. Series: Earth and Environmental Science **328** (2019) 012060 doi:10.1088/1755-1315/328/1/012060

or every five days, allow for equality of territory or village. Each village has a chance to gain easy access to fulfill their daily needs and market their products directly in the market, and every village has a chance to be vital. With this market rotation pattern, it can alleviate the vitality dichotomy between the mother village and other surrounding villages. Each has the same opportunity to become a rotating central-place.

Those can be a benchmark for classical location theories that tend to see one location as a central location for the surrounding areas. Some theories and concepts were, among others, formulated by Von Thunen who envisioned the achievement of minimal economic costs through the creation of a growth center in one city with the concentrated supply of agricultural areas of the city concentrated around him, Christaller, with the Central Place Theory, formulated the hexagonal form as a concept illustration Hierarchy of space based on the hierarchy of economic service level. Friedmann with his core-periphery theory which explained the existence of a central region with some peripheral regions. The application of the eurocentric theories in regional development, especially in developing countries, has implications for the Desakota phenomenon. The city as a central-place became a location that syphon resources from the periphery, giving stress to the environment and societal rural communities. Orientation of economic vitality to the city always places the surrounding area under the hierarchy of the city economy.

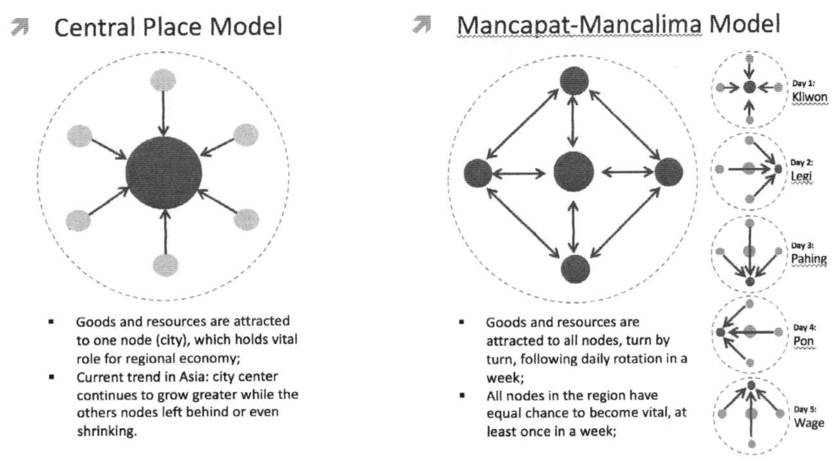

Figure 3. Comparison the Central Place Model and *Mancapat-Mancalima* Model

Critical reviews of these theories mainly concern the relevance of their application in Indonesian territory to the diversity of geographic and social conditions of society. Simplification is related to geographical issues, distance, and human behavior. Thus, the failure of the theory's application in regional development practice can be identified by the problem of spatial inequality and dependence. This situation is, of course, contrary to the traditional Javanese concept application that seeks to harmonize and balance life.

The market rotation system based on the concept of "market" not only creates the balance and dynamics of space, but furthermore, it triggers the socio-cultural dynamics of society in every village. The rotation of trading activity according to market day, encourages the movement of traders from one village to another. Therefore, in social schemes, the community groups of economic actors are intertwined not only internally but also externally with market mechanisms outside the village. Traders from a village will meet traders from different villages. In addition, this encounter also occurs between the merchant and certain commodity producers and buyers from different villages. The routine of the encounter not only creates, but also strengthens the social ties between regions. In a cultural scheme, the same external fabric allows for the exchange of information, ideas, ideas and cultures. The market no longer means economic, but social and cultural.

In the context of the present, in the midst of rapid economic activity, the existence of economic space or traditional markets can survive because it is closely related to the social fabric in it. The loss of existence of one of them will affect other existence. The "go-to" rotation that is applied in the tradition of the economic system of ancient Javanese society, enables the sustainability of market, village, and socio- cultural community existence. In this case, the market is not merely present in the economic life of society, but it exists to maintain its existence at that time and develop in the future. The distribution pattern of bounded reciprocal commodities described by meetings between traders / producers from different villages in the border region to conduct transactions and down-the-line trade patterns, where hopefully in the future it will lead to new small markets. Through this way, buy-sell transactions usually occured at crossroads or in open ground. The meeting triggered by the need for commodity exchange among traders created wider economic activity, i.e. between traders and the people living around the transit area.

5. Conclusion

In Ancient Javanese civilizations, religious-spiritual beliefs proved not merely as a manifest in ceremonial worship activities. From the textual study, it can be learned that the beliefs become the basis of attitude in every fabric of Javanese human life. Concept of "*mancapat-mancalima*" and the concept of classification of time "*pancawarna* and *kesasaran*" is a proof of the role of cosmological beliefs embraced by ancient Mataram people in the arrangement of space and livelihood activities. Meanwhile, the concept of traditional Java market rotation showed that the condition that space, time, human, and activity cannot be separated. These four aspects cannot be understood as single entities, but must be understood as a coherent and holistic entities.

Market rotation system is one of the component in the ancient Mataram market economy system in addition to production, distribution, transportation, and transactions. In spatial context, this rotation system established inter and intra-territorial networks, especially villages involving community interconnection. In a socio-cultural context, the system proved capable of creating a wider rural community knit. Knitted social community stems from the entanglement of economic activity actors internally and externally beyond the administrative boundary of the area, grown extensively to form the fabric of society producers, traders, consumers, authorities, and ordinary people.

Five patterns of commodity distribution in traditional Javanese market based on the "target" cycle enable the expansion and the strengthening of social fabric, as well as the survival and development of market existence itself. This is possible because through the system, there are guaranteed availability of traded commodities, the fabric of economic actors, social fabric with the wider public. In addition, transactions that occurred due to the movement of traders from one village to another, resulted the embryo formation of new markets that can be found until today. The market is no longer just a meaningful economic space, but further, the market can be a social and cultural space, i.e. a space that is enabled by economic activity. Therefore, the disappearance of one of these values will threaten the existence of not only the market as an economic space, but also a village. The concept of "*sepasaran*" that manifests itself in a rotational system market, is the spirit of vitality generating a fair and sustainable space at the time of the ancient Java spatial order.

References

[1] Mitchell, N., M. Rössler, P.Tricaud. 2009. *World Heritage Cultural Landscapes A: Handbook for Conservation and Management.* (France: UNESCO)
[2] Campolo, D. et.al. 2016. Cultural Landscape and cultural routes: infrastructur role and indigenous knowledge for sustainable development of inland. *Proceedia Social and Behavioral Sciences*, 223:576-582.
[3] Young, Greg. 2008. *Reshaping Planning with Culture*. Hampshire: Ashgate Publishing Limited.
[4] McGee, Terry G. (2002), "Reconstructing ›The Southeast Asian City‹ in an Era of Volatile Globalization", Critical Reflections on Cities in Southeast Asia, pp. 33–34
[5] Djojodipuro, M. 1992. *Teori Lokasi*. Jakarta: Lembaga Penerbit Fakultas Ekonomi Universitas

Indonesia.
[6] McGee. 2002. *Reconstructing The Southeast Asian City*, pp. 47–48
[7] Cairns, Stephen. 2002. "Troubling Real-estate: Reflecting on Urban Form in Southeast Asia", Critical Reflections on Cities in Southeast Asia, p. 117
[8] Sui, Daniel Z.; Zeng, Hui (2001), "Modeling the dynamics of landscape structure in Asia's emerging desakota regions: A case study in Shenzhen", Landscape and Urban Planning, 53 (1–4): 37–52, doi:10.1016/s0169- 2046(00)00136-5
[9] McGee (1991), The Emergence of Desakota Regions in Asia., pp. 6–7
[10] Chia, Lin Sien; Perry, Martin (2003), "Introduction", Southeast Asia Transformed: A Geography of Change, Singapur: Institute of Southeast Asian Studies, p. 12, ISBN 981-230-119-4
[11] Cooper, E., P. Ghazi, K. Holmes, K. Kichnak, Y. Kura, A. LaVina, D. Prager, P. Steele, W. Vanasselt, and J. Virdin. 2005. World Resources 2005: The Wealth of the Poor— Managing Ecosystems to Fight Poverty. World Resources Institute in collaboration with United Nations Development Programme, United Nations Environment Programme, and World Bank, Washington DC.
[12] Allen, A., J. D. Davila, and P. Hofmann. 2006. Governance of Water and Sanitation Services for the Peri-urban Poor: A Framework for Understanding and Action in Metropolitan Regions. Development Planning Unit, University College London, London.
[13] Borner, J., A. Mendoza, and S. A. Vosti. 2007. Ecosystem services, agriculture, and rural poverty in the Eastern Brazilian Amazon: Interrelationships and policy prescriptions. Ecological Economics 64:356-373.
[14] Bailey, K., 1994, Methods of Social Research, Fourth Edition, New York: The Free Press.
[15] Payne, G. and Payne, J., 2004, Key Concepts in Social Research, London: Sage Publications.
[16] Scott, J., 1990, A Matter of Record, Documentary Sources in Social Research, Cambridge: Polity Pressl
[17] Coe, M.D. 1961. Social Typology and The Tropical Forest Civilization. Dalam Studies in Society and History. 4(1): 65-85.
[18] Kumar, A. (2005). Vaastu: The Art And Science Of Living. New Delhi: Sterling Publishers Private Limited.
[19] Van Ossenbruggen. 1975. Asal-usul konsep Jawa tentang mancapat dalam hubungannya dengan sistim-sistim klasifikasi primitif. Jakarta: Bhratara
[20] Junianto. 2017. Konsep Mancapat-Mancalima dalam Struktur Kota Kerajaan Mataram Islam Periode Kerajaan Pajang sampai dengan Surakarta. SEMINAR NASIONAL SPACE #3: Membingkai Multikultur dalam Kearifan Lokal Melalui Perencanaan Wilayah dan Kota.
[21] Wurjantoro, E. 1981. Wanua i Tpi Siring. Seri Penerbitan Ilmiah. No. 7. Fakultas Sastra Universitas Indonesia. pp: 79-109.
[22] Rahardjo, S. (2011). Peradaban Jawa Dari Mataram Kuno Sampai Majapahit Akhir. Jakarta: Komunitas Bambu.
[23] Jacobs, M and Stern, B.J. 1955. General Anthropology. A Brief Survey of Physical Cultural, and Social Antrophology. New York: Barnes & Noble, Inc.
[24] Theodorson, G.A. and Achilles G.T. 1970. Modern Dictionary of Sociology. The Concepts and Terminology of Sociology and Related Disciplines. New York: Thomas Y. Crowell Company.
[25] Nastiti, T.S. 2003. Pasar-Pasar Di Jawa Masa Matara Kuno Abad VIII – XI Masehi. Bandung: Pustaka Jaya.
[26] Yunus, H.S. 1987. Beberapa Determinan Perkembangan Permukiman Kota: Dampak dan Pengelolaanya. Yogyakarta: Fakultas Geografi UGM.

The Influence of Physical Quality on Service Scale and Frequency of Visits in "Alun-Alun Malang" City Square Park

E B Kurniawan[1*], W D Purnamasari[1], D A Setyono[1], W. Astrini[2]

[1]Department of Urban and Regional Planning, Faculty of Engineering, Universitas Brawijaya, Malang, Indonesia
[2]Department of Architecture, Faculty of Engineering, Universitas Brawijaya, Malang, Indonesia

*Email: eddi@ub.ac.id

Abstract. Malang City is one of the cities with rapid physical growth, but the availability of green open space (RTH) is inadequate. The construction of RTH as a solution to fulfill 20% of public RTH raises new problems of: the occurrence of inequality in the use of city parks. There are city parks that are not functioned optimally, judging from the intensity (number and frequency). Based on these conditions, the study aims to assess the performance, service scale and frequency of visits in Alun-Alun Malang Park, Klojen District, Malang City. The method of analysis utilizes park quality analysis, park service scale analysis, frequency of park visits analysis, and multiple linear regression analysis. Based on the results of the regression test, it is apparent that there is an influence between the park service scale and the services quality of Alun-Alun Malang Park, including amenities (X_1), easy access (X_4), and facilities completeness (X_5). Based on the regression analysis using the successive intervals method, it is obvious that variables which significantly influence the frequency of parks visits, are amenities (X_1), security and safety (X_2), parks identity (X_3), easy access (X_4), facilities completeness (X_5), park beautiness (X_6), activity diversity (X_7), user diversity (X_8).

Keywords: park quality, service scale, frequency of parks visits.

1. Introduction

Open space or public space is an important element in planning a city. Public space has an important role that is not only concerned with environmental, economic and social aspects but also plays a role in the quality of life and sustainability aspects of a city [1,2]. Good public space is a public space for all groups of society, without the dominance of certain groups, which is a safe and friendly for all societies including women [3]. City parks are defined as open land with social and aesthetic functions as a means of recreational activities, education or other activities at the city level [4]. City parks and open spaces are considered able to support the welfare of the community, promoting lifestyle changes, and contributing to the development of the social capital and healthy cities [5]. Planning an open space must consider various attributes such as: the needs of users, population, population density, people's preferences, and accessibility to reach open spaces. The quality of public space is not only seen from the number of people's visit, diverse activities, spatial planning that has aesthetic value, but is seen from a quality public space, which is easily accessed by the public [3].

Malang City is one of the cities with rapid physical growth. The total population of Malang City on the Year 2017 was 861,414 people [6]. However, the rapid development of Malang City's population is not matched by the availability of green open space (RTH). The requirement for providing green open space (RTH) in a city is 30% of the total city area consisting of 20% public green open space (RTH) and 10% private green open space (RTH) [7]. Whereas, based on existing conditions the percentage of public green open space (RTH) in Malang City, it is still less than 20%. The width of public open space in Malang City is 1,760.15 hectares or 15.92% of the total area of Malang City [8]. The percentage of public green open space (RTH) is dominated by the border of the river reaching 11.41%, while the city park and urban forest had a percentage of 1.82% and 0.35% of the total area of the city.

Malang City Government through The Housing and The Settlement Area Agency (Disperkim) supports the efforts to fulfill 30% of green open space (RTH). New development, as well as the revitalization of green open space (RTH), is being actively carried out in Malang City. The development of green open space (RTH) is more focused on the construction of active parks that can be used as a place of interaction and socialization of the community. The development and design of open green space must pay attention to the needs of its users including the provision of supporting facilities. However, there are imbalances in the utilization of city parks in Malang City. There are city parks that are not functioned optimally, judging from the intensity (number and frequency) of the small number of visitors, and visitors who come only from the vicinity of the park. However, there is a city park that is crowded with visitors and even parks visitors from outside the Malang City. Based on the initial observations of the researchers, *Alun-Alun Tugu* and *Alun-Alun* Malang are city parks that are often visited by visitors who come from outside Malang City, especially during weekends. While city parks such as *Kunang-Kunang* Park and Ronggowarsito Park, in addition to having fewer visitors, visitors come to these parks only from around the park's location.

Therefore, it is estimated that there are factors that influence the utilization of each city park. Therefore, a study was conducted on the relationship between the influence of the city parks quality on service scale and the frequency of city park visits in *Alun-Alun* Malang Park, in Klojen District, at Malang City.

2. Methods

The analytical methods used in the study were park quality analysis, park service scale analysis, frequency of park visits analysis, and multiple linear regression analysis. Variables and sub-variables used in the study are sourced from previous research and literature review (Table 1).

Table 1. Research Variable

Aim	Variable	Sub Variable	Analysis Techniques
Identify the park physical quality characteristics in Alun-Alun Malang	Site [9]	• Size and area arrangement • Regional link • Legal description • Natural physical features • Artificial physical features • Circulation • Utilities • The five senses • Climate	Park physical quality analysis with descriptive methods
Identify the service scale in Alun-Alun Malang Park	The visitor radius service scale		Park service scale analysis with descriptive-quantitative methods
Identify the frequency of visits to Alun-Alun Malang Park	The frequency of visits by visitors		Frequency of park visits analysis with descriptive-quantitative methods
Identify the influence of physical quality on service scale and frequency of visits	The influence of physical	• Amenities • Security and Safety • Park Identity	Multiple linear regression analysis

Aim	Variable	Sub Variable	Analysis Techniques
	quality on service scale and frequency of visits	• Easy Access • Facilities Completeness • Park Beautiness • Diversity of Activities • Diversity of Users • Park Hospitality	

3. Result and Discussion

3.1. Physical Quality Characteristics of Alun-Alun Malang Park

The physical characteristics of Alun-Alun Malang Park are seen in ten aspects which mention information in site planning seen from the size and territorial, regional links, legal descriptions, natural physical features, artificial physical features, circulation, utilities, five sensories, humans and culture, and climate [8]. The quality of Alun-Alun Malang Park is based on nine variables. Each variable consists of 5 ranges of values from very bad to very good. The evaluation of the Alun-Alun Malang quality is based on the sample of visitors to Alun-Alun Malang (Figure 1).

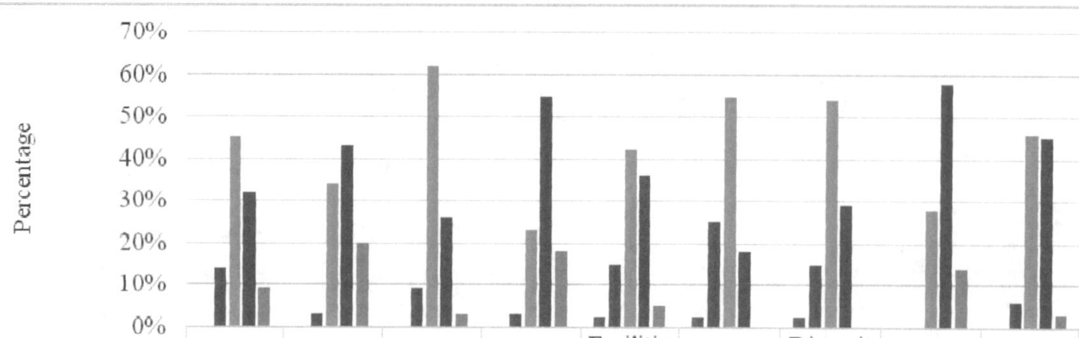

	Amenities	Security and Safety	Park Identity	Easy Access	Facilities Completeness	Park Beautiness	Diversity of Activities	Diversity of Users	Park Hospitality
Very bad	0%	0%	0%	0%	2%	2%	2%	0%	0%
Bad	14%	3%	9%	3%	15%	25%	15%	0%	6%
Good enough	45%	34%	62%	23%	42%	55%	54%	28%	46%
Good	32%	43%	26%	55%	36%	18%	29%	58%	45%
Very good	9%	20%	3%	18%	5%	0%	0%	14%	3%

Figure 1. Quality Chart of Alun-Alun Malang by Park Visitors Perception

3.2. Service Scale of Alun-Alun Malang Park

In this research, the service scale is based on the reach of the park radius. The service scale for city parks in Malang City has a different range of radius for each park. *Alun-Alun* Malang Park has a reach radius up to the outside of Malang city with a range radius of 13.5 km and 11.2 km (seen in Table 2).

Table 2. Service Scale of Alun-Alun Malang Park

Farthestest Service Scale		Nearest Service Scale		Average of Service Scale for Each Park
Distance	Visitors Origin	Distance	Origin of Visitors	
13,5 km	Pagentan, Singosari	1,8 km	Bareng	6,81 km

3.3. Frequency of Park Visits Analysis in Alun-Alun Malang Park

The characteristics are based on the frequency of the visit including the number of visits or the number of times when visitors visited the city park. The number of visits is calculated based on the number of visits made by visitors before the survey activities conducted by researchers, until the visit was made when filling out the research questionnaire. Figure 2 reports that as many as 17% of respondents first visited Alun-Alun Malang. The highest number of respondent visits to Alun-Alun Malang is 2-3 times of the frequency of visits (33%). However, the respondents visiting Alun-Alun Malang in > 7 times are only 17%.

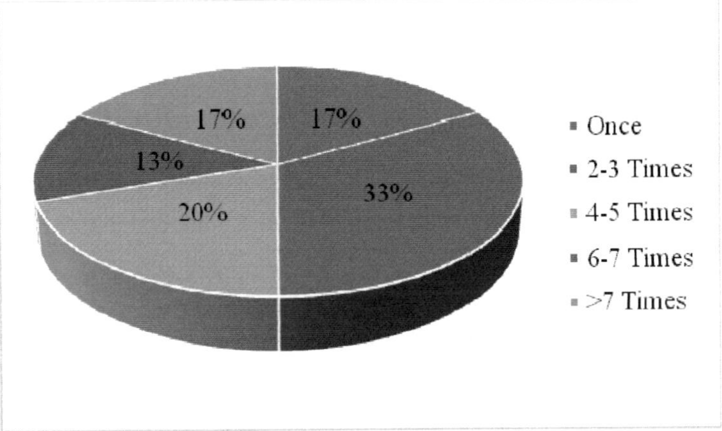

Figure 2. Percentage of City Park Visitors Characteristics Based on Visit Frequency

3.4. The Influence of Physical Quality on Service Scale and Frequency of Visits

3.4.1 The Influence of Physical Quality (X) on Service Scale (Y)

Linear regression analysis of the city parks quality with a coverage radius of the park was carried out by using SPSS with the enter method. The results of the Ftest (Table 3) show that the Fcount value in the research model is 20.902 with a significance level of 0.000. A significance value of less than 0.05 indicates that the data will produce a good model.

Table 3. ANOVA

Model		Sum of Squares	df	Mean Square	F	Sig.
1	Regression	105.420	9	11.713	20.902	.000[a]
	Residual	78.453	140	.560		
	Total	183.873	149			

a. Predictors: (Constant), Amenities, Security and Safety, Park Identity, Easy Access, Facilities Completeness, Park Beautiness, Diversity of Activities, Diversity of Users, Park Hospitality.

b. Dependent Variable: Coverage Radius

Table 4. Model Summary

Model	R	R.Square	Adjusted R Square	Std. Error of the Estimate
1	.757[a]	.573	.546	0.748586

The goodness of fit test aims to see the suitability of the model, or the ability of the independent variable to explain the dependent variable variance. Table 4 presents an R-value of 0.757 and the value of R-square is 0.573, indicating that the ability of the independent variable to explain the variance of the dependent variable is 57.3%.

Table 5. Coefficients

		Unstandardized Coefficients		Standardized Coefficients		
Model		B	Std. Error	Beta	t	Sig.
1	(Constant)	-.750	.425		-1.764	.080
	Amenities	.228	.088	.192	2.589	.011
	Security and Safety	.006	.071	.005	.091	.928
	Park Identity	-.044	.072	-.036	-.608	.544
	Easy Access	.200	.076	.164	2.646	.009
	Facilities Completeness	.547	.101	.465	5.436	.000
	Park Beautiness	-.028	.104	-.019	-.268	.789
	Diversity of Activities	.97	.095	.069	1.021	.309
	Diversity of Users	-.012	.087	-.009	-.133	.895
	Park Hospitality	.102	.084	.083	1.210	.228

T-test (partial) aims to see the influence of the variables that most influence the dependent variable. The independent variable that has the most influence on the dependent variable is based on the significance value of <0.05. T-test results in Table 5 indicate the three independent variables that have a significance value of <0.05including variable amenities, easy access, and facilities completeness. Based on the results in Table 5, the regression equation can be arranged as follows:

$$Y = -0.750 + 0.228\ X1 + 0.200\ X4 + 0.547\ X5$$

Information:
Y = Park coverage radius
X1 = Amenities
X4 = Easy access
X5 = Facilities completeness

3.4.2 The Influence of Physical Quality (X) on Visit Frequency (Y)

The independent variable forming the regression model is considered to have a significant influence together on the dependent variable if Fcount >Ftable or significant <α=0.05. Based on the results of regression models test (Table 6), the obtained Fcount is 22.677 (Sig F=0.000). Ftable at the 5%indicates level with independent degrees of 10 and 189 at 1.88. Because Fcount>Ftable (22.677>1.88) and Sig F<5% (0.000<0.05) then Ho is rejected which means that the variables such as amenities (X1), security and safety (X2), park identity (X3), ease of access (X4), facilities completeness (X5), park beautiness (X6), diversity of activities (X7), diversity of users (X8), park hospitality (X9) have a significant influence on visit frequency variables (Y).

Table 6. F Test

Model	df	Fcount	Ftable	Sig.
Regression	10	22.677	1.88	0.000
Residual	189			

Partial regression models test is used to determine whether each of the independent variables forming the regression model individually with a significant influence on the dependent variable or not. The independent variable forming the regression model is assumed to have a significant effect if Tcount>Ttable or significant <α=0.05. The value of Ttable is obtained by using Ttable (df 189) and significant level of 0.05 that is 1.973. After obtaining T-table value, the next step is to compare the t - value on each variable. The independent variable will have a significant effect on the dependent variable if the value of Tcount>Ttable, indicating 8 independent variables that affect the dependent variable. Variables that influence are variables such as: amenities (X1), security and safety (X2), park identity (

X3), ease of access (X4), facilities completeness (X5), park beautiness (X6), diversity of activities (X7), and diversity of users (X8).

Table 7. T-test

Independent Variable	T_{count}	Sig. t	T_{table}	Information
Amenities (X1)	3.602	0.001	1.973	Significant
Security and Safety (X2)	3.230	0.001	1.973	Significant
Park Identity (X3)	2.659	0.008	1.973	Significant
Easy Access (X4)	2.881	0.001	1.973	Significant
Facilities Completeness (X5)	2.367	0.030	1.973	Significant
Park Beautiness (X6)	4.445	0.000	1.973	Significant
Diversity of Activities (X7)	2.106	0.028	1.973	Significant
Diversity of Users (X8)	2.950	0.002	1.973	Significant
Park Hospitality (X9)	0.587	0.664	1.973	Not significant

Independent variable included in model regression is an independent variable that takes significant influence corresponding with T-test results that have been carried out. Table 8 presents the constant value and coefficient value of the independent variable included in the regression equation.

Table 8. Results of Multiple Regression Calculation

	Unstandardized Coefficients		Standardized Coefficients
	B	Std. Error	Beta
(Constant)	-1.943	0.330	
Amenities (X1)	0.200	0.056	0.202
Security and Safety (X2)	0.177	0.055	0.171
Park Identity (X3)	0.138	0.052	0.137
Easy Access (X4)	0.156	0.054	0.154
Facilities Completeness (X5)	0.126	0.053	0.125
Park Beautiness (X6)	0.264	0.059	0.257
Diversity of Activities (X7)	0.116	0.055	0.114
Diversity of Users (X8)	0.154	0.052	0.155

The coefficient values included in the regression model are taken from the B value (beta) in the Unstandardized Coefficients table, where the regression model is:

$$Y = -1.943 + 0.177\ X2 + 0.200\ X1 + 0.154\ X8 + 0.116\ X7 + 0.156\ X4 + 0.138\ X3 + 0.264\ X6 + 0.126\ X5$$

Information:

Y = Frequency of Park Visit
X1 = Amenities
X2 = Security and Safety
X3 = Park Identity
X4 = Easy access
X5 = Facilities completeness
X6 = Park Beautiness
X7 = Diversity of Activities
X8 = Diversity of Users

4. Conclusion

Based on the results of the analysis that has been done, it can be concluded:

1. *Alun-Alun* Malang visitors are very diverse not only from within the city but also from outside the city. The city park services scale in Klojen District, at Malang City can reach visitors from outside

IOP Conf. Series: Earth and Environmental Science **328** (2019) 012061 doi:10.1088/1755-1315/328/1/012061

the Malang City. *Alun-Alun* Malang has the highest service scale which is 13.5 km, if compared to the average city park services scale in Klojen District which is 6.34 km.

2. Out of a total of respondents, 17% respondents make the first visit to *Alun-Alun* Malang. The highest number of respondent visits to Alun-Alun Malang is 2-3 times the frequency of visits from 33% of respondents. Whereas, the respondents who visited the Alun-Alun Malang about >7 times are only 17% respondents.
3. Multiple linear regression results in Park Service Scale is = -0.750 + 0.228, Amenities is+ 0.200 Easy Access is + 0.547 Facilities Completeness. It is known that there is an influence between the park service scale and the services quality of *Alun-Alun* Malang Park, including amenities (X1), easy access (X4), and facilities completeness (X5). Based on the regression analysis using the successive intervals method, it is revealed that variables that significantly influence the frequency of parks visits, are amenities (X1), security and safety (X2), parks identity (X3), easy access (X4), facilities completeness (X5), park beautiness (X6), activity diversity (X7), user diversity (X8), with the regression model of Y= -1.943+ 0.177 X2+ 0.200 X1 + 0.154 X8 + 0.116 X7 + 0.156 X4 + 0.138 X3 + 0.264 X6 + 0.126 X5.

References

[1] Cheisura, A. 2004. The Role of Urban Parks for The Sustainability City. *Journal Landscape and Urban Planning*. 129-38
[2] Era, T.R. 2012. Improving Pedestrian Accessibility to Public Space Through Space Syntax Analysis. *Proceeding Eight International Space Syntax Symposium.* Spain: Environmental Planning Laboratory, Department of Urbanism and Spatial Planning, University of Granada.
[3] Handbook Project for Public Space. 2011. Placemaking and The Future of Cities. UN-HABITAT Sustainable Urban Development Network (SUD-Net).
[4] General Director of Spatial Planning, Ministry of Public Works. 2008. Minister of Public Works Regulation Number 5 of 2008 concerning Provisioning Guidelines and Utilization of Green Open Space in Urban Areas. Jakarta: General Directorate of Spatial Planning Department of Public Works (Direktoral Jenderal Penataan Ruang Departemen Pekerjaan Umum. 2008. Peraturan Menteri Pekerjaan Umum Nomor 5 Tahun 2008 tentang Pedoman Penyediaan dan Pemanfaatan Ruang Terbuka Hijau di Kawasan Perkotaan. Jakarta: Direktoral Jenderal Penataan Ruang Departemen Pekerjaan Umum)
[5] Wang, D. et al. 2013. Rethinking Accesibility in Planning of Urban Space Using an Integrative Theoretical Framework. *Final Paper.* Brisbane: University of Queensland
[6] Statistics Central Agency of Malang Municipal. 2018. Malang City in Figures Year 2018. Malang: Statistics Central Agency of Malang Municipal (Badan Pusat Statistik Kota Malang. 2018. Kota Malang Dalam Angka Tahun 2018. Malang: Badan Pusat Statistik Kota Malang)
[7] The Republic of Indonesia. 2007. Undang-Undang Number 26 of 2007 concerning Spatial Planning. Jakarta: Secretariat of State (Republik Indonesia. 2007. Undang-Undang Nomor 26 Tahun 2007 tentang Penataan Ruang. Jakarta: Sekretariat Negara)
[8] Sanitation and Landscaping Agency of Malang Municipal. Masterplan of Malang City Open Space 2012-2032 Malang: Sanitation and Landscaping Agency of Malang Municipal (Dinas Kebersihan dan Pertamanan Kota Malang. Masterplan Ruang Terbuka Hijau Kota Malang Tahun 2012-2032. Malang: Dinas Kebersihan dan Pertamanan Kota Malang)
[9] Edwart T. White . 1985 . Site Analysis: Making Information Diagrams for Architectural Design (Edwart T. White. 1985. Analisis Tapak: Pembuatan Diagram Informasi bagi Perancangan Arsitektur)

Study of Performance-Based Contract Application in Building Maintenance

E J Malelak[1], J U D Hatmoko[2], and M A Wibowo[3]

[123]Department of Civil Engineering, Faculty of Engineering, Diponegoro University, Jl. Prof. Soedarto, Semarang 50275, Indonesia

*Email: erasthus.j.malelak@gmail.com

Abstract. Some buildings have no clear and structured maintenance. Currently building maintenance is mostly done when there has been damage or system instability attached to the building. Besides that, there are still not many buildings that have data about the age of use of each particular facility, for example plumbing, lights, faucets, telephones, and past data related to the changes that have been made. In one hand, a good maintenance system has been implemented in road projects because it uses Performance Based Contracts (PBCs). Therefore, the purpose of this study is to examine the application of Performance Based Contracts in buildings. The research method is in quantitative and qualitative ways to analyze the opportunities for the implementation of Performance Based Contracts in building maintenance by way of case studies. The object of research is government and private building. The results of this study are the compilation of the initial concept of building maintenance based on the Performance Based Contract (PBC). The implication of this research is the potential for making performance-based contract standard documents for more structured buildings so that they can effectively manage building maintenance.

1. Introduction

Building construction in Indonesia is increasing rapidly every year, both commercial buildings and buildings for public services. But the rapid construction of buildings is often not accompanied by increased maintenance activities. Maintenance is a very important activity in preventing damage [6]. In addition, building maintenance will also reduce the risk of greater repair costs in the future [5].

In general, building construction uses a traditional contract system both with a system of unit price payments and lump sum [8]. Currently there has been an innovative form of contract that is being used, namely performance-based contracts (PBCs). Performance-based contracts are multi-year contracts, the responsibility for design, construction and maintenance is fully left to the contractor [3]. One of the fundamental differences between traditional contracts and performance-based contracts is: at the maintenance stage, in the traditional contract the contractor is not responsible for maintenance after the project is completed, while in performance-based contracts, the contractor is the partly responsible [2].

The implementation of performance-based contracts is currently only used in the construction of road infrastructure that requires service providers to participate in supervising the road so that the factors that cause road damage can be minimized [1]. Even though risk and waste occur in all stages of the project life cycle of all types of projects [7], it becomes an interesting discussion if maintenance such as the Performance Based Contract can be applied to other types of projects.

Poor building maintenance activities are currently caused by ineffective maintenance strategies (Zawawi et al. 2010). Therefore, to improve the efficiency of building maintenance systems, the authors want to conduct research on the implementation of a performance-based contract system in building infrastructure. Implementation of performance-based contracts is expected to provide effective results in building maintenance activities so that there is guaranteed quality planning and construction as well as operational guarantees so that they can meet the age of the building.

The purpose of this study is to examine the application of performance-based contracts in buildings. This is due to the lack of serious maintenance in the building. Therefore, it is necessary to examine what factors can be the drivers and constraints in the implementation of performance-based contracts in buildings. In addition, it will be examined whether performance-based contracts can be applied as alternative contracts for building maintenance.

2. Research Methods

This research was conducted on government and private buildings. The main location of this research was conducted in one of the Building Buildings located in Kudus Regency - Central Java. Based on its nature, this study includes qualitative research. Qualitative research is systematic scientific research on parts and phenomena and their relationships [4].

The first step is to do a literature study. Literature studies are grouped into several sections according to the research objectives. In reviewing the reasons for implementing the PBC on the maintenance of related literature study buildings in the form of: the application of PBC in other countries, the application of PBC in Indonesia, and the reasons for the PBC being applied. To identify constraints and assess the magnitude of the constraints, we study the related literature in the form of: constraints to the application of PBC, laws and policies that form the basis for managing building maintenance.

There are several stages of research carried out, namely:

1. Data collection
2. The data collection techniques used in this study include:
   a. Questionnaire
      Data collection is done by distributing questionnaires to respondents by being given directly and re-collecting questionnaires that have been distributed and filled out by respondents.
3. Processing data
   a. The interview data is assessed using the assessment measures that have been developed.
   b. The results of the questionnaire data are used as input to calculate the importance of the existing constraints.
4. Discussion
   a. From the results of the assessment of the constraints that exist in implementing the PBC in the maintenance of buildings, it will be known the magnitude of the constraints at each level of assessment.
   b. This value will then be analyzed to determine the strengths and weaknesses of the assessment.

3. Results and Analysis

PT. XXX as a case study is one of the national cigarette companies located in Kudus City. The area is in one of the factory locations of the company of PT. XXX with 82 Ha, which is divided into open and closed areas. To support the company's business processes, the role of building facilities needs to be maintained and maintenance of factory buildings, warehouses, public facilities, parks and road and channel infrastructure. Targets achieved in the process of maintaining and maintaining these building facilities so that the building is safe, healthy, comfortable and convenient.

To achieve this goal, PT. XXX in conducting maintenance and maintenance of building facilities divides 2 (two) Maintenance and Maintenance Division of Building Facilities which consists of (1) Division of Maintenance and Maintenance of Security Systems and Security Systems, (2) Division of

Maintenance and Maintenance of Building, Roads and Parks. The maintenance and maintenance division of the utility and security system is called Utility Engineering (EU), while the maintenance and maintenance division of buildings, roads and parks is called the General Service Building Operation and Maintenance (GS).

Every time building maintenance is done , the GS team must make an estimation of its cost budget every year. The estimated budget is carried out by referring to the calculation of the maintenance budget that has been carried out in previous years, based on the cost of previous development, as well as routine maintenance costs carried out in line with the operation of the building.

The distribution of costs to total maintenance costs consists of vehicle maintenance, maintenance of safety and health equipment, landscape maintenance, maintenance of furniture and office equipment, AC maintenance, photocopy usage and maintenance, cleaning tools & materials, external cleaning with each cost value (in billion) seen in Figure 1.

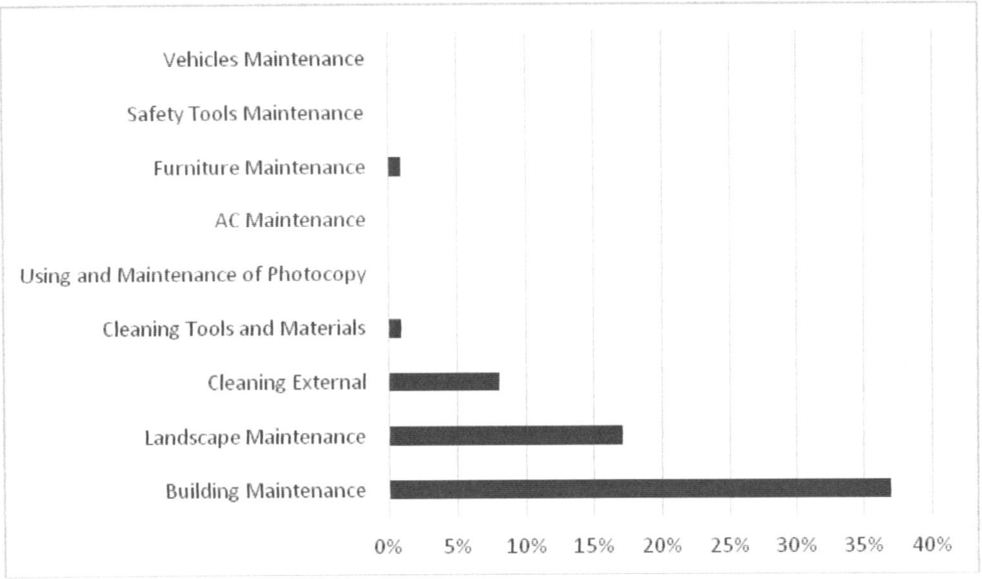

Figure 1. Percentage of Maintenance Costs

Based on Figure 1, the criteria included in building maintenance and the environment can be seen from the maintenance group from the top to the bottom, namely the cost percentage is building maintenance (37%), landscape maintenance (17%), external cleaning (8%) and cleaning tools & materials (1%) of the total maintenance group costs.

The building maintenance group is again grouped into two categories of building maintenance, namely the maintenance of buildings that can be predicted and maintenance of buildings that cannot be predicted. The characteristics / types of work related to building maintenance that are predictable and unpredictable can be seen in Table 1 below.

Table 1. Characteristics of Building Building Maintenance Works

Type of Maintenance	Predicted	Unpredictable
Periodic maintenance	√	-
Maintenance for prevention	√	-
Repair due to operational damage	-	√
Repair due to natural disasters and human errors	-	√

The predicted process of estimated costs for building maintenance uses a reference from the cost of building maintenance in previous years. How to calculate the estimated maintenance budget for this building is still using the traditional method. In the process of calculating predictable maintenance costs, this is more used.

As a comparison of studies, patterns and procedures for maintenance and maintenance of buildings in Kudus Regency Government, it is only through submission / request every year (by order). They cited the maintenance of the building walls in the form of painting the outer and inner walls of the building, if the paint color was really dull and dirty, then repainting would be submitted with the system of calculating the work volume multiplied by the work unit price set by the Government. Another example is if there is a ceiling that is damaged or collapsed, then repairs are immediately carried out at the expense of local governance.

3.1. Character Building Maintenance Related to Procurement of Implementing Parties

The process of handling building construction should be separated from the contract system and the type to be used, related to the procurement of implementing parties. The following is a summary of building maintenance related to alternative handling, funding sources, and types of contracts.

Table 2. Alternative Handling, Funding Sources and Types of Contracts

Handling Category	Nature of Financing	Contract System	Handling Characteristics	Handling Description	Budget Percentage
Construction	Investation	DBB	New construction	Steel construction	0.26%
				Building construction	6.55%
				Pond making in the east of OASIS	15.70%
				Garden making	0.24%
Total Construction					22.76%
Enhancement	Investation	DB	Improved building functions	Building improvement	1.06%
		DBB	Improved building functions	Building improvement	3.17%
			Garden function improvement	Garden improvement	0.23%
Total Construction					4.46%
Rehabilitation	Preservation	DB	Periodic maintenance	Utility maintenance	0.09%
				Repair of buildings	0.18%
				Furniture repairs	0.23%
		DBB	Periodic maintenance	Washing biomass walls	0.04%
				Painting new walls/ceilings	0.40%
				New floor painting	0.04%
				Painting of iron/ steel	0.35%
				Repainting walls/ ceilings	7.06%
				Painting the floor again	27.37%
				Repainting markings	0.80%
				Repair of buildings	5.08%
				Garden improvement	0.04%

Handling Category	Nature of Financing	Contract System	Handling Characteristics	Handling Description	Budget Percentage
Total rehabilitation					41.66%
Maintenance	Preservation	DBB	Periodic maintenance	Garden improvement	0.99%
			Routine maintenance	Repair of buildings	2.94%
		PBC	Periodic maintenance	Building maintenance	0.44%
			Routine maintenance	Building maintenance	9.46%
				Garden improvement	17.29%
Total Maintenance					31.12%
Grand Total					100.00%

Note: DBB: Design-Bid-Build; DB: Design-Build; PPP: Public Private Partnership; UP: Unit Price; LS: Lumpsum; BOT: Build, Operate, Transfer; APBN: State Expenditure Budget; PBC: Performance Based Contract; BOM: Build-Operate-Maintenance.

Based on Table 2 above it can be seen, the character of building maintenance is the category of handling rehabilitation and maintenance with a preservation financing system, DBB, DB and PBC contract systems with values that tend to be relatively small. In the maintenance of performance based contract, there are two types of maintenance, namely periodic maintenance and routine maintenance. Periodic maintenance with a budget percentage of 0.44% while routine maintenance is 9.46%.

Routine maintenance is carried out by self-management, while periodic maintenance is done by contracting. Constraints that arise in routine maintenance with self-management, namely sometimes the owner does not carry out maintenance activities even tend to extend maintenance intervals to reduce maintenance costs. Periodic maintenance is by scheduling maintenance from the start of the building, for example, periodically in daily, monthly, and yearly.

3.2. Building Maintenance Performance Parameters

The basis for determining building maintenance and maintenance parameters, which are maintenance indicators can be seen in the table below. The table will be compared with how the hygiene standards that have been done at PT XXX as a case study.

Table 3. Adoption Standards for Adoption of Minister of Public Works Regulation No. 24/PRT/M/2008

Maintenance object		Cleanliness standards	Comparison with the cleanliness standards of PT XXX
Diffuser/Grill	:	Clean, immaculate, no cobwebs, not dusty	Perform routine cleaning on the diffuser/grille
Plafond	:	Free from dirty, no stains, not dusty, no spiderwebs	The ceiling is cleaned regularly
Glass	:	Clean, clear, clear, no stains, no dirt, no dust, clean glass frames	The glass is cleaned every dirty
Horizontal Blind	:	Clean, not dirty, not dusty, neat	Horizontal blinds are cleaned regularly
Switch and Socket	:	Not dusty, stained	The switch and socket are cleaned periodically and checked for function
Furniture	:	Clean, not dusty, immaculate, if rubbed does not imprint, no rubbish, no spiderwebs	Furniture is cleaned regularly
Floor	:	Clean, not dusty, immaculate, not blurry, not wet, no smell, clean floor	The floor is cleaned every day

Maintenance object		Cleanliness standards	Comparison with the cleanliness standards of PT XXX
		grout	
Carpet	:	Clean, not dusty, immaculate, not smelly, not wet, neatly combed	The carpet is cleaned regularly
Toilet			
Room	:	No smell: piss, fishy, rancid	Cleaned twice a day
Mirror	:	Clear, bright, not dull, not stained, not wet	Mirror glass is cleaned using a glass cleaner
Toilet	:	Smooth flowing, no stains, no water spots around it, no smell	The toilet water flows smoothly and is cleaned twice a day
Faucet	:	Not rusty, not wet, not dull	The faucet is cleaned and replaced if it has rust and is dull
Floor	:	Clean, dry, no stains, no trash	The toilet floor is cleaned regularly and marked before it is dry
Door handle	:	Clean, no stains, shiny (as original)	Door handles are cleaned regularly
Urinoir	:	Clean, no stains, no smell, no rust	Urinal is cleaned twice a day and always check the water can flow
Mat	:	Not dusty, not wet, no garbage, no smell	Doormat cleaned and replaced regularly
Faucet	:	Not rusty, not wet, not dull	The faucet is cleaned and checked for function
Stairs			
Railing	:	Odorless, immaculate, if rubbed does not imprint	Railing stairs are cleaned regularly
Bordes	:	Not dusty, no garbage, not wet, no smell	Bordes stairs are cleaned regularly
Iron	:	Odorless, immaculate, no spots	Iron stairs are cleaned regularly
Garden	:	Fertile, clean, neat, beautiful	The park is kept clean and tidied up to look beautiful
Asphalt floor	:	Clean no garbage, no flooding, no dirt	Asphalt floors are cleaned of dust and garbage from company activities

In general, the cleaning standards at PT XXX in this case study have been done well. When compared with the standards in the regulation, almost all meet the hygiene standards. The building maintenance parameters are quite complex by considering many aspects. Maintenance of building refers to the scope of handling components of the General Service (GS) team building which includes structural, architectural, mechanical, outdoor, and housekeeping components.

3.3. Opportunities for the Implementation of Performance-Based Contracts in Building Maintenance

Based on the characteristics of building maintenance and the problems that often occur in buildings, a performance standard based on references can be made that has been set by the government related to the procedure for calculating buildings. The first is the adoption of a contract system. The contract system commonly used in building construction is a traditional contract that separates the person in charge at each stage of the project cycle. Design stage by planning consultants, implementation by contractors, and maintenance by building owners. The adoption of Performance Based Contracts (PBCs) that integrate the design, implementation and maintenance phases is an opportunity to be used in building projects. The first step is to map or design the activities carried out by the responsible person (stakeholder) at each stage of the project cycle.

The second is a maintenance component that is an opportunity to be applied to buildings based on analysis, which is building maintenance, landscape maintenance, external cleaning, and cleaning tools & materials.

3.4. Effectiveness of Building at PBC

The effectiveness of Performance Based Contracts (PBCs) in buildings is very potential to be applied but will face several differences when compared to the implementation of road projects. In road projects, the type of maintenance is not as much as in buildings. In the road project most of the maintenance is to ensure the road is not hollow, cracked, and deflated for a number of years in accordance with the contract document. But in buildings, which must be maintained by the contractor very much includes the main structure, complementary buildings, and contents in the building.

Therefore it is necessary to arrange in a detailed contract document to implement a Performance Based Contract (PBC) in a building which is the responsibility of the building owner (owner), which is the authority of the contractor. So far, building maintenance is still the responsibility of building owners. In the case study of this study at PT XXX Kudus Regency, a good and structured maintenance was carried out. This is because the owner is a private party whose policies and budgets can be managed independently. In contrast to government projects that have more rules.

4. Conclusion

The conclusion that can be obtained from this study is the maintenance of a Performance Based Contract showing that there are two types of maintenance, namely periodic maintenance and routine maintenance. In the case study it was found periodic maintenance with a budget percentage of 0.44% while routine maintenance amounted to 9.46%.

Parameters of building maintenance performance by considering many aspects. Maintenance of building refers to the scope of handling components of the General Service (GS) team building which includes structural, architectural, mechanical, outdoor, and housekeeping components.

The opportunity to implement PBCs in buildings is the adoption of PBC contracts and maintenance components, namely maintenance of buildings, maintenance of landscape, external cleaning, and cleaning tools & materials due to the highest percentage of maintenance costs among other components.

The pattern of building maintenance after the Final Hand Over (FHO) on building maintenance will need to be mentioned in the contract document. The age of the building is based on the duration of the minimum service standard of the building.

The effectiveness of Performance Based Contracts (PBCs) on building maintenance is very potential to be applied but will face some differences when compared to the implementation of road projects. Maintenance of building includes structural, architectural, mechanical, outer space and housekeeping aspects while road projects include structures and supporting facilities.

References

[1] Catur, Stefanus et al. 2016. "Model Sistem Peringkat Pada Proyek Konstruksi Jalan." 16(3): 213–22.

[2] Chan Edwin H. W., and Raymond Y. C. Tse. 2003. "Cultural Considerations in International Construction Contracts." Journal of Construction Engineering and Management 129(4): 375–81.

[3] Guajardo, Jose, Morris A. Cohen, Sang-Hyun Kim, and Serguei Netessine. 2011. "Impact of Performance-Based Contracting on Product Reliability: An Empirical Analysis." Ssrn (September 2016).

[4] Gunawan. 2016. "Kualitatif Imam Gunawan."

[5] Rawis, Thresia Deisy. 2016. "Perencanaan Biaya Keselamatan Dan Kesehatan Kerja (K3) Pada Proyek Konstruksi Bangunan (Studi Kasus : Sekolah St. Ursula Kotamobagu)." 4(4): 241–52.

[6] Usman, Kristianto, and Restita Winandi. 2009. "Kajian Manajeman Pemeliharaan Gedung (Building Maintenance) Di Universitas Lampung." Jurnal Rekayasa 13(2): 157–66.

[7] Wibowo, Mochammad Agung, Naniek Utami Handayani, Asri Nurdiana, and Moh Nur Sholeh. 2017. "The Identification of Waste Construction at Construction Project Life Cycle." Advanced Science Letters 23(3): 2633–35.

[8] Wirabakti, Deden Matri, Rahman Abdullah, and Andi Maddeppungeng. 2014. "Studi Faktor-Faktor Penyebab Keterlambatan Proyek Konstruksi Bangunan Gedung." Teknik Sipil Universitas Agung Tirtayasa Vol. 6: 15–29.

[9] Zawawi, Emma Marinie Ahmad, Syahrul Nizam Kamaruzzaman, Azlan Shah Ali, and Raha Sulaiman. 2010. "Assessment of Building Maintenance Management in Malaysia: Resolving Using a Solution Diagram." Journal of Retail and Leisure Property 9(4): 349–56.

IOP Conf. Series: Earth and Environmental Science **328** (2019) 012063 doi:10.1088/1755-1315/328/1/012063

Improving Urban Environmental Quality and Human Well-Being towards Healthy Urbanism: Case Study in Soetomo General Hospital, Surabaya, Indonesia

E T Sunarti[1], R P Kurniasanti[2], R D Pratowo[2], P B R Anggraeni[2], F A Wibowo[2]

[1]Professor in Architectural and Urban Morphology, Head of Urban Design Laboratory, Department of Architecture Faculty of Architecture, Design and Planning, Sepuluh Nopember Institute of Technology
[2]Graduated Student, Department of Architecture Faculty of Architecture, Design and Planning, Sepuluh Nopember Institute of Technology

*Email: endar@arch.its.ac.id

Abstract. Nowadays, there is overwhelming evidence that regular physical activity has important and wide-ranging health benefits. The health risk and mortality factors were indirectly influenced by physical inactivity, the sedentary lifestyles developed by the car-oriented urban approach. That evidence supports the conclusion that physical inactivity becomes one of the most important public health problems of the 21st century, and may even be the most important. In order to encourage healthy urbanism, focusing on cultural and environmental aspects become necessary to Indonesian people. Soetomo General Hospital is a provincial health centre to provide a sense of security, comfort and convenience which are equal for all kinds of people, especially to walk. Therefore, environmental quality is a very important component to be considered. This paper utilizes character appraisal to find Soetomo General Hospital area's physical condition. The result concludes design criteria for the built environment of the area. These design criteria are used as the main concept for elaborating the concept design of the place to improve its quality of environment towards healthy urbanism.

Keywords: *Environmental Quality; Human Well-being; Public Space; Healthy Urbanism.*

1. Introduction

Understanding of empirical facts through the analysis of three road segments in RSUD dr. Soetomo can conclude that land use in all three road segments is quite diverse having supported Soetomo General Hospital as the center of activity. In the general sphere, each road has its own characteristics and needs with different handling. It means as a case, the part of the city's area of the hospital area becomes something unique to illustrate the linkage between urban face changes influenced by the meaning of an environment according to Shultz's statement of the wider meaning of placement in an environment not only related to occupancy, but also to reach shelter or neighborhood [1].

This research still refers to the arrangement that prioritizes the needs of each road corridor. The arrangement in question is to rearrange the area, especially the corridor of the road experiencing a decrease in its condition and even the transfer of function. This is conducted by paying attention to the

Sustainability Principle which requires to wisely choosing a more cost-effective alternative to natural resources, using urban area development patterns with new functions that provide better appeal. Attention here includes the thoughts on the conditions of achievement of the Sustainable Development Process.

In general, the first phase is more focused on physical conditions. The form of road connectivity, free from obstacles, the quality of the area for walking and the need for a map can provide important building location information. However, there is also a very important non-physical factor to improve the need for a place to gather and sit comfortably; as well as the need for supervision and stern action on the behavior of motorists who are potentially harmful and take pedestrian rights. Therefore, there needs to be a review of the needs of the road users as well as the priorities of design development and design direction in the future.

1.1. Literature Review

Environmental physical design can affect the level of public health. In relations between humans and the surrounding environment, the challenges in healthy city design deal with developing good communication between the medical community and land use [2]. Then, physical activity, obesity, and other diseases, are closely related to the lack of land use management, such as low density development, single use development, and road networks that are not well-connected [3]. Regarding parameters related to community activities and their health, land use is an important factor in forming a healthy space in a city that can have a direct impact on the community [4].

Land use as a reason related to public health, is also mentioned by several earlier studies where land use becomes a reason related to public health as well as inherited community characteristics, lifestyle and social economic variables [5] [6]. It all reflects a broad view of health as a positive experience of well-being and not just absence of disease [7]. However, despite the symbiotic relationship between land use planning and health, this connection has in practice been forgotten to date [8]. Regarding land use matters, the two interrelated sets of theories underlie this diagram: about urban ecosystems and about health determinants [9]. Each model layer has an impact on health and well-being, and changes in land use affect each layer. An artificial environment layer is a field where land use planning has a direct impact. Changes in the built environment structure change the natural, the social and economic environment [8].

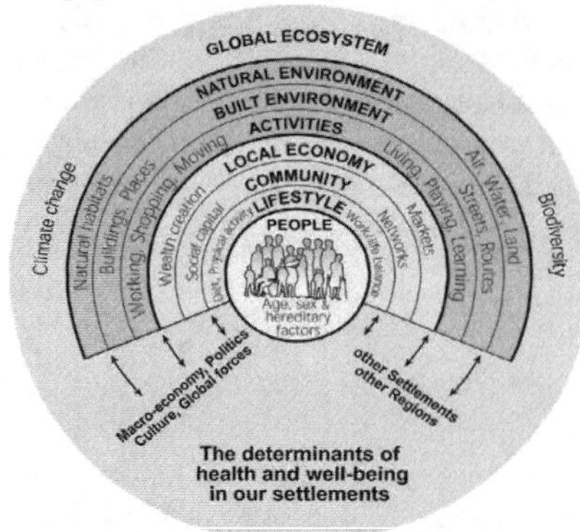

Figure 1. The built environment as part of healthy urbanism
Source: Barton and Grant, 2006

Physical boundaries and territorial boundaries and access in the area influence each other higher density environments allow better access and increased use of local facilities, although access to services for people living in areas far from the city center is also related to get transportation services [10].

Type of land use. The mixed impact of land use on health has to do with increasing levels of physical activity with health benefits that have resulted in including reduced obesity rates, identifying types of environments or environmental aspects that will ease a more active lifestyle [11] [12] [13].

Environmental profile. This highlights a number of physical characteristics that have a positive impact on health and well-being, including: an attractive and well-maintained environment; bright and well-built sidewalks for pedestrians; road patterns that offer opportunities for informal contact among residents [14].

The above description provides a solid summary of the relationship between the concept of Healthy Urbanism and land use. The community in this concept plays an important role in providing active participation and gaining experience with communication between activities that actively support health and land use. The use of land that supports the concept of Healthy Urbanism must be able to provide an active response hindering people for choosing the second option other than walking in a movement towards the destination such as using public transportation or choosing not to move at all.

2. Material and Methods

The research method in this study employs character appraisal [15] to find Soetomo General Hospital area's physical condition. The method is used to obtain information about the functional form and to record activities to list the particular characteristics that occur in this area. The result will conclude design criteria for the built environment of the area. These design criteria used as the main concept to elaborate the concept design of the place to improve its quality of environment towards healthy urbanism.

The condition and quality of the corridors around the hospital will be the basis of this research. The corridors are Dharmawangsa Street, Prof. Dr. Moestopo Street and Airlangga Street. The following is a map that explains the location of the study site.

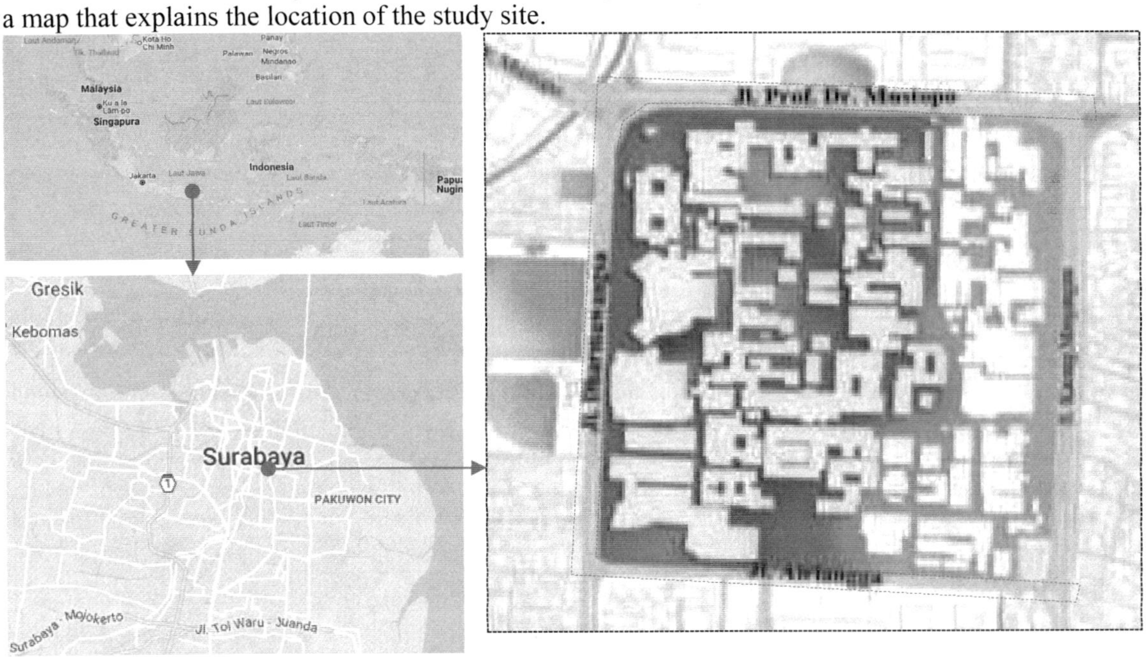

Figure 2. Key-map of research study

3. Result and Discussion

3.1. Physical Boundaries

The physical boundary becomes one aspect of environmental quality that affects comfort in an area. The appraisal character of the three corridors shows the condition and situation of each physical boundaries problem.

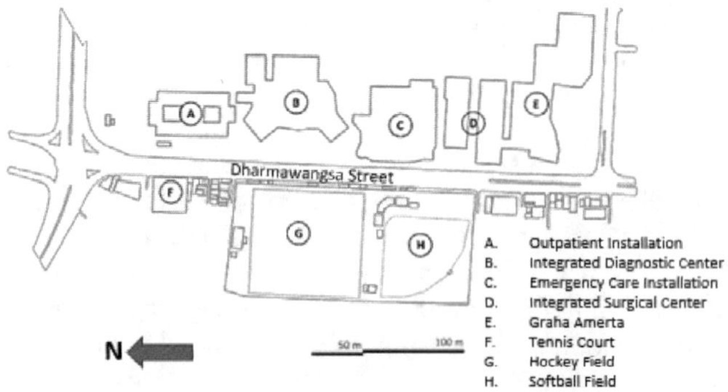

Figure 3. Dharmawangsa Street Layout

On Dharmawangasa Street, the building borderline for the east side segment perimeter area is 12 m, while for the west side commercial land is 10 m. There are some misuses of the pedestrian lane as a place for passing motorbikes, bicycles and PKL carts selling rather than using highway shoulder segments. On the edge of the highway, there was no clear marker of the location of parking vehicles resulting in cars and motorbikes parked in any place neglecting the time and crowds of the road.

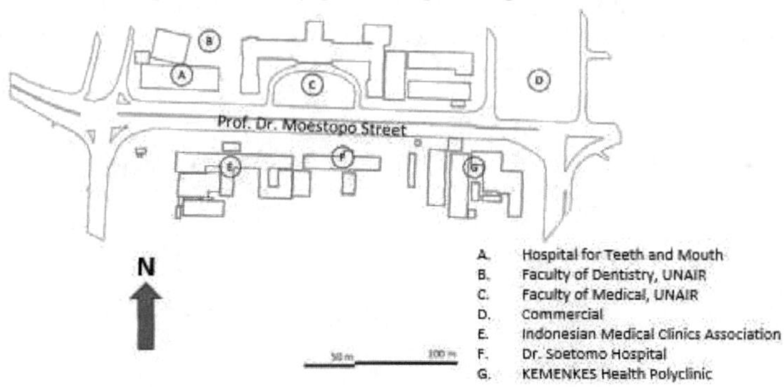

Figure 4. Prof. Dr. Moestopo Street Layout

Prof. Dr. Mustopo street is included in the classification of secondary arterial roads which have a width of 25 meters. Each side has 3 lanes of vehicles which are usually used for parking the vehicle. The borderline of the north and south side of the building in this corridor is 8 meters, each with priority for pedestrians.

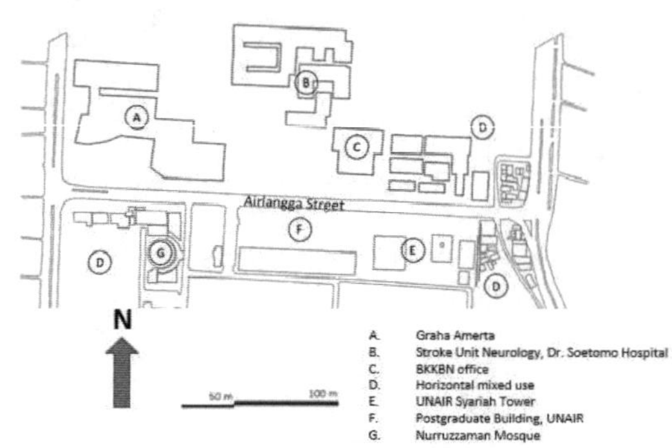

Figure 5. Airlangga Street Layout

Airlangga street is included in the classification of secondary collector roads that have a meter road width. Each side has 3 lanes of vehicles which are usually used for parking the vehicle. The border lines of the north and south sides of the corridor are 5 meters, each with pedestrian priority.

3.2. Type of Land Use

The main land-use in Dharmawangsa Street is oriented towards the health industry because there are several hospitals or health facilities, education and community housing in the development area. Health facility building that supports Dr. Soetomo on this road is on the east side of the road. This area also contains other public facilities such as Tennis Court, Hockey Field, and Softball Field. On the west side, most are filled with shops that sell daily equipment, food, and medicine.

Land use in the corridor of Prof. Dr. Moestopo street is dominated by health and education facilities. The functions of the buildings found are health facilities in the form of dental and oral hospitals, education in the form of Airlangga University campus A, and commercial functions in the form of small stalls which are business enterprises owned by residents.

Land use in the Airlangga Road corridor is mainly for public facilities and some buildings for services and commercial trade. On the north side, several health facility buildings are located such as Graha Amerta and RSUD Dr. Soetomo, as well as offices such as the BKKBN Center in East Java Province.
Some small and medium-sized commercial buildings also appear to fill the corridor both on the north and south, including restaurants, photocopies, jewelry stores, and posyandu. In existing conditions, there is land use for informal traders using portable places to sell.

3.3. Environmental Profile

Table 1. Environmental Profile

Character Appraisal - Environmental Profile
Dharmawangsa Street Corridor
East Side 
West Side 
The east side of the Dharmawangsa road is filled with high-rise hospital health facilities. Some of them have a height of 4 floors or more, thus giving the character of a corridor with high buildings on this road. While the west side looks more grounded with pedestrian lane users because it manages their front to be easily accessible without a fence. The character of Dharmawangsa street neighborhood is formed by the profile of the environment of buildings in the style of modern architecture. The building's façade is displayed in the form of vertical and horizontal lines formed by concrete structures and other ornaments.
Prof. Dr. Moestopo Street Corridor
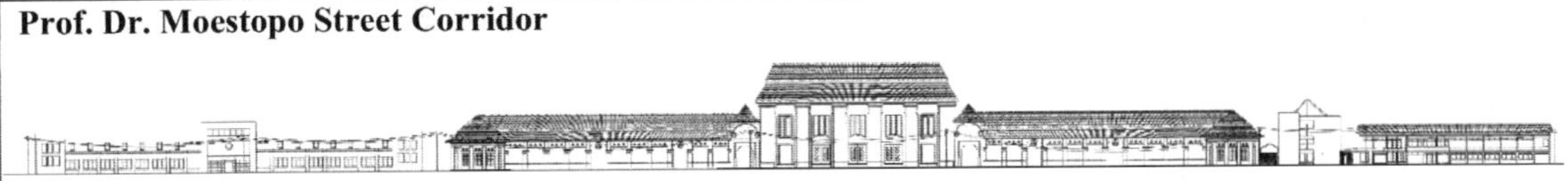
North Side South Side 

Character Appraisal - Environmental Profile
The corridor of Prof. Dr. Mustopo gives the impression of a monotonous facade, especially on the south side where the dividing gate stretches from one side to another on the road. The number of offices in this area makes this area seem more closed and private for pedestrians. Environmental characteristics in Prof. Dr. Moestopo street is formed by the profile of the buildings in the style of colonial architecture or the *indische empire style* using a shield or *limasan type roof* and the use of columns in buildings without any cover ornament.
Airlangga Street Corridor 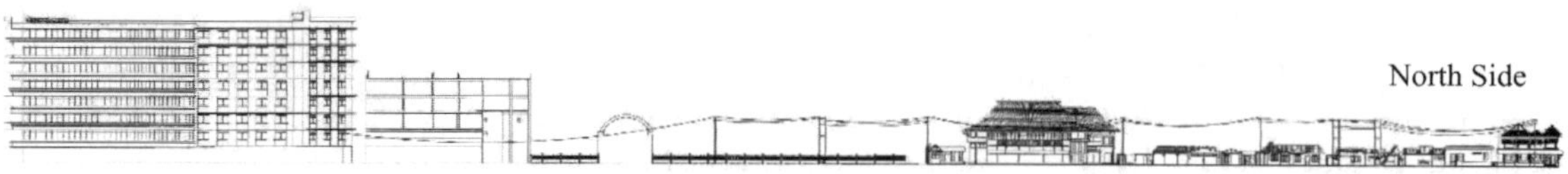 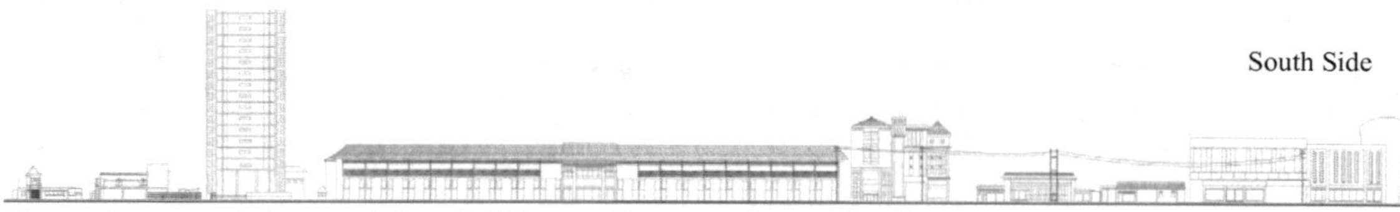
The environmental characteristics of Airlangga Street are formed by diverse environmental profiles. There are buildings that use the colonial architectural style or the Indische empire style. There are also buildings that are stylish in modern architecture, especially in new buildings, as well as buildings that have Mediterranean architectural style, especially in residential buildings.

4. Conclusion

Soetomo General Hospital in an urban context and environmentally-conscious region should basically consider the following several things. The first consideration is setting the skyline of buildings around the hospital area to create a balance of the neighborhood environment. The use of zoning in the wider corridor as a commercial area for informal traders must be encouraged. Then, utilization of spaces on pedestrian lanes as well as corridors as passive public spaces is also required. Design of pedestrian pathways is equipped with barrier and passive public spaces as a barrier between pedestrian circulation area and motorized circulation. All corridors need to add pedestrian maps presenting location of buildings around the area.

The mentioned items above are integrated into a schematic design proposal with the aim to actualize Dr. Soetomo General Hospital's sustainability and the characteristics of each corridor within the area.

Table 2. Schematic Design

Concept Visualization		
	Align the facade of the building by equalizing the position of the banner in the vertical part of the partition. 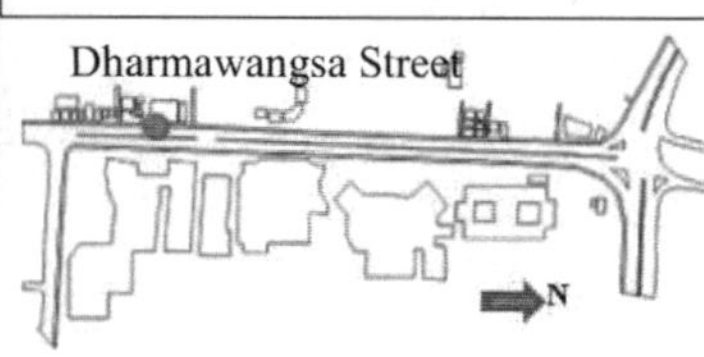	

Concept Visualization

The addition of *overstek* in front of the shopping area and stalls that provide harmony between facades of buildings.

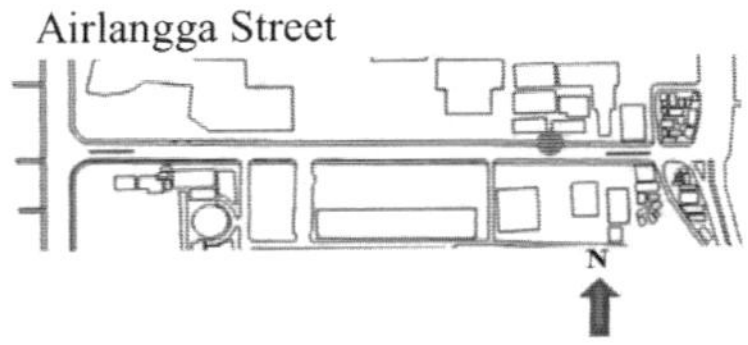

Addition of street furniture such as zebra cross line, pelican cross, pedestrian maps, and tactile paving

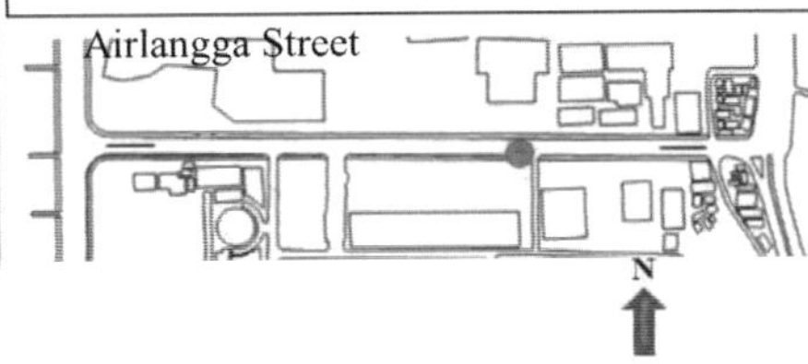

Seats and trash bins are placed repetitively on the edge of the pedestrian lane as a separator of the

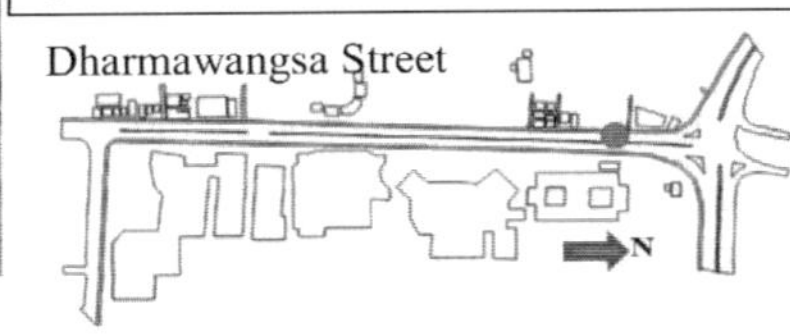

Plantswith low shrub plants to optimize vegetation and improve the aesthetic quality of the corridor.

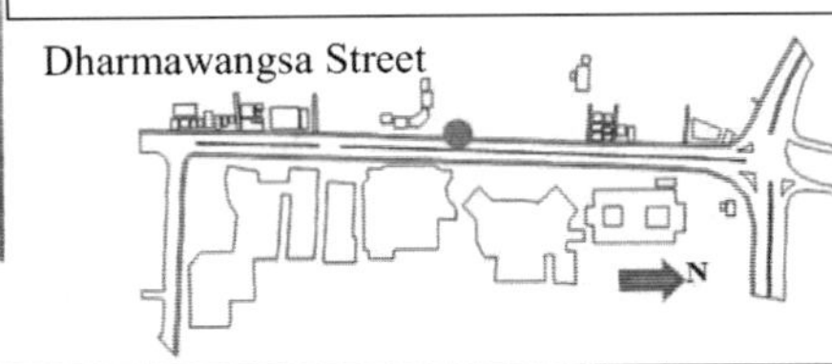

Structuring the zoning of commercial areas of informal traders with the concept of open space to give an open

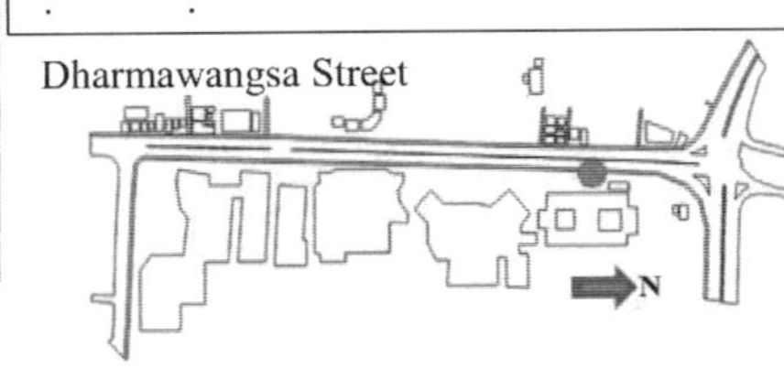

Acknowledgement
This publication is part of research, conducted in 2017-2018, entitled Designing Soetomo General Hospital in the context of Urban Environmental Development. This research is funded by "Ministry of Research, Technology and Higher Education Republic of Indonesia".

References
[1] Shultz, John L. 1975. *The Vocational Gudance Quaterly*. Willey One Library. Vol. 23. https://doi.org/10.1002/j.2164-585X.1975.tb00915.x. p 271-274
[2] Eitler, T. W., McMahon, E. T, Thoerig, T.C. 2013. *Ten Principles for Building Healthy Places*. Urban Land Institute, Washington.
[3] Ewing R, Schmid T, Killingsworth R, et al. 2003. *Relationship Between Urban Sprawl and Physical Activity, Obesity and Morbidity*. American Journal of Health Promotion 18(1), Page 47–57.
[4] Cubukcu, E. 2013. *Walking for Sustainable Living*. Procedia - Social and Behavioral Sciences, 85,33-42.
[5] Lalonde, M. 1974. *A New Perspective on the Health of Canadians*. Health and Welfare, Canada, Ottawa.
[6] Whitehead, M., Dahlgren, G. 1991. *What Can Be Done About Inequalities In Health?*. The Lancet 338, 1059–1063.
[7] WHO, 2009
[8] Barton, Hugh. 2009. *Land Use Planning and Health and Well-Being*. Land Use Policy 26S S115–S123
[9] Barton, H. and Grant, M. 2006. *A health map for the local human habitat*. The Journal for the Royal Society for the Promotion of Health, 126 (6). pp. 252-253
[10] Bramley. G., Dempsey, N., Power, S., Brown, C. and Watkins, D. (2009). *Social sustainability and urban form: evidence from five British cities*. Environment and Planning A 41 pp.2125-2142.
[11] Papas, M. A., Alberg, A. J., Ewing, R., Helzlsouer, K. J., Gary, T. L., & Klassen, A. C. (2007). *The built environment and obesity*. Epidemiologic Reviews, 29, 129-143.
[12] Feng, J., Glass, T. A., Curriero, F. C., Stewart, W. F., & Schwartz, B. S. 2010. *The built environment and obesity: a systematic review of the epidemiologic evidence*. Health & Place, 16(2), 175-190.
[13] Rodriguez, D. A., Khattak, A. J., & Evenson, K. R. 2006. *Can new urbanism encourage physical activity? Comparing a new urbanist neighborhood with conventional suburbs*. Journal of the American Planning Association, 72(1), 43-54.
[14] McDonald, J., Wise, M and Harris, P. (2008) *The health impacts of the urban form: a review of the reviews*. Sydney, Centre for Health Equity Training, Research and Evaluation.
[15] Ministy for the Environment. 2006. *Urban Design Toolkit*. New Zealand : Third Edition. ISBN: 978-0-478-33153-0

Urban Population Growth and Their Implication to Agricultural Land in the Process of Metropolitanization: The case of *Kabupaten* Sukoharjo, in Metropolitan Surakarta

F H Mardiansjah[1], A Sugiri[2], G P Sari[3]

[123] University of Diponegoro, Prof. Soedarto St., SH, Tembalang, Semarang, Central Java, Indonesia

*Email: fadjar.mardiansjah@pwk.undip.ac.id

Abstract. Extended urban spatial formation has been a major phenomena in many cities' development, in Indonesia, including big, medium and smaller size of cities. Using the case which occurred in the *Kabupaten* Sukoharjo, an adjacent kabupaten of the City of Surakarta, this paper discusses the process of population growth, land expansion and their impact on the agricultural land outside the city's perimeter, in the metropolitanization process of Surakarta as a secondary metropolitan growth in Java based on a medium city size as the core of the metropolitan. The analysis applies time-serial statistical data gathered from the *kabupaten*'s monographs. It also applies comparative analysis among the *kecamatan* (districts) in the *kabupaten* that are classified into two types of *kecamatan*: the adjacent and the non-adjacent. The analysis indicates a dramatic changing of urban land, which confirmed the impact of uncontrolled urbanization into agricultural land in the peripheries in terms of losing of fertile agricultural land. It also indicates that the process leads to the needs to develop other agricultural land to substitute the loss of fertile agricultural land. Therefore, the analysis indicates the double impacts of metropolitanization process into agricultural land, such as: the loss of fertile agricultural land in its peri-urban and the implication of efforts in producing fertile land in other areas.

Keywords: *land conversion, metropolitanization, peri-urban, population growth, Sukoharjo, Surakarta.*

1. Introduction

As happening in many developing countries, Indonesia has also experienced a tremendous urbanization process that brings many challenges and problems in the management of its cities. According to a report in CICRED Series, The Demography Institute of the University of Indonesia [1], it is predicted that urban population in Indonesia was about 14.4 million in 1961. It was about 14.8% of the total population [1]. The number has increased to 85.24 million in 2000 that represented about 41.9% of the total population [2]. Later, Indonesia has achieved the key milestone of its urban population growth by exceeding the 50 per cents of the percentage of population living in urban areas for the first time in history in 2012 [3]. Further, Indonesia is predicted to have about 135.6 million urban populations, amounting to 53.1% of Indonesian population, according to the result of Intercensal Population Survey in 2015. Higher growth of urban population occurred since 1970s, after Indonesia entered into a new scheme of development efforts that was called as the New Order Era. The number of urban population in 2015 has become about six and half times of its number in 1971.In this identifiable time, Indonesia had nearly three million additional urban populations annually.

However, similar to the total population, urban population has not been distributed evenly in Indonesia. Java, as the smallest of the five biggest islands in Indonesia, which only has 7% of the national land area, accommodates 56.8% of the country's population. It also becomes the home for 67% of its urban population. The huge number of population lives in Java has made the island as the most densely populated in the country. In 2015, the density of Java was 1.120 inhabitants per square kilometer. It is much denser than Bali and Nusa Tenggara, the second most densely populated region, which has density to 193 inhabitants per square kilometer, and the average density of the country covers 130 inhabitants per square kilometer.

Table 1. Distribution of urban and total populations among regions in Indonesia, 2015

Regions	Area (km2)	Area Share	Urban Population	Urb. Pop Share	Total Population	Total Pop. Share	% Urban
Java	129,438	7.0%	90,825,696	67.0%	145,013,573	56.8%	62.60%
Sumatra	480,793	26.2%	22,848,775	16.8%	55,198,752	21.6%	41.40%
Sulawesi	188,522	10.3%	6,892,367	5.1%	18,702,298	7.3%	36.90%
Kalimantan	468,682	25.5%	6,864,938	5.1%	15,320,017	6.0%	44.80%
Bali & Nusa Tenggara	73,070	4.0%	5,998,627	4.4%	14,091,466	5.5%	42.60%
Papua	418,708	22.8%	1,191,056	0.9%	4,011,907	1.6%	29.70%
Maluku	78,897	4.3%	991,627	0.7%	2,844,131	1.1%	34.90%
Indonesia	***1,838,110***	***100.0%***	***135,613,086***	***100.0%***	***255,182,144***	***100.0%***	***53.10%***

Sources: Calculated from SUPAS data, 2015.

With the high density of population, Java becomes a fertile land for urban population development. Two-thirds of the national population increase occurs in Java as two-thirds of the national urban population living on Java. In 1990s, urban population growth in Java in 1980-1990 was concentrated dominantly in north coast of Java [4], especially in the corridor between Jakarta to Semarang. In 2016, the distribution has become more dispersed as the occurrence of in-situ urbanization creating the gradual change of many Javanese villages from non-urban villages to urban villages [5]. The spatial distribution of urban population growth has also been characterized by the continuing process of mega-urbanization as the high growth of urban population in the outskirt of the Javanese big cities with more than one million populations such as in Jakarta, Surabaya, Bandung and Semarang [5].

The extended urban formation in the urbanization process in Java has not occurred in its big cities but also in its much smaller cities [6]. The case of Cirebon Urban Area of West Java, a metropolitan area that is formed by Municipality Cirebon, where a 300 thousand population city size as the core area, and *Kabupaten* Cirebon as the peripheries of the metropolitan has been one of the evidences [6]. The growth of urban population in Java has not only occurred in big city based urban regions, but also in many urban regions based on smaller cities like Malang and Tegal [7]. In fact, urban population growth has not only occurred in the Javanese cities (*daerah kota* or urban districts) but also in many Javanese *kabupaten* (non-urban districts), especially in the outskirts of big, medium and small cities in Java, including in some *kabupaten* that are not neighboring to any city in Java [7].

There are many studies and analyzes that have been carried out on urbanization and metropolitanization focusing on large metropolitan areas in Indonesia, such as on Metropolitan Jakarta [8] [9] [10] and [11] and Metropolitan Bandung [12] [11] and [13]. However, there are still very few studies conducted on smaller urban regions in Indonesia. There is a study on the Cirebon Urban Region, attempting to explore the urbanization process in much smaller city in Java [6].

This article aims to discuss population growth in periurban area of medium size metropolitan in Indonesia, using the case of the periurban of Metropolitan Surakarta, a medium size metropolitan in Java. It is expected to contribute to the discourses of urbanization and metropolitanization in Java. The small-and-medium-city-size-based urbanization process in Java is very important to be understood. In addition to its condition that is still under-explored, the importance raises also because small and medium cities are the dominant type of cities in Java playing important roles in Javanese regional development as well as improving linkages to the agricultural and non-urban activities.

This article will be presented in several sections that elaborate the impact of population growth in Sukoharjo Regency, the implications of the development in built land and its effect on the availability of agricultural land. Following this introduction, an analysis of the population growth of the district will be carried out. Further, the analysis on the growth of urban land will be presented, followed by the analysis on the reduction on agricultural land, especially in paddy fields. Later, after a general discussion on the topic, the last part will be a conclusion and recommendation.

2. Data and Methods

This research utilizes the case of urban growth that occurs in *Kabupaten* Sukoharjo, one of three bordering *kabupaten* in the outskirt of Surakarta City, which has the longest border with the city compared to the other two. The *kabupaten* bordering on eight *kecamatan* (sub-districts) in the south of the city. The elaboration applies the secondary data to explain the process of population growth in the suburbs of the metropolitan as well as its relation to the growing built-up area and the conversion of agricultural land into the built-up ones. Data used in this research are compiled from monographs of the *kabupaten*, especially on population and land uses, such as: in built-up and agricultural land (paddy fields and farm land). Data analysis is gathered in the *kecamatan* (sub-district) units, classified into two groups of: the adjacent *kecamatan* for those are bordering to Surakarta City, and the non-adjacent *kecamatan* for the rest. The research employs a time serial analysis in order to perform spatial-temporal comparisons.

3. Result and Discussion

3.1. Population Growth in Kabupaten Sukoharjo

Population of the *kabupaten* has grown from 695,732 in 1990 to 875,917 in 2015. Meanwhile, urban population of this *kabupaten* has grown more rapid, from 126,811 in 1980 to 632,367 in 2010 [15]. Population growth before 2000 has been higher than after, and the period from 1995 to 2000 has the highest population growth among other periods. The more rapid urban population growth has led the *kabupaten* into an urbanized *kabupaten* in 1990s with a jump in the level of urbanization from only 21.3% in 1980 into 76.7% in 2010 [15]. Eight of its *kecamatan* have been classified as urbanized *kecamatan* in 2010. All of the *kecamatan* in adjacent areas have become urbanized *kecamatan*, and only *Kecamatan* Gatak that has started to be an urbanized *kecamatan* since 2000, while the others have been urbanized since 1990 or before. *Kecamatan* Sukoharjo, Bulu and Bendosari are the three kecamatan of the non-adjacent areas that also have been urbanized since 2000.

The *kabupaten* had about 10,755 additional populations annually in which *Kecamatan* Kertasura and Grogol had more than 2,500 annual additional populations each. The population growth rate has been different among areas in the *kabupaten*. Although they have less population, the adjacent areas faced more rapid population growth than the other areas. *Kecamatan* Grogol, Kertasura and Mojolaban had been always the top three *kecamatan* in receiving additional populations annually. Two third additional populations of the *kabupaten* occur in the adjacent areas, with the *Kecamatan* of Grogol, Kertasura and Mojolaban are the place where the growths are more concentrated. In temporal pattern, population growth in the adjacent areas has very similar pattern to the *kabupaten*. The different was only in the magnitude that was higher than that of the *kabupaten*. Even, population growth rates of these areas have been higher than that of *Kecamatan* Sukoharjo, the capital of the *kabupaten*. The more far areas from the border with Surakarta City have lower population growth.

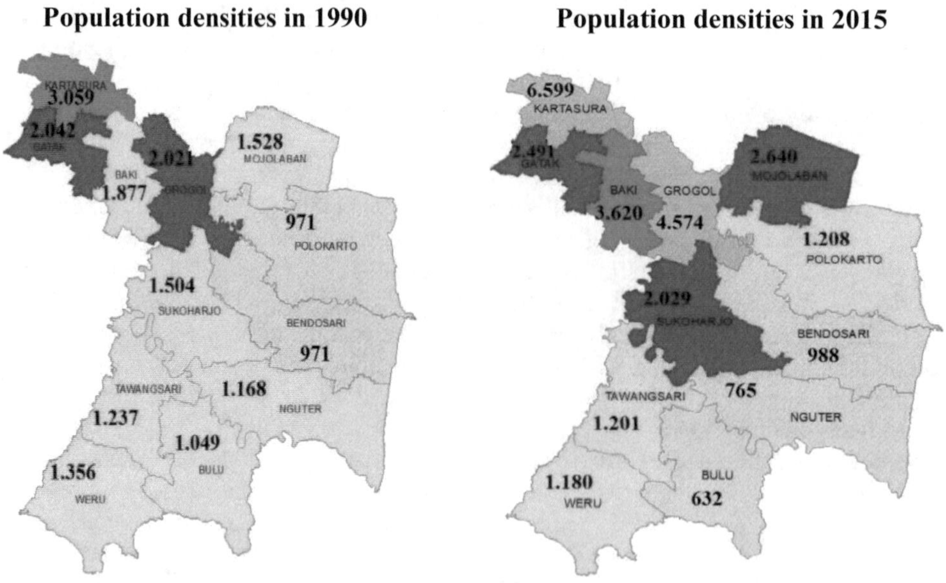

Figure 1. Change in population density of the *kecamatan* 1990 – 2015 (in pop./sq.km.).

The concentration of population growth in the adjacent areas has increased population density of all *kecamatan* in the areas, so their density are much higher than those of non-adjacent ones (see Figure 1). Even, *Kecamatan* Kertasura and Grogol have very big leap from 1990 to 2015 that has made their density double than before. Meanwhile all *kecamatan* in the non-adjacent areas had still less than 1500 population per square kilometer. *Kecamatan* Sukoharjo and Kecamatan Polokarto of these areas, which are located not too far from Surakarta, still have significant increases, but some *kecamatan*, like Bulu, Nguter, Weru and Tawangsari that are located more far than the others have decreasing population density along the observed period.

3.2. The growing built-up areas in Kabupaten Sukoharjo

The growing population in the *kabupaten* has also increased its built-up areas from 14.285 hectares in 1990 to 16.676 hectares in 2015, so the *kabupaten* had an increase of 2,391 hectares of built-up areas to accommodate 180,185 additional populations in the period. The increase was a huge since the additional built-up land was more than half of the area of Surakarta City, the core of the metropolitan area, which has about 500 thousand populations. This indicates that urban growth in the *kabupaten* was extensive although the average built-up land consumption per population in the *kabupaten* has decreased from 205 m^2 per person in 1990 to 190 m^2 per person in 2015, which still more than three times of those in Surakarta that has decreased from 65.2 m^2 in 1990 to 56.2 m^2 in 2015.

The densification process in the adjacent areas has occurred in a more intense than in the non-adjacent ones. Figure 2 shows that *Kecamatan* Grogol of the adjacent areas became the *kecamatan* that faced highest increase with 672 hectares increase along the period to accommodate 39,396 additional populations. *Kecamatan* Nguter of the non-adjacent areas faced the second highest increase of 409 hectares although its population increase was very little (only 96 additional population) as in 1995 to 2000 this *kecamatan* faced a significant decreasing population. However, it is interesting to know that most of all *kecamatan* in the adjacent areas have big increase of built-up land, while the *kecamatan* in the non-adjacent areas have much less increase. *Kecamatan* Grogol, *Kecamatan* Kertasura and

Mojolaban have more than 250 hectares of increase each, while Baki and Gatak have 200 and 107 hectares of increase respectively. These situations indicate that the adjacent areas have become the main area of urban growth in the *kabupaten*.

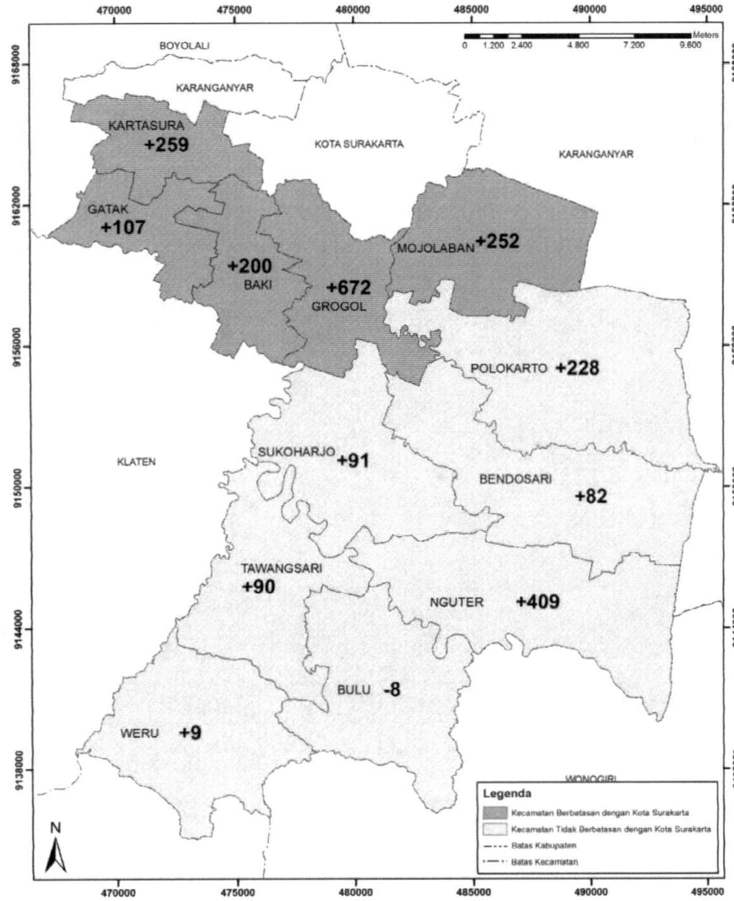

Figure 2. Increase of built-up land in every *kecamatan* 1990 – 2015 (in hectares).

The highest annual increase of built-up land occurred in 2010 to 2015 with an increase of 123.6 hectares of built-up land per year, although the additional population in this period has become more moderate to less than 6000 population annually. It was different with that of the annual increase of population that was in 1995-2000, with an average of 10,755 populations per year. The two classification areas have different time of the highest average increase of built-up land. The adjacent areas had the highest increase in 1990 to 2005 when they got about 70.9 hectares additional built-up land per year with *Kecamatan* Grogol as the main area of the increase with an average of 35.5 hectares per year. The increase in the adjacent areas has been slowing down in the next periods, even though they were still high as they have 69.2 hectares annually in the period of 2010-2015. The non-adjacent areas had different pattern that shows an increasing annual increase according the time. Therefore the latest period, 2010-2015, was the period with the highest annual increase in the non-adjacent areas with *Kecamatan* Nguter has become the main area of the increase.

3.3. Reduction of paddy fields in Kabupaten Sukoharjo between 1990 into 2015

The increasing in built-up in the *kabupaten* has been followed by the reduction of paddy fields and farmland significantly. The *kabupaten* lost at least 1,732.0 hectares of paddy fields in the periods. However, since the *kabupaten* has also developed some new paddy fields, especially in the non-adjacent areas, the reduction of paddy fields was ‘only’ 1,093.0 hectares. Within the loss of paddy fields, there are include the loss of 1,407.0.0 hectares of paddy fields with technical irrigation systems, the best agricultural land to raise paddy rice since they usually could be cultivated three times per year as they could arrange their irrigation.

Table 2. Changes of land uses in *Kabupaten* Sukoharjo from 1990 to 2015 (in hectares)

Kecamatan	A	B	C	D	E	F	G
Kartasura	259.0	-266.0	-266.0	0.0	0.0	0.0	0.0
Gatak	107.0	-83.0	-83.0	0.0	0.0	0.0	-2.0
Baki	200.0	-190.0	444.0	-616.0	0.0	-2.0	-3.0
Grogol	672.0	-509.0	-476.0	-110.0	273.0	-73.0	21.0
Mojolaban	252.0	-197.0	-153.0	-28.0	0.0	-9.0	-16.0
Adjacent Areas	1,490.0	-1,245.0	-534.0	-754.0	273.0	-84.0	0.0
Sukoharjo	91.0	-328.0	-328.0	0.0	0.0	0.0	-28.0
Bendosari	82.0	-124.0	-101.0	294.0	0.0	-317.0	-48.0
Polokarto	228.0	22.0	28.0	-10.0	318.0	-143.0	-203.0
Tawangsari	90.0	34.0	1,484.0	0.0	165.0	-1,303.0	-371.0
Nguter	409.0	33.0	416.0	-258.0	669.0	-457.0	-437.0
Weru	-8.0	550.0	1,082.0	-10.0	511.0	-225.0	-755.0
Bulu	9.0	-35.0	385.0	-193.0	8.0	-183.0	-519.0
Non-Adjacent Areas	901.0	152.0	2,966.0	-177.0	1,671.0	-2,628.0	-2,361.0
Kabupaten Sukoharjo	2,391.0	-1,093.0	2,432.0	-931.0	1,944.0	-2,712.0	-2,361.0

Sources: Results of the analysis, 2019.
Notes: A : built-up land.
B : paddy fields.
C : paddy fields with technical irrigation systems.
D : paddy fields with half-technical irrigation systems.
E : paddy fields with simple irrigation systems.
F : paddy fields without any irrigation systems.
G : farm land

The adjacent areas was the place where the loss of technical irrigation system of paddy fields has dominantly occurred, as the areas has not many farmland as well as other low quality paddy field anymore since 1990, so most agricultural land conversion in these areas occurred by consuming their paddy fields. It was about 978.0 hectares in the adjacent areas and another 429.0 hectares in the non-adjacent ones. However, since the *kabupaten* has also did some efforts in transforming other paddy fields to create 3,395 hectares of paddy fields with technical irrigation systems, mostly in the non-adjacent areas, the *kabupaten* has not experienced reduction in its paddy fields with technical irrigation systems. Therefore, it could be said that the *kabupaten* had a double impacts in these phenomena, in which it had some loss of investments with the loss of its paddy fields with the technical irrigation systems in the adjacent areas, and it has to invest a lot more again to create new paddy fields with technical irrigation systems in the non-adjacent areas to replace the previous loss.

Another land use change pattern occurred in the non-adjacent areas. The increase of built-up land in these areas was not followed by reduction in the paddy fields. Instead of deceasing, the paddy fields in these areas have also been increasing, including the paddy fields with technical irrigation systems. The big creation of new paddy fields including those with irrigation systems, especially in *Kecamatan* Tawangsari and Weru, and also in *Kecamatan* Baki, Nguter and Bulu, has replaced the loss of paddy fields in the *kabupaten*, so the *kabupaten* does not seem to have some loss of their paddy fields, including those with the technical irrigation systems (see Table 2.).

Another interesting phenomenon occurred in the *kabupaten* is the facts that the increasing paddy fields with irrigation systems is not only followed by decreasing of other type of paddy fields especially those without irrigation systems, the lowest quality of paddy fields since they are highly depend on the rain water as the source of their irrigation, but also is followed by very significant decrease in their farmland. The very big decrease of paddy fields without any irrigation systems and farmland in these areas has coincided with the very significant increase in the paddy fields with technical irrigation systems and the paddy fields with simple irrigation systems. These phenomena indicate that the big loss of paddy fields with technical irrigation systems in the *kabupaten* has been responded by the improvements of other agricultural lands in the context of creating new paddy fields in order to maintain the rice production, the main staple food of the nation.

3.4. Implication of population growth in periurban areas in Kabupaten Sukoharjo

The phenomena occurred in *Kabupaten* Sukoharjo shows that population growth in a medium metropolitan's peripheries leads to influence the availability of agricultural land in several ways. The population growth into the peripheries that could be said as an extension of the urban growth to the peri-urban areas, the extension of urban growth become a predator to agricultural land, including their best quality like the paddy fields with technical irrigation systems. The conversion patterns in the peri-urban areas occurred in two pattern. The first is the agriculture-to-urban land conversions, stimulated by urban growth that consumed agricultural land, including the best paddy fields, to create into built-up lands as the place for the growing urban activities. The second conversion is the agriculture-to-agriculture land conversion stimulated by the losses of best agricultural land (paddy fields including those with technical irrigation systems) that convert lower quality agricultural land into better ones, which occurred in two patterns: the conversion of low quality of irrigated paddy fields into the technical irrigation systems paddy fields; and the conversion of farmlands and paddy fields without irrigation systems into better quality of paddy fields including those with technical irrigation systems. If the first conversion pattern could be said as the implication of meeting the need of population growth, the second conversion is the implication of maintaining the areas' capacities in producing food by improving their available agricultural lands to become better quality agricultural land in producing food.

4. Conclusion

In conclusion, the case of population growth in the peri-urban of metropolitan areas in *Kabupaten* Sukoharjo presents an interesting interaction process between urban population growths and land use conversions. The growth of population in the agricultural peri-urban areas leads to the consumption of agricultural land resources, including the most valuable agricultural land, which later leads to the situation in which the peri-urban region needs to find the way to replace the lost agricultural land resources in other parts of the region.

In the growth management point of view, these situations could be considered an indication of weak growth management system in which the peri-urban region tend to eliminate its important resources somewhere in meeting the needs of their growing population, which stimulate to increase their needs to develop their remaining important agricultural land resources in order to maintain their capacity in food production. Therefore, in this context, peri-urban growth or the urbanization process in the fringe areas needs to have better and comprehensive approaches in managing their lanf resources.

Acknowledgement

We thank the Faculty of Engineering, Universitas Diponegoro, for the funding of this research through the Competitive Grants on Strategic Research of RKAT Funds in the Budget Year of 2018 (Contract no. 2687/UN7.3.3/PG/2018).

References

[1] CICRED 1974 *The Population of Indonesia*: 1974 World Population Year Available from: http://www.cicred.org/Eng/Publications/pdf/c-c24.pdf

[2] BPS 2001 Population of Indonesia: Result of the 2000 Populstion Census Series: L2.2 Available from: https://www.bps.go.id/publication/download.html?nrbvfeve=YjgzZTQwODI2OGYwYmFmMDQ0ZGQ4MjJi&xzmn=aHR0cHM6Ly93d3cuYnBzLmdvLmlkL3B1YmxpY2F0aW9uLzIwMDEvMTIvMTcvYjgzZTQwODI2OGYwYmFmMDQ0ZGQ4MjJiL3BlbmR1ZHVrLW1uZG9uZXNpYS1oYXNpbC1zZW5zdXMtcGVuZHVkdWstMjAwMC1zZXJpLWwyLTIuaHRtbA%3D%3D&twoadfnoarfeauf=MjAxOS0wMi0wMyAyMDo0MToxNg%3D%3D

[3] Jones and Mulyana 2015 Urbanization in Indonesia: UNFPA Indonesia Monograph Series No.4 Available from: https://indonesia.unfpa.org/sites/default/files/pub-pdf/FA_Isi_BUKU_Monograph_No.4_URBANIZATION_.pdf

[4] Rahayu P and Mardiansjah F.H. 2018 Characteristics of peri-urbanization of a secondary city: a challenge in recent urban development *IOP Conference Series: Earth and Environmental Science, Volume 126, conference 1.*

[5] Firman T 1992 The Spatial Pattern of Urban Population Growth in Java, 1980–1990, *Bulletin of Indonesian Economic Studies*, 28:2, pp. 95-109, DOI: 10.1080/00074919212331336234

[6] Firman T 2016 The urbanisation of Java, 2000–2010: towards 'the island of mega-urban regions', *Asian Population Studies*, DOI:10.1080/17441730.2016.1247587

[7] Fahmi F Z, Hudalah D, Rahayu P, and Woltjer J 2014 Extended urbanization in small and medium-sized cities: the case of Cirebon, Indonesia Habitat International 42, pp. 1-10 DOI: 10.1016/j.habitatint.2013.10.003

[8] Mardiansjah, F. H. (2013). *Urbanisation durable des territoires et politiques de développement urbain en Indonésie: Étude de trois kabupaten en voie d'urbanisation rapide dans l'île de Java*. University of Paris-Est.

[9] Wahyudi A, Liu Y, and Corcoran J 2018 Combining Landsat and landscape metrics to analyse large-scale urban land cover change: a case study in the Jakarta Metropolitan Area. *Journal of Spatial Science*, 1–20. doi:10.1080/14498596.2018.1443849.

[10] Firman T and Fahmi F Z 2017 The privatization of metropolitan Jakarta's (Jabodetabek) urban fringes: The early stages of "post-suburbanization" in Indonesia. *Journal of the American Planning Association*, 83(1), 68–79. doi:10.1080/01944363.2016.1249010

[11] Winarso H, Hudalah D, and Firman T 2015 Peri-urban transformation in the Jakarta metropolitan area. *Habitat International*, 49, 221–229. doi:10.1016/j.habitatint.2015.05.024.

[12] Dorodjatoen A M H 2009 The emergence of Jakarta-Bandung mega-urban region and its future challenges. *Jurnal Perencanaan Wilayah Dan Kota*, *20*(1), 15–33.

[13] Budiyantini Y and Pratiwi V 2016 Peri-urban typology of Bandung Metropolitan Area. *Procedia-Social and Behavioral Sciences*, 227(November 2015), 833–837. doi:10.1016/j.sbspro.2016.06.152.

[14] Firman T 2009 The continuity and change in mega-urbanization in Indonesia: A survey of Jakarta-Bandung Region (JBR) development. *Habitat International*, 33(4), 327–339. doi:10.1016/j.habitatint.2008.08.005

[15] Mardiansjah F. H., Hndayani W, and Setyono J.S. 2018 Pertumbuhan penduduk perkotaan dan perkembangan pola distribusinya pada Kawasan Metropolitan Surakarta. *Jurnal Wilayah dan Lingkungan*, 6(3), 215-233. doi:10.14710/jwl.6.3.215-233

IOP Conf. Series: Earth and Environmental Science **328** (2019) 012065 doi:10.1088/1755-1315/328/1/012065

Understanding the Characteristics of "Ekistics" Elements in Determining Factors of Urban Settlement Growth

I A Farizkha[1], **R Alfiah**[1], **R Novi Listyawati**[1], **R S Aji**[1]
[1]Urban and Regional Planning Department, Faculty of Engineering, University of Jember, Indonesia

*Email: ivanagustafariska@gmail.com

Abstract. The development of urban settlements causes fast flow of urbanization. These condition leads to the growth of uncontrolled settlements which indicates Urban Sprawl Phenomenon. The phenomenon of urban sprawl begins with the need of the community to obtain better living standard, especially for residential functions. That residential function associated with urban sprawl is driven by settlement growth in certain locations (sporadic growth), and it is difficult to control (the impacts on the decline in environmental quality due to changes in certain land functions). By looking at these conditions, the understanding of the phenomenon of settlement growth must be conducted through the right approach. The approach views from an ecological element by examining a settlement in one large frame, total settlement patterns with a two-way classification scheme [1]. This research was conducted to understand the condition of settlement growth in order to find out the existing growth factors. The research area is located in the urban Situbondo with growth characteristics influenced by coastal and agriculture condition becoming the trigger of growth. Based on the results of the AHP calculation in Situbondo Regency, it is known that the most influential factor on settlement growth is natural factors with an assessment value of 0.243. Based on the morphological approach, the influence of natural factors shaping the pattern of ribbon settlement growth a small observation scale and the octopus pattern in the big range scale observation

Keyword : *Ekistic, Urban Sprawl, Urban Settlement*

1. Introduction

The growth of urban settlements due to the high flow of urbanization often leads to uncontrolled growth of urban settlements. The uncontrolled growth of urban settlements becomes an indication of the phenomenon of urban sprawl [1]. This phenomenon begins with the need for the community to occupy empty space, especially for residential functions. The occurrence of residential function associated with urban sprawl is driven by the growth of settlements in certain locations (sporadic growth), and it is generally uncontrolled. The growth of settlements with the characteristics of urban sprawl growth is not accompanied by the provision of basic infrastructures; thus, it often creates urban problems (problems of waste, sanitation, drainage, etc.) [2].

As an example of the phenomenon of urban sprawl is development of settlements in the metropolitan area of Indonesia, such as the Jabodetabek and Bandung which tend to lead to suburban areas. On the suburban areas of Jabodetabek, especially in Tangerang Regency, there are many large-scale housing developers. In total approximately 60 developers who have assets of housing development. Bandung suburbs such as Parongpong, Lembang, Cimenyan, Cilengkrang, Cileunyi, Bojongsoang, Dayeuhkolot, Margahayu and Margaasih also needed 51 formal housing developments with 93 locations and 2,382.13 hectares of conversion [1]. As a result of developing suburban areas there have been many city, physical, economic and social problems such as congestion due to inefficient mobility leading to workplaces, transportation, land, air, reduced suburban land for agriculture, land conversion, emergence of slums, etc. [3].

Based on the location of the settlements, the most common location of the settlements there is on the path to human movement and street. However, if viewed based on the characteristics of each sub-district, the location of these settlements has certain characteristics. For example Situbondo, Panji, and Panarukan Sub districts in addition to following the district road lane, settlement growth occurs following the knots of trade and service activities. For Mangaran sub-districts, the location of settlements tends to spread because of the attraction of natural resources in the form of fertile soil potential and potential marine resources related to agricultural activities of the population. Housing tends to develop to the north, which is located in the coastal area of Situbondo Regency. Growth in coastal areas has characteristics which become dense and slum areas. Therefore, if there is no growth control of government regulation, the development of settlements will lead to Urban Sprawl.

By examining these background conditions, an understanding of the phenomenon of settlement growth must be established. The elemental approach to understand the characteristics of residential growth is one of the suitable approaches. Ekistics views a settlement in one large frame, total settlements patterns with a two-way classification scheme [4]. The first classification relates to settlement hierarchies based on scale. The second classification relates to elements of settlements, namely nature, arthropods, society, shells, and networks [5]. This research was conducted to comprehend the condition of settlement growth in order to find out the occurring growth factors. The research area is situated in urban Situbondo with its growth characteristics influenced by coastal presence. This research aims to understand the characteristics of ecological elements by comparing the direction of spatial settlement growth. There are two objectives to be achieved: (1) Understanding settlement growth factors based on the characteristics of the forming element (the approach); and (2) Analyzing the spatial pattern of settlement development. This study employs quantitative exploration method which is supported by qualitative methods. The stages of analysis carried out in this study include: (1) Analyzing the factors influencing the development of settlements; and (2) Analyzing the periodic trend with periodic analysis of regional image. Factors which influence the development of settlements are analyzed using the AHP technique to see the weight of the influence of each variable forming a settlement on the existing development of settlements.

1.1 Characteristics of Settlements

An understanding of the characteristics of settlements in this study broadly refers to several sources, namely Turner related to the characteristics of settlements based on the nature and Doxiadis related to settlement characteristics seen from the forming elements and units.

According to Turner (1976) [6] in his book Housing By People, there are 2 types of settlements based on their nature, namely formal and informal settlements. Formal settlements are settlements built by the formal sector referring to housing development which is built on several development regulations and through legal procedures. On the other hand, informal settlements are built by the informal sector referring to development without regulation and without legal procedures.

Based on its constituent elements, settlements consist of humans, their culture and the place where humans live (the container) which includes natural and artificial elements (home and networks) [7]. All of these forming elements combine to form a residential unit scale that is divided into groups based on the scale from the smallest to the largest unit. Settlement unit groups based on the theory by Doxiadis are:

- Non-Permanent Settlements (Temporary Human Settlement)
- Villages

- City (Polises)
- Metropolises
- Megalopolises
- National System
- International System

1.2 Settlement Growth

Growth of settlements can be understood by making observations on components from their forming aspects. Settlements are part of sustainable development that has three forming pillars namely Economic, Social and Environmental [7]. Balanced interaction between these pillars in the development of settlements will lead to sustainable development. The concept of sustainable settlement is a strong global concept that is expressed and actualized locally. Sustainable settlements must have a strong economy, a harmonious environment, equal and full levels of justice, high levels of community participation, and well-controlled energy conservation.

On the other hand according to Doxiadis (1976) [8], settlements can be analogous to living things that have body parts as formers. When one part of the body does not function properly it will affect the other part. Physically, for example with regard to buildings and the availability of infrastructure, if these physical parts are not fulfill properly, the quality of the environment formed will also not be good and will lead to a decline in the quality of life of the community.

To find out the factors that influence the growth of settlements, two major theories regarding sustainable development and characteristics of the elements forming settlements are closely related. If it is spelled out between the pillars of sustainable development and elements forming settlements, the economic and social aspects are related to human and community elements (the content) while the environment is related to natural elements, housing / housing conditions, and networks (the container). The results of balanced interaction complement each other between the elements forming settlements are adjusted to the pillars of sustainable development will form a sustainable and ideal settlement with good aspects, both physically and non-physically.

1.3 Ekistic

Human settlement is a place that is inhabited by humans that includes content elements and container elements. Elements of ekistics are divided into five parts, namely nature (human), human (antrophos), society (society), reflection (shells) and networks (networks)[8]. Ekistics theory is a theory to explain settlement (human settlement) developed by doxiadis.

a. Natural
Theoretically, it can be said that settlements are part of nature. The research was carried out by looking at natural elements related to physical appearance of the earth, availability and ability of the environment.
b. Human
Humans initially started by changing nature by building huts. After that, began to have expertise in the agricultural revolution which subsequently created various types of houses. In research, the human element examined is related to the conditions, numbers, and relationships between individuals.
c. Community
This study examines the elements of society through sub-variables in the form of social conditions that exist in the scope of settlements (such as livelihoods, income, education level, economic conditions and cultural characteristics that exist in the community in the scope of the research room)
d. Protection
Protection is divided into several categories, namely education, health, administration, security, industry, storage and others [10].

e. Network
The choice of a comfortable dwelling will take into account the availability of the network, especially the transportation network that contributes to the access of occupancy to centers of activities and clean water networks as a means of basic human needs.

2. Methodology

2.1 Research Variable

The research variables used in this study are residential building elements that describe the characteristics of settlements comprehensively. This variable is obtained from a literature review related to theories about settlements.

Table 1 Research Variable

INDICATORS	VARIABLES	SUB VARIABLES
Physical : • Nature • Houses / housing • network	Nature	1. Availability of natural resources 2. Land capability
	Housing	1. Type of housing (permanent or non-permanent) 2. Housing model 3. Location of settlement 4. Density level of residential area 5. Housing facilities (completeness and distribution) 6. Land use
	Settlement infrastructure and facilities	1. Clean water infrastructure 2. Electricity infrastructure 3. Road infrastructure 4. Waste and waste processing infrastructure 5. Telecommunications infrastructure
Non physical • Humans • community	Social	1. population 2. Livelihoods 3. Education level 4. Existence of social institutions 5. Human growth
	Economic	1. Level income distribution 2. Existence of economic institutions
	Culture	1. Habits 2. Activity

2.2 Source of data and information

1. Determination of the Analysis Unit
The unit of analysis in this study is urban residential environmental units which are recently developed in the urban area of Situbondo Regency.
2. Determination of Observation Units:
The observation unit of this study is a variable that indicates the characteristics of settlements based on settlement elements in residential areas.
3. Determination of Information Units:
The information unit of this study is government stakeholders in residential areas (on a district, sub-district and village scale) as influential and understanding parties to the conditions of settlement development in Situbondo and also the developers (formal sector) as formal housing providers who contribute to the development of settlements.

2.3 Stages of Analysis

Table 2 Stage of Analysis

No	The Analysis Stage	Data Input	Analysis Tools	Output
1.	Mapping the conditions of development of settlements	Periodic image area	Analysis of regional images with GIS devices	Knowing the direction of trends in residential growth
2.	Identification of factors causing settlement growth	Variables were obtained from literature review	AHP Analysis supported by stakeholder descriptions	Knowing the factors that cause the growth of settlements

- Stage 1: Mapping the conditions of development of settlements

  Mapping the conditions of development of settlements is done by analyzing the regional image to see the form of development and direction of settlement growth. This is done by comparing the built up area and not being built in each year period. The development of periodically built and non-built areas can be known with the help of ArcGIS software on remote sensing tools.

- Stage 2. Identification of factors causing settlement growth

  The analysis technique used at this stage is Analytical Hierarchy Process (AHP). AHP is used to determine the magnitude of the priority of each criteria variable that affects the growth of settlements based on the studies that have been conducted. In process, each of the indicated variables has an influence on the development of settlements in the study area compared to each other and then assessed by respondents (expert stakeholders who understand the condition of the research area).

- AHP analysis stages:
  1. Compilation of hierarchies
     Hierarchy preparation is done by elaborating the influence of each variable that has been determined on the growth of settlements. This hierarchy is based on understanding according to the study of the theory that the growth of settlements is influenced by the characteristics of settlements and the aspects of sustainable development that are described by their constituent elements
  2. Distribution of questionnaires
     Questionnaire that contains a comparison between the determinants that influence the growth of settlements which will later be used as a reference in controlling the development of settlements.
  3. Processing with Pairwise Comparisons)
     The results of the comparison questionnaire between the criteria were processed with pairwise matrices (Pairwise Comparisons), where processing with this paired matrix to find out the comparison values of the importance of a criterion relative to other criteria which are then processed to determine relative rank / priority of all criteria.
  4. Comparison to get priority (Normality Test)
  5. Consistency Test
  6. Merging Respondents' Opinions

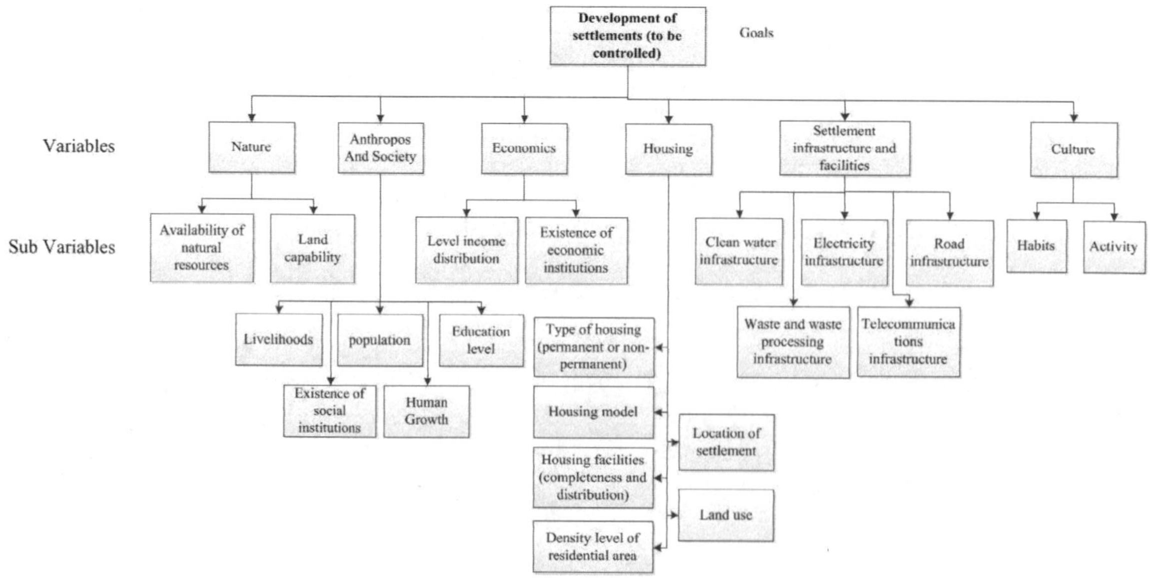

Figure 1 Hierarchical structure of criteria in AHP (factors that influence the development of settlements based on forming elements)

The output of the AHP analysis is to know the weight of the influence of each variable and sub-variables forming the settlements on the phenomenon of residential growth in the research area. The description of output is supported by an explanation of the results in deep determined stakeholder interviews into a sample. The results of in deep interview are in the form of understanding the phenomenon of settlement areas from the point of view of several related stakeholders.

3. Result and Discussion

Based on the results of the AHP calculation in Situbondo Regency, it is identified that the most influential factor on settlement growth is natural factor with an assessment value of 0.243. The condition of natural resource presents a significant effect because of the urban location of Situbondo which is affected by coastal areas. The tendency of body settlement growth takes form of a ribbon settlement pattern, extending along the road close to sea access. The condition of the pattern is related to human activities in Situbondo which are related to coastal areas. This condition is corroborated by the results of the assessment of the second most influential factor on residential growth, namely community factors (eco-culture) with a value of 0.238. The highest variable is in the livelihood variable.

Figure 2 Results of AHP Calculation

If seen in a spatial pattern, the growth of settlements in the urban area of Situbondo can be divided into two classification zones, the central zone of the urban center growing extensively from east to west and the northern zone approaching the coastal areas growing from 2003 in Tanjung Chinatown.

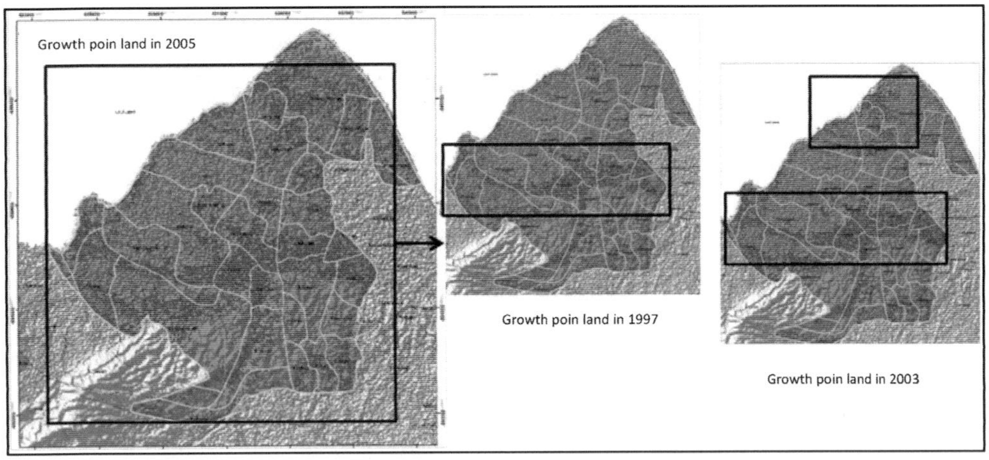

Figure 3 Growth Point from 1997 to 2005

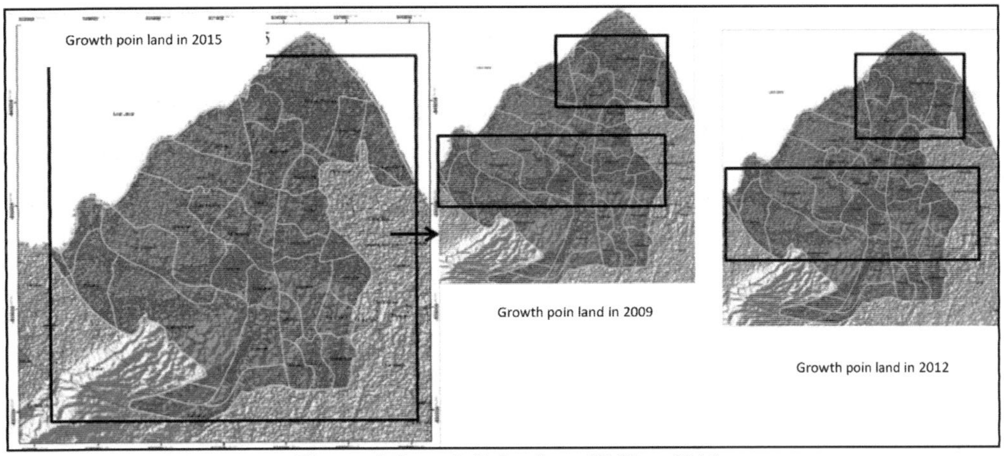

Figure 4 Growth Point from 2009 to 2015

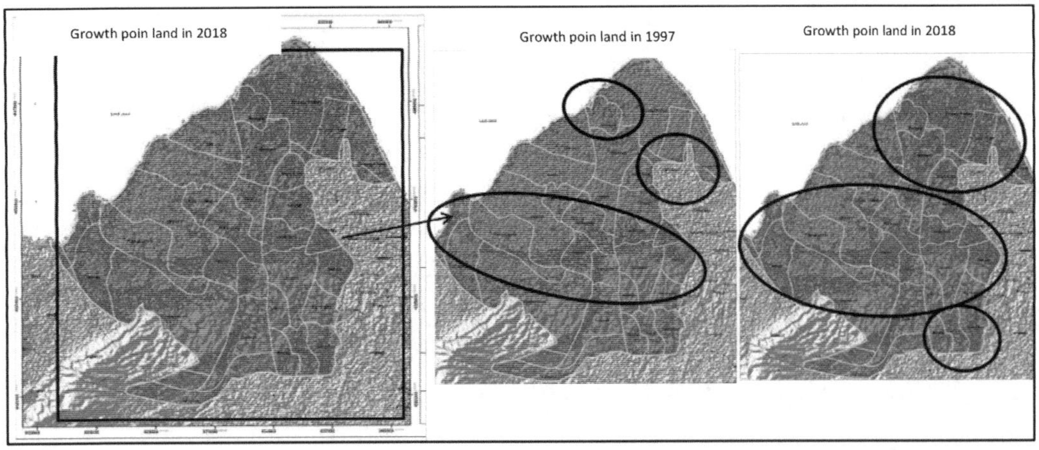

Figure 5 Growth Point from 1997 to 2018

The settlement growth with a ribbon settlement pattern, extending along the road close to sea access. The condition of the pattern is related to human activities in Situbondo which are related to coastal areas.

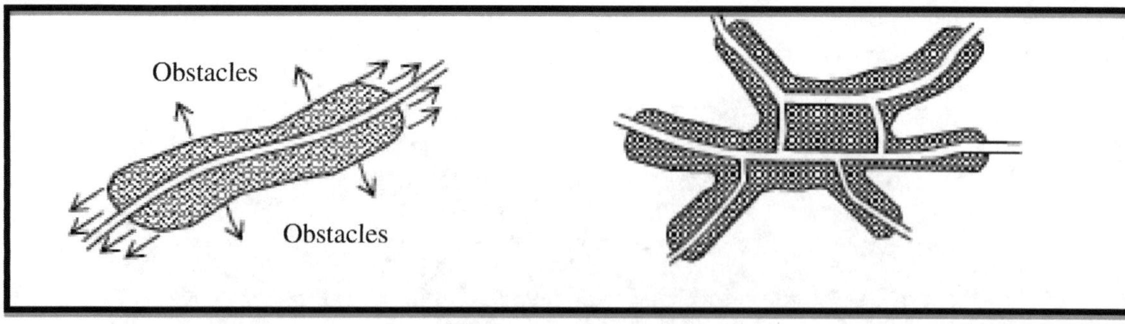

Figure 6 Octopus Pattern

The development of settlements at the beginning of the research year in 1997 showed that there were concentrations of settlements in the city center, namely Situbondo sub-District, Panji sub-District with horizontal forms from west to east following arterial roads. But in the period 2009 to 2012 the development of settlements began to spread southward sporadically, but the highest density concentrations remained in the city center. In the physical organization of space, the development of settlements forms the octopus pattern (Octopus / Star Shape Cities) where in this octopus pattern there are several dominant transportation routes

4. Conclusion

Based on the results of the AHP calculation in Situbondo Regency, it can be seen that the most spatial influential factor in the development of settlements in Situbondo Regency is a natural factor with an assessment value of 0.243. Natural factors consisting of natural conditions and availability of natural resources are important factors in the growth of settlements in Situbondo Regency. This physical factor greatly affects the development of residence in Situbondo because the existence of houses and settlements will not be separated from the conditions of land and natural resources from the location occupied. These natural factors will later provide a pattern for expanding settlements in Situbondo Regency with a value of 0.238. Moreover, the community factors consist of community quality, income distribution, livelihoods and culture

References

[1] Salvatia, L. (2012). Low-density settlements and land use changes in a Mediterranean urban region. Landscape and Urban Planning vol 105.
[2] Moitra, M. K. (1991). "Environmental improvement of slums The Calcutta experience." pergamon press Vol 26 (Building and Environment): 253-257.
[3] Hakim, Ikhwan. (2010). Struktur Ruang dan isu Keberlanjutan Perkotaan di Jabodetabek.JURNAL perencanaan pembangunan edisi 1.
[4] Doxiadis, C. A. 1976. Ekistic Elements, Action for Human Settlements: 11-27.
[5] Turner, John, (1976), Housing By People: Towards Autonomy in Buildings Environtment: Pantheon Books, New York
[6] IUCN. (2006). The Future of Sustainability: Re-thinking Environment and Development in the Twenty-first Century. Report of the IUCN Renowned Thinkers Meeting, 29-31 January 2006. www.iucn.org
[7] Doxiadis, C. A. 1968. Ekistics : an introduction to the science of human settlements. New York: Oxford University Press

Defining Gap between Housing Land Value and the Facilities Provision in Urban Housing Development by Private Developers in Malang City, Indonesia

I N S Wijaya[1]

[1] Department of Urban and Regional Planning, Brawijaya University, Malang 65145, Indonesia

*Email : nyoman@ub.ac.id

Abstract. One of Indonesian government strategy in fulfilling the national housing target is by improving private developer's roles in housing provision. Due to the provision of housing facilities, obligated by Indonesian government for every private developer since 1987, the practical implementation shows opposite circumstances for the private residential area. This research aims at describing conflict between the fulfilment of urban housing facilities and the housing land price in Malang City. Hence, comparison between the existed land prices to the availability of facilities is conducted to identify the gap. Since the price-contour is resulted by the land price mapping, this paper compares between the price and the existed facilities in order to show the practical gaps between the prices and the facilities. Finally, this research indicates: 1). The residential areas with unfulfilled facilities are located at the urban area with high prices, 2) Theefficiency of housing prices becomes the main motivation of the developers in building the facilities, without much consideration on environmental quality of the residential areas.
***Keywords**: land value, housing facilities, gap, and urban areas*

1. Introduction

The 11th target of SGD's is actualizing sustainable cities and communities. For many city governments, this target is concerned to the development of urban inclusivity. It may be appreciated by all of city's government as giving adequate opportunity for all people to live in urban areas. In other word, city government should assure their urban areas become adequate and affordable shelter and the facilities for all people. It becomes a complicated work for city governments in many developing countries where urban agglomeration, population and land are in unfavourable circumstances in this past decade, especially in Indonesia.

Indonesian government, since 1987, had motivated private developers in taking over the government role as housing provider instead of enabler that is still on government hand. Ministerial Regulation No 34/Permen/M/2006 (Permen PU 34/2006) replaces the government provider roles for housing facilities development to private developer. Private developers are given with broad opportunity in investing their money to housing development. They are obligated to provide adequate housing facilities for all their invested residential areas. Corresponding to the regulation; in 2007, many Indonesian cities face the boost of private housing development. Thus, urban population grow in great speed followed by the demand of housing development. It is characterized as fast growth housing development with high price; high marketable house located in high value urban areas. As the most valuable area in Von Thunen's perspective, these areas should become the best-serviced area for

IOP Conf. Series: Earth and Environmental Science **328** (2019) 012066 doi:10.1088/1755-1315/328/1/012066

housing facilities compared to the other urban areas [1]. However, this high value may be lead to the unwillingness of the developers in providing the facilities. Additionally, the use-value of the land tends to be the crucial factors in determining the price [2], [3], [4].

However, along with the trend of the development, the value of land (read: the price) is also rising up. Private developer responds to this situation by focusing their investment merely to houses development. As a result, housing price becomes irrational when the demand is inflated. Irrational means the price is extremely high comparing to the facilities that the buyers can get. It is apparent that with the high price, government and private cannot assure the availability of adequate housing facilities. The number of houses development significantly increases; whereas, the infrastructure is suffered from serious insufficiencies.

In order to build comprehensive knowledge, this paper describes the practical gap between urban housing land prices and the availability of facilities. It is crucial in building government comprehension towards private developer's attitudes in managing the value of their project. Further, the resulted discussion should be the government basic consideration in controlling the situation. Therefore, research towards the availability of urban housing facilities has broadly conducted by many researchers for the land value. However, towards the comprehensive gap identification between them might not be much discussed. This research thus presents the novelty in such area.

In building the comprehensive description, these paper overlays the evaluated availability of housing facilities to the practical circumstance of urban land prices. In this case, Malang City is one case study of the discussion, as one of big cities of East Java that, hardly suffered from housing and the facilities development. Firstly, this paper identifies the distribution of land price in Malang city to draw the contour of land prices and define the highest price in the city. The contour of land prices is developed based upon the price of the land corresponding to NJOP. Secondly, this paper evaluates the availability of housing facilities based on Ministerial Regulation No 34/Permen/M/2006. The result of evaluation demonstrates the available percentage of the facilities to the regulation. Thirdly, the result of evaluation is overlaid to the resulted land prices contour in order to identify the gap between prices and the facilities provided. Finally, this paper discusses and concludes the gap completed by some perspectives of the developer in defining their market price.

2. Method

This research overlays between maps that are a contour map of land prices to the map of facilities availability, described in the following steps. Firstly, in order to build the contour map, the price of land is utilized as the result of investigation towards practical market land prices. It reflects real price of land in many terms, such as the hedonic terms of land based on Indonesian Formal tax Value of Land (*Nilai Jual Object Pajak* – NJOP). The resulted map gives indication about the position of each observed residential area in the land price contour. Table 1 presents the land prices of 9 samples, categorized as private residential areas as Cluster A by Malang Municipal Government.

Table 1. Land Prices (existing)

No	Housing/Residential Name	Land price Existing (rupiah)
1.	Bukit Cemara Tidar (*Tidar Pines Hills*)	5.000.000
2.	Griyashanta Eksekutif (*Griyashanta Executive*)	5.000.000
3.	Istana Gajayana (*Gajahyana Palace*)	6.500.000
4.	Pondok Blimbing Indah (*Blimbing Indah Estate*)	5.000.000
5.	Riverside (*Riverside*)	4.000.000
6.	Puncak Buring Indah (*Beautifull Buring Hills*)	7.000.000
7.	De Cassablanca (*De Cassablanca*)	3.500.000

8	De Rumah (*De Rumah*)	8.500.000
9	Ijen Nirwana Residence (*Ijen Nirwana Residence*)	9.000.000

Source: Survey, 2016

Secondly, the map of available facilities is built as the result of evaluation towards the availability, consisting of supporting facilities and basic facilities as detailed by Table 1. Further, for each private residential area, the availability of housing facility is valued based on Ministerial Regulation No 34/Permen/M/2006. **Table 2** gives list of type of facilities that are observed in 9 samples of residential areas.

Table 2. Type of Observed Facilities

Supporting	**Basic Facilities**
Public services and government offices	Roads
Education	Drainage
Health	Fresh Water Instalation
Mosque, Church, Temple, etc	Grey Water Instalation
Shopping Facilities	Garbage
Culture and Recreation	Electrical
Open Green Spaces, Park, Sports Park	Communication

Source: Analysis, 2016

3. Result and Discussion

3.1 Indicated Value

A. Land Prices Distribution

Table 1 depicts 9 samples from 105 samples of land prices in Malang City surveyed by this research. Based on these samples, the range of land prices is defined. These ranges are utilized in drawing the contour of land price in Figure 1. The ranges of the actual prices are shown as follows:

a. IDR 2.000.000 – 3.400.000
b. IDR 3.400.000 – 4.800.000
c. IDR 4.800.000 – 6.200.000
d. IDR 6.200.000 - 7.000.000
e. IDR 7.000.000 – 9.000.000

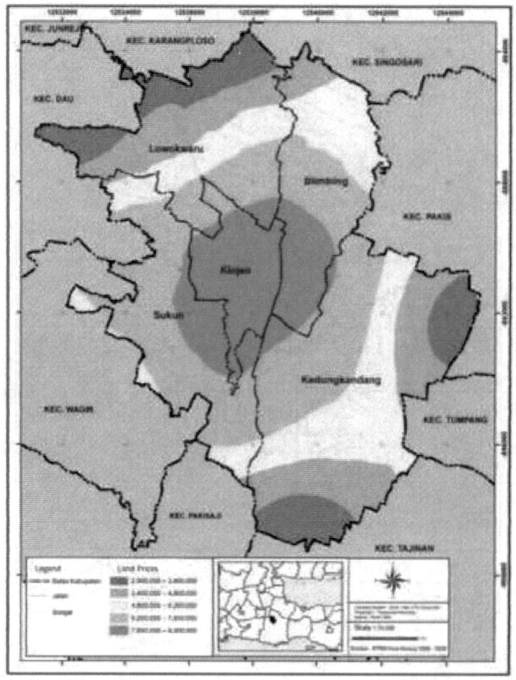

Figure 1. Resulted Land Price Contour

Figure 1describes smooth distribution of land prices for whole area of Malang City indicating that the highest price is located in centre area of the City. The contour, gradually, changes from the highest in the centre to the lowest in the peripheral areas. Logically, the highest price reflects the highest value of the land corresponded to the availability of the facilities. Therefore, the practical situation is clarified by the further discussion.

3.2 Housing Facilities Availability

Table 3. Percentage of Availability

Private Residential Areas	**Facilities**	**Percentage Availability(%)**
Bukit Cemara Tidar	Supporting	5
	Basic	57
De Cassablanca	Supporting	6
	Basic	57
Istana Gajayana	Supporting	0
	Basic	71
Pondok Blimbing Indah	Supporting	10
	Basic	57
Puncak Buring Indah	Supporting	6
	Basic	71
Riverside	Supporting	24
	Basic	57
Griyashanta Eksekutif	Supporting	0
	Basic	71
De Rumah	Supporting	0

	Basic	71
Ijen Nirwana	Supportin	0
	Basic	7

Source: Analysis 2016

For each area, **Table 3** presents specific result of assessment for the facilities availability. Based on Table 3, mostly, the percentage availability values are similar among 9 samples. All private residential areas have no significant differences in their provision of basic facilities. Meanwhile, for supporting facilities, significant differences are clearly indicated by the high differences percentage. Hence, Riverside shows highest value of availability of the supporting facilities. The other two (Griyashanta Executive and Istana Gajahyana) similarly have low percentage of the supporting facilities availability. Further, Figure 2 describes the map of availability housing facilities of private residential areas in Malang City.

Generally, based on two resulted values on **Table 3**, there is no specific pattern to draw specific type of the availability of housing facilities in Malang City. The above situation indicates that distribution of supporting facilities which significantly varies among the private residential areas in Malang City.

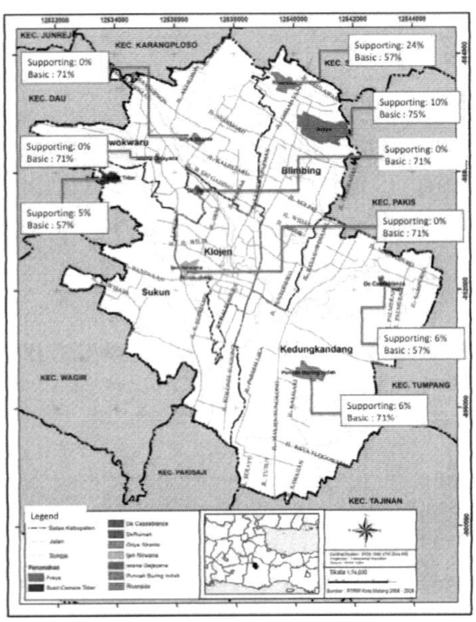

Figure 2. The availability of housing facilities for each

3.3. Defining the Gaps

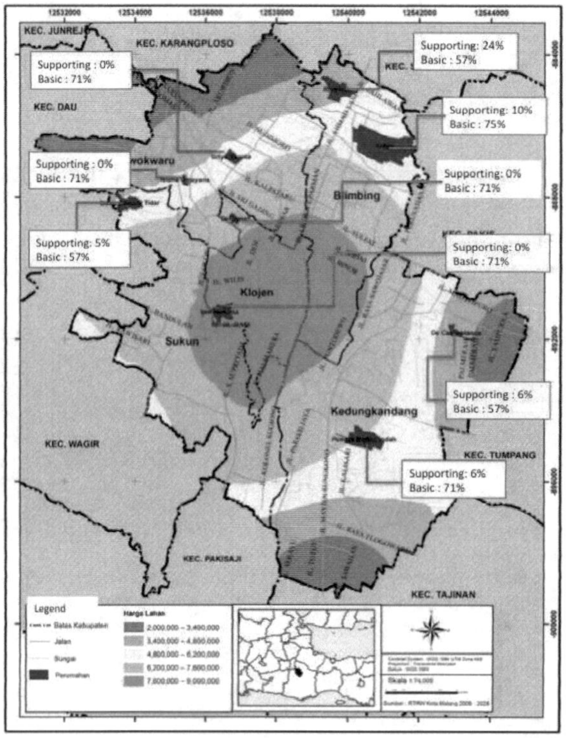

Figure 3. Overlaid Land Price Contour and Availability of Facilities

By overlying the map of facilities availability (Figure2) to the map of land prices contour (**Figure 1**), the gap between the prices and value of availability can be clearly described. **Figure 3** describes the overlaid map as well as the real situation, marking one residential area (Ijen Nirwana) located in the fifth/highest contour of land prices. There is also one residential sample (De Rumah) located in the fourth contour of land prices. Four samples are located in area of the third contour, such as Bukit Cemara Tidar, Griyashanta Executive, istana Gajahyana, and Pondok Blimbing Indah (Araya). In addition, there are two samples (De Casablanca and Riverside) located in area of the second contour, indicating Ijen Nirwana as the one located in fifth contour having average basic facilities (71%) and 0% (zero) supporting facilities. This lowest availability percentage of supporting facilities is also founded in the fourth and the third contour. Meanwhile, Figure 3 does not present the lowest availability of supporting service in the fourth area.

Based on **Figure 3**, irregular pattern of availability percentage does not correspond to the pattern of the contour. There is no specific pattern to draw specific type of the availability of housing facilities in Malang City concerning to the location of residential areas. Figure 3 indicates the distributions of availability percentage of housing facilities which are randomly spread over the land price contour. It is assumed that the highest contour does not mean that the area for residential with highest availability percentage of housing facilities. Also, the lowest contour is not the area with lowest availability percentage. This circumstance is contradicted to the preposition that the highest contour should be an urban area with highest land price with best-serviced facilities. Concerning to the percentage, the highest availability percentages are founded in the third and fourth contour of land prices.

Interestingly, this research also reveals two important matters corresponded to the private developer's behaviour in treating their project of residential areas. First, for basic facilities, 9 samples of residential area have more than 50% availability. However, there is no maximum availability percentage for all private residential areas. Although it may not be seen as a serious urban problem, the above circumstance means that compatibility of practical provision of urban housing facilities to the national standard is low. Without any concern to this phenomena, the aimed of 'role sharing' between government and private investors cannot achieve the desired result.

The second matter is related to availability percentage of the supporting facilities. The highest percentages are founded in the third and fourth contour of land prices. Two residential areas (Riverside and Pondok Blimbing Indah (Araya)) respectively have 24% and 10% availability. Moreover, these two residential areas (Riverside and Pondok Blimbing Indah) also indicate high percentage availability of basic housing facilities compared to others. Pondok Blimbing Indah has highest percentage (75%) and Riverside has 57% availability. Concerning to the developer's behaviour in treating their project, it may be reasonable that lowest contour of land prices gives much opportunity for the developer in fulfilling the facilities, especially the supporting one. In this situation, the best-serviced facilities escalate the market appreciation and the value of the project.

In spite of high valued land in Thunen's location theory [1], the above circumstance shows some contradiction. The result of this research presented in Figure 3 does not reveal high valued urban area reflected by the high accessibility to the urban centre and urban facilities. High valued urban land that should be located in the fifth contour does not conform to the completeness of the facilities in servicing the inhabitant. It also means that high price of urban land does not reflect the actual value of the land in Thunen's perspective. Moreover, Malang case tends to indicate that land prices for residential area are not merely determined by the location. It is closely related to the Eckert's preposition that perspective of provider (private developer) and buyer towards the use of land is the most determining factors to decide the price [5]. In this research, it is revealed by the provider's lower opportunity in escalating the availability percentage of housing facilities for their residential area in highest contour of urban land price.

4. Conclusion

By this research, there are some concluding remarks which are:

1. There is no specific pattern in the relation between the land prices contour and the availability of housing service.
2. The availability percentage of basic facilities is relatively similar among the samples of residential areas.
3. Significant variance of percentage is revealed on the availability percentage of supporting facilities although it is not clearly correspond to the pattern of land price contour
4. Unavailable maximum percentage of basic facilities leads to the fact that the compatibility of practical housing facilities to the National standard is low.

In order to develop comprehensive understanding about the practical trend of Indonesian housing price, developer's motivation, deliberation, and decision in treating the land related to land market system, it should be the future research's discussion. Indeed, more case studies about some similar cities should be the observed cases in building the complete figures of the practical trend and model of housing facilities provision in Indonesia.

References

[1] Von Thunen, J. H; Wartenberg, C. M; Hall, P. Isolated State (an english version Oxford). London. Pergamon press (1966).
[2] Sujarto, D. Beberapa Pengertian tentang Perencanaan Fisik. Jakarta. Brata Karya Aksara (1985).
[3] Wolcott, Richard C. The Appraisal of Real Estate American Institute of Real Estate Appraiser. North Michigan, Chicago Illinois. p. 22-63 (1987).
[4] Sutawijaya, A. Analisis Faktor-Faktor yang Mempengaruhi Nilai Tanah Sebagai Dasar Penilaian Nilai Jual Obyek Pajak (NJOP) PBB di Kota Semarang. Jurnal Ekonomi Pembangunan, 76. (2004).
[5] Eckert, J.K. Property Appraisal and Assessment Administration. IAAO, Chicago. Illinois. p. 151-180. (1990).

The 4rd International Conference in Planning in the 2019 Era of Uncertainty IOP Publishing
IOP Conf. Series: Earth and Environmental Science **328** (2019) 012067 doi:10.1088/1755-1315/328/1/012067

Architecture *Stasiungebouw* Medium Type (Case Study : Ngabean Station Yogyakarta)

M A A Riza[1], S W Firmandhani[2]
[1,2]Universitas Diponegoro, Semarang, Indonesia

*eEmail: arifah@arsitektur.undip.ac.id

Abstract. Ngabean Station was built by the *Nederlandsch-Indische Spoorweg* (the first Railway company in Indonesia) in 1895-1919. The location of Ngabean Station is on Wahid Hasyim St., of Yogyakarta. Ngabean Station has been out of operation since the 1970s. Ngabean Station is included in the type typology of *Stasiungebouw* medium type. This article aims to determine the architectural characters of Ngabean Station (spatial aspects, architectural aspects, structural aspects). This research utilizes qualitative research methods with a rationalistic paradigm approach. From the results of the study, the researchers found that Ngabean Station has a type of passenger emplacement as well as luggage emplacement. Spatial space of Ngabean Station is divided into 3 zones. First, the passenger public zone is shown by the *wachtkamer*; second, administrative zones are indicated by *kantoor / kartjees*; and third, technical operational zones of the train are indicated by the *magazine*. From an architectural aspect, this station creates a balance of facades formed from a symmetrical facade system between the opposite facade walls. The facade of the front is a reflection of the rear facade, as well as the facade on the side, as indicated from the laying of door and window openings. The results of this reflection form a symmetrical layout of the openings in each facade creating the balance of the shape in the building. The building structure of the Ngabean Station is a closed frame structure which is a combination of the frame system with the construction of a buffer wall.

.

1. Introduction

Indonesia is a country with rail transport technology since 1864. Railway transportation is marked by the construction of the first train in Kemijen Village, Semarang [1]. Railway technology in Indonesia is even ahead of Japan. The first railroad in Japan was made in 1870 with the Tokyo line to Yokohama [2]. However in Indonesia, especially in Yogyakarta, there has been a reduction in rail transport. One of them is Line of Yogyakarta-Sewugalur which has 12 stations in the beginning, but in 2017 there are only 5 stations [3]. Five stations no longer serve as stations.

One of the five stations is Ngabean Station, which was constructed by *Nederlandsch-Indische Spoorweg* (the first railway company in Indonesia) in 1895-1919. Ngabean Station is located at Wahid Hasyim St., of Yogyakarta. Ngabean Station has been in operation since the 1970s used as a tourism car to support office managed by Yogyakarta Transport Agency. The station architecture in Java is generally divided into types of building stations, such as *haltegebouw* and *stationgebouw* [4].

According to Ikaputra, *a stationgebouw type is* divided into: small type, medium type, and big type (main station) (presented in Figure 1).

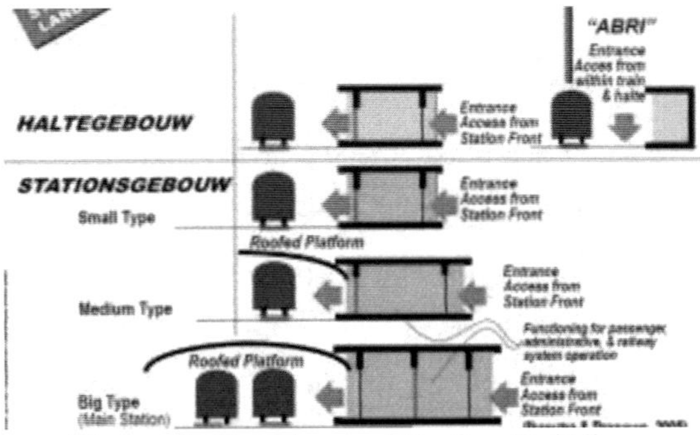

Figure 1. Typology of Station Building.

From the results of the previous study, it was found that Ngabean Station (Figure 2) has a type of Stasiungebouw [5]. This study is a follow-up study about character research of Ngabean Station (spatial aspect, architectural aspect, structural aspect). The study aims to reveal the characteristic architecture of the Indish station in 18 as a reference material for maintaining the preservation of historical buildings in Indonesia.

Figure 2. Ngabean Staion

2. Methods

This research applies qualitative research method with rationalistic paradigm approach. Analysis of descriptive research results creates an overview and exposure to explore the existence of physical changes in the style of buildings on the Ngabean station. The data analysis in this study is performed by descriptive method to explain and provide an overview of the building style character at each station. Exposure is conducted by analyzing historical and environmental contexts at the station. Building information is collected with building modelling to understand the physical characteristics of the station building. Then the connection will be made to the colonial style that affects the physical characteristics of the building.

3. Discussion

3.1. Spatial System

Ngabean Station has parallel type station, indicating that the location of the station is parallel to the railroad tracks on one side; thus, the mass building extends from the north-south to the direction of the railroad tracks. The orientation of the station building faces from west to east. Ngabean Station is located in the middle of Yogyakarta city and is directly accessed to the highway. Around the station building, there are several official houses both owned by station officials and by sugar factory officers.

Ngabean Station has a type of passenger and baggage emplacements. Ngabean Station as well as train passengers also serves to distribute sugar industry production. Despite the placement of these stations to not transport factory produce, there is activity in the distribution of industrial products. This station serves to advance the sugar industry production to urban residents and to continue production to areas outside the city of Yogyakarta by passing to Tugu station passed through the railway line.

The space at Ngabean Station is arranged linearly. Therefore, the station forms a cubic building with a dominant horizontal appearance. The circulation of the building is carried out outside the building, which is on the terrace of the building facing the railway track. The terrace is also used as a platform for passengers waiting for the train. This creates the rigidity of lines arranged linearly due to the outdoor circulation space. By arranging space in parallel way, it makes the space function to be clearly defined. Different space functions will be indicated by different spaces. The layout of the placeholders is presented in the following designs (Figure 3a).

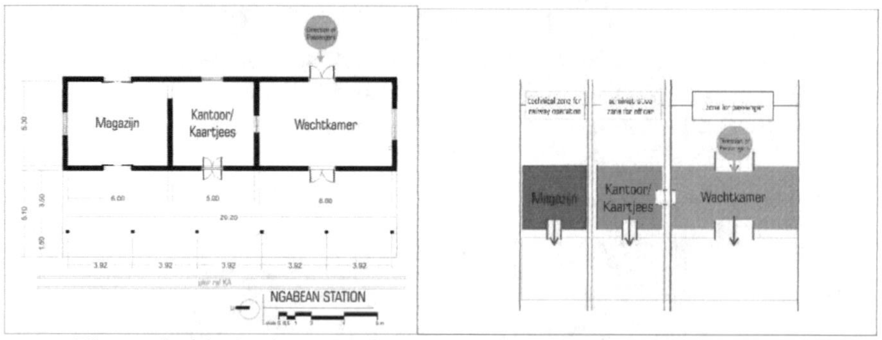

Figure 3. **(a)** Ngabean Station Plan, **(b)** 3 zones at Ngaben Station

The parallel space composition forms a linear space organization by following the direction of the train. The use of linear organizations can emphasize the existence of movements linking linear organization to the building due to the function of the building centred on the direction of the railroad movement. Every activity and function of space that occurs inside and outside the station depends on the movement of the train. This activity is divided into 3 zones (Figure 3b), including public zone for passenger, administrative zone for officer, technical zone for railway operation. Public passenger zones are indicated by *wachtkamer* (platform). The administrative zone of the officer is represented by the *kantoor / kartjees* (office). Train in the technical zone of the operation is shown by *magazijn* (baggage space).

Circulation inside Ngabean Station by Ching [6] is categorized by using passage through space to create flexible route configuration with the integrity of the retained space. Circulatory paths use the core / platform that runs along the station building. This circulatory line is located on the edge of the building to the west of the building. From this circular, the orientation of the room overlooking the train tracks is indicated by the entrance of each room on the side of the corridor. The form of circulation space inside Ngabean Station is a form of open space on one side. This kind of circulation conveys spatial and visual continuity with the connected space, enabling the arrival and departure activities of the train, and the associated space in this corridor are clearly visible (seen in Figure 4).

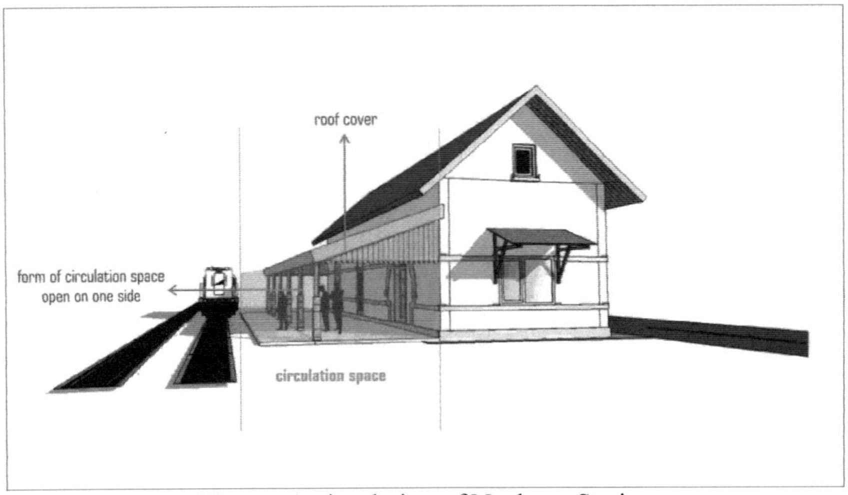

Figure 4. circulation of Ngabean Station

3.2. Physical System

Ngabean Station has a simple plan with rectangular shape with a linear layout, shaping the whole building as a cube (Figure %). The length of the cubic structure of the building has a rectangular base extending along the rail track. Ngabean Station has a length of 20 m and 5 m wide, with a roof height of about 2.5 m. From observation, it is found that the proportion of Ngabean stations is proportional to the proportion of building height H = W (Figure 6). The balance of Ngabean Station's facade is formed from a symmetrical facade system in the opposite facade wall. The front facade is a rear facade mirror and facade on the side, as seen from the laying of doors and windows. The results of this reflection form a symmetrical layout of openings in each facade and create a balance in the shape of the building.

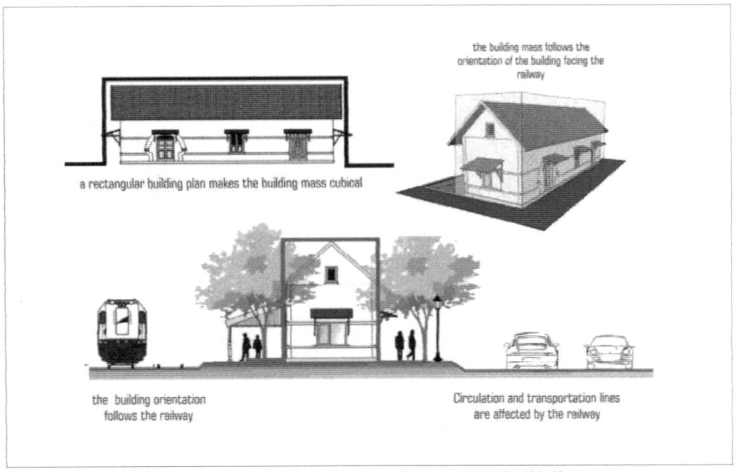

Figure 5. Ngabean Station mass building .

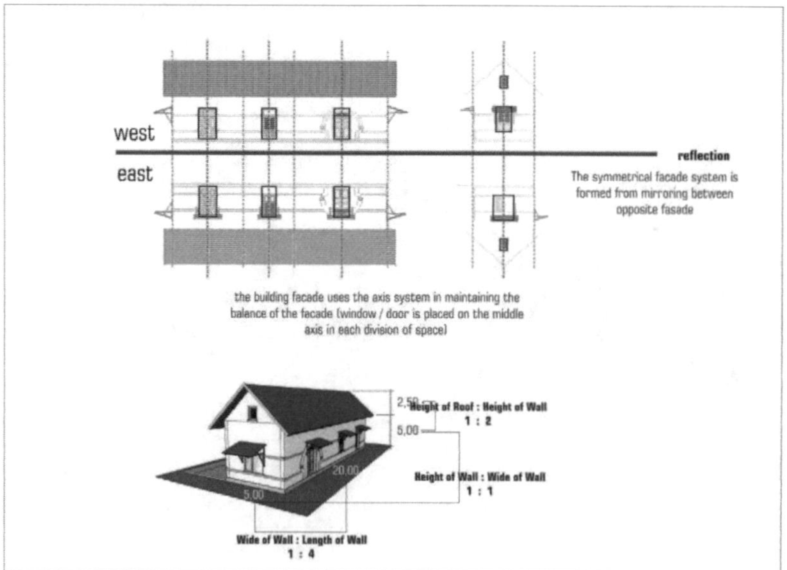

Figure 6. (a & b) Ngabean Station propotion.

Ngabean Station has a saddle roof with a height of 2.5 m. The roof slope level ranges from 41.5 degrees. Based on the observation, the roof formation is addressed in a response from the climate in Indonesia which has a high rainfall. For the part of the roof, there is a rather suspended hanging over 50-75 cm to prevent disturbance of rainwater. However, there is still an additional roof at each opening that serves as a truss, because the main roof is not enough to protect against rain. The materials used are wooden horses with clay tile. Clay materials on tiles are not soaked as the weather in Indonesia tends to dry. Extra roof is made of zinc material as it does not work to prevent heat from rain. In addition, the selection of lightweight materials is due to the easy roof console structure. At the top of the boarding house, it is easy to find the air inside the building (Figure 7).

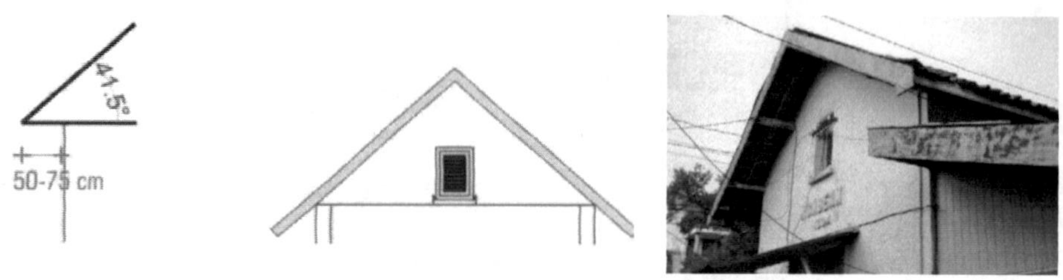

Figure 7. (a, b, & c) Roof Detail of Ngabean Station

The walls are white with simple ornaments that is light blue / gray. The thickness of the Ngabean Station wall is 30-40 cm which aims to drain the roof from the roof. On the surface, there is a simple wall decoration in the form of a horizontal field with a thickness of 2-5 cm and a diameter of 15-30 cm. This horizontal decoration connects the boundary line between openings to each other, presenting a line of relief in maintaining every opening height. The doors and windows openings mostly use wood materials with a combination of glass technology. The colour of this building has a tendency to cool colour tone, which is in the form of white, light blue and dark chocolate. These colours are available from the roof, walls, and openings, to wall ornaments. The use of cool colours was influenced by the style of colonial style at that time

The Ngabean Station facility combines the rectangular formations (at doors and windows) with gilded decoration (above the decorative wall). Both are connected to horizontal lines online by adding it to the wall ornaments, portraying a perception of the line on the wall to connect the edges of the openings with gates, where harmony is formed on the face of the earth. This ornament is a half-circle (arch) that runs from the bottom of the visitor's door. The presence of arch ornament gives a contrast to the Ngabean Station facade.

3.3. Structure System

The building structure of Ngabean Station is a closed frame structure that is a combination of frame system with construction of buffer wall. The supporting building at Ngabean Station is a thick wall between 30-40 cm from a pair of coated bricks. Thick walls are developed to carry loads from the roof to the foundation so that the wall also functions as a bearing wall. For the foundation, exact structure and materials are unidentified.

The roof of the building uses a gable structure with a structural frame in the form of teak wood and roof cover in the form of brick tiles. Ngabean station roofing ranges from 38 to 45 degrees which is also the adjustment of colonial architecture building on climate in Indonesia having tropical weather (in the corners of the roof slope of colonial building) measuring 30 degrees. In the roof of Ngabean Station, the ornament decoration has never been found. The station roof is supported by a wooden horse connected to a ring on the surface of the wall and arranged in lines, ribs, and battens to be placed on the roof as a roof cover. In the roof structure, there is a trinity of 0.5-0.75 meters from the outer wall. The purpose of this triangular production is to prevent the influx of rainwater into the building. There is an additional roof structure on a platform terrace separated from the main roof structure. This additional roof structure is a horse-drawn structure with support on several wooden pillars at several ends.

At Ngabean Station, there are variations in the use of building materials, local materials such as wood, roof tile or modern materials such as brick, glass, tile and ceramic are pasted. The use of wood materials is on the frame and the roof easel. The brick materials are seen on the walls and on columns of the building. The roof of the building uses tile material made of clay. Building floors use ceramic tiles arranged according to room size. In the openings, there is a variation in the use of materials between wood and glass. The use of this building material is as the impact of the transition architecture development in which the construction of buildings is in the form of bricks and wood, but has penetrated the technology of glass at door and window openings. The door and window laying system is by symmetric reflection causing cross ventilation allowing air to enter and exit easily to make the comfortable temperature inside the room.

4. Conclusion

Ngabean Station has a type of passenger emplacement as well as luggage emplacement. Spatial space of Ngabean Station is divided into 3 zones. First, the passenger public zone is shown by the *wachtkamer*; second, administrative zones are indicated by *kantoor / kartjees*; and third, technical operational zones of the train are indicated by the *magajine*. From an architectural aspect, this station creates a balance of facades formed from a symmetrical facade system between the opposite facade walls. The facade of the front is a reflection of the rear facade, as well as the facade on the side. This is seen from the laying of door and window openings. The results of this reflection form a symmetrical layout of the openings in each facade and make the balance of the shape of the building. The building structure of the Ngabean Station is a closed frame structure which is a combination of the frame system with the construction of a buffer wall. The door and window laying system is equipped by symmetric reflection causing cross ventilation allowing air to enter and exit easily to make the comfortable temperature inside the room.

References

[1] Ratnawati Y 2015 PERKEMBANGAN PERKERETAAPIAN PADA MASA KOLONIAL DI SEMARANG TAHUN 1867-1901 *J. Indones. Hist.* **3** 65–9

[2] Aoki E 1994 Dawn of Japanese Railways *Japan Railw. Transp. Rev.* 28–30

[3] Riza M, Ikaputra and Wihardyanto D 2017 Perubahan Era Gaya Arsitektur Stasiun Jalur Yogyakarta- Bantul pp 47–54

[4] Lehner E, Doubrawa I, Ikaputra., IVA-ICRA I for C R in A (Vienna). and Gadjah Mada University. Department of Architecture and Planning. 2013 *Insular diversity : architecture, culture, identity in Indonesia* (IVA-ICRA, Institute for Comparative Research in Architecture)

[5] Riza M A A, Ikaputra and Wihardyanto D 2017 *KARAKTERISTIK ARSITEKTUR BANGUNAN STASIUN NIS JALUR YOGYAKARTA-BANTUL* (Universitas Gadjah Mada)

[6] Ching F D K 2014 *Architecture: Form, space, and order* (John Wiley & Sons)

IOP Conf. Series: Earth and Environmental Science **328** (2019) 012068 doi:10.1088/1755-1315/328/1/012068

Contesting Historic Urban Space: Dispute Over Heritage Conservation of Colonial *Alun-Alun* in Serang, Indonesia

M A A Syahid[1], **K R Kurniawan**[2]

[1]Department of Architecture, Universitas Indonesia
[2]Department of Architecture, Universitas Indonesia

*E-mail: kemas.ridwan@gmail.com

Abstract. Urban preservation program in many cities has been challenged by particular interests and heterogeneous perspectives towards the idea of building a modern city. The dispute often rises over uncertain controversial spatial planning to erect any new large structure on historic urban space. In the case of colonial *alun-alun* of Serang, the issue occurred when the city ex-mayor designed to build a new *Masjid Agung* (Great Mosque) on the remaining empty spaces of *alun-alun*, amidst the pros and cons amongst urban activists, religious groups, and community. Contrary to all expectations, due to the political regime changes, the new government deals with the former idea by collaborating together with university team by justifying in academic attitude to assess the feasibility of the *alun-alun* to bear such project. The results instead indicated the un-feasibility of it, primarily due to its registered later status as *cagar budaya* (cultural heritage). It certainly triggered more disputes following the unsatisfactory comments, heated debates and rebuttals during the open presentation session with the government and public citizens of Serang city. This article hence is the proceeding of authors' experiences after conducting the feasibility study for 2 (two) months, by discussing and encouraging extensive views about the dynamics of power relations involving academics, bureaucrats, and urban citizens in contesting the historic urban space as urban heritage beyond the 'scientific' study. Therefore, besides putting efforts on mere analysis the importance of *alun-alun* of Serang and its heritage values, this article also inclines to underline the intentions of local strategies in utilizing presevation issues as tools throughout political decisions for succeeding multiple aims. The research method then investigates the prescribed regulations, law, urban policies and its implementations in regards with the spatial planning and preservation actions in the level of city. The result of this research is also expected to exemplify the conservation process of colonial *alun-alun* as an architectural element, urban space, and cultural heritage in local cities of Indonesia.

1. Introduction

The city of Serang in Banten Province has a slogan known as the 'Kota Serang Madani', refers to the name origin of Madinah (or Arabic as *mudunun*, a city) to be an example of prominent civilized Islamic city throughout history. The enthusiasm for this slogan is probably identified with the character and nuances of the Islamic image and perception in the city, given the long history of the city development, large populations and the role of muslim communities as majority who have been forming their spaces until today's contemporary times. The mosque building primarily becomes a concrete physical evidence which is considered representing the characteristic of the 'Madani City'.

Besides, the rising populations following the need for extending volume and number of prayer spaces, where until now, Serang does not yet have a so-called modern grandeur mosque anew in its city centre. The municipal government of Serang in 2018 was planning the construction of a new mosque which can become the central mosque for Serang. The construction of the Great Mosque of Serang City refers to the spirit of building the order of Islamic city space in the direction of its vision and identity. Serang as the capital city of Banten Province is ought to have a represented worship building facility with adequate infrastructure.

Figure 1. The bird-eye view images of the masterplan of Grand Mosque of Serang (source: *Dinas Pekerjaan Umum dan Penataan Ruang Kota Serang*, 2018)

However, obstacles and the on-going dynamic debates emerge since the early phase of appointing the site plan. The problem was raised a lot especially when the ex-mayor of Serang city, Haerul Jaman, chose the site exactly on the empty space of *alun-alun*, which is alleged to contain numerous values of historical, cultural, social and environmental. The square of *alun-alun* also functions a public open space that is important element in community's urban spaces. It have played an important role with its relationship to the city development since the precolonial sultanate, colonial administrative, until the postcolonial period in Indonesia. It was one of the results of early modern urban space development carried out by the Dutch colonies since the late 19th century. On the other hand, several people who agreed to implement the project have presented thoughts that the situated *alun-alun* in the city centre reflects importance of displaying the very first impression of city image, and the construction of mosque was assumed to strengthen the identity on city centre following the slogan of 'Kota Serang Madani'. [1]

The dispute often rises over uncertain controversial spatial planning to erect any new large structure on historic urban space. This also indicates that urban preservation program in many cities has been challenged by particular interests and heterogeneous perspectives on how a modern city should be built. Looking at the dynamic views above, it is a need to properly have a justifiable action together with the form of a feasibility study as a scientific consideration to assist the stakeholders (Municipal Government of Serang City and Bappeda; *Badan Perencanaan Pembangunan Daerah*) in creating fair and responsible and feasibly decide spatial policies in regards with plan to build the mosque. The data findings and analysis of information are performed in a multidisciplinary approach, involving quantitative and qualitative aspects in studying *alun-alun* as site for construction of new *Masjid Agung*.

2. Methods

This article is the proceeding of authors' experiences after conducting the feasibility study for about 2 (two) months, which will discuss and encourage extensive views about the dynamics of power relations involving academics, bureaucrats, and urban citizens in contesting the historic urban space as urban heritage beyond the 'scientific' study. Therefore, besides only putting efforts on mere analysis why *alun-alun* and its heritage values should be preserved, this article is also inclined to underline the intentions of local strategies in using presevation issues as tools throughout political decisions for succeeding multiple aims. The research method then investigates the prescribed regulations, law policy and its implementations regarding spatial planning and preservation actions in the level of city, regional and national as well. It is also expected to exemplify the conservation process of colonial *alun-alun* as an architectural element, urban space, and cultural heritage in local cities of Indonesia. [2]

The composition of team members came from the Department of Architecture and Department of Civil and Environmental Engineering, Faculty of Engineering Universitas Indonesia. This research team consists of members who have diverse educational backgrounds, experience in practice, and expertise in various related fields of science that completes and assists in the realization of ideas, findings, analysis and justifications that are scientific, comprehensive and accountable. The research method of this study using a mixed method of quantitative and qualitative approach by involving scientific views in a multidisciplinary field of sciences. This study will accommodate a variety of perspectives especially urban aspects, architecture, civil and environmental engineering, and socio-cultural aspects so as to enrich arguments and conclusions about the feasibility of the site.

The collection of primary data and secondary data is gathered before analytical actions. Primary data is generated by conducting direct observations through surveys and site visits to the area of *alun-alun* and its surroundings, correspondence and in-depth discussion to related parties who understand and having insight and knowledge into cultural preservation regulations in Serang City and Banten Province, such as the BPCB Banten and the Regional Development Planning Agency (Bappeda) of Serang. The secondary data is obtained through literature studies of written sources both printed and electronic. The team collected topographic maps as well as cartography records since colonial to post-colonial modern period from various official websites of institutions providing open access of public data.

As the research team studied of local regulations such as the Serang City Spatial Plan (*Rencana Tata Ruang Wilayah*, abbreviated as RTRW) for year 2010-2030 as stipulated in the Regional Regulation (*Perda*) of Serang City No. 6 of 2011, Spatial Detail Plan (*Rencana Detail Tata Ruang*, abbreviated as RDTR) of Serang and Cipocokjaya District Zoning for year 2013-2033 in Perda No. 9 of 2014. In addition to observing the physical condition of the site, the study also underlines the concern of status of *alun-alun* as a cultural heritage structure in Serang registered in the List of Cultural Heritage Objects by Banten Cultural Heritage Preservation Centre (*Balai Pelestarian Cagar Budaya Banten*, or BPCB Banten), protected by the Law of the Republic of Indonesia No. 11 / 2010.

In addition, the research team also included various opinions, proposals, suggestions, criticisms and recommendations from the audience during the open presentation of the feasibility study results since the Initial Report stage (13 November 2018 in the Department of Architecture of FTUI, Depok), Intermediate Report (30 November 2018 in the Serang City Government Complex, Serang), up to the Final Report (13 December 2018 at Puri Kayana Hotel, Serang) including bureaucrats, religious leaders, community leaders and the general public to enrich the arguments of the study. Contrary to all expectations due to the political regime changes, the new government in opposite dealing that former idea by collaborating together with university team, asking them to justify in academic attitude to assess how feasible the *alun-alun* to bear such project. The results instead showed the un-feasibility of it, primarily due to its registered later status as *cagar budaya* (cultural heritage). It certainly raised more dispute following the unsatisfactory comments, heated debates and rebuttals during the open presentation session with the government and public citizens of Serang city.

3. Results and Discussions

Most of the historical buildings in Serang tend to be centred around two regions, namely Old Town of Banten (*Banten Lama*) and city centre of Serang. Some alleged cultural heritage objects in Serang City mainly include the colonial *Alun-alun* (covering East Square and West Square), Office of Regent (*Bupati*) Daerah Tingkat II Serang, railway station, and Class II B Serang Detention House. In 2017, there were at least 58 total documented cultural heritage in Serang by BPCB Banten consisting of 37

IOP Conf. Series: Earth and Environmental Science **328** (2019) 012068 doi:10.1088/1755-1315/328/1/012068

buildings, 16 structures, and 5 sites. Based on a list of inventory, *Alun-alun* object is noted in number 50 with details No. Old Inventory 416.01.01.04; 074.01.01.04.04, and No. New Inventory 054.01.07.04.08, and falls into the category of Structure [4]. The definition of the 'Structure of Cultural Heritage' itself is well regulated in Article 1 Paragraph (41) point 4 of Law No. 11 / 2010, which explains that, "Cultural Heritage Structure is a built structure made of natural objects and / or man-made objects to meet the space needs of activities that are integrated with nature, facilities and infrastructure to accommodate human needs." [5]

Table 1. Data Inventory of *Alun-alun kota Serang* as the city cultural heritage [6]

DATA OBJEK CAGAR BUDAYA	
ID Pendaftaran Objek	PO2018091900001
Tanggal Daftar	19 September 2018
Nama Pendaftar	Balai Pelestarian Cagar Budaya (BPCB) Banten Hendri Prasetyo
Kategori Cagar Budaya	Struktur
Nama Objek	Alun-alun Kota Serang
Kategori Objek	Struktur
Periode/Masa	Islam/Kolonial
Letak Administratif	
Alamat	Jl. Ki Mas Jong
Desa/Kelurahan	Cipare
Kecamatan	Serang
Kabupaten/Kota	Kota Serang
Provinsi	Banten
Kode Pos	42117
Letak Geografis	
Koordinat UTM/Zona UTM	48
X	627413
Y	9323686
Koordinat Degrees Minutes Seconds	
Latitude	6' 7' 2.47 " LS
Longitude	106' 9' 5 29 " BT
Koordinat Decimal Degree	
Latitude	-6.117353
Longitude	106.151469
Ketinggian	30 m dpl
Ukuran	-
Batas-Batas	
Utara	Jl. Veteran
Selatan	Jl. Ki Mas Jong
Timur	Jl. Pangeran Diponegoro
Barat	Jl. Bridgen KH. Sam'un
Kondisi Saat Ini	Terawat
Sejarah	Alun-alun merupakan peninggalan masa kolonial Belanda
Deskripsi	Alun-alun Kota Serang merupakan tinggalan kolonial Belanda yang terbagi menjadi dua bagian, yaitu alun-alun bagian barat dan bagian timur. Saat ini bagian timur digunakan untuk aktivitas olahraga, sedangkan alun-alun sebelah timur digunakan sebagai aktivitas kegiatan upacara kenegaraan.
Pemilik	Pemkot Serang
Riwayat Kepemilikan	-
Pengelola	Pemkot Serang

(source: *Balai Pelestarian Cagar Budaya Banten*, 2018)

Both Old and New Inventory Number BPCB Banten appeared due to reference of the old inventory to Law Number 5 of 1992 concerning Cultural Heritage Objects (*Benda Cagar Budaya*) not in accordance with developments and legal current needs of Indonesian societies, while the new inventories have referred to in Law Number 11 / 2010. As it also relates with the substances of *Perda* of Serang City regarding the RTRW and RDTR that several articles and verses might be applied in considering the feasibility of building new mosque in *alun-alun*. [7] [8]

Unfortunately, the city of Serang does not yet have any policy that specifically regulates the management of preservation of cultural heritage, especially Cultural Heritage Buildings and Heritage Sites, let alone the expert team of cultural heritage TACB (*Tim Ahli Cagar Budaya*) every municipalities in Indonesia should form. The *Perda* No. 5 of 2017 of RTRW classifies cultural heritage as 'part of the protected space pattern areas' (Article 40 point a juncto Article 41 Paragraph (1) point d), among other scientific knowledges to covering the preservation of cultural environments in Banten [7]. In addition, *Perda* No. 9 of 2014 concerning RDTR of Serang and Cipocok Jaya Subdistricts have the Protected Area Patterns Plan, described in Article 22 Paragraph (1) point c including cultural and scientific reserves. [8]

In the discussion about determining cultural heritage, an understanding of cultural significance is required to be categorized as a cultural heritage. Cultural heritage as a cultural resource has been a fragile, unique, rare, limited, and not up-to-date nature, as described in the Explanation of the Republic of Indonesia Law Number 11 / 2010, in paragraph 5. *Alun-alun of Serang* clearly has the characteristics as mentioned above. When the research team compared the existing situated map (2018) with the colonial map of 1915 of *alun-alun* of Serang as shown in Figure 2, a general information drawn is that the present morphological patterns of Serang City are still strongly influenced by the past morphological pattern of Serang City since the Dutch colonial period [9]. This conclusion is withdrawn from interesting findings in the process of comparing the two maps, which are:

(1) The shape of *alun-alun* from the very beginning of its presence was designed and built by the presence of a *lane* that separates the field between West Square and East Square, to arrange a unique pattern existing in its original model currently in general.

(2) There are several similarities in the city pattern including the colonial buildings that still continuing its functions like offices/pendopo (*Regentschapskantoor*) and regent houses, post offices, detention house (*Gevangenis*), old market, stations, and school. As for some colonial buildings that their functions have been adjusted, among others are the O.S.V.I.A school which has now shifted functions to the police resort office. There are also sites undergone changes in shape but still retain similar functions, such as church, post offices, mosques (Ats-Tsauroh the Great Mosque), and cemeteries complexes scattered in several places around the city centre.

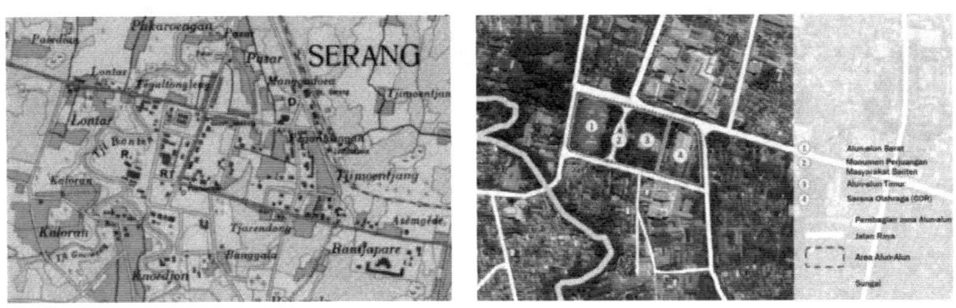

Figure. 2. City centre of Serang *afdeeling* in 1915 (left) and 2018 (right)
(sources: *Topographische Inrichting, Batavia 1915 Blad 12 a* & Google Maps, edited)

Even though the centre of Serang City and its surroundings have not been designated as Cultural Heritage Sites or Regions with contained buildings and historic structures, it should be underlined that the treatment of objects presumed as Cultural Heritage must be the same as the legitimate Cultural Heritage. Therefore, they must be preserved by referring to statutory provisions according to the law No. 11 / 2010. Maintaining the original characteristics of the Cultural Heritage Structure as well as the original characteristics of the cultural landscape and/or land surface is required before making adaptations. The construction of the Great Mosque of Serang does not guarantee the authenticity of the city image or *vista*, and it clearly risks destroying the physical values of the structure of *alun-alun*. [5]

The provincial and municipality governments in this case are responsible for the duties and authorities of Cultural Heritage as well as the alleged Cultural Heritage, such as the management of National Register of Cultural Heritage in the area (Article 40 Paragraph (2) juncto Article 95 Paragraph (1) and (2) Chapter VIII Duties and Authorities of Law No. 11 / 2010. [5] Besides, there are several assumptions and aspirations among the people who consider that the construction of the grand mosque in the *alun-alun* is based on historical arguments which refer to the model of urban centre space in the kingdom/sultanate era in the archipelago. The argument may not be fully in accordance with the context of *alun-alun* of Serang, due to these following reasons.

IOP Conf. Series: Earth and Environmental Science **328** (2019) 012068 doi:10.1088/1755-1315/328/1/012068

The first reason is that *alun-alun* was built during the Dutch East Indies (or *Nederlands Indies*) colonial state period between the moving centre of power of Old Banten to Serang City (around the 1800s), and before the beginning of the 20th century. *Alun-alun* here was a form 'new city centre' which were built later as a physical consequence after the kingdom and sultanate demise in Old Banten. This square less or more has a different space configuration with the precolonial *alun-alun* model in Nusantara, thus made it distinctive and unique urban element in local cities of Java. The similarity between the precolonial and colonial *alun-alun* was the office or pavilion of the regent as local authorities still represents the presence of the palace as central power in the precolonial city model. Many elements of the city found were not in accordance with the principles or types of Islamic sultanate square, such as the absence of mosque building, and even was replaced by the church as the official religious symbol of Dutch colonial state. [9]

Even though the *alun-alun* was built as referred to the precolonial type, the siteplan of *Masjid Agung* should be chosen and built around the *alun-alun* and would be side by side with the Serang Regent's office and market too, not *on* the very existing space volume of the *alun-alun*. Building a new structure/building on top of the square is conducted by removing the square area. Thus, a firm policy to guarantee and maintain the existence of cultural heritage from the threat of physical development in urban areas is absolutely necessary, when *alun-alun* as a 'living monument' governed by its utilization.

In additon to being a Cultural Heritage Structure, *alun-alun* of Serang also functions as a green open space in Serang City which is regulated by Regional Regulation (*Perda*) of Serang City No. 5 of 2017 concerning the Banten Province RTRW, which can also be found in Article 44 paragraph (1) point f to regulate the local protected area, juncto Article 41 Paragraph (1) point c which covers urban Green Open Space (*Ruang Terbuka Hijau*, abbreviated as RTH). [7] This is in line with the *Perda* of Serang City No. 9 of 2014 concerning RDTR and Zonation Regulations of Serang & Cipocok Jaya District in 2013-2033, located in Article 22 Paragraph (1) regulating Protected Area Pattern Plans include cultural preservation and RTH. [8]

The total volume of RTH in Serang City is regulated in *Perda* No. 9 of 2014 in "Chapter III Objectives, Policies, and Strategies". In Article 6 point p, the amount of RTH has been set at 30% (thirty percent) of the total area of city. The Article 7 Paragraph (16) point a and point b further regulate the strategy for determining the amount, by maintaining the existing open space of approximately 39 km^2, or about 9-14% of the total area of Serang. The above strategy is also supported by Article 7 Paragraph (16) point d by keeping open space in regards with its function, to sustain the maintenance of the amount of RTH remains effective and stable. The article that lists *alun-alun* of Serang as part of RTH of Serang City can easily be found in "Chapter V Regional Space Pattern Plans" in the same RDTR Regulation in Paragraph 7 Green Open Space, Article 29 Paragraph (2) point g including an area of approximately 20 Ha, such as Maulana Yusuf Stadium, Brimob Stadium, and *Alun-alun*. [8]

Therefore, it is clear that *alun-alun* of Serang City contributes to the amount of RTH in the city. By continuing the construction plan on the site, it will automatically reduce the amount of existing green open space, thus, the number of RTH decreases before it even reaches 30%, referring to the amount of 14% listed in *Perda* No. 9 of 2014 of RDTR. *Perda* No. 9 of 2014, in Paragraph 7 about Disaster Mitigation Facilities and Infrastructure System especially in Article 19 Paragraph (2) letter b, it regulates the direction of development of disaster mitigation facilities and infrastructure including the Earthquake Disaster Mitigation Space to the Maulana Yusuf Stadium, West and East Square of *alun-alun* and any vacant land closest to community housing. Specifically, the empty space on the *alun-alun* also serves as a community gathering place in order to mitigate natural disasters. [8]

As another consideration, the research team tried to look at the Draft of Revisions of *Perda* of RTRW of Serawng City that had not been approved as an additional reference. By referring at Article 50 paragraph (5) letter a, a similar development may be carried out on cultural heritage that are in the

IOP Conf. Series: Earth and Environmental Science **328** (2019) 012068 doi:10.1088/1755-1315/328/1/012068

zoning of public and social facilities by maintaining the following matters, and construction of places of worship on the *alun-alun* in other sense is performed with the following not, that of: (1) not eliminating the original function, in regards with Article 7 Paragraph (11), namely its use as a community and open space is kept well maintained, (2) meeting the requirements and intensity standards of supporting zones and developers in cultural heritage areas, as stated in Article 50 letter b, and (3) meeting the requirements of establishing a worship house in accordance with the Joint Regulation of the Minister of Religion and Minister of Home Affairs Number 8 and 9 of 2006. [10]

4. Conclusions

The strict protection and preservation efforts against the Cultural Heritage (*Cagar Budaya*) and Green Open Space (*Ruang Terbuka Hijau*) regulated by the formal regulations and legislations reinforce to the conclusion that there is a need to reconsider the feasibility of colonial *alun-alun* (both West and East Square) including its surrounding historic environments intended as a site for the construction of the *Masjid Agung*. Some recommendations were given by the research team as a series of alternative solutions to resolve the presented problems as input that might be implemented at the next stage of the assessment and development planning. This reconsideration does not mean declaring a plain rejection of the plan in building a new grand mosque in Serang, but as a special statement regarding the location plan. The authors continue to support the Municipal Government and the people of Serang City in holding the construction, yet with some recommendations provided as next better steps to be implemented by both Municipal Government and the general public after the activity of feasibility study.

Figure 3. Historical mosques surrounding *alun-alun*, left: Masjid Kaujon, right: Masjid Raya Ats-Tsauroh (source: Private collection, 2018)

The first conclusion is not recommending the *alun-alun* of Serang, which strongly has many historical buildings surrounding, to serve as the construction site for the Grand Mosque erection or any similar massive structures, with due consideration referring to the legal basis and as for implementing cultural preservation management efforts in Serang. Encouraging the Regional Government of Serang City is another effort to submit a city centre consisting of squares and various historical buildings around it into the Cultural Heritage National Register System (*Sistem Register Nasional Cagar Budaya*), especially buildings and other structures that have not been registered and are still alleged as Cultural Heritage.

Other suggestion is encouraging any authorized stakeholders to conduct intensive and comprehensive publications and socializations to the people of Serang City regarding Cultural Heritage laws, its historic knowledge, and the importance of distinctive, unique, rare and valuable cultural significance. Optimizing the use of *alun-alun* as a Cultural Heritage Structure as facility of public infrastructure and Green Open Space (RTH) by inviting the general public to hold various activities in it for the interests of religion, social, education, science, technology, culture and tourism, as stipulated in Article 85 Paragraph (1) of Law No. 11 / 2010. [5] Reflecting on various religious activities in recent years, the concerned efforts have often been carried out in the National Monument (Monas) Structure in DKI Jakarta, *alun-alun* of Serang can also potentially be utilized as a facility for religious activities such as *tabligh akbar* congregations, *zikr* assembly, and other religious activities.

IOP Conf. Series: Earth and Environmental Science **328** (2019) 012068 doi:10.1088/1755-1315/328/1/012068

Organizing efforts to optimize and develop other main types of worship areas and buildings around the city centre, such as the Great Mosque of Kaujon and the Great Mosque of Ats-Tsauroh which are both historical buildings that have historical links and physical connectivities in the location. The problem of spatial need for prayers space can be adjusted to the scheme of developing the area around the already existing built mosque(s), especially the Ats-Tsauroh Great Mosque which is an old mosque but not included yet in the Cultural Heritage List of BPCB Banten, therefore, development can be perform according the community's need towards adequate worship facilities in the city centre. Finally, the research also suggests to be more considerate in selecting sites to hold the possibility of increasing land property values in other areas instead of city centre, so that the construction of Grand Mosque would give benefits to potential lands to open new centre of religious, economic, social and open spaces too.

However, the use and abuse of the political decision towards heritage conservation were eventually immense, such as documentation and registering actions of colonial *alun-alun* were only taken into account when the changes of political regimes happened, probably done for any particular partisan's interests. The issues on heritage and/or the need for open space of *alun-alun* in the city become a merely political tool to legitimate political decisions rather than as quintessential fields to take into account for building a better city planning [11]. Finally, the urge for creating the expert team of cultural heritage (*Tim Ahli Cagar Budaya*) in recommending law and regulations towards cultural heritage (*cagar budaya*) should be immediately decided by local municipalities, together with the educational programme in to encourage, introduce and familiarize the values of local unique cultural heritage in city.

References

[1] Pusat Penelitian Sains dan Teknologi (PPST) Fakultas Teknik Universtas Indonesia (2018) *Laporan akhir studi kelayakan (feasibility study) alun-alun kota Serang sebagai tapak pembangunan masjid agung kota Serang*, Badan Perencanaan Pembangunan Daerah, Serang.

[2] Silverman, ed. (2011) *Contested cultural heritage: religion, nationalism, erasure, and exclusion in a global world*. Springer, New York.

[3] Karlstrom, A. (2013) "Local heritage and the problem with conservation", Brockwell, O'Connor & Byrne, eds., *Transcending the culture—Nature divide in cultural heritage, views from the Asia-pacific region*. ANU Press.

[4] Balai Pelestarian Cagar Budaya (BPCB) Banten (2017) *Daftar keseluruhan benda/situs cagar budaya provinsi Banten tahun 2017*, Direktoral Jenderal Kebudayaan, Kementerian Pendidikan dan Kebudayaan Republik Indonesia.

[5] *Undang-Undang Republik Indonesia Nomor 11 Tahun 2010 tentang Cagar Budaya.*

[6] Balai Pelestarian Cagar Budaya (BPCB) Banten (2018) *Data objek cagar budaya: Alun-alun kota Serang*, Direktoral Jenderal Kebudayaan, Kementerian Pendidikan dan Kebudayaan Republik Indonesia.

[7] *Peraturan Daerah Kota Serang Nomor 6 Tahun 2011 tentang Rencana Tata Ruang Wilayah (RTRW) Provinsi Banten.*

[8] *Peraturan Daerah Kota Serang Nomor 11 Tahun 2014 tentang Rencana Detail Tata Ruang (RDTR) Kecamatan Cipocokjaya dan Kecamatan Serang.*

[9] Knaap, Gerrit (2007) *Grote Atlas van de Verenigde Oost-Indische Compagnie, II Java en Madoera* ('Comprehensive Atlas of the Dutch United East India Company'). Royal Dutch Geographical Society, Utrecht; Nationaal Archief, The Hague.

[10] *Rancangan Peraturan Daerah (Ranperda) sebagai Perubahan atas Peraturan Daerah Kota Serang Nomor 6 Tahun 2011 tentang Rencana Tata Ruang Wilayah Provinsi Banten.*

[11] Lim & Padawangi (2008) "Contesting alun-alun: power relations, identities and the production of urban space in Bandung, Indonesia", *IDPR*, 30 (3), p. 307-326.

3rd World Development Model in Indonesia: How to deal with the issues of decreasing public finance and overcoming the demands for public participation in Industrial Revolution 4.0 era?

M Janah[1] **A Y S Rahayu**[2]
[1]Master Student of Local Government Studies, Faculty of Administrative Science, University of Indonesia (UI).
Coordinating Ministry for Human Development and Cultural Affairs Republic of Indonesia (Kemenko-PMK)
[2]Professor of Public Policy and Administration at University of Indonesia (UI)

*Email: amy_soeroso@yahoo.com
miftahusein5@gmail.com

Abstract. This work consists of discussions as well as practical examples about potentials and challenges for regional managers in Indonesia en-routing to the 3rd World Development Model. The model emerged as a critique to the old schemes of government collaborations and partnerships in which the participation of the end-users is not proportionally given. This work aims to understand the relationships among actors, the divisions of their roles, the pattern of participation as well as the advantages of "end-users" participation in the development project. This concept is a Constructive Model of Collaborative, a meeting point between Participatory and Collaborative Governance concept. This study is conducted through post-positivistic approach and qualitative methods by using cases and theoretical studies. Furthermore, through this article, the authors focused on the two major issues in the development nowadays. Firstly, the issue of decreasing public finance, particularly in the case of financial deficit in national health insurance program known as JKN-KIS. Secondly, the issue of public participation in regional and urban development planning through case of Bandar Kayangan City Development Project in North Lombok, West Nusa Tenggara. Hence, this work drives a finding which offers solutions for the challenges on how to run the development project without depending on State or Local Budget at the same time to leverage the public supports and participation in government's development agenda. Finally, it is concluded that there are needs to understand the tendencies of conflicting visions and goals, power or influence asymmetric and resources imbalance among the involved actors in the new model of collaborative governance.

Keywords: collaborative governance, urban development, decreasing public finance, development model

1. Introduction

Health become one of the basic criteria for a country to be classified as a developed and developing country. The other one, amongst others, is the citizen-oriented infrastructure development. Both issues have become the focus of attention in this 4.0 industrial revolution era. Based on the criteria arranged on the official website of the United Nation Development Program (UNDP), the main criteria used as a benchmark in the classification of countries recognized by the United Nations (UN) are based on the value of the Human Development Index (HDI). The assessment considers all forms of multifaceted nature. The Human Development Index (HDI) is a comparative measurement of life expectancy, education, and living standards. HDI is used as an indicator to assess the quality aspects of development and to classify whether a country includes developed countries, developing countries, or underdeveloped countries and also to measure the influence of economic policies on quality of life (Directorate of Analysis and Development of BPS Statistics, 2015).

The HDI was first introduced by one of the institutions of the United Nations in 1990 and regularly published in the annual Human Development Report (HDR). In various literatures related to the process of polarization and classification of countries, one of them is closely related to the level of public health and public infrastructure development reflected in the direction of a country's national strategic policies. A good country is defined as a country that thoroughly guarantees the quality of life of its people, both the quality of their health and homes.

On the other hand, external demands that globally apply to each country such as the ideals of the SDGs enable the government of a country to formulate appropriate policies and encourage new innovations in the development of bureaucracy. In an effort to fulfill a good score in HDI, the government of a country is currently required to be able to carry out governance that involves community participation excluding the dependence on the unsettled state budget. Thus, there is a need for a joint governance model that involves many stakeholders both the private sector and the community in matters of public service provision. In the current scientific development of governance, several constructs of collaborative governance are established. Through the last three decades, scholars have developed more definitions of collaboration emphasizing the preconditions, process, or outcomes of the relationship.

[8] argued that collaboration is a negotiating process of interdependently organizational actors as a response to find the solution to the shared concerns (pp.12-13). The next decades, [4] defined collaboration as the efforts to leverage the public value by working together through joint partnerships of two or more agencies (p.8). In addition, [1] added the definition of collaboration based on the value of reciprocity arranged to solve problems that cannot be solved by single organizations. They argued that collaboration is a multi-sector and multi-actor relationships particularly working across boundaries in the effort to achieve common goals. Furthermore, the dynamics and characteristics of the collaboration among actors or agencies in governance system are now widely known in many shapes and forms. In some ideals in the literature and practices, the term of Collaborative Governance is formerly used, which according to [2] is a collectively direct meeting between state and non-state stakeholders through a formal, deliberative, and consensus-oriented decision-making process in order to achieve a multilaterally implemented public-policy and mutually managed public programs or assets.

Moreover, this paper argues that the framework and constructive model of collaborative governance applicable to the analysis of cooperation in health service and urban development. This paper has four parts. After this introduction, the key approaches are discussed by examining the collaborative governance forms and argue that studies on collaborative governance offer a useful analytical framework for this study. The next section overviews the two major cases through the concept analysis by presenting a collaborative framework to the intergovernmental interactions. The final part briefly concludes the key impacts of the constructive model towards the regional development and health insurance issues recently.

Table 1.1 The Collaborative Governance forms

Collaborative Governance	PPP	4Ps
• Formal and Collective direct meeting • Achieve decision-making consensus • Represent consensus between public agencies and non-state stakeholders • Function to create public policy and consensus decision-making • Joint ownership of decision • Bottom-up decision making • Interdependencies among actors	▪ Formal contracts of Cooperation to function ▪ Achieve Cooperation between governments agencies and private actors ▪ Represent agreement /contracts between public and private actors ▪ Function to deliver certain service or perform certain tasks ▪ Joint ownership of the risks ▪ Top-down planning ▪ Dependent among actors	• Formal and informal Participation • Achieve Collaboration among governments actors, private actors and people (end-users). • Represent commitment to active and positive participation of all actors. • Function to formulating the needs and wants of the end-users (people) and sustained the whole A-to-Z process • Joint ownerships of the process • Bottom-up planning • Interdependencies among actors

Source: Literature review; Modified by the authors

1.1. Problem Statement and Research Goals

A. An overview of financial deficit case on National Health Insurance program (JKN-KIS)
The JKN program is an aspiration set in the 5th Nawa Cita, which is improving the Quality of Indonesian Human Life and is one of the national strategic programs where the implemented system is a mandatory social health insurance system. The program aims to provide a guarantee of protection for socio-economic risks that afflict participants and / or family members. The organizers of the health social security system are in accordance with the mandate of Law Number 40 Year 2004 and Law Number 24 Year 2011 is the Health Social Security Organizing Agency (BPJS *Kesehatan*) and the form of participation is through the *Kartu Indonesia Sehat* (KIS).

One of the biggest problems faced by BPJS *Kesehatan* concerns with the income that comes from participant contributions which is not proportional to the amount of funds spent to pay capitation and claims for hospital payments, or in other words BPJS has a budget deficit. Furthermore, Law Number 24 of 2011 concerning the Health Insurance Organizing Agency has regulated the BPJS funding mechanism. However, the funding arrangement has not explained detail about the flows and sanctions for participants who do not meet the rules of the BPJS membership. In the BPJS annual financial report, it shows that in 2014 BPJS had a deficit IDR. 3.8 Trillion, then in the following year, the 2015 budget deficit returned as much as IDR. 5.9 Trillion, as presented in the following table. 2.1:

Table 1.2 Number of Budget Deficits in BPJS *Kesehatan*

No	Year	Budget Deficit (IDR)
1	2014	3,8 Trillion

IOP Conf. Series: Earth and Environmental Science **328** (2019) 012069 doi:10.1088/1755-1315/328/1/012069

2	2015	5,9 Trillion
3	2016	9,7 Trillion
4	2017	10,2 Trillion
5	2018	8 Trillion (Projection)

Source: Processed from the BPJS *Kesehatan* Report, 2017

Throughout 2017, the contribution from National Health Insurance (JKN) program amounted to only IDR. 74.25 trillion, while the claims reached IDR. 84 trillion. As a result, BPJS Kesehatan had to bear a financial deficit of up to IDR 10.25 trillion, a greater amount than the previous year. Even 2018 report did not present profitable scheme. Referring to the BPJS Health Performance and Annual Budget, in 2018 the company targeted revenues from IDR.79.77 trillion with health financing reaching IDR. 87.80 trillion. Certainly, in 2018 BPJS *Kesehatan* is projected to have a deficit of around IDR. 8.03 trillion.

B. An Overview of people participation in regional development planning in Bandar Kayangan Global Hub Project

The implementation of local autonomy through UU No. 22 1999, replaced by UU No. 32 2004 and revised by UU No. 23 2014 about local government, has open the access for citizens widely to participate in the process of development. In this case, the provincial government of West Nusa Tenggara realized that the society plays an important role in the development of the region. The governor optimistically announced that it is expected to accelerate the economic growth in the region and be more adaptive to meets the demands of the current and future inhabitants in current and future era. To respond to the smart and sustainable city development challenges, among the three new cities are targeted by the government. The NTB Government in the development project of Bandar Kayangan brings innovation in the principle of development that involves the private sector as well as engaging community involvement. This innovation concept distinguishes the principle of Kota Bandar Kayangan's development with Maja and Sofifi and becomes a "point of interest" for researchers to unfold the concept.

C. Research Goals

This article explores the independent provisions of health services on state or local budget. Moreover, the issue creating an inclusive, safe, resilient and sustainable city is also conveyed by applying the Public-Private-People-Partnerships and other forms of collaborations as one of the most successful mechanisms in urban planning. Thus, several aspects are required to be analyzed such as: a) the real interest of all actors in the development project through 4Ps concept of urban/regional planning as well as the power gaps among actors in public health service provisions, b) the real influence of the public administration on the entire project, in accrodance with a theory-based on coordination and control related to both cases.

1.2. Conceptual Review

An explosion of new developments in the area of Collaborative Governance has been debatable concerning its in the latest few decades. [7] stated that "*Collaboration generally involves a higher degree of mutual planning and management among peers; the conscious alignment of goals, strategies, agendas, resources and activities; an equitable commitment of investment and capacities; and the sharing of risks, liabilities and benefits. . . . Collaboration, therefore, suggests something less than authoritative coordination and something more than tacit cooperation.*"

Lately, there are large number of countries, among others Scandinavian countries, for instance "Baltic Urban Lab - Integrated Planning and Partnership Model for Brownfield Development - Project" in Finland has started to developing and testing new integrated models of Public-Private-People-Partnership (4Ps) progressing to grow as a developing concept or subject of discussions in the urban development planning and health services (Nordregio, 2016). This concept offers a platform of fruitful partnerships that is theoretically and practically more accommodative towards the real needs of people or the end-users as "connoisseur" of development in the future. 4Ps concept in urban planning developed by Majamaa Wisa presented in the following Diagram 3.1:

Diagram 3.1 The 4th P – People- in Urban Development base on Public-Private-People-Partnership

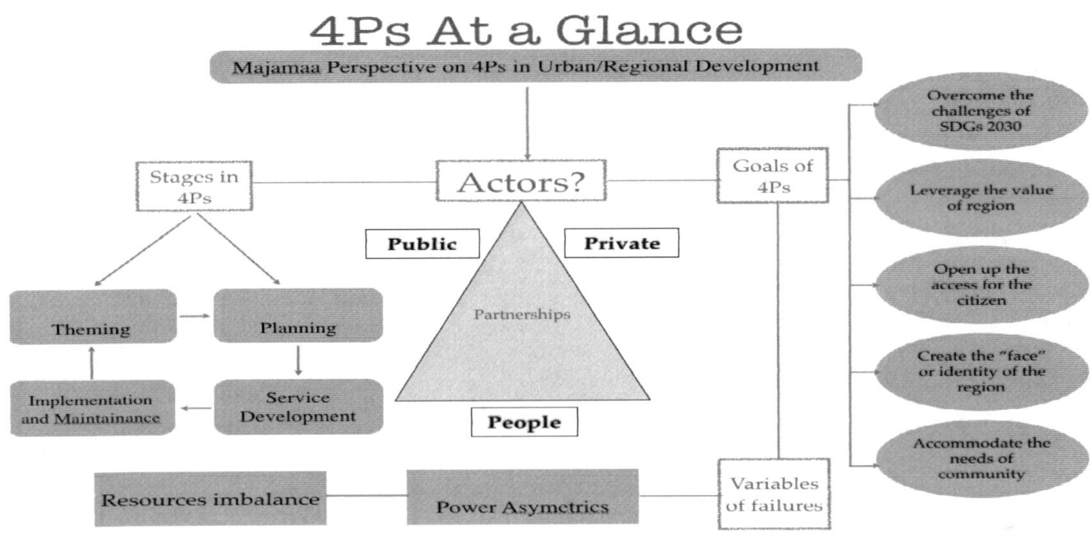

Source: Wisa Risto Majamaa, 2008

Public-Private-People Partnership (4Ps) provides a new principle in urban development planning and public health services. This principle develops new ways of engaging public / government actors, private actors, local people or communities, NGOs, and civil society actors in various development planning and public service processes. To understand the 4P concept, it is important to understand the background and foundations of the Public-Private-Partnerships (PPPs) principles. Public-Private-People-Partnerships (4Ps) came as a result of the dominant role that the government play in the PPPs model of partnerships, where the government used to centralize all the authorities and to take large amount of control in the decision-making process. Moreover, the growing demand of undesirable project failures and various development problems have been encountered in previous PPPs-based project, enabling governments to involve people in the partnership in order to avoid issues and obstacles (S.T. Ng et al., 2013, pp. 370). Hence, S.T. Ng et al (2013) added that the governments develop the 4Ps for the same purpose they use in PPPs for, which is to upgrade the quality of public services, particularly the infrastructure of services. He identified the Public-Private-People-Partnership (4Ps) meaning as;

"process framework embraces the bottom-up participative strategies which bring the public engagement clearly visible for infrastructure planning and policy making. With this newly developed framework and associated engagement strategies, decision-making power can deviate

IOP Conf. Series: Earth and Environmental Science **328** (2019) 012069 doi:10.1088/1755-1315/328/1/012069

from policy makers, who are traditionally holding the ultimate decision authority, towards the citizens through proactive engagement" (S.T. Ng et al., 2013, pp. 370).

Non-governmental organizations, local communities, and civil societies all represent the people's participation. Zhang et al. (2015) explained people's participation as the community-based policy formulation and implementation from the very beginning step to achieving the desired goals" (Zhang et al., 2015, pp. 409).

2. Research Method

This research utilizes the post-positivitism approach by using two main cases with the same concept as a research analysis framework which is Public-Private-People-Partnerships (4Ps). Case study becomes the main research method by collecting and compiling descriptive data derived from some key informants. Case study utilized to describe field phenomena. Therefore, perception and valuation of researcher(s) in vieweing phenomena must be framed by a research guidance to keep research focused on the goal(s). The research site is conducted in two places which are City development project in North Lombok, West Nusa Tenggara and another issue in BPJS *Kesehatan* collaborated with National health insurance council (DJSN) under coordinating ministry of human development and cultural (Kemenko-PMK). Both cases are chosen because the authors argued that both issue are relevant to the approach and constructive concept of collaboration, where the citizen active participation could help governance dealing with the issue of decreasing public finance and overcoming the demands of public participation through the third world development model known as 4Ps.

3. Research Findings and Discussion

3.1. Issue of Budget Deficit in BPJS Kesehatan

Basically, collaborative efforts in dealing with budget deficit problems in the JKN-KIS program have been carried out by involving many actors who have produced policy options that are currently in the review process to find a joint solution. Director of Expansion and Services of BPJS Health Participants, Andayani Budi Lestari (2018) stated that the support and participation of the Regional Government was very strategic and decisive in optimizing the JKN-KIS Program, where there were at least 3 important roles of regional governments including expanding the scope of participation to encourage Universal Health Coverage (UHC), improving service quality, and increasing compliance for non-contribution recipients (non-PBI) and participants. In this case, the model of 4Ps is needed in order to commonly overcome the financial deficit problem.

In this era, the state budget is limited and citizen participation is much required. Through 4Ps model of collaboration; participatory budgeting is not a second option. If citizen, in this case the non-PBI participants is aware towards their contribution and the financial deficit is preventable. Bardach (1998) argued to increase public value by working together through joint partnerships of two or more institutions. Joint efforts to address budget deficit problems have been addressed with meetings and coordination between stakeholders.

To overcome this issue, local governments are encouraged not only to act preventively and promotively, but to be more active in addressing this problem. Local governments are asked to allocate regional budgets to the health sector in accordance with the rules of the central government. The budget from the central (APBN) should be five percents for the allocation of health in the region, where the local government can allocate a minimum of ten percents of the APBD.

[6] explained that the key to collaborative governance is to provide wisdom carefully and strategically to private entities, whether for profit or non-profit, which simultaneously motivate and empower them to create public value. The essence of collaboration is generally related to multi-sector and multi-actor relations that specifically work across borders in an effort to achieve common goals.

3.2. Issue of Citizen Participation in Bandar Kayangan Global Hub Regional Development Project

This development project had been started without mentioning the end-users involvement in the early bird processes. The Governor of NTB explained that the development of Kota Kayangan will apply the principles of "Public Private People Partnership" (4P), which was started by establishing a joint company, whose shares will be owned by West Nusa Tenggara Provincial Government, North Lombok Regency Government, investors and professional management, and land owners. The scheme is intended to apply the principles of sustainable development, ensuring future stability, where people will have a sense of belonging to development and will also benefit the development of social justice.

A company called Bandar Kayangan International is trusted by local governments as executor working with investors and land-owners/end-users (current-future inhabitants). In this situation, the control is in the hand of that leading company. It is mentioned that there will be a sharing benefits of the regional development with all actors to gain the Return on Investment in certain percentage. In detail, it is not, whether the government, companies as well as people (end-users), ever been highlighted have conducted a formal meeting and discussion in the planning process.

In urban development processes, participatory planning is applied to involve stakeholders in the planning and development processes. However in this case, the initiator to run the development project is in the hand of pointed private company. It is important to be aware of the knowledge of the current inhabitants or people about their region. They know more about their land. Moreover, it is need a continual redefinitions on the type actors , their roles, relationships pattern and influences. All those aspectrs are necessary to build and design 4Ps concept. Van Herzele (2004) mentioned based on rational planning theories, participatory planning will emphasizes more future seeking actions than future defining. It seems obvious that the vital source of knowledge is in the existing inhabitants so that providing the hope for future inhabitants overlooked in 4Ps.

4. Conclusion

There are two aspects to be underlined in the conclusions, which are:

1) There are no concepts that are really compatible at all time and conditions, including 4Ps concept referring to the older concepts as well. In the future, this concept will be diminished and changed by a new concept regarding the demands of the time and technology. The success of development is not solely caused by the concept adopted, but more possibly influenced by the leader in playing their roles and responsibility. It is unquestionable that a leader One thing for sure, a leader should have future-oriented or long-term projection.
2) The specification of the role of all actors seems to be very limited. The end-users must be included in all steps in the public policy formulation process. Moreover, the power gaps between government and all stakeholders are amticipated to avoid monopoly of control as well as resources by certain actors.

References

[1] Agranoff, R. and McGuire, M. (2003). *Collaborative Public Management.* DCGeorgetown University Press.

[2] Ansell, C. and Gash, A. (2007). *Collaborative Governance in Theory and Practice.* Journal of Public Administration Research and Theory, 18(4), pp.543-571.

[3] Bach, D. (1999). Regionalism Versus Regional Integration: The Emergence of a New Paradigm in Africa in J. Grugel and W. Hout (eds), Regionalism Across the North-South Divide: State Strategies and Globalization. London: Rontledge.

[4] Bardach, E.(1998). Getting Agencies to Work Together: The Practice and Theory ofManagerial Craftsmanship. Washington, DC: Brookings Institution.

[5] Creswell, John W. 2014. Research Design: Qualitative, Quantitative, and Mixed Methods Approaches, Fourth Edition. Thousand Oaks – California: SAGE Publications Incorporations.

[6] Donahoe, John D & Zeckhauser, Rrichard J.(2011).Collaborative Private Roles For Public

Goals In Turbulent Times Governance. Princeton University Press.
[7] Fosler, R. Scott. (2002). Working Better Together: How Government, Business, and Nonprofit Organizations Can Achieve Public Purposes through Cross-Sector Collaboration, Alliances, and Partnerships. Washington, DC: Independent Sector.
[8] Gray, B. (1989). Collaborating: Finding Common Ground for Multiparty Problems. San Francisco: CAJossey-Bass
[9] Klijn, E.H., G.R. Teisman. (2003). Institutional and Strategic Barriers to Public-Private Partnership: an Analysis of Dutch cases. Public money and Management, 23 (3): 137-146
[10] Lewis, D. (2001). *Soul of the New Consumer.* London: Publishing.
[11] Lizzie, Crowley. (2011). *Streets Ahead: what makes a city innovative?.* The Work Foundation.
[12] O'Leary, R. and Vij, N. (2012). *Collaborative Public Management: Where Have We Been and Where Are We Going?,* The American Review of Public Administration, 42(5), pp.507-522.
[13] Osborne, S. P. (Ed.). (2010). The new public governance? Emerging perspectives on the theory and practice of public governance. London: Routledge.
[14] Majamaa, Wisa et al, (2008). The 4th P -People- in urban development based on Public-Private-People Partnerships, Helsinki University of Technology.
[15] Montanheiro, L., et al. (1998). *Public and Private Sector Partnerships: Fostering Enterprise.*Sheffield: Sheffield Hallam University Press.

IOP Conf. Series: Earth and Environmental Science **328** (2019) 012070 doi:10.1088/1755-1315/328/1/012070

Bauhäusle as a Cohousing Project

M Nasution[1] and B K Napitupulu[1]

[1]Master of Infrastructure Planning, University of Stuttgart, Germany

E-mail: bonikasih@gmail.com

Abstract. In recent years, housing prices in many countries constantly increased followed by limited availability of land. How to meet the demand for development with increasing number of population and city growth, particularly rapid development in cities, become one of main issues of housing supply. The effects lead to several problems, one of which is adequate housing with a strong social interaction. To fulfil the demand and improve social well-being in the neighbourhood can be achieved by applying cohousing concept in the neighbourhood. This study aims to analyse whether cohousing can be such a coherent solution in tackling issue of housing problems, furthermore, it is also a way to promote social inclusion. Bauhäusle is located on campus in Vaihingen, Stuttgart. It is designed with ecological construction concept which uses readily available raw materials generally donated by DIY firms, companies and also discarded materials. In disguise, the concept of Bauhäusle also initiated other similar projects for example: Construction of the student dormitory ESA in Kaiserslautern or the mound houses in Stuttgart-Hohenheim. This concept applied according to social relationship which can fulfil the aims of sustainability through an inclusive community. Method used in this research are literature reviews, on-site visitation, interview, and descriptive analysis. The results of this research is to give advantageous and disadvantageous Bauhäusle cohousing by providing stakeholder analysis, action plan, SWOT analysis, and potential period of implementation to upscale the project.

1. Introduction

Sustainability development agenda is now commonly evolved in urban development. This concept concerns to improve inclusiveness, sustainable and resilient future for development. Sustainability elements should be promoted for reducing inequalities, increasing standards of living, maintaining social development and inclusion [8]. However, how to meet the demand for development with increasing number of population and city growth, especially the rapid development in cities, become one of the main issues to implement sustainability. The effects lead to several problems such as adequate housing, basic infrastructures, transportation, and also affect to social relation. To fulfil the demand and improve social well-being in the neighbourhood can be achieved by applying cohousing concept in the neighbourhood. Cohousing is one of the solutions which offer people not only to have an access and share the ownership, but also create a strong bond and build great responsibility within a community [9]. Cohousing was established in the middle 1960s by a Denmark architect and it started to spread to other countries. This concept applied according to social relationship which can fulfil the aims of sustainability through an inclusive community.

Inclusiveness is a way to enhance participation within development processes and goals [5]. These are elements which are taken into account for enhancing inclusion.

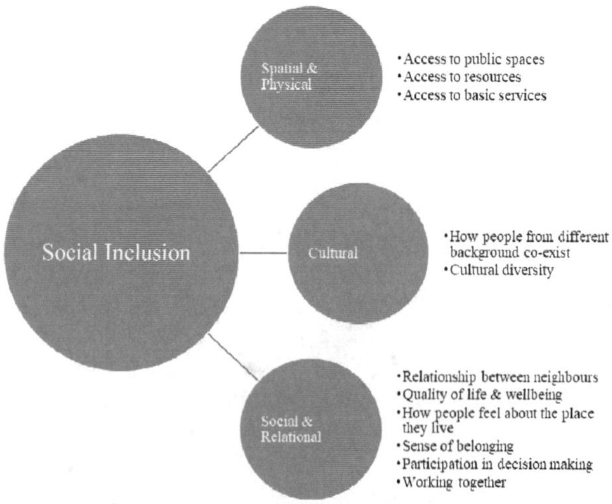

Figure 1. Elements of Social Exclusion and its Indicator. Source: DESA, 2009

According to the [9], Bauhäusle can be assumed that it is applied cohousing concept particularly refer to resident-led model. This model include the residents' involvement in designing, managing, and producing processes of the project, together with forming the community. They may get help from professionals in designing, legal forms, and technical aspects. Bauhäusle was built under the supervision of the great minds Peter Sulzer and Peter Hübner. It is the most beautiful, untraditional and at the same time the cheapest dormitory in the state of Baden-Württemberg. The construction of Bauhäusle began in summer 1981-1983. The Bauhäusle is located on campus in Vaihingen, Stuttgart. It is designed with ecological construction concept which uses readily available raw materials generally donated by DIY firms, companies and also discarded materials. In disguise, the concept of Bauhäusle also initiated other similar projects for example: Construction of the student dormitory ESA in Kaiserslautern or the mound houses in Stuttgart-Hohenheim. As the research purpose, this paper aims to analyse whether cohousing can be such a coherent solution in tackling issue of housing problems such as limited land and land price, furthermore, it is also a way to promote social inclusion.

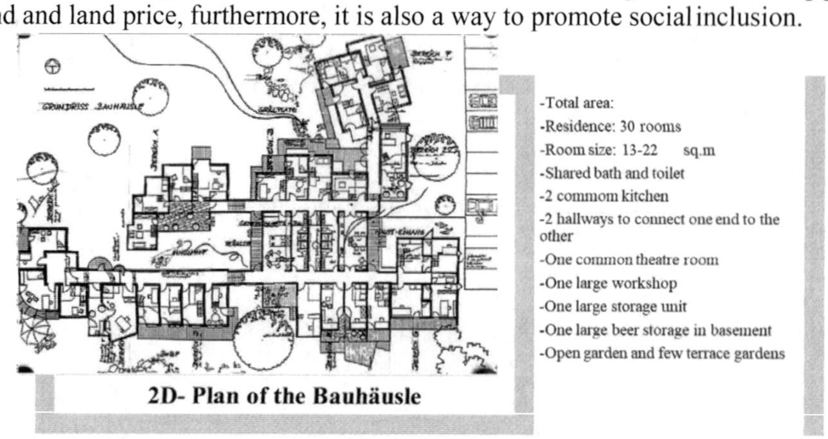

Figure 2. Bauhäusle 2D Plan, Source: Sack, 1983

2. Method

This research was conducted by doing several research methods, such are: 1) Interviewing students who live in the Bauhäusle to get information regarding activities and inclusivity in Bauhäusle, 2) literature reviews which are taken from websites and books. Those two methods then to be processed with qualitative descriptive analysis which included SWOT analysis and stakeholder mapping for discussion and conclusion.

3. Results

3.1. SWOT analysis

Inclusiveness in Bauhäusle in this research can be analysed using SWOT analysis, as a way to represent the internal and external factors to compare the ends results with the objectives.

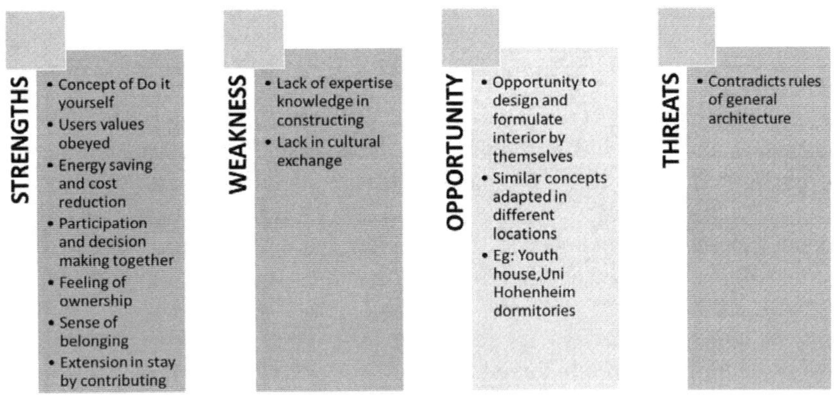

Figure 3. SWOT Analysis

Bauhäusle as a cohousing in Stuttgart has prominent strength in energy saving and cost reduction. Since it is a cohousing project, Bauhäusle promote values, sense of belonging, participation, and feeling of ownership as parts of social inclusion. Inclusiveness in Bauhäusle will increase opportunity to formulate interior design of houses by people who live inside it. On the contrary, it will contradicts rules of general architecture. In the end, it is basically depends on the dwellers in making decision of interior design. Rules of general architecture can be discarded if all dwellers feel comfortable in designing their houses as what they want.

3.2. Stakeholder Mapping

In Bauhäusle, there are several stakeholders involved in order to manage and strengthen social relation among them. Table 1 shows the roles of each stakeholders involved in Bauhäusle.

Table 1. Stakeholder Mapping

Stakeholders	Roles	Note
Students	• Conceptualization, designing, and implementing. • Managing daily needs in the cohousing. • Forming and strengthening social relationship among the students.	• Still needs professional helps to promote such activities. • The bonds and social relationship among students who live in Bauhäusle are strong. • The residents have less social relation with people outside Bauhäusle.
Hausmeister	Managing and maintaining daily needs in the cohousing.	-
Management (Studierendenwerk)	• Manager of the land and the building. • Allocating the cohousing to the tenants.	Lack of consideration to assign multicultural students in the Bauhäusle.
University department	• In the beginning of the concept, the professors were the initiator and assisted the students from designing until constructing phase of the housing project.	The university less participates to the Bauhäusle particularly from architecture and civil department to assist the students if they seek of assistance and professional help.

Private company	a) Providing raw materials which are used for constructing the cohousing and supporting the fund (in the beginning of the project)	After Bauhäusle had already constructed, the private company do not supply more the raw materials. However, if the students need materials to maintain or construct new facilities, they will find it by their own effort or it will be provided by *Studierendenwerk.*

3.3. Vision and Objectives
After analysing Bauhausle as the experimental project and refer to SWOT analysis, it is a good practice for enhancing social inclusion and idea of cohousing. The vision is to upscale the idea and concepts of sharing and cohousing such as sharing space, facilities, responsibility, and activities to a community led housing. This idea will bring more cohesion by creating social links and networks within a community, empowering people through engagement in the decision making and led residents to be more active, creative and participative regarding their communal space and society. To achieve the vision, the project focuses on strategies with two main objectives and sub-objectives.

1. Enhancing Inclusiveness

a) Inclusive design and support shared facilities
- Creating design concept to promote inclusiveness
- Including user involvement in design process
- Creating common sport centres and Green space/ open space
- Collaborating with private company which can supply raw materials for construction and maintenance of the cohousing project.
- Strengthening the relationship and interaction of the community

b) Promote cultural diversity
- Organizing more events and parties to strengthen the residence bond.
- Promoting cultural activity and intergenerational meetings.
- Forming a community association within the community

c) Strengthening participation and community engagement in planning and decision-making process.
- Demonstrating visible leadership through associations within the community to provide equality, diversity, and inclusion matters.
- Ensuring everyone understand their personal responsibilities to promote equality and inclusive practice and remove any unfair barriers.

2. Promote Cohousing

a) Introducing new stakeholders and strengthening the coordination among stakeholders
- Involving influential parties
- Holding stakeholder's meetings in a regular basis
- Improving partnership (private public partnership)

b) Regulating housing policy
- Introducing Subsidies and incentives

3.4. Action Plan
Based on the objectives and strategies, an action plan is required to measure the respective indicators of each objective. An action plan is necessary to provide an overview of the implementation in order to achieve the vision.

Table 2. Stakeholder Mapping

Wider Objective:	Indicators of Progress	Indicator Measurements	Assumptions and Risks
Enhancing sharing concept in Bauhäusle	• Shared facilities and common areas, such as common sport centres and green space/ open space. • Increasing social inclusion.	• Number of common space survey • Interviewing Stakeholders	Residents are interested in sharing concept
Specific Project Objective:	Indicators of Progress:	Indicator Measurements	Assumptions and Risks
Improving relationship among stakeholders as influential parties to the project.	• Meetings and discussions • Coordination among stakeholders • Sharing responsibilities • Stakeholders participation	• Number of meetings and discussion • Number of stakeholders who participate • Interview each stakeholders regarding their responsibilities	Stakeholders participate in discussion and sharing responsibilities
Specific Project Objective:	Indicators of Progress:	Indicator Measurements	Assumptions and Risks
Strengthening the interaction between the residents	Organising common events	Number of events organized (Parties, Cultural events, seminars/workshops, Sport events)	Residents participate in the events
Output	Indicators of Output	Output	Assumptions and Risks
• **Community association** • **Activity centre** • **Common facilities**	• Members of association • Physical structure • Shared facilities	• The number of members • Building design and the physical structure • The number of shared facilities	• Resident representatives engage in association • Privacy concern
Outcomes	Indicators of Outcomes	Outcomes	Assumptions and Risks
• **Sharing ownership** • **Sense of belonging** • **Volunteers in taking initiatives for betterment**	• Responsibility • Maintenance system	• Interview the residents regarding the responsibility and maintenance system • The number of resident who participate and initiate for betterment	Volunteers are interested in taking part in the project
Activities	Input	Indicator Measurements	Assumptions and Risks
• **Physical structures** • **Community events**	• Community agreement • Funds for construction • Sponsors for conducting community events. • Involvement from institution for learning events	• Number of funds • Number of sponsorships • Number of participants and institution • Number of events conducted	Private companies support the concept of cohousing

4. Conclusion

Bauhäusle set itself as a good example for a co-housing project. Based on the current trend of sharing concepts and the SWOT analysis, it leads to an opportunity to upscale the project on higher grounds like community led housings. SWOT, stakeholder mapping, vision and objectives clearly shown how cohousing project can be possibly applied to cities or region which have problems with limited land and land price. It can also be applied to refugees housing or *Studentenwohnheim (*students' *dorm).* To upscale the project, it needs strategies which can enhance inclusiveness, for example providing more public facilities and community events, then, most importantly promote the project by involving more stakeholders which has more influencing power, but not forgetting people's participation in making decision towards the cohousing, since they are the person who will live in it. In a nutshell, projects like community led housing is a key influential factor in the growing economy. However, this research has a limitation regarding growing economy in Bauhäusle. Nevertheless, with rapid urbanization, such projects can overcome scarcity of housing issues and likely improve social inclusion to have a sustainable society.

In such a contribution towards Indonesia, this concept might be applicable to cities which have land issues but in need to provide the inhabitants places to live. Government, local actors, stakeholders, and citizens should participate in arranging policy to ease people in getting houses with affordable price and can still maintain social relationship to each citizens. Government and stakeholders could promote available land which can be offered to create cohousing and targeted specific citizen who are willingly to live in a cohousing. In regards of criminality, to reduce the risk of criminality, citizen who are motivated to be in part of cohousing should provide full information about themselves and give it to developers and other citizens who want to participate in the same cohousing, so they will know each other. This is such a simple solution to reduce criminality internally. They can also provide security guard if it is necessary.

Finally, at the end of this paper, we would like to thank University of Stuttgart and *Lembaga Pengelola Dana Pendidikan (*Indonesia Endowment Fund*)* in helping us to realize this paper. We hope this paper could be used as development towards cohousing in Indonesia.

5. References

[1] Bauhäusle. (2018). *Am Anfang: Hör Mal Wer Da Hämmert*. Retrieved from Bauhäusle: http://www.bauhaeusle.de/about.html

[2] Colantonio, A. (2009). Social Sustainability: A Review and Critique of Traditional Versus Emerging Themes and Assessment Methods. In A. P. M. Horner, *Second International Conference on Whole Life Urban Sustainability and its Assessment* (pp. 865-885). Loughborough: SUE-MoT.

[3] DESA. (2009). *Creating an Inclusive Society: Practical Strategies to Promote Social Integration.*

[4] Emas, R. (2015). *The Concept of Sustainable Development: Definition and Defining.* Florida: Florida International University.

[5] Joyeeta Gupta, C. V. (2016). Sustainable development goals and inclusive development. *Int Environ Agreements*, 433-448.

[6] Morelli, J. (2011). Environmental Sustainability: A Definition for Environmental Professionals. *Journal of Environmental Sustainability*, 1-9.

[7] Natasha Palich, A. E. (2013). Social Sustainability: Creating Places and Participatory Processes that Perform Well for People. *Environment Design Guide*.

[8] Nation, U. (2018, May 31). *United Nation*. Retrieved from United Nation: https://www.un.org/sustainabledevelopment/development-agenda/

[9] Norhaslina Jumadi, N. A. (2016). An Introspective view of Sustainable Cohousing with The Malaysian Malaysian Housing Concept. *MATEC Web of Conferences 66*, 1-9.

[10] Sack, M. (1983, June 24). *Zeit Online*. Retrieved from Zeit Online: https://www.zeit.de/1983/26/mit-dem-kopf-und-mit-den-haenden

[11] *Zeit Online*. (1992, April 3). Retrieved from Zeit Online: https://www.zeit.de/1992/15/meister-unordentlicher-haeuser

Impacts of Urban Consolidation Centres for Sustainable City Logistics Using Adaptive Dynamic Programming Based Multi-Agent Simulation

N Firdausiyah[1]*, E Taniguchi[2], A G Qureshi[3]

[1]Urban and Regional Planning Department, Brawijaya University, Malang
[2]Resilience Research Unit, Kyoto University, Japan
[3]Department of Urban Management, Kyoto University, Japan

*email: firdausiyah@ub.ac.id

Abstract. This paper aims at evaluating the impacts of Urban Consolidation Centers (UCC) for sustainable city logistics using Adaptive Dynamic Programming (ADP) based multi-agent simulation (MAS-ADP). Economic efficiency and environment friendliness criteria were used to evaluate the sustainability of UCC. The results proved that the implementation of UCC as a sustainable city logistics scheme is efficient in reducing 8% of the total delivery cost for freight carrier and reducing 36% of the total emissions released to the environment. It is also showed that the use of learning agents is essential to demonstrate the successful implementation of the UCC, as it is only in the learning-based simulation, UCC operator could get a profit. Our simulation analysis also confirmed that compared to widely used reinforced learning algorithms (Q-learning), MAS-ADP brought increased accuracy to the evaluations' outcomes of UCC.
Keywords: sustainability, city logistics, urban consolidation centre, adaptive dynamic programming, multi-agent system

1. Introduction

Sustainable city logistics has become an important issue in urban and transportation planning due to high population density in urban areas as well as due to the social, economic, and environmental problems associated with it. City logistics is defined as the process of fully optimizing the logistics and transport activities with the support of advanced information systems in urban areas considering the traffic environment, the traffic congestion, the traffic safety, and the energy savings within the framework of a market economy [1]. The harmonization of economic efficiency and environmental friendliness in city logistics is essential for ensuring sustainable development in urban areas [2], which faces two difficult problems. First is the efficiency of goods delivery within the uncertain environment (due to the parking issues, traffic congestion, and other restrictions in the urban area) that directly effects the operational cost as well as the action selection in presence of optional solutions or policies. The second issue is the involvement of multiple agents in city logistics system, such as freight carriers, shippers, customers, and administrator. All these key stakeholders in urban freight transport have their own specific objectives and tend to behave in a different manner to any urban freight policy [3]. These stakeholders also interact and influence each other in the city logistics environment, which makes the environment unpredictable. Therefore, the main challenge for the city logistics is to provide a sustainable urban freight transportation while considering multi-agent problems within the uncertain environment.

In order to achieve these aims, numerous city logistics initiatives have been proposed and implemented in several cities, including the Urban Consolidation Centers (UCC) [4]. It is important to evaluate the city logistics policies before they can be effectively deployed due to their manifold implications on different city logistics stakeholders [5]. For that purpose, decision support tools (DST) are needed to help public decision makers and practitioners make decisions acceptable to all parties. These DSTs are mainly based on modeling, optimization, simulation, and evaluation procedures.

There have been many attempts to develop multi-agent simulations to analyze decision-making process of various stakeholders in city logistics, but almost all of them rely on Q-learning [6], [7], [8]. However, based on previous research experiences, which will be described in more detail in the next section, it has been found that ADP-based learning performed better in the accuracy of the outcomes when agents need to interact in uncertain environment, such as city logistics. Therefore, in order to have an accurate evaluation of the UCC, an ADP-based multi-agent simulation has also been developed, which can be used as a DST to achieve better outcomes in the decision process of designing and implementing sustainable city logistics policies.

1.1. Evaluation models for evaluating city logistics measures

Multi-agent systems (MAS) based on the reinforced learning (RL) algorithms have been used for evaluating the behavior of stakeholders, who are affected by the implementation of a city logistics policy. In MAS environment, multiple agents come together and interact, cooperate, coordinate, and negotiate with each other to reach their intended objectives. Various other city logistics policies have been evaluated using MAS with Q-learning such as load factor control and road pricing [6], e-Commerce [7], truck ban and motorway toll discounts [8]. These researches used Q-learning to model evolving behavior of the key stakeholders, namely the carriers, shippers, administrator, and residents relating to urban freight transport. The MAS with Q-learning algorithms have also been used to evaluate the dynamic usage of UCC [9]. The results indicated that the main policy measures that contributed to the successful functioning of the UCC are road pricing, operational subsidies, and the application of time windows. Another study evaluated the Joint Delivery System (JDS) with parking restrictions using MAS with Q-learning [10] and the results showed that JDS with UCC and car parking management have the potential for improving environmental issues related with the urban freight. The results indicated that implementing a truck ban in environmentally damaged areas and discounting motorway tolls in the urban motorway network will have a large environmental impact, resulting in an acceptable environment for all stakeholders.

It can be observed that most of the MAS research in city logistics use Q-learning to represent the decision making of the agents. Some comparative studies modeled the freight carriers and UCC operator behavior in JDS with UCC using on policy algorithm named MAS-ADP based RL [11], [12]. Their result showed that MAS-ADP based RL have better capability than MAS-Q-learning to adjust and adapt the uncertain environment, thus improving the accuracy of decision-making. This study, therefore, would like to apply the MAS-ADP based RL for evaluating the impacts of JDS with UCC for sustainable city logistics.

1.2. MAS-ADP for evaluating city logistics measures

ADP is a learning model in RL part that can be used in the simulation field and optimal control field. As an optimal control tool, [13] has described that ADP scheme is suitable for applications to the systems with strong coupling, strong nonlinearity, and high complexity. It has also been concluded that the ADP is capable to deal with uncertainty [14]. ADP has been widely implemented at the confluence of control problem (15), intelligence traffic systems [16], and robotics [17]. However, none of these previous researches has used the ADP in the multi-agent simulation field, particularly in the area of city logistics, which represents a highly uncertain environment. Therefore, this study will use the ADP based multi-agent simulation (MAS-ADP) to evaluate the UCC for sustainable city logistics.

2. Methods

2.1. Objective of the learning agents

Recognizing agent's objective is important in modeling its learning behavior. The freight carrier's objective is minimizing the total cost of delivering goods to customers. These costs include fixed vehicle utilization cost, the travel cost, and the penalty cost, which are calculated by the Vehicle Routing Problem with Soft Time Windows (VRPSTW) [18].

The UCC operator has been considered as a private or public company that consolidates and delivers the goods from the UCC to customers. Therefore, the UCC operator's objective is to maximize the profit, which is the difference between revenue (obtained by multiplying the UCC fee with the total demand (parcels) that the freight carrier gives to the UCC) and delivery cost (calculated by the VRPSTW).

2.2. Actions and rewards of the learning agents

The freight carrier can choose two possible actions, viz., direct delivery (DD) or Joint Delivery System (JDS). The operational cost and the total additional parking cost are the immediate rewards for a freight carrier associated with DD action. While UCC fee is the consequence of choosing the JDS. The second learning agent, the UCC operator has three options of managing the UCC fee, which are price going up, flat price, and price going down. The UCC operator will receive the immediate profits associated with the selected action.

2.3. Model for the learning agents

This research applied the Multi-agent-, Simulation-Adaptive-, Dynamic-Programming-based Reinforcement Learning (MAS-ADP based RL) to evaluate the impacts of Joint Delivery Systems (JDS) measured with Urban Consolidation Center. For more details on the MAS-ADP based RL formulation, readers are referred to Firdausiyah et al., [11], [12]. As mentioned earlier, Q-learning-based MAS has been extensively used in the evaluation of the city logistics policies (including the UCC) and because one of the objectives of this research is to come up with a more accurate evaluation tool (i.e. the ADP-based MAS), this research compared the MAS-ADP with the Q-learning [19].

In addition, this research used the VRPTW (Vehicle routing problem with soft-time window) model to calculate the transportation cost either for freight carriers or for UCC operator. For more details on the VRPSTW formulation and solution algorithms, readers are referred to Qureshi et al., [18].

2.4. Environmental emissions model

To evaluate the benefit of using UCC for the environment, this research calculated the emissions using qualitative appraisal index for the basic unit of carbon dioxide (CO_2), oxides of nitrogen (NO_x) and suspended particulate matter (SPM), produced by the freight trucks. For more details on the formulation, readers are referred to NILIM [20].

3. Results and Discussions

A square topology-based, hypothetical network (Figure 1) was used for the evaluation of the UCC based on the simulations using ADP and Q-learning models within MAS. Four carriers (A, B, C, D), one UCC, and 20 customers were involved in this network. The agents will make decisions as actions by considering the fluctuating parking cost, UCC fees, and demand received from their customer every day.

IOP Conf. Series: Earth and Environmental Science **328** (2019) 012071 doi:10.1088/1755-1315/328/1/012071

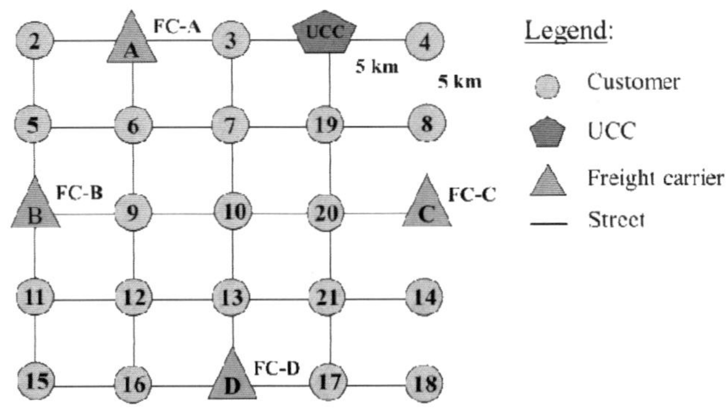

Figure 1. Test road network

This research used some assumptions as listed in **Table 1**;

Table 1. Simulation Assumptions

Item	Value
Working time	8 AM to 8 PM
Time window	60 minutes per customer
Capacity of the truck	200 parcels/ truck
Waiting charge (c_e) for early arrival	I Yen/ minute
Penalty charge (c_l) for late arrival	5 Yen/ minute

All simulations are done using MATLAB. The learning rate and the discount factor for ADP used were 0.2 and 0.6, respectively; whereas, the learning rate and the discount factor for Q-learning have been set as 0.2 and 0.8, respectively. These values were based on the results of a sensitivity analysis that has been done prior to the case study. In this study, this research performed two separate simulations using ADP and Q-learning to evaluate the impacts of UCC for sustainable city logistics using the following criteria; 1) economics' efficiency for each learning agent, i.e., cost saving for the freight carrier, and profitability for the UCC operator; 2) environmentally friendliness.

3.1. *Accuracy of the learning models*

Accuracy of the outcomes obtained in the MAS-ADP and MAS-Q-learning is important for the learning models to evaluate a sustainable city logistics scheme (such as UCC in this study). The "accuracy" refers to the closeness of the gap between the expected value obtained in the MAS-ADP and MAS-Q-learning based simulations to the corresponding value experienced by the learning agent. Smaller gap between these two costs means more accurate method.

Figure 2(a) shows that the percentage gap between expected cost and the experienced cost in the ADP-based simulation is lower (39.6%) than the Q-learning based simulation (47.7%) for freight carrier A; similar pattern were obtained for freight carriers B, C, and D.

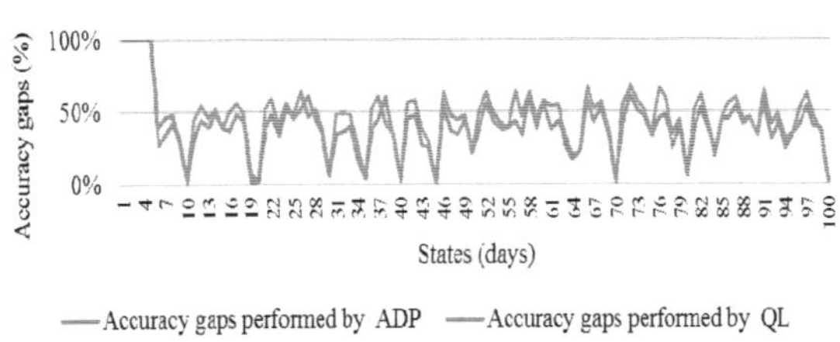

(a) Accuracy gaps (%) in case of freight carrier A

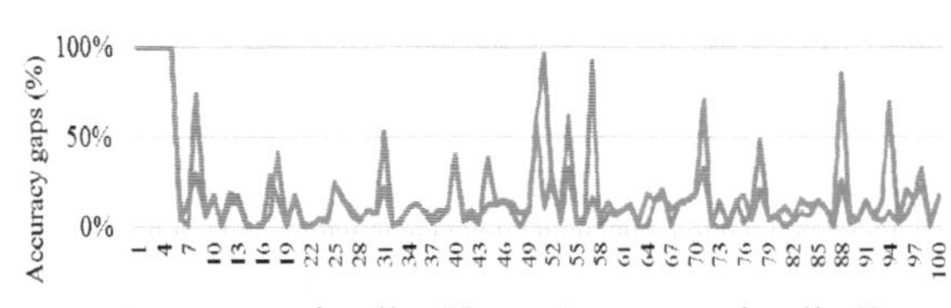

(b) Accuracy gaps (%) in case of UCC operator

Figure 2. Accuracy gaps (%) in ADP and Q-learning based simulations

Similarly, in the case of the UCC operator, the percentage gap between the expected profit and the experienced profits in the ADP-based simulation was also lower (46.4%) than the Q-learning-based simulation (51.9%) as shown in **Figure 2**(b). It proved that the ADP-based learning can improve the accuracy of the simulation, thereby improving the quality of simulation.

3.2. Sustainability criteria: economics' efficiency

3.2.1. Freight carriers

The meaning of efficiency for freight carrier is the delivery of goods at lower cost to the customers. To calculate the efficiency for freight carrier, this research compared the difference in the experienced cost for the freight carrier with UCC and without UCC. The existence of UCC provided more alternatives of goods' delivery for freight carrier, which are direct delivery, and JDS with UCC. The MAS-ADP and MAS-Q-learning as the learning models will suggest the actions based on the reward received from the environment.

Both ADP and Q-learning-based simulation resulted in a lower delivery costs experienced by freight carriers in the case of UCC than the one without UCC case (Figure 3(a)). The delivery cost with UCC resulted from the fact that ADP-based simulation was 8.4% lower on average compared to the experienced delivery cost without UCC. Corresponding figures for the Q-learning-based simulation was 6.7%. Figure 3(b) illustrates more details with the cumulative experienced delivery cost for a freight along the simulation. It means that implementing the UCC as a sustainable city logistics policy is efficient to minimize the delivery costs for a freight carrier. Moreover, using ADP as the freight carriers' behavior learning model is better than using the Q-learning, as the former choice can further save almost 1.7%, on average, of the total delivery costs; thereby improving the quality of the city logistics simulation i.e. ADP-based simulation provided more favorable comparison with the no-policy (without UCC) case.

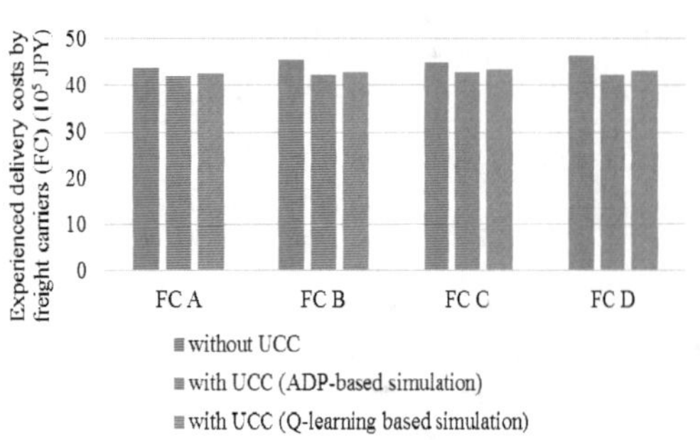

(a) Experienced delivery costs for freight carriers (FC) with UCC and without UCC

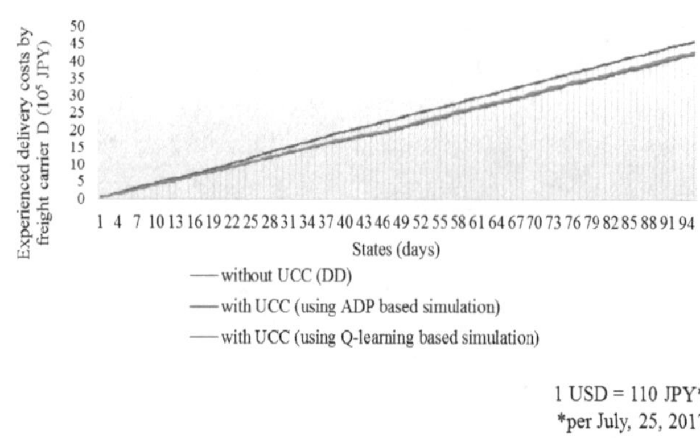

(b) Simulation results of cumulative experienced delivery cost for the freight carrier D

Figure 3. Experienced delivery costs for freight carriers (FC)

3.2.2. UCC operator

The economic efficiency for UCC operator meant higher profits at lower UCC fee offered to freight carriers to foster more demand. To calculate the profitability, we compared the difference in the experienced profits for UCC operator under the learning environment with freight carrier and without learning. In the learning environment, we assumed the UCC operator will update the UCC fee every day by learning the reward received (profits) based on the business provided by the freight carriers. Without learning (α=0), the UCC operator will not consider the current information from the environment. Therefore, the UCC operator is assumed to offer fixed UCC fee (150 JPY/ parcel) to the freight carrier every day.

Figure 4 clearly shows that the UCC would fail dramatically if a fixed UCC fee policy is followed without learning from the environment, as the cumulative experienced profit received by the UCC operator without learning is way lower than the cumulative experienced profits resulted from learning using either of ADP or Q-learning. A dip in the cumulative experienced profit curve shows a negative profit obtained in that episode. The negative profit means the UCC failed to cover the downstream delivery cost based on the business (demand) received from the freight carriers. If the UCC operator is modelled as a learning agent, it learns from these negative reward values to adjust the UCC fee (may be able to attract more demand) and becomes profitable again. It showed the importance of the MAS-based simulations in the evaluation of the city logistics policies where a stakeholder's action/behavior can seriously impact the other. It is important for the UCC operator to learn from the behavior of the freight carriers (refusing to join UCC due to high fees) in order to become profitable.

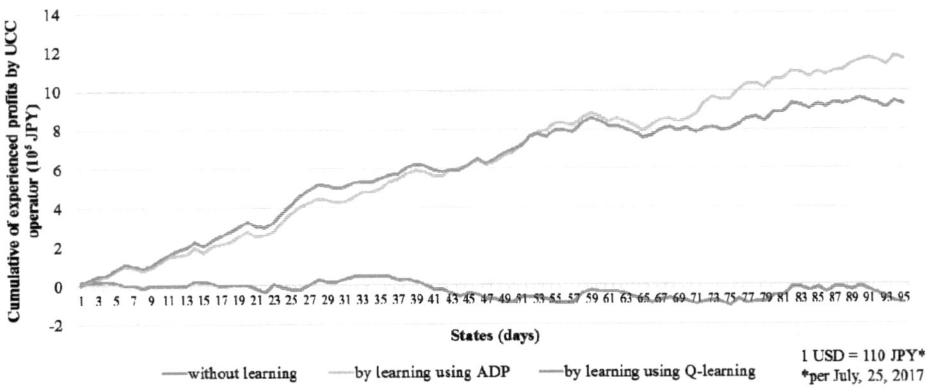

Figure 4. Experienced profits with and without learning by the UCC operator

Moreover, **Figure 5** shows that using ADP as the UCC operator's behavior learning model is better than the Q-learning. The two simulations suggest different actions of managing the UCC fee level by the UCC operator, which resulted in the difference of the experienced profits. The UCC fee suggested by the ADP is always 7% lower, on average, than the UCC fee suggested by Q-learning (**Figure 5**(a)). It increases the possibility of choosing JDS with UCC by freight carrier than the direct delivery under the ADP-based learning. The impact is also evident in the profits received by UCC operator, which were 3.7% when using the UCC fee suggested by ADP (130 JPY/ parcel on the average) as compared to 2.1% when using the Q-learning (140 JPY/ parcel on the average) (**Figure 5**(b)).

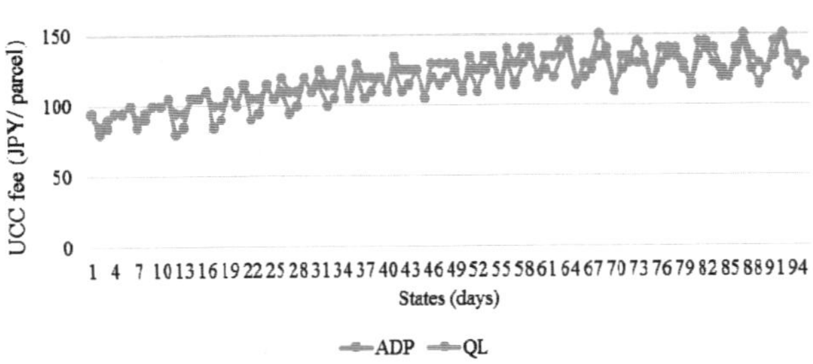

(a) UCC fee per parcel offered by UCC operator

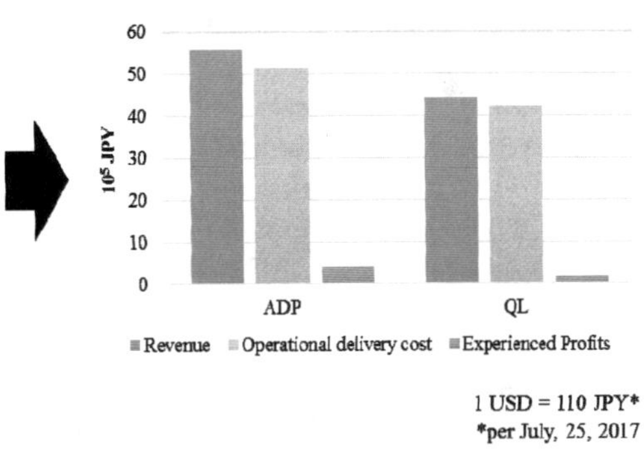

(b) experienced profits received by UCC operator

Figure 5. UCC fee per parcel (a) and experienced profits by UCC operator (b)

IOP Conf. Series: Earth and Environmental Science **328** (2019) 012071 doi:10.1088/1755-1315/328/1/012071

3.3. Sustainability criteria: environmentally friendliness

We evaluated the impacts of UCC by calculating the total emissions (CO_2, NO_x, and SPM) from the delivery activities made by freight carriers and the UCC operator with and without UCC using ADP and Q-learning. The existence of UCC will reduce 36% (ADP) and 31% (Q-learning) of the total emissions as compared to the condition without UCC (**Figure 6**). As explained earlier, the differences of the result between these two simulations arose from the different action selection suggested by the learning model. The experienced emission level of CO_2 (**Figure 6**(a)), NO_x (**Figure 6**(b)), and SMP (**Figure 6**(c)), obtained under ADP-based simulation is 5% lower than the Q-learning based simulation. It means that using ADP as the learning model for both agents is better than using the Q-learning, as the former choice can reduce 36% of the total emissions released to the environment; thereby improving the quality of the city logistics simulation, i.e. ADP-based simulation provides better comparison with the no-policy (without UCC) case.

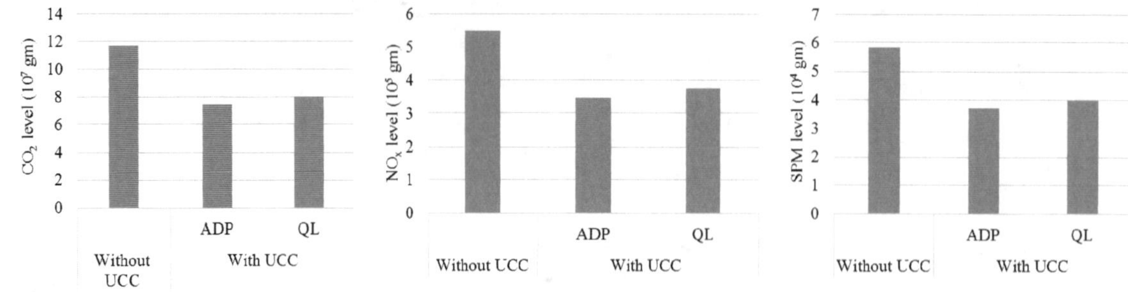

(b) NO_X emissions (c) SPM emissions

Figure 6. Total emissions for 24 episodes (120 days) of simulation

4. Conclusions

This paper developed the ADP models to evaluate the UCC as a sustainable city logistics policy. Economic efficiency and environmentally friendliness criteria were used to evaluate the sustainability of UCC. The results showed that the implementation of UCC as a sustainable city logistics scheme is efficient by reducing 8% of the total delivery cost for the freight carrier and reducing 36% of the total emissions released to the environment. It was also showed that the use of learning agents is essential to demonstrate the successful implementation of the UCC, as it is capable to increase the experienced profit gained.

In addition, simulations should accommodate the agent's objective. It was observed that simulation using ADP resulted in a further 1.7% less experienced cost as compared to the simulation done using Q-learning. In case of UCC operator, the ADP satisfied its objective by getting higher profits than the Q-learning based simulation. The differences of the results between these two simulations arise from the different action selection suggested by the learning model. Therefore, the accuracy of the outcomes is also very important for the learning models, especially within the uncertain environment. It was found that ADP-based simulation improved the accuracy of the expected delivery costs for the freight carrier compared to the Q-learning based simulation. The accuracy of the expected profits received by the UCC operator in the ADP-based simulation was also much better.

References

[1] Taniguchi, E., and Thompson, R. G. City Logistics I. Institute of Systems Science Research. Kyoto., Japan, 1999.

[2] Taniguchi, E., Thompson, R. G., Yamada, T. Logistics systems for sustainable cities: Visions for city logistics. Elsevier Ltd., UK, 2004.

[3] Taniguchi, E. & Tamagawa, D. Evaluating city logistics measures considering the behavior of several stakeholders. Journal of the Eastern Asia Society for Transportation Studies, Volume 6, 2005, pp. 3062-3076

[4] Browne, M., Allen, J., & Leonardi, J. Evaluating the use of an urban consolidation centre and electric vehicles in central London. IATSS Research 35, 2011, Page 1-6.

[5] Taniguchi, E., and Thompson, Russell G. City Logistics: Mapping the Future. CRC Press Taylor and Francis Group., UK, 2015.

[6] Teo, J. S. E., Taniguchi, E. & Qureshi, A. G Evaluation of Load Factor Control and Urban Freight Road Pricing Joint Schemes with Multi-agent Systems Learning Models. Procedia Social and Behavioral Sciences, vol. 125, 2014, pp. 62 – 74.

[7] Teo, J. S. E., Taniguchi, E., and Qureshi, A. G. Evaluation of Distance-Based and Cordon-Based Urban Freight Road Pricing in E-Commerce Environment with Multiagent Model. In Transportation Research Record: Journal of the Transportation Research Board, No. 2269, Transportation Research Board of the National Academies, Washington, D.C., 2012, pp. 127–134.

[8] Tamagawa, D., Taniguchi, E., and Yamada, T. Evaluating City Logistics Measures Using a Multi-agent Model. Procedia Social and Behavioral Sciences, Vol. 2, 2010, pp. 6002–6012.

[9] van Duin, J. H. R., van Kolck, A., Anand, N. and Tavasszy, L. A. Towards an agent-based modeling approach for the evaluation of dynamic usage of urban distribution centers. Procedia Social and Behavioral Sciences, Volume 39, 2012, pp. 333-348.

[10] Wangapisit, O., Taniguchi, E., Teo, J. S. E. & Qureshi, A. G. Multi-Agent systems modelling for Evaluating Joint Delivery Systems. Procedia-Social and Behavior Sciences, Volume 125C, 2014, pp. 472-483.

[11] Firdausiyah, N., Taniguchi, E., and Qureshi, A. G., (2017). Multi-agent Simulation using Adaptive Dynamic Programming in The Existence of Urban Consolidation Centers. *Journal of Japan Society of Civil Engineers, Ser. D3 (Infrastructure Planning and Management)*, 73(5), I_835-I_846.

[12] Firdausiyah, N., Taniguchi, E., Qureshi, A. G., (2019). Modeling City Logistics using Adaptive Dynamic Programming Based Multi-Agent Simulation. *Transportation Research Part E: Logistics and Transportation Review*. 125, 2019, pp. 74-96.

[13] Zhang, Hua-Guang, Xin Zhang, Yan-Hong Luo, and Jun Yang. An overview of research on adaptive dynamic programming. Acta Automatica Sinica, volume 39(4), 2013, pp. 303–311.

[14] Godfrey, G. & Powell, W. B. An adaptive, dynamic programming algorithm for stochastic resource allocation problems I: Single period travel times. Transportation Science, volume 36(1), 2002, pp. 21–39.

[15] Venayagamoorthy G K, Harley R G, Wunsch D C. Dual heuristic programming excitation neuro control for generators in a multimachine power system. IEEE Transactions on Industry Applications, volume 39(2), 2003, pp. 382–394.

[16] Fagan, D., and Meier, R. Dynamic Multi-Agent Reinforcement Learning for Control Optimization. Fifth International Conference on Intelligent System, Modelling and Simulation, 2014.

[17] Karthikeyan, R., SheelaRani, B., and Renganathan, K. An Instant Path Planning Algorithm for Indoor Mobile A robot Using Adaptive Dynamic Programming and Reinforcement Learning. International Journal of Engineering and Technology, volume 6, no.2, 2014, pp. 1224-1231.

[18] Qureshi, A.G., Taniguchi, E., and Yamada, T. Hybrid insertion heuristics for vehicle routing problem with soft time windows. Journal of the Eastern Asia Society for Transportation Studies, Volume 8, 2010, pp. 827–841.

[19] Watkins, C.J.C.H and Dayyan, P. Q.Learning. Machine Learning, 8, 1992, pp. 279-292.

[20] NILIM. Qualitative appraisal index calculations used for basic unit for computation of CO2, NOx and SPM (in Japanese), 2003.

IOP Conf. Series: Earth and Environmental Science **328** (2019) 012072 doi:10.1088/1755-1315/328/1/012072

Carrying Capacity of Grompol Watershed Settlements, Central Java

Nasrudin[1], C Muryani[2], Y Yusuf[3]

[1]Master Program of Geography Education of Studies Sebelas Maret University, Surakarta, Indonesia
[2]Departement of Geography Education, Sebelas Maret University, Surakarta, Indonesia
[3]Departement of Geography Education, Sebelas Maret University, Surakarta, Indonesia

*Email: nazruroeddy33@gmail.com

Abstract. Settlements within the watershed will be damaged by natural disasters if the carrying capacity has been exceeded, so research is needed to determine the limits of the carrying capacity of settlements in the watershed so that natural disasters can be minimized. The purpose of this study is to determine the limit of the carrying capacity for settlements in the Grompol watershed. Data on rainfall, soil type, and contours were collected from the BPDAS Solo and BIG. The data obtained were then analyzed into rainfall intensity using the Spline method, erodibility based on BPN soil types and slope classes using the Surface Analysis with Slope method, then the three criteria were overlaid based on Decree of Minister of Agriculture Number 837 / KPTS / UM / II / 1980, so that the region use of the settlement area can be obtained. After that, the settlement suited region will be analysed based on disaster-prone areas. Settlement carrying capacity is calculated from the settlement suited region divided by population and space / capita needs. The result of calculation of settlement carrying capacity is obtained DDPm = 1.23. This condition shows that the carrying capacity for settlements is still high and can support the growth of settlements.

1. Introduction

Watershed is an ecosystem unit- where organisms and their environment interact dynamically and have a dependence on each other in their composition [1]. Watershed becomes an area whose spatial activities must be managed thoroughly in accordance with the principle of sustainable development based on environmental carrying capacity, so as not to experience land degradation. The problem of watershed land management starts from land use that is not in accordance with the carrying capacity of the land and is followed by the application of cultivation techniques that are not in accordance with conservation rules, thus causing degradation of watershed land.

Development must be carried out optimally according to the carrying capacity of the environment, namely by processing land that is comparable to environmental capabilities. The ability of the environment to support human life, living things, and balance between them are called environmental carrying capacity. Regional development and spatial planning need to examine the carrying capacity of the environment [2]. Study of environmental carrying capacity is very important to do to avoid environmental damage.

1.1 Literature Review

Carrying capacity has a great value as a concept around which to organize development. The very nature of carrying capacity requires long-range planning, a feature so notably lacking from most of the development plans laid for rainforest areas [3]. Development of settlements must indeed be done, but there is a need for spatial planning that is able to manage the environment so that it goes hand in hand. Spatial planning is done to get the benefits of available resources- optimally by not ignoring environmental sustainability and the intended use does not exceed the carrying capacity of the environment [4].

The quality of the environment must be maintained properly. A settlement must take into account the existence of nature, 1/3 - 2/3 of all spatial plans to be managed or changed, must be allowed to develop naturally [4].

2. Methods

2.1. Region Use Analysis

This analysis was conducted to find out how much the area functions to be able to support efforts to utilize settlements. The analysis begins analyzing the limiting aspects of regional functions based on the Decree of Minister of Agriculture Number 837 / KPTS / UM / II / 1980, namely slope class, soil sensitivity, and rain intensity.

The three criteria above are slope class, soil sensitivity, and rainfall intensity overlay analysis and total score scoring using GIS. A total score of more than 175 is a protected area, the total score between 125-17 has a buffer zone function, a total score of less than 125 with a slope of less than 15% is an area of annual crop cultivation, and a total score of less than 125 with a slope of less than 8% is a function of seasonal crop cultivation and settlement areas. Thus, the distribution of regional functions in the research area is obtained.

2.1.1. Slope Class Analysis. Analysis was carried out by means of contour map processing obtained from Geospasial Information Agency (*Badan Informasi Gespasial*) or (BIG) using Topo to Raster extensions, and it was then analyzed using the Surface Analysis with Slope in ArcGIS.

2.1.2. Erodibility Analysis. The analysis was carried out by determining soil sensitivity based on BPN soil type data obtained from Solo BPDAS.

2.1.3. Rainfall Intensity Analysis. The analysis was carried out by processing rainfall data obtained from Water and Human Resources Agency Karanganyar and Sragen using the Interpolation Spline extension in ArcGIS.

2.2. Settlement Suited Region Analysis

This analysis shows the ability of land in the study area for settlements, taking into account the criteria of the function of the area and being prone to disasters. The analysis is carried out by overlaying the two criteria and calculations using the formula :

$$LPm = LW - (LKL + LKP + LKBTT + LKBTS + LKRB)$$

Note :

LPm = Settlement suited region
LW = Area of the watershed
LKL = Protected area
LKP = Buffer zone
LKBTT = Annual crop cultivation area
LKBTS = Seasonal crop cultivation area
LKRB = Disaster-prone area

2.3. Settlement Carrying Capacity Analysis

This analysis is intended to find out or provide information on the size of land resource capacity that is still available and can be utilized for settlements. The analysis is done by the formula:

$$DDPm = \frac{LPm / Jp}{\alpha}$$

Note :
DDPm = Settlement carrying capacity
LPm = Settlement suited region
Jp = Total population
α = Coefficient of space need per capita (m^2/ capita)

3. Result

Slope class classification based on the Decree of The Minister of Agriculture Number 837 / KPTS / UM / II / 1980 and the following results are obtained:

Table 1. Slope Class of Grompol Watershed

No	Slope Class Classification	Information	Area (ha)	%
1.	0-8	Flat	15,123.56	84.80
2.	8-15	Sloping	1,376.88	7.72
3.	15-25	Slightly Steep	540.52	3.03
4.	25-40	Steep	596.18	3.34
5.	>40	Very Steep	197.78	1.11
	Total		17.834,90	100

Soil sensitivity classification in the study area consisted of insensitivity, a little sensitive, insensitive, sensitive and very sensitive :

Table 2. Soil Type of Grompol Watershed

No	Soil Type Group	Information	Area (ha)	%
1.	Alluvial	Insensitivity	3,176.26	17.81
2.	Latosol	a Little Sensitive	4,851.13	27.20
3.	Mediteran	Insensitive	4,936.07	27.67
4.	Grumosol + Andosol	Sensitive	4,870.36	27.30
5.	Andosol + Litosol	Very Sensitive	2.97	0.02
	Total		17,834.90	100

Based on the recording of rainfall observation stations, Grompol watershed can be divided into five rain fall intensity classifications :

Table 3. Rainfall Intensity of Grompol Watershed

No	Classification of Rainfall Intensity (mm / rainy day)	Information	Area (ha)	%
1.	0 – 13,6	Very Low	1,817.94	10
2.	13,6 – 20,7	Low	2,504.41	14
3.	20,7 – 27,7	Medium	8,866.13	50
4.	27,7 – 34,8	High	3,904.06	22
5.	> 34,8	Very High	742.36	4
	Total		17.834,90	100

Based on the above criteria, the region use of Grompol watershed area can be determined :

Table 4. Region Use of Grompol Watershed

No	Region Use	Area (ha)	%
1.	Protected	402.92	2.26 %
2.	Buffer	1,096.79	6.15 %
3.	Annual Plant Cultivation	1,409.06	7.90 %
4.	Seasonal Cultivation	9,599.05	53.59 %
5.	Settlements	5,326.41	30.10 %
	Total	17,834.90	100 %

Further analysis is carried out by looking for the area of disaster-prone settlement cultivation areas (LKBRB). The analysis was carried out by overlaying the map of settlement cultivation areas with disaster-prone maps so that LKBRB obtained 1,851.94 ha.

Based on the results of the overlay analysis and the results of calculations using the formula, the ability of the study area for settlements can be determined by the ability of the land in the form of settlement suited region.

$$LPm = LW - (LKL + LKP + LKBTT + LKBTS + LKRB)$$

$$LPm = 17{,}834.90 - (402.92 + 1{,}096.79 + 1{,}409.06 + 9{,}599.05 + 1{,}851.94)$$

$$LPm = 17{,}834.90 - 14{,}360.43$$

$$LPm = 3{,}474.47\ ha$$

The analysis of the population of Grompol watershed is done by finding the percentage of the area of village settlements that enter the watershed, after that it is multiplied by the total population per village. The results of the analysis of the population of Grompol watershed were 211,961.

Based on the Regulation of the Minister of Public Housing No. 11 / PERMEN / M / 2008 concerning Guidelines for Harmony of Housing and Settlement Areas, the needs of the maximum space of the population vary more according to the zone of the region. Determination of the coefficient value (α) of the total space requirements per capita can be determined by looking over the density of the population of Grompol watershed. The value of population density in Grompol watershed is calculated by means of the population entering the watershed divided by the watershed area.

$$\text{Population Density} = \frac{\text{Total Population}}{\text{Area}}$$

$$PD = \frac{211{,}961}{17{,}834.90}$$

$$\text{PD} = 11.88 \text{ inhabitants/ hectare}$$

The results of the calculation of population density amounted to 11.88 inhabitants per hectare, so based on Regulation of the Minister of Public Housing No. II / PERMEN / M / 2008, the area coefficient of space needs per capita in Grompol watershed is $\alpha = 133\ m^2$ / capita.

Finally, the analysis of settlement carrying capacity is obtained from, the area of settlement suited region per total population of the watershed divided by the coefficient of space/capita needs (m^2 / capita), so the carrying capacity of the environment for settlements in Grompol watershed is:

$$\text{DDPm} = \frac{\text{LPm}(\text{m}^2)/\text{Jp}}{\alpha}$$

$$DDPm = \frac{34{,}744{,}663.82 / 211{,}961}{133}$$

$$DDPm = 1.23$$

Based on the results of the analysis, Grompol watershed has a total carrying capacity of settlements of 1.23 which mean high and residential development can still be carried out based on the optimal total population (*JPo*) with the formula:

$$JPo = DDPm \times Total\ Population$$

$$JPo = 1.23 \times 211{,}961$$

$$JPo = 261{,}238$$

and the optimal area of suitable for settlement (*LPmo)* with the formula:

$$LPmo = \frac{1}{DDPm} \times LPm$$

$$LPmo = \frac{1}{1.23} \times 3{,}474.47\ ha$$

$$LPmo = 2{,}819.08\ ha$$

4. Conclusion

Based on the analysis using the Settlement Carrying Capacity (DDPm) formula above, the environmental carrying capacity for settlements in the study area is 1.23. Therefore, Grompol watershed has a high carrying capacity of settlements. JPo value = 261,238, means that the current population can still accommodate 49,277 people to get the optimal population. Whereas LPmo = 2,819.08 ha, means that the size of residential settlements can still be added to 655.38 ha in order to be eligible for optimal settlements. The calculation is based on the Decree of Minister of State of Population/ Head of National Population and Family Planning Board (BKKBN) Number: Kep-03 / MEN / MENEG. K / 2/1998 which states that environmental quality will be well maintained if humans manage to carry capacity based on the minimum and optimum limits (30-70%). If the land is used more than 70%, it will result in decreased environmental quality [5].

The ideal direction for settlements based on the analysis of environmental carrying capacity is to place the location of the settlement according to its designation. Settlement areas should be developed as much as possible on land with settlement suited region- according to several criteria above. However, with the limited number of ideal location, priority is needed in the plan for utilizing further settlements- along with efforts to improve the quality of land in less ideal areas.

References

[1] Asdak, C. 2010. *Hidrologi dan Pengelolaan Daerah Aliran Sungai.* (Yogyakarta : Gadjah Mada University Press)

[2] Muta'ali, L. 2011. Environmental Carrying Capacity Based on Spatial Planning. *Indonesian Journal of Geography* **43(2):** 142-155.

[3] Fearnside, P.M. 1984. Brazil's Amazon Settlement Schemes: Conflicting Objectives and Huma Carrying Capacity (Habitat International **8(1)**: 45-61.

[4] Riyanto, A. 2003. Kajian Kemampuan Lahan untuk Arahan Kegiatan Permukiman Berdasarkan Aspek Fisik Dasar. Unpublished thesis. Diponegoro University, Semarang.

[5] Muta'ali, L. 2012. *Daya Dukung Lingkungan untuk Perencanaan Pengembangan Wilayah.* (Yogyakarta : Badan Penerbit Fakultas Geografi (BPFG) Universitas Gadjah Mada)

IOP Conf. Series: Earth and Environmental Science **328** (2019) 012073 doi:10.1088/1755-1315/328/1/012073

Morphological Study of Planned Human Settlement Affected by Service Trade Activities

R Aditantri[1], N A Krisnaputri[2], A Oktaviyani[3]

[123]Universitas Agung Podomoro

*Email: rahmatyas.aditantri@podomorouniversity.ac.id

Abstract. A planned human settlement is a settlement which is planned and controlled by government. In its development, this human settlement needs support from other areas, for example trade and service areas. However, this service trading area will give effect to residential areas if it's not controlled, especially in its morphological aspects. There are many residential areas which are close to the service trade area and affected by this activity, for example in this case is Tanjung Duren Selatan, West Jakarta. This paper aims to study morphological aspect of planned human settlement affected by service and trading activities. This paper used qualitative method with mapping and parameter assessment based on aspects in human settlement morphology. The study shows that the trade and service sector in Tanjung Duren Selatan give the impact for housing morphology, such as increasing of land use for housing, solid formation dominates more than the formation of voids, morphological formation of figure ground theory, linkage and place that adjusts the impact.
Keywords: settlement; trade service; morphology, study

1. Introduction

A dynamic developing city cannot be separated from the development of the service trade sector. The service trade sector that grows in a region reflects the development of the region itself. On the other hand, a city that is developing dynamically will lead to increased demand for space, especially the need for residential area[1].

In urban areas, service trade areas are often seen alongside residential areas. The existence of this service trade area, more or less, have an impact on settlements, both from the physical aspects of the environment and the impact in the form of socio-economic. Furthermore, this raises social changes and these changes are manifested in the physical form of the region. Products of social change in the physical area are known as morphology[2].

Morphology is the physical appearance of the area in terms of the structure that forms the shape of a particular appearance. Physical appearance of morphology is not only a form but also the existence of relations between areas.

Tanjung Duren Selatan is an urban village located in Grogol Petamburan District, West Jakarta. In this region there are service trade superblocks, namely Podomoro City and Taman Anggrek Mall. The existence of the service trade area in Tanjung Duren Selatan is adjacent to the residential area that has been there before.

IOP Conf. Series: Earth and Environmental Science **328** (2019) 012073 doi:10.1088/1755-1315/328/1/012073

Existing land use in this location study is mixed land use which consists of settlements and trade services. This service trade is also supported for the existence of the Podomoro City superblock area and Taman Anggrek Mall. This area has an activity character which is a mix used development [3]. Many land use functions have sprung up, for example the emergence of eating places, houses that have turned function into kiosks / stalls and varied land uses in this area. Retail presence is also seen in this region. Therefore, this random land use will affect the growth and development of the region in the future.

Morphology is defined as knowledge to study the physical form of a city. Morphology is a form of science or approach to understanding the development of a city that continues to experience changes during its development process through the form and pattern of urban spatial structure [4].

The physical appearance of residential areas in Tanjung Duren Selatan developed with the function of settlements. Continuous development without control results in random shapes, making it difficult to predict future needs. The existence of the existing service trade area also has an impact on the physical appearance of the existing area. Thus, research will be carried out to examine the physical appearance of the Tanjung Duren Selatan residential area due to existing trade services.

1.1. Planned Human Settlement

In the physical form of cities, there are several types of settlements which can be classified as planned and unplanned settlements [5]. Planned settlements are settlements whose overall planned development includes road facilities, drainage, open space, mass management patterns, and so forth. Unplanned settlements or spontaneous settlements built in stages from units by each owner with unplanned environmental conditions such as roads and drainage. According to Yunus [6], continuously uncontrolled settlement development will become a building-dense settlement, unregulated and a trigger for taudification or the process of forming slums.

1.2. Morphology

In designing a city, Trancik in Yunus [7] assumed that cities can be approached from 3 approaches / theories, namely Figure Ground Theory, Linkage Theory and Place Theory. These three theories will give an idea of the morphological characteristics of a city. Furthermore, these three theories will be useful for analysing potential problems that exist in the study location.

1.2.1. Figure Ground Theory. This theory emphasizes the relationship between the form that is built (building mass) and open space (open space) from an area / urban environment. In theory, what is meant by Figure is mass that is built in an area. Usually the map is shown in black. Meanwhile, the Ground is an open space in an area or all the space outside the built mass. Usually in maps it is shown in white. This theory can be used to analyse the texture and patterns of urban spatial planning and identify problems related to the mass / urban space.

1.2.2. Linkage Theory. Linkage means a pseudo line that connects between one element with another element, node one with another node or district one with another district. This line can be in the form of road networks, pedestrian lines, open spaces in the form of lines and so on. Linkage theory is an analysis that is used to pay attention to and confirm the relationships and movements of an urban fabric. This theory sees cities as dynamic and urban dynamics can be used as generators / generators of urban activities.

The most important thing about Linkage Theory is the integration of 2 regions visually and connecting two regions with priority to one area. An example is a linkage to see the relationship between the core city and its satellite city. The relationship of a place with others can be seen through a visual, structural or collective form of acceleration.

1.2.3. Place Theory. This theory is related to space / space which lies in understanding or understanding of culture and human characteristics of physical space. Space is a void that lives and has a physical connection. This space will be a place if given a contextual meaning from the cultural content or local content potential. Place Theory sees the design of cities approached from how big the interests of urban places are open to history, culture and socialization. Place analysis is a tool for giving understanding of urban space through signs of urban life and giving understanding of urban space. The nature of place theory in spatial design lies in understanding culture and human characteristics in its place. An architect in designing a city / region based on the place theory approach, must pay attention to the history of society and the location, culture and historical monuments that may exist as a basis for design so that the direction of the development of the city can be determined.

2. Research Method

This study is used mapping analysis and qualitative descriptive analysis to gain potential and problems that exist in the study location. In this study, observations of existing settlement morphology were carried out by referring to the three previously mentioned theories, namely Figure Ground Theory, Linkage Theory and Place Theory. The morphological formation observed was changed in land function and physical formation of the area. The results of the observations are then transformed into regional maps and analysed based on the theory.

3. Discussion and Result

Tanjung Duren Selatan is located in Grogol Petamburan District, West Jakarta. The study location has an area of 1.77 km^2, with residential land use of 1.66 km^2 and the use of service trade 0.11 km^2(8). The study location can be seen on figure 1.

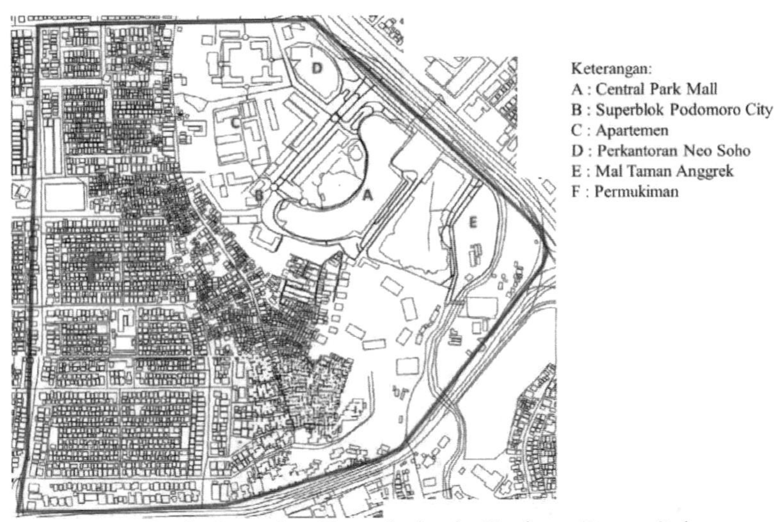

Figure 1. Study Location Boundaries in Tanjung Duren Selatan

Land Use Conversion

The development of trade and service activities in a region spatially has had an effect, such as the addition of a wide area to the built due to similar activities and supporting activities that arise. Expansion of built-up areas such as commercial land use, trade in services, and public facilities occurs by occupying areas not built so as to reduce the area of vacant land [9].

The existence of Central Park mall, which was established around 2009 - 2010, also had an impact on land changing in Tanjung Duren Selatan Village, especially for residential land use and service trade. The percentage of land changes that occur is presented in the following charts as seen on chart 1 and 2.

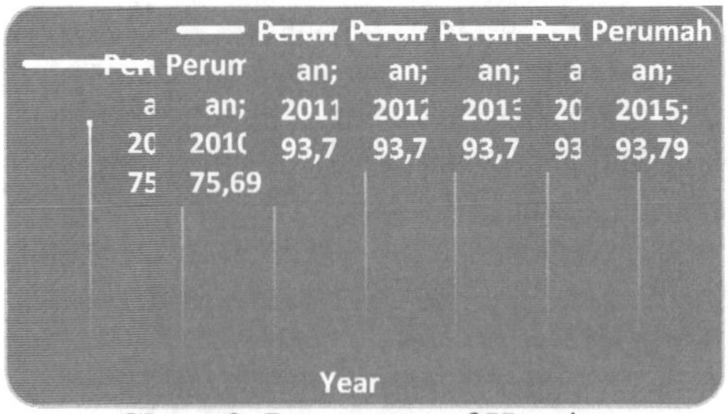

Chart 2. Percentage of Housing Land Use in 2009 - 2015

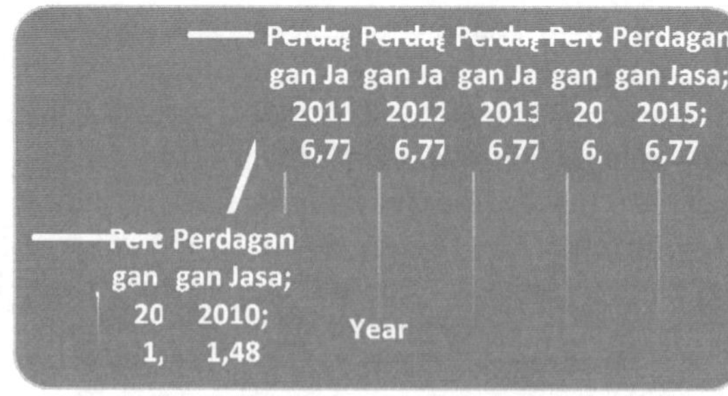

Chart 1. Percentage of Service and Trade Land Use in 2009 - 2015

Chart 1 shows that the percentage of land use for housing in 2009 - 2010 was 75.69% of the total land area or 1.34 km^2 and increased to 93.79% of the total land area in 2011 or 1.66 km^2. In graph 2 it also shows that the percentage of land use for services trade in 2009 - 2010 was 1.48% of the total land area or an area of 0.026 km^2 and increased to 6.77% of the total land area in 2011 or an area of 0.11 km^2.

The increasing of land use for housing was affected by increasing the service trade activities in the Podomoro City area. Apartments that are part of housing contribute to this increase. The market share for vertical and landed house housing in Tanjung Duren Selatan is increasing due to the ease of access to Podomoro City as a one stop living center. People who want to work in the Podomoro City area can choose to live in Tanjung Duren Selatan. The community does not need long time to arrive at the Podomoro City service trade area or in the Central Park mall, and both daily and secondary needs can be easily found in this region. This trend is a factor in the increase in land use for housing in Tanjung Duren Selatan

Housing Morphology

Tanjung Duren Selatan has 2 types of building mass typology (solid), blocks seen from several areas such as the Podomoro City superblock area and solid formed from residential settlements that are parallel to road formations. For void typology (space) is seen with a linear open system because it follows the shape of the residential environment which is a main road. There are also a number of closed systems that focus on several points in the residential area as well as in the superblock area. The solid void typology of study location can be seen on figure 2.

Figure 2. Solid Void Typology in Tanjung Duren Selatan

From this solid void it is also seen that solid formation dominates more than the formation of voids that exist. This is happened as the impact of the high intensity of land use in Tanjung Duren Selatan.

While in terms of urban texture, overall urban textures in Tanjung Duren Selatan region are heterogeneous, visible from several scattered masses and different shapes and heights so that they look

IOP Conf. Series: Earth and Environmental Science **328** (2019) 012073 doi:10.1088/1755-1315/328/1/012073

like unregulated areas and can adversely affect the city's image. The urban texture in Tanjung Duren Selatan can be seen on figure 3.

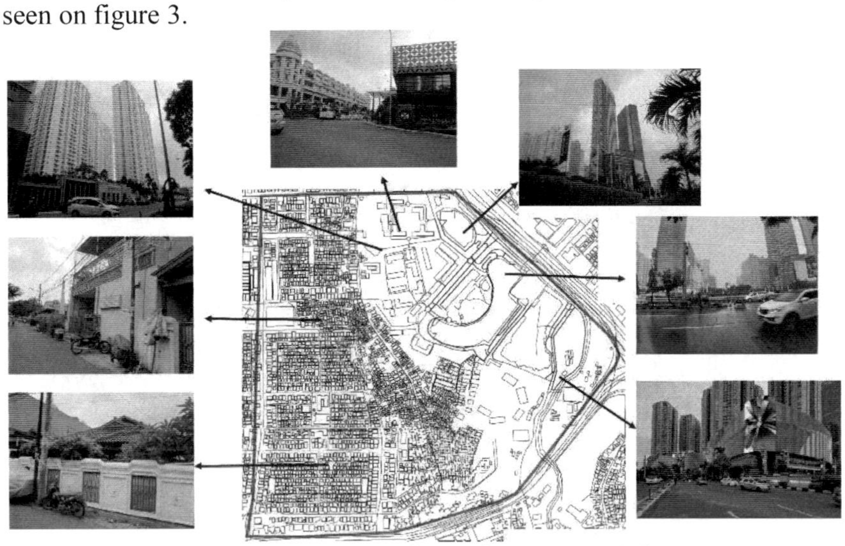

Figure 3. Urban Texture in Tanjung Duren Selatan

For the linkage aspect, the linkage that appears is a visual linkage. The visual linkage in the area is marked by the axis of the road, namely Tanjung Duren Timur Raya road which is neutral or serves as the main lane connecting between regions and is characterized by rows of trees and buildings around the road so as to form line elements. The linkage in study location can be seen on figure 4 as follows.

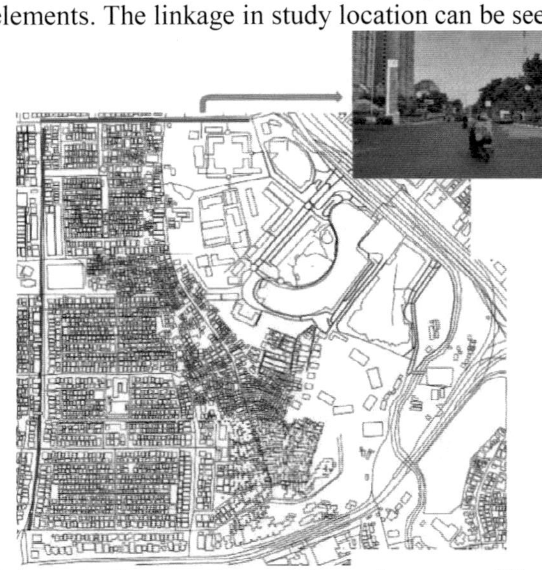

Figure 4. Linkage in Tanjung Duren Selatan Appears as Visual Linkage

Next element is path. As a connector for human and vehicle mobility, path also can make urban structure. There are 2 main path in this area. First, Jalan Tanjung Duren Selatan which is connecting settlement in Tanjung Duren Selatan and Jl. Letjend S. Parman (beyond area). Secondly, Boulevard Podomoro City which is connecting superblock Podomoro City with Jl. Letjend S. Parman. As a connector for 2 district/area, these 2 paths are always crowd by vehicle. Beside, retails are seen alongside Jalan Tanjung Duren Selatan. Figure 5 shows path element in location study.

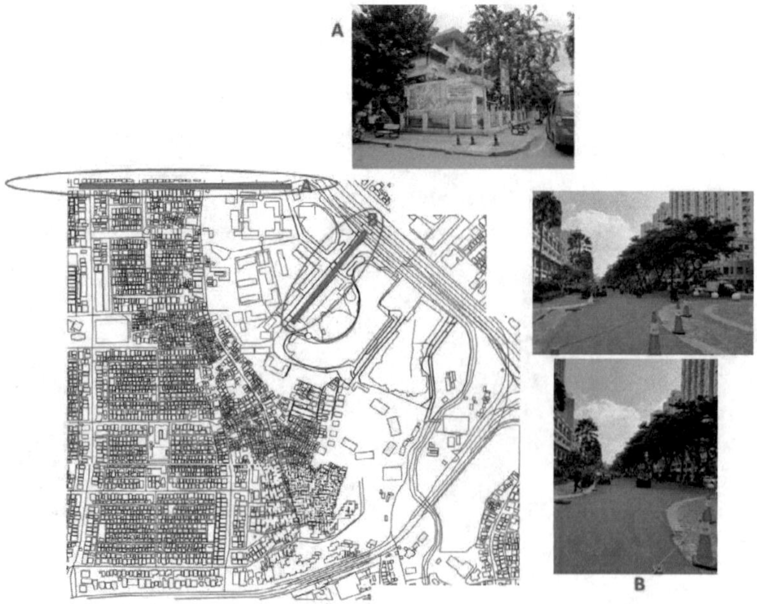

Figure 5. Path Element in Tanjung Duren Selatan

Edges are terminations of a particular district or region. Formation of building masses forms and limits a particular functional area with other functional areas within the city. Grogol River is a very clear edge to the east of the area, but in the north, east and west it still does not stand out because it is a building with less clear mass.

District is an integration of various functional activities which generally focus on certain areas in the city. This concentration is based on the main orientation, interests and role within a city. Figure 5 as follow, shows the district element in study location.

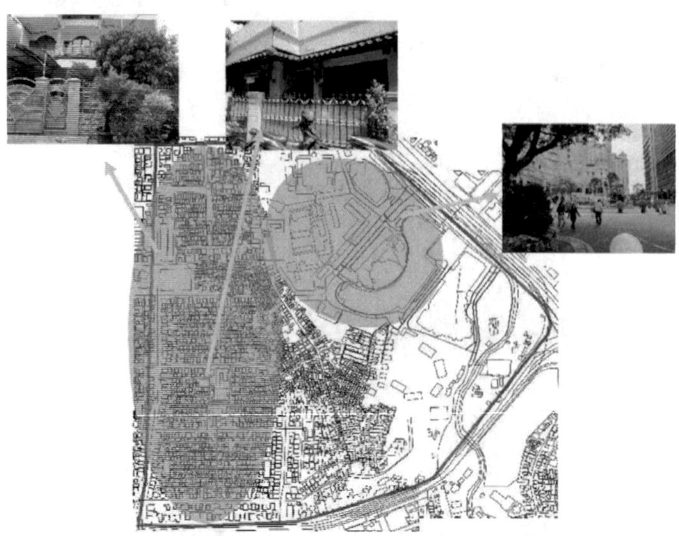

Figure 6. District Element in Tanjung Duren Selatan

As a result of trade in service activities in Podomoro City superblock, it has resulted in 2 dominant district formations, namely districts in the form of residential areas that dominate the region and districts in the form of markets and retails spread in several points in Tanjung Duren Selatan.

Nodes are a meeting between several roads / aisles in the city, thus forming a space that is usually the center of certain functional activities. Each node has different characteristics, both the formation of space and the pattern of general activities that occur. Buildings located at these nodes are often specifically designed to provide a certain image or space identity.

Figure 7. Nodes Element in Tanjung Duren Selatan

The increasing trade service activities in Tanjung Duren Selatan makes Kopro Market as regional nodes. It is no longer a settlement that has become nodes in Tanjung Duren Selatan.

Landmark is one of the elements that also enriches urban space. Landmark is buildings that provide certain images so that they are easily known and remembered and can also provide orientation for people and vehicles to circulate. A good landmark is a physical structure that stands out from the surrounding environment but is still a harmonious part of the whole environment.

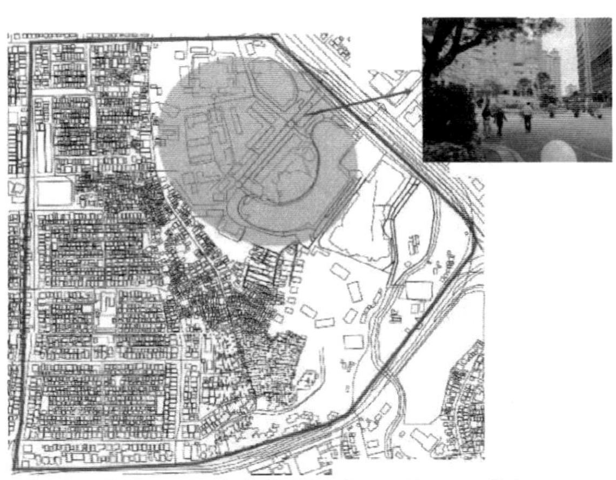

Figure 8. Landmark in Tanjung Duren Selatan

The increasing of service trade activities in TDS make Podomoro City a landmark for the region. It is no longer a settlement but is a landmark in TDS.

From this discussion on settlement morphology in Tanjung Duren Selatan, the typical formations of existing settlements are less prominent. The existing form of housing has no difference with other regions. The influence of the trade and service area indirectly took over the formation of morphology in Tanjung Duren Selatan settlement. If there is no appropriate utilization control, the existing settlement will lose its identity.

IOP Conf. Series: Earth and Environmental Science **328** (2019) 012073 doi:10.1088/1755-1315/328/1/012073

4. Conclusion

Regarding the analysis, the study shows that the trade and service sector in Tanjung Duren Selatan give the impact for housing morphology such as increasing of land use for housing, solid formation dominates more than the formation of voids, morphological formation of figure ground theory, linkage and place that adjusts the impact.

References

[1] Soetomo S. Urbanisasi dan Morfologi. Graha Ilmu; 2009.
[2] Putri MA, Rahayu MJ, Putri RA. Bentuk Morfologi Kawasan Permukiman Urban Fringe Selatan Kota Surakarta. Jurnal Pengemb Kota. 2016;4(2):120. Available from: http://ejournal2.undip.ac.id/index.php/jpk/article/view/720
[3] Krisnaputri NAC. Penataan Ruang Terbuka Publik (PS) dan Fasad Depan Komersial (CFF) di Koridor Utama Jl. Letjen S.Parman Jakarta Barat. In: Seminar Nasional dan Konferensi Sosiologi Perkotaan 2018. 2018. p. 157–67.
[4] Zahnd M. Model Baru Perancangan Kota yang Kontekstual. Yogyakarta: Kanisius; 2008.
[5] Tilaar, Sonny;Rogi, Octavianus H.A.; Tinangon AJ. Kajian Tipomorfologi Kawasan Permukiman Terencana Di Kota Manado. 2012;9(3):82–103.
[6] Yunus HS. Manajemen Kota: Perspektif Spasial. Yogyakarta: Pustaka Pelajar; 2005.
[7] Yunus HS. Struktur Tata Ruang Kota. Yogyakarta: Pustaka Pelajar; 2000.
[8] BPS. Grogol Petamburan Dalam Angka 2015. 2015. 66 p.
[9] Arifia Soedwiwahjono; Utomo, Rizon Pamardhi DS. Pengaruh Perkembangan Kegiatan Perdagangan dan Jasa terhadap Perubahan Penggunaan Lahan di Kawasan Solo Baru.Arsitektura. 2017;(Vol 15, No 1 (2017)):1–9.

IOP Conf. Series: Earth and Environmental Science **328** (2019) 012074 doi:10.1088/1755-1315/328/1/012074

Clean Water Network Planning in Ba'engas Village, Labang Sub-District, Bangkalan Regency, Indonesia

R D Lufira[1], M J Ismoyo[2], D Chandrasasi[3], A Firmandika[4]
[1234]University of Brawijaya, Malang, Indonesia

*Email: rahmahdara@ub.ac.id

Abstract. Access to clean water supply systems have not been fully available to all people, especially those living in rural areas. Ba'engas Village, located in Labang Sub-District, Bangkalan Regency, has not possessed clean water facilities to fulfil its people's needs. The elevation difference and the distance of about 9 km between water resources causing difficulties in water provision. This research discusses clean water network planning in Ba'engas Village by utilizing the Pocong River, at a rate of 3,236 l/sec. The objective of this research is to investigate the provision of water availability for consumption up to the year 2035, simulate clean water networks and find out its results, and calculate planned budget. The modelling and simulation of clean water networks are developed utilizing WaterCAD V.8.i software. This research employs three alternatives, with different numbers of pumps and tanks each. The result of clean water projection is a rate of 4,936 l/sec. Based on the simulation results, alternative 1 was chosen, using 1 pump and 1 tank, according to the permitted planning criteria of speed between 0.1-2.5 m/sec, head loss gradient from 0 to 15 m/km, and pressure from 0.5 to 8 atm. The cost for clean water network planning with alternative 1 is estimated IDR 8,021,618,000.

1. Introduction

Water becomes one of the primary needs for survival. As such, the provision of clean water plays a very important role to improve their quality of living. Based on data obtained from the Regional Drinking Water Company (PDAM, *Perusahaan Daerah Air Minum*) of Bangkalan Regency, many villages have not yet obtained clean water services, particularly those located in inland or peripheral areas. One of the efforts taken to resolve this issue is by creating a good plan for clean water networks, making use of distribution system of pipe networks.

Ba'engas Village is one of the villages located in Labang Sub-District, Bangkalan Regency. This village suffers from the lack of clean water facilities, resolving by utilizing Pocong River, with discharge level of 3.236 l/sec. Distant location of the water source as well as difference in elevation with low-lying residential areas complicate water provision. Thus, this study discusses an effective planning for clean water networks to solve the problems at the study location, by considering the utilization of pumps and tanks. Pumps are utilized to increase water pressure and elevate water from a low elevation region to a higher elevation area; meanwhile, tanks function to store water for distribution to the service area or residential locations.

The aim of this study is to investigate clean water provision for the residents of Ba'engas village up to the year 2035, simulate clean water network planning for the study location, and to calculate the size of the Planned Budget (RAB, *Rencana Anggaran Biaya*) for building clean water networks in Ba'engas Village in Labang Sub-District, Bangkalan Regency.

This study will benefit in providing inputs or information to the Regional Drinking Water Company of Bangkalan Regency to improve services and the provision of clean water, developing planning of clean water networks for residents of Ba'engas Village in Labang Sub-District of Bangkalan Regency, and adding further insights to the field of clean water network planning.

2. Materials and Methods

The location of the study area, Ba'engas Village in Labang Sub-District, Bangkalan Regency, is depicted in Figure3. The village is located in the southern part of Bangkalan Regency at elevations ranging from 20 to 50 m above sea level and geographically positioned between 112° and 113° BT and between 6° and 7° LS, bordered by these administrative regions of:

- North : Bringin Village
- South : West Sukolilo Village
- East : Bunajih Village
- West : Morkepek Village

The required supporting analytic data for the pipe network system in this planning study cover:

- Data of water availability, utilizing the discharge flow of Pocong River;
- Data of population, comprising the number of residents of Ba'engas Village spreading out among 6 hamlets; and
- Data of topography, consisting of the elevation data for water sources, locations of pump, placement of the tank bases, and the targeted areas or regions served.

The steps involved in the data processing for the completion of this planning study are as follows:

- Collecting primary and secondary data consisting of technical and other supporting data used in the analysis of clean water network systems
- Processing the population and amount of service data
- Calculating clean water needs
- Establishing the planned networks at service up to the year 2035
- Calculating the Planned Budget for building the plans

2.1. Major Losses

Fluid flowing into the pipe will experience shear stress and the velocity gradient across the field due to the kinematic viscosity. The shear stress will cause power loss during drainage (Triatmodjo II, 1993: 25). Shear stress occurs at the pipe wall becomes the main cause of energy loss in a flow line (major losses); in addition, this stress depends on the type of pipe. High loss equation by major press Hazen – Williams (Webber 1971:121) is as follows:

$$H_f = k.Q^{1,85} \quad (1)$$

$$k = \frac{10,675.L}{C_{hw}^{\ 1,85}.D^{4,87}} \quad (2)$$

with:

Hf = height major loss press (m)
k = coefficient of pipe characteristics

IOP Conf. Series: Earth and Environmental Science **328** (2019) 012074 doi:10.1088/1755-1315/328/1/012074

Q = flow in pipe (m3/sec)
D = diameter of pipe (m)
L = length of pipe (m)
Chw = Hazen - Williams roughness coefficient

Table 1. Coefficient Characteristics of Pipe According to Hazen – Williams

Nilai C_{rw}	Type of Pipes
140	extremely smooth pipe
130	smooth pipes, cement, cast iron new
120	new welded steel pipe
110	new riveted steel pipe
100	old cast iron pipes
95	old riveted steel pipe
60-80	the old pipe

Sources: Triatmodjo II, 1993: 51

2.2. Minor Losses

Various contributors of minor losses include: sudden narrowing or widening of pipes, bends, joints, and the presence of valves in pipes.

If a pipeline consists of pipes of different sizes connected to the same diameter, the pipe is in a series relationship, leading to several high losses (Priyantoro, 1991: 49)

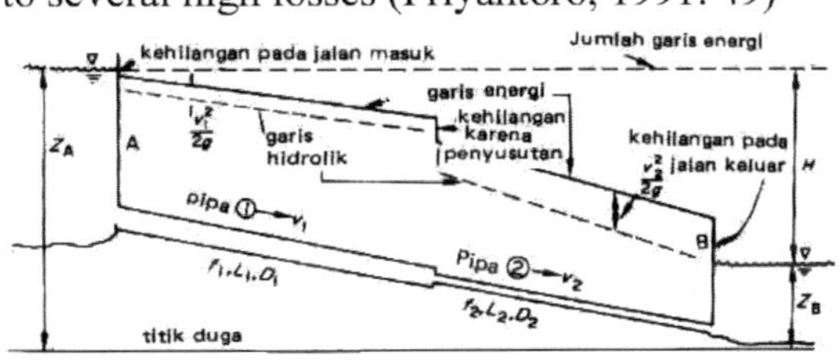

Figure 1. Series Pipe Line: Dake (1958 : 78)

Equation of series pipe continuity:
$Q = Q1 = Q_2$ with:
Q = total discharge on installed pipeline (m^3/sec)
Q1, Q2 = discharge in pipes 1 and 2 (m^3/sec)
Total pressure loss in pipes installed in series (Triatmodjo, 1996: 74):
$H = hl_1 + hl_2$
with:
H = total loss of press on the pipe installed in series (m)
hl_1, hl_2, = loss of pressure per pipe (m)

2.3. Criteria for Clean Water Pipes

Plumbing network planning must meet certain criteria in accordance with the existing standards. The criteria for pipelines are presented in the following table:

Table 2. Criteria of Pipelines

Criteria	Alteration

Criteria	Alteration
1. Velocity (0,1 - 2,5 m/sec)	- Speed less than 0.3m / sec a. Reduced pipe diameter b. Added pump c. Higher upstream pipe elevation (adjusted in the field) - Speed of more than 4.5 m / sec a. Enlarged diameter b. Higher elevation in the upstream pipe compared to the downstream
2. Headloss Gradient (0 – 15 m/km)	- Head loss Gradient of more than 15 m / km a. Enlarged pipe diameter b. Larger upstream pipe elevation compared to the downstream pipe
3. Pressure (0,5 – 8 atm)	- Pressure of less than 0.5 atm a. Enlarged diameter of the pipe is enlarged b. Added pump c. Installed second pipe at the top, part or all of the length of the pipe - Pressure of more than 8 atm a. Reduced pipe diameter b. Established press release building - Pressure Reducer Valve (PRV) Installation

Source: Minister of Public Works Regulation on the Implementation of SPAM Development, 2007

2.4. Budget plan (RAB)

RAB is an estimated cost before the building or project is carried out. The basic calculation of the RAB is principally obtained from the sum of all volumes from each type of work which is available at each unit price.

3. Results and Discussion

The projection of the population growth of Ba'engas Village was calculated using three methods of projection, Arithmetic Method, Geometric Method, and Exponential Method. The projection of the population in this study was conducted up to 20 years in the future.

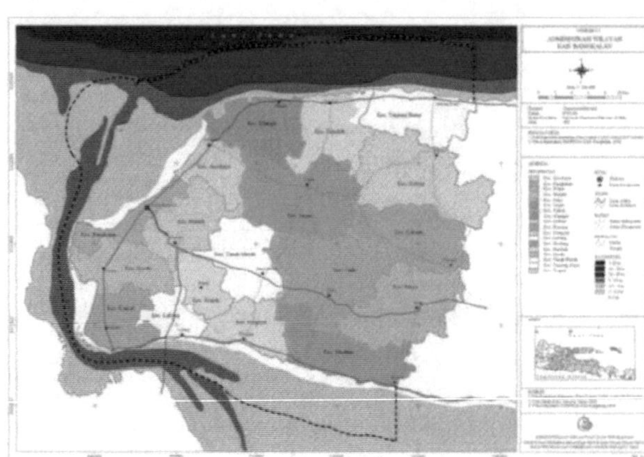

Figure 2. Administrative Map of Bangkalan Regency

The planning for the next 20 years was adjusted in accordance to the SPAM (*Sistem Penyediaan Air Minum*, Drinking Water Supply System) planning criteria. According to the composition of primary plans for SPAM development, there is a criterion stipulating between 15-20 years in the planning horizon portion. To determine which method should be best employed to project the size of the population from the three methods of projection above, the appropriateness was tested using statistical testing using standard deviations and correlation coefficients. From the results of the conducted appropriateness testing for the projection methods, arithmetic method was selected as the projection method measuring the population growth. Arithmetic method was selected because it has the smallest standard deviation values. Meanwhile, the resulting correlation coefficient values were nearly equal; and thus, the selection only used the standard deviation appropriateness testing. The calculation results for the three methods of projection as well as the test of appropriateness for the standard deviations are presented in Table 3.

Table 3. Results of Standard Deviations

Method	Hamlet		
	Sogah	T. Agung	Ba'engas
Arithmetic	47,532	117,103	69,6028
Geometric	55,486	135,853	80,6434
Exponential	55,997	137,047	81,3453
Method	Hamlet		
	Tengginah	Kolpoh	Pangloros
Arithmetic	50,661	82,482	34,947
Geometric	58,648	97,439	41,214
Exponential	59,155	98,416	41,623

Source: Calculation Results

After calculating the projections for the population, the projection of average water needs for the people of Ba'engas Village was afterwards calculated, with the assumption of water needs of 60 liters/person/day (SPAM, 2007, p.59).

The results of calculated projections for the average clean water needs of Ba'engas Village residents in 2035 are presented in Table 4. The table informs the total average clean water needs for Ba'engas Village population in 2035 will be 4.936 l/sec. This total average clean water needs used in the planning is modeled and simulated using the WaterCAD V.8.i software. The modeling and simulation for the planning of clean water networks used three alternative plans.

Alternative 1 utilizes 1 pump and 1 tank; alternative 2 utilizes 2 pumps and 2 tanks; and alternative 3 utilizes 3 pumps and 2 tanks.

Table 4. Projection of Clean Water Needs

No	Hamlet	Unit	Year
			2035
1	Sogah	kl/s	0.580
2	T. Agung	kl/s	1.473
3	Ba'engas	kl/s	0.881
4	Tengginah	kl/s	0.644
5	Kolpoh	kl/s	0.952
6	Pangloros	kl/s	0.407
	Total		4.936

Source: Calculation Results

For this modeling and simulation of clean water networks, alternative 1 was opted because the simulation results for this alternative satisfied the permitted planning criteria for the planned networks comprising speed, head loss gradient, and pressure. The planning criteria for the planned pipe networks refer to the Regulation of the Public Works Minister PU on SPAM development, which states that the criteria for the parameter of speed is between 0.1 to 2.5 m/sec, the parameter of head loss gradient is from 0 to 15 m/km, and the parameter of pressure is from 0.5 to 8 atm (SPAM, 2007, p.55).

In the modeling and simulation, the pump components used in the planning with alternative 1 are rotodynamic pumps of the centrifugal type with the pump catalog number of 65 x 50 B2 – 5 3.7 (Sularso & Tahara, 2000:p.53). The calculation of pump selection specification was adjusted with the calculation of rotodynamic pumps of the centrifugal type specifications as well as with pump selection flowchart. The scheme of clean water network planning with alternative 1 is illustrated in Figure 4. The pump is simulated for 24 hours with a discharge capacity of 7 l/sec and a pump head of size 24.79 m. The time operating pattern is arranged so that the pumps are turned off and on automatically or by leveling.

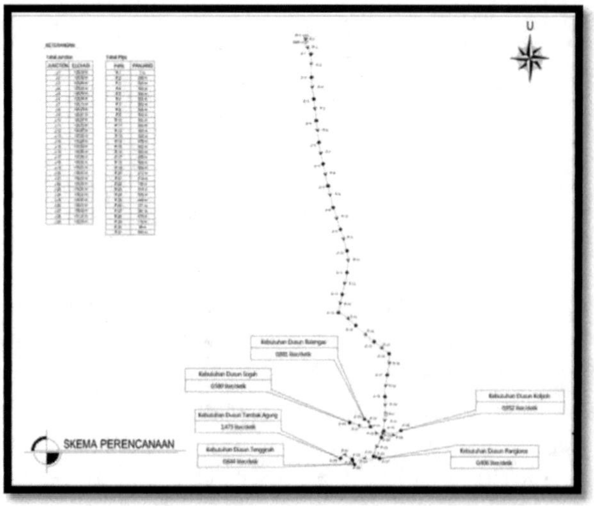

Figure 3. Scheme of Clean Water Network

The modeling and simulation of the tank components were conducted within a period of 24 hours with the tanks designed to never be in an empty or full state (leveling). From the simulation of the tanks, the resulting tank dimension was found to store water volume of 36.000 liters.
The design of the tanks for this clean water network planning stipulates tanks with dimensions of 4 m in length, 3 m in width, and 3 m in height.

The results of modeling and simulation with pipes of HDPE material are presented in Table 5. The following results offer pipe simulation at 8 AM, as the peak hour of water usage.

Table 5. Results of Pipe Simulation at 8 AM

Pipe No.	Diameter (inch)	Speed (m/s)	Head loss Gradient (m/km)	Pipe No.	Diameter (inch)	Speed (m/s)	Head loss Gradient (m/km)
P-1	8	0.23	0.30	P-17	8	0.23	0.30
P-2	8	0.23	0.30	P-18	8	0.23	0.30
P-3	8	0.23	0.30	P-19	8	0.23	0.30
P-4	8	0.23	0.30	P-20	8	0.23	0.30
P-5	8	0.23	0.30	P-21	8	0.23	0.30
P-6	8	0.23	0.30	P-22	6	0.32	0.81

Pipe No.	Diameter	Speed	Head loss Gradient	Pipe No.	Diameter	Speed	Head loss Gradient
	(inch)	(m/s)	(m/km)		(inch)	(m/s)	(m/km)
P-7	8	0.23	0.30	P-23	6	0.21	0.35
P-8	8	0.23	0.30	P-24	4	0.39	1.80
P-9	8	0.23	0.30	P-25	4	0.27	0.91
P-10	8	0.23	0.30	P-26	3	0.29	1.44
P-11	8	0.23	0.30	P-27	3	0.19	0.67
P-12	8	0.23	0.30	P-28	3	0.31	1.67
P-13	8	0.23	0.30	P-29	3	0.13	0.34
P-14	8	0.23	0.30	P-30	3	0.21	0.81
P-15	8	0.23	0.30	P-31	3	0.48	3.74
P-16	8	0.23	0.30				

Source: Calculation Results

As informed by the previous table, the speed and head loss gradient have met the permitted planning criteria, the speed parameter ranging from 0.1 to 2.5 m/sec and the parameter of head loss gradient ranging from 0 to 5 m/km.

From the results of modeling and simulation of junction points, it was found that the pressure value meets the permitted criteria for pipe network planning, ranging from 0.5 to 8 atm. The results of simulated junction points are presented in Table 6.

Table 6. Results of Junction Simulation at 8 AM

Label	Demand	Pressure	Notes	Label	Demand	Pressure	Notes
	(L/s)	(atm)			(L/s)	(atm)	
J-1	0	2.28	Satisfactory	J-16	0	0.96	Satisfactory
J-2	0	2.25	Satisfactory	J-17	0	0.81	Satisfactory
J-3	0	2.22	Satisfactory	J-18	0	0.65	Satisfactory
J-4	0	2.19	Satisfactory	J-19	0	0.67	Satisfactory
J-5	0	2.16	Satisfactory	J-20	0	0.76	Satisfactory
J-6	0	2.13	Satisfactory	J-21	0	0.94	Satisfactory
J-7	0	2.10	Satisfactory	J-22	0	1.12	Satisfactory
J-8	0	2.07	Satisfactory	J-23	0	0.86	Satisfactory
J-9	0	2.04	Satisfactory	J-24	1.31	0.97	Satisfactory
J-10	0	2.01	Satisfactory	J-25	0.86	1.02	Satisfactory
J-11	0	1.98	Satisfactory	J-26	1.41	0.69	Satisfactory
J-12	0	1.95	Satisfactory	J-27	0.60	1.03	Satisfactory
J-13	0	1.93	Satisfactory	J-28	0.96	1.31	Satisfactory
J-14	0	1.59	Satisfactory	J-29	2.19	1.02	Satisfactory
J-15	0	1.27	Satisfactory				

Source: Calculation Results

The calculation of the Planned Budget refers to the Analysis of Labor Unit Costs of the Department of Public Works and Public Housing and Development in 2016 and the list of standard units for building materials, equipment, and labor costs for Bangkalan Regency in 2017. The formula used to calculate the Planned Budget consists of multiplying the work volume by the price of the work unit.

IOP Conf. Series: Earth and Environmental Science **328** (2019) 012074 doi:10.1088/1755-1315/328/1/012074

In calculating the construction budget, there are three differentiated kinds of analysis for the materials, labor, and work unit price (Ibrahim, 1993, p.138). The recapitulated results of the calculated Planned Budget are presented in Table 7.

Table 7. Planned Budget

Description	Price for Alternative 1 (IDR)
Pipe Work	6,982,466,700.44
Pump Work	141,385,417.17
Tank Work	168,527,672.84
Total	7,292,379,790.45
Tax	729,237,979,04
Total + Tax	8,021,617,769.49
Rounding	8,021,618,000.00

Source: Calculation Results

4. Conclusion

Based on the results of the analysis carried out in this planning study, the conclusions of the study are as follows:

1. The projected average clean water needs for Ba'engas Village in Labang Sub-District, Bangkalan Regency in 2035 is 4.936 l/sec.
2. The results of modeling and simulation of the clean water network planning with alternative 1 show that the parameters of speed, head loss gradient, and pressure meet the standard planning criteria.
3. The Planned Budget needed for the development of clean water networks in Ba'engas Village in Labang Sub-District, Bangkalan Regency is IDR 8,021,618,000.00.

The suggestions based on the discussed results of this planning study are as follows:

1. The local government of Ba'engas Village in Labang Sub-District, Bangkalan Regency is encouraged to realize the proposed plans for clean water networks by establishing facilities to provide clean water which sufficiently meets the needs of its residents.
2. The local communities is encouraged to cooperate in developing, maintaining, and operating clean water networks in order to supply and meet their clean water needs on daily basis improving the quality of living.

References

[1] Ibrahim, Bachtiar. 1993. *Rencana Dan Estimate Real Of Cost* [Planned and Estimated Real Costs]. Jakarta: Bumi Aksara.

[2] Menteri Pekerjaan Umum. 2007. *Penyelenggaraan Pengembangan Sistem Penyediaan Air Minum (SPAM)* [The Provision and Development of Drinking Water Supply Systems]. Jakarta: Departemen Pekerjaan Umum.

[3] Menteri Pekerjaan Umum dan Perumahan Rakyat. 2016. Analisa Harga Satuan Pekerjaan [Analysis of Work Unit Prices]. Jakarta: Departemen Pekerjaan Umum.

[4] Sularso dan Tahara, Haruo. 2000. *Pompa dan Kompresor* [Pumps and Compressors]. Jakarta: PT Pradnya Paramita

[5] Webber, N. B. 1971. *Fluid Mechanics For Civil Engineering, S. I Edition*. London: Chapman and Hall Ltd.

Study of Physical Arrangement with M3K Concept Based On Element Urban Design in Gemblakan Bawah Village Kota Yogyakarta

R Nurrokhmi[1] and R Widjajanti[2]

[12]Urban and Regional Planning Department, Diponegoro University, Semarang, Indonesia

*Email[1]: revi.armi@gmail.com
*Email[2]: nyo2_wien@yahoo.com

Abstract. The river boundary is a buffer zone between the river and terrestrial aquatic ecosystems. The community's lack understanding of river border functions causes the use of river boundary as residential areas. This condition occurs in the River Code boundary, Yogyakarta. To restore the river border, Pamerti Kali Code (a community engaged in the Code River observer) initiated the concept of M3K (Mundur Munggah Madhep Kali). The physical arrangement of settlements with the M3K concept has been realized in the slum village in Gemblakan Bawah Village located on the banks of the Code River. This village is the 1st segment of 4 segments that are planned for structuring. After the M3K concept was implemented, this village was transformed into a tourist village. In the urban design context, every city must pay attention to element urban design so that the city has clear characteristics. So structuring settlements with M3K needs to be assessed based on the elements of urban design. The purpose of this paper is to identify the application of elements urban design in physical arrangement with the concept of M3K in Kampung Gemblakan Bawah. This research used quantitative methods with quantitative descriptive and frequency distribution analysis. The author obtained data by conducting surveys and questionnaires to 78 families. The analysis was carried out by analyzing the physical condition of the Gemblakan Bawah Village after applying M3K concept based on urban design elements (land use, building and massing, circulation and parking, open space, signage, pedestrian ways, and activity support). The results of the study prove that settlement arrangements with the M3K concept make urban design elements in residential areas increase. Among them are improvements in the form of the regularity of buildings, improvement of river bank circulation into non-motorized lanes as well as new open spaces and the existence of new spaces, support activities in the form of tourism activities that result in improved environmental quality in Gemblakan Bawah Village.

Keywords : *urban design, settlement, stream buffer, urban*

1. Introduction

Urbanization is one of the external physical factors that influence the development of the city [1]. The urbanization process in urban areas due to the encouragement factors is contained in the social-psychological aspects of immigrants, the attraction factor because of the more attractive business, economic and socio-cultural opportunities compared to the village [2]. Population growth continues to have spatial consequences, namely the demands of space for housing needs, residential houses or trade and services. The limitation of land causes a high density of buildings with irregular distribution directions which cause pressure on the surrounding area, especially in the riverbank area [3].

Settlements on river banks are dense settlements that occupy land on the river banks that can contaminate the river, which lead to flooding [4]. As a result of the increasing number of people living on rivers banks, rivers lose morphological stability and hydraulic components [5]. So, the arrangement of the waterfront area needs to be done. The arrangement of the riverfront area is related to the physical and non-physical elements of the area, namely the characteristics of the activities. The structuring model of a residential area along the river includes vitalization, revitalization, redevelopment, renewal, intensification of development, rehabilitation of the area, and improvement of environmental quality through improved facilities and infrastructure [5]. Yogyakarta is one of the cities that implements a river bank settlement arrangement model

Yogyakarta city is divided by 3 rivers, namely the Winongo, Code and Gajahwong, where the majority of the river border areas undergo changes in function to become a cultivation function. Settlements Problems in river banks also occur in the Yogyakarta city. River boundary settlements in the city of Yogyakarta have a variety of different problems, especially in terms of land ownership status. There are two land ownership statuses (legal and sultan ground) in settlements located on the Stream buffer in Yogyakarta city. The people living in the sultan ground area on the stream buffer will find it easier to arrange their settlements compared to the people who have certificates of ownership rights to the land they occupy. This is because the people who live in the sultan ground area will surely obey the Sultan's command to arrange their settlements. Until now the Yogyakarta Palace has a role that remains important in the dynamics of the life of the people of Yogyakarta [6]. The condition of the stream buffer settlement in Yogyakarta city is a unique condition that does not occur in other cities in Indonesia.

The Code River is an undulating river and is located in the central Yogyakarta. Code River area administratively passes 8 (eight) districts and 14 sub-districts. Code River area is part of an urban area in the middle of the urban area of Yogyakarta. River Code is in the middle of Yogyakarta City which is the center of economic, political, social and cultural activities. The stream buffer located in or across the central part of the urban area of Yogyakarta is interesting to study because as an urban area, some of the lands have been used for building construction. This is reinforced by the results of the study of the most dominant types of spatial utilization in stream buffer of Code River is informal settlements(65%) [7].

The Code River problem is like the classic problem of urban settlements on stream buffer, some of the Code dikes are directly adjacent to the back wall of the house [8]. The settlement of the stream buffer Code in the central area of Yogyakarta City has similar characteristics, river boundaries are lost for occupancy, BCR (Building Coverage Ratio) is very high, lack of green space, difficult accessibility and building orientation back to the river. The disordered settlement causes slum [9]. Most of the problems mentioned have occurred in Gemblakan Bawah which fall into the category of slums.

One of the efforts sought to deal with the slum area issue was launched through 'Mundur Munggah Madhep Kali' or M3K concept. Mundur means moving away from the river, madhep means the building faces the river munggah increasing the intensity (floor) of the building. Through this concept, stream buffer is expanded according to standards and make the river a front yard to encourage residents to care of the river [10]. The standard width of stream buffer in urban areas is three meters [11]. The application of the M3K concept was carried out since 2014 at several points in the Code River. Applied to the M3K concept, one of the benefits is open space [12].

In the context of design, every city design must pay attention to the existing design elements so that the city has clear characteristics. So structuring settlements with M3K needs to be assessed based on urban design elements. This study aims to identify the application of urban design elements in physical arrangement with the M3K concept in Gemblakan Bawah Village.

2. Methods

IOP Conf. Series: Earth and Environmental Science **328** (2019) 012075 doi:10.1088/1755-1315/328/1/012075

The research method used is a quantitative method. Data were obtained through surveys and questionnaires. In determining the research sample the writer using stratified proportional sampling and the research respondents were households in Gemblakan Bawah, and analysis method used was descriptive quantitative.

3. Research Result

This analysis was conducted to identify the application of elements urban design in physical arrangement with the M3K concept in Gemblakan Bawah Village. Aspects that will be studied include land use, building form and mass, circulation and parking, open space, pedestrian area, signage, and activity support.

3.1 Land Use

Land use in Kampung Gemblakan Bawah has settlement functions. The current function is in accordance with the directives in PERDA Yogyakarta No. 1 Tahun 2015, as a residential area. The application of the M3K concept resulted in the width of the stream buffer reaching 3 meters. This condition is in accordance with the rules about the width of the undulating river in urban areas of 3 meters [11]. Visually, settlement arrangement results can be seen through imagery, taking into account the comparison of river border lines (GSS) after the arrangement.

▭: the width of the stream buffer reaching 3 meters

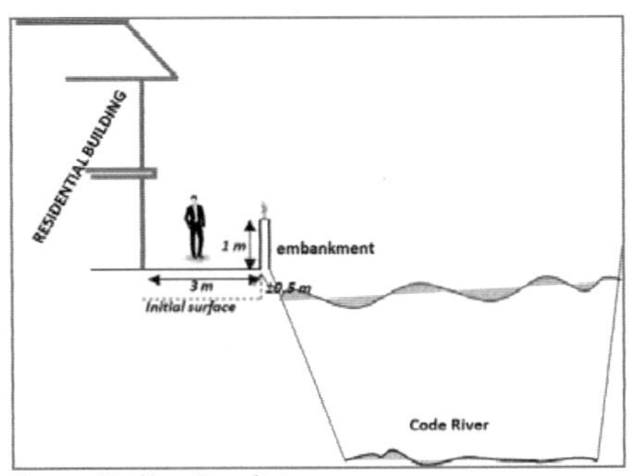

Strem Buffer Conditions

Figure 1. Overview of the Conditions of River Border Lines

The arrangement of stream buffer settlements is done by raising the river embankment and widening the road in the river border. The hope is that when the water discharge on the Code River rises, the water from the Code River no longer inundates the residents' settlements. After structuring, the Gemblakan Bawah Village is no longer experiencing flooding even though the Code River is flooded (Questionnaire Results, 2018). The existence of a wide enough circulation in the river boundary makes the evacuation path easier. At present Gemblakan Bawah Village has flood mitigation efforts. The flood disaster mitigation system includes making safe distances between rivers and settlements, not disposing garbage / or hard objects into rivers, not planting trees in river areas, maintaining and caring for facilities built as a means of prevention and handling of floods. Efforts that have been made in flood mitigation have been synergized between physical development and non-physical development. This synergy is expected to provide optimal results in flood disaster mitigation [13].

The settlement arrangement efforts carried out have been carried out in accordance with the settlement arrangement theory on the banks of the river. Efforts made in structuring settlements on the banks of the River Code have been carried out in stages to free the river border from buildings. The arrangement carried out is in accordance with efforts to control settlements to restore conservation land by Maryono [11]. The theory of structuring riverbank settlements can be implemented in Gemblakan Bawah Village.

IOP Conf. Series: Earth and Environmental Science **328** (2019) 012075 doi:10.1088/1755-1315/328/1/012075

3.2 Building and Massing

The settlement arrangement that is applied only affects changes in buildings along the river and some buildings behind it due to budget constraints (Figure 2). The condition of the building is dominated by permanent buildings with a BCR of 100%, the distance between buildings is <1m, and there are no building boundaries. Riverbanks have wide front limits, ranging from 2.5 to 3 meters (increasing from 1 to 1.5 meters from the previous scale). Steps to regulate the layout of buildings from river boundaries can be adjusted to the theory of building layout rules on the road lines in the concept of provision regarding the placement and placement of buildings.

Delineation of the arranged building *Building Condition In-Stream Buffer*

Figure 2. Building and Massing in Gemblakan Bawah Village

The arrangement of buildings with M3K is in accordance with the theory of building layout regulation, the arrangement of continuity of building faces along the road, the retreat of the ground floor of the building, the retreat of the corners of the building, but did not carry out the retreat of the top of the building. This is because the results of community consultations that have two-level buildings on the banks of the river do not agree to be supported, ideal conditions for building boundaries cannot be fulfilled. So there is a river border section that is less than 3 meters wide.

After structuring, the facade of the riverbank building is changed. The face of the riverbank building is dominated by buildings that have been plastered. The arrangement of riverbank settlements also affects changes in building functions, namely residential buildings (2nd floor) and stalls (1st floor). However, the addition of building functions did not occur in all buildings affected by the M3K program.

The orientation of the riverbank building is facing the river [14]. The river can be cleaner because the river is used as the front yard of the house. The implementation of the M3K concept results in the regularity of buildings that cause sunlight and wind to enter residential areas so that the residential environment is not humid, comfort, health, and freshness can be felt in homes. All of these changes result in an increase in the visual quality of the environment, and the creation of an impression of relief in balancing building height and building density.

Efforts to rebuild buildings affected by the M3K program were not immediately resolved by the community because of limited funding and personal funds. There was one building that was not completed by the building owner because they thought the entire rebuilding was borne by the government. Based on this fact, it can be indicated that the need for more in-depth socialization by the facilitator team relates to the funding and technical implementation of the program to minimize problems during program execution.

Regarding the conditions of settlements in Gemblakan Bawah which are included in fire-prone areas, currently the village has hydrants as a fire disaster mitigation system i.e. Fire Hydrant Bar - dry. The use of dry hydrants is efficient enough to help outages during fires in densely populated villages.

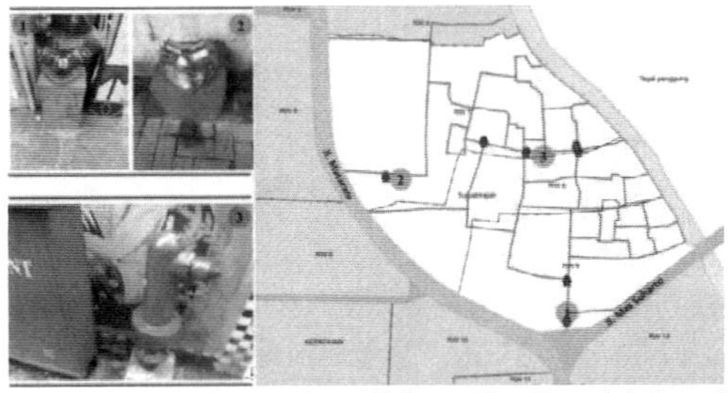

Figure 3. Fire Hydrant Conditions In Gemblakan Bawah

3.3 Circulation and Parking

Gemblakan Bawah village can be accessed through 4 main door points, 3 points are on Jl. Mataram and 1 point are on Jl. Mas Suharto. The three points are on Jl. Mataram has a width of 1 - 2.5 m and can be passed by motorized, while one main point is on Jl. Mas Suharto has a width of 2 m, can only be passed on foot. The width of the Gemblakan Bawah road consists of <1.5 meters and 2 meters. Road pavement in the form of paving, cement, and stone temples are in good condition. Changes in the pavement from paving/cement in some parts of the village road to stone temples have created new problems. The community complained that the change of pavement into stone temples made the village road more slippery. So that if the road is wet due to rain, the community must be more carefully walking or driving on the road that uses temple pavement so as not to slip.

Figure 4. Variety of Road Pavement

Circulation in the village area does not distinguish between human circulation and vehicles. Based on the description of the circulation of space in the Gemblakan Bawah Village, circulation patterns can be produced. The existing circulation pattern develops in all directions, does not have a center of space, creates a square or rectangular space area. Based on the description of the condition, the circulation pattern of the Gemblakan Bawah Village space is a grid. Environmental management only changes the dimensions and functions of circulation along the stream buffer river and the alley in the village through the road at the stream buffer. The arrangement with the M3K concept resulted in a circulation width on the edge of the river to 2-3 meters and widening the alley into the settlement through a border area almost 1 meter wide. The stream buffer width is in accordance with the stream buffer width standard. Circulation in the border area is only used for pedestrians. One of the efforts made to prevent vehicles from entering the border area of the river is to make different heights between the entrance alley and the road at the river border (Figure 5). After the establishment of the riverbank circulation path as a pedestrian pathway, Gemblakan Bawah Village looks more comfortable.

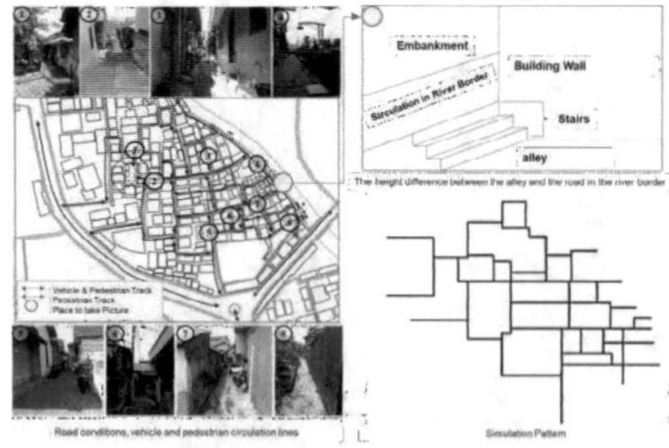

Figure 5. Circulation Kampung Gemblakan Bawah

Villagers parked vehicles along the road around his house because of limited space for parking vehicles. This condition can still be tolerated by the community, even though it results in reduced dimensions of road width. People who live close to the sports field will park their vehicles on the edge of the field and free of charge.

Parking for vehicles around the field and on the road

Figure 6. Community Parking Habits

Circulation conditions after the settlement arrangement program is then improved. Circulation of circulation is done by widening the circulation and adding special circulation for pedestrians. Whereas in terms of parking, settlement arrangement only affects the changes in the parking habits of the riverbank community because of the function of pedestrian-specific river bank circulation.

3.4 Open Space

Open space in Gemblakan Bawah Village concerns landscaping; hard elements (hardscape) which include sports fields, pedestrian lanes, circulation paths (linear space) and space/hall of community associations. The sports field in RW 7 has street furniture in the form of lights, seating, and plants with pavement in the form of stone temples. The field in RW 7 is a private field owned by people who are freed in general use to the community. Apart from being used as a sports facility, this field is used as a parking bag for the people who live around the field and also as a socialization space for citizens (Questionnaire Results, 2018).

Based on the provisions of open space regulated in SNI 03-1733-2004 [15], Gemblakan Village must have at least one open space which functions as a sports room and children's playground. The physical arrangement with M3K produces new open spaces in the river border area that are used as a space for interaction between the people who live on the banks of the river and places for organizing community events. In addition, in order to have community space amid limited land, the community elevates the MCK building and utilizes the second floor as a community room. After structuring settlements, open spaces in the river border area make residents feel more comfortable because space is cleaner, more beautiful and green. The addition of children's play space in the new open space makes parents not worried because children no longer play on the river. The wide dimensions of this linear open space are 2-3 m with stone pavement temples which are also equipped with.

Outdoor playground facilities and benches, lights and plants as street furniture in open spaces

Figure 7. Open Space in Gemblakan Bawah Village

Based on the depiction of open space in the Lower Gemblakan Village after structuring with the M3K concept it can be concluded that there are many new open spaces equipped with supporting facilities. The availability of open space is in accordance with SNI 03-1733-2004. New open spaces need more attention to hygiene conditions so that more space is maintained.

3.5 Pedestrian Ways

The arrangement of settlements on the banks of the river produces additional dimensions of the circulation path in the border zone. The addition of dimensions in the circulation path in the border zone is designed to be used as a pedestrian path (3 meters). Regional development plans are directed into the concept of tourism. Current pedestrian elements are pedestrian path elements (temple stone), and pedestrian path support elements (lighting, vegetation, trash cans, and signs).

Pedestrian lanes formed in Gemblakan Bawah not only function as pedestrian circulation paths but also function as a space for social interaction and add to the beauty and comfort of the city. The benefits obtained from the planning and implementation of pedestrian lanes are to foster healthy activities so as to reduce the vulnerability of crime, stimulate various economic activities that can support the development of the region, event promotion/exhibition activities, and can be used for social activities for the village community.

All the people of Gemblakan Bawah Village agreed that riverbank arrangements have made the image of a slum village a healthy village (Questionnaire Results, 2018). The existence of a pedestrian path makes people feel more comfortable, clean and beautiful. As a result, the pedestrian area is often used by the local community as a lounge, meeting room, and children's playroom. Based on these conditions, it can be seen that the planning of pedestrian paths in Kampung Gemblakan Bawah not only emphasizes quality and quantity, but also environmental management [16].

The compatibility of the use of temple stone material as a pavement of pedestrian path elements and supporting elements of pedestrian lines such as greenery, street lights, and benches & Pedestrian Path Width

Figure 8. Pedestrian Condition

The pedestrian route also managed to attract some visitors from outside the village to take pictures in the pedestrian area. Currently, the street furniture available in the pedestrian area includes gazebos, lighting, benches, plants in pots, trash cans, and spot taking pictures. All road furniture in the pedestrian area can be used and maintained. The arrangement of street furniture, the use of lighting arrangements with a typical atmosphere of pedestrians, has increased the physical value of the area. The results of the implementation of the pedestrian route planning have paid attention to compatibility, the scale of infrastructure material, and dimensions so that pedestrians feel comfortable walking. The development of pedestrian lines along the Code River can have a positive impact on environmental and economic improvements. Pedestrian lanes in the border area of the river can be an alternative destination for tourists visiting the city of Yogyakarta, especially Malioboro

3.6 Signage

After the arrangement of settlements, the walls of the pedestrian section are decorated with signages that provide educational information to the villagers. Signage was placed on the wall of the building so that it can be classified in the type of wall sign. The nature of signage communication on pedestrian lines is indirect, meaning that the information conveyed has nothing to do with the building where the marker is placed. Signages in pedestrian areas contain educational information such as appeals to love the environment, stay away from drugs, and harmony in life. Design signages along the pedestrian path have been synergized with views of the pedestrian lane. So that pedestrians can feel the match between the marker and the pedestrian path condition. This is in accordance with the design criteria proposed by Lestariningsih [17] regarding the synergy between marker design and the visual area of the city. In addition, mounting markers on pedestrian lines has fulfilled the technical guidelines for mounting signages, ie markers have reflected adequate and regular regional, distance and size characteristics, and markers of harmony with architectural buildings around the location [18].

However, the flood evacuation route marker cannot be clearly seen. The marking of the flood evacuation route is only a sticker that is affixed to the wall or building glass. Marking the flood evacuation route is important to note so that the flood evacuation route needs to be clearly known by villagers and tourists. Given the history of the Lower Gemblakan Village which is prone to flooding from the River Code runoff, the flood evacuation route can be installed along the evacuation route with. Whereas in the river border area a flood evacuation route map can be installed so that information can be read and clearly understood by the village community and tourists.

Figure 9. Signage in Gemblakan BawahVillage

3.7 Activity Support.

Activity Support in the Gemblakan Bawah Village after making arrangements is a tourist activity. The goal is to be able to have an impact on the economy of the people in Gemblakan Bawah Village. In fact, the economic impact is felt by people who work as traders. This condition is in accordance with the main function of supporting activities, namely moving the function of the main activities of the city to be more alive, continuous, and peaceful [19]. In addition, the arrangement of settlements carried out in Gemblakan Bawah Village attracts many academics/institutions to visit to conduct studies.

Gemblakan Bawah Village provides tourism support facilities such as benches, places to take pictures (selfie locations), and bins. The existence of tourist facilities in the form of places that can be used to take pictures of visitors is used as a way to attract tourists. The decoration of the spot selfie is on the river bank and settlement walls.

Figure 10. Tourism Facilities in Gemblakan BawahVillage

4 Conclusion

The results of the study prove that settlement arrangements with the M3K concept make the urban elements of the city experience improvements in quality and quantity. Improving the quality and quantity of urban design elements makes the quality of the village environment better. Improving the quality and quantity of urban design elements includes:

- Creation of river border spaces that are free from buildings

- Improving the visual quality of the environment and building security due to the regularity of buildings and fire disaster mitigation
- Structuring and widening the circulation of river banks into non-motorized lanes and new open spaces
- Increased number of open spaces equipped with children's play facilities
- Availability of pedestrian that fosters regional vitality and improves regional identity
- Availability of educative markers that reflect regional characteristics and harmony with the architecture of surrounding buildings. However, the design and size of the flood evacuation route markers need to be improved so that they can be read clearly.
- The emergence of new activities in the form of tourism activities that provide economic and social impacts. However, the diversity and intensity of supporting activities need to have more attention to attract tourists to come to the Lower Gemblakan Village

References

[1] I. Khambali, *Model Perencanaan Vegetasi Hutan Kota*, 1st ed. Yogyakarta: Andi Publisher, 2017.
[2] B. Martopo, Sugeng & Mitchell, *Bali : balancing environment, economy, and culture.* Australia: University of Waterloo. Department of Geography, 1995.
[3] B. K. Mokodongan, R. L. E. Sela, and H. H. Karongkong, "Identifikasi Pemanfaatan Kawasan Bantaran Sungai Dayanan Di Kotamobagu," *Sabua*, vol. 6, no. 3, pp. 273–283, 2014.
[4] E. Poedjioetami, "Penataan Ulang Kawasan Bantaran Sungai dengan Menghadirkan Sentra Ekonomi dan Rekreasi Kota: Studi Kasus Kawasan Dinoyo Tenun, Surabaya," *J. Rekayasa Perenc.*, vol. 4, no. 3, 2008.
[5] D. K. Rahmadi, "Permukiman Bantaran Sungai.pdf." p. 2, 2009.
[6] S. A. H. Al Munawar, "Khasanah Budaya Keraton Yogyakarta II," 2001.
[7] Y. I. Winarto, "Identifikasi Variasi, Kondisi, Dan Permasalahan Pemanfaatan Ruang Di Sempadan Sungai Code, Kota Yogyakarta," 2015.
[8] T. A. Fitria, "Revitalisasi Permukiman di Tepi Sungai Dengan Pendekatan Lansekap Berkelanjutan untuk Meningkatkan Kesehatan Lingkungan," vol. 1, no. 1, pp. 195–197, 2017.
[9] N. Amri, "Karakteristik Lingkungan Permukiman Kumuh Tepian Sungai Kecamatan Kolaka, Sulawesi Tenggara," *Univ. Hasanuddin*, no. 1927, 2009.
[10] Z. Dzulfia, "Bangun Rumah 'Madhep Kali', Cara Warga Yogyakarta Bersahabat dengan Sungai," Jakarta, 2017.
[11] A. Maryono, *Pengelolaan Kawasan Sempadan Sungai Dengan Pendekatan Integral : Peraturan, Kelembagaan, Tata Ruang, Sosial, Morfologi, Ekologi, Hidrologi, dan Keteknikan*. Yogyakarta: Gadjah Mada University Press, 2014.
[12] Superman, "Mengelola Koridor Sungai," 2015.
[13] H. Mulyandari, "Upaya pengelolaan lahan bangunan pada bantaran sungai berbasis lingkungan di kabupaten sleman diy," pp. 31–40, 2011.
[14] Y. A. Birwono Joga, *Bahasa Pohon Selamatkan Bumi*. Jakarta: PT Gramedia Pustaka Utama, 2013.
[15] National Standardization Agency of Indonesia, *SNI 03-1733-2004 about Housing Planning Procedures in Urban Areas (Tata Cara Perencanaan Lingkungan Perumahan di Perkotaan)*. 2004.
[16] J. T. Rochadi, Mochammad Tri, Titien Woro Murtini, Dwi Suci Sri Lestari, "Linkage System," in *Perencanaan Urban*, Bandung: ITB, 1991.
[17] D. J. Lestariningsih, "PENGARUH SIGNAGE TERHADAP ESTETIKA VISUAL (KORIDOR KOMERSIAL) JALAN AGUS SALIM SEMARANG." Universitas Diponegoro, Semarang, 2002.
[18] E. Darmawan, *Teori dan Implementasi Perancaangan Kota*, I. Semarang: Badan Penerbit

Universitas Diponegoro, 2003.
[19] H. Shirvani, *The Urban Design Process*. New York: Van Nostrand Rainhold Company, 1985.

A Comparative Study on Resident Satisfaction of Formal Housing and Self-help Housing in Cemorokandang Urban Village, Malang

W Sasongko[1], A N Fadhillah[2], and T A Rachmawati[3]

[1] Departement of Urban and Regional Planning, Faculty of Engineering, Brawijaya University, Malang, East Java Indonesia.

[2] Departement of Regional Planning, Goverment of Malang City, East Java, Indonesia.

[3] Departement of Urban and Regional Planning, Faculty of Engineering, Brawijaya University, Malang, East Java Indonesia.

*E-mail: wisnusasongko72@gmail.com

Abstract. Existing housing in Cemorokandang Urban Village can be classified into 2: formal housing built by developers and self-help housing built by the community's own initiative. There were 4,802 houses that have been built and there are 884 houses uninhabited in Cemorokandang Urban Village. The majority of uninhabitable houses were in housing built by the developer (the formal housing) while only 8% of which were self-help housing. There are significant differences in the number of empty houses in formal and self-help housing in Cemorokandang Urban Village. The research was conducted to find out whether there are differences related to the satisfaction of the residents in the formal housing and self-help housing in Cemorokandang Urban Village. There are some factors that influence the satisfaction of residents, where in this study factors influencing satisfaction focused on the variables of accessibility, physical attributes and the quality of the environment. The results of the comparison of the formal and self-help housing that variables of accessibility and environmental quality are higher in self-help housing, while the variable of physical attributes is higher in formal housing.

Keywords: Housing satisfaction, accessibility, physical atributes, environment quality

1. Introduction

The development of city can be seen from the physical characteristics such as population growth, increased building area, and the availability of city facilities that support social and economic activities of the city. Along with the increase in the number of inhabitants in a city it will bring effect to the increasing needs of a home as one of the basic human needs as part of the quality of life and well-being of the people. 1 Execution of development in fact experiences a variety of limitations of the level of social development, organization of housing development, the area of settlements and the provision of land and infrastructure of public facilities, and it is often found that the settlement conditions are not qualified. 2 Regional regulation number 4 year 2011 about the Spatial Plan of the city of Malang year 2010-2030 states that the territory of Kedungkandang includes as a region of high complexity. The region of Kedungkandang has one of the primary functions as housing. 3 According to BPS data from the twelve existing wards in Kedungkandang district, in 2014-2015 Cemorokadang Village has the highest growth rate i.e. 2.64. 4

The problems encountered due to the increasing population are high demand on food, energy, and housing. In Kelurahan Cemorokandang problems occurred related to conditions of housing. In the Village of Cemorokandang there were 4,802 houses that have been built and there were 884 uninhabited houses. The majority wereformal housing and only 8 percent of them were self-help housing. In addition, the provision and development of infrastructure in the Sub-District of Kedungkandang is still not optimal. 5 The preliminary survey (2017) mentions that the self-help housing infrastructure in the village of Cemorokandang in particular in RW 6 is still not optimal. In Kedungkandang subdistrict criminal cases that often occur are theft, and from 142 reported cases only 47 cases were successfully resolved by the police. 4 At the village of Cemorokandang, criminal cases most often occurs in formal residential area. Kedungkandang subdistrict has low accessibility because the distance between the center of the city and the villageis quite far 6 and public transporthas inefficient headway that results in a decrease in the number of passengers and the number of public transport. 7 The purpose of the study is to know the comparison of the variables of accessibility, physical attributes, and environmental quality in the formal housing and self-help housing in Cemorokandang Village.

The formal housing that is built has clear rules with a regular pattern, while the self-help houses are organized based on the initiative and efforts of the community. 8 The satisfaction of housing is used to evaluate the perceptions and feelings of the citizens about the home and the environment. There are reviews of various studies that discuss an assessment about the satisfaction of residents. The study refers to an assessment of how big the occupants are satisfied with their homes where the comfort of staying is a product of complacency and the neighborhood is able to meet the needs of residents on affordablility and convenient facilities, and can make the residents feel welcome. 9 The satisfaction of living is affected by accessibility, physical attributes, environmental health, and public facilities. 10

This research uses 3 variables that affect satisfaction, namely accessibility, physical attributes, and fulfillment of environment. The accessibility can be judged from the availability of public transport, road conditions and the distance to the Center. 10 Physical attributes include two criteria, namely the design and construction of buildings. The design of the building is determined by the owner of the house and building construction is determined by the selection of good material, the method of installation and compliance with the environment. Environmental quality in housing is affected by residential location, completeness, thoroughness of the infrastructure facility, the level of security and social relations with neighbours. 9

2. Methods

Comparative research is a research that compares two or more symptoms. Comparative research can be either descriptive or comparative. Comparative descriptive is comparative correlational, comparing the same variable for different samples. 11 The research is using the comparative descriptive. Analysis of descriptive statistics are used to express the image of the descriptive data by way of exposing the results tabulated late processing frequency in order to unveil the trend nominal data and the description of empirical data, and analysis results are useful to support descriptive interpretation against the results of the analysis with other techniques. 12 The descriptive statistics analysis is used to explain the results of the questionnaire of the respondents between formal housing residents and self-help housing, explain the difference between the formal housing resident satisfaction and self-help housing, and interpret chart, diagram and results analysis. Discussion of the value of the variables of satisfaction, namely physical attributes, accessibility and environmental quality, done by weighing the results of the questionnaire and then calculating the average value so it can be matched with the table of satisfaction criteria. According to Hariyanto the criterion of satisfaction obtained from the difference between the highest score minus the lowest score of the answers to the questionnaire and divided by the number of classes 13 Therefore, the class interval is (5-1)/5 = 0.8. Note the class interval of 0.8 is then compiled with an average assessment criteria of the respondent's answers.

IOP Conf. Series: Earth and Environmental Science **328** (2019) 012076 doi:10.1088/1755-1315/328/1/012076

3. Results and Discussion

3.1.Characteristics of the housing in Cemorokandang Village

3.1.1. The Characteristics Of Accessibility

Accessibility characteristic is seen from the indicators of the availability of public transport, the distance from home to work, education, health, commerce and recreation. Assessment of public transport in terms of the Division of respondents based on formal housing and self-help housing tells us that respondents in the formal housing and non-governmental housing in majority say that the availability of public transport is difficult, in formal housing 63 respondents and in self-help housing 41 respondents. The reasons respondents why they say the availability of public transportation is difficult because the public prefer to use their private vehicle for personal transportation because it is more flexible and almost every household in Kelurahan of Cemorokandang has a private vehicle.

In the formal housing, the distance from home to work is not the main determinant in choosing a place to live. Having a place to live far away from work is not a barrier, whereas in self-help housing respondents tend to choose the place of residence near the location of work

According to the survey results in the self-help housing, 72 respondents said the distance from home to educational facility is near and in the village Cemorokandang. The number of respondents who states that self-help housing is near because they choose the educational facility located in Cemorokandang Village, while formal housing dwellers fall into the category of not very close because they choose the means of education outside Cemorokandang village.

Respondents residing in formal residential in Kelurahan Cemorokandang as many as 58 people stated that the distance from home to health facility is within a radius of nearby, that is between 1 to 3 kilometers, while those who stated that it is not near are 3 persons by reason of using health facilities outside Cemorokandang village. In contrast, respondents on self-help housing who stated that the distance from home to health facility is near are 55 people and there is no one who stated that the distance from home to health facility is not very near or over 7 kilometres since the respondents in self-help housing are using health facilities in Cemorokandang Village. As many as 67% state that the distance from the house to the means of trade and services is between one to three kilometers from their home. It indicates that the public in Kelurahan Cemorokandang meet the needs of their life by doing shopping activities and getting the services which are not far from their home. There are indications there are respondents doing the shopping activity outside the Cemorokandang village which is over 7 kilometers.

As many as 42% of respondents use a nearby recreational facility within a short distance from the house and outside the village Cemorokandang. Respondents tend togo sightseeing in Wendit recreational park or Velodrom. meanwhile, only 2% of respondents who stated that the distance from home torecreation facility is near because the recreational field is utilizing IE sport around their home.

3.1.2. The characteristic of physical attributes

Building area in the most formal housing includes in the category of undecided (33 respondents), whereas in non residential building area of the majority is not broad categories, i.e. as many as 36 respondents. Differences in formal residential building area and self help is due to the different ability to purchase land and build a house on a formal society and community based organizations. Respondents in the formal housing has a greater income than residents in self-help housing, so the purchasing power to own a house is higher as the financial ability is higher.

If divided based on formal housing groups and self-help, formal housing respondents with the number of 52 people build houses with quality building materials. The buildings built in the formal housing is permanent, and the vast majority of the respondents do a home repair maximum ≤ 3 times in 5 years in the form of roof repair roof tiles. As for self-help housing the quality of building material most widely chosen by as many as 43 respondents are unconvincing. In self-help housing the majority is permanent house, the quality of this type of building materials is lower than formal housing because of the frequent occurrence

IOP Conf. Series: Earth and Environmental Science **328** (2019) 012076 doi:10.1088/1755-1315/328/1/012076

of damage such as cracked walls and leaking roofs or flaking, and the repair of damage to the House is done ≤ 5 times a year.

Regarding the comparison between formal and self-help housing, respondents most widely agreed to build the design of the house in accordance with the wishes and the environment. In the formal housing 44 respondents agreed about their house and the building design while in self-help housing 68 respondents agreed with the design of the building of their house. In formal residential home building erected by the developer, although some residents have renovated the building of their house so that it is in accordance with the wishes and the environment. There is no residents who disagree with the design of their house building.

3.1.3. The Characteristics of environmental quality

42 respondents in the formal housing judge agreed that the location of their home were in accordance with the wishes and the environment, but in the highest scoring non-governmental housing there is hesitancy in the category ofhome location, i.e. as many as 38 respondents. There are similarities among the respondents of formal and self-help housing. They say that the suitability of the location of the House is good because before buying the House, occupants have been considering the location of the house.

In comparison between the assessment of completeness of formal housing and infrastructures of self-help housing, all respondents in formal housing have the highest scoring (63 persons) because the infrastructure in formal housing is readily available, for example roads with good conditions, clean water, drainage, waste management and the public could use those infrastructures according to their needs. As for self-help housing only 30 respondents say that the infrastructure is complete while 34 respondents disagree with this because the respondents consider the existing infrastructure is not yet able to meet the comfort in the area of residence as some of the roads are not paved and would hamper transportation. , Regarding waste management, some residents still use the method of burning garbage as there is no garbage collector.

43% of the society States that in Kelurahan Cemorokandang the facilities are complete. There are 218 facilities starting from the means of Government and public service, education, health facilities, a means of trade and services, means of expression, means the industrial and warehousing, security means, the means of open space and sports, as well as the means of cemetery. 47 the respondent stated that the facilities in Kelurahan Cemorokandang are incomplete because they thought that the completeness of the facility have been met but it spreads unevenly, so that in some areas several facilities are unavailable.

The level of security that is in the village of Cemorokandang has started to improve in line with the growth of the housing development and the rise of the level of awareness of society to jointly maintain the security of the environment. The occurrence of criminal cases particularly theft can be reduced by having the system of regular patrol at night, a security guard who watches over the entrance, and one-gate access. However, there is still the chance of theft, especially in areas prone to theft like on the road, the path of Perdana Kusuma Airport Abdurahman Saleh, around housing in RW 9, RW 8, and RW 6 (formal housing).

The social circumstances in the neighbouring Village of Cemorokandang are rated good by 81 respondents or 45%. This is supported because in some neighborhoods such as RW and RW 6 social interaction is well maintained because there is still a relative or relatives. At the another site social relations are maintained by holding activities that involve citizens in it. As many as 1% of the community stated social relationships with neighbor isn't very good and 1% of the community stated social relationships with neighbor is not goodbecause there are still some people who have a high degree of individualism. Respondents in the formal housing are more hesitant to choose social relations i.e. neighborly as many as 46 people, whereas in self-help housing clumped on the criteria respondents assessment of good and very good. Only 2 people who are doubtfulwhile 61 people judge that the condition of social relationship with neighbor is good. Assessment of good neighborly relations to social housing conditions are supported by the non-governmental interaction between neighboring social activities goes by and large.

Characteristics of satisfaction in Cemorokandang: 44% community declared that they are satisfied with their house. It is detached from the House is large or small, fancy or simple. The reason why the respondents are satisfied because there is a sense of comfort residing in the house and the majority is because they are the owner of the house (not rent house), that is why the respondents feel satisfied. Meanwhile, as many as 18 respondents said they are less satisfied with their house because the house is not large enough. There is a difference between the assessment of satisfaction on the formal housing and self-help housing. Most respondents from formal housing are satisfied with their house 49 people respondents choose satisfied because conditions in the existing house is in accordance with their wish. As many as 48 respondents argued that their house are not good enough when compared to existing houses in residential homes and the land is less spacious, while those who say . satisfied are persons. However, there is no self-help housing residents who feel very dissatisfied because the house is livable and they can adapt to the place of residence they have.

Related to satisfaction of the environment for the most formal housing 39 respondents are in doubt because of the location of the house is facilitated with good infrastructure but less secure home environment, while for self-help housing as many as 73 respondents are satisfied. The feeling of satisfied from self-help housing residents is bigger because they have been adapting to environmental conditions and the possibility to move is small. Self-help housing residents do not have the option to stay in place so they try to feel comfortable. In contrast the formal housing residents if they feel dissatisfied they can move. There are parallels between formal and self-help housing in assessing the satisfaction of the environment i.e. no respondents who state very dissatisfied whether in formal and in self-help housing.

Comparative analysis of physical attributes, accessibility, and the quality of the environment in the formal housing and self-help housing in Kelurahan Cemorokandang is based on a questionnaire that has been filled by 90 people from formal housing and 90 respondents respondents from self-help housing. The result of the filling of the questionnaire is weighed to get the value of each answer on each indicator. After each value is known to every answer on each indicator, then the results are aggregated so that the total value per indicator is known, then the average assessment per indicator is calculated by dividing by the number of respondents. The average assessment per indicator is needed to determine criteria for the satisfaction of the respondents. The respondent satisfaction criteria is derived from the difference between the highest score minus the lowest score of the answers to the questionnaire divided by the number of classes. Therefore, the class interval is (5-1)/5 = 0.8. Interval class 0.8 is then drafted criteria for the assessment of the average answer of the repondents: The criteria for customer satisfaction: Very unsatisfied (1.00-1.8), not satisfied (1.81-2.6), enaugh/undecided (2.61-3.4), satisfied (3.41-4.2), very satisfied (4.21-5.00).

The results of the average of each indicator is then matched with the satisfaction criteria table,so the retrieved data about satisfaction of each indicator is shown in table 1. Table 1 aims at knowing the average value per indicator, where the average is a number that represents a set of data. On the formal housing accessibility variable, indicators that get the most satisfaction assessment based on average assessment indicators is an indicator of the healthwith the average value per indicator of 3.48 which means satisfied. Based on existing conditions, the majority of the respondents have felt satisfied due to the proximity of the distance from home to health facility. On the formal housing from the 6 indicators, indicator 2 get assessment of not satisfied because it is difficult to get the transportation facility as well as the distance to home recreation, 3 indicators get enough assessment because of the distance from the means of education, trade and office are varied. Unlike the self-help housing, there are 4 indicators of accessibility variables that get an average per indicator which falls into the category of satisfied namely, work, education, health, and trade with the highest average rating is on educational indicators i.e. 3.96 because the distance from the house into education facility is between 1-3 km (near). On the variable of accessibility, there is a difference of the highest average value indicator. 58 of respondents in the formal housing assess the distance from home to health facility is close and they use health facilities that are located in Cemorokandang village. For self-help housing when converted in intervals of satisfaction falls into the category of satisfied with the education facility. The means of education from elementary to junior high school are in within the environment and can be utilized by the respondents of self-help housing. 72 respondents rate the distance

IOP Conf. Series: Earth and Environmental Science **328** (2019) 012076 doi:10.1088/1755-1315/328/1/012076

from home toeducation facility is near (1-3 km) and using the means of education in Cemorokandang village.

Variable physical attributes has three indicators, namely building area, type of material and design, where formal and self-help housing on indicators that have the highest average is the reason of repondents because according to the design of the house. The design is an indicator that most easily tailored to the desires of the self help housing, on average the residents do not have sufficient charge for using this type of building materials of the best quality, but they still could have a house with a desirable design in accordance with the cost owned. On the variable of physical attributes, there is a difference in valuation of the kind of formal housing materials and self-help, self-help housing which fall into the category of formal housing while on the regular entry in the category are satisfied.

Table 1. The Comparison of the results of W eighting Indicator formal housing and self help housing

Variable	Indicator	Formal			self-help Housing		
		total value (indicator)	average (indicator)	Description	total value (indicator)	average (indicator)	Description
access	Public transport	179	1,99	not satisfied	247	2,74	enough
	place of work	252	2,8	enough	341	3,79	satisfied
	school	300	3,33	enough	356	3,96	satisfied
	Health place	313	3,48	satisfied	331	3,68	satisfied
	store/market	305	3,39	enough	325	3,61	satisfied
	recreation areas	208	2,31	not satisfied	198	2,2	enough
Physical attributes	building area	286	3,18	enough	247	2,74	enough
	types of building materials	319	3,54	satisfied	295	3,28	enough
	model	346	3,84	satisfied	335	3,72	satisfied
Environmental quality	Residential location	350	3,89	satisfied	298	3,31	enough
	the completeness of the infrastructure	347	3,86	satisfied	289	3,21	enough
	the completeness of the utility	289	3,21	enough	284	3,16	enough
	security	279	3,1	enough	324	3,6	satisfied
	social communities	286	3,18	enough	385	4,28	satisfied

Residential location indicator on the quality of the environment variables in the formal housing had the highest median value i.e. 3.89 in the category of satisfied because the respondents contended with the location that suits their wishes and followed by the suitability of the means and infrastructure. In contrast to the formal housing, self-help housing has averaged the highest social indicators are neighbors. Social indicators of the neighbors have an average of 4.28 and belongs to the category of satisfied. The average value of the neighboring social indicators in self-help housing is the highest value among the average value of the indicators in the variable of environmental quality according to the respondents. The self-help housing indicators of social capital is the neighborhood. It is important that in the neighborhood, like the example with social satisfaction, the environment is more secure because there is a togetherness in safeguarding the environment. After that, the average value of each indicator is then concluded for the average assessment per variable. The average value per variable obtained by way of summing the average value of each indicator in one variable and then divided by the number of indicators. Table 2 describes the results of the average value of the variable with a description of the level of satisfaction.

Table 2 The comparison of Formal Housing and Self-help Housing

Formal Housing		Variables	Self-help Housing	
average	description		average	description
2,89	enough	access <	3,33	enough
3,52	satisfied	physical attributes >	3,25	enough
3,45	satisfied	environment quality <	3,51	satisfied

Table 2 shows that when compared between formal and self-help housing based on variables of accessibility, physical attributes, and the quality of the environment, on housing access formal variable and non-governmental alike fall into the category enough that is not too satisfied or not satisfied. However, the self-help housing has the average value of the variable access to higher (3.33) than formal housing (2.89) because it is supported by indicators of transport, workplace, education, health, trade value has higher satisfaction rate than formal housing. On the variable of physical attributes , formal housing has an average rating of higher i.e. 3.52 in category satisfied compared to self-help housing has a value of 3.25 and belongs to the category fairly, in accordance with the existing conditions of the respondents of formal housing has a building area of more than self-help housing, has the type of building and the designdesign is better than self help housing. The variable of review of the quality of the environment, the average value of the variables of formal housing is lower than the average value of the variables of self help housing although both entered category satisfied , the average value of formal housing was only 3.45 while self-help housing higher i.e. 3.51 because the security assessments and social communities in self-help housing are more superior than in formal housing.

4. Conclusion

Self-help housing has an average value of variables and the variable of access environmental quality higher than formal housing. On the variables of physical attributes of, formal housing has the average value that is higher than the value of self-help housing.

References

[1] Hilman, Maman. 2004. Perkembangan Lokasi Perumahan di Wilayah Gedebage Kota Bandung Akibat Pemekaran Kota. Jurnal Dimensi Teknik Arsitektur Vol. 32, No. 2, Desember 2004: 157 - 160

[2] Bappeda. 2012. Naskah Akademis dan Rancangan Peraturan Daerah tentang Rencana Pembangunan dan Pengembangan Perumahan dan Kawasan Permukiman di Kota Malang (Malang: Bappeda)

[3] DPR Kota Malang. 2010. Peraturan Daerah Nomor 4 Tahun 2011 Tentang Rencana Tata Ruang Wilayah Kota Malang Tahun 2010-2030

[4] BPS Kota Malang. 2017. Kedungkandang Dalam Angka 2016 (Malang: Bappeda)

[5] Febriyani, Risna et al.2014. Pengembangan Wilayah Hinterland di Kecamatan Kedung Kandang Sebagai Upaya Penigkatan Pelayanan Publik. Jurnal Administrasi Publik (JAP), Vol. 2, No. 3, Hal. 440-446
[6] Mahendra, Yusril Ihza. 2016. Transformasi Spasial di Kawasan Peri Urban Kota Malang. Jurnal Pembangunan Wilayah & Kota Vol. 12(1):112-126 Maret 2016
[7] Batlayeri, Liberatus. 2015. Evaluasi Kinerja Angkutan Umum Penumpang Kota Malang Pada Jalur Ckl (Cemoro Kandang - Kedawung – Landungsari). Skripsi Sarjana Teknik Universitas Muhammadiyah Malang. Malang.
[8] Rachman, Hamzah. 2010. Kajian Pola Spasial Pertumbuhan Kawasan Perumahan dan Permukiman di Kecamatan Limboto Kabupaten Gorontalo. Semarang
[9] Widiastomo. 2014. Pengaruh Kualitas Rumah dan Lingkungan terhadap Kepuasan Penghuni dan Kecenderungan Berpindah di Perumnas Bukit Sendangmulyo Semarang. Jurnal Pembangunan Wilayah dan Kota Vol. 10
[10] Primaningtyas. 2012. Pengaruh aksesibilitas, Atribut fisik, Kesehatan Lingkungan dan Fasilitas Publik Terhadap Kepuasan Bermukim (Studi Kasus pada PT. Armada Hada Graha Magelang)
[11] Mardhiyah, Khusna. 2012. Studi Komparasi Kualitas Pelayanan Pendidikan di Sekolah Menengah Atas Negeri 2 Rembang dan Madrasah Aliyah Negeri Rembang. Tesis Program Pascasarjana Undip.Semarang.
[12] Ujianto. 2004. Analisis Faktor-Faktor yang Menimbulkan Kecenderungan Minat Beli Konsumen Sarung (Studi Perilaku Konsumen Sarung di Jawa Timur). Jurnal Manajemen & Kewirausahaan Vol. 6, No. 1, Maret 2004: 34 - 53
[13] Harianto, Feri & Prasetyo, Fendy .2010. Analisis Tingkat Kepuasan Penghuni Rumah di Perumahan Puri Surya Jaya Gedangan Sidoarjo. Seminar Nasional VI 2010 Teknik Sipil ITS Surabaya Pengembangan Infrastruktur Dalam Menunjang Pembangunan Ekonomi Nasional

AUTHOR INDEX

AUTHOR INDEX

Institute of Physics
Dirac House, Temple Back
Bristol BS1 6BE UK

ISSN: 1755-1307
ISBN 978-1-7138-1299-9

9 781713 812999

CURRAN ASSOCIATES INC.
proceedings.com